12th Edition

Elementary Technical Mathematics

Dale Ewen
Parkland Community College

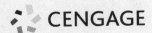

Australia • Brazil • Canada • Mexico • Singapore • United Kingdom • United States

Elementary Technical Mathematics,
Twelfth Edition
Dale Ewen

Product Director: Terry Boyle

Product Manager: Rita Lombard

Content Developer: Powell Vacha

Product Assistant: Abby DeVeuve

Marketing Manager: Ana Albinson

Content Project Manager: Corinna Dibble

Manufacturing Planner: Rebecca Cross

IP Analyst: Ann Hoffman

IP Project Manager: Erika Mugavin

Production Service and Compositor: MPS Limited

Art Director: Vernon Boes

Text and Cover Designer: Terri Wright

Cover Image: Hero Images/Getty Images, Masterfile, Lisa F. Young/Shutterstock.com, Monkey Business Images/Shutterstock.com, kurhan/Shutterstock.com, SantiPhotoSS/Shutterstock.com, anyaivanova/Shutterstock.com, Nolanberg11/Shutterstock.com

© 2019, 2015, 2011 Cengage Learning, Inc.

All items are copyright of Cengage unless otherwise noted.

ALL RIGHTS RESERVED. No part of this work covered by the copyright herein may be reproduced or distributed in any form or by any means, except as permitted by U.S. copyright law, without the prior written permission of the copyright owner.

> For product information and technology assistance, contact us at
> **Cengage Customer & Sales Support, 1-800-354-9706**
> **or support.cengage.com.**
>
> For permission to use material from this text or product, submit all requests online at **www.cengage.com/permissions.**

Library of Congress Control Number: 2017938035

Student Edition:
ISBN: 978-1-337-63058-0

Loose-leaf Edition:
ISBN: 978-1-337-63065-8

Cengage
200 Pier 4 Boulevard
Boston, MA 02210
USA

Cengage is a leading provider of customized learning solutions with employees residing in nearly 40 different countries and sales in more than 125 countries around the world. Find your local representative at: **www.cengage.com.**

To learn more about Cengage platforms and services, register or access your online learning solution, or purchase materials for your course, visit **www.cengage.com.**

Printed at CLDPC, USA, 11-21

CONTENTS

List of Applications   viii
Preface   xii

CHAPTER 1   Basic Concepts   1

UNIT 1A   REVIEW OF OPERATIONS WITH WHOLE NUMBERS   2
- **1.1**   Review of Basic Operations   2
- **1.2**   Order of Operations   11
- **1.3**   Area and Volume   13
- **1.4**   Formulas   18
- **1.5**   Prime Factorization   20
   UNIT 1A: REVIEW   23

UNIT 1B   REVIEW OF OPERATIONS WITH FRACTIONS   24
- **1.6**   Introduction to Fractions   24
- **1.7**   Addition and Subtraction of Fractions   29
- **1.8**   Multiplication and Division of Fractions   41
- **1.9**   The U.S. System of Weights and Measures   48
   UNIT 1B: REVIEW   52

UNIT 1C   REVIEW OF OPERATIONS WITH DECIMAL FRACTIONS AND PERCENT   53
- **1.10**   Addition and Subtraction of Decimal Fractions   53
- **1.11**   Rounding Numbers   61
- **1.12**   Multiplication and Division of Decimal Fractions   64
- **1.13**   Percent   71
- **1.14**   Rate, Base, and Part   76
- **1.15**   Powers and Roots   83
- **1.16**   Applications Involving Percent: Business and Personal Finance   86
   UNIT 1C: REVIEW   92
   CHAPTER 1: SUMMARY   93
   CHAPTER 1: REVIEW   96
   CHAPTER 1: TEST   98

CHAPTER 2   Signed Numbers and Powers of 10   100

- **2.1**   Addition of Signed Numbers   101
- **2.2**   Subtraction of Signed Numbers   105
- **2.3**   Multiplication and Division of Signed Numbers   107
- **2.4**   Signed Fractions   110
- **2.5**   Powers of 10   114
- **2.6**   Scientific Notation   118
- **2.7**   Engineering Notation   123
   CHAPTER 2: SUMMARY   126

CHAPTER 2: REVIEW   127
CHAPTER 2: TEST   128
CHAPTERS 1–2: CUMULATIVE REVIEW   129

CHAPTER 3 The Metric System   130

3.1 Introduction to the Metric System   131
3.2 Length   133
3.3 Mass and Weight   136
3.4 Volume and Area   138
3.5 Time, Current, and Other Units   142
3.6 Temperature   144
3.7 Metric and U.S. Conversion   146
CHAPTER 3: SUMMARY   150
CHAPTER 3: REVIEW   151
CHAPTER 3: TEST   152

CHAPTER 4 Measurement   153

4.1 Approximate Numbers and Accuracy   154
4.2 Precision and Greatest Possible Error   157
4.3 The Vernier Caliper   161
4.4 The Micrometer Caliper   167
4.5 Addition and Subtraction of Measurements   174
4.6 Multiplication and Division of Measurements   178
4.7 Relative Error and Percent of Error   182
4.8 Color Code of Electrical Resistors   185
4.9 Reading Scales   189
CHAPTER 4: SUMMARY   194
CHAPTER 4: REVIEW   196
CHAPTER 4: TEST   197
CHAPTERS 1–4: CUMULATIVE REVIEW   198

CHAPTER 5 An Introduction to Algebra   199

5.1 Fundamental Operations   200
5.2 Simplifying Algebraic Expressions   202
5.3 Addition and Subtraction of Polynomials   206
5.4 Multiplication of Monomials   209
5.5 Multiplication of Polynomials   211
5.6 Division by a Monomial   213
5.7 Division by a Polynomial   215
CHAPTER 5: SUMMARY   218
CHAPTER 5: REVIEW   219
CHAPTER 5: TEST   220

CHAPTER 6 Equations and Formulas   221

6.1 Equations   222
6.2 Equations with Variables in Both Members   226
6.3 Equations with Parentheses   228
6.4 Equations with Fractions   230
6.5 Translating Words into Algebraic Symbols   235

- **6.6** Applications Involving Equations 236
- **6.7** Formulas 240
- **6.8** Substituting Data into Formulas 244
- **6.9** Reciprocal Formulas Using a Calculator 247
  - **CHAPTER 6: SUMMARY** 250
  - **CHAPTER 6: REVIEW** 251
  - **CHAPTER 6: TEST** 251
  - **CHAPTERS 1–6: CUMULATIVE REVIEW** 252

CHAPTER 7 Ratio and Proportion   253

- **7.1** Ratio 254
- **7.2** Proportion 257
- **7.3** Direct Variation 265
- **7.4** Inverse Variation 271
  - **CHAPTER 7: SUMMARY** 274
  - **CHAPTER 7: REVIEW** 275
  - **CHAPTER 7: TEST** 275

CHAPTER 8 Graphing Linear Equations   277

- **8.1** Linear Equations with Two Variables 278
- **8.2** Graphing Linear Equations 283
- **8.3** The Slope of a Line 289
- **8.4** The Equation of a Line 295
  - **CHAPTER 8: SUMMARY** 300
  - **CHAPTER 8: REVIEW** 301
  - **CHAPTER 8: TEST** 302
  - **CHAPTERS 1–8: CUMULATIVE REVIEW** 303

CHAPTER 9 Systems of Linear Equations   304

- **9.1** Solving Pairs of Linear Equations by Graphing 305
- **9.2** Solving Pairs of Linear Equations by Addition 310
- **9.3** Solving Pairs of Linear Equations by Substitution 316
- **9.4** Applications Involving Pairs of Linear Equations 317
  - **CHAPTER 9: SUMMARY** 323
  - **CHAPTER 9: REVIEW** 324
  - **CHAPTER 9: TEST** 324

CHAPTER 10 Factoring Algebraic Expressions   325

- **10.1** Finding Monomial Factors 326
- **10.2** Finding the Product of Two Binomials Mentally 327
- **10.3** Finding Binomial Factors 330
- **10.4** Special Products 332
- **10.5** Finding Factors of Special Products 334
- **10.6** Factoring General Trinomials 336
  - **CHAPTER 10: SUMMARY** 339
  - **CHAPTER 10: REVIEW** 339
  - **CHAPTER 10: TEST** 340
  - **CHAPTERS 1–10: CUMULATIVE REVIEW** 340

CHAPTER 11   Quadratic Equations   342

- **11.1** Solving Quadratic Equations by Factoring   343
- **11.2** The Quadratic Formula   345
- **11.3** Applications Involving Quadratic Equations   348
- **11.4** Graphs of Quadratic Equations   352
- **11.5** Imaginary Numbers   356
  - **CHAPTER 11: SUMMARY**   359
  - **CHAPTER 11: REVIEW**   360
  - **CHAPTER 11: TEST**   361

CHAPTER 12   Geometry   362

- **12.1** Angles and Polygons   363
- **12.2** Quadrilaterals   369
- **12.3** Triangles   373
- **12.4** Similar Polygons   381
- **12.5** Circles   385
- **12.6** Radian Measure   392
- **12.7** Prisms   397
- **12.8** Cylinders   402
- **12.9** Pyramids and Cones   407
- **12.10** Spheres   414
  - **CHAPTER 12: SUMMARY**   416
  - **CHAPTER 12: REVIEW**   419
  - **CHAPTER 12: TEST**   421
  - **CHAPTERS 1–12: CUMULATIVE REVIEW**   421

CHAPTER 13   Right Triangle Trigonometry   423

- **13.1** Trigonometric Ratios   424
- **13.2** Using Trigonometric Ratios to Find Angles   428
- **13.3** Using Trigonometric Ratios to Find Sides   430
- **13.4** Solving Right Triangles   432
- **13.5** Applications Involving Trigonometric Ratios   434
  - **CHAPTER 13: SUMMARY**   441
  - **CHAPTER 13: REVIEW**   442
  - **CHAPTER 13: TEST**   443

CHAPTER 14   Trigonometry with Any Angle   444

- **14.1** Sine and Cosine Graphs   445
- **14.2** Period and Phase Shift   451
- **14.3** Solving Oblique Triangles: Law of Sines   454
- **14.4** Law of Sines: The Ambiguous Case   457
- **14.5** Solving Oblique Triangles: Law of Cosines   462
  - **CHAPTER 14: SUMMARY**   467
  - **CHAPTER 14: REVIEW**   468

CHAPTER 14: TEST 469
CHAPTERS 1–14: CUMULATIVE REVIEW 469

CHAPTER 15 Basic Statistics   471

- **15.1** Bar Graphs 472
- **15.2** Circle Graphs 474
- **15.3** Line Graphs 477
- **15.4** Other Graphs 480
- **15.5** Mean Measurement 481
- **15.6** Other Average Measurements and Percentiles 483
- **15.7** Range and Standard Deviation 485
- **15.8** Grouped Data 488
- **15.9** Standard Deviation for Grouped Data 494
- **15.10** Statistical Process Control 496
- **15.11** Other Graphs for Statistical Data 499
- **15.12** Normal Distribution 502
- **15.13** Probability 505
- **15.14** Independent Events 507
  - CHAPTER 15: SUMMARY 508
  - CHAPTER 15: REVIEW 509
  - CHAPTER 15: TEST 510

CHAPTER 16 Binary and Hexadecimal Numbers   512

- **16.1** Introduction to Binary Numbers 513
- **16.2** Addition of Binary Numbers 514
- **16.3** Subtraction of Binary Numbers 516
- **16.4** Multiplication of Binary Numbers 517
- **16.5** Conversion from Decimal to Binary System 518
- **16.6** Conversion from Binary to Decimal System 519
- **16.7** Hexadecimal System 520
- **16.8** Addition and Subtraction of Hexadecimal Numbers 522
- **16.9** Binary to Hexadecimal Conversion 525
  - CHAPTER 16: SUMMARY 527
  - CHAPTER 16: REVIEW 528
  - CHAPTER 16: TEST 528
  - CHAPTERS 1–16: CUMULATIVE REVIEW 529

APPENDIXES

- **A** Exponential Equations 530
- **B** Simple Inequalities 535
- **C** Answers to Odd-Numbered Exercises and All Chapter Review and Cumulative Review Exercises 540

Index 569

LIST OF APPLICATIONS

Agriculture and Horticulture
Corn storage, 9
Crop yields, 9, 47, 150
Railroad freight cars needed, 9
Feed consumption, 9
Weight of hay, 9
Weight of cotton, 9, 47
Tractor depreciation, 9
Pesticide application, 9, 47
Planting daylilies, 9
Mulch for flowerbed, 17
Placing plant containers, 17
Herbicide application, 47, 70
Concrete feed lot, 47
Weight of feed mixture, 51
Fertilizer cost, 61
Insecticide application, 70
Feeder cattle weight gain, 70
Ranch herd loss, 80
Chemical active ingredients, 81
Protein in soybeans, 81
Butterfat in milk, 81
Lawn seed, 81
Percent of plants that lived, 81
Grain contract delivery sheets, 83
Land purchase, 91
Combine purchase, 92
Tractor purchase, 92
Hay dry matter percent, 97
Acres and hectares, 150
Planting seed corn, 150
Volume of storage bin cylinder, 181
Difference of yield, 181
Mixing two types of milk, 240
Weight of grain ratio of pounds per bushel, 256
Crop yield, 256, 263
Rate of gallon per acre, 256
Herbicide rate per acre, 256
Sand & topsoil mixture, 256
Yellow & red peppers planted ratio, 256
Pesticide mixture, 263
Chemical for field, 263
Pounds of N, P, K removed per acre of use, 263
Yield of apples per tree and income from sales of apples, 263

Fertilizer needed for lawn, 263
Percent of live hog that is carcass, 263
Percent of fat in beef and number of pounds in a carcass, 263
Percent of antifreeze in tractor radiator, 264
Protein mixture for feed, 321
Butterfat mixture, 321
Corn and soybean sales, 321
Pesticide mix, 321
Grass seed mix, 321
Border width around rectangular garden, 351
Ranch acreage, 372
Corn yield, 372
Percent of lot that is lawn, 373
Using a wheel to measure length of field, 389
Diameter of circular silo, 389
Number of smaller pipes needed to approximate one larger pipe, 389
Area of cross section of pipe, 389
Volume of wagon box, 401
Volume of gravity bin, 401
Painting cylindrical silo, 406
Sheet metal in trough, 406
Feeding bin capacity, 412

Allied Health
Fluid input & output, 8
Amount of orange juice, 9
Medicine dosage, 9, 10, 47, 70
Alcohol percentage, 47
Weight of baby, 47
Weight loss of a newborn, 47
Number of doses of medicine doses from bottle, 47
Total ounces of daily medication, 47
Number of teaspoons of medicine, 47
Number of mg of medicine, 70
Amount of medicine in one dose, 70
Liquid solutions, 81
Ratio of g/mL of dextrose, 257
Rate of intravenous solution, 257, 322
Number of drops to set up IV, 257

Time to infuse IV, 257
IV flow rate, 257
Length of time for IV, 257
mL needed for a given dose, 263
Number of mL of pure ingredient to prepare a solution, 264
Number of grams of pure ingredient to prepare a solution, 264
Preparing a saline solution, 322
IV solution administer times, 322
Number of vials of two medications, 322
Area of X-ray film, 372
Placing hospital beds in wards, 372
Placing respirator units in storeroom, 372

Auto/Diesel Service
Distance traveled on tank of gas, 8
Piston displacement, 8, 69, 70, 247
Hourly labor cost, 8
Miles per gallon, 8, 69
Kilometres per litre, 8
Tire cost, 8, 69
Area an automobile occupies, 15
Volume of oil pan, 17
Oil used, 38
Auto service time, 38, 46
Tool length, 40
Copper tubing length, 40
Heater hose length, 46
Time to detail autos, 46
Time to change tires, 46
Convert gallons to quarts and pints, 51
Tire tread, 59, 97
Length of socket, 60
Piston ring wear, 60
Length of valve stem, 60
Length of crankshaft, 69
Overtime hours, 69
Brake pad wear, 70
Cost of set of tires, 81
Percent of oil in filter, 81
Cooling system leak, 178
Total miles on trip, 178
Engine horsepower, 181, 263
Area of windshield, 181

Vehicle mileage, 181
Volume of auto trunk, 181
Cost of batteries, 239
Strengthening antifreeze mixture in radiator, 240
Mixing two types of gasoline, 240
Length of cylinder, 247
Alternator-to-engine drive ratio, 256
Oil flow rate, 256
Flywheel-drive gear ratio, 256
Ratio of secondary voltage to primary voltage in auto coil, 263
Amount of fuel required, 263
Fuel pump fuel delivery, 263
Tire pressure, 263
Fuel tank capacity, 263
Small engine testing time, 321
Hybrid engine fuel testing, 321
Engine testing time, 321
Mixing parts of cleaning solution, 321
Area of rear view mirror, 372
Similar fan belt arrangements, 384
Similar side mirrors on trucks, 384
Circumference of wheel, 389
Volume of oil filter, 405
Volume of air filter, 405
Total piston displacement, 405
Cylindrical bore increase, 406
Cylindrical bore lateral surface area, 406
Radius of crankshaft journal, 438
Piston movement distance, 439
Distance from driver's side front tire to passenger's side rear tire after accident, 466
Distance from front tip of seat cushion to tip of head rest, 466

Aviation
Certificate flight time, 8
Flight distance, 8
Plane climb rate, 8
Area of runway, 15
Area of military operating zone, 15
Fuel used, 37, 38, 181, 263
Plane speed, 45, 69
Search time, 45

viii

LIST OF APPLICATIONS

Plane altitude, 51
Runway length, 51
Flying time, 59
Flight mileage, 59
Cost of fuel, 69
Nautical miles flown, 70
Plane rental, 81
Cross-country trip, 81
Baggage weight, 178
Draining fuel tank, 178
Hours of flying lessons, 181
Area formed by flight pattern, 181
Ratio of flight time for single engine rating to commercial rating, 257
Airplane rental, 322
Wing dimensions, 350
Area from chart used for aviation navigation, 371
Flight distance, 371, 380
Angle in flight diagram, 380
Similar hospital helicopter landing pads, 384
Similar runways, 384
Area of side of tire, 389
Helicopter baggage compartment volume, 400
Lateral surface area of nose of airplane, 412
Surface area of hemispherical cockpit, 415
Ground length of flight, 438
Straight-line distance back to base airport, 466
Taxiway length, 466

Business & Personal Finance
Rate of interest on loan, 80
Salary increase, 80
Sale price of discounted items, 80
Decrease in house value, 82
Salary decrease, 82
Final sale price, 82
Family loan, 91
Savings account interest, 91
Money owed on loan, 91, 92, 98
Savings account amount accumulation, 91
Investing money, 91, 92
House payment on home loan, 91
Payment on new truck, 91
Auto financing, 91, 92, 98
Effective annual rate of interest for value of discount, 92
Effective rate of interest on early payment, 92
Effective rate of return, 92

Commercial space rental, 149
Money distribution, 239
Number of boards purchased, 239
Number of hours worked, 240
Amount borrowed from bank, 240
Amount invested to earn interest, 240
Amount needed to generate given return, 240
Country club dues, 240
Siding replacement cost, 257
Rate of pay per hour, 257
Paint coverage, 257
Unit cost of material, 263
Commission, 263
Percent of reduction of list price, 263
Percent of pay increase, 264
Carpet sales, 322
Apartment rentals, 322
Types of snorkels sold, 322
Bond investments, 323
Display floor space, 372
Cost of rectangular pieces of canvas, 372
Cost of fencing business lot, 372
Holes drilled in circular plate, 389
Maximum number of boxes shipped, 402

CAD/Drafting
Difference in output of drawings, 9
Shopping center design, 17
Shipping box design, 17
Packaging, 17
Distance between points, 38
Length of shaft, 39
Length and width of steel strip, 40
Channel dimensions, 47
Tank capacity, 51
Internal diameter of tube, 59
Height needed for riser, 69
Number of windows per code, 81
Dimensions of embankment, 81
Catwalk dimensions, 81
Length of drawing dimensions, 178
Dimensions of barn model, 264
Dimensions of plot, 322
Original room dimensions, 322
Dimensions of walkway, 322
Original building dimensions, 322
Increase in door area, 352
Bay window area added to room, 400
Triangular display pedestal design, 401
Concrete tube design, 401
Cardboard box design, 401

Volume of air in room of Victorian building, 402
Scuppers needed in swimming pool design, 402
Cylindrical tank design, 405
Volume and weight of steel plate, 405
Gallons of water in cooling tank, 406
Concrete forming paper tube design, 406
Hemispherical dome house design, 415
Angles for rafters, 438
Distance across corners of hex bolt, 440
Hydraulic control valve dimensions, 440
Locating a benchmark, 440
Mating part design, 440

Culinary Arts
Maximum seating, 10
Total loin end cut servings possible, 10
Number of items delivered to kitchen, 10
Least number of servers needed, 10
Dividing tips at end of day, 10
Amount of butter used, 40
Remaining pie, 40
Remaining flour, 40
Remaining lettuce, 40
Remaining French fries, 41
Potatoes in kitchen when new order needed, 41
Scoops of sugar needed, 48
Number of pie crusts from pie dough, 48
Number of steaks cut from a loin, 48
Edible portion of watermelon, 48
Cooking oil available, 48
Short loin available for soup, 48
tsp needed in recipe, 52
Quarts of fruit juice, 52
Number of servings from container, 52
Soup recipe in gallons, 52
Volume of punch from recipe, 61
Amount of cooking oil, 61
Number of ounces in drink of the day, 61
Weight in pounds of ingredients in recipe, 61
Syrup for ice cream, 70
Wedding mints, 71
Pasta salad purchase, 71

Food costs determine menu prices, 82
Beef shrinkage, 82
Octoberfest brats purchase, 150
Soup in 1-litre containers, 150
Table top requirements, 240
Diluting chicken soup, 240
Cost using two types of ground beef, 240
Tomato paste recipe ratio, 257
Volume of water to beef broth ratio, 257
Cost per pound of pork loin, 257
Ratio of amount of potatoes per person, 257
Pork : beef ratio for meat loaf, 264
Number of bone-in prime rib cuts from same number of beef loins, 264
Amounts of ingredients to make given recipes, 264, 265
Amounts of ingredients to serve given number of people, 264
Cups of ingredients to make given number of servings, 265
Kitchen ratio, 265
Mixing different types of ground beef, 323
Seating of guests at tables, 323
Selling cups and bowls of chili, 323
Difference in area of banquet and dinner plates, 392
Wedding reception dinner seating, 392
Cookies ordered for special event, 392
Area of slice of pizza, 396
Batter and icing needed for sheet cakes, 402
Cylindrical stock pot capacity, 407
Batter needed for round wedding cake, 407

Electronics
Total resistance in series circuit, 8, 60
Ohm's Law, 9
Total current in parallel circuit, 38, 60
Load in circuit, 46
Voltage of electric iron, 46
Power used in drill, 46
Cable for wiring, 46
Current needed, 47
Length of wire needed, 47, 322
Outlet spacing, 47
Total resistance in a parallel circuit, 48, 247

LIST OF APPLICATIONS

Resistance in copper wire, 51
Voltage of source, 60
Inductive reactance in circuit, 70
Power in circuit, 70
Current in circuit, 70, 322
Resistance in flashlight bulb, 70
Resistance in lamp, 70
Current in heating element, 70
Line voltage surge, 81
Electronics parts invoice, 83
Electronics business overhead, 97
Current through one branch of parallel circuit, 177
Current draw in a drill, 247
Resistance in flashlight bulb, 247
Transformer voltage, 256
Ratio of voltage drops across resistors, 256
Transformer coil ratio, 256
Voltage drop in resistor, 256, 263
Resistance in copper wire, 263
Ratio of secondary turns to primary turns in transformer, 263
Size of two types of capacitors, 321
Batteries in series, 321
Current in branches of parallel circuit, 321
Electrolyte solution, 321
Size of two resistors, 322
Variable current, 350
Variable voltage, 350
Applied voltage, voltage across a coil, and voltage across a resistance in a circuit, 379
Total current, coil current, and resistor current, 379
Impedance, reactance, and resistance of a circuit, 379
Conduit length and angle, 437
Right triangle relationships in electrical circuits, 438, 439
Frequency of radar waves, 451
Wavelength of radio waves, 451

HVAC
Ductwork replacement cost, 8
Volume of circulated air, 17
Duct volume, 17
Furnace filter volume, 17
Cost of heating a building, 17
Duct length, 38, 46
Cooling requirements, 38
Pieces of duct, 46
Airflow in cubic feet per second, 51
Duct cost, 59, 69
Percent of moisture removed, 81
Air flow through duct, 81
Gas used over given period, 178
Airflow supply of unit, 178
Ventilation requirement CFM, 181
Furnace space, 181
Sections of duct for furnace, 181
Ratio of the BTU of two air conditioners, 257
Metal duct cost, 263
Current needed for compressors and air-handling units, 321
Flow of two air ducts, 322
Building dimensions, 322
Height and area of rectangular metal duct, 371
Length of sides of triangular duct, 380
Similar heater filter sizes, 384
Similar ducts, 384
Diameter of round metal duct, 389
Joining metal ducts, 413
Volume of coolant canister, 415
Duct length along stairs, 438
Lengths of ducts in kite shaped room, 466
Angles for placing air handlers, 466

Industrial/Construction Trades
Number of studs, 8
Cutting pipe, 8
Number of boards in order, 9
Space between walls and windows, 9, 10
Concrete blocks needed for wall, 10
Tiles needed for wall, 16
Number of ceiling tiles, 16
Gallons of paint needed, 16
Pieces of drywall needed, 16
Insurance for replacement cost, 16
Weight of cement floor, 17
Distance between floor joists, 39
Tap drill size, 39
Reducing diameter of shaft, 40
Difference in plate thickness, 40
Distance of house from sides of lot, 40
Thickness of plate after lathe pass, 40
Board feet of lumber, 46
Length of steel pipes, 46
Inside diameter of pipe, 46
Distance between rivets, 46
Distance between centers of circles, 46
Vent dimensions, 46
Volume of concrete pad, 46
Cutting a bar, 47
Weight of iron rods, 51
Mixing chemicals, 51
Difference of diameter ends of taper, 60
Thickness of pipe wall, 60
Cutting cable, 69
Building floor space, 69
Cost of excavation, 69
Number of days to complete job, 70
Increase in floor space, 82
Plumbing supplier invoice, 82
Lumberyard invoice, 83
Thickness of hole, 98
Shipping box design, 98
Sidewalk cost, 149
House lot in acres, 150
Thickness of sheets of metal, 177
Bookshelves construction, 239
Length of cut boards, 239, 321
Types of light fixtures, 239
Yard dimensions, 239
Mixing concrete, 239
Cutting a beam to meet specifications, 239
R value of insulation, 247
Copper tubing cost, 256
Ratio of wall area to window area, 256
Cost of home, 257, 263
Ratio of volume of concrete to volume of cement, 256
Amount of sand to make concrete, 263
Pitch of roof, 263
Number of bricks for wall, 263
Percent of volume of dry concrete mix of cement, sand, and gravel, 264
Capacity of two trucks, 321
Contractor testing pumps, 321
Working time of two bricklayers, 321
Number of each type of ceiling tiles, 321
Material for concrete, 322
Cutting squares of corners of rectangular material to form rectangular container, 351
Size of square sheet of aluminum to form rectangular container, 351
Increase length and width of lot with given increase in area, 351
Dimensions of warehouse to give maximum floor space, 352
Dimensions of storage building to minimize the outside walls, 352
Area of sheet metal, 372
Number of squares of shingles for roof, 372
Number of ceiling suspension panels, 372
Cost of painting a house, 372
Number of bricks needed for wall, 372
Length of support braces, 378
Depth of keyway cut, 378
Mill round stock into square stock, 378
Length of rafter, 379
Offset distance, 379
Length of conduit, 379
Length of ladder to reach window, 379
Area of hole cut in steel plate, 380
Braces for inclined ramp, 383
Dimensions of finished stock, 383
Length of bookcase crosspiece, 384
Length of tower guy wires, 384
Width of insulation wrapped around circular pipe, 389
Area of metal after circular holes are removed, 390
Length of strapping needed for pipe, 390
Satellite bracket design, 391
Area and volume of various parts of building, 400
Weight of rectangular piece of steel, 401
Volume of rectangular lead sleeve, 401
Volume of cylindrical tank, 405
Height of cylindrical tank, 405
Volume of cylindrical piece of steel, 405
Volume of refrigerant in copper tubing, 405
Sheet metal needed for sides of cylindrical tank, 406
Volume of lead in "pig", 406
Metal in cans, 406
Weight of circular tank, 412
Volume of gravel, 412
Plastic resin pellets hopper design, 413
Round stock tapered to cone, 413
Diameter of shut off ball float, 415
Gallons of water in spherical tank, 415
Ratio of surface area to volume in spherical tank, 415
Conveyor length, 437
Safety height of ladder, 437
Width of river, 437
Roadway inclination, 437
Length of guy wire, 437
Drilling holes in metal plate, 437
Height of TV relay tower, 438

LIST OF APPLICATIONS

Litres of liquid in right circular conical tank, 438
Distance between adjacent drilled holes, 439
Check dimension of dovetail, 439
Head angle of metal screw, 439
Length of roofline, 440
Height of building, 440
Lengths and angles in framing a roof, 465, 466

Manufacturing
Linear feet of pipe in inventory, 9
Distance between hazard stripes, 10
Drums of oil needed, 10
Length of shaft, 39, 40, 59
Distance between holes, 39
Length of rod, 39, 46, 69
Diameter of largest part, 40
Number of pins after cuts, 46
Lathe turn time, 46
Length of side of hexagon, 59
Distance of hole from end, 59
Find missing dimension, 60
Pitch of screw, 69
Sheet metal stack height, 69
Number of metal sheets, 69
Number of cuts needed to turn down metal stock, 69
Amperage requirement, 70
Weight of steel plate, 70
Number of defective tires in plant, 80
Defective resistors, 80
Hydraulic pressure, 81
Machinist pay increase, 82
Length of drying booth, 264
Diameter of pulleys, 322
Length and cost of fiberglass, 372
Floor area and cost of garage, 372
Machinist building a screen around shop area, 372
Area needed that is unavailable for manufacturing, 373
Fertilizer needed in shrub garden, 380
Material needed for water trough, 381
Manufacturing canisters to fit inside each other, 384
Reducing mold to scale, 385

Work station design, 390
Boiler placement in corner of room, 390
Length of pulley, 390
Central angle of equally spaced holes in metal plate, 390
Cardboard box design, 402
Volume of trash can, 402
Capacity of parts washer, 407
Cylindrical steel tanks capacity, 407
Cost cutting material, 413
Centers of equally spaced bolt holes in metal, 439
Length of antenna guy wire, 467

Natural Resources
Cruising timber, 10
Tilapia feed, 10
Volume of cordwood, 17
Volume of settling tank at wastewater plant, 17
Product weight, 38
Cords of firewood burned, 40
Homeowner lawn, 40
Hiking distance, 40
Allowance for kerf, 48
Crossing plants, 48
Tree harvested for firewood, 48
Truckloads of fish, 52
Convert lawn area to acres, 52
Using Biltmore stick to measure height of tree, 52
Population increase, 60
Fertilizer cost, 61
Petroleum reserves, 61
Municipal solid waste (MSW), 70, 82
Capacity of silo, 70
Volume of rick of firewood, 70
Weight of firewood, 82
Fish catch, 82
Survival rate of flock of ducks (sord), 82
Deer population, 82
Deer density, 82
Weight of trash for a week, 178
Weight of fish, 178
Water in shopping center parking lot, 181
CO_2 level in atmosphere, 181
Food waste compost, 181
Cubic miles of water in Cayuga Lake, 181

Collecting sea salt, 240
Deer and elk population control, 240
Pressure at bottom of lake, 247
Ratio of cougars per living area, 247
Fish farming feed-to-weight-gain ratio, 247
Salt contained in sea water, 264
Amount of N-P-K applied, 264
Gear ratio of fishing reel, 264
Amount of one inch of water over one acre, 264
Length of boards, 322
Difference in height of two waterfalls, 322
Mixing two types of grain for animal feed, 322
Dimensions of sod area, 352
Dimensions of forest plot, 352
Cross-sectional area of water in canal, 373
Area in game preserve, 373
Slope of hill, 381
Hiking distance, 381
Rock climbers estimate height of cliff, 385
Similar cat stretching posts, 385
Windmill blades, 391
Water sprinkler use, 391
Volume of swimming pool, 402
Fish tank design, 402
Oil pipeline volume, 407
Cylindrical silo capacity, 407
Wastewater treatment plant sediment tank capacity, 407
Obelisk design, 413
Volume of weather balloon, 416
Volume of air balloon, 416
Solar panels position, 441
Lean-to shelter design, 441
Width of jaw opening of snake, 467
Distance of kite from a person, 467
U.S. coal production, 482
Tree ring mean growth, 482
Tree ring thickness, 485
Worldwide coal production, 485

Welding
Length of welded pipe, 8, 37, 59
Argon gas used, 8
Volume of welded container, 15, 17
Length of welded piece, 37

Difference in diameter of welding rods, 37
Total length of weld, 45, 47
Cutting pieces of pipe, 45
Area of piece of sheet metal, 51
Total length of steel angle weld, 51
Steel angle divided into equal parts, 69
I-beam divided into equal parts, 69
Percent of welds completed, 81
Number of high quality welds, 81
Length of steel angle welds, 178
Weight of scrap metal, 178
Rods used in welds, 181
Volume of storage bin, 181
Ratio of steel angle pieces, 257
Ratio of welding rods, 257
Cost of welding rods, 263
Hours of work for each welder, 321
Earnings of experienced and beginning welders, 322
Dimensions of sheet metal to patch hole in large metal tank, 350
Area of side of welded metal storage bin, 371
Area of triangular gusset, 380
Similar support gussets, 384
Similar pieces of steel, 384
Area of lid in welded circular metal tank, 389
Radius of hole in metal, 389
Volume of gusset, 400
Metal duct volume and lateral surface area, 400
Volume of pyramid, 412
Fabricating storage compartments, 413
Volume of pan in shape of hemisphere, 415
Sheet metal trough capacity, 421
Length of support for a conveyor belt, 437
Angle of taper, 439
Measure of angles in triangular metal sheet, 465, 466

PREFACE

Elementary Technical Mathematics, Twelfth Edition, is intended for technical, trade, allied health, or Tech Prep programs. This book was written for students who plan to learn a technical skill, but who have minimal background in mathematics or need considerable review. To become proficient in most technical programs, students must learn basic mathematical skills. To that end, Chapters 1 through 4 cover basic arithmetic operations, fractions, decimals, percent, the metric system, and numbers as measurements. Chapters 5 through 11 present essential algebra needed in technical and trade programs. The essentials of geometry—relationships and formulas for the most common two- and three-dimensional figures—are given in detail in Chapter 12. Chapters 13 and 14 present a short but intensive study of trigonometry that includes right-triangle trigonometry as well as oblique triangles and graphing. The concepts of statistics that are most important to technical fields are discussed in Chapter 15. An introduction to binary and hexadecimal numbers is found in Chapter 16 for those who requested this material.

This text is written to match the reading level of most technical students. Visual images engage these readers and stimulate the problem-solving process. These skills are essential for success in technical courses. This text is written to be as flexible as possible for the wide range of student backgrounds and technical program needs. Sections may be easily combined for the better prepared class of students.

The following important text features have been retained from previous editions:

♦ A large number of applications are used from a wide variety of technical areas, including agriculture and horticulture, allied health, auto/diesel service, aviation, business and personal finance, CAD/drafting, culinary arts, electronics, HVAC, industrial and construction trades, manufacturing, natural resources, and welding.

♦ Chapter 1 reviews basic concepts in such a way that individuals, groups of students, or the entire class can easily study only those sections they need to review.

♦ A comprehensive introduction to basic algebra is presented for those students who need it as a prerequisite to more advanced algebra courses. However, the book has been written to allow the omission of selected sections or chapters without loss of continuity, to meet the needs of specific students.

♦ More than 6500 exercises assist student learning of skills and concepts.

♦ More than 750 detailed, well-illustrated examples, many with step-by-step comments, support student understanding of skills and concepts.

♦ Learning objectives are listed with each Chapter Opener to give a clear outline of topics covered in the chapter. This serves as a reference for students when completing homework assignments or studying for exams, and it also helps with lesson and assessment preparation for instructors.

- A chapter summary with a glossary of basic terms, a chapter review, and a chapter test appear at the end of each chapter as aids for students in preparing for quizzes and exams. Each chapter test is designed to be completed by an average student in no more than approximately 50 minutes.

REVIEW | CHAPTER 3

Give the metric prefix for each value:

1. 0.001
2. 1000

Give the SI abbreviation for each prefix:

3. mega
4. micro

Write the SI abbreviation for each quantity:

5. 42 millilitres
6. 8.3 nanoseconds

Write the SI unit for each abbreviation:

7. 18 km
8. 350 mA
9. 50 µs

Which is larger?

10. 1 L or 1 mL
11. 1 kW or 1 MW
12. 1 km² or 1 ha
13. 1 m³ or 1 L

Fill in each blank:

14. 650 m = _____ km
15. 750 mL = _____ L
16. 6.1 kg = _____ g
17. 4.2 A = _____ µA
27. Water boils at _____ °C.
28. 180 lb = _____ kg
29. 126 ft = _____ m
30. 360 cm = _____ in.
31. 275 in² = _____ cm²
32. 18 yd² = _____ ft²
33. 5 m³ = _____ ft³
34. 15.0 acres = _____ ha

Choose the most reasonable quantity:

35. Jorge and Maria drive **a.** 1600 cm, **b.** 470 m, **c.** 12 km, or **d.** 2400 mm to college each day.
36. Chuck's mass is **a.** 80 kg, **b.** 175 kg, **c.** 14 µg, or **d.** 160 Mg.
37. An automobile's fuel tank holds **a.** 18 L, **b.** 15 kL, **c.** 240 mL, or **d.** 60 L of gasoline.
38. Jamilla, being of average height, is **a.** 5.5 m, **b.** 325 mm, **c.** 55 cm, or **d.** 165 cm tall.
39. An automobile's average fuel consumption is **a.** 320 km/L, **b.** 15 km/L, **c.** 35 km/L, or **d.** 0.75 km/L.

TEST | CHAPTER 3

1. Give the metric prefix for 1000.
2. Give the metric prefix for 0.01.
3. Which is larger, 200 mg or 1 g?
4. Write the SI unit for the abbreviation 240 µL.
5. Write the abbreviation for 30 hectograms.
6. Which is longer, 1 km or 25 cm?

Fill in each blank:

7. 4.25 km = _____ m
8. 7.28 mm = _____ µm
9. 72 m = _____ mm
10. 256 hm = _____ cm
11. 12 dg = _____ mg
12. 16.2 g = _____ mg
13. 7.236 metric tons = _____ kg
14. 310 g = _____ cg
15. 72 hg = _____ mg
16. 1.52 dL = _____ L
17. 175 L = _____ m³
18. 2.7 m³ = _____ cm³
19. 400 ha = _____ km²
20. 0.2 L = _____ mL
21. What is the basic SI unit of time?
22. Write the abbreviation for 25 kilowatts.

Fill in each blank:

23. 280 W = _____ kW
24. 13.9 mA = _____ A
25. 720 ps = _____ ns
26. What is the basic SI unit for temperature?
27. What is the freezing temperature of water on the Celsius scale?

Fill in each blank, rounding each result to three significant digits when necessary:

28. 25°C = _____ °F
29. 28°F = _____ °C
30. 98.6°F = _____ °C
31. 100 km = _____ mi
32. 200 cm = _____ in.
33. 1.8 ft³ = _____ in³
34. 37.8 ha = _____ acres
35. 80.2 kg = _____ lb

- The text design and second color help to make the text more easily understood, highlight important concepts, and enhance the art presentation.
- A reference of useful, frequently referenced information—such as metric system prefixes, U.S. weights and measures, metric and U.S. conversion, and formulas from geometry—is printed on the inside covers.

xiv   PREFACE

- The use of a basic scientific calculator has been integrated in an easy-to-use format with calculator flowcharts and displays throughout the text to reflect its nearly universal use in technical classes and on the job. The instructor should inform the students when *not* to use a calculator.

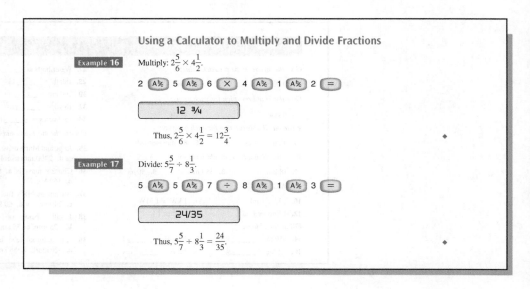

- Cumulative reviews are provided at the end of every even-numbered chapter to help students review for comprehensive exams.

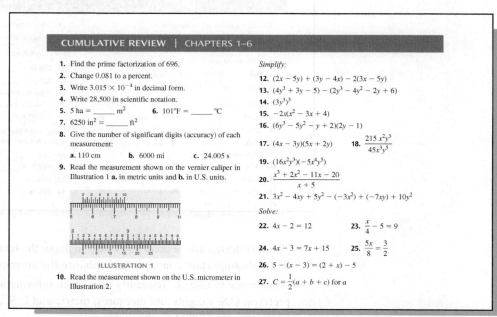

PREFACE

- Studies show that current students will experience several career changes during their working lives. The chapter-opening pages illustrate various career paths for students to consider as their careers, technology, and the workplace evolve. The basic information provided in the chapter openers about a technical career is explored in further detail on the Cengage book companion website at www.cengage.com/mathematics/ewen.

Mathematics at Work

Electronics technicians perform a variety of jobs. Electronic engineering technicians apply electrical and electronic theory and knowledge to design, build, test, repair, and modify experimental and production electrical equipment in industrial or commercial plants for use by engineering personnel in making engineering design and evaluation decisions.

Other electronics technicians repair electronic equipment such as industrial controls, telemetering and missile control systems, radar systems, and transmitters and antennas using testing instruments. Industrial controls automatically monitor and direct production processes on the factory floor. Transmitters and antennas provide communications links for many organizations. The federal government uses radar and missile control systems for national defense as well as other applications.

Electricians install, maintain, and repair wiring, equipment, and fixtures and ensure that work is in accordance with relevant codes. They also travel to locations to repair equipment and perform preventive maintenance on a regular basis. They use schematics and manufacturers' specifications that show connections and provide instructions on how to locate problems. They also use software programs and testing equipment to diagnose malfunctions. For more information, please visit **www.cengage.com** and access the Student Online Resources for this text.

Electronics Technician
Electronics technician checking a fuse box

- Special application exercises in the areas of agriculture and horticulture, allied health, auto/diesel service, aviation, business and personal finance, CAD/drafting, culinary arts, electronics, HVAC, industrial and construction trades, manufacturing, natural resources, and welding have been submitted by faculty in these technical areas and are marked with related icons.

44. Find the total piston displacement of a six-cylinder engine if each piston displaces 0.9 litres (L).

45. A four-cylinder engine has a total displacement of 2.0 L. Find the displacement of each piston.

46. An eight-cylinder engine has a total displacement of 318 in^3. Find the displacement of each piston.

47. New front brake pads are 0.375 in. thick. The average wear rate of these pads in a particular vehicle is 0.062 in. per 15,000 mi. Determine **a.** the expected wear after 45,000 mi and **b.** the expected pad thickness after 60,000 mi.

48. A certain job requires 500 person-hours to complete. How many days will it take for five people working 8 hours per day to complete the job?

49. How many gallons of herbicide are needed for 150 acres of soybeans if 1.6 gal/acre are applied?

50. Suppose 10 gal of water and 1.7 lb of pesticide are to be applied per acre. **a.** How much pesticide would you put in a 300-gal spray tank? **b.** How many acres can be covered with one tankful? (Assume the pesticide dissolves in the water and has no volume.)

51. A cattle feeder buys some feeder cattle, which average 550 lb at $145/hundredweight (that is, $145 per hundred pounds, or $1.45/lb). The price he receives when he sells them as slaughter cattle is $120/hundredweight. If he plans to make a profit of $150 per head, what will be his cost per pound for a 500-lb weight gain?

52. An insecticide is to be applied at a rate of 2 pt/100 gal of water. How many pints are needed for a tank that holds 20 gal? 60 gal? 150 gal? 350 gal? (Assume that the insecticide dissolves in the water and has no volume.)

59. A lamp that requires 0.84 A of current is connected to a 115-V source. What is the lamp's resistance? (Resistance equals voltage divided by current.)

60. A heating element operates on a 115-V line. If it has a resistance of 18 Ω, what current does it draw? (Current equals voltage divided by resistance.)

61. A patient takes 3 tablets of clonidine hydrochloride, containing 0.1 mg each. How many milligrams are taken?

62. One dose of ampicillin for a patient with bronchitis is 2 tablets each containing 0.25 g of medication. How many grams are in one dose?

63. An order reads 0.5 mg of digitalis, and each tablet contains 0.1 mg. How many tablets should be given?

64. An order reads 1.25 mg of digoxin, and the tablets on hand are 0.25 mg. How many tablets should be given?

65. A statute mile is 5280 ft. A nautical mile used in aviation is 6080.6 ft. This gives the conversion 1 statute mile = 0.868 nautical miles. If a plane flew 350 statute miles, how many nautical miles were flown?

66. Five lathes and four milling machines are to be on one circuit. If each lathe uses 16.0 A and each milling machine uses 13.8 A, what is the amperage requirement for this circuit?

67. A steel plate 1.00 in. thick weighs 40.32 lb/ft^2. Find the weight of a 4.00 ft × 8.00 ft sheet.

68. Municipal solid waste (MSW) consists basically of trash and recycle that is produced by nonindustrial and nonagricultural sources. According to Environmental Protection Agency estimates, as of 2014, each person in the United States generated an average of 4.44 lb of MSW each day. If you are an average American, how

- Group activity projects have been moved to the Instructor Companion website.
- An instructor's edition that includes all the answers to exercises is available.

Significant changes in the twelfth edition include the following:

- New and revised applications with the help and expertise of professionals in the areas of industrial and construction trades, electronics, and CAD/drafting.
- All areas have been reviewed and updated with current information and data.
- The material on measurement has been reorganized and revised to provide better rationale for measurement accuracy and precision and for calculations with measurements. Single versus multiple measurements are compared, and the concept of random and systematic errors have been introduced.
- Major effort was made to contain cost to students by having a more space-efficient page design, reviewing art size and placement, moving Group Activities from the end of each chapter to the Instructor Companion website, and deleting dial indicator material from Section 4.9 that seemingly was not being used.
- More than 140 exercises have been updated, replaced, or improved.

Useful ancillaries available to qualified adopters of this text include the following:

WEBASSIGN From Cengage www.webassign.com/cengage

- WebAssign from Cengage *Elementary Technical Mathematics*, Twelfth Edition, is a fully customizable online solution for STEM disciplines that empowers you to help your students learn, not just do homework. Insightful tools save you time and highlight exactly where your students are struggling. Decide when and what type of help students can access while working on assignments—and incentivize independent work so help features aren't abused. Meanwhile, your students get an engaging experience, instant feedback, and better outcomes. A total win-win!

  To try a sample assignment, learn about LMS integration or connect with our digital course support; visit www.webassign.com/cengage

- **Instructor's Edition** The Instructor's Edition features an appendix containing the answers to all problems in the book. (978-1-337-63059-7)
- **Instructor Solutions Manual** (ISBN: 978-1-337-63063-4): This guide contains solutions to every exercise in the book. You can download the solutions manual from the Instructor Companion Site.
- **Instructor Companion Website:** Everything you need for your course in one place! Access the Instructor Solutions Manual, full lecture PowerPoints, Group Projects, and other support materials. This collection of book-specific lecture and class tools is available via www.cengage.com/login

Student Resources:

WEBASSIGN From Cengage www.webassign.com

- Prepare for class with confidence using WebAssign from Cengage *Elementary Technical Mathematics*, Twelfth Edition. This online learning platform fuels practice, so you truly absorb what you learn—and are better prepared come test time. Videos and tutorials walk you through concepts and deliver instant feedback and grading, so you always know where you stand in class. Focus your study time and get extra practice where you need it most. Study smarter with WebAssign!

  Ask your instructor today how you can get access to WebAssign, or learn about self-study options at www.webassign.com

◆ **Student Solutions Manual**
Author: James Lapp
(ISBN: 978-1-337-63060-3)
The Student Solutions Manual provides worked-out solutions to all of the odd-numbered exercises in the text, as well as solutions to all chapter review and cumulative review exercises.

I am grateful for the courtesy of the L. S. Starrett Company in allowing the use of photographs of their instruments in Chapter 4. A special thank you to Sarah Alamilla, Waukesha County Technical College, and Taylor Moore, Joliet Junior College, for lending their professional expertise in reviewing and updating the applications.

I also thank the many faculty members who used earlier editions and who offered suggestions. In particular, I thank Sarah Alamilla, Waukesha County Technical College; Yelda Aydin-Mullen, Parkland College; Adebayo Badmos, Black Hawk College; Royetta Ealba, Henry Ford Community College; Ben Falero, Central Carolina Community College–Sanford Campus; Jared Harvey, Kennebec Valley Technical College; Vanessa Hill, Springfield Technical Community College; and Taylor Moore, Joliet Junior College.

Anyone wishing to correspond regarding suggestions or questions should contact Dale Ewen through the publisher.

For all their help, I thank our Product Manager Rita Lombard, Content Developer Powell Vacha, and Product Assistant Abby DeVeuve. I am especially grateful to Senior Content Project Manager Corinna Dibble and Project Manager Lori Hazzard of MPS Limited for their professional commitment to quality, to James Lapp for his thorough work authoring the solutions manuals, and to Scott Barnett for his outstanding work and attention to the details of accuracy checking and proofreading.

Finally, I thank my friend and colleague of many years C. Robert Nelson for his work on all of the previous editions and wish him the very best.

Dale Ewen

CENGAGE.com

Cengage.com is the smart move when it comes to getting the right stuff on time, every time. Whether you rent or buy, we'll save you time, money, and frustration.

- **You've Got Options:** Convenient digital solutions and textbooks the way you want them — to buy or rent.

- **You Get Access:** Anytime, anywhere access of digital products, eBooks, and eChapters, on your desktop, laptop, or phone.

- **You Get Free Stuff:** Free 14-day eBook access, free shipping on orders of $25 or more, free study tools like flashcards and quizzes, and a free trial period for most digital products.

Look, we get it. You've got a full schedule — we've got your back(pack). Get what you need to get the grades at Cengage.com

Basic Concepts

CHAPTER 1

OBJECTIVES

- Add, subtract, multiply, and divide whole numbers.
- Add, subtract, multiply, and divide whole numbers with a basic scientific calculator.
- Apply the rules for order of operations.
- Find the area and volume of geometric figures.
- Evaluate formulas.
- Find the prime factorization of whole numbers.
- Add, subtract, multiply, and divide fractions.
- Add, subtract, multiply, and divide fractions with a basic scientific calculator.
- Use conversion factors to change from one unit to another within the U.S. system of weights and measures.
- Add, subtract, multiply, and divide decimal fractions.
- Add, subtract, multiply, and divide decimal fractions with a basic scientific calculator.
- Round numbers to a particular place value.
- Apply the percent concept; change a percent to a decimal, a decimal to a percent, a fraction to a percent, and a percent to a fraction.
- Solve application problems involving the addition, subtraction, multiplication, and division of whole numbers, fractions, and decimal fractions and percents.
- Find powers and roots of numbers using a scientific calculator.
- Solve personal finance problems involving percent.

Mathematics at Work

Modern manufacturing companies require a wide variety of technology specialists for their operations. Manufacturing technology specialists set up, operate, and maintain industrial and manufacturing equipment as well as computer-numeric-controlled (CNC) and other automated equipment that make a large variety of products according to controlled specifications. Some focus on systematic equipment maintenance and repair. Others specialize in materials transportation and distribution; that is, they are responsible for moving and distributing the products to the sales locations and/or consumers after they are manufactured. Other key team members include designers, engineers, draftspersons, and quality control specialists. Training and education for these careers are available at many community colleges and trade schools. Some require a bachelor's degree. For more information, please visit **www.cengage.com** and access the Student Online Resources for this text.

Manufacturing Technology Specialist
Technician working with numerically controlled milling machine

Dmitry Kalinovsky/Shutterstock.com

UNIT 1A Review of Operations with Whole Numbers

1.1 Review of Basic Operations

The **positive integers** are the numbers 1, 2, 3, 4, 5, 6, and so on. They can also be written as +1, +2, +3, and so on, but usually the *positive* (+) sign is omitted. The **whole numbers** are the numbers 0, 1, 2, 3, 4, 5, 6, and so on. That is, the whole numbers consist of the positive integers and zero.

The value of any digit in a number is determined by its place in the particular number. Each place represents a certain power of 10. By powers of 10, we mean the following:

$10^0 = 1$
$10^1 = 10$
$10^2 = 10 \times 10 = 100$ (the second power of 10)
$10^3 = 10 \times 10 \times 10 = 1000$ (the third power of 10)
$10^4 = 10 \times 10 \times 10 \times 10 = 10,000$ (the fourth power of 10) and so on.

NOTE: A small superscript number (such as the 2 in 10^2) is called an *exponent*.

The number 2354 means 2 thousands plus 3 hundreds plus 5 tens plus 4 ones.

In the number 236,895,174, each digit has been multiplied by some power of 10, as shown below.

	(ten millions)		(hundred thousands)		(thousands)		(tens)	
	10^7		10^5		10^3		10^1	
2	3	6,	8	9	5,	1	7	4
10^8		10^6		10^4		10^2		10^0
(hundred millions)		(millions)		(ten thousands)		(hundreds)		(units)

The "+" (plus) symbol is the sign for addition, as in the expression $5 + 7$. The result of adding the numbers (in this case, 12) is called the **sum**. Integers are added in columns with the digits representing like powers of 10 in the same vertical line. (*Vertical* means up and down.)

Example 1 Add: $238 + 15 + 9 + 3564$.

$$\begin{array}{r} 238 \\ 15 \\ 9 \\ \underline{3564} \\ 3826 \end{array}$$

♦

Subtraction is the inverse operation of addition. Therefore, subtraction can be thought of in terms of addition. The "−" (minus) sign is the symbol for subtraction. The quantity $5 − 3$ can be thought of as "what number added to 3 gives 5?" The result of subtraction is called the **difference**.

To check a subtraction, add the difference to the second number. If the sum is equal to the first number, the subtraction has been done correctly.

Example 2 Subtract: 2843 − 1928.

Subtract:
```
 2843    first number
-1928    second number
  915    difference
```

Check:
```
 1928    second number
+ 915    difference
 2843    This sum equals the first number, so
         915 is the correct difference.
```

Next, let's study some applications. To communicate about problems in electricity, technicians have developed a "language" of their own. It is a picture language that uses symbols and diagrams. The symbols used most often are listed in Table 2 of Appendix A. An electric circuit is a conducting loop in which electrons carrying electric energy may be transferred from a source to do useful work and returned to the source. The circuit diagram is the most common and useful way to show an electric circuit. Note how each component (part) of the picture (Figure 1.1a) is represented by its symbol in the circuit diagram (Figure 1.1b) in the same relative position.

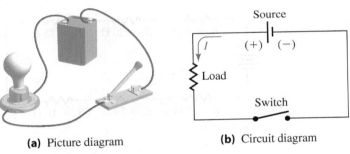

(a) Picture diagram (b) Circuit diagram

Figure 1.1
Components in an electric circuit

The light bulb may be represented as a resistance. Then the circuit diagram in Figure 1.1b would appear as in Figure 1.2, where

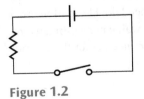

Figure 1.2

—⋀⋀⋀— represents the resistor 
—o o— represents the switch 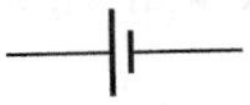
—|⊢— represents the source. The short line represents the negative terminal of a battery, and the long line represents the positive terminal. The current flows from positive to negative.

Energy is stored in the battery. When the switch is closed, energy is transferred to the light, and the light glows.

NOTE: In this book assume that the charge carriers are positive and draw current arrows in the direction that a positive charge would flow. This is a common practice used by most technicians and engineers. However, you may find the negative-charge–current-flow convention is also used in some books. Regardless of the convention used, the formulas and results are the same.

4   CHAPTER 1 ♦ Basic Concepts

There are two basic types of electric circuits: series and parallel. An electric circuit with only one path for the current, *I*, to flow is called a *series* circuit (Figure 1.3a). An electric circuit with more than one path for the current to flow is called a *parallel* circuit (Figure 1.3b). A circuit breaker or fuse in a house is wired in series with its outlets. The outlets themselves are wired in parallel.

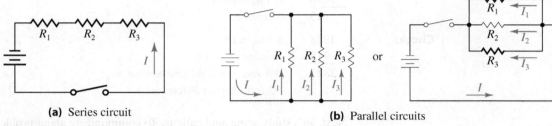

(a) Series circuit   (b) Parallel circuits

Figure 1.3
Two basic types of electric circuits

Example 3   In a series circuit, the total resistance equals the sum of all the resistances in the circuit. Find the total resistance in the series circuit in Figure 1.4. Resistance is measured in ohms, Ω.

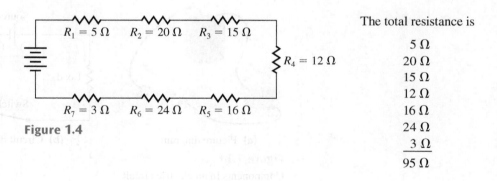

Figure 1.4

The total resistance is

5 Ω
20 Ω
15 Ω
12 Ω
16 Ω
24 Ω
 3 Ω
95 Ω

♦

Example 4   Studs are upright wooden or metal pieces in the walls of a building, to which siding, insulation panels, drywall, or decorative paneling is attached. (A wall portion with seven studs is shown in Figure 1.5.) Studs are normally placed 16 in. on center and are placed double at all internal and external corners of a building. The number of studs needed in a wall can be estimated by finding the number of linear feet (ft) of the wall. How many studs are needed for the exterior walls of the building in Figure 1.6?

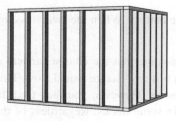   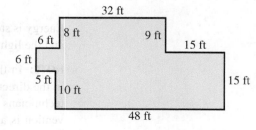

Figure 1.5   **Figure 1.6**

The outside perimeter of the building is the sum of the lengths of the sides of the building:

48 ft
15 ft
15 ft
9 ft
32 ft
8 ft
6 ft
6 ft
5 ft
10 ft
154 ft

Therefore, approximately 154 studs are needed in the outside wall. ◆

Repeated addition of the same number can be shortened by multiplication. The "×" (times) and the "·" (raised dot) are used to indicate multiplication. When adding the lengths of five pipes, each 7 ft long, we have 7 ft + 7 ft + 7 ft + 7 ft + 7 ft = 35 ft of pipe. In multiplication, this would be 5 × 7 ft = 35 ft. The 5 and 7 are called *factors*. The result of multiplying numbers (in this case, 35) is called the **product**. Computing areas, volumes, forces, and distances requires skills in multiplication.

Example 5 Multiply: 358 × 18.

$$\begin{array}{r} 358 \\ \times 18 \\ \hline 2864 \\ 358 \\ \hline 6444 \end{array}$$

◆

Division is the inverse operation of multiplication. The following symbols are used to show division: 15 ÷ 5, $5\overline{)15}$, 15/5, and $\frac{15}{5}$. The quantity 15 ÷ 5 can also be thought of as "what number times 5 gives 15?" The answer to this question is 3, which is 15 divided by 5. The result of dividing numbers (in this case, 3) is called the **quotient**. The number to be divided, 15, is called the *dividend*. The number you divide by, 5, is called the *divisor*.

Example 6 Divide: 84 ÷ 6.

$$\begin{array}{r} 14 \leftarrow \text{quotient} \\ 6\overline{)84} \leftarrow \text{dividend} \\ \underline{6} \\ 24 \\ \underline{24} \\ 0 \leftarrow \text{remainder} \end{array}$$

divisor ↑

Example 7 Divide: 115 ÷ 7.

$$\begin{array}{r} 16 \leftarrow \text{quotient} \\ 7\overline{)115} \leftarrow \text{dividend} \\ \underline{7} \\ 45 \\ \underline{42} \\ 3 \leftarrow \text{remainder} \end{array}$$

divisor ↑

◆

The *remainder* (when not 0) is usually written in one of two ways: with an "r" preceding it or with the remainder written over the divisor as a fraction, as shown in Example 8. (Fractions are discussed in Unit 1B.)

Example 8 Divide: 534 ÷ 24.

$$24\overline{)534} \quad \begin{array}{r} 22\,r\,6 \\ 48 \\ \hline 54 \\ 48 \\ \hline 6 \end{array} \quad \text{or} \quad 22\tfrac{6}{24} \qquad \text{This quotient may be written } 22\,r\,6 \text{ or } 22\tfrac{6}{24}.$$

Example 9 Ohm's law states that in a simple electric circuit, the current I (measured in amps, A) equals the voltage E (measured in volts, V) divided by the resistance R (measured in ohms, Ω). Find the current in the circuit of Figure 1.7.

$$\text{The current } I = \frac{E}{R} = \frac{110}{22} = 5 \text{ A}.$$

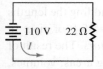

Figure 1.7

Example 10 A 16-row corn planter costs $128,500. It has a 10-year life and a salvage value of $10,000. What is the annual depreciation? (Use the straight-line depreciation method.)

The straight-line depreciation method means that the difference between the cost and the salvage value is divided evenly over the life of the item. In this case, the difference between the cost and the salvage value is

$128,500	cost
−$10,000	salvage
$118,500	difference

This difference divided by 10, the life of the item, is $11,850. This is the annual depreciation.

Example 11 Restaurants purchase potatoes to use for baked potatoes. The potatoes are often called bakers and can come in cases containing 90, 120, and so on, potatoes. If 3 cases of bakers with 90 potatoes per case are ordered plus 2 cases of bakers with 120 potatoes per case, how many total individual bakers are ordered?

3 cases × 90 potatoes/case = 270 potatoes
2 cases × 120 potatoes/case = <u>240 potatoes</u>
Total   510 potatoes

Using a Scientific Calculator

Use of a scientific calculator is integrated throughout this text. To demonstrate how to use a common scientific calculator, we show which keys to use and the order in which they are pushed. We have chosen to illustrate the most common types of algebraic logic calculators. Yours may differ. If so, consult your manual.

NOTE: Your calculator should be cleared before you begin any calculation.

1.1 ♦ Review of Basic Operations   7

Use a calculator to add, subtract, multiply, and divide as shown in the following examples.

Example 12 Add: 9463
               125
                 9
                80

9463 [+] 125 [+] 9 [+] 80 [=]

[9677]

The sum is 9677.

Example 13 Subtract: 3500
                          1628

3500 [−] 1628 [=]

[1872]

The result is 1872.

Example 14 Multiply: 125×68.

125 [×] 68 [=]

[8500]

The product is 8500.

Example 15 Divide: $8700 \div 15$.

8700 [÷] 15 [=]

[580]

The quotient is 580.

NOTE: Your instructor will indicate which exercises should be completed using a calculator.

EXERCISES 1.1

Add:

1. $832 + 9 + 56 + 2358$

2. $324 + 973 + 66 + 9430$

3. 384
     291
     147
     632

4. 78
     107
      45
     217
       9
     123

5. $197 + 1072 + 10{,}877 + 15{,}532 + 768{,}098$

6. $160{,}000 + 19{,}000 + 4{,}160{,}000 + 506{,}000$

Subtract and check:

7. 7561
     2397

8. 4000
      702

9. $98{,}405 - 72{,}397$

10. $417{,}286 - 287{,}156$

11. 4000
      1180

12. 60,000
       9,876

Find the total resistance in each series circuit:

13. $R_1 = 460\,\Omega$, $R_2 = 825\,\Omega$, $R_3 = 750\,\Omega$, $R_4 = 1500\,\Omega$, $R_5 = 650\,\Omega$, $R_6 = 10\,\Omega$

14. $R_1 = 3600\,\Omega$, $R_2 = 560\,\Omega$, $R_3 = 75\,\Omega$, $R_4 = 100\,\Omega$, $R_5 = 1200\,\Omega$, $R_6 = 575\,\Omega$, $R_7 = 5\,\Omega$, $R_8 = 2500\,\Omega$

15. Approximately how many studs are needed for the exterior walls in the building shown in Illustration 1? (See Example 4.)

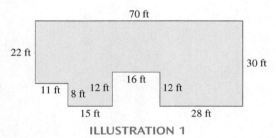

ILLUSTRATION 1

16. A pipe 24 ft long is cut into four pieces: the first 4 ft long, the second 5 ft long, and the third 7 ft long. What is the length of the remaining piece? (Assume no waste from cutting.)

17. A welder needs to weld together pipes of lengths 10 ft, 15 ft, and 14 ft. What is the total length of the new pipe?

18. A welder ordered a 125-ft^3 cylinder of argon gas, a semi-inert shielding gas for TIG welding. After a few days, only 78 ft^3 remained. How much argon was used?

19. Find the total input and output (I-O) in cubic centimetres (cm^3)* for a patient. By how much does the input of fluids exceed the output?

Input: 300 cm^3, 550 cm^3, 150 cm^3, 75 cm^3, 150 cm^3, 450 cm^3, 250 cm^3

Output: 325 cm^3, 150 cm^3, 525 cm^3, 250 cm^3, 175 cm^3

20. A student pilot must complete 40 h of total flight time as required for her private pilot certificate. She had already entered 31 h of flight time in her logbook. Monday she logged another 2 h, then Wednesday she logged another 3 h, and Friday she logged yet another 2 h. If she can fly 3 h more on Saturday, will she have enough total time as required for the certificate?

Multiply:

21. 567 × 48

22. 8374 × 203

23. 71,263 × 255

24. 1520 × 320

25. 6800 × 5200

26. 30,010 × 4080

Divide (use the remainder form with r):

27. $4\overline{)7236}$

28. $5\overline{)308{,}736}$

29. 4668 ÷ 12

30. 15,648 ÷ 36

31. 67,560 ÷ 80

32. $\dfrac{188{,}000}{120}$

33. An automobile uses gasoline at the rate of 31 miles per gallon (mi/gal or mpg) and has a 16-gallon tank. How far can it travel on one tank of gas?

34. An automobile uses gasoline at a rate of 12 kilometres per litre (km/L) and has a 65-litre tank. How far can it travel on one tank of gas?

35. A four-cylinder engine has a total displacement of 1300 cm^3. Find the displacement of each piston.

36. An automobile travels 1274 mi and uses 49 gal of gasoline. Find its mileage in miles per gallon.

37. An automobile travels 2340 km and uses 180 L of gasoline. Find its fuel consumption in kilometres per litre.

38. To replace some damaged ductwork, 20 linear feet of 8-in. × 16-in. duct is needed. The cost is $13 per 4 linear feet. What is the cost of replacement?

39. The bill for a new transmission was received. The total cost for labor was $516. If the car was serviced for 6 h, find the cost of labor per hour.

40. The cost for a set of four tires is $596. What is the cost of each tire?

41. A small Cessna aircraft has enough fuel to fly for 4 h. If the aircraft cruises at a ground speed of 125 miles per hour (mi/h or mph), how many miles can the aircraft fly in the 4 h?

42. A small plane takes off and climbs at a rate of 500 ft/min. If the plane levels off after 15 min, how high is the plane?

---

*Although cm^3 is the "official" metric abbreviation for cubic centimetres and will be used throughout this book, some readers may be more familiar with the abbreviation "cc," which is still used in some medical and allied health areas.

43. Inventory shows the following lengths of 3-inch steel pipe:

5 pieces 18 ft long
42 pieces 15 ft long
158 pieces 12 ft long
105 pieces 10 ft long
79 pieces 8 ft long
87 pieces 6 ft long

What is the total linear feet of pipe in inventory?

44. An order of lumber contains 36 boards 12 ft long, 28 boards 10 ft long, 36 boards 8 ft long, and 12 boards 16 ft long. How many boards are contained in the order? How many linear feet of lumber are contained in the order?

45. Two draftspersons, operating the same computer plotter, each work 8 hours per day. One produces 80 drawings per hour; the other produces 120 drawings per hour. What is the difference in their outputs after 30 work days?

46. A shipment contains a total of 5232 linear feet of steel pipe. Each piece of pipe is 12 ft long. How many pieces should be expected?

47. The wall is 10 ft high and the vertical length of the window is 54 in. The center of the window needs to be at a distance of 5/8 of the height of the wall above the floor (to meet the special Fibonacci ratio criteria). How should a window 75 in. wide be horizontally placed so that it is centered on a wall 17 ft 5 in. wide? How high is the bottom of the window above the floor?

48. A farmer expects a yield of 165 bushels per acre (bu/acre) from 260 acres of corn. If the corn is stored, how many bushels of storage are needed?

49. A farmer harvests 6864 bushels (bu) of soybeans from 156 acres. What is his yield per acre?

50. A railroad freight car can hold 2035 bu of corn. How many freight cars are needed to haul the expected 12,000,000 bu from a local grain elevator?

51. On a given day, eight steers weighed 856 lb, 754 lb, 1044 lb, 928 lb, 888 lb, 734 lb, 953 lb, and 891 lb. **a.** What is the average weight? **b.** In 36 days, 4320 lb of feed is consumed. What is the average feed consumption per day per steer?

52. What is the weight (in tons) of a stack of hay bales 6 bales wide, 110 bales long, and 15 bales high? The average weight of each bale is 80 lb. (1 ton = 2000 lb.)

53. From a 34-acre field, 92,480 lb of oats are harvested. Find the yield in bushels per acre. (1 bu of oats weighs 32 lb.)

54. A standard bale of cotton weighs approximately 500 lb. How many bales are contained in 15 tons of cotton?

55. A tractor costs $175,000. It has a 10-year life and a salvage value of $3000. What is the annual depreciation? (Use the straight-line depreciation method. See Example 10.)

56. How much pesticide powder would you put in a 400-gal spray tank if 10 gal of spray, containing 2 lb of pesticide, are applied per acre?

57. Daylilies are to be planted along one side of a 30-ft walk in front of a house. The daylilies are planted 5 in. from each end and 10 in. apart along the walk. How many daylilies are needed?

58. A potato patch has 7 rows with 75 hills of potatoes per row. If each potato hill yields 3 lb of marketable potatoes, how many pounds of marketable potatoes were produced?

Using Ohm's law, find the current I in amps (A) in each electric circuit (see Example 9):

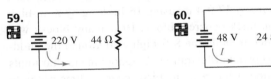

59. 220 V, 44 Ω

60. 48 V, 24 Ω

Ohm's law, in another form, states that in a simple circuit the voltage E (measured in volts, V) equals the current I (measured in amps, A) times the resistance R (measured in ohms, Ω). Find the voltage E measured in volts (V) in each electric circuit:

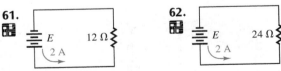

61. E, 12 Ω, 2 A

62. E, 24 Ω, 2 A

63. A hospital dietitian determines that each patient needs 4 ounces (oz) of orange juice. How many ounces of orange juice must be prepared for 220 patients?

64. During 24 hours, a patient is to receive three 60-mg doses of phenobarbital. Each tablet contains 30 mg of phenobarbital. How many milligrams of phenobarbital does the patient receive altogether in 24 hours? How many pills does the patient take in 24 hours?

65. To give 800 mg of quinine sulfate from 200-mg tablets, how many tablets would you use?

66. A nurse used two 5-g tablets of potassium permanganate in the preparation of a medication. How much potassium permanganate was used?

67. A sun room addition to a home has a wall 14 ft 6 in. long measured from inside wall to inside wall. Four windows are to be equally spaced from each other in this wall. The windows are 2 ft 6 in. wide including the inside window molding. What is the space between the wall and windows shown in Illustration 2?

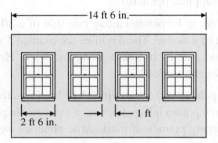

ILLUSTRATION 2

68. A solid concrete block wall is being built around a rectangular storage building with outside dimensions 12 ft 8 in. by 17 ft 4 in. using 16-in.-long by 8-in.-high by 4-in.-thick concrete block. How many blocks will be needed to build the 8-ft-high wall around the building as shown in Illustration 3? (Ignore the mortar joints and a door.) Suggestion: First, change dimensions to inches.

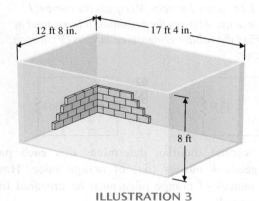

ILLUSTRATION 3

69. A sheet of plywood 8 ft long is painted with three equally spaced stripes to mark off a hazardous area as shown in Illustration 4. If each stripe is 10 in. wide, what is the space between the end of the plywood and the first stripe?

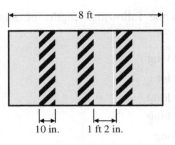

ILLUSTRATION 4

70. In a small machine shop, eight 5-gallon drums of oil are on hand. If 2 gallons are used each day and the owner wants a 30-day supply on hand, how many drums should be ordered?

71. Using a process called "cruising timber," foresters can estimate the amount of lumber in board feet in trees before they are cut down. In a stand of 1000 trees, a forester selects a representative sample of 100 trees and estimates that the sample contains 8540 board feet of lumber. If the entire stand containing 2500 trees is harvested, how many board feet would the landowner expect to harvest?

72. In tilapia aquaculture production, a feed conversion ratio of 2 lb of high-protein pelleted feed per pound of weight gain, after death losses, is not unusual. At that rate of feed conversion, if fish food costs $520 per ton (2000 lb), what would be the feed cost per pound of live fish produced?

73. A banquet facility has section areas separated with folding walls. Section A has a total of 50 seats, Section B has a total of 125 seats, Section C has a total of 110 seats, and Section D has a total of 35 seats. What is the maximum number of guests who could be seated using all sections?

74. Each beef export loin will yield 11 prime rib servings. **a.** How many beef export loins are needed to serve prime rib to 125 guests? **b.** How many total end cut servings are possible for these guests?

75. A delivery to the kitchen includes the following: 2 boxes of bakers (potatoes, 90 per box), 3 boxes of beef roast 109s (prime ribs, 4 per box), and 2 boxes of pork loins (pork, 4 per box). How many total individual items are in all the boxes?

76. The Sun Rise Restaurant has 10 tables that seat 6 people each plus 12 tables that seat 4 people each. Each server is assigned at most 6 tables. What is the least number of servers needed when the restaurant is full of customers?

77. The wait staff decides to pool (combine and divide evenly) their tips for the evening shift. The three servers have rounded tips of: $131, $152, and $128. **a.** What is the total amount to be divided? **b.** How much does each server receive?

1.2 Order of Operations

The expression 5^3 means to use 5 as a factor 3 times. We say that 5^3 is the third *power* of 5, where 5 is called the *base* and 3 is called the *exponent*. Here, 5^3 means $5 \times 5 \times 5 = 125$. The expression 2^4 means that 2 is used as a factor 4 times; that is, $2^4 = 2 \times 2 \times 2 \times 2 = 16$. Here, 2^4 is the fourth power of 2.

Just as we use periods, commas, and other punctuation marks to help make sentences more readable, we use **grouping symbols** in mathematics such as *parentheses* "( )" and *brackets* "[ ]" to help clarify the meaning of mathematical expressions. Parentheses not only give an expression a particular meaning, they also specify the order to be followed in evaluating and simplifying expressions.

What is the value of $8 - 3 \cdot 2$? Is it 10? Is it 2? Or is it some other number? It is very important that each mathematical expression have only one value. For this to happen, we all must not only perform the exact *same operations* in a given mathematical expression or problem but also perform them in exactly the *same order*. The following order of operations is followed by all:

Order of Operations

1. Always do the operations within parentheses or other grouping symbols first.

2. Then evaluate each power, if any. Examples:

$$4 \times 3^2 = 4 \times (3 \times 3) = 4 \times 9 = 36$$
$$5^2 \times 6 = (5 \times 5) \times 6 = 25 \times 6 = 150$$
$$\frac{5^3}{6^2} = \frac{5 \times 5 \times 5}{6 \times 6} = \frac{125}{36}$$

3. Next, perform multiplications and divisions in the order in which they appear as you read from left to right. For example,

$$60 \times 5 \div 4 \div 3 \times 2$$
$$= 300 \div 4 \div 3 \times 2$$
$$= 75 \div 3 \times 2$$
$$= 25 \times 2$$
$$= 50$$

4. Finally, perform additions and subtractions in the order in which they appear as you read from left to right.

NOTE: If two parentheses or a number and a parenthesis occur next to one another without any sign between them, multiplication is indicated.

By using the above procedure, we find that $8 - 3 \cdot 2 = 8 - 6 = 2$.

Example 1 Evaluate: $2 + 5(7 + 6)$.

	$= 2 + 5(13)$	Add within parentheses.
	$= 2 + 65$	Multiply.
	$= 67$	Add.

NOTE: A number next to parentheses indicates multiplication. In Example 1, 5(13) means 5×13. Adjacent parentheses also indicate multiplication: (5)(13) also means 5×13. ◆

Example 2 Evaluate: $(9 + 4) \times 16 + 8$.

$$= 13 \times 16 + 8 \quad \text{Add within parentheses.}$$
$$= 208 + 8 \quad \text{Multiply.}$$
$$= 216 \quad \text{Add.}$$

♦

Example 3 Evaluate: $(6 + 1) \times 3 + (4 + 5)$.

$$= 7 \times 3 + 9 \quad \text{Add within parentheses.}$$
$$= 21 + 9 \quad \text{Multiply.}$$
$$= 30 \quad \text{Add.}$$

♦

Example 4 Evaluate: $4(16 - 4) + \dfrac{14}{7} - 8$.

$$= 4(12) + \dfrac{14}{7} - 8 \quad \text{Subtract within parentheses.}$$
$$= 48 + 2 - 8 \quad \text{Multiply and divide.}$$
$$= 42 \quad \text{Add and subtract.}$$

♦

Example 5 Evaluate: $7 + (6 - 2)^2$.

$$= 7 + 4^2 \quad \text{Subtract within parentheses.}$$
$$= 7 + 16 \quad \text{Evaluate the power.}$$
$$= 23 \quad \text{Add.}$$

♦

Example 6 Evaluate: $25 - 3 \cdot 2^3$.

$$= 25 - 3 \cdot 8 \quad \text{Evaluate the power.}$$
$$= 25 - 24 \quad \text{Multiply.}$$
$$= 1 \quad \text{Subtract.}$$

♦

Example 7 Evaluate: $\dfrac{6^2 \cdot 4 - 2 \cdot 3^3}{5^2 + 5 \cdot 2^2}$

$$= \dfrac{36 \cdot 4 - 2 \cdot 27}{25 + 5 \cdot 4} \quad \text{Evaluate each power.}$$
$$= \dfrac{144 - 54}{25 + 20} \quad \text{Multiply.}$$
$$= \dfrac{90}{45} \quad \text{Subtract in the numerator and add in the denominator.}$$
$$= 2 \quad \text{Divide.}$$

NOTE: You must evaluate the numerator and the denominator separately before you divide in the last step.

♦

If pairs of parentheses are nested (parentheses within parentheses, or within brackets), work from the innermost pair of parentheses to the outermost pair. That is, remove the innermost parentheses first, remove the next innermost parentheses second, and so on.

Example 8 Evaluate: $6 \times 2 + 3[7 + 4(8 - 6)]$.

$$= 6 \times 2 + 3[7 + 4(2)] \quad \text{Subtract within parentheses.}$$
$$= 6 \times 2 + 3[7 + 8] \quad \text{Multiply.}$$
$$= 6 \times 2 + 3[15] \quad \text{Add within brackets.}$$
$$= 12 + 45 \quad \text{Multiply.}$$
$$= 57 \quad \text{Add.}$$

♦

EXERCISES 1.2

Evaluate each expression:

1. $8 - 3(4 - 2)$
2. $(8 + 6)4 + 8$
3. $(8 + 6) - (7 - 3)$
4. $4 \times (2 \times 6) + (6 + 2) \div 4$
5. $2(9 + 5) - 6 \times (13 + 2) \div 9$
6. $5(8 \times 9) + (13 + 7) \div 4$
7. $27 + 13 \times (7 - 3)(12 + 6) \div 9$
8. $123 - 3(8 + 9) + 17$
9. $16 + 4(7 + 8) - 3$
10. $(18 + 17)(12 + 9) - (7 \times 16)(4 + 2)$
11. $9 - 2(17 - 15) + 18$
12. $(9 + 7)5 + 13$
13. $(39 - 18) - (23 - 18)$
14. $5(3 \times 7) + (8 + 4) \div 3$
15. $3(8 + 6) - 7(13 + 3) \div 14$
16. $6(4 \times 5) + (15 + 9) \div 6$
17. $42 + 12(9 - 3)(12 + 13) \div 30$
18. $228 - 4 \times (7 + 6) - 8(6 - 2)$
19. $38 + 9 \times (8 + 4) - 3(5 - 2)$
20. $(19 + 8)(4 + 3) \div 21 + (8 \times 15) \div (4 \times 3)$
21. $27 - 2 \times (18 - 9) - 3 + 8(43 - 15)$
22. $6 \times 8 \div 2 \times 9 \div 12 + 6$
23. $12 \times 9 \div 18 \times 64 \div 8 + 7$
24. $18 \div 6 \times 24 \div 4 \div 6$
25. $7 + 6(3 + 2) - 7 - 5(4 + 2)$
26. $5 + 3(7 \times 7) - 6 - 2(4 + 7)$
27. $3 + 17(2 \times 2) - 67$
28. $8 - 3(9 - 2) \div 21 - 7$
29. $28 - 4(2 \times 3) + 4 - (16 \times 8) \div (4 \times 4)$
30. $6 + 4(9 + 6) + 8 - 2(7 + 3) - (3 \times 12) \div 9$
31. $24/(6 - 2) + 4 \times 3 - 15/3$
32. $(36 - 6)/(5 + 10) + (16 - 1)/3$
33. $3 \times 15 \div 9 + (13 - 5)/2 \times 4 - 2$
34. $28/2 \times 7 - (6 + 10)/(6 - 2)$
35. $10 + 4^2$
36. $4 + 2 \cdot 3^2$
37. $\dfrac{3 \cdot 5 + 6 \cdot 8}{53 - 2 \cdot 5^2}$
38. $\dfrac{3 \cdot 4 + 2 \cdot 3}{42 - 20 \cdot 2 \cdot 1}$
39. $\dfrac{20 + (2 \cdot 3)^2}{7 \cdot 2^3}$
40. $\dfrac{(20 - 2 \cdot 5)^2}{3^3 - 2}$
41. $6[3 + 2(2 + 5)]$
42. $5((4 + 6) + 2(5 - 2))$
43. $5 \times 2 + 3[2(5 - 3) + 4(4 + 2) - 3]$
44. $3(10 + 2(1 + 3(2 + 6(4 - 2))))$

1.3 Area and Volume

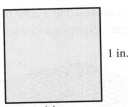

1 square inch (in²)

1 square centimetre (cm²)

Figure 1.8
Square units

To measure the length of an object, you must first select a suitable standard unit of length. To measure short lengths, choose a unit such as centimetres or millimetres in the metric system, or inches in the United States or, as it is still sometimes called, the English system. For long distances, choose metres or kilometres in the metric system, or yards or miles in the U.S. system.

Area

The **area** of a plane geometric figure is the number of square units of measure it contains. To measure the surface area of an object, first select a standard unit of area suitable to the object to be measured. Standard units of area are based on the square and are called square units. For example, a square inch (in²) is the amount of surface area within a square that measures one inch on a side. A square centimetre (cm²) is the amount of surface area within a square that is 1 cm on a side. (See Figure 1.8.)

Example 1 What is the area of a rectangle measuring 4 cm by 3 cm?

Each square in Figure 1.9 represents 1 cm². By simply counting the number of squares, you find that the area of the rectangle is 12 cm².

14 CHAPTER 1 ♦ Basic Concepts

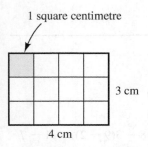

Figure 1.9

You can also find the area by multiplying the length times the width:

Area = $l \times w$
　　　= 4 cm × 3 cm = 12 cm^2　　Note: cm × cm = cm^2
　　　(length)　(width)

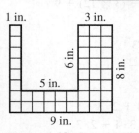

Figure 1.10

Example 2 What is the area of the metal plate represented in Figure 1.10?

Each square represents 1 square inch. By simply counting the number of squares, we find that the area of the metal plate is 42 in^2.

Another way to find the area of the figure is to find the areas of two rectangles and then find their difference, as in Figure 1.11.

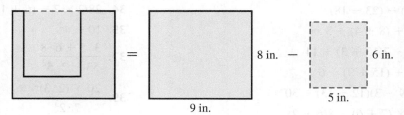

Figure 1.11

Area of outer rectangle: 9 in. × 8 in. = 72 in^2
Area of inner rectangle: 5 in. × 6 in. = 30 in^2
Area of metal plate:　　　　　　　　= 42 in^2　　Subtract. ♦

Volume

The **volume** of a solid geometric figure is the number of cubic units of measure it contains. In area measurement, the standard units are based on the square and called square units. For volume measurement, the standard units are based on the cube and called cubic units. For example, a cubic inch (in^3) is the amount of space contained in a cube that measures 1 in. on each edge. A cubic centimetre (cm^3) is the amount of space contained in a cube that measures 1 cm on each edge. A cubic foot (ft^3) is the amount of space contained in a cube that measures 1 ft (or 12 in.) on each edge. (See Figure 1.12.)

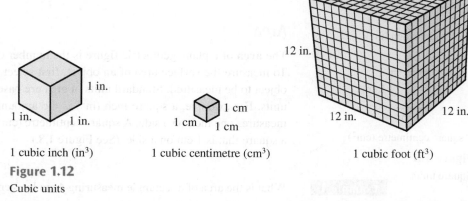

Figure 1.12
Cubic units

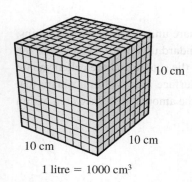

Figure 1.13
Litre

Figure 1.13 shows that the cubic decimetre (litre) is made up of 10 layers, each containing 100 cm^3, for a total of 1000 cm^3.

1.3 ♦ Area and Volume

Example 3 Find the volume of a rectangular box 8 cm long, 4 cm wide, and 6 cm high.

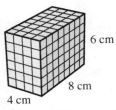
Figure 1.14

Suppose you placed one-centimetre cubes in the box, as in Figure 1.14. On the bottom layer, there would be 8×4, or 32, one-cm cubes. In all, there are six such layers, or $6 \times 32 = 192$ one-cm cubes. Therefore, the volume is 192 cm³.

You can also find the volume of a rectangular solid by multiplying the length times the width times the height:

$$V = l \times w \times h$$
$$= 8 \text{ cm} \times 4 \text{ cm} \times 6 \text{ cm}$$
$$= 192 \text{ cm}^3$$

Note: cm × cm × cm = cm³ ♦

Example 4 How many cubic inches are in one cubic foot?

The bottom layer of Figure 1.15 contains 12×12, or 144, one-inch cubes. There are 12 such layers, or $12 \times 144 = 1728$ one-inch cubes. Therefore, 1 ft³ = 1728 in³. ♦

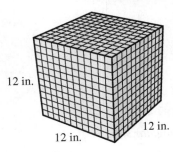
Figure 1.15
Cubic foot

EXERCISES 1.3

1. How many square yards (yd²) are contained in a rectangle 12 yd long and 8 yd wide?
2. How many square metres (m²) are contained in a rectangle 12 m long and 8 m wide?
3. At a small airport, Runway 11-29 is 4100 ft long and 75 ft wide. What is the area of the runway?
4. A small rectangular military operating zone has dimensions 12 mi by 22 mi. What is its area?
5. An automobile measures 191 in. by 73 in. Find the area it occupies.
6. Five pieces of sheet metal have been cut to form a container. The five pieces are of sizes 27 by 15, 15 by 18, 27 by 18, 27 by 18, and 15 by 18 (all in inches). What is the total area of all five pieces?

In exercises assume that corners are square and that like measurements are not repeated because the figures have consistent lengths. Note: All three of the following mean that the length of a side is 3 cm:

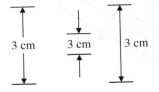

Find the area of each figure:

7.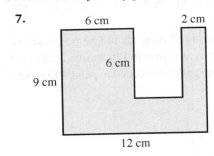

8.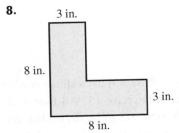

9.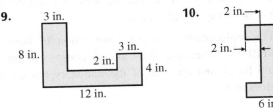

10.

11. 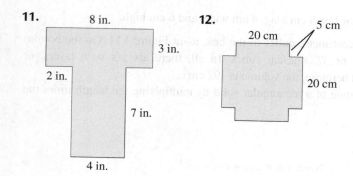 **12.**

13. How many tiles 4 in. on a side should be used to cover a portion of a wall 48 in. long by 36 in. high? (See Illustration 1.)

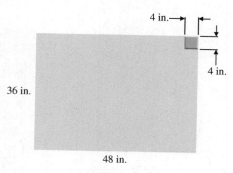

ILLUSTRATION 1

14. How many ceiling tiles 2 ft by 4 ft are needed to tile a ceiling that is 24 ft by 26 ft? (Be careful how you arrange the tiles.)

15. How many gallons of paint should be purchased to paint 20 motel rooms as shown in Illustration 2? (Do not paint the floor.) One gallon is needed to paint 400 square feet (ft²).

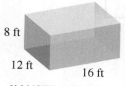

ILLUSTRATION 2

16. How many pieces of 4-ft by 8-ft drywall are needed for the 20 motel rooms in Exercise 15? All four walls and the ceiling in each room are to be drywalled. Assume that the drywall cut out for windows and doors cannot be salvaged and used again.

17. The replacement cost for construction of houses is $110/ft². Determine how much house insurance should be carried on each of the one-story houses in Illustration 3.

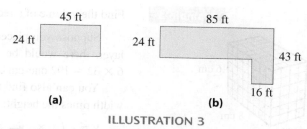

ILLUSTRATION 3

18. The replacement cost for construction of the building in Illustration 4 is $90/ft². Determine how much insurance should be carried for full replacement.

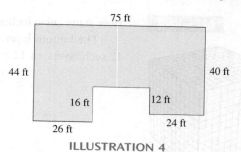

ILLUSTRATION 4

Find the volume of each rectangular solid:

19. **20.**

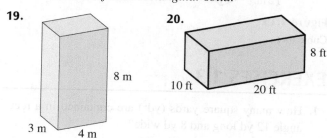

21.

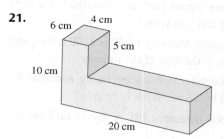

22.

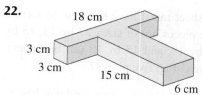

23.

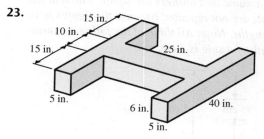

24.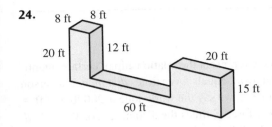

25. Find the volume of a rectangular box 10 cm by 12 cm by 5 cm.

26. 🏠 A mountain cabin has a single room 20 ft by 10 ft by 8 ft high. What is the total volume of air in the room that will be circulated through the central ventilating fan?

27. 🏠 Common house duct is 8-in. by 20-in. rectangular metal duct. If the length of a piece of duct is 72 in., what is its volume?

28. 🏠 A furnace filter measures 16 in. by 20 in. by 1 in. What is its volume?

29. ✦ A large rectangular tank is to be made of sheet metal as follows: 3 ft by 5 ft for the top and the base, two sides consisting of 2 ft by 3 ft, and two sides consisting of 2 ft by 5 ft. Find the volume of this container.

30. 🛢 Suppose an oil pan has the rectangular dimensions 14 in. by 16 in. by 4 in. Find its volume.

31. 🧱 Find the weight of a cement floor that is 15 ft by 12 ft by 2 ft if 1 ft^3 of cement weighs 193 lb.

32. A trailer 5 ft by 6 ft by 5 ft is filled with coal. Given that 1 ft^3 of coal weighs 40 lb and 1 ton = 2000 lb, how many tons of coal are in the trailer?

33. A rectangular tank is 8 ft long by 5 ft wide by 6 ft high. Water weighs approximately 62 lb/ft^3. Find the weight of water if the tank is full.

34. A rectangular tank is 9 ft by 6 ft by 4 ft. Gasoline weighs approximately 42 lb/ft^3. Find the weight of gasoline if the tank is full.

35. 🏠 A building is 100 ft long, 50 ft wide, and 10 ft high. Estimate the cost of heating it at the rate of $55 per 1000 ft^3.

36. 💻 A single-story shopping center is being designed to be 483 ft long by 90 ft deep. Two stores have been preleased. One occupies 160 linear feet and the other will occupy 210 linear feet. The owner is trying to decide how to divide the remaining space, knowing that the smallest possible space should be 4000 ft^2. How many stores can occupy the remaining area as shown in Illustration 5?

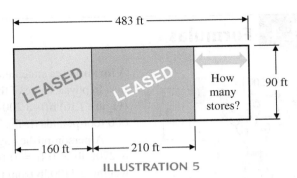

ILLUSTRATION 5

37. 💻 A trophy company needs a shipping box for a trophy 15 in. high with an 8-in.-square base. The box company is drawing the die to cut the cardboard for this box. **a.** What are the dimensions of the smallest rectangular sheet of cardboard needed to make one box that allows 1 in. for packing and 1 in. for a glue edge as shown in Illustration 6? **b.** What is the area of the rectangular cardboard to be mass produced? **c.** What is the total area of the cardboad to be removed?

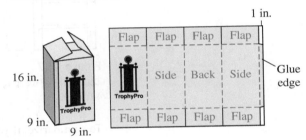

ILLUSTRATION 6

38. 💻 Styrofoam "peanuts" will be used to pack the trophy in the box in Illustration 6 to prevent the trophy from being broken during shipment. Ignoring the box wall thickness, how many cubic inches of peanuts including air space will be used for each package if the volume of the trophy is 450 in^3?

39. ✦ A standard cord of wood measures 4 ft by 4 ft by 8 ft. What is the volume in cubic feet of a cord of wood?

40. ♻ A municipal wastewater treatment plant has a settling tank that is 125 ft long and 24 ft wide with an effective depth of 12 ft. What is the surface area of the liquid in the tank and what is the volume of sewerage that the settling tank will hold?

41. 🌱 Three inches of mulch need to be applied to a rectangular flower bed 8 ft by 24 ft between a house and a walk. How many cubic feet of mulch are needed?

42. 🌱 How many 4 × 4 inch plant containers can be placed in a greenhouse on a table 4 ft wide and 8 ft long?

1.4 Formulas

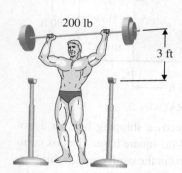

Figure 1.16
600 ft-lb of work is done moving this 200-lb weight a distance of 3 ft.

A **formula** is a statement of a rule using letters to represent the relationship of certain quantities. In physics, one of the basic rules states that *work* equals *force* times *distance*. If a person (Figure 1.16) lifts a 200-lb weight a distance of 3 ft, we say the work done is 200 lb × 3 ft = 600 foot-pounds (ft-lb). The work, W, equals the force, f, times the distance, d, or $W = f \times d$.

A person pushes against a car weighing 2700 lb but does not move it. The work done is 2700 lb × 0 ft = 0 ft-lb. An automotive technician (Figure 1.17) moves a diesel engine weighing 1100 lb from the floor to a workbench 4 ft high. The work done in moving the engine is 1100 lb × 4 ft = 4400 ft-lb.

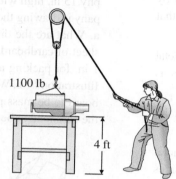

Figure 1.17
4400 ft-lb of work is done moving this 1100-lb diesel engine from the floor to this workbench 4 ft high.

To summarize, if you know the amount of force and the distance the force is applied, the work can be found by simply multiplying the force and distance. The formula $W = f \times d$ is often written $W = f \cdot d$, or simply $W = fd$. Whenever there is no symbol between a number and a letter or between two letters, it is assumed that the operation to be performed is multiplication.

Example 1 If $W = fd$, $f = 10$, and $d = 16$, find W.

$$W = fd$$
$$W = (10)(16)$$
$$W = 160 \quad \text{Multiply.}$$

Example 2 If $I = \dfrac{E}{R}$, $E = 450$, and $R = 15$, find I.

$$I = \frac{E}{R}$$
$$I = \frac{450}{15}$$
$$I = 30 \quad \text{Divide.}$$

Example 3 If $P = I^2 R$, $I = 3$, and $R = 600$, find P.

$$P = I^2 R$$
$$P = (3)^2(600)$$
$$P = (9)(600) \quad \text{Evaluate the power.}$$
$$P = 5400 \quad \text{Multiply.}$$

There are many other formulas used in science and technology. Some examples are given here:

a. $d = vt$ c. $f = ma$ e. $I = \dfrac{E}{R}$

b. $W = IEt$ d. $P = IE$ f. $P = \dfrac{V^2}{R}$

Formulas from Geometry

The area of a triangle is given by the formula $A = \frac{1}{2}bh$, where b is the length of the base and h, the height, is the length of the altitude to the base (Figure 1.18). (An altitude of a triangle is a line from a vertex perpendicular to the opposite side, as shown in (a), or to the opposite side extended, as shown in (b).)

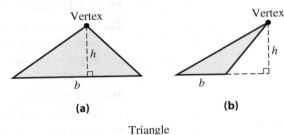

Triangle

Figure 1.18

Example 4 Find the area of a triangle whose base is 18 in. and whose height is 10 in.

$A = \dfrac{1}{2}bh$

$A = \dfrac{1}{2}(18 \text{ in.})(10 \text{ in.})$

$\quad = 90 \text{ in}^2$ Note: (in.)(in.) = in^2

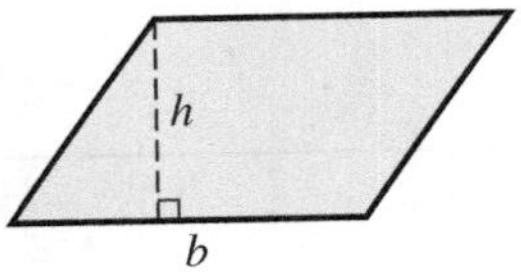

The area of a *parallelogram* (a four-sided figure whose opposite sides are parallel) is given by the formula $A = bh$, where b is the length of the base and h is the perpendicular distance between the base and its opposite side (Figure 1.19).

Parallelogram
Figure 1.19

Example 5 Find the area of a parallelogram with base 24 cm and height 10 cm.

$A = bh$

$A = (24 \text{ cm})(10 \text{ cm})$

$\quad = 240 \text{ cm}^2$ Note: (cm)(cm) = cm^2

The area of a *trapezoid* (a four-sided figure with one pair of parallel sides) is given by the formula $A = \left(\dfrac{a+b}{2}\right)h$, where a and b are the lengths of the parallel sides (called *bases*), and h is the perpendicular distance between the bases (Figure 1.20).

Trapezoid
Figure 1.20

Example 6 Find the area of the trapezoid in Figure 1.21.

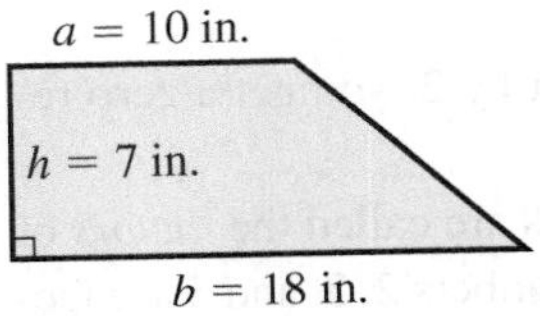

$A = \left(\dfrac{a+b}{2}\right)h$

$A = \left(\dfrac{10 \text{ in.} + 18 \text{ in.}}{2}\right)(7 \text{ in.})$

$\quad = \left(\dfrac{28 \text{ in.}}{2}\right)(7 \text{ in.})$ Add within parentheses.

$\quad = (14 \text{ in.})(7 \text{ in.})$ Divide.

$\quad = 98 \text{ in}^2$ Multiply.

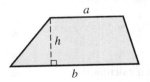

$a = 10$ in.

$h = 7$ in.

$b = 18$ in.

Figure 1.21

EXERCISES 1.4

Use the formula $W = fd$, where f represents a force and d represents the distance that the force is applied. Find the work done, W:

1. $f = 30, d = 20$
2. $f = 17, d = 9$
3. $f = 1125, d = 10$
4. $f = 203, d = 27$
5. $f = 176, d = 326$
6. $f = 2400, d = 120$

*From formulas **a.** to **f.** on page 19, choose one that contains all the given letters. Then use the formula to find the unknown letter:*

7. If $m = 1600$ and $a = 24$, find f.
8. If $V = 120$ and $R = 24$, find P.
9. If $E = 120$ and $R = 15$, find I.
10. If $v = 372$ and $t = 18$, find d.
11. If $I = 29$ and $E = 173$, find P.
12. If $I = 11, E = 95$, and $t = 46$, find W.

Find the area of each triangle:

13. $b = 10$ in., $h = 8$ in.
14. $b = 36$ cm, $h = 20$ cm
15. $b = 54$ ft, $h = 30$ ft
16. $b = 188$ m, $h = 220$ m

Find the area of each rectangle:

17. $b = 8$ m, $h = 7$ m
18. $b = 24$ in., $h = 15$ in.
19. $b = 36$ ft, $h = 18$ ft
20. $b = 250$ cm, $h = 120$ cm

Find the area of each trapezoid:

21. $a = 7$ ft, $b = 9$ ft, $h = 4$ ft
22. $a = 30$ in., $b = 50$ in., $h = 24$ in.
23. $a = 96$ cm, $b = 24$ cm, $h = 30$ cm
24. $a = 450$ m, $b = 750$ m, $h = 250$ m

25. The volume of a rectangular solid is given by $V = lwh$, where l is the length, w is the width, and h is the height of the solid. Find V if $l = 25$ cm, $w = 15$ cm, and $h = 12$ cm.

26. Find the volume of a box with dimensions $l = 48$ in., $w = 24$ in., and $h = 96$ in.

27. Given $v = v_0 + gt$, $v_0 = 12$, $g = 32$, and $t = 5$, find v.

28. Given $Q = CV$, $C = 12$, and $V = 2500$, find Q.

29. Given $I = \dfrac{E}{Z}$, $E = 240$, and $Z = 15$, find I.

30. Given $P = I^2 R$, $I = 4$, and $R = 2000$, find P.

1.5 Prime Factorization

Divisibility

A number is **divisible** by a second number if, when you divide the first number by the second number, you get a zero remainder. That is, the second number *divides* the first number.

Example 1 12 is divisible by 3, because 3 divides 12. ♦

Example 2 124 is not divisible by 7, because 7 does not divide 124. Check with a calculator. ♦

There are many ways of classifying the positive integers. They can be classified as even or odd, as divisible by 3 or not divisible by 3, as larger than 10 or smaller than 10, and so on. One of the most important classifications involves the concept of a **prime number**: an integer greater than 1 that has no divisors except itself and 1. The first ten prime numbers are 2, 3, 5, 7, 11, 13, 17, 19, 23, and 29.

An integer is **even** if it is divisible by 2; that is, if you divide it by 2, you get a zero remainder. An integer is **odd** if it is not divisible by 2.

In multiplying two or more positive integers, the positive integers are called the *factors* of the product. Thus, 2 and 5 are factors of 10, since $2 \times 5 = 10$. The numbers 2, 3, and 5 are factors of 30, since $2 \times 3 \times 5 = 30$. Those prime numbers whose product equals the given integer are called **prime factors**. The process of finding the prime factors of a positive integer is called **prime factorization**. The prime factorization of a given number is the product of its prime factors. That is, each of the factors is prime, and their product equals the given number. One of the most useful applications of prime factorization is in finding the least common denominator (LCD) when adding and subtracting fractions. This application is found in Section 1.7.

Example 3 Factor 28 into prime factors.

a. $28 = 4 \cdot 7$
    $= 2 \cdot 2 \cdot 7$

b. $28 = 7 \cdot 4$
    $= 7 \cdot 2 \cdot 2$

c. $28 = 2 \cdot 14$
    $= 2 \cdot 7 \cdot 2$

In each case, you have three prime factors of 28; one factor is 7, the other two are 2's. The factors may be written in any order, but we usually list all the factors in order from smallest to largest. It would not be correct in the examples above to leave $7 \cdot 4$, $4 \cdot 7$, or $2 \cdot 14$ as factors of 28, since 4 and 14 are not prime numbers. ♦

Short division, a condensed form of long division, is a helpful way to find prime factors. Find a prime number that divides the given number. Divide, using short division. Then find a second prime number that divides the result. Divide, using short division. Keep repeating this process of stacking the quotients and divisors (as shown below) until the final quotient is also prime. The prime factors will be the product of the divisors and the final quotient of the repeated short divisions.

Example 4 Find the prime factorization of 45.

$3\underline{|45}$ Divide by 3.
$3\underline{|15}$ Divide by 3.
5

The prime factorization of 45 is $3 \cdot 3 \cdot 5$. ♦

Example 5 Find the prime factorization of 60.

$2\underline{|60}$ Divide by 2.
$2\underline{|30}$ Divide by 2.
$3\underline{|15}$ Divide by 3.
5

The prime factorization of 60 is $2 \cdot 2 \cdot 3 \cdot 5$. ♦

Example 6 Find the prime factorization of 17.

17 has no factors except for itself and 1. Thus, 17 is a prime number. When asked for factors of a prime number, write "prime" as your answer. ♦

Divisibility Tests

To eliminate some of the guesswork involved in finding prime factors, divisibility tests can be used. Such tests determine whether or not a particular positive integer divides another integer without carrying out the division. Divisibility tests and prime factorization are used to reduce fractions to lowest terms and to find the lowest common denominator. (See Unit 1B.)

The following divisibility tests for certain positive integers are most helpful.

Divisibility by 2

If a number ends with an even digit, then the number is divisible by 2.

NOTE: Zero is even.

Example 7 Is 4258 divisible by 2?

Yes; since 8, the last digit of the number, is even, 4258 is divisible by 2.

NOTE: Check each example with a calculator.

Example 8 Is 215,517 divisible by 2?

Since 7 (the last digit) is odd, 215,517 is not divisible by 2.

Divisibility by 3

If the sum of the digits of a number is divisible by 3, then the number itself is divisible by 3.

Example 9 Is 531 divisible by 3?

The sum of the digits $5 + 3 + 1 = 9$. Since 9 is divisible by 3, then 531 is divisible by 3.

Example 10 Is 551 divisible by 3?

The sum of the digits $5 + 5 + 1 = 11$. Since 11 is not divisible by 3, then 551 is not divisible by 3.

Divisibility by 5

If a number has 0 or 5 as its last digit, then the number is divisible by 5.

Example 11 Is 2372 divisible by 5?

The last digit of 2372 is neither 0 nor 5, so 2372 is not divisible by 5.

Example 12 Is 3210 divisible by 5?

The last digit of 3210 is 0, so 3210 is divisible by 5.

Example 13 Find the prime factorization of 204.

$2\underline{|204}$ Last digit is even.
$2\underline{|102}$ Last digit is even.
$3\underline{|\ 51}$ Sum of digits is divisible by 3.
$\ \ \ \ 17$

The prime factorization of 204 is $2 \cdot 2 \cdot 3 \cdot 17$.

Example 14 Find the prime factorization of 630.

$2\underline{|630}$ Last digit is even.
$3\underline{|315}$ Sum of digits is divisible by 3.
$3\underline{|105}$ Sum of digits is divisible by 3.
$5\underline{|\ 35}$ Last digit is 5.
$\ \ \ \ 7$

The prime factorization of 630 is $2 \cdot 3 \cdot 3 \cdot 5 \cdot 7$.

NOTE: As a general rule of thumb:

1. Keep dividing by 2 until the quotient is not even.
2. Keep dividing by 3 until the quotient's sum of digits is not divisible by 3.
3. Keep dividing by 5 until the quotient does not end in 0 or 5.

That is, if you divide out all the factors of 2, 3, and 5, the remaining factors, if any, will be much smaller and easier to work with, and perhaps prime.

NOTE: Some people prefer to use the divisibility tests for 2 and 5 before using the divisibility test for 3.

EXERCISES 1.5

Which numbers are divisible a. by 3 and b. by 4?

1. 15
2. 28
3. 96
4. 172
5. 78
6. 675

Classify each number as prime or not prime:

7. 53
8. 57
9. 93
10. 121
11. 16
12. 123
13. 39
14. 87

Test for divisibility by 2:

15. 458
16. 12,746
17. 315,817
18. 877,778
19. 1367
20. 1205

Test for divisibility by 3 and check your results with a calculator:

21. 387
22. 1254
23. 453,128
24. 178,213
25. 218,745
26. 15,690

Test for divisibility by 5 and check your results with a calculator:

27. 70
28. 145
29. 366
30. 56,665
31. 63,227
32. 14,601

Test the divisibility of each first number by the second number:

33. 56; 2
34. 42; 3
35. 218; 3
36. 375; 5
37. 528; 5
38. 2184; 3
39. 198; 3
40. 2236; 3
41. 1,820,670; 2
42. 2,817,638; 2
43. 7,215,720; 5
44. 5,275,343; 3

Find the prime factorization of each number (use divisibility tests where helpful):

45. 20
46. 18
47. 66
48. 30
49. 36
50. 25
51. 27
52. 59
53. 51
54. 56
55. 42
56. 63
57. 120
58. 72
59. 171
60. 360
61. 105
62. 78
63. 252
64. 444

UNIT 1A REVIEW

1. Add: 33 + 104 + 75 + 29
2. Subtract: 2301 − 506
3. Multiply: 3709 × 731
4. Divide: 9300 ÷ 15
5. Josh has the following lengths of 3-inch plastic pipe:

   3 pieces 12 ft long
   8 pieces 8 ft long
   9 pieces 10 ft long
   12 pieces 6 ft long

   Find the total length of pipe on hand.
6. If one bushel of corn weighs 56 lb, how many bushels are contained in 14,224 lb of corn?

Evaluate each expression:

7. 6 + 2(5 × 4 − 2)
8. $3^2 + 12 \div 3 - 2 \times 3$
9. 12 + 2[3(8 − 2) − 2(3 + 1)]
10. In Illustration 1, find the area.

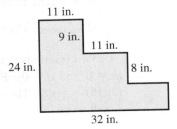

ILLUSTRATION 1

11. In Illustration 2, find the volume.

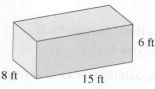

ILLUSTRATION 2

12. If $d = vt$, $v = 45$, and $t = 4$, find d.

13. If $I = \dfrac{E}{R}$, $E = 120$, and $R = 12$, find I.

14. If $A = \dfrac{1}{2}bh$, $b = 40$, and $h = 15$, find A.

Classify each number as prime or not prime:

15. 51   **16.** 47

Test for divisibility of each first number by the second number:

17. 195; 3   **18.** 821; 5

Find the prime factorization of each number:

19. 40   **20.** 135

UNIT 1B Review of Operations with Fractions

1.6 Introduction to Fractions

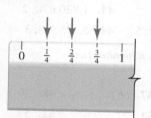

Figure 1.22

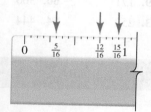

Figure 1.23

The U.S. system of measurement, which is derived from and sometimes called the English system, is a system whose units are often expressed as common fractions and mixed numbers. The metric system of measurement is a system whose units are expressed as decimal fractions and powers of 10. We must feel comfortable working with both systems of measurement and working with both fractions and decimals. The metric system is developed in Chapter 3.

A **common fraction** may be defined as the ratio or quotient of two integers (say, a and b) in the form $\dfrac{a}{b}$ (where $b \neq 0$). Examples are $\dfrac{1}{2}$, $\dfrac{7}{11}$, $\dfrac{3}{8}$, and $\dfrac{37}{22}$. The integer below the line is called the *denominator*. It gives the denomination (size) of equal parts into which the fraction unit is divided. The integer above the line is called the *numerator*. It numerates (counts) the number of times the denominator is used. Look at one inch on a ruler, as shown in Figures 1.22 and 1.23.

$\dfrac{1}{4}$ in. means 1 of 4 equal parts of an inch.

$\dfrac{2}{4}$ in. means 2 of 4 equal parts of an inch.

$\dfrac{3}{4}$ in. means 3 of 4 equal parts of an inch.

$\dfrac{5}{16}$ in. means 5 of 16 equal parts of an inch.

$\dfrac{12}{16}$ in. means 12 of 16 equal parts of an inch.

$\dfrac{15}{16}$ in. means 15 of 16 equal parts of an inch.

Two fractions $\dfrac{a}{b}$ and $\dfrac{c}{d}$ are *equal* or *equivalent* if $ad = bc$. That is, $\dfrac{a}{b} = \dfrac{c}{d}$ if $ad = bc$ ($b \neq 0$ and $d \neq 0$). For example, $\dfrac{2}{4}$ and $\dfrac{8}{16}$ are names for the same fraction, because $(2)(16) = (4)(8)$. There are many other names for this same fraction, such as $\dfrac{1}{2}$, $\dfrac{9}{18}$, $\dfrac{10}{20}$, $\dfrac{5}{10}$, $\dfrac{3}{6}$, and so on.

$\dfrac{2}{4} = \dfrac{1}{2}$, because $(2)(2) = (4)(1)$ $\dfrac{2}{4} = \dfrac{5}{10}$, because $(2)(10) = (4)(5)$

Figure 1.24 may help you visualize the relative sizes of fractions. Note that $\frac{1}{2} = \frac{2}{4} = \frac{4}{8} = \frac{8}{16}$, $\frac{7}{8}$ is greater than $\frac{3}{4}$, and $\frac{3}{16}$ is less than $\frac{1}{4}$.

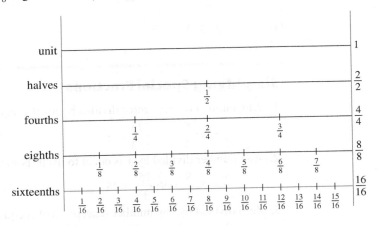

Figure 1.24
Relative sizes of fractions

We have two rules for finding equal (or *equivalent*) fractions:

Equal or Equivalent Fractions

1. The numerator and denominator of any fraction may be *multiplied* by the same number (except zero) without changing the value of the given fraction, thus producing an equivalent fraction. For example, $\frac{4}{9} = \frac{4 \cdot 5}{9 \cdot 5} = \frac{20}{45}$.

2. The numerator and denominator of any fraction may be *divided* by the same number (except zero) without changing the value of the given fraction. For example, $\frac{6}{10} = \frac{6 \div 2}{10 \div 2} = \frac{3}{5}$.

We use these rules for equivalent fractions (a) to reduce a fraction to lowest terms and (b) to change a fraction to higher terms when adding and subtracting fractions with different denominators.

To *simplify* a fraction means to find an equivalent fraction whose numerator and denominator are *relatively prime*—that is, a fraction whose numerator and denominator have no common divisor. This is also called **reducing a fraction to lowest terms**.

To reduce a fraction to lowest terms, write the prime factorization of the numerator and the denominator. Then divide (cancel) numerator and denominator by any pair of common factors. You may find it helpful to use the divisibility tests in Section 1.5 to write the prime factorizations.

Example 1 Simplify: $\dfrac{35}{50}$.

$$\frac{35}{50} = \frac{\cancel{5} \cdot 7}{2 \cdot \cancel{5} \cdot 5} = \frac{7}{10}$$ Note the use of the divisibility test for 5. A last digit of 0 or 5 indicates the number is divisible by 5. ◆

Example 2 Simplify: $\dfrac{63}{99}$.

$$\frac{63}{99} = \frac{\cancel{3} \cdot \cancel{3} \cdot 7}{\cancel{3} \cdot \cancel{3} \cdot 11} = \frac{7}{11}$$ Note the use of the divisibility test for 3 twice. ◆

Example 3 Simplify: $\frac{84}{300}$.

$$\frac{84}{300} = \frac{\cancel{2}\cdot\cancel{2}\cdot\cancel{3}\cdot 7}{\cancel{2}\cdot\cancel{2}\cdot\cancel{3}\cdot 5\cdot 5} = \frac{7}{25}$$

> ### Simplifying Special Fractions
>
> 1. Any number (except zero) divided by itself is equal to 1. For example,
> $$\frac{3}{3} = 1; \frac{5}{5} = 1; \frac{173}{173} = 1$$
>
> 2. Any number divided by 1 is equal to itself. For example,
> $$\frac{5}{1} = 5; \frac{9}{1} = 9; \frac{25}{1} = 25$$
>
> 3. Zero divided by any number (except zero) is equal to zero. For example,
> $$\frac{0}{6} = 0; \frac{0}{13} = 0; \frac{0}{8} = 0$$
>
> 4. Any number *divided by zero* is not meaningful and is called *undefined*. For example, $\frac{4}{0}$ is undefined.

A **proper fraction** is a fraction whose numerator is less than its denominator. Examples of proper fractions are $\frac{2}{3}$, $\frac{5}{14}$, and $\frac{3}{8}$. An **improper fraction** is a fraction whose numerator is greater than or equal to its denominator. Examples of improper fractions are $\frac{7}{5}$, $\frac{11}{11}$, and $\frac{9}{4}$.

A **mixed number** is an integer plus a proper fraction. Examples of mixed numbers are $1\frac{3}{4}$ $\left(1 + \frac{3}{4}\right)$, $14\frac{1}{9}$, and $5\frac{2}{15}$.

> ### Changing an Improper Fraction to a Mixed Number
>
> To change an improper fraction to a mixed number, divide the numerator by the denominator. The quotient is the whole-number part. The remainder over the divisor is the proper fraction part of the mixed number.

Example 4 Change $\frac{17}{3}$ to a mixed number.

$$\frac{17}{3} = 17 \div 3 = 3\overline{)17} = 5\frac{2}{3}$$
$$\phantom{\frac{17}{3} = 17 \div 3 = 3\overline{)17}} 5\text{ r }2$$

Example 5 Change $\frac{78}{7}$ to a mixed number.

$$\frac{78}{7} = 78 \div 7 = 7\overline{)78} = 11\frac{1}{7}$$
$$\phantom{\frac{78}{7} = 78 \div 7 = 7\overline{)78}} 11\text{ r }1$$

If the improper fraction is not reduced to lowest terms, you may find it easier to reduce it before changing it to a mixed number. Of course, you may reduce the proper fraction after the division if you prefer.

1.6 ♦ Introduction to Fractions

Example 6 Change $\dfrac{324}{48}$ to a mixed number and simplify.

Method 1: Reduce the improper fraction to lowest terms first.

$$\frac{324}{48} = \frac{\cancel{2}\cdot\cancel{2}\cdot 3\cdot 3\cdot 3\cdot \cancel{3}}{\cancel{2}\cdot\cancel{2}\cdot 2\cdot 2\cdot \cancel{3}} = \frac{27}{4} = 4\overline{)27}\ \ 6\text{ r }3 = 6\frac{3}{4}$$

Method 2: Change the improper fraction to a mixed number first.

$$\frac{324}{48} = 48\overline{)324}^{\ 6\text{ r }36}\ \ \underline{288}\ \ 36 = 6\frac{36}{48} = 6\frac{\cancel{2}\cdot\cancel{2}\cdot 3\cdot \cancel{3}}{\cancel{2}\cdot\cancel{2}\cdot 2\cdot 2\cdot \cancel{3}} = 6\frac{3}{4}$$

One way to change a mixed number to an improper fraction is to multiply the integer by the denominator of the fraction and then add the numerator of the fraction. Then place this sum over the original denominator.

Example 7 Change $2\dfrac{1}{3}$ to an improper fraction.

$$2\frac{1}{3} = \frac{(2\times 3)+1}{3} = \frac{7}{3}$$

Example 8 Change $5\dfrac{3}{8}$ to an improper fraction.

$$5\frac{3}{8} = \frac{(5\times 8)+3}{8} = \frac{43}{8}$$

Example 9 Change $10\dfrac{5}{9}$ to an improper fraction.

$$10\frac{5}{9} = \frac{(10\times 9)+5}{9} = \frac{95}{9}$$

A number containing an integer and an improper fraction may be simplified as follows.

Example 10 Change $3\dfrac{8}{5}$ **a.** to an improper fraction and then **b.** to a mixed number in simplest form.

a. $3\dfrac{8}{5} = \dfrac{(3\times 5)+8}{5} = \dfrac{23}{5}$

b. $\dfrac{23}{5} = 23 \div 5 = 5\overline{)23}\ \ 4\text{ r }3 = 4\dfrac{3}{5}$

A calculator with a fraction key may be used to simplify fractions as follows. The fraction key often looks similar to $\boxed{A\%}$.

Example 11 Reduce $\frac{108}{144}$ to lowest terms.

108 [A%] 144 [=]

[3/4]

Thus, $\frac{108}{144} = \frac{3}{4}$ in lowest terms.

A calculator with a fraction key may be used to change an improper fraction to a mixed number, as follows.

Example 12 Change $\frac{25}{6}$ to a mixed number.

25 [A%] 6 [=]

[4 1/6]

Thus, $\frac{25}{6} = 4\frac{1}{6}$.

A calculator with a fraction key may be used to change a mixed number to an improper fraction as follows. The improper fraction key often looks similar to .

NOTE: Most scientific calculators are programmed so that several keys will perform more than one function. These calculators have what is called a *second function key*. To access this function, press the second function key first.

Example 13 Change $6\frac{3}{5}$ to an improper fraction.

6 [A%] 3 [A%] 5 [%] [=]

[33/5]

Thus, $6\frac{3}{5} = \frac{33}{5}$.

♦

EXERCISES 1.6

Simplify:

1. $\frac{12}{28}$ 2. $\frac{9}{12}$ 3. $\frac{36}{42}$ 4. $\frac{12}{18}$ 9. $\frac{48}{60}$ 10. $\frac{72}{96}$ 11. $\frac{9}{9}$ 12. $\frac{15}{1}$

5. $\frac{9}{48}$ 6. $\frac{8}{10}$ 7. $\frac{13}{39}$ 8. $\frac{24}{36}$ 13. $\frac{0}{8}$ 14. $\frac{6}{6}$ 15. $\frac{9}{0}$ 16. $\frac{6}{8}$

17. $\dfrac{14}{16}$ 18. $\dfrac{7}{28}$ 19. $\dfrac{27}{36}$ 20. $\dfrac{15}{18}$

21. $\dfrac{12}{16}$ 22. $\dfrac{9}{18}$ 23. $\dfrac{20}{25}$ 24. $\dfrac{12}{36}$

25. $\dfrac{12}{40}$ 26. $\dfrac{54}{72}$ 27. $\dfrac{112}{128}$ 28. $\dfrac{330}{360}$

29. $\dfrac{112}{144}$ 30. $\dfrac{525}{1155}$

Change each fraction to a mixed number in simplest form:

31. $\dfrac{78}{5}$ 32. $\dfrac{11}{4}$ 33. $\dfrac{28}{3}$ 34. $\dfrac{21}{3}$

35. $\dfrac{45}{36}$ 36. $\dfrac{67}{16}$ 37. $\dfrac{57}{6}$ 38. $\dfrac{84}{9}$

39. $5\dfrac{15}{12}$ 40. $2\dfrac{70}{16}$

Change each mixed number to an improper fraction:

41. $3\dfrac{5}{6}$ 42. $6\dfrac{3}{4}$ 43. $2\dfrac{1}{8}$ 44. $5\dfrac{2}{3}$

45. $1\dfrac{7}{16}$ 46. $4\dfrac{1}{2}$ 47. $6\dfrac{7}{8}$ 48. $8\dfrac{1}{5}$

49. $10\dfrac{3}{5}$ 50. $12\dfrac{5}{6}$

51. Pies in a restaurant are cut into 6 pieces each. Twenty-eight pieces would be equivalent to $\dfrac{28}{6}$ pies. Change this improper fraction to a mixed number reduced to lowest terms.

52. The chef asks the prep cook to convert each of the following as indicated: **a.** $1\dfrac{1}{3}$ cups of butter, to an improper fraction. **b.** $\dfrac{15}{4}$ cups of milk, to a mixed number. **c.** $\dfrac{3}{2}$ pints, to a mixed number.

1.7 Addition and Subtraction of Fractions

Technicians must be able to compute fractions accurately, because mistakes on the job can be quite costly. Also, many shop drawing dimensions contain fractions.

> **Adding Fractions**
>
> $$\dfrac{a}{c} + \dfrac{b}{c} = \dfrac{a+b}{c} \quad (c \neq 0)$$
>
> That is, to add two or more fractions with the same denominator, first add their numerators. Then place this sum over the common denominator and simplify.

Example 1   Add: $\dfrac{1}{8} + \dfrac{3}{8}$.

$$\dfrac{1}{8} + \dfrac{3}{8} = \dfrac{1+3}{8} = \dfrac{4}{8} = \dfrac{1}{2} \quad \text{Add the numerators and simplify.}$$ ◆

Example 2   Add: $\dfrac{2}{16} + \dfrac{5}{16}$.

$$\dfrac{2}{16} + \dfrac{5}{16} = \dfrac{2+5}{16} = \dfrac{7}{16} \quad \text{Add the numerators.}$$ ◆

Example 3   Add: $\dfrac{2}{31} + \dfrac{7}{31} + \dfrac{15}{31}$.

$$\dfrac{2}{31} + \dfrac{7}{31} + \dfrac{15}{31} = \dfrac{2+7+15}{31} = \dfrac{24}{31} \quad \text{Add the numerators.}$$ ◆

To add fractions with different denominators, we first need to find a common denominator. When reducing fractions to lowest terms, we *divide* both numerator and denominator by the same nonzero number, which does not change the value of the fraction. Similarly, we can *multiply* both numerator and denominator by the same nonzero number without changing the value of the fraction. It is customary to find the **least common denominator (LCD)** for fractions with unlike denominators. The LCD is the smallest positive integer that has all the denominators as divisors. Then, multiply both numerator and denominator of each fraction by a number that makes the denominator of the given fraction the same as the LCD.

> To find the least common denominator (LCD) of a set of fractions:
> 1. Factor each denominator into its prime factors.
> 2. Write each prime factor the number of times it appears *most* in any *one* denominator in Step 1. The LCD is the product of these prime factors.

Example 4 Find the LCD of the following fractions: $\frac{1}{6}, \frac{1}{8}$, and $\frac{1}{18}$.

STEP 1 Factor each denominator into prime factors. (Prime factorization may be reviewed in Section 1.5.)

$$6 = 2 \cdot 3$$
$$8 = 2 \cdot 2 \cdot 2$$
$$18 = 2 \cdot 3 \cdot 3$$

STEP 2 Write each prime factor the number of times it appears *most* in any *one* denominator in Step 1. The LCD is the product of these prime factors.

Here, 2 appears once as a factor of 6, three times as a factor of 8, and once as a factor of 18. So 2 appears at most *three* times in any one denominator. Therefore, you have $2 \cdot 2 \cdot 2$ as factors of the LCD. The factor 3 appears at most twice in any one denominator, so you have $3 \cdot 3$ as factors of the LCD. Now 2 and 3 are the only factors of the three given denominators. The LCD for $\frac{1}{6}, \frac{1}{8}$, and $\frac{1}{18}$ must be $2 \cdot 2 \cdot 2 \cdot 3 \cdot 3 = 72$. Note that 72 does have divisors 6, 8, and 18. This procedure is shown in Table 1.1.

Table 1.1

Prime factor	Denominator	Number of times the prime factor appears in given denominator	*most* in any one denominator
2	$6 = 2 \cdot 3$	once	three times
	$8 = 2 \cdot 2 \cdot 2$	three times	
	$18 = 2 \cdot 3 \cdot 3$	once	
3	$6 = 2 \cdot 3$	once	twice
	$8 = 2 \cdot 2 \cdot 2$	none	
	$18 = 2 \cdot 3 \cdot 3$	twice	

From the table, we see that the LCD contains the factor 2 three times and the factor 3 two times. Thus, LCD $= 2 \cdot 2 \cdot 2 \cdot 3 \cdot 3 = 72$.

1.7 ♦ Addition and Subtraction of Fractions

Example 5 Find the LCD of $\frac{3}{4}$, $\frac{3}{8}$, and $\frac{3}{16}$.

$$4 = 2 \cdot 2$$
$$8 = 2 \cdot 2 \cdot 2$$
$$16 = 2 \cdot 2 \cdot 2 \cdot 2$$

The factor 2 appears at most *four* times in any one denominator, so the LCD is $2 \cdot 2 \cdot 2 \cdot 2 = 16$. Note that 16 does have divisors 4, 8, and 16. ♦

Example 6 Find the LCD of $\frac{2}{5}$, $\frac{4}{15}$, and $\frac{3}{20}$.

$$5 = 5$$
$$15 = 3 \cdot 5$$
$$20 = 2 \cdot 2 \cdot 5$$

The LCD is $2 \cdot 2 \cdot 3 \cdot 5 = 60$. ♦

Of course, if you can find the LCD by inspection, you need not go through the method shown in the examples.

Example 7 Find the LCD of $\frac{3}{8}$ and $\frac{5}{16}$.

By inspection, the LCD is 16, because 16 is the smallest number that has divisors 8 and 16. ♦

After finding the LCD of the fractions you wish to add, change each of the original fractions to a fraction of equal value, with the LCD as its denominator.

Example 8 Add: $\frac{3}{8} + \frac{5}{16}$.

First, find the LCD of $\frac{3}{8}$ and $\frac{5}{16}$. The LCD is 16. Now change $\frac{3}{8}$ to a fraction of equal value with a denominator of 16.

Write: $\frac{3}{8} = \frac{?}{16}$.   Think: $8 \times ? = 16$.

Since $8 \times 2 = 16$, we multiply both the numerator and the denominator by 2. The numerator is 6, and the denominator is 16.

$$\frac{3}{8} \times \frac{2}{2} = \frac{6}{16}$$

Now, using the rule for adding fractions, we have

$$\frac{3}{8} + \frac{5}{16} = \frac{6}{16} + \frac{5}{16} = \frac{6 + 5}{16} = \frac{11}{16}$$ ♦

Now try adding some fractions for which the LCD is more difficult to find.

Example 9 Add: $\frac{3}{4} + \frac{1}{6} + \frac{5}{16} + \frac{7}{12}$.

First, find the LCD.

$$4 = 2 \cdot 2$$
$$6 = 2 \cdot 3$$
$$16 = 2 \cdot 2 \cdot 2 \cdot 2$$
$$12 = 2 \cdot 2 \cdot 3$$

Note that 2 is used as a factor at most four times in any one denominator and 3 as a factor at most once. Thus, the LCD = $2 \cdot 2 \cdot 2 \cdot 2 \cdot 3 = 48$.

Second, change each fraction to an equivalent fraction with 48 as its denominator.

$$\frac{3}{4} = \frac{?}{48} \qquad \frac{3 \times 12}{4 \times 12} = \frac{36}{48}$$

$$\frac{1}{6} = \frac{?}{48} \qquad \frac{1 \times 8}{6 \times 8} = \frac{8}{48}$$

$$\frac{5}{16} = \frac{?}{48} \qquad \frac{5 \times 3}{16 \times 3} = \frac{15}{48}$$

$$\frac{7}{12} = \frac{?}{48} \qquad \frac{7 \times 4}{12 \times 4} = \frac{28}{48}$$

$$\frac{3}{4} + \frac{1}{6} + \frac{5}{16} + \frac{7}{12} = \frac{36}{48} + \frac{8}{48} + \frac{15}{48} + \frac{28}{48}$$

$$= \frac{36 + 8 + 15 + 28}{48}$$

$$= \frac{87}{48}$$

Simplifying, we have

$$\frac{87}{48} = 1\frac{39}{48} = 1\frac{\cancel{3} \cdot 13}{\cancel{3} \cdot 16} = 1\frac{13}{16}$$

Subtracting Fractions

$$\frac{a}{c} - \frac{b}{c} = \frac{a - b}{c} \qquad (c \neq 0)$$

That is, to subtract two (or more) fractions with the same denominator, first subtract their numerators. Then place the difference over the common denominator and simplify.

Example 10 Subtract: $\frac{3}{5} - \frac{2}{5}$.

$$\frac{3}{5} - \frac{2}{5} = \frac{3 - 2}{5} = \frac{1}{5} \qquad \text{Subtract the numerators.}$$

Example 11 Subtract: $\frac{5}{8} - \frac{3}{8}$.

$$\frac{5}{8} - \frac{3}{8} = \frac{5 - 3}{8} = \frac{2}{8} = \frac{\cancel{2} \cdot 1}{\cancel{2} \cdot 4} = \frac{1}{4} \qquad \text{Subtract the numerators and simplify.}$$

1.7 ♦ Addition and Subtraction of Fractions

To subtract two fractions that have different denominators, first find the LCD. Then express each fraction as an equivalent fraction using the LCD, and subtract the numerators.

Example 12 Subtract: $\dfrac{5}{6} - \dfrac{4}{15}$.

$\dfrac{5}{6} - \dfrac{4}{15} = \dfrac{25}{30} - \dfrac{8}{30}$ First, change the fractions to the LCD, 30.

$\phantom{\dfrac{5}{6} - \dfrac{4}{15}} = \dfrac{25 - 8}{30} = \dfrac{17}{30}$ Subtract the numerators.

Adding Mixed Numbers

To add mixed numbers, find the LCD of the fractions. Add the fractions, then add the whole numbers. Finally, add these two results and simplify.

Example 13 Add: $2\dfrac{1}{2}$ and $3\dfrac{3}{5}$.

$2\dfrac{1}{2} = 2\dfrac{5}{10}$ First, change the proper fractions to the LCD, 10.

$3\dfrac{3}{5} = 3\dfrac{6}{10}$

$5\dfrac{11}{10} = 5 + \dfrac{11}{10} = 5 + 1\dfrac{1}{10} = 6\dfrac{1}{10}$

Subtracting Mixed Numbers

To subtract mixed numbers, find the LCD of the fractions. Subtract the fractions, then subtract the whole numbers and simplify.

Example 14 Subtract: $8\dfrac{2}{3}$ from $13\dfrac{3}{4}$.

$13\dfrac{3}{4} = 13\dfrac{9}{12}$ First, change the proper fractions to the LCD, 12.

$8\dfrac{2}{3} = 8\dfrac{8}{12}$

$\phantom{8\dfrac{2}{3} = 1}5\dfrac{1}{12}$

If the larger of the two mixed numbers does not also have the larger proper fraction, borrow 1 from the whole number. Then add it to the proper fraction before subtracting.

Example 15 Subtract: $2\dfrac{3}{5}$ from $4\dfrac{1}{2}$.

$4\dfrac{1}{2} = 4\dfrac{5}{10} = 3\dfrac{15}{10}$ First, change the proper fractions to the LCD, 10. Then, borrow 1 from 4 and add $\dfrac{10}{10}$ to $\dfrac{5}{10}$.

$2\dfrac{3}{5} = 2\dfrac{6}{10} = 2\dfrac{6}{10}$

$\phantom{2\dfrac{3}{5} = 2\dfrac{6}{10} =\ }1\dfrac{9}{10}$

34 CHAPTER 1 ♦ Basic Concepts

Example 16 Subtract: $2\dfrac{3}{7}$ from $8\dfrac{1}{4}$.

$$8\dfrac{1}{4} = 8\dfrac{7}{28} = 7\dfrac{35}{28}$$

First, change the proper fractions to the LCD, 28. Then, borrow 1 from 8 and add $\dfrac{28}{28}$ to $\dfrac{7}{28}$.

$$2\dfrac{3}{7} = 2\dfrac{12}{28} = 2\dfrac{12}{28}$$

$$5\dfrac{23}{28}$$

♦

Example 17 Subtract: $12 - 4\dfrac{3}{8}$.

$$12 = 11\dfrac{8}{8}$$

First, borrow 1 from 12. Then, rewrite it in terms of the LCD, 8, as shown.

$$4\dfrac{3}{8} = 4\dfrac{3}{8}$$

$$7\dfrac{5}{8}$$

♦

Applications Involving Addition and Subtraction of Fractions

An electrical circuit with more than one path for the current to flow is called a *parallel circuit*. See Figure 1.25. The current in a parallel circuit is divided among the branches in the circuit. How it is divided depends on the resistance in each branch. Since the current is divided among the branches, the total current (I_T) of the circuit is the same as the current from the source. This equals the sum of the currents through the individual branches of the circuit. That is, $I_T = I_1 + I_2 + I_3 + \cdots$.

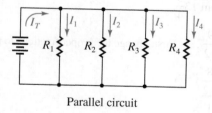

Parallel circuit

Figure 1.25

Example 18 Find the total current in the parallel circuit in Figure 1.26.

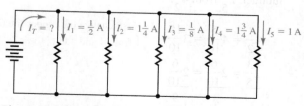

Figure 1.26

1.7 ♦ Addition and Subtraction of Fractions

$$I_T = I_1 + I_2 + I_3 + I_4 + I_5$$

$\dfrac{1}{2} A = \dfrac{4}{8} A$ First, change the proper fractions to the LCD, 8.

$1\dfrac{1}{4} A = 1\dfrac{2}{8} A$

$\dfrac{1}{8} A = \dfrac{1}{8} A$

$1\dfrac{3}{4} A = 1\dfrac{6}{8} A$

$\underline{1\ \ A = 1\ \ A}$

$3\dfrac{13}{8} A = 4\dfrac{5}{8} A$ ♦

Example 19   Find the missing dimension in Figure 1.27.

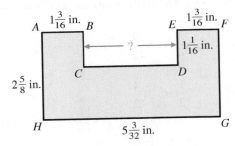

Figure 1.27

First, note that the length of side *HG* equals the sum of the lengths of sides *AB*, *CD*, and *EF*. To find the length of the missing dimension, subtract the sum of side *AB* and side *EF* from side *HG*.

That is, first add *AB* + *EF*:   Then subtract *HG* − (*AB* + *EF*):

AB: $1\dfrac{3}{16}$ in.   *HG*: $5\dfrac{3}{32}$ in. = $5\dfrac{3}{32}$ in. = $4\dfrac{35}{32}$ in.

EF: $\underline{1\dfrac{3}{16}\text{ in.}}$ *AB* + *EF*: $2\dfrac{3}{8}$ in. = $2\dfrac{12}{32}$ in. = $\underline{2\dfrac{12}{32}\text{ in.}}$

AB + *EF*: $2\dfrac{6}{16}$ in., or $2\dfrac{3}{8}$ in.   = $2\dfrac{23}{32}$ in.

Therefore, the missing dimension is $2\dfrac{23}{32}$ in. ♦

The **perimeter** of a geometric figure is the sum of the lengths of its sides.

Example 20   Find the perimeter of Figure 1.27.

AB: $1\dfrac{3}{16}$ in. = $1\dfrac{6}{32}$ in.

BC: $1\dfrac{1}{16}$ in. = $1\dfrac{2}{32}$ in.

CD: $2\dfrac{23}{32}$ in. = $2\dfrac{23}{32}$ in.

DE: $1\frac{1}{16}$ in. = $1\frac{2}{32}$ in.

EF: $1\frac{3}{16}$ in. = $1\frac{6}{32}$ in.

FG: $2\frac{5}{8}$ in. = $2\frac{20}{32}$ in.

GH: $5\frac{3}{32}$ in. = $5\frac{3}{32}$ in.

HA: $2\frac{5}{8}$ in. = $2\frac{20}{32}$ in.

Perimeter: $15\frac{82}{32}$ in. = $15 + 2\frac{18}{32}$ in. = $17\frac{9}{16}$ in.

Example 21 The kitchen staff is doing monthly inventory and determines there are $5\frac{3}{4}$ lb of beef patties, $7\frac{1}{3}$ lb of ground beef, and $13\frac{5}{6}$ lb of frozen beef. What is the total amount of beef?

Beef patties: $5\frac{3}{4} = 5\frac{9}{12}$ First, change the proper fractions to the LCD, 12.

Ground beef: $7\frac{1}{3} = 7\frac{4}{12}$

Frozen beef: $13\frac{5}{6} = 13\frac{10}{12}$

Total: $25\frac{23}{12} = 26\frac{11}{12}$ Add the whole numbers, add the fractions, change the improper fraction to a mixed number, and combine.

Using a Calculator to Add and Subtract Fractions

Example 22 Add: $\frac{2}{3} + \frac{7}{12}$.

2 [A^b/$_c$] 3 [+] 7 [A^b/$_c$] 12 [=]

[1 ¼]

Thus, $\frac{2}{3} + \frac{7}{12} = 1\frac{1}{4}$.

Example 23 Add: $7\frac{3}{4} + 5\frac{11}{12}$.

7 [A^b/$_c$] 3 [A^b/$_c$] 4 [+] 5 [A^b/$_c$] 11 [A^b/$_c$] 12 [=]

[13 ⅔]

Thus, $7\frac{3}{4} + 5\frac{11}{12} = 13\frac{2}{3}$.

1.7 ♦ Addition and Subtraction of Fractions

Example 24   Subtract: $8\frac{1}{4} - 3\frac{5}{12}$.

8 [A b/c] 1 [A b/c] 4 [−] 3 [A b/c] 5 [A b/c] 12 [=]

```
4 5/6
```

Thus, $8\frac{1}{4} - 3\frac{5}{12} = 4\frac{5}{6}$.

EXERCISES 1.7

Find the LCD of each set of fractions:

1. $\frac{1}{2}, \frac{1}{8}, \frac{1}{16}$
2. $\frac{1}{3}, \frac{2}{5}, \frac{3}{7}$
3. $\frac{1}{6}, \frac{3}{10}, \frac{1}{14}$
4. $\frac{1}{9}, \frac{1}{15}, \frac{5}{21}$
5. $\frac{1}{3}, \frac{1}{16}, \frac{7}{8}$
6. $\frac{1}{5}, \frac{3}{14}, \frac{4}{35}$

Perform the indicated operations and simplify:

7. $\frac{2}{3} + \frac{1}{6}$
8. $\frac{1}{2} + \frac{3}{8}$
9. $\frac{1}{16} + \frac{3}{32}$
10. $\frac{5}{6} + \frac{1}{18}$
11. $\frac{2}{7} + \frac{3}{28}$
12. $\frac{1}{9} + \frac{2}{45}$
13. $\frac{3}{8} + \frac{5}{64}$
14. $\frac{3}{10} + \frac{7}{100}$
15. $\frac{1}{5} + \frac{3}{20}$
16. $\frac{3}{4} + \frac{3}{16}$
17. $\frac{4}{5} + \frac{1}{2}$
18. $\frac{2}{3} + \frac{4}{9}$
19. $\frac{1}{3} + \frac{1}{6} + \frac{3}{16} + \frac{1}{12}$
20. $\frac{3}{16} + \frac{1}{8} + \frac{1}{3} + \frac{1}{4}$
21. $\frac{1}{20} + \frac{1}{30} + \frac{1}{40}$
22. $\frac{1}{14} + \frac{1}{15} + \frac{1}{6}$
23. $\frac{3}{10} + \frac{1}{14} + \frac{4}{15}$
24. $\frac{5}{36} + \frac{11}{72} + \frac{5}{6}$
25. $\frac{7}{8} - \frac{3}{4}$
26. $\frac{9}{64} - \frac{2}{128}$
27. $\frac{4}{5} - \frac{3}{10}$
28. $\frac{7}{16} - \frac{1}{3}$
29. $\frac{9}{14} - \frac{3}{42}$
30. $\frac{8}{9} - \frac{5}{24}$
31. $\frac{9}{16} - \frac{13}{32} - \frac{1}{8}$
32. $\frac{7}{8} - \frac{2}{9} - \frac{1}{12}$
33. $2\frac{1}{2} + 4\frac{3}{4}$
34. $3\frac{5}{8} + 5\frac{3}{4}$
35. $3 - \frac{3}{8}$
36. $8 - 5\frac{3}{4}$
37. $8\frac{3}{16} - 3\frac{7}{16}$
38. $5\frac{3}{8} + 2\frac{3}{4}$
39. $7\frac{3}{16} - 4\frac{7}{8}$
40. $8\frac{1}{4} - 4\frac{7}{16}$
41. $3\frac{4}{5} + 9\frac{8}{9}$
42. $4\frac{5}{12} + 6\frac{17}{20}$
43. $3\frac{9}{16} + 4\frac{7}{12} + 3\frac{1}{6}$
44. $5\frac{2}{5} + 3\frac{7}{10} + 4\frac{7}{15}$
45. $16\frac{5}{8} - 4\frac{7}{12} - 2\frac{1}{2}$
46. $12\frac{9}{16} - 3\frac{1}{6} + 2\frac{1}{4}$

47. Find the perimeter of the triangular plot in Illustration 1.

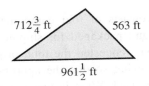

ILLUSTRATION 1

48. ✷ A welder has four pieces of scrap steel angle of lengths $3\frac{1}{4}$ ft, $2\frac{3}{8}$ ft, $3\frac{1}{8}$ ft, and $4\frac{3}{16}$ ft. If they are welded together, how long is the welded piece?

49. ✷ A welder has two pieces of half-inch pipe, one of length $2\frac{3}{8}$ ft and another of length $3\frac{7}{8}$ ft. **a.** What is the total length of the two welded together? **b.** If she needs a total length of $4\frac{3}{4}$ ft, how much must be cut off?

50. ✷ What is the difference in the size (diameter) of 6011 welding rods of diameter $\frac{1}{8}$ in. and Super Strength 100 rods of diameter $\frac{3}{32}$ in.?

51. ✈ A pilot flies a small plane on a cross-country trip to two cities and begins with a full tank of fuel from the home airport. Upon arrival at the first location, the plane required $13\frac{3}{4}$ gal of aviation fuel. At the next stop, the plane required $11\frac{2}{5}$ gal. Upon return to the home airport, the plane took $10\frac{2}{5}$ gal to fill the tank. How much total fuel was used on the trip?

38 CHAPTER 1 ♦ Basic Concepts

52. A Piper Warrior holds 50 gal of aviation fuel. A pilot takes off and lands at another airport and fills up the tank, which takes $17\frac{1}{2}$ gal. The pilot then flies to a second airport, which requires $20\frac{3}{8}$ gal. Had the pilot made the trip to the two airports without refueling each time, how much fuel would have been left in the tank?

53. A pilot flies to an island off the coast of North Carolina and uses $25\frac{1}{4}$ gal of fuel. The return trip only uses $23\frac{3}{4}$ gal. The difference is due to the wind. What is the difference in the fuel used?

54. Oil is changed in three automobiles. They hold $4\frac{1}{2}$ qt, $4\frac{1}{4}$ qt, and $4\frac{3}{8}$ qt. How much total oil is used?

55. An automotive technician spent $\frac{1}{3}$ h changing spark plugs, $\frac{1}{4}$ h changing an air filter, and $\frac{1}{4}$ h changing the oil and oil filter. How much total time was spent servicing this car?

56. A heating and cooling specialist needs two pieces of duct $3\frac{3}{4}$ ft and $2\frac{1}{4}$ ft in length. There are two pieces in stock that are each 4 ft long. After these two lengths are cut off, what excess will be left?

57. The cooling requirements for the three separate incubation rooms are $\frac{1}{3}$ ton, $\frac{3}{4}$ ton, and $\frac{9}{16}$ ton. If a central HVAC unit will be installed, how many tons are required?

58. A finished product consists of four components that will be assembled and packaged for shipment. The box manufacturer has requested the total product weight be on the drawing so that the appropriate strength cardboard is used. What is the product weight? (1 lb = 16 oz)

Part	Weight each
1	$3\frac{1}{2}$ oz
2	$33\frac{1}{8}$ oz
3	6 lb
4	$10\frac{1}{3}$ oz

59. What is the distance between points A and B in Illustration 2?

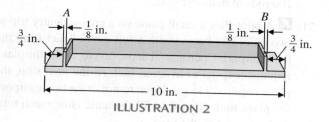

ILLUSTRATION 2

Find **a.** *the length of the missing dimension and* **b.** *the perimeter of each figure.*

60.
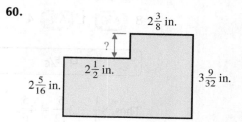

61.
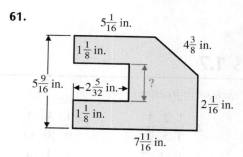

62.
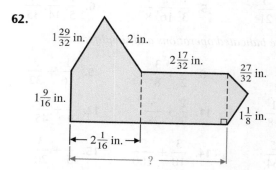

63.
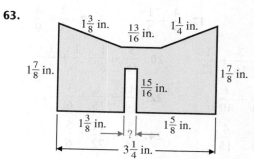

64. The perimeter of a triangle is $59\frac{9}{32}$ in. One side is $19\frac{5}{8}$ in., and a second side is $17\frac{13}{16}$ in. How long is the remaining side?

Find the total current in each parallel circuit:

65.
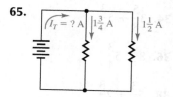

66.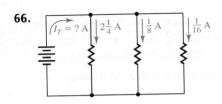

67.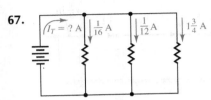

68.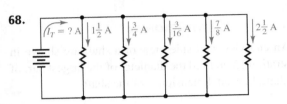

69. Find the length of the shaft in Illustration 3.

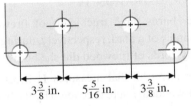

ILLUSTRATION 3

70. Find the distance between the centers of the two end-holes of the plate in Illustration 4.

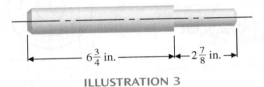

ILLUSTRATION 4

71. In Illustration 5, find **a.** the length of the tool and **b.** the length of diameter A.

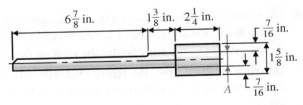

ILLUSTRATION 5

72. A rod $13\frac{13}{16}$ in. long has been cut as shown in Illustration 6. Assume that the waste in each cut is $\frac{1}{16}$ in. What is the length of the remaining piece?

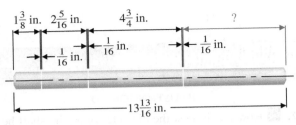

ILLUSTRATION 6

73. Find **a.** the length and **b.** the diameter of the shaft in Illustration 7.

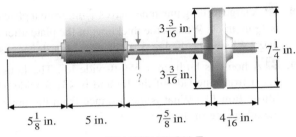

ILLUSTRATION 7

74. Find the missing dimension of the shaft in Illustration 8.

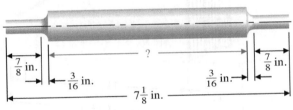

ILLUSTRATION 8

75. Floor joists are spaced 16 in. OC (on center) and are $1\frac{5}{8}$ in. thick, as shown in Illustration 9. What is the distance between them?

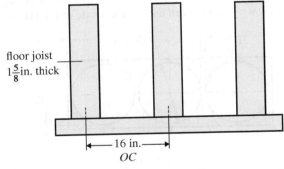

ILLUSTRATION 9

76. If no tap drill chart is available, the correct drill size (TDS) can be found by using the formula $TDS = OD - P$, where OD is the outside diameter and P is the pitch of the thread (the distance between successive threads). See Illustration 10. Find the tap drill size for a $\frac{3}{8}$-in. outside diameter if the pitch is $\frac{1}{16}$ in.

ILLUSTRATION 10

77. How much must the diameter of a $\frac{7}{8}$-in. shaft be reduced so that its diameter will be $\frac{51}{64}$ in.?

78. What is the difference in thickness between a $\frac{7}{16}$-in. steel plate and a $\frac{5}{8}$-in. steel plate?

79. A large milling machine shaves $\frac{3}{32}$-in. from a plate that is $1\frac{7}{8}$ in. thick. What is the thickness of the plate after one pass? What is the thickness of the plate after three passes?

80. A home is built on a $65\frac{3}{4}$-ft-wide lot. The house is $5\frac{5}{12}$ ft from one side of the lot and is $43\frac{5}{6}$ ft wide. (See Illustration 11.) What is the distance from the house to the other side of the lot?

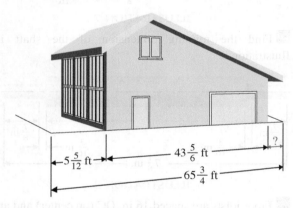

ILLUSTRATION 11

81. See Illustration 12. What width and length steel strip is needed in order to drill three holes of diameter $3\frac{5}{16}$ in.? Allow $\frac{7}{32}$ in. between and on each side of the holes.

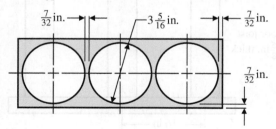

ILLUSTRATION 12

82. Find length x in Illustration 13.

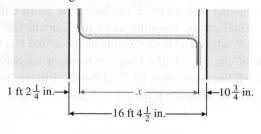

ILLUSTRATION 13

83. An automotive technician needs the following lengths of $\frac{3}{8}$-in. copper tubing: $15\frac{3}{8}$ in., $7\frac{3}{4}$ in., $11\frac{1}{2}$ in., $7\frac{7}{32}$ in., and $10\frac{5}{16}$ in. What is the total length of tubing needed?

84. Find the total length of the shaft in Illustration 14.

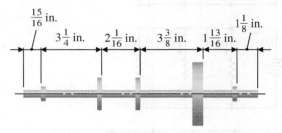

ILLUSTRATION 14

85. An end view and side view of a shaft are shown in Illustration 15. **a.** Find the diameter of the largest part of the shaft. **b.** Find dimension A of the shaft.

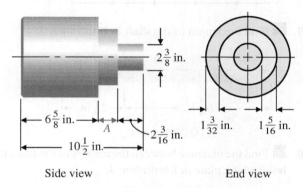

Side view   End view

ILLUSTRATION 15

86. A homeowner burns three truckloads of firewood that contain $\frac{2}{3}$, $\frac{3}{4}$, and $\frac{2}{3}$ of a cord, respectively, during the winter. How many cords of firewood did she burn?

87. A homeowner has a lot that measures $1\frac{1}{2}$ acres. Of that area, $\frac{1}{2}$ acre is wooded, $\frac{1}{6}$ acre is covered by the house and shrubbery, and $\frac{1}{3}$ acre is covered by a driveway and groundcover. The rest is planted in grass for lawns all around the house. How much lawn does the homeowner have?

88. A recreational hiker walks a trail with signs that indicate the following distances between points: $1\frac{1}{2}$ mi, $2\frac{3}{4}$ mi, $\frac{3}{4}$ mi, and $\frac{1}{2}$ mi. How far did she walk on the trail?

89. While making a pie, the pastry chef has $\frac{3}{4}$ stick of butter and $\frac{1}{2}$ stick of butter. How much is the total?

90. How many servings of pie remain if there are $15\frac{3}{8}$ pies in the restaurant and $12\frac{1}{2}$ pies are used?

91. Chef Alex has $3\frac{3}{8}$ cups of flour but needs only $2\frac{1}{4}$ cups of flour. How much flour is left?

92. A salad person has $5\frac{1}{2}$ heads of lettuce. If $1\frac{1}{2}$ heads of lettuce are used for sandwiches and $2\frac{3}{4}$ heads of lettuce are used for salads, how many heads of lettuce remain?

93. ✖ Some French fries come in 5-lb bags. Today $1\frac{1}{2}$ bags of fries are available at the start of the shift and 3 additional bags are brought from the freezer. Sandwich orders use $1\frac{3}{4}$ bags of fries, individual orders of fries use $2\frac{1}{2}$ bags, and $\frac{1}{8}$ bag is wasted. How many French fries remain at the end of the shift?

94. ✖ The kitchen wait staff keeps track of inventory and orders baker potatoes when the case is $\frac{3}{8}$ full. An additional $\frac{5}{16}$ of a case is used before the 2 cases ordered are received. How many potatoes (in cases or parts of cases) are in the kitchen when the 2 ordered cases arrive?

1.8 Multiplication and Division of Fractions

> **Multiplying Fractions**
>
> $$\frac{a}{b} \times \frac{c}{d} = \frac{a \cdot c}{b \cdot d} \qquad (b \neq 0, d \neq 0)$$
>
> To multiply fractions, multiply the numerators and multiply the denominators. Then reduce the resulting fraction to lowest terms.

Example 1 Multiply: $\frac{5}{9} \times \frac{3}{10}$.

$$\frac{5}{9} \times \frac{3}{10} = \frac{5 \cdot 3}{9 \cdot 10} = \frac{15}{90} = \frac{\cancel{15} \cdot 1}{\cancel{15} \cdot 6} = \frac{1}{6}$$

To simplify the work, consider the following alternative method:

$$\frac{\cancel{5}^1}{\cancel{9}_3} \times \frac{\cancel{3}^1}{\cancel{10}_2} = \frac{1 \cdot 1}{3 \cdot 2} = \frac{1}{6}$$

This method divides the numerator by 15, or (5 × 3), and the denominator by 15, or (5 × 3). It does not change the value of the fraction. ◆

Example 2 Multiply: $\frac{18}{25} \times \frac{7}{27}$.

As a shortcut, divide a numerator by 9 and a denominator by 9. Then multiply:

$$\frac{\cancel{18}^2}{25} \times \frac{7}{\cancel{27}_3} = \frac{2 \cdot 7}{25 \cdot 3} = \frac{14}{75}$$
◆

Any mixed number must be replaced by an equivalent improper fraction before multiplying or dividing two or more fractions.

Example 3 Multiply: $8 \times 3\frac{3}{4}$.

$$8 \times 3\frac{3}{4} = \frac{\cancel{8}^2}{1} \times \frac{15}{\cancel{4}_1} = \frac{30}{1} = 30 \qquad \text{First, change } 3\frac{3}{4} \text{ to an improper fraction.}$$
◆

Example 4 Multiply: $\frac{9}{16} \times \frac{5}{22} \times \frac{4}{7} \times 3\frac{2}{3}$.

$$\frac{9}{16} \times \frac{5}{22} \times \frac{4}{7} \times 3\frac{2}{3} = \frac{\cancel{9}^3}{\cancel{16}_4} \times \frac{5}{\cancel{22}_2} \times \frac{\cancel{4}^1}{7} \times \frac{\cancel{11}^1}{\cancel{3}_1} = \frac{15}{56}$$
◆

CHAPTER 1 ♦ Basic Concepts

NOTE: Whenever you multiply several fractions, you may simplify the computation by dividing *any* numerator and *any* denominator by the same number.

Dividing Fractions

$$\frac{a}{b} \div \frac{c}{d} = \frac{a}{b} \times \frac{d}{c} = \frac{a \cdot d}{b \cdot c} \qquad (b \neq 0, c \neq 0, d \neq 0)$$

To divide a fraction by a fraction, invert the fraction (interchange numerator and denominator) that follows the division sign ($\div$). Then multiply the resulting fractions.

Example 5 Divide: $\frac{5}{6} \div \frac{2}{3}$.

$$\frac{5}{6} \div \frac{2}{3} = \frac{5}{\underset{2}{\cancel{6}}} \times \frac{\overset{1}{\cancel{3}}}{2} = \frac{5 \cdot 1}{2 \cdot 2} = \frac{5}{4} \text{ or } 1\frac{1}{4} \qquad \text{Invert } \tfrac{2}{3} \text{ and multiply.}$$

Example 6 Divide: $7 \div \frac{2}{5}$.

$$7 \div \frac{2}{5} = \frac{7}{1} \times \frac{5}{2} = \frac{35}{2} \text{ or } 17\frac{1}{2} \qquad \text{Invert } \tfrac{2}{5} \text{ and multiply.}$$

Example 7 Divide: $\frac{3}{5} \div 4$.

$$\frac{3}{5} \div 4 = \frac{3}{5} \div \frac{4}{1} = \frac{3}{5} \times \frac{1}{4} = \frac{3}{20} \qquad \text{Invert 4 and multiply.}$$

Example 8 Divide: $\frac{9}{10} \div 5\frac{2}{5}$.

$$\frac{9}{10} \div 5\frac{2}{5} = \frac{9}{10} \div \frac{27}{5} = \frac{\overset{1}{\cancel{9}}}{\underset{2}{\cancel{10}}} \times \frac{\overset{1}{\cancel{5}}}{\underset{3}{\cancel{27}}} = \frac{1}{6} \qquad \text{Invert } \tfrac{27}{5} \text{ and multiply.}$$

When both multiplication and division of fractions occur, invert only the first fraction that follows a division sign ($\div$). Then proceed according to the rules for multiplying fractions.

Example 9 Perform the indicated operations and simplify: $\frac{2}{5} \times \frac{1}{3} \div \frac{3}{4}$.

$$\frac{2}{5} \times \frac{1}{3} \div \frac{3}{4} = \frac{2}{5} \times \frac{1}{3} \times \frac{4}{3} = \frac{2 \cdot 1 \cdot 4}{5 \cdot 3 \cdot 3} = \frac{8}{45}$$

Example 10 Perform the indicated operations and simplify: $7\frac{1}{3} \div 4 \times 2$.

$$7\frac{1}{3} \div 4 \times 2 = \frac{\overset{11}{\cancel{22}}}{3} \times \frac{1}{\underset{1}{\cancel{4}}_{2}} \times \frac{\overset{1}{\cancel{2}}}{1} = \frac{11}{3} \text{ or } 3\frac{2}{3}$$

Applications Involving Multiplication and Division of Fractions

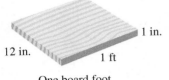

Figure 1.28 One board foot

Lumber is usually measured in board feet. One *board foot* is the amount of wood contained in a piece of wood that measures one inch thick and one square foot in area, or its equivalent. (See Figure 1.28.) The number of board feet in lumber may be found by the formula

$$\text{bd ft} = \frac{\text{number of boards} \times \text{thickness (in in.)} \times \text{width (in in.)} \times \text{length (in ft)}}{12}$$

The 12 in the denominator comes from the fact that the simplest form of one board foot can be thought of as a board that is 1 in. thick × 12 in. wide × 1 ft long.

Lumber is either rough or finished. *Rough stock* is lumber that is not planed or dressed; *finished stock* is planed on one or more sides. When measuring lumber, we compute the full size. That is, we compute the measure of the rough stock that is required to make the desired finished piece. When lumber is finished or planed, $\frac{1}{16}$ in. is taken off each side when the lumber is less than $1\frac{1}{2}$ in. thick. If the lumber is $1\frac{1}{2}$ in. or more in thickness, $\frac{1}{8}$ in. is taken off each side. (**Note:** Lumber for framing houses usually measures $\frac{1}{2}$ in. less than the name that we call the piece. For example, a "two-by-four," a piece 2 in. by 4 in., actually measures $1\frac{1}{2}$ in. by $3\frac{1}{2}$ in.)

Example 11 Find the number of board feet contained in 6 pieces of lumber 2 in. × 8 in. × 16 ft (Figure 1.29).

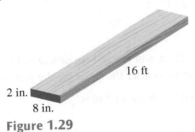

Figure 1.29

$$\text{bd ft} = \frac{\text{number of boards} \times \text{thickness (in in.)} \times \text{width (in in.)} \times \text{length (in ft)}}{12}$$

$$= \frac{6 \times 2 \times 8 \times 16}{12} = 128 \text{ bd ft}$$

◆

Example 12 Energy in the form of electric power is used by industry and consumers alike. Power (in watts, W) equals the voltage (in volts, V) times the current (in amperes, or amps, A). A soldering iron draws a current of $7\frac{1}{2}$ A on a 110-V circuit. Find the wattage, or power, rating of this soldering iron.

$$\text{Power} = (\text{voltage}) \times (\text{current})$$
$$= 110 \times 7\frac{1}{2}$$
$$= 110 \times \frac{15}{2}$$
$$= 825 \text{ W}$$

Power may also be found by computing the product of the square of the current (in amps, A) and the resistance (in ohms, Ω).

◆

44 CHAPTER 1 ♦ Basic Concepts

Example 13 A baby measures $19\frac{1}{4}$ in. long at birth. At the 12-month checkup, the baby is $1\frac{1}{2}$ times as long as it was at birth. How long is the baby at its 12-month checkup?

$$19\frac{1}{4} \text{ in.} \times 1\frac{1}{2} = \frac{77}{4}\text{in.} \times \frac{3}{2}$$

$$= \frac{231}{8}\text{in.}$$

$$= 28\frac{7}{8}\text{in.}$$

♦

Example 14 One form of Ohm's law states that the current I (in amps, A) in a simple circuit equals the voltage E (in volts, V) divided by the resistance R (in ohms, Ω). What current is required for a heating element with a resistance of $7\frac{1}{2}\,\Omega$ operating in a 12-V circuit?

$$\text{Current} = (\text{voltage}) \div (\text{resistance})$$

$$= 12 \div 7\frac{1}{2}$$

$$= 12 \div \frac{15}{2}$$

$$= \cancel{12}^{4} \times \frac{2}{\cancel{15}_{5}}$$

$$= \frac{8}{5} \text{ or } 1\frac{3}{5}\text{ A}$$

♦

Example 15 A pastry chef has $12\frac{1}{2}$ cups of sugar. Five pies are to be made, each requiring $1\frac{1}{3}$ cups of sugar. How much sugar remains after the sugar for the pies is removed?

First, find the amount of sugar in 5 pies:

$$5 \times 1\frac{1}{3} = 5 \times \frac{4}{3} \qquad \text{Change to an improper fraction.}$$

$$= \frac{20}{3}$$

$$= 6\frac{2}{3}\text{ cups} \qquad \text{Change to a mixed number}$$

Then, subtract this amount from $12\frac{1}{2}$ cups. (The LCD is 6.)

$$12\frac{1}{2} = 12\frac{3}{6} = 11\frac{9}{6} \qquad \text{Change the proper fractions to the LCD, 6. Then borrow 1 from 12 and add } \frac{6}{6} \text{ to } \frac{3}{6} \text{ and subtract.}$$

$$\underline{6\frac{2}{3}} = \underline{6\frac{4}{6}} = \underline{6\frac{4}{6}}$$

$$5\frac{5}{6}\text{ cups remain}$$

♦

Using a Calculator to Multiply and Divide Fractions

Example 16 Multiply: $2\frac{5}{6} \times 4\frac{1}{2}$.

1.8 ♦ Multiplication and Division of Fractions

2 [A%] 5 [A%] 6 [×] 4 [A%] 1 [A%] 2 [=]

[12 ¾]

Thus, $2\frac{5}{6} \times 4\frac{1}{2} = 12\frac{3}{4}$.

Example 17 Divide: $5\frac{5}{7} \div 8\frac{1}{3}$.

5 [A%] 5 [A%] 7 [÷] 8 [A%] 1 [A%] 3 [=]

[24/35]

Thus, $5\frac{5}{7} \div 8\frac{1}{3} = \frac{24}{35}$.

EXERCISES 1.8

Perform the indicated operations and simplify:

1. $\frac{2}{3} \times 18$
2. $8 \times \frac{1}{2}$
3. $\frac{3}{4} \times 12$
4. $3\frac{1}{2} \times \frac{2}{5}$
5. $1\frac{3}{4} \times \frac{5}{16}$
6. $\frac{1}{3} \times \frac{1}{3} \times \frac{1}{3}$
7. $\frac{16}{21} \times \frac{7}{8}$
8. $\frac{7}{12} \times \frac{45}{56}$
9. $\frac{2}{7} \times 35$
10. $\frac{9}{16} \times \frac{2}{3} \times 1\frac{6}{15}$
11. $\frac{5}{8} \times \frac{7}{10} \times \frac{2}{7}$
12. $\frac{9}{16} \times \frac{5}{9} \times \frac{4}{25}$
13. $2\frac{1}{3} \times \frac{5}{8} \times \frac{6}{7}$
14. $\frac{5}{28} \times \frac{3}{5} \times \frac{2}{3} \times \frac{2}{9}$
15. $\frac{6}{11} \times \frac{26}{35} \times 1\frac{9}{13} \times \frac{7}{12}$
16. $\frac{3}{8} \div \frac{1}{4}$
17. $\frac{3}{5} \div \frac{10}{12}$
18. $\frac{10}{12} \div \frac{3}{5}$
19. $4\frac{1}{2} \div \frac{1}{4}$
20. $18\frac{2}{3} \div 6$
21. $15 \div \frac{3}{8}$
22. $\frac{77}{6} \div 6$
23. $\frac{7}{11} \div \frac{3}{5}$
24. $7 \div 3\frac{1}{8}$
25. $\frac{2}{5} \times 3\frac{2}{3} \div \frac{3}{4}$
26. $\frac{7}{8} \times \frac{1}{2} \div \frac{2}{7}$
27. $\frac{16}{5} \times \frac{3}{2} \times \frac{10}{4} \div 5\frac{1}{3}$
28. $6 \times 6 \times \frac{21}{7} \div 48$
29. $\frac{7}{9} \times \frac{3}{8} \div \frac{28}{81}$
30. $2\frac{1}{3} \times \frac{5}{8} \div \frac{10}{4}$
31. $\frac{2}{7} \times \frac{5}{9} \times \frac{3}{10} \div 6$
32. $\frac{9}{4} \times \frac{9}{4} \times \frac{21}{7} \div 81$
33. $\frac{7}{16} \div \frac{3}{8} \times \frac{1}{2}$
34. $\frac{5}{8} \div \frac{25}{64} \times \frac{5}{6}$

35. A barrel has a capacity of 42 gal. How many gallons does it contain when it is $\frac{3}{4}$ full?
36. a. Find the area of a rectangle with length $6\frac{1}{3}$ ft and width $3\frac{3}{4}$ ft. (Area = length × width.)
    b. Find its perimeter.
37. ❋ A welder uses seven 6011 welding rods to weld two metal slabs. If each rod makes a $6\frac{1}{2}$-in. weld, find the total length of the weld.
38. ❋ A welder has $6\frac{3}{4}$ ft of $\frac{1}{2}$-in. pipe. How many pieces of pipe, each of length $1\frac{3}{4}$ ft, can be obtained from the original pipe if each cut removes $\frac{1}{16}$ in. of pipe? How long is the remaining piece of pipe?
39. ✈ A small aircraft flew a total of $684\frac{1}{4}$ mi. If it took the plane $5\frac{2}{3}$ h to make the trip, how fast was the aircraft flying?
40. ✈ On a Civil Air Patrol mission, five search planes searched for $3\frac{1}{4}$ h for a missing aircraft. How many total hours did they search?

41. Nine pieces of 8-in. × 12-in. duct that is $3\frac{2}{3}$ ft in length is needed for a building. What is the total length needed?

42. The HVAC supply duct is 17 ft long. Our truck bed can only carry ducts $4\frac{1}{2}$ ft long. How many pieces must the 17-ft duct be cut into and how long is each piece, assuming most of the pieces will be $4\frac{1}{2}$ ft?

Find the number of board feet in each quantity of lumber:

43. 10 pieces 2 in. × 4 in. × 12 ft
44. 24 pieces 4 in. × 4 in. × 16 ft
45. 175 pieces 1 in. × 8 in. × 14 ft
46. Find the total length of eight pieces of steel each $5\frac{3}{4}$ in. long.
47. The outside diameter (OD) of a pipe is $4\frac{9}{32}$ in. The walls are $\frac{7}{32}$ in. thick. Find the inside diameter. (See Illustration 1.)

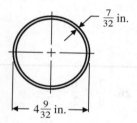

ILLUSTRATION 1

48. Two strips of metal are riveted together in a straight line, with nine rivets equally spaced $2\frac{5}{16}$ in. apart center to center. What is the distance between the first and last rivet?

49. Two metal strips are riveted together in a straight line, with 16 equally spaced rivets. The distance between the first and last rivet is $28\frac{1}{8}$ in. center to center. Find the distance between any two consecutive rivets.

50. Find length x, the distance between centers, in Illustration 2.

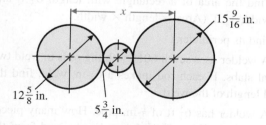

ILLUSTRATION 2

51. From a steel rod 36 in. long, the following pieces are cut:

    3 pieces $2\frac{1}{8}$ in. long
    2 pieces $5\frac{3}{4}$ in. long
    6 pieces $\frac{7}{8}$ in. long
    1 piece $3\frac{1}{2}$ in. long

Assume $\frac{1}{16}$ in. of waste in each cut. Find the length of the remaining piece.

52. A piece of drill rod 2 ft 6 in. long is to be cut into pins, each $2\frac{1}{2}$ in. long.
    a. Assume no loss of material in cutting. How many pins will you get?
    b. Assume $\frac{1}{16}$ in. waste in each cut. How many pins will you get?

53. The cutting tool on a lathe moves $\frac{3}{128}$ in. along the piece being turned for each revolution of work. The piece revolves at 45 revolutions per minute. How long will it take to turn a length of $9\frac{9}{64}$-in. stock in one operation?

54. Three vents are equally spaced along a wall 26 ft 6 in. (318 in.) long, as shown in Illustration 3. Find dimension d.

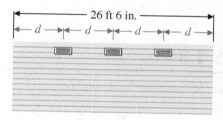

ILLUSTRATION 3

55. A concrete pad for mounting a condensing unit is 4 ft long, $2\frac{2}{3}$ ft wide, and 3 in. ($\frac{1}{4}$ ft) thick. Find its volume in cubic feet.

56. How many lengths of heater hose, each $5\frac{1}{4}$ in. long, can be cut from a 6-ft roll?

57. An automobile dealership received six cars that needed to be detailed. If the service staff took $7\frac{1}{2}$ h to detail the automobiles, how long did each car take?

58. Four tires can be replaced on an automobile in $\frac{3}{4}$ h. If 11 cars needed their tires changed, how long would it take?

59. Find the load of a circuit (power in watts) that takes $12\frac{1}{2}$ A at 220 V. (See Example 12.)

60. An electric iron requires $4\frac{1}{4}$ A and has a resistance of $24\frac{1}{2}$ Ω. What voltage does it require to operate? ($V = IR$.)

61. An electric hand drill draws $3\frac{3}{4}$ A and has a resistance of $5\frac{1}{3}$ Ω. What power does it use? ($P = I^2R$.)

62. A wiring job requires the following lengths of BX cable:

    12 pieces $8\frac{1}{2}$ ft long
    7 pieces $18\frac{1}{2}$ ft long
    24 pieces $1\frac{3}{4}$ ft long
    12 pieces $6\frac{1}{2}$ ft long
    2 pieces $34\frac{1}{4}$ ft long

How much cable is needed to complete the job?

63. What current is required for a heating element with a resistance of $10\frac{1}{2}$ Ω operating in a 24-V circuit? (See Example 14.)

64. How many lengths of wire, each $3\frac{7}{8}$ in. long, can be cut from a 25-ft roll? What length remains?

65. Install 7 equally-spaced wall outlets in a straight line in a hallway so that the distance between the centers of the two end outlets is $43\frac{1}{2}$ ft. **a.** What is the distance, center to center, between two adjacent outlets? **b.** If each outlet is 3 in. wide, what is the total distance between the outside edges of the two most outer outlets? **c.** What is the distance from the left edge of the most left outlet to the center of the fourth outlet from the left?

66. Tom needs to apply $1\frac{3}{4}$ gal of herbicide per acre of soybeans. How many gallons of herbicide are needed for 120 acres?

67. An airplane sprayer tank holds 60 gal. If $\frac{3}{4}$ gal of water and $\frac{1}{2}$ lb of pesticide are applied per acre, how much pesticide powder is needed per tankful?

68. If 1 ft³ of cotton weighs $22\frac{1}{2}$ lb, how many cubic feet are contained in a bale of cotton weighing 500 lb? In 15 tons of cotton?

69. A test plot of $\frac{1}{20}$ acre produces 448 lb of shelled corn. Find the yield in bushels per acre. (1 bu of shelled corn weighs 56 lb.)

70. A farmer wishes to concrete his rectangular feed lot, which measures 120 ft by 180 ft. He wants to have a base of 4 in. of gravel covered with $3\frac{1}{2}$ in. of concrete.

   a. How many cubic yards of each material must he purchase?

   b. What is his total materials cost rounded to the nearest dollar? Concrete costs \$94/yd³ delivered, and gravel costs \$14/ton delivered. (1 yd³ of gravel weighs approximately 2500 lb.)

71. A bottle holds $2\frac{1}{2}$ oz of medicine of which is $\frac{1}{5}$ alcohol. How many ounces of alcohol does the bottle contain?

72. The doctor orders 45 mg of prednisone. Each tablet contains 10 mg. How many tablets are given to the patient?

73. To give 15 mg of codeine sulfate from 30-mg scored tablets, how many tablets should be given?

74. To give 45 mg of codeine sulfate from 30-mg scored tablets, how many tablets should be given?

75. A baby weighs $7\frac{1}{4}$ lb at birth. At its 3-month checkup, the nurse notes that the baby has doubled its birth weight. Calculate the baby's weight at 3 months.

76. A newborn baby is expected to lose $\frac{1}{20}$ of its body weight during the first week after birth. How much weight would a newborn weighing $7\frac{1}{2}$ lb be expected to lose?

77. How many $\frac{1}{2}$-ounce doses of cough syrup can be given from a 12-ounce bottle?

78. How many ounces of medication will a patient take daily if the prescription states: $2\frac{1}{2}$ oz taken 3 times per day?

79. The dosage of cough medicine for a child 4 to 6 years old is $\frac{1}{2}$ tsp every 4 hours. If a child receives 5 doses of medicine over a period of 2 days, how many teaspoons has the child received?

80. Six pieces of pipe are to be welded together with a flat $\frac{1}{4}$-in. plate between them to form guides. The pipes are each $6\frac{1}{8}$ in. long. What would be the overall length l of the assembly shown in Illustration 4?

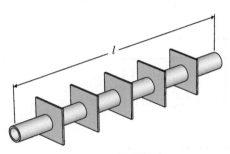

ILLUSTRATION 4

81. A drawing is lacking dimension A. It is critical to make the channel 1 ft long with a cross section as shown in Illustration 5. Find **a.** dimension A and **b.** the volume of metal in the channel.

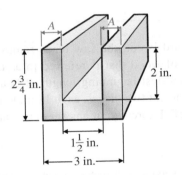

ILLUSTRATION 5

82. A CAD drawing in Illustration 6 shows a bar 2 in. by 4 in. and 36 in. long. How many $3\frac{1}{8}$-in. pieces can be cut if each saw cut is $\frac{1}{8}$-in. scrap?

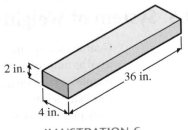

ILLUSTRATION 6

48 CHAPTER 1 ♦ Basic Concepts

*In a parallel circuit, the total resistance (R_T) is given by the formula**

$$R_T = \frac{1}{\frac{1}{R_1} + \frac{1}{R_2} + \frac{1}{R_3} + \cdots}$$

NOTE: The three dots mean that you should use as many fractions in the denominator as there are resistances in the circuit.

Find the total resistance in each parallel circuit:

83.

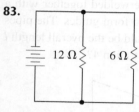

84.

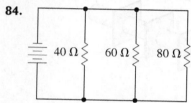

85.

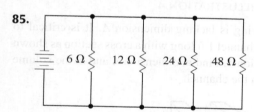

86. ✣ Each time a board is cut, the saw blade removes an additional $\frac{1}{8}$ in. of wood that is referred to as the *kerf*. What is the shortest beginning length of board that will allow you to cut five pieces, each 18 in. long, allowing for the kerf? Express your answer in feet and inches.

---

*A calculator approach to working with such equations is shown in Section 6.9.

87. ✣ A horticulturist crosses two plants of the same species: one has a red flower and the other has a white flower. He knows from genetics that $\frac{1}{4}$ of the resulting seeds will produce plants with white flowers and $\frac{3}{4}$ will produce plants with red flowers. He harvests seeds and grows out 300 plants. How many of those would he expect to produce white flowers? Red flowers?

88. ✣ A tree harvested for firewood has a trunk that is 27 ft long. The woodcutter plans to cut it into $1\frac{1}{2}$-ft lengths before splitting. How many lengths will result?

89. ✖ A recipe needs $1\frac{1}{2}$ cups of sugar. If a $\frac{1}{4}$ cup scoop is used to measure, how many scoops are needed?

90. ✖ A recipe makes 12 lb of pie dough. If $\frac{1}{4}$ lb of dough is needed for the bottom crust and $\frac{1}{8}$ lb of dough is needed for the top crust, how many pies can be made from the 12 lb of dough?

91. ✖ A $16\frac{1}{4}$-lb beef short loin is to be trimmed and then cut into 14-oz steaks. If $5\frac{1}{2}$ lb of trimmings are removed, how many 14-oz steaks can be cut from the trimmed short loin? (16 oz = 1 lb)

92. ✖ A watermelon is cut and prepared into individual servings. The watermelon weighs 12 lb. The melon rind is removed and found to weigh 28 oz. Express the edible portion as a fraction of the whole.

93. ✖ Three deep fryers each require $2\frac{1}{2}$ gal of cooking oil. If there are $10\frac{1}{3}$ gal of cooking oil available, how much cooking oil remains after the 3 empty fryers are filled?

94. ✖ A restaurant uses $\frac{5}{8}$ of each entire short loin for steaks, $\frac{1}{4}$ of each loin for beef sandwiches, and the remainder of each loin for soup. If a total of 3 short loins are used, how much short loin is available for soup?

1.9 The U.S. System of Weights and Measures

Centuries ago, the thumb, hand, foot, and length from nose to outstretched fingers were used as units of measurement. These methods, of course, were not very satisfactory because people's sizes varied. In the 14th century, King Edward II proclaimed the length of the English inch to be the same as three barleycorn grains laid end to end. (See Figure 1.30.) This proclamation helped some, but it did not eliminate disputes over the length of the English inch.

1.9 ♦ The U.S. System of Weights and Measures

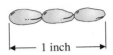

(a) One old way to define one yard.

(b) Three barley corns laid end to end used to define one inch.

Figure 1.30
Old English units

Each of these methods provides rough estimations of measurements. Actually, **measurement** is the comparison of an *observed* quantity with a *standard unit* quantity. In the estimation above, there is no one standard unit. A standard unit that is constant, accurate, and accepted by all is needed for technical measurements. Today, nations have bureaus to set national standards for all measures.

The U.S. system of weights and measures, which is derived from and sometimes called the English system, is a combination of makeshift units of Anglo-Saxon, Roman, and French-Norman weights and measures. The metric system, which is now used by international industry and business, all major U.S. industries, and most federal agencies, is presented in Chapter 3.

The U.S. system requires us to understand and be able to use fractions in everyday life. One advantage of the metric system is that the importance of fractional computations is greatly reduced.

Take a moment to review the table of U.S. weights and measures inside the front cover. Become familiar with this table, because you will use it when changing units.

Example 1

Change 5 ft 9 in. to inches.

$$1 \text{ ft} = 12 \text{ in., so } 5 \text{ ft} = 5 \times 12 \text{ in.} = 60 \text{ in.}$$
$$5 \text{ ft } 9 \text{ in.} = 60 \text{ in.} + 9 \text{ in.} = 69 \text{ in.}$$

♦

To change from one unit or set of units to another, we use what is commonly called a conversion factor. We know that we can multiply any number or quantity by 1 (one) without changing its value. We also know that any fraction whose numerator and denominator are the same is equal to 1. For example, $\frac{5}{5} = 1$, $\frac{16}{16} = 1$, and $\frac{7 \text{ ft}}{7 \text{ ft}} = 1$. Also, since 12 in. = 1 ft, $\frac{12 \text{ in.}}{1 \text{ ft}} = 1$, and likewise, $\frac{1 \text{ ft}}{12 \text{ in.}} = 1$, because the numerator equals the denominator. We call such names for 1 **conversion factors** (or *unit conversion factors*). The information necessary for forming conversion factors is found in tables, many of which are provided inside the front and back covers of this book.

Choosing Conversion Factors

The correct choice for a given conversion factor is the one in which the old units are in the numerator of the original expression and in the denominator of the conversion factor, or the old units are in the denominator of the original expression and in the numerator of the conversion factor. That is, set up the conversion factor so that the old units cancel each other.

50 CHAPTER 1 ♦ Basic Concepts

Example 2 Change 19 ft to inches.

Since 1 ft = 12 in., the two possible conversion factors are $\frac{1 \text{ ft}}{12 \text{ in.}} = 1$ and $\frac{12 \text{ in.}}{1 \text{ ft}} = 1$. We want to choose the one whose numerator is expressed in the new units (in.) and whose denominator is expressed in the old units (ft); that is, $\frac{12 \text{ in.}}{1 \text{ ft}}$. Therefore,

$$19 \text{ ft} \times \frac{12 \text{ in.}}{1 \text{ ft}} = 19 \times 12 \text{ in.} = 228 \text{ in.}$$

— conversion factor

Example 3 Change 8 yd to feet.

3 ft = 1 yd,

so $8 \text{ yd} \times \frac{3 \text{ ft}}{1 \text{ yd}} = 8 \times 3 \text{ ft} = 24 \text{ ft}$

— conversion factor

Example 4 Change 76 oz to pounds.

$$76 \text{ oz} \times \frac{1 \text{ lb}}{16 \text{ oz}} = \frac{76}{16} \text{ lb} = \frac{19}{4} \text{ lb} = 4\frac{3}{4} \text{ lb}$$

— conversion factor

Sometimes it is necessary to use more than one conversion factor.

Example 5 Change 6 mi to yards.

In the table on your reference card, there is no expression equating miles with yards. Thus, it is necessary to use two conversion factors.

$$6 \text{ mi} \times \frac{5280 \text{ ft}}{1 \text{ mi}} \times \frac{1 \text{ yd}}{3 \text{ ft}} = \frac{6 \times 5280}{3} \text{ yd} = 10{,}560 \text{ yd}$$

— conversion factors

Example 6 How could a technician mixing chemicals express 4800 fluid ounces (fl oz) in gallons?

No conversions between fluid ounces and gallons are given in the tables. Use the conversion factors for (a) fluid ounces to pints (pt); (b) pints to quarts (qt); and (c) quarts to gallons (gal).

$$4800 \text{ fl oz} \times \frac{1 \text{ pt}}{16 \text{ fl oz}} \times \frac{1 \text{ qt}}{2 \text{ pt}} \times \frac{1 \text{ gal}}{4 \text{ qt}} = \frac{4800}{16 \times 2 \times 4} \text{ gal} = 37.5 \text{ gal}$$

conversion factors for (a) (b) (c)

The use of a conversion factor is especially helpful for units with which you are unfamiliar, such as rods, chains, or fathoms.

Example 7 Change 561 ft to rods.

Given 1 rod = 16.5 ft, proceed as follows:

$$561 \text{ ft} \times \frac{1 \text{ rod}}{16.5 \text{ ft}} = 34 \text{ rods}$$

Example 8

Change 320 ft/min to ft/h.

Here, choose the conversion factor whose denominator is expressed in the new units (hours) and whose numerator is expressed in the old units (minutes).

$$320 \frac{\text{ft}}{\text{min}} \times \frac{60 \text{ min}}{1 \text{ h}} = 19{,}200 \text{ ft/h}$$

The following example shows how to use multiple conversion factors in more complex units.

Example 9

Change 60 mi/h to ft/s.

This requires a series of conversions as follows: (a) from hours to minutes; (b) from minutes to seconds; and (c) from miles to feet.

$$60 \frac{\text{mi}}{\text{h}} \times \underbrace{\frac{1 \text{ h}}{60 \text{ min}}}_{(a)} \times \underbrace{\frac{1 \text{ min}}{60 \text{ s}}}_{(b)} \times \underbrace{\frac{5280 \text{ ft}}{1 \text{ mi}}}_{(c)} = 88 \text{ ft/s}$$

conversion factors for

EXERCISES 1.9

Fill in each blank:

1. 3 ft 7 in. = _____ in.
2. 6 yd 4 ft = _____ ft
3. 5 lb 3 oz = _____ oz
4. 7 yd 3 ft 6 in. = _____ in.
5. 4 qt 1 pt = _____ pt
6. 6 gal 3 qt = _____ pt
7. 3 tbs = _____ tsp
8. 2 gal = _____ pt
9. 8 ft = _____ in.
10. 5 yd = _____ ft
11. 3 qt = _____ pt
12. 4 mi = _____ ft
13. 96 in. = _____ ft
14. 72 ft = _____ yd
15. 10 pt = _____ qt
16. 54 in. = _____ ft
17. 88 oz = _____ lb
18. 32 fl oz = _____ pt
19. 14 qt = _____ gal
20. 3 bu = _____ pk
21. 56 fl oz = _____ pt
22. 7040 ft = _____ mi
23. 92 ft = _____ yd
24. 9000 lb = _____ tons
25. 2 mi = _____ yd
26. 6000 fl oz = _____ gal
27. 500 fl oz = _____ qt
28. 3 mi = _____ rods

29. A door is 80 in. in height. Find its height in feet and inches.

30. A plane is flying at 22,000 ft. How many miles high is it?

31. Change the length of a shaft $12\frac{3}{4}$ ft long to inches.

32. A machinist has 15 wrought-iron rods to mill. Each rod weighs 24 oz. What is the total weight of the rods in pounds?

33. The instructions on a carton of chemicals call for mixing 144 fl oz of water, 24 fl oz of chemical No. 1, and 56 fl oz of chemical No. 2. How many quarts are contained in the final mixture?

34. The resistance of 1 ft of No. 32-gauge copper wire is $\frac{4}{25}$ Ω. What is the resistance of 15 yd of this wire?

35. A farmer wishes to wire a shed that is 1 mi from the electricity source in his barn. He uses No. 0-gauge copper wire, which has a resistance of $\frac{1}{10}$ ohm (Ω) per 1000 ft. What is the resistance for the mile of wire?

36. To mix an order of feed, the following quantities of feed are combined: 4200 lb, 600 lb, 5800 lb, 1300 lb, and 2100 lb. How many tons are in the final mixture?

37. A piece of sheet metal has dimensions $3\frac{3}{4}$ ft $\times$ $4\frac{2}{3}$ ft. What is the area in square inches?

38. Three pieces of steel angle of lengths 72 in., 68 in., and 82 in. are welded together. **a.** What is the total length in feet? **b.** Find the total length in yards.

39. An airport runway is 2 mi long. How long is it in **a.** feet and **b.** yards? (1 mi = 5280 ft)

40. A given automobile holds $17\frac{1}{2}$ gal of gasoline. How many **a.** quarts and **b.** pints is this?

41. A small window air conditioner is charged with 3 lb of refrigerant. How many ounces is this?

42. Air flows through a metal duct at 2200 cubic feet per minute (CFM). Find this airflow in cubic feet per second.

43. A CAD drawing survey sheet shows a property road frontage as 153 ft. How many yards is this?

44. A tank that exists on a property is 3 ft by 6 ft by 4 ft deep. How many gallons of water will this tank hold? (Water weighs 62.4 lb/ft^3; 1 gal of water weighs 8.34 lb.)

52 CHAPTER 1 ♦ Basic Concepts

45. Given 1 chain = 66 ft, change 561 ft to chains.
46. Given 1 fathom = 6 ft, change 12 fathoms to feet.
47. Given 1 dram = $27\frac{17}{50}$ grains, change 15 drams to grains.
48. Given 1 ounce = 8 drams, change 96 drams to ounces.
49. Change 4500 ft/h to ft/min.
50. Change 28 ft/s to ft/min.
51. Change $1\frac{1}{5}$ mi/s to mi/min.
52. Change 7200 ft/min to ft/s.
53. Change 40 mi/h to ft/s.
54. Change 64 ft/s to mi/h.
55. Change 24 in./s to ft/min.
56. Change 36 in./s to mi/h.
57. Add: 6 yd 2 ft 11 in.
    2 yd 1 ft 8 in.
    5 yd 2 ft 9 in.
    1 yd   6 in.
58. Subtract: 8 yd 1 ft 3 in.
              2 yd 2 ft 6 in.

59. ✺ A fish farmer sells three truckloads of fish that weigh an average of 1.5 tons each. How many pounds of fish did she sell?
60. ✺ A homeowner estimates that she has 34,850 ft² of lawn. Convert that area of lawn into acres.
61. ✺ To estimate the height of a tree using a device called a Biltmore stick, a forester must stand 4 rods away from the base of the tree. If the forester has an average pace length of 3 ft, how many paces must he walk from the tree to be at that approximate distance from the tree?
62. ✖ A recipe calls for 3 tbs (tablespoons) of honey. Only a tsp (teaspoon) measure is available. How many tsp of honey are needed?
63. ✖ Fruit juice used to make punch is purchased in quarts. To use 7 gal of fruit juice for punch, how many quarts are needed?
64. ✖ How many $1\frac{1}{3}$-oz servings of salad dressing are in a $1\frac{1}{4}$-gal salad dressing container?
65. ✖ The daily soup in a restaurant contains 2 gal of beef stock, 2 qt of beef, 3 pt of vegetables, and $\frac{1}{2}$ gal of tomatoes. How many gallons is the finished soup?

UNIT 1B REVIEW

Simplify:

1. $\dfrac{9}{15}$
2. $\dfrac{48}{54}$
3. Change $\dfrac{27}{6}$ to a mixed number in simplest form.
4. Change $3\dfrac{2}{5}$ to an improper fraction.

Perform the indicated operations and simplify:

5. $\dfrac{5}{6} + \dfrac{2}{3}$
6. $5\dfrac{3}{8} - 2\dfrac{5}{12}$
7. $\dfrac{5}{12} \times \dfrac{16}{25}$
8. $\dfrac{3}{4} \div 1\dfrac{5}{8}$
9. $1\dfrac{2}{3} + 3\dfrac{5}{6} - 2\dfrac{1}{4}$
10. $4\dfrac{2}{3} \div 3\dfrac{1}{2} \times 1\dfrac{1}{2}$

11. Find the missing dimension in Illustration 1.

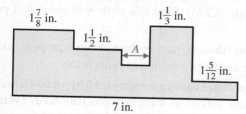

ILLUSTRATION 1

12. A pipe is 72 in. long. Cut three pieces of the following lengths from the pipe: $16\dfrac{3}{4}$ in., $24\dfrac{7}{8}$ in., and $12\dfrac{5}{16}$ in. Assume $\dfrac{1}{16}$ in. waste in each cut. What length of pipe is left?
13. Find the perimeter of a rectangle with length $6\dfrac{1}{4}$ in. and width $2\dfrac{2}{3}$ in.
14. Find the area of a rectangle with length $6\dfrac{1}{4}$ in. and width $2\dfrac{2}{3}$ in.
15. Change 4 ft to inches.
16. Change 24 ft to yards.
17. Change 3 lb to ounces.
18. Change 20 qt to gallons.
19. Change 60 mi/h to ft/s.
20. Subtract: 14 ft 4 in.
              8 ft 8 in.

UNIT 1C Review of Operations with Decimal Fractions and Percent

1.10 Addition and Subtraction of Decimal Fractions

Introduction to Decimals

A fraction whose denominator is 10, 100, 1000, or any power of 10 is called a **decimal fraction**. Decimal calculations and measuring instruments calibrated in decimals are the basic tools for measurement in the metric system. The common use of the calculator makes a basic understanding of decimal principles necessary.

Recall the place values of the digits of a whole number from Section 1.1. Each digit to the left of the decimal point represents a multiple of a power of 10. Each digit to the right of the decimal point represents a multiple of a power of $\frac{1}{10}$. Study Table 1.2, which shows place values for decimals.

Note that $10^0 = 1$. (See Section 2.5.)

Table 1.2 Place Values for Decimals

Number	Words	Product form	Exponential form
1,000,000	One million	$10 \times 10 \times 10 \times 10 \times 10 \times 10$	10^6
100,000	One hundred thousand	$10 \times 10 \times 10 \times 10 \times 10$	10^5
10,000	Ten thousand	$10 \times 10 \times 10 \times 10$	10^4
1000	One thousand	$10 \times 10 \times 10$	10^3
100	One hundred	10×10	10^2
10	Ten	10	10^1
1	One	1	10^0
0.1	One tenth	$\frac{1}{10}$	$\left(\frac{1}{10}\right)^1$ or 10^{-1}
0.01	One hundredth	$\frac{1}{10} \times \frac{1}{10}$	$\left(\frac{1}{10}\right)^2$ or 10^{-2}
0.001	One thousandth	$\frac{1}{10} \times \frac{1}{10} \times \frac{1}{10}$	$\left(\frac{1}{10}\right)^3$ or 10^{-3}
0.0001	One ten-thousandth	$\frac{1}{10} \times \frac{1}{10} \times \frac{1}{10} \times \frac{1}{10}$	$\left(\frac{1}{10}\right)^4$ or 10^{-4}
0.00001	One hundred-thousandth	$\frac{1}{10} \times \frac{1}{10} \times \frac{1}{10} \times \frac{1}{10} \times \frac{1}{10}$	$\left(\frac{1}{10}\right)^5$ or 10^{-5}
0.000001	One millionth	$\frac{1}{10} \times \frac{1}{10} \times \frac{1}{10} \times \frac{1}{10} \times \frac{1}{10} \times \frac{1}{10}$	$\left(\frac{1}{10}\right)^6$ or 10^{-6}

Example 1 In the number 123.456, find the place value of each digit and the number it represents.

Digit	Place value	Number represented
1	Hundreds	1×10^2
2	Tens	2×10^1
3	Ones or units	3×10^0 or 3×1
4	Tenths	$4 \times \left(\frac{1}{10}\right)^1$ or 4×10^{-1}
5	Hundredths	$5 \times \left(\frac{1}{10}\right)^2$ or 5×10^{-2}
6	Thousandths	$6 \times \left(\frac{1}{10}\right)^3$ or 6×10^{-3}

Recall that place values to the left of the decimal point are powers of 10 and place values to the right of the decimal point are powers of $\frac{1}{10}$.

Example 2 Write each decimal in words: 0.05; 0.0006; 24.41; 234.001207.

Decimal	Word form
0.05	Five hundredths
0.0006	Six ten-thousandths
24.41	Twenty-four *and* forty-one hundredths
234.001207	Two hundred thirty-four *and* one thousand two hundred seven millionths

Note that the decimal point is read *and*.

Example 3 Write each number as a decimal and as a common fraction.

Number	Decimal	Common fraction
One hundred four and seventeen hundredths	104.17	$104\frac{17}{100}$
Fifty and three thousandths	50.003	$50\frac{3}{1000}$
Five hundred eleven hundred-thousandths	0.00511*	$\frac{511}{100,000}$

*This book follows the common practice of writing a zero before the decimal point in a decimal smaller than 1.

Often, common fractions are easier to use if they are expressed as decimal equivalents. Every common fraction can be expressed as a decimal. A *repeating decimal* is one in which a digit or a group of digits repeats again and again; it may be written as a common fraction.

A bar over a digit or group of digits means that this digit or group of digits is repeated without ending. Each of the following numbers is a repeating decimal:

$0.33333\ldots$ is written $0.\overline{3}$

$72.64444\ldots$ is written $72.6\overline{4}$

$0.21212121\ldots$ is written $0.\overline{21}$

$6.33120120120\ldots$ is written $6.33\overline{120}$

A *terminating decimal* is a decimal number with a given number of digits. Examples are 0.75, 12.505, and 0.000612.

Changing a Common Fraction to a Decimal

To change a common fraction to a decimal, divide the numerator of the fraction by the denominator.

Example 4 Change $\dfrac{3}{4}$ to a decimal.

$$\begin{array}{r} 0.75 \\ 4\overline{)3.00} \\ \underline{2\ 8} \\ 20 \\ \underline{20} \end{array}$$ Divide the numerator by the denominator.

$\dfrac{3}{4} = 0.75$ (a terminating decimal)

Example 5 Change $\dfrac{8}{15}$ to a decimal.

$$\begin{array}{r} 0.5333 \\ 15\overline{)8.0000} \\ \underline{7\ 5} \\ 50 \\ \underline{45} \\ 50 \\ \underline{45} \\ 50 \\ \underline{45} \\ 5 \end{array}$$ Divide the numerator by the denominator.

$\dfrac{8}{15} = 0.5333\ldots = 0.5\overline{3}$ (a repeating decimal)

The result could be written $0.53\overline{3}$ or $0.5\overline{3}$. It is not necessary to continue the division once it has been established that the quotient has begun to repeat.

Since a decimal fraction can be written as a common fraction with a denominator that is a power of 10, it is easy to change a decimal fraction to a common fraction. Simply use the digits that appear to the right of the decimal point (disregarding beginning zeros) as the numerator. Use the place value of the last digit as the denominator. Any digits to the left of the decimal point will be the whole-number part of the resulting mixed number.

Example 6 Change each decimal to a common fraction or a mixed number.

Decimal	Common fraction or mixed number
a. 0.3	$\dfrac{3}{10}$
b. 0.17	$\dfrac{17}{100}$
c. 0.25	$\dfrac{25}{100} = \dfrac{1}{4}$
d. 0.125	$\dfrac{125}{1000} = \dfrac{1}{8}$
e. 0.86	$\dfrac{86}{100} = \dfrac{43}{50}$
f. 8.1	$8\dfrac{1}{10}$
g. 13.64	$13\dfrac{64}{100} = 13\dfrac{16}{25}$
h. 5.034	$5\dfrac{34}{1000} = 5\dfrac{17}{500}$

♦

In on-the-job situations, it is often more convenient to add, subtract, multiply, and divide measurements that are in decimal form rather than in fractional form. Except for the placement of the decimal point, the four arithmetic operations are the same for decimal fractions as they are for whole numbers.

Example 7 Add 13.2, 8.42, and 120.1.

a. Using decimal fractions:

$$\begin{array}{r} 13.2 \\ 8.42 \\ \underline{120.1} \\ 141.72 \end{array}$$

b. Using common fractions:

$$13\dfrac{2}{10} = 13\dfrac{20}{100}$$
$$8\dfrac{42}{100} = 8\dfrac{42}{100}$$
$$120\dfrac{1}{10} = 120\dfrac{10}{100}$$
$$141\dfrac{72}{100} = 141.72$$

♦

Adding or Subtracting Decimal Fractions

STEP 1 Write the decimals so that the digits having the same place value are in vertical columns. (Make certain that the decimal points are also lined up vertically.)

STEP 2 Add or subtract as with whole numbers.

STEP 3 Place the decimal point between the ones digit and the tenths digit of the sum or the difference. (Be certain the decimal point is in the same vertical line as the other decimal points.)

1.10 ♦ Addition and Subtraction of Decimal Fractions

Example 8 Subtract 1.28 from 17.9.

```
  17.90
   1.28
  16.62
```
Zeros can be supplied after the last digit at the right of the decimal point without changing the value of a number. Therefore, 17.9 = 17.90.

Example 9 Add 24.1, 26, and 37.02.

```
  24.10
  26.00
  37.02
  87.12
```
A decimal point can be placed at the right of any whole number, and zeros can be supplied without changing the value of the number.

Example 10 Perform the indicated operations: $51.6 - 2.45 + 7.3 - 14.92$.

```
   51.60
 -  2.45
   49.15    difference
 +  7.30
   56.45    sum
 - 14.92
   41.53    final difference
```

Example 11 Find the missing dimension in Figure 1.31.

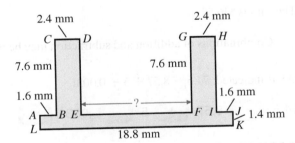

Figure 1.31

The missing dimension EF equals the sum of the lengths AB, CD, GH, and IJ subtracted from the length LK. That is, add

AB:	1.6 mm
CD:	2.4 mm
GH:	2.4 mm
IJ:	1.6 mm
	8.0 mm

Then subtract

LK:	18.8 mm
	−8.0 mm
	10.8 mm

That is, length $EF = 10.8$ mm.

Example 12 As we saw in Unit 1B, the total current in a parallel circuit equals the sum of the currents in the branches of the circuit. Find the total current in the parallel circuit shown in Figure 1.32.

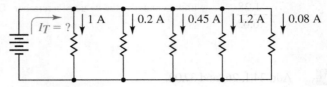

Figure 1.32

```
  1   A
  0.2 A
  0.45 A
  1.2 A
  0.08 A
─────────
  2.93 A
```

Using a Calculator to Add and Subtract Decimal Fractions

Example 13 Add: $14.62 + $0.78 + $1.40 + $0.05.

14.62 [+] .78 [+] 1.4 [+] .05 [=]

[16.85]

The sum is $16.85.

Combinations of addition and subtraction may be performed on a calculator as follows.

Example 14 Do as indicated: $74.6 - 8.57 + 5 - 0.0031$.

74.6 [−] 8.57 [+] 5 [−] .0031 [=]

[71.0269]

The result is 71.0269.

EXERCISES 1.10

Write each decimal in words:

1. 0.004
2. 0.021
3. 0.0005
4. 7.1
5. 1.00421
6. 1042.007
7. 6.092
8. 8.1461

11. Seventy-one and twenty-one ten-thousandths
12. Sixty-five thousandths
13. Forty-three and one hundred one ten-thousandths
14. Five hundred sixty-three millionths

Write each number both as a decimal and as a common fraction or mixed number:

9. Five and two hundredths
10. One hundred twenty-three and six thousandths

Change each common fraction to a decimal:

15. $\dfrac{3}{8}$
16. $\dfrac{16}{25}$
17. $\dfrac{11}{15}$
18. $\dfrac{2}{5}$

1.10 ♦ Addition and Subtraction of Decimal Fractions

19. $\dfrac{17}{50}$ **20.** $\dfrac{11}{9}$ **21.** $\dfrac{14}{11}$ **22.** $\dfrac{128}{25}$

23. $\dfrac{128}{7}$ **24.** $\dfrac{603}{24}$ **25.** $\dfrac{308}{9}$ **26.** $\dfrac{230}{6}$

Change each decimal to a common fraction or a mixed number:

27. 0.7 **28.** 0.6 **29.** 0.11
30. 0.75 **31.** 0.8425 **32.** 3.14
33. 10.76 **34.** 148.255

Find each sum:

35. 137.64
 7.14
 0.008
 6.1

36. 63
 4.7
 19.45
 120.015

37. $147.49 + 7.31 + 0.004 + 8.4$
38. $47 + 6.3 + 20.71 + 170.027$

Subtract:

39. 72.4 from 159 **40.** 3.12 from 4.7
41. $64.718 - 49.41$ **42.** $140 - 16.412$

Perform the indicated operations:

43. $18.4 - 13.72 + 4$ **44.** $34.14 - 8.7 - 16.5$
45. $0.37 + 4.5 - 0.008$
46. $51.7 - 1.11 - 4.6 + 84.1$
47. $1.511 + 14.714 - 6.1743$
48. $0.0056 + 0.023 - 0.00456 + 0.9005$

49. A piece of $\frac{1}{4}$-in. flat steel is 6.25 ft long by 4.2 ft wide. If you cut off two equal pieces of length 2.4 ft and width 4.2 ft, what size piece will be left?

50. A welder needs to weld together pipes of lengths 10.25 ft, 15.4 ft, and 14.1 ft. What is the total length of the new pipe?

51. A crop duster flew 2.3 h on Monday, 3.1 h on Wednesday, and 5.4 h on Friday and Saturday combined. What was the total flying time for the week?

52. An ultralight aircraft flew 125.5 mi to a small airport, then another 110.3 mi to another airport. After spending the night, it flew 97.8 mi to yet another airport. What was the total mileage for the trip?

53. An automobile needs new tires. The tread on the old tires measures $\frac{1}{16}$ in. If the new tires have a tread of $\frac{3}{8}$ in., what is the difference in the tread written as a decimal?

54. What is the total cost for one piece of 8-in. by 16-in. metal duct at $17.33 and two pieces of 8-in. by 12-in. metal duct at $11.58?

55. Find the missing dimensions in Illustration 1.
56. Find the perimeter of the figure in Illustration 1.

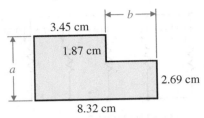

ILLUSTRATION 1

57. Find the length of the shaft shown in Illustration 2.

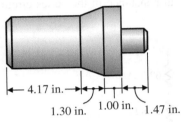

ILLUSTRATION 2

58. The perimeter of the hexagon in Illustration 3 is 6.573 in. Find the length of side *x*.

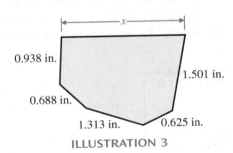

ILLUSTRATION 3

59. A steel axle is being designed and drawn as in Illustration 4. It has a $\frac{1}{8}$-in.-diameter hole drilled in the center of its length. If the axle is 9.625 in. long, how far from the end should the center of the hole be dimensioned?

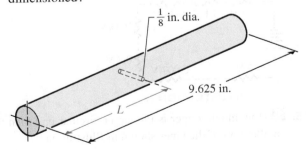

ILLUSTRATION 4

60. What is the internal diameter of a circular tube having an outside diameter (OD) of 1.125 in. and a wall thickness of 0.046 in.?

60 CHAPTER 1 ♦ Basic Concepts

61. Find the total current in the parallel circuit in Illustration 5.

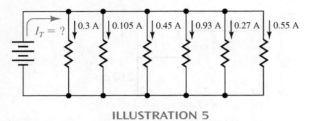

ILLUSTRATION 5

62. As we saw in Section 1.1, the total resistance in a series circuit is equal to the sum of the resistances in the circuit. Find the total resistance in the series circuit in Illustration 6.

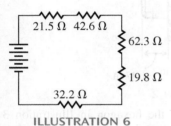

ILLUSTRATION 6

63. Find the total resistance in the series circuit in Illustration 7. The total resistance in a series circuit equals the sum of the resistances in the circuit (Section 1.1).

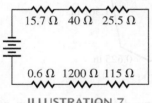

ILLUSTRATION 7

64. In a series circuit, the voltage of the source equals the sum of the separate voltage drops in the circuit. Find the voltage of the source in the circuit in Illustration 8.

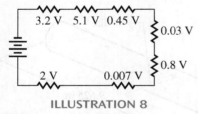

ILLUSTRATION 8

65. How much longer is the larger circular end than the smaller end of the taper shown in Illustration 9?

ILLUSTRATION 9

66. Find the missing dimension in each figure in Illustration 10.

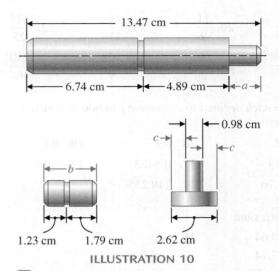

ILLUSTRATION 10

67. Find the wall thickness of the pipe in Illustration 11.

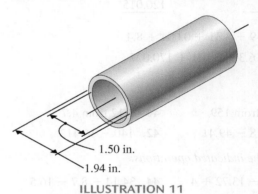

ILLUSTRATION 11

68. Find the length, l, of the socket in Illustration 12. Also, find its diameter A.

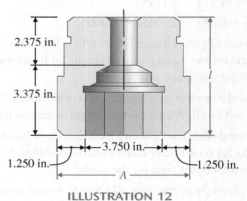

ILLUSTRATION 12

69. During a valve grinding operation, 0.007 in. was removed from the valve stem tip. If the original length of the valve was 4.125 in., what is the new length after the valve stem tip was serviced?

70. The standard width of a new piston ring is 0.2573 in. The used ring measures 0.2476 in. How much has it worn?

71. According to the United Nations, the human population in year 2016 was 7.4 billion and it was expected to be about 11.2 billion by the year 2100. If the estimate is accurate, by how much will the human population increase during the rest of this century?

72. A homeowner fertilizes his lawn in March at a cost of $114.57, in June at a cost of $145.36, and in September at a cost of $99.21. How much was spent on lawn fertilizer over the course of the year?

73. According to the *CIA World Factbook*, as of January 2015 the top ten countries with the most proven oil reserves in billions of barrels (bbl) from known reservoirs were as follows:

Venezuela	298.4 bbl
Saudi Arabia	268.3 bbl
Canada	171.0 bbl
Iran	157.8 bbl
Iraq	144.2 bbl
Kuwait	104.0 bbl
Russia	103.2 bbl
United Arab Emirates	97.8 bbl
Libya	48.4 bbl
Nigeria	37.1 bbl

a. Find the total oil reserves of these ten countries. b. If the total world oil reserves estimate of 1697.6 bbl is accurate, how much oil reserves are in the rest of the world?

74. The recipe for a popular punch for spring includes $2\frac{1}{3}$ qt of pineapple juice, $1\frac{1}{6}$ qt of orange juice, and $3\frac{1}{4}$ qt of club soda. How much punch will this recipe make? Express the answer as a decimal.

75. The fry cook has 4 partial containers of cooking oil to combine. Express the total amount of cooking oil as a decimal. The amounts are: $1\frac{3}{4}$ gal, 0.4 gal, 0.75 gal, and 0.5 gal.

76. The drink of the day in a restaurant contains 0.75 oz of pineapple juice, 1.3 oz of orange juice, 2.5 oz of soda water, and 0.1 oz of coconut extract. How many ounces is this drink?

77. Chili ingredients are 2.5 lb of cooked ground beef, 12 oz of ketchup, 1.5 oz of seasoning, 0.7 lb of beans, 14 oz of elbow macaroni, and 18 oz of tomato juice. How many pounds (rounded to the nearest pound) are in the completed recipe?

1.11 Rounding Numbers

We often need to make an estimate of a number or a measurement. When a truck driver makes a delivery from one side of a city to another, he or she can only estimate the time it will take to make the trip. An automobile technician must estimate the cost of a repair job and the number of mechanics to assign to that job. On such occasions, estimates are *rounded*.

Earlier, you found that $\frac{1}{3} = 0.33\overline{3}$. There is no exact decimal value to use in a calculation. You must round $0.33\overline{3}$ to a certain number of decimal places, depending on the accuracy needed in a given situation.

There are many rounding procedures in general use today. Some are complicated, and others are simple. We use one of the simplest methods, which will be outlined in the next examples and then stated in the form of a rule.

Example 1 Round 25,348 to the nearest thousand.

Note that 25,348 is more than 25,000 and less than 26,000.

26,000
25,900
25,800
25,700
25,600
25,500
25,400
25,348   As you can see, 25,348 is closer to 25,000 than to 26,000. Therefore,
25,300   25,348 rounded to the nearest thousand is 25,000.
25,200
25,100
25,000

Example 2 Round 2.5271 to the nearest hundredth.

Note that 2.5271 is more than 2.5200 but less than 2.5300.

2.5300 = 2.53
2.5290
2.5280
2.5271   2.5271 is nearer to 2.53 than to 2.52. Therefore, 2.5271 rounded to the nearest hundredth is 2.53.
2.5270
2.5260
2.5250
2.5240
2.5230
2.5220
2.5210
2.5200 = 2.52

NOTE: If a number is exactly halfway between two numbers, round up to the larger number. ♦

Rounding Numbers to a Particular Place Value

To round a number to a particular place value:

1. If the digit in the next place to the right is less than 5, drop that digit and all other following digits. Use zeros to replace any whole-number places dropped.
2. If the digit in the next place to the right is 5 or greater, add 1 to the digit in the place to which you are rounding. Drop all other following digits. Use zeros to replace any whole-number places dropped.

Example 3 Round each number in the left column to the place indicated in each of the other columns.

Number	Hundred	Ten	Unit	Tenth	Hundredth	Thousandth
a. 158.6147	200	160	159	158.6	158.61	158.615
b. 4562.7155	4600	4560	4563	4562.7	4562.72	4562.716
c. 7.12579	0	10	7	7.1	7.13	7.126
d. 63,576.15	63,600	63,580	63,576	63,576.2	63,576.15	—
e. 845.9981	800	850	846	846.0	846.00	845.998

Instead of rounding a number to a particular place value, we often need to round a number to a given number of significant digits. **Significant digits** are those digits in a number we are reasonably sure of being able to rely on in a measurement. Here we present a brief introduction to significant digits. (An in-depth discussion of accuracy and significant digits is given in Section 4.1.)

Significant Digits

The following digits in a number are significant:

- All nonzero digits (258 has three significant digits)
- All zeros between significant digits (2007 has four significant digits)
- All zeros at the end of a decimal number (2.000 and 0.09500 have four significant digits)

The following digits in a number are not significant:

- All zeros at the beginning of a decimal number less than 1 (0.00775 has three significant digits)
- All zeros at the end of a whole number (36,000 has two significant digits)

Rounding Numbers to a Given Number of Significant Digits

To round a number to a given number of significant digits:

1. Count the given number of significant digits from left to right, starting with the first nonzero digit.
2. If the next digit to the right is less than 5, drop that digit and all other following digits. Use zeros to replace any whole-number places dropped.
3. If the next digit to the right is 5 or greater, add 1 to the digit in the place to which you are rounding. Drop all other following digits. Use zeros to replace any whole-number places dropped.

Example 4 Round each number to three significant digits.

a. 74,123   Count three digits from left to right, which is the digit 1. Since the next digit to its right is less than 5, replace the next two digits with zeros. Thus, 74,123 rounded to three significant digits is 74,100.

b. 0.002976401   Count three nonzero digits from left to right, which is the digit 7. Since the next digit to its right is greater than 5, increase the digit 7 by 1 and drop the next four digits. Thus, 0.002976401 rounded to three significant digits is 0.00298.

Example 5 Round each number to the given number of significant digits.

Number	Given number of significant digits	Rounded number
a. 2571.88	3	2570
b. 2571.88	4	2572
c. 345,175	2	350,000
d. 345,175	4	345,200
e. 0.0030162	2	0.0030
f. 0.0030162	3	0.00302
g. 24.00055	3	24.0
h. 24.00055	5	24.001

EXERCISES 1.11

*Round each number to **a.** the nearest hundred and **b.** the nearest ten:*

1. 1652
2. 1760
3. 3125.4
4. 73.82
5. 18,675
6. 5968

*Round each number to **a.** the nearest tenth and **b.** the nearest thousandth:*

7. 3.1416
8. 0.161616
9. 0.05731
10. 0.9836
11. 0.07046
12. 3.7654

Round each number in the left column to the place indicated in each of the other columns:

	Number	Hundred	Ten	Unit	Tenth	Hundredth	Thousandth
13.	636.1825						
14.	1451.5254						
15.	17,159.1666						
16.	8.171717						
17.	1543,679						
18.	41,892.1565						
19.	10,649.83						
20.	84.00659						
21.	649.8995						
22.	147.99545						

Round each number to three significant digits:

23. 236,534
24. 202.505
25. 0.03275

Round each number to two significant digits:

26. 63,914
27. 71.613
28. 0.03275

Round each number to four significant digits:

29. 1,462,304
30. 23.2347
31. 0.000337567

Round each number to three significant digits:

32. 20,714
33. 1.00782
34. 0.00118952

1.12 Multiplication and Division of Decimal Fractions

Multiplying Two Decimal Fractions

To multiply two decimal fractions:
1. Multiply the numbers as you would whole numbers.
2. Count the total number of digits to the right of the decimal points in the two numbers being multiplied. Then place the decimal in the product so that it has that same total number of digits to the right of the decimal point.

1.12 ♦ Multiplication and Division of Decimal Fractions

Example 1 Multiply: 42.6×1.73.

$$
\begin{array}{r}
42.6 \\
\underline{1.73} \\
12\ 78 \\
298\ 2 \\
\underline{426} \\
73.698
\end{array}
$$

Note that 42.6 has one digit to the *right* of the decimal point and 1.73 has two digits to the *right* of the decimal point. The product has three digits to the *right* of the decimal point. ♦

Example 2 Multiply: 30.6×4200.

$$
\begin{array}{r}
30.6 \\
\underline{4200} \\
61\ 200 \\
\underline{1224} \\
1285\ 20.0
\end{array}
$$

This product has one digit to the *right* of the decimal point. ♦

Dividing Two Decimal Fractions

To divide two decimal fractions:

STEP 1 Use the same form as in dividing two whole numbers.

STEP 2 Multiply both the dividend and the divisor (numerator and denominator) by a power of 10 that makes the divisor a whole number.

STEP 3 Divide as you would with whole numbers, and place the decimal point in the quotient directly above the decimal point in the dividend.

Example 3 Divide 24.32 by 6.4.

Method 1:

$$\frac{24.32}{6.4} \times \frac{10}{10} = \frac{243.2}{64}$$

$$
\begin{array}{r}
3.8 \\
64\overline{)243.2} \\
\underline{192} \\
51\ 2 \\
\underline{51\ 2}
\end{array}
$$

Method 2:

$$
\begin{array}{r}
3.8 \\
6.4\overline{)24.3.2} \\
\underline{19\ 2} \\
5\ 1\ 2 \\
\underline{5\ 1\ 2}
\end{array}
$$

Moving the decimal point one place to the right in both the dividend and divisor here is the same as multiplying numerator and denominator by 10 in Method 1. ♦

Example 4 Divide 75.1 by 1.62 and round to the nearest hundredth.

To round to the nearest hundredth, you must carry the division out to the thousandths place and then round to hundredths. We show two methods.

CHAPTER 1 ♦ Basic Concepts

Method 1:

$$\frac{75.1}{1.62} \times \frac{100}{100} = \frac{7510}{162}$$

```
         46.358
    162)7510.000
         648
         ———
         1030
          972
         ———
          580
          486
         ———
           940
           810
          ———
           1300
           1296
          ————
              4
```

Method 2:

```
            46.358
    1.62.)75.10.000
     →       →
            648
            ———
            1030
             972
            ———
             580
             486
            ———
              940
              810
             ———
              1300
              1296
             ————
                 4
```

Moving the decimal point two places to the right in both the dividend and divisor here is the same as multiplying both numerator and denominator by 100 in Method 1.

In both methods, you need to add zeros after the decimal point and carry the division out to the thousandths place. Then round to the nearest hundredth. This gives 46.36 as the result. ♦

Example 5 A gasoline station is leased for $1155 per month. How much gasoline must be sold each month to make the cost of the lease equal to 3.5¢ ($0.035) per gallon?

Divide the cost of the lease per gallon into the cost of the lease per month.

```
                33 000.
    0.035.)1155.000.
      →        →
             105
             ———
              105
              105
             ———
              105
              105
```

That is, 33,000 gal of gasoline must be sold each month. ♦

Example 6 A sprayer tank holds 350 gal. Suppose 20 gal of water and 1.25 gal of pesticide are applied to each acre.

  a. How many acres can be treated on one tankful?
  b. How much pesticide is needed per tankful?

  a. To find the number of acres treated on one tankful, divide the number of gallons of water *and* pesticide into the number of gallons of a full tank.

```
              16.4          or approximately 16 acres/tankful
    21.25.)350.00.0
       →      →
            212 5
            —————
            137 50
            127 50
            ——————
             10 00 0
              8 50 0
             ————————
              1 50 0
```

  b. To find the amount of pesticide needed per tankful, multiply the number of gallons of pesticide applied per acre times the number of acres treated on one tankful.

```
      1.25
       16
      ───
      7 50
      12 5
      ─────
      20.00   or approximately 20 gal/tankful.
```

Using a Calculator to Multiply and Divide Decimal Fractions

Example 7 Multiply: $8.23 \times 65 \times 0.4$.

8.23 [×] 65 [×] .4 [=]

[213.98]

The product is 213.98.

To divide numbers using a calculator, use the steps in the following example.

Example 8 Divide: $3.69 \div 8.2$.

3.69 [÷] 8.2 [=]

[0.45]

The quotient is 0.45.

Now we expand the order of operations from Section 1.2 to include how to treat the fraction bar.

Example 9 Evaluate $\dfrac{(145)(0.64) - 12.5}{16.3 + (9.6)(3.2)}$ using a calculator rounded to three significant digits.

You will need to separate both the numerator and the denominator of the fraction with parentheses and a ÷ sign as follows:

[(] 145 [×] 0.64 [−] 12.5 [)] [÷] [(] 16.3 [+] 9.6 [×] 3.2 [)] [=]

[17.07783922]

Thus, the result is 17.1 rounded to three significant digits.

Example 10 Evaluate: $\dfrac{4 + (9 - 3)^2}{4^3 - 2 \cdot 12}$

$= \dfrac{4 + (6)^2}{4^3 - 2 \cdot 12}$ Subtract within parentheses.

$= \dfrac{4 + 36}{64 - 2 \cdot 12}$ Evaluate the powers.

68 CHAPTER 1 ♦ Basic Concepts

$$= \frac{4+36}{64-24} \quad \text{Multiply.}$$

$$= \frac{40}{40} \quad \text{Add the numbers in the numerator and subtract the numbers in the denominator. When a problem contains a fraction bar, treat the numerator and the denominator separately before dividing.}$$

$$= 1. \quad \text{Divide.} \quad \blacklozenge$$

Example 11 The inductive reactance (in ohms, Ω) in an ac circuit equals the product of 2π times the frequency (in hertz, Hz, that is, cycles/second) times the inductance (in henries, H). Find the inductive reactance in the ac circuit in Figure 1.33. (Use the π key on your calculator, or use $\pi = 3.14$.)

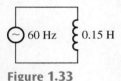

Figure 1.33

The inductive reactance is

$$2\pi \times \text{frequency} \times \text{inductance}$$
$$2 \times \pi \times \quad 60 \quad \times \quad 0.15 \quad = 56.5 \, \Omega \text{ (rounded to 3 significant digits)} \quad \blacklozenge$$

Example 12 The effect of both resistance and inductance in a circuit is called *impedance*. Ohm's law for an ac circuit states that the current (in amps, A) equals the voltage (in volts, V) divided by the impedance (in ohms, Ω). Find the current in a 110-V ac circuit that has an impedance of 65 Ω.

$$\text{current} = \text{voltage} \div \text{impedance}$$
$$= 110 \quad \div \quad 65$$
$$= 1.69 \text{ A (rounded to 3 significant digits)} \quad \blacklozenge$$

Example 13 Bulk salad dressing can come in $\frac{1}{2}$ gal containers. How many ounces of salad dressing are left if a full container is used to fill 15 salad dressing cups that each hold $1\frac{1}{4}$ oz?

First, determine the number of ounces in $\frac{1}{2}$ gallon:

$$\frac{1}{2}\cancel{\text{gal}} \times \frac{4 \cancel{\text{qt}}}{1 \cancel{\text{gal}}} \times \frac{2 \cancel{\text{pt}}}{1 \cancel{\text{qt}}} \times \frac{2 \cancel{\text{cups}}}{1 \cancel{\text{pt}}} \times \frac{8 \text{ oz}}{1 \cancel{\text{cup}}} = 64 \text{ oz}$$

Then, multiply 15 times $1\frac{1}{4}$ oz to find the number of ounces used to fill the cups.

$$15 \times 1\tfrac{1}{4} = 15 \times \frac{5}{4} = \frac{75}{4} = 18\tfrac{3}{4} \text{ oz}$$

Subtract to find the amount left.

$$64 - 18\tfrac{3}{4} = 45\tfrac{1}{4} \text{ oz} \quad \blacklozenge$$

EXERCISES 1.12

Multiply:

1. 3.7 × 0.15
2. 14.1 × 1.7
3. 25.03 × 0.42
4. 4000 × 6.75
5. 5800 × 1600
6. 90,000 × 0.00705

Divide:

7. $36 \div 1.2$
8. $5.1 \div 1.7$
9. $0.6 \div 0.04$
10. $14.356 \div 0.74$

Divide and round to the nearest hundredth:

11. $17,500 \div 70.5$
12. $7900 \div 1.52$
13. $75,000 \div 20.4$
14. $1850 \div 0.75$

Use a calculator to find each result rounded to three significant digits.

15. $\dfrac{27.5 + 136.2}{0.48 + 20.5}$

16. $\dfrac{18.7 - 6.6}{28.4 - 16.1}$

17. $\dfrac{(8.5)(4.7) - (1.4)(2.6)}{6 - (0.28)(1.75)}$

18. $\dfrac{(16.5)(1.95)(12.4) + (6.3)(0.75)}{(125)(0.05) - (0.15)(4.7)}$

Evaluate each expression following the order of operations:

19. $\dfrac{8^2 - 6^2}{4 \cdot 8 + (7 + 9)}$ 20. $\dfrac{148 - 3 \cdot 4^2}{5^3 - 2 \cdot 5^2}$

21. $\dfrac{4 \cdot 5 \cdot 6 - 5 \cdot 2^3}{4^2 \cdot 5 + 5 \cdot 2^2}$ 22. $\dfrac{2^3 + (2 + 3 \cdot 6)^2}{(2 \cdot 5 - 4)^2 + 3 \cdot 5}$

23. A 3.6-ft piece of steel angle is to be divided into 3 equal parts. What is the length of each piece?
24. A 7-ft I-beam is to be divided into 4 equal parts. What is the length of each piece?
25. A small plane flew 321.3 mi in 2.7 h. How fast did the plane fly?
26. To fill the tanks of a small plane after a trip cost $104.06. If it took 24.2 gal of gas, find the price per gallon.
27. An automobile can travel 475 mi on a full tank of fuel. If the automobile holds 17.12 gal of fuel, how many miles per gallon can it get?
28. A set of four new tires cost $565.40. What was the price per tire?
29. A stair detail has 12 risers of $8\tfrac{7}{8}$ in. each. The owner wants only 11 risers, but the total height must be as before. What is the new riser height dimension (in a decimal fraction) for the drawing in Illustration 1?

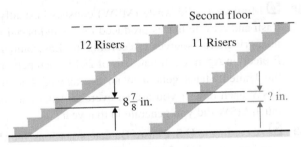

ILLUSTRATION 1

30. If the cost of 8-in. by 20-in. metal duct is $24.96 for 4 ft, how much is it per foot? Per inch?
31. In Illustration 2, find **a.** the perimeter of the outside square and **b.** the length of the center line, l.

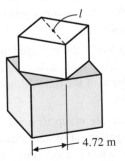

ILLUSTRATION 2

32. In Illustration 3, find the perimeter of the octagon, which has eight equal sides.

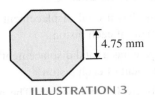

ILLUSTRATION 3

33. The pitch p of a screw is the reciprocal of the number of threads per inch n; that is, $p = \dfrac{1}{n}$. If the pitch is 0.0125, find the number of threads per inch.
34. A 78-ft cable is to be cut into 3.25-ft lengths. How many such lengths need to be cut?
35. A steel rod 32.63 in. long is to be cut into 8 pieces. Each piece is 3.56 in. long. Each cut wastes 0.15 in. of rod in shavings. How many inches of the rod are left?
36. How high is a pile of 32 metal sheets if each sheet is 0.045 in. thick?
37. How many metal sheets are in a stack that measures 18 in. high if each sheet is 0.0060 in. thick?
38. A building measures 45 ft 3 in. by 64 ft 6 in. inside. How many square feet of possible floor space does it contain?
39. Find the cost of excavating a space 87 ft long, 42 ft wide, and 9 ft deep at a cost of $39/yd^3.
40. Each cut on a lathe is 0.018 in. deep. How many cuts would be needed to turn down 2.640-in. metal stock to 2.388 in.?
41. Find the total length of the crankshaft shown in Illustration 4.

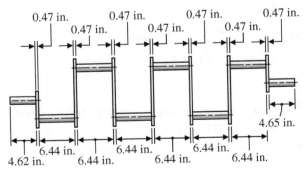

ILLUSTRATION 4

42. A garage service manager has authorized $595 to be spent on overtime to complete a job. If the overtime pay is $29.75/h, how many hours of overtime were authorized?
43. Find the total piston displacement of a six-cylinder engine if each piston displaces a volume of 56.25 in^3.

44. Find the total piston displacement of a six-cylinder engine if each piston displaces 0.9 litres (L).

45. A four-cylinder engine has a total displacement of 2.0 L. Find the displacement of each piston.

46. An eight-cylinder engine has a total displacement of 318 in³. Find the displacement of each piston.

47. New front brake pads are 0.375 in. thick. The average wear rate of these pads in a particular vehicle is 0.062 in. per 15,000 mi. Determine **a.** the expected wear after 45,000 mi and **b.** the expected pad thickness after 60,000 mi.

48. A certain job requires 500 person-hours to complete. How many days will it take for five people working 8 hours per day to complete the job?

49. How many gallons of herbicide are needed for 150 acres of soybeans if 1.6 gal/acre are applied?

50. Suppose 10 gal of water and 1.7 lb of pesticide are to be applied per acre. **a.** How much pesticide would you put in a 300-gal spray tank? **b.** How many acres can be covered with one tankful? (Assume the pesticide dissolves in the water and has no volume.)

51. A cattle feeder buys some feeder cattle, which average 550 lb at $145/hundredweight (that is, $145 per hundred pounds, or $1.45/lb). The price he receives when he sells them as slaughter cattle is $120/hundredweight. If he plans to make a profit of $150 per head, what will be his cost per pound for a 500-lb weight gain?

52. An insecticide is to be applied at a rate of 2 pt/100 gal of water. How many pints are needed for a tank that holds 20 gal? 60 gal? 150 gal? 350 gal? (Assume that the insecticide dissolves in the water and has no volume.)

Find the inductive reactance in each ac circuit (see Example 11):

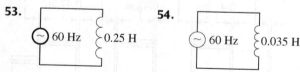

53.   54.

Power (in watts, W) equals voltage times current. Find the power in each circuit:

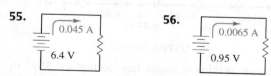

55.   56.

57. Find the current in a 220-V ac circuit with impedance 35.5 Ω. (See Example 12.)

58. A flashlight bulb is connected to a 1.5-V dry cell. If it draws 0.25 A, what is its resistance? (Resistance equals voltage divided by current.)

59. A lamp that requires 0.84 A of current is connected to a 115-V source. What is the lamp's resistance? (Resistance equals voltage divided by current.)

60. A heating element operates on a 115-V line. If it has a resistance of 18 Ω, what current does it draw? (Current equals voltage divided by resistance.)

61. A patient takes 3 tablets of clonidine hydrochloride, containing 0.1 mg each. How many milligrams are taken?

62. One dose of ampicillin for a patient with bronchitis is 2 tablets each containing 0.25 g of medication. How many grams are in one dose?

63. An order reads 0.5 mg of digitalis, and each tablet contains 0.1 mg. How many tablets should be given?

64. An order reads 1.25 mg of digoxin, and the tablets on hand are 0.25 mg. How many tablets should be given?

65. A statute mile is 5280 ft. A nautical mile used in aviation is 6080.6 ft. This gives the conversion 1 statute mile = 0.868 nautical miles. If a plane flew 350 statute miles, how many nautical miles were flown?

66. Five lathes and four milling machines are to be on one circuit. If each lathe uses 16.0 A and each milling machine uses 13.8 A, what is the amperage requirement for this circuit?

67. A steel plate 1.00 in. thick weighs 40.32 lb/ft². Find the weight of a 4.00 ft × 8.00 ft sheet.

68. Municipal solid waste (MSW) consists basically of trash and recycle that is produced by nonindustrial and nonagricultural sources. According to Environmental Protection Agency estimates, as of 2014, each person in the United States generated an average of 4.44 lb of MSW each day. If you are an average American, how much MSW did you generate in that year?

69. According to U.S. Census Bureau projections, the U.S. population was 316,128,839 on December 31, 2013. If that is accurate and if the average person in the United States actually generated 4.4 lb of MSW on that day, how many tons of MSW (rounded to the nearest thousand) were generated in the United States on December 31, 2013?

70. One U.S. bushel contains 1.2445 ft³. A silo that has a capacity of 10,240 ft³ can store how many bushels of corn? Round to the nearest bushel.

71. A rick of firewood is 4 ft by 8 ft by the average length of the stick of firewood. If the firewood in a rick is cut to 16-in. lengths, what is the volume of the rick? Round to the nearest tenth of a cubic foot.

72. An ice cream shop uses simple syrup to make the toppings for sundaes. Each gallon of water requires 0.25 gal of sugar. If 3.25 gal of water is used, how much sugar is needed to make the simple syrup?

73. ✖ A wedding planner provides a memento for each guest. The memento is a 1.5-oz container of mints. Mints are purchased in 5-lb bags. How many bags of mints are needed for 200 guests?

74. ✖ A cafeteria serves 2.2 oz of pasta salad with each sandwich order. **a.** If 110 sandwich orders are expected, how much pasta salad will be needed? **b.** If pasta salad comes in 5.5-lb containers, how many containers need to be purchased? **c.** If the correct number of containers of pasta salad is purchased, how much pasta salad is left after 110 sandwich orders?

1.13 Percent

Percent is the comparison of any number of parts to 100 parts. The word *percent* means "per hundred." The symbol for percent is %.

You wish to put milk in a pitcher so that it is 25% "full" (Figure 1.34a). First, imagine a line drawn down the side of the pitcher. Then imagine the line divided into 100 equal parts. Each mark shows 1%: that is, each mark shows 1 out of 100 parts.

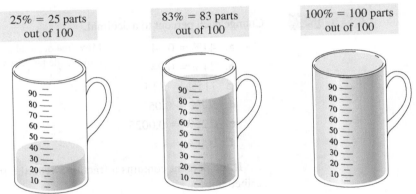

(a) This pitcher is 25% full. (b) This pitcher is 83% full. (c) This pitcher is 100% full.

Figure 1.34
How full is each pitcher?

Finally, count 25 marks from the bottom. The amount of milk below the line is 25% of what the pitcher will hold. Note that 100% is a full, or one whole, pitcher of milk.

One dollar equals 100 cents or 100 pennies. Then, 36% of one dollar equals 36 of 100 parts, or 36 cents or 36 pennies. (See Figure 1.35.)

To save 10% of your salary, you would have to save $10 out of each $100 earned.

When the U.S. government spends 6% of its budget on its debt, interest payments are taking $6 out of every $100 the government collects without reducing the debt.

A salesperson who earns a commission of 8% receives $8 out of each $100 of goods he or she sells.

A car's radiator holds a mixture that is 25% antifreeze. That is, in each hundred parts of mixture, there are 25 parts of pure antifreeze.

A state charges a 5% sales tax. That is, for each $100 of goods that you buy, a tax of $5 is added to your bill. The $5, a 5% tax, is then paid to the state.

Just remember: *percent* means "per hundred."

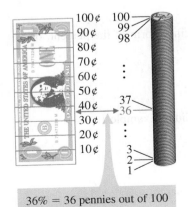

36% = 36 pennies out of 100

Figure 1.35

Changing a Percent to a Decimal

Percent means the number of parts per 100 parts. Any percent can be written as a fraction with 100 as the denominator.

Example 1   Change each percent to a fraction and then to a decimal.

a. $75\% = \dfrac{75}{100} = 0.75$ 75 hundredths

b. $45\% = \dfrac{45}{100} = 0.45$ 45 hundredths

c. $16\% = \dfrac{16}{100} = 0.16$ 16 hundredths

d. $7\% = \dfrac{7}{100} = 0.07$ 7 hundredths

Changing a Percent to a Decimal

To change a percent to a decimal, move the decimal point two places to the *left* (divide by 100). Then remove the percent sign (%).

Example 2   Change each percent to a decimal.

a. $44\% = 0.44$ Move the decimal point two places to the *left* and remove the percent sign (%).
b. $24\% = 0.24$
c. $115\% = 1.15$
d. $5.7\% = 0.057$
e. $0.25\% = 0.0025$
f. $100\% = 1$

If the percent contains a fraction, write the fraction as a decimal. Then proceed as described above.

Example 3   Change each percent to a decimal.

a. $12\dfrac{1}{2}\% = 12.5\% = 0.125$ Write the fraction part as a decimal and then change the percent to a decimal.

b. $6\dfrac{3}{4}\% = 6.75\% = 0.0675$

c. $165\dfrac{1}{4}\% = 165.25\% = 1.6525$

d. $\dfrac{3}{5}\% = 0.6\% = 0.006$

For problems involving percents, we must use the decimal form of the percent, or its equivalent fractional form.

Changing a Decimal to a Percent

Changing a decimal to a percent is the reverse of what we did in Example 1.

Example 4   Write 0.75 as a percent.

$0.75 = \dfrac{75}{100}$ 75 hundredths

      $= 75\%$ *hundredths* means percent

Changing a Decimal to a Percent

To change a decimal to a percent, move the decimal point two places to the *right* (multiply by 100). Write the percent sign (%) after the number.

Example 5 Change each decimal to a percent.

a. $0.38 = 38\%$ Move the decimal point two places to the *right*. Write the percent sign (%) after the number.
b. $0.42 = 42\%$
c. $0.08 = 8\%$
d. $0.195 = 19.5\%$
e. $1.25 = 125\%$
f. $2 = 200\%$

Changing a Fraction to a Percent

In some problems, we need to change a fraction to a percent.

Changing a Fraction to a Percent
1. First, change the fraction to a decimal.
2. Then, change this decimal to a percent.

Example 6 Change $\frac{3}{5}$ to a percent.

$$\begin{array}{r} 0.6 \\ 5\overline{)3.0} \\ \underline{3\ 0} \end{array}$$

First, change $\frac{3}{5}$ to a decimal by dividing the numerator by the denominator.

$0.6 = 60\%$ Then, change 0.6 to a percent by moving the decimal point two places to the right. Write the percent sign (%) after the number.

So $\frac{3}{5} = 0.6 = 60\%$.

Example 7 Change $\frac{3}{8}$ to a percent.

$$\begin{array}{r} 0.375 \\ 8\overline{)3.000} \\ \underline{2\ 4} \\ 60 \\ \underline{56} \\ 40 \\ \underline{40} \end{array}$$

First, change $\frac{3}{8}$ to a decimal.

$0.375 = 37.5\%$ Then, change 0.375 to a percent.

So $\frac{3}{8} = 0.375 = 37.5\%$.

CHAPTER 1 ♦ Basic Concepts

Example 8 Change $\frac{5}{6}$ to a percent.

$$\begin{array}{r} 0.83 \text{ r } 2 \text{ or } 0.83\frac{2}{6} = 0.83\frac{1}{3} \\ 6\overline{)5.00} \\ \underline{4\ 8} \\ 20 \\ \underline{18} \\ 2 \end{array}$$

First, change $\frac{5}{6}$ to a decimal.

NOTE: When the division is carried out to the hundredths place and the remainder is not zero, write the remainder in fraction form, with the remainder over the divisor.

$0.83\frac{1}{3} = 83\frac{1}{3}\%$ Then, change $0.83\frac{1}{3}$ to a percent.

So $\frac{5}{6} = 0.83\frac{1}{3} = 83\frac{1}{3}\%$.

Example 9 Change $1\frac{2}{3}$ to a percent.

$$\begin{array}{r} 0.66 \text{ r } 2 \text{ or } 0.66\frac{2}{3} \\ 3\overline{)2.00} \\ \underline{1\ 8} \\ 20 \\ \underline{18} \\ 2 \end{array}$$

First, change $1\frac{2}{3}$ to a decimal.

That is, $1\frac{2}{3} = 1.66\frac{2}{3}$.

$1.66\frac{2}{3} = 166\frac{2}{3}\%$ Then, change $1.66\frac{2}{3}$ to a percent.

So $1\frac{2}{3} = 1.66\frac{2}{3} = 166\frac{2}{3}\%$.

Changing a Percent to a Fraction

Changing a Percent to a Fraction

1. Change the percent to a decimal.
2. Then change the decimal to a fraction in lowest terms.

Example 10 Change 25% to a fraction in lowest terms.

$25\% = 0.25$ First, change 25% to a decimal by moving the decimal point two places to the *left*. Remove the percent sign (%).

$0.25 = \frac{25}{100} = \frac{1}{4}$ Then, change 0.25 to a fraction. Reduce it to lowest terms.

So $25\% = 0.25 = \frac{1}{4}$.

Example 11 Change 215% to a mixed number.

$$215\% = 2.15$$ First, change 215% to a decimal.

$$2.15 = 2\frac{15}{100} = 2\frac{3}{20}$$ Then, change 2.15 to a mixed number in lowest terms.

So $215\% = 2.15 = 2\frac{3}{20}$.

Changing a Percent That Contains a Mixed Number to a Fraction

1. Change the mixed number to an improper fraction.
2. Then multiply this result by $\frac{1}{100}$* and remove the percent sign (%).

*Multiplying by $\frac{1}{100}$ is the same as dividing by 100. This is what we do to change a percent to a decimal.

Example 12 Change $33\frac{1}{3}\%$ to a fraction.

$$33\frac{1}{3}\% = \frac{100}{3}\%$$ First, change the mixed number to an improper fraction.

$$\frac{100}{3}\% \times \frac{1}{100} = \frac{\cancel{100}^{1}}{3} \times \frac{1}{\cancel{100}_{1}} = \frac{1}{3}$$ Then, multiply this result by $\frac{1}{100}$ and remove the percent sign (%).

So $33\frac{1}{3}\% = \frac{1}{3}$.

Example 13 Change $83\frac{1}{3}\%$ to a fraction.

First, $83\frac{1}{3}\% = \frac{250}{3}\%$.

Then $\frac{250}{3}\% \times \frac{1}{100} = \frac{\cancel{250}^{5}}{3} \times \frac{1}{\cancel{100}_{2}} = \frac{5}{6}$.

So $83\frac{1}{3}\% = \frac{5}{6}$.

EXERCISES 1.13

Change each percent to a decimal:

1. 27%
2. 15%
3. 6%
4. 5%
5. 156%
6. 232%
7. 29.2%
8. 36.2%
9. 8.7%
10. 128.7%
11. 947.8%
12. 68.29%
13. 0.28%
14. 0.78%
15. 0.068%
16. 0.0093%
17. $4\frac{1}{4}\%$
18. $9\frac{1}{2}\%$
19. $\frac{3}{8}\%$
20. $50\frac{1}{3}\%$

Change each decimal to a percent:

21. 0.54
22. 0.25
23. 0.08
24. 0.02
25. 0.62
26. 0.79

27. 2.17   **28.** 0.345   **29.** 4.35
30. 0.225 **31.** 0.185   **32.** 6.25
33. 0.297 **34.** 7.11   **35.** 5.19
36. 0.815 **37.** 0.0187 **38.** 0.0342
39. 0.0029 **40.** 0.00062

Change each percent to a fraction or a mixed number in lowest terms:

61. 75%   **62.** 45%   **63.** 16%
64. 80%   **65.** 60%   **66.** 15%
67. 93%   **68.** 32%   **69.** 275%
70. 325% **71.** 125% **72.** 150%
73. $10\frac{3}{4}$% **74.** $13\frac{2}{5}$% **75.** $10\frac{7}{10}$%
76. $40\frac{7}{20}$% **77.** $17\frac{1}{4}$% **78.** $6\frac{1}{3}$%
79. $16\frac{1}{6}$%   **80.** $72\frac{1}{8}$%

Change each fraction to a percent:

41. $\frac{4}{5}$ **42.** $\frac{3}{4}$ **43.** $\frac{1}{8}$ **44.** $\frac{2}{5}$
45. $\frac{1}{6}$ **46.** $\frac{1}{3}$ **47.** $\frac{4}{9}$ **48.** $\frac{3}{7}$
49. $\frac{3}{5}$ **50.** $\frac{5}{6}$ **51.** $\frac{13}{40}$ **52.** $\frac{17}{50}$
53. $\frac{7}{16}$ **54.** $\frac{15}{16}$ **55.** $\frac{96}{40}$ **56.** $\frac{100}{16}$
57. $1\frac{3}{4}$ **58.** $2\frac{1}{3}$ **59.** $2\frac{5}{12}$ **60.** $5\frac{3}{8}$

81. Complete the following table with the equivalents:

Fraction	Decimal	Percent
$\frac{3}{8}$		
	0.45	
		18%
$1\frac{2}{5}$		
	1.08	
		$16\frac{3}{4}$%

1.14 Rate, Base, and Part

Any percent problem calls for finding one of three things:

1. the rate (percent),
2. the base, or
3. the part.

Such problems are solved using one of three percent formulas. In these formulas, we let

R = the rate (percent)
B = the base
P = the part or amount (sometimes called the percentage)

The following may help you identify which letter stands for each given number and the unknown in a problem:

1. The rate, R, usually has either a percent sign (%) or the word *percent* with it.
2. The base, B, is usually the whole (or entire) amount. The base is often the number that follows the word *of*.
3. The part, P, is usually some fractional part of the base, B. If you identify R and B first, then P will be the number that is not R or B.

NOTE: The base and the part should have the same unit(s) of measure.

Example 1   Given: 25% of $80 is $20. Identify R, B, and P.

R is 25%.   25 is the number with a percent sign. Remember to change 25% to the decimal 0.25 for use in a formula.

B is $80. $80 is the whole amount. It also follows the word *of*.

P is $20. $20 is the part. It is also the number that is not R or B.

Example 2

Given: 72% of the 75 students who took this course last year are now working; find how many are now working. Identify R, B, and P.

R is 72%.	72 is the number with a percent sign.
B is 75 students.	75 is the whole amount. It also follows the word *of*.
P is the unknown.	The unknown is the number that is some fractional part of the base. It is also the number that is not R or B.

Percent Problems: Finding the Part

After you have determined which two numbers are known, you find the third or unknown number by using one of three formulas:

Formulas for Finding Part, Base, and Rate

1. $P = BR$ Use to find the part.

2. $B = \dfrac{P}{R}$ Use to find the base.

3. $R = \dfrac{P}{B}$ Use to find the rate or percent.

NOTE: After you have studied algebra later in the text, you will need to remember only the first formula. Another option for solving percent problems is using proportions, which is shown in Section 7.2.

Example 3

Find 75% of 180.

$R = 75\% = 0.75$ 75 is the number with a percent sign.
$B = 180$ 180 is the whole amount and follows the word *of*.
$P =$ the unknown   Use Formula 1.
$P = BR$
$P = (180)(0.75)$
$ = 135$

Example 4

$45 is $9\dfrac{3}{4}\%$ of what amount?

$R = 9\dfrac{3}{4}\% = 9.75\% = 0.0975$ $9\dfrac{3}{4}\%$ is the number with a percent sign.

$B =$ the unknown   Use Formula 2.
$P = \$45$ $45 is the part.

$B = \dfrac{P}{R}$

$B = \dfrac{\$45}{0.0975}$

$ = \$461.54$ (rounded to the nearest cent)

78 CHAPTER 1 ♦ Basic Concepts

Example 5 What percent of 20 metres is 5 metres?

R = the unknown   Use Formula 3.
B = 20 m   20 m is the whole amount and follows the word *of*.
P = 5 m   5 m is the part.

$$R = \frac{P}{B}$$

$$R = \frac{5 \text{ m}}{20 \text{ m}}$$

$$= 0.25 = 25\%$$

Example 6 Aluminum is 12% of the mass of a given car. This car has 186 kg of aluminum in it. What is the total mass of the car?

R = 12% = 0.12   12 is the number with a percent sign.
B = the unknown   Use Formula 2.
P = 186 kg   186 kg is the part.

$$B = \frac{P}{R}$$

$$B = \frac{186 \text{ kg}}{0.12}$$

$$= 1550 \text{ kg}$$

Example 7 A fuse is a safety device with a core. When too much current flows, the core melts and breaks the circuit. The size of a fuse is the number of amperes of current the fuse can safely carry. A given 50-amp (50-A) fuse blows at 20% overload. What is the maximum current the fuse will carry?

First, find the amount of current overload:

R = 20% = 0.20   20 is the number with a percent sign.
B = 50 A   50 A is the base.
P = the unknown   Use Formula 1.
$P = BR$
$P = (50 \text{ A})(0.20)$
$= 10 \text{ A}$

The maximum current the fuse will carry is the normal current plus the overload:

$$50 \text{ A} + 10 \text{ A} = 60 \text{ A}$$

Example 8 Georgia's salary was $600 per week. Then she was given a raise of $50 per week. What percent raise did she get?

R = the unknown   Use Formula 3.
B = $600 $600 is the base.
P = $50 $50 is the part.

$$R = \frac{P}{B}$$

$$R = \frac{\$50}{\$600}$$

$$= 0.08\frac{1}{3} = 8\frac{1}{3}\%$$

1.14 ♦ Rate, Base, and Part

Example 9 How many millilitres of boric acid are in 1000 mL of a 40% boric acid antiseptic solution?

$R = 40\%$ or 0.40 40 is the number with a percent sign.
$B = 1000$ mL   1000 mL is the base.
$P = $ the unknown   Use Formula 1.
$P = BR$
$P = (1000 \text{ mL})(0.40)$
$ = 400$ mL

There are 400 mL of boric acid in a 40% boric acid antiseptic solution. ♦

Example 10 Castings are listed at $9.50 each. A 12% discount is given if 50 or more are bought at one time. We buy 60 castings.

  a. What is the discount on one casting?
  b. What is the cost of one casting?
  c. What is the total cost?

  a. Discount equals 12% of $9.50.

$R = 12\% = 0.12$
$B = \$9.50$
$P = $ the unknown (the discount)
$P = BR$
$P = (\$9.50)(0.12)$
$ = \$1.14$ (the discount on one casting)

  b. Cost (of one casting) $= $ list $-$ discount
$\phantom{\text{Cost (of one casting)}} = \$9.50 - \$1.14$
$\phantom{\text{Cost (of one casting)}} = \$8.36$

  c. Total cost $=$ cost of one casting times the number of castings
$\phantom{\text{Total cost}} = (\$8.36)(60)$
$\phantom{\text{Total cost}} = \$501.60$ ♦

You may also need to find the percent increase or decrease in a given quantity.

Example 11 Mary's hourly wages changed from $18.40 to $19.55. Find the percent increase in her wages.

First, let's find the change in her wages.

$\$19.55 - \$18.40 = \$1.15$

Then, this change is what percent of her original wage?

$R = \dfrac{P}{B} = \dfrac{\$1.15}{\$18.40} = 0.0625 = 6.25\%$ ♦

The process of finding the percent increase or percent decrease may be summarized by the following formula:

$$\text{percent increase (or percent decrease)} = \dfrac{\text{the change}}{\text{the original value}} \times 100\%$$

CHAPTER 1 ♦ Basic Concepts

Example 12 Normal ac line voltage is 115 volts (V). Find the percent decrease if the line voltage drops to 109 V.

$$\text{percent decrease} = \frac{\text{the change}}{\text{the original value}} \times 100\%$$

$$= \frac{115 \text{ V} - 109 \text{ V}}{115 \text{ V}} \times 100\%$$

$$= 5.2\% \quad \text{(rounded to the nearest tenth of a percent)} \quad ♦$$

Example 13 A banquet hall calculates an 18% tip of the bill before tax for the waitstaff. The waitstaff gives 15% of the tips they receive to the bus help (those who help clear tables, get drinks, etc.). **a.** If the banquet bill before tax and tip is $1348.24, how much is the waitstaff tip before paying the bus help? **b.** How much tip does the bus help receive from the waitstaff? **c.** If sales tax is $7\frac{3}{4}\%$, what is the total bill?

a. The waitstaff receives $0.18 \times \$1348.24 = \$242.68$.
b. The bus help receives $0.15 \times \$242.68 = \$36.40$.
c. Sales tax is $0.0775 \times \$1348.24 = \$104.49$.

So, the total bill is $\$1348.24 + \$242.68 + \$104.49 = \$1695.41$. ♦

The triangle in Figure 1.36 can be used to help you remember the three percent formulas, as follows:

1. $P = BR$ To find the part, cover P; B and R are next to each other on the same line, as in multiplication.
2. $B = \dfrac{P}{R}$ To find the base, cover B; P is over R, as in division.
3. $R = \dfrac{P}{B}$ To find the rate, cover R; P is over B, as in division.

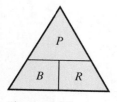

Figure 1.36

EXERCISES 1.14

Identify the rate (R), the base (B), and the part (P) in each statement 1–10 (do not solve the problem):

1. 60 is 25% of 240.
2. $33\frac{1}{3}\%$ of $300 is $100.
3. 40% of 270 is 108.
4. 72 is 15% of 480.
5. At plant A, 4% of the tires made were defective. Plant A made 28,000 tires. How many tires were defective?
6. On the last test, 25 of the 28 students earned passing grades. What percent of students passed?
7. A girls' volleyball team won 60% of its games. The team won 21 games. How many games did it play?
8. A rancher usually loses 10% of his herd every winter due to weather. He has a herd of 15,000. How many does he expect to lose this winter?
9. An electronics firm finds that 6% of the resistors it makes are defective. There were 2050 defective resistors. How many resistors were made?
10. The interest on a $500 loan is $40. What is the rate of interest?
11. Monica receives an annual salary of $32,500 and receives an 8% raise. What is her new annual salary?
12. Jose receives a monthly salary of $2870. His annual salary is raised by 6%. What is his new annual salary?
13. A store has a sale offering 10% off items priced less than $10, 20% off items priced $10–$25, and 30% off items priced more than $25. You purchase items with the following original price tags: $5.49, $3.28, $7.22, $2.12, $12.57, $22.12, $38.42, $40.12, $35.18, $17.88. **a.** What is the total sale price before any sales tax is included? **b.** What is the total sale price if a 6.25% sales tax is added?

When finding the percent, round to the nearest tenth of a percent when necessary:

14. The number 2040 is 7.5% of what number?
15. What percent of 5280 ft is 880 yd?
16. 0.35 mi is 4% of what amount?
17. $72 is 4.5% of what amount?
18. What percent of 7.15 is 3.5?
19. Find 235% of 48.

20. What percent of $\frac{1}{8}$ is $\frac{1}{15}$?
21. Find 28% of 32 volts (V).
22. A customer wants to buy four tires that cost $159.95 each. There is a sales tax of 7.5% on the tires plus an EPA fee of $2.50 for each tire and a disposal fee of $3.69 for each tire. What is the total bill?
23. A welder needs to complete 130 welds. If 97 have been completed so far, what is the percent completed?
24. A welder makes high-quality welds 92% of the time. Out of 115 welds, how many are expected to be of high quality?
25. A small airport has a Cessna 172 rental plane. In one month, 24 h of the 65 total rental hours were for lessons. What percent of the total rental time was the plane rented for lessons?
26. On a cross-country trip, 1.5 h were flown under VFR (Visual Flight Rules), and 0.4 h was flown under IFR (Instrument Flight Rules). What percent of the trip was flown under IFR?
27. An automobile oil filter holds 0.3 qt of oil. The automobile holds 4.5 qt of oil including the filter. What percent of the oil is in the filter?
28. Air enters an air conditioner at the rate of 75 lb/h, and the unit can remove 1.5 lb/h of moisture. If the air entering contains 2 lb/h moisture, what percent of the moisture is removed?
29. Air flows through a duct at 2400 cubic feet per minute (CFM). After several feet and a few vents, the airflow decreases to 1920 CFM. What is the percent drop that has occurred?
30. A building being designed will have fixed windows. Including the frame, the windows are 2 ft wide and 6 ft high. The south wall is 78 ft 6 in. wide by 12 ft 2 in. high. Local codes allow only 20% window area on south walls. How many windows can you draw on this wall?
31. The embankment leading to a bridge must have a maximum 3% slope. The change in elevation shown in Illustration 1 must be dimensioned to meet these criteria. Find dimension A to complete the drawing.

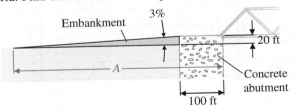

ILLUSTRATION 1

32. A rectangular tank is being designed with an internal catwalk around the inside as shown in Illustration 2. For functional reasons, the walkway cannot be more than 3 in. above the liquid. The liquid level in the tank will be maintained $\frac{3}{4}$ full at all times. What height dimension for the walkway should be put on the drawing?

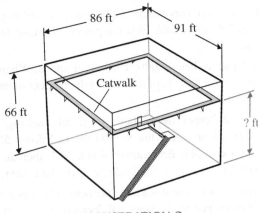

ILLUSTRATION 2

33. The application rate of a chemical is $2\frac{3}{4}$ lb/acre. How many pounds are needed for 160 acres of corn? If the chemical contains 80% active ingredients by weight, how many pounds of active ingredients will be applied? How many pounds of inert ingredients will be applied?
34. U.S. soybeans average 39% protein. A bushel of soybeans weighs 60 lb. How many pounds of protein are in a bushel? A 120-acre field yields 45 bu/acre. How many pounds of protein does that field yield?
35. A dairy cow produced 7310 lb of milk in a year. A gallon of milk weighs 8.6 lb. How many gallons of milk did the cow produce? The milk tested at 4.2% butterfat. How many gallons of butterfat did the cow produce?
36. A lawn has 18,400 ft² of grass. A 60% portion of the lawn is seeded with 2 lb of a lawn seed mix per 1000 ft². How many pounds of seed are needed?
37. A new homeowner planted 150 plants. Because of a hot dry summer, 39 plants died. What percent of the plants lived?
38. How many millilitres of alcohol are in 500 mL of a 15% alcohol solution?
39. How many millilitres of hydrogen peroxide are in 250 mL of a 3% solution?
40. You need 0.15% of 2000 mL. How many millilitres do you need?
41. A 1000-mL solution of acetic acid contains 25 mL of pure acetic acid. Find the percent concentration of this solution.
42. During a line voltage surge, the normal ac voltage increased from 115 V to 128 V. Find the percent increase.
43. During manufacturing, the pressure in a hydraulic line increases from 75 lb/in² to 115 lb/in². What is the percent increase in pressure?

44. The value of Caroline's house decreased from $98,500 to $79,400 when the area's major employer closed the local plant. Find the percent decrease in the value of her house.

45. Due to wage concessions, Bill's hourly wages dropped from $25.50 to $21.88. Find the percent decrease in his wages.

46. A building has 28,000 ft^2 of floor space. When an addition of 6500 ft^2 is built, what is the percent increase in floor space?

47. Two different items both originally selling for $100.00 are on sale. One item is marked down 55%. The second item is first marked down 40%, then an additional 15%. Find the final sale price for each item.

48. A machinist is hired at $22.15 per hour. After a 6-month probationary period, the wage will increase by 32%. If the machinist successfully completes the apprenticeship, what will the pay be per hour?

49. A homeowner harvests a tree to use for firewood. The entire tree weighs 1640 lb. Of that, 95% is cut and split into sticks of firewood. The rest is leaves and branches too small to use for firewood. How much did the firewood weigh?

50. A fisherman catches a total of 125 lb of fish. When the fish are cleaned, 59 lb of fillet remain and the rest is discarded. What percent of the fish was usable as fillets?

51. A flock of mallards (ducks) is called a sord. One particular mallard sord has 250 live mallards when the last hatchling emerges in the spring. At the end of the following winter, the sord has 187 birds remaining before the first egg hatches. What was the survival rate for the sord?

52. In a local community, wildlife biologists estimate a deer population of 135 on January 1. Over the following 12 months, there are 42 live births of deer fawn, 7 are killed by vehicles on the highway, 3 fawns are killed by dogs, 5 are killed by hunters, and 10 die of disease or other injury. As of December 31, what was the deer population and what was the percentage change?

53. Populations of any organism increase when births exceed deaths. In a suburban area in the upper Midwest, the large number of deer was becoming a problem. A Deer Task Force survey suggested that 20 deer per square mile might be an acceptable population level for the citizens in that area. Assume a current population density of 25 deer per square mile and a population growth rate (births minus deaths) of 40% per year. **a.** If there are no significant predators and hunting is not allowed, how many deer can the town planners expect per square mile in the following year? **b.** How many deer can the town planners expect per square mile in the following second year with the same population growth rate?

54. A community has a goal to decrease its municipal solid waste (MSW) by 25% over a 5-year period. Assuming the community has 75,000 residents who each average 4.6 lb of MSW each day, **a.** how much MSW would each resident average each day if the goal were met? **b.** How many tons of MSW would the community generate annually if the goal were met? **c.** Another nearby larger community had decreased its annual MSW by 30% to 73,500 tons. How much was its previous annual amount of MSW?

55. Food cost is often used to determine the price of menu items. Food costs are the total price paid for the food served. An individual steak dinner has the following food costs: steak ($3.96), potato ($0.24), salad ($0.96), vegetable ($0.27), roll ($0.19), and pat of butter ($0.04). If the total food cost is to be 34% of the selling price, what should be the selling price of the steak dinner?

56. The shrinkage of beef in the cooking process is 17%. If the chef starts with 70 lb of beef, how many pounds remain after the cooking process?

57. An *invoice* is an itemized list of goods and services specifying the price and terms of sale. Illustration 3

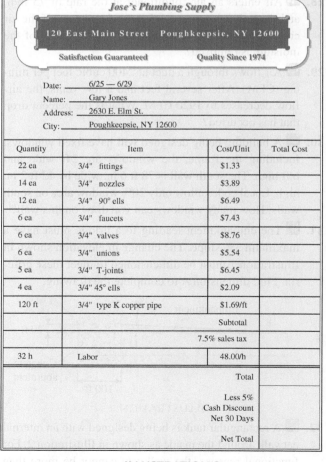

ILLUSTRATION 3

shows an invoice for parts and labor for an addition to a home for the week indicated. Complete the invoice.

58. Delivery sheets are used to document the delivery and itemize the settlement of grain contracts.

a. Find the net amount to be paid to the farmer for the delivery of soybeans on contract as shown in Illustration 4:

Contract:	Delivery Sheet:	Bushels	Price	Amount
S04091803	10750908	1495.47	$9.52	_____

Truck Freight: −$224.67
Grain Assessment: −$72.26
Moisture: −$27.76
Net amount _____

ILLUSTRATION 4

b. Find the net amount to be paid to the farmer for the delivery of corn on contract as shown in Illustration 5:

Contract:	Delivery Sheet:	Bushels	Price	Amount
S0389701	10860301	2398.88	$4.23	_____
S0394456	10860302	1755.89	$4.18	_____
S0398827	10860303	977.60	$3.98	_____

Truck Freight: −$740.67
Grain Assessment: −$55.92
Moisture: −$320.75
Damage: −$337.54
Net amount _____

ILLUSTRATION 5

59. Many lumberyards write invoices for their lumber by the piece. (See Illustration 6.) Complete the invoice, which is for materials delivered to a construction site.

KURT'S LUMBER   400 WEST OAK   AKRON, OHIO 44300

SOLD TO: Robert Bennett 32 Park Pl E. Akron 44305   DATE: 5/16

QUANTITY	DESCRIPTION	UNIT PRICE	TOTAL
66	2" x 4" x 16', fir	$ 7.97	
30	2" x 4" x 10', fir	3.95	
14	2" x 4" x 8', fir	3.39	
17	2" x 6" x 12', fir	6.59	
9	2" x 6" x 10', fir	5.39	
7	2" x 4" x 12', fir	4.97	
10	2" x 8" x 12', fir	11.97	
6	2" x 8" x 16', fir	16.89	
15	4' x 8' x 3/4", T & G plywood	24.25	
80	4' x 8' x 1/2", roof decking	17.29	
250	precut fir studs	2.18	

Subtotal _____
Less 2% cash discount _____
Subtotal _____
5 3/4% sales tax _____
NET TOTAL _____

ILLUSTRATION 6

60. Complete the electronics parts invoice shown in Illustration 7.

APPLIANCE DISTRIBUTORS INCORPORATED
1400 West Elm Street St. Louis, Missouri 63100

Sold to: Maria's Appliance Repair   Date: 9/26
1915 W. Main, Florissant, MO 63031

Quantity	Description	Unit price	Discount	Net amount
3	67A76-1	$ 18.58	40%	
5	A8934-1	65.10	25%	
5	A8935-1	73.95	25%	
8	A8922-2	43.90	25%	
2	A8919-2X	124.60	20%	
5	700A256	18.80	15%	

SUBTOTAL _____
Less 5% if paid in 30 days _____
TOTAL _____

ILLUSTRATION 7

1.15 Powers and Roots

The **square of a number** is the product of that number times itself. The square of 3 is $3 \cdot 3$ or 3^2 or 9. The square of a number may be found with a calculator as follows.

Example 1

Find 73.6^2 rounded to three significant digits.

73.6 x^2 $=$

5416.96

Thus, $73.6^2 = 5420$ rounded to three significant digits.

Example 2 Find 0.135^2 rounded to three significant digits.

.135

0.018225

Thus, $0.135^2 = 0.0182$ rounded to three significant digits. ◆

The **square root of a number** is that positive number which, when multiplied by itself, gives the original number. The square root of 25 is 5 and is written as $\sqrt{25}$. The symbol $\sqrt{}$ is called a *radical*.

Example 3 Find the square roots of **a.** 16, **b.** 64, **c.** 100, and **d.** 144.

  a. $\sqrt{16} = 4$ because $4 \cdot 4 = 16$
  b. $\sqrt{64} = 8$ because $8 \cdot 8 = 64$
  c. $\sqrt{100} = 10$ because $10 \cdot 10 = 100$
  d. $\sqrt{144} = 12$ because $12 \cdot 12 = 144$

Numbers whose square roots are whole numbers are called **perfect squares**. For example, 1, 4, 9, 16, 25, 36, 49, and 64 are perfect squares. ◆

The square root of a number may be found with a calculator as follows.

Example 4 Find $\sqrt{21.4}$ rounded to three significant digits.

 21.4

4.626013402

Thus, $\sqrt{21.4} = 4.63$ rounded to three significant digits. ◆

Example 5 Find $\sqrt{0.000594}$ rounded to three significant digits.

$\sqrt{}$ .000594 =

0.024372115

Thus, $\sqrt{0.000594} = 0.0244$ rounded to three significant digits. ◆

The **cube of a number** is the product of that number times itself three times. The cube of 5 is $5 \cdot 5 \cdot 5$ or 5^3 or 125.

Example 6 Find the cubes of **a.** 2, **b.** 3, **c.** 4, and **d.** 10.

  a. $2^3 = 2 \cdot 2 \cdot 2 = 8$
  b. $3^3 = 3 \cdot 3 \cdot 3 = 27$
  c. $4^3 = 4 \cdot 4 \cdot 4 = 64$
  d. $10^3 = 10 \cdot 10 \cdot 10 = 1000$ ◆

The cube of a number may be found with a calculator as follows.

Example 7   Find 12^3.

12 [∧] *3 [=]

[1728]

Thus, $12^3 = 1728$.

Example 8   Find 4.25^3 rounded to three significant digits.

4.25 [∧] 3 [=]

[76.765625]

Thus, $4.25^3 = 76.8$ rounded to three significant digits.

The **cube root of a number** is that number which, when multiplied by itself three times, gives the original number. The cube root of 8 is 2 and is written as $\sqrt[3]{8}$. (**Note:** $2 \cdot 2 \cdot 2 = 8$. The small ^{3} in the radical is called the *index*.)

Example 9   Find the cube roots of **a.** 8, **b.** 27, and **c.** 125.

  a. $\sqrt[3]{8} = 2$ because $2 \cdot 2 \cdot 2 = 8$
  b. $\sqrt[3]{27} = 3$ because $3 \cdot 3 \cdot 3 = 27$
  c. $\sqrt[3]{125} = 5$ because $5 \cdot 5 \cdot 5 = 125$

Numbers whose cube roots are whole numbers are called **perfect cubes**. For example, 1, 8, 27, 64, 125, and 216 are perfect cubes.

The cube root of a number may be found with a calculator as follows.

Example 10   Find $\sqrt[3]{512}$.

  a. If your calculator has a $\sqrt[3]{}$ button,

[∛] 512 [=]

[8]

Thus, $\sqrt[3]{512} = 8$.

  b. If your calculator has a $\sqrt[x]{}$ button,

3** [ˣ√] 512 [=]

[8]

Thus, $\sqrt[3]{512} = 8$.

---

*Some calculators use the [y^x] button to find a power.

**Here you need to enter the index of the root first.

86 CHAPTER 1 ◆ Basic Concepts

Example 11 Find $\sqrt[3]{4532}$ rounded to three significant digits.

a. If your calculator has a $\sqrt[3]{}$ button,

$\boxed{\sqrt[3]{}}$ 4532 $\boxed{=}$

$\boxed{16.5486778}$

Thus, $\sqrt[3]{4532} = 16.5$ rounded to three significant digits.

b. If your calculator has a $\sqrt[x]{}$ button,

3 $\boxed{\sqrt[x]{}}$ 4532 $\boxed{=}$

$\boxed{16.5486778}$

Thus, $\sqrt[3]{4532} = 16.5$ rounded to three significant digits. ◆

In general, in a **power** of a number, the *exponent* indicates the number of times the *base* is used as a factor. For example, the 4th power of 3 is written 3^4, which means that 3 is used as a factor 4 times ($3^4 = 3 \cdot 3 \cdot 3 \cdot 3 = 81$).

Example 12 Find 2.24^5 rounded to three significant digits.

2.24 $\boxed{\wedge}$ *5 $\boxed{=}$

$\boxed{56.39493386}$

Thus, $2.24^5 = 56.4$ rounded to three significant digits. ◆

EXERCISES 1.15

Find each power rounded to three significant digits:

1. 15^2
2. 25^2
3. 14.9^2
4. 0.0279^2
5. 0.00257^2
6. $54,200^2$
7. 9^3
8. 14^3
9. 8.25^3
10. 0.0225^3
11. 0.169^3
12. 24.8^3
13. 2.75^5
14. 3.5^{10}

Find each root rounded to three significant digits:

15. $\sqrt{8.75}$
16. $\sqrt{12,500}$
17. $\sqrt{4750}$
18. $\sqrt{0.0065}$
19. $\sqrt[3]{75,975}$
20. $\sqrt[3]{9.59}$
21. $\sqrt[3]{0.00777}$
22. $\sqrt[3]{675.88}$

1.16 Applications Involving Percent: Business and Personal Finance

When money is loaned, the borrower pays a fee for using the money to the lender; this fee is called *interest*. Similarly, when someone deposits money into a savings account, the bank or other financial institution pays an interest fee for using the money to the depositor. There are several types of interest problems. We will start with simple interest, the

---

*Some calculators use the $\boxed{y^x}$ button to find a power.

1.16 ♦ Applications Involving Percent: Business and Personal Finance

interest paid only on the original principal. Simple interest problems are based on the following formula:

Simple Interest

$$i = prt$$

where i = the amount of interest paid
p = the principal, which is the original amount of money borrowed or deposited
r = the interest rate written as a decimal
t = the time in years the money is being used

Example 1 A student obtains a simple interest loan to purchase a computer for $1200 at 12.5% over 2 years. How much interest is paid?

Here, $p = \$1200$
$r = 12.5\% = 0.125$, written as a decimal
$t = 2$ years

Then, $i = prt$
$i = (\$1200)(0.125)(2)$
$= \$300$

Thus, the student pays an extra $300 fee to be able to use the computer while the money is being paid. ♦

Savings accounts paid with simple interest are calculated similarly.

When money is borrowed, the total amount to be paid back is the amount borrowed (the principal) plus the interest. The money is usually paid back in regular monthly or weekly installments. The following formula is used to determine the regular payment amount:

Payment Amount

$$\text{payment amount} = \frac{\text{principal} + \text{interest}}{\text{loan period in months or weeks}}$$

Example 2 Find the monthly payment amount for the computer purchase in Example 1.

$$\text{payment amount} = \frac{\text{principal} + \text{interest}}{\text{loan period in months or weeks}}$$

$$\text{payment amount} = \frac{\$1200 + \$300}{24} \quad \text{(2 years} \times \text{12 months)}$$

$$= \$62.50 \text{ per month}$$ ♦

You have probably heard the term *compound interest* used. What is the difference between simple interest and compound interest? *Simple interest* is the amount of money paid or earned on the principal (initial amount) without the interest added to the initial amount. For example, if you accepted a loan of $1000 with 1% simple interest per month, you would owe

$1000 + ($1000)(0.01) = $1000 + $10 = 1010 at the end of one month,
$1000 + $10 + ($1000)(0.01) = $1000 + $10 + $10 = 1020 at the end of two months,
$1000 + $20 + ($1000)(0.01) = $1000 + $20 + $10 = 1030 at the end of three months, etc.

Compound interest is the amount of money paid or earned on the accumulated interest plus the principal with the accumulated interest for the given period then added to the principal. For example, if you accepted a loan of $1000 with 1% interest compounded monthly, you would owe

$1000 + ($1000)(0.01) = $1000 + $10 = 1010 at the end of one month,
$1010 + ($1010)(0.01) = $1010 + $10.10 = 1020.10 at the end of two months,
$1020.10 + ($1020.10)(0.01) = $1020.10 + $10.20 = 1030.30 at the end of three months, etc.

The following formula is used to compute compound interest for any given number of times the interest is compounded annually. Basic examples would include finding the amount of money owed on a loan after a given period of time with no payments made or the amount of money in a savings account after a given period of time with no additional deposits made.

Compound Interest

$$A = P\left(1 + \frac{r}{n}\right)^{nt}$$

where A = the amount after time t
       P = the principal or initial amount
       r = the annual interest rate written as a decimal
       n = the number of times the interest is compounded each year
       t = the number of years

Example 3 Find the amount of money owed at the end of 3 years if $10,000 is borrowed at 8% per year compounded quarterly and no payments are made on the loan.

Here, $P = $10,000$
       $r = 8\% = 0.08$
       $n = 4$ (compounded quarterly means at the end of each of four quarters)
       $t = 3$ years

$$A = P\left(1 + \frac{r}{n}\right)^{nt}$$

$$= $10,000\left(1 + \frac{0.08}{4}\right)^{(4)(3)}$$

$$= $10,000\ (1.02)^{12}$$

$$= $12,682.42$$

Using a scientific calculator, we have

Thus, the answer is $12,682.42 rounded to the nearest cent.

1.16 ♦ Applications Involving Percent: Business and Personal Finance

A variety of formulas can be used to calculate interest. You are advised to always ask how your interest is calculated.

A *mortgage* is a legal document that pledges a house or other real estate as security for the repayment of a loan. This enables a person to buy and use property without having the funds on hand to pay for the house and property outright; a significant down payment is almost always required before a loan is approved. If the borrower fails to repay the loan, the lender may foreclose, which means that the house and property would be sold so that the lender can recover the amount of the loan remaining. A car loan works similarly. If the buyer fails to repay the car loan, the car is reclaimed so that the lender can recover the amount of the car loan remaining. Most mortgage and car loans are calculated with compound interest.

Amortization is the distribution of a single given amount of money into smaller regular installment loan repayments. Each installment consists of both principal and interest. The following amortization formula is used to calculate the periodic (often monthly) installment payment amount for a given loan amount. Although the amount of each installment payment remains the same, some goes toward the interest and some goes toward the principal, with early payments loaded toward interest and later payments loaded toward principal.

Amortization Formula

$$A = P\left(\frac{i(1+i)^n}{(1+i)^n - 1}\right)$$

where A = the periodic payment amount
P = the amount of the principal
i = the interest rate as an annual percentage rate (APR) as a decimal divided by the number of annual interest payments (divide by 12 if monthly or by 52 if weekly)
n = the total number of payments

NOTE: Interest rate i must be given in terms of APR, annual percentage rate, and not APY, annual percentage yield.

Example 4 You need a home loan of $120,000 after your down payment. How much would your monthly house payment be if the bank charges 6% APR for a loan of 30 years?

Here, $P = \$120{,}000$
$i = 0.06/12 = 0.005$
$n = 12$ payments/year $\times$ 30 years $= 360$ monthly payments

$$A = P\left(\frac{i(1+i)^n}{(1+i)^n - 1}\right)$$

$$= \$120{,}000\left(\frac{0.005(1 + 0.005)^{360}}{(1 + 0.005)^{360} - 1}\right)$$

$$= \$120{,}000\left(\frac{0.005(1.005)^{360}}{(1.005)^{360} - 1}\right)$$

$$= \$719.46$$

Using a scientific calculator, we have

120000 ⊠ .005 ⊠ 1.005 ∧ 360 ÷ (1.005 ∧ 360 − 1) =

[719.4606302]

That is, you would pay $719.46 rounded to the nearest cent each month for 30 years to repay this loan. ◆

Example 5 You need an automobile loan of $18,000 after the trade-in and down payment. You are trying to decide between two choices. (a) Receive a loan and pay 1.5% APR for 36 months through the dealership or (b) pay cash to receive a $2500 discount and obtain a loan for the difference at a bank at 6% APR for 36 months.

a. Here, $P = \$18{,}000$
$i = 0.015/12 = 0.00125$
$n = 36$
$$A = P\left(\frac{i(1+i)^n}{(1+i)^n - 1}\right)$$
$$= \$18{,}000\left(\frac{0.00125(1+0.00125)^{36}}{(1+0.00125)^{36} - 1}\right)$$
$$= \$511.65/\text{month} \text{ or } \$511.65 \times 36 = \$18{,}419.40 \text{ total}$$

b. Here, $P = \$18{,}000 - \$2500 = \$15{,}500$
$i = 0.06/12 = 0.005$
$n = 36$
$$A = P\left(\frac{i(1+i)^n}{(1+i)^n - 1}\right)$$
$$= \$15{,}500\left(\frac{0.005(1+0.005)^{36}}{(1+0.005)^{36} - 1}\right)$$
$$= \$471.54/\text{month or } \$471.54 \times 36 = \$16{,}975.44 \text{ total}$$

It pays to do some good financial research! ◆

Many businesses offer cash discounts to encourage customers to buy or use their services and also as an incentive for prompt payment of invoices. These discounts are often called prompt discounts and are also called $\frac{2}{10}$ net 30, which means a 2% discount if paid within 10 days and total amount due in 30 days. Whether you are the seller or the buyer, you need to understand how cash discounts work and be able to determine their cost or benefit to you. The value of the discount may be expressed as an effective annual rate of interest using the following formula:

Value of Cash Discount as an Effective Annual Rate of Interest

$$\text{Interest} = \frac{\text{Discount amount}}{\text{Invoice amount} - \text{Discount amount}} \times \frac{\text{Number of days in year}}{\text{Number of days paid early}^*}$$

---

*This is the number of days between the final discount date and the date when the full payment is due. For example, if the invoice states that if you pay within 10 days of the date on the invoice, you can subtract a cash discount of a given percent of the invoice amount; otherwise the full amount of the invoice is due on the stated date (often 30 days from the invoice date) on the invoice. Here the number of days paid early is 20 days.

1.16 ♦ Applications Involving Percent: Business and Personal Finance

Example 6 You purchased materials and are invoiced $600. The invoice states that if you pay within 10 days, you can subtract a cash discount of 2%; otherwise the full $600 is due within 30 days of the invoice date. Find the effective annual rate of interest for the value of the discount.

The discount amount is (0.02)($600) = $12.

The number of days between the final discount date and the date when the full payment is due is 20 days.

$$\text{Interest} = \frac{\text{Discount amount}}{\text{Invoice amount} - \text{Discount amount}} \times \frac{\text{Number of days in year}}{\text{Number of days paid early}}$$

$$= \frac{\$12}{\$600 - \$12} \times \frac{365 \text{ days}}{20 \text{ days}}$$

$$= 0.372 = 37.2\%$$

As the buyer, this discount gives you a 37.2% effective annual rate of return on your $600. You are not likely to get a higher rate of return on your money elsewhere. You may also want to consider borrowing, especially if the invoice amount is significant, because you would be ahead by the difference between 37.2% and any loan rate you would receive. Prompt payment could also earn you preferred customer status as well as a good credit rating.

As the seller in this example, this discount would cost you 37.2%, and you need to understand the many factors involved of such discounts; for example, they are common and expected in your business area, are needed to help promote sales, or help keep a favorable cash flow. ♦

EXERCISES 1.16

1. One family member loans $2000 to another family member for 3 years at 5% simple interest. **a.** How much interest is paid? **b.** Find the monthly payment.

2. Kim deposits $2500 in a savings account that pays 4.5% simple interest. How much interest does she earn in 2 years?

3. Find the amount of money owed at the end of 4 years if $7500 is borrowed at 6.5% per year compounded quarterly and no payments are made on the loan.

4. Find the amount of money owed at the end of 6 years if $10,500 is borrowed at 5.75% per year compounded semiannually and no payments are made on the loan.

5. Mary Lou received $15,000 from her grandparents for her college education 8 years prior to her enrolling in college. Mary Lou invested the money at 5.5% compounded semiannually. How much money would she have in her savings account when she is ready to enroll in college?

6. Ted invests $6000 at 7.5% compounded quarterly for 5 years. How much money does he have at the end of 5 years?

7. You need a home loan of $150,000 after your down payment. How much would your monthly house payment be if the bank charges 6.5% APR for a loan of 30 years?

8. You need a home loan of $75,000 after your down payment. How much would your monthly house payment be if the bank charges 6.25% APR for a loan of 15 years?

9. A farmer purchased 275 acres of land for $4100/acre. He paid 25% down and obtained a loan for the balance at 6.75% APR over a 20-year period. How much is the annual payment?

10. Denny purchased a new truck for $45,500. He received a rebate of $4500 and paid a sales tax of 6.5% after the rebate. He paid a down payment of 20% and obtained a loan for the balance at 7.25% APR over a 5-year period. How much is the monthly payment?

11. You need an automobile loan of $24,000 after the trade-in and down payment. You are trying to decide between two choices. **a.** Receive a loan and pay 0.75% APR for 36 months through the dealership. Find the monthly payment and the total amount paid. **b.** Pay cash to receive a $1500 discount and obtain a loan for the difference at a bank at 8.5% APR for 36 months. Find the monthly payment and the total amount paid.

12. You need an automobile loan of $19,500 after the trade-in and down payment. You are trying to decide between two choices. **a.** Receive a loan and pay 1.75% APR for 36 months through the dealership. Find the

monthly payment and the total amount paid. **b.** Pay cash to receive a $2500 discount and obtain a loan for the difference at a bank at 6.5% APR for 36 months. Find the monthly payment and the total amount paid.

13. Larry purchased a new combine that cost $220,500 less a rebate of $4500, a trade-in of $9500, and a down payment of $8000. He takes out a loan for the balance at 8% APR over 4 years. Find the annual payment.

14. An agricultural equipment dealer bought a tractor for $150,500 and then includes markups of 3.5% to cover expenses and 0.95% for profit. Fred buys this tractor less a trade-in of $7500 and a down payment of $10,000. He takes out a loan for the balance at 7.25% APR over 5 years. Find the annual payment.

15. Find the amount of money owed at the end of 3 years if $30,000 is borrowed at 5% per year compounded annually and no payments are made on the loan.

16. Find the amount of money owed at the end of 3 years if $30,000 is borrowed at 5% per year compounded monthly and no payments are made on the loan.

17. Find the amount of money owed at the end of 3 years if $30,000 is borrowed at 5% per year compounded daily and no payments are made on the loan.

18. Find the amount of money owed at the end of 3 years if $30,000 is borrowed at 5% per year compounded weekly and no payments are made on the loan.

19. Find the amount of money owed at the end of 5 years if $8400 is borrowed at 3.5% per year compounded monthly and no payments are made on the loan.

20. Nola invests $4000 at 5.5% compounded weekly for 4 years. How much money does she have at the end of 4 years if she makes no withdrawals?

21. Smitty purchased a new automobile for $37,500. The sales tax was 6%. He made a down payment of 10% of the purchase price. He then obtained a loan at 4.2% APR over a 3-year period on the difference. How much is his monthly payment?

22. A dealer offers a 2% discount on an order of $12,000 if paid within 10 days and the total amount due in 30 days. Find the effective annual rate of interest for the value of the discount.

23. A golf shop has an order from a golf club manufacturer to restock the inventory. The order amounts to $15,870. The terms are $\frac{3}{10}$ net 30. What is the effective rate of return on the early payment?

24. You purchased materials and are invoiced $3000. The invoice states that if you pay within 12 days, you can subtract a cash discount of 2%; otherwise the full $3000 is due within 30 days of the invoice date. Find the effective annual rate of interest for the value of the discount.

25. A retailer ordered merchandise totaling $129,115.23 with terms 2.5%/10 net 30. What is the effective rate of return?

26. A homeowner has the siding replaced on his home. The siding contractor offered to do the work for $22,000 with $\frac{2}{30}$ net 60. What is the effective rate of return?

27. An auto dealer offers a discount on a $21,500 automobile if the customer pays cash with terms $\frac{1}{10}$ net 20. What is the effective rate of return?

28. You purchased materials and are invoiced $16,000. The invoice states that if you pay within 10 days, you can subtract a cash discount of 1.5%; otherwise the full $16,000 is due within 30 days of the invoice date. Find the annual rate of interest for the value of the discount.

UNIT 1C REVIEW

1. Change $1\frac{5}{8}$ to a decimal.
2. Change 0.45 to a common fraction in lowest terms.

Perform the indicated operations:

3. $4.206 + 0.023 + 5.9$
4. $120 - 3.065$
5. $12.1 - 6.25 + 0.004$
6. Find the missing dimension in the figure in Illustration 1.
7. Find the perimeter of the figure in Illustration 1.

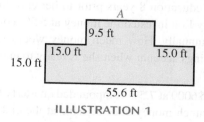

ILLUSTRATION 1

8. Round 45.0649 to **a.** the nearest tenth and **b.** the nearest hundredth.
9. Round 45.0649 to **a.** three significant digits and **b.** four significant digits.

10. Multiply: 4.606 × 0.025
11. Divide: 45.24 ÷ 2.4
12. A cable 18.5 in. long is to be cut into lengths of 2.75 in. each. How many cables of this length can be cut? How much of the cable is left?
13. Change 25% to a decimal.
14. Change 0.724 to a percent.
15. Find 16.5% of 420.
16. 240 is 12% of what number?
17. What percent of 240 yd is 96 yd?
18. Jean makes $16.50/h. If she receives a raise of 6%, find her new wage.

Find each power or root rounded to three significant digits:

19. 45.9^2
20. $\sqrt[3]{831}$

SUMMARY | CHAPTER 1

Glossary of Basic Terms

Area. The number of square units of measure contained in a plane geometric figure. (p. 13)

Common fraction. The ratio or quotient of two integers in the form $\frac{a}{b}$, where $b \neq 0$. The integer above the line is called the numerator; the integer below the line is called the denominator. (p. 24)

Conversion factor. A fraction whose numerator equals its denominator (equal to 1) but with different units to change from one unit or set of units to another. (p. 49)

Cube of a number. The product of that number times itself three times. (p. 84)

Cube root of a number. That number which, when multiplied by itself three times, gives the original number. (p. 85)

Decimal fraction. A fraction whose denominator is 10, 100, 1000, or any power of 10. (p. 53)

Difference. The result of subtracting numbers. (p. 2)

Divisible. One number is divisible by a second number if, when you divide the first number by the second number, you get a zero remainder. (p. 20)

Even integer. An integer divisible by 2. (p. 20)

Formula. A statement of a rule using letters to represent the relationship of certain quantities. (p. 18)

Fraction reduced to lowest terms. A fraction whose numerator and denominator have no common factors. (p. 25)

Grouping symbols. Often parentheses ( ) or brackets [ ] that help to clarify the meaning of mathematical expressions. (p. 11)

Improper fraction. A fraction whose numerator is greater than or equal to its denominator. (p. 26)

Least common denominator (LCD). The smallest positive integer that has all the denominators as divisors. (p. 30)

Measurement. The comparison of an observed quantity with a standard unit quantity. (p. 49)

Mixed number. An integer plus a proper fraction. (p. 26)

Odd integer. An integer that is not divisible by 2. (p. 20)

Percent. The comparison of any number of parts to 100 parts; *percent* means "per hundred." (p. 71)

Perfect cubes. Numbers whose cube roots are whole numbers. (p. 85)

Perfect squares. Numbers whose square roots are whole numbers. (p. 84)

Perimeter. The sum of the lengths of the sides of a geometric figure. (p. 35)

Positive integers. The numbers 1, 2, 3, ... (p. 2)

Power. A number, called the base, and an exponent, which indicates the number of times the base is used as a factor. (p. 86)

Prime factorization. The process of finding the prime factors of a positive integer. (p. 20)

Prime factors of a positive integer. Those prime numbers whose product equals the given positive integer. (p. 20)

Prime number. An integer greater than 1 that has no divisors except itself and 1; the first ten prime numbers are 2, 3, 5, 7, 11, 13, 17, 19, 23, and 29. (p. 20)

Product. The result of multiplying numbers. (p. 5)

Proper fraction. A fraction whose numerator is less than its denominator. (p. 26)

Quotient. The result of dividing numbers. (p. 5)

Significant digits. Those digits in a number we are reasonably sure of being able to rely on in a measurement. (p. 62)

Square of a number. The product of that number times itself. (p. 83)

Square root of a number. That positive number which, when multiplied by itself, gives the original number. (p. 84)

Sum. The result of adding numbers. (p. 2)

Volume. The number of cubic units of measure contained in a solid geometric figure. (p. 14)

Whole numbers. The numbers 0, 1, 2, 3, ... (p. 2)

1.2 Order of Operations

1. **Order of Operations:**
    a. Always do the operations within parentheses or other grouping symbols first.
    b. Then evaluate each power, if any.
    c. Next, perform multiplications and divisions in the order in which they appear as you read from left to right.

CHAPTER 1 ♦ Basic Concepts

    d. Finally, perform additions and subtractions in the order in which they appear as you read from left to right. (p. 11)

1.3 Area and Volume

1. **Area of a rectangle:** $A = lw$ (p. 14)
2. **Volume of a rectangular solid:** $V = lwh$ (p. 15)

1.4 Formulas

1. **Formulas from geometry:**
   a. Area of a triangle: $A = \frac{1}{2}bh$ (p. 19)
   b. Area of a parallelogram: $A = bh$ (p. 19)
   c. Area of a trapezoid: $A = \left(\frac{a+b}{2}\right)h$ (p. 19)

1.5 Prime Factorization

1. Divisibility tests:
   a. *Divisibility by 2:* If a number ends with an even digit, then the number is *divisible by 2*. (p. 21)
   b. *Divisibility by 3:* If the sum of the digits of a number is divisible by 3, then the number itself is *divisible by 3*. (p. 22)
   c. *Divisibility by 5:* If a number has 0 or 5 as its last digit, then the number is *divisible by 5*. (p. 22)

1.6 Introduction to Fractions

1. **Equal or equivalent fractions:**
   a. The numerator and denominator of any fraction may be multiplied or divided by the same number (except zero) without changing the value of the fraction.
   b. Two fractions $\frac{a}{b}$ and $\frac{c}{d}$ are equal or equivalent if $ad = bc$, where $b \neq 0$ and $d \neq 0$. (p. 24)
2. **Simplifying special fractions:**
   a. Any number (except zero) divided by itself equals 1.
   b. Any number divided by 1 equals itself.
   c. Zero divided by any number (except zero) equals zero.
   d. Any number divided by zero is not meaningful and is called *undefined*. (p. 26)
3. **Changing an improper fraction to a mixed number:** To change an improper fraction to a mixed number, divide the numerator by the denominator. The quotient is the whole-number part. The remainder over the divisor is the proper fraction part of the mixed number. (p. 26)

1.7 Addition and Subtraction of Fractions

1. **Finding the least common denominator:** To find the least common denominator (LCD) of a set of fractions:
   a. Factor each denominator into its prime factors.
   b. Write each prime factor the number of times it appears *most* in any *one* denominator in step (a). The LCD is the product of these prime factors. (p. 30)
2. **Adding fractions:** To add two or more fractions with the same denominator, first add the numerators. Then place the sum over the common denominator and simplify. (p. 29)
3. **Subtracting fractions:** To subtract two (or more) fractions with the same denominator, first subtract their numerators. Then place the difference over the common denominator and simplify. (p. 32)
4. **Adding mixed numbers:** To add mixed numbers, find the LCD of the fractions. Add the fractions, then add the whole numbers. Finally, add these two results and simplify. (p. 33)
5. **Subtracting mixed numbers:** To subtract mixed numbers, find the LCD of the fractions. Subtract the fractions, then subtract the whole numbers and simplify. (p. 33)

1.8 Multiplication and Division of Fractions

1. **Multiplying fractions:** To multiply fractions, multiply the numerators and multiply the denominators. Then reduce the resulting fraction to lowest terms. (p. 41)
2. **Dividing fractions:** To divide a fraction by a fraction, invert the fraction that follows the division sign. Then multiply the resulting fractions as described above. (p. 42)

1.9 The U.S. System of Weights and Measures

1. **Choosing conversion factors:** The correct choice for a given conversion factor is the one in which the old units are in the numerator of the original expression and in the denominator of the conversion factor or the old units are in the denominator of the original expression and in the numerator of the conversion factor. That is, set up the conversion factor so that the old units cancel each other. (p. 49)

1.10 Addition and Subtraction of Decimal Fractions

1. **Place values for decimals:** Review Table 1.2. (p. 53)
2. **Changing a common fraction to a decimal:** To change a common fraction to a decimal, divide the numerator of the fraction by the denominator. (p. 55)
3. **Adding or subtracting decimal fractions:** To add or subtract decimal fractions:
   a. Write the decimals so that the digits having the same place value are in vertical columns. (Make certain that the decimal points are also lined up vertically.)

Summary

b. Add or subtract as with whole numbers.
c. Place the decimal point between the ones digit and the tenths digit of the sum or the difference. (Be certain that the decimal point is in the same vertical line as the other decimal points.) (p. 56)

1.11 Rounding Numbers

1. **Rounding numbers to a particular place value:** To round a number to a particular place value:
   a. If the digit in the next place to the right is less than 5, drop that digit and all other following digits. Use zeros to replace any whole-number places dropped.
   b. If the digit in the next place is 5 or greater, add 1 to the digit in the place to which you are rounding. Drop all other following digits. Use zeros to replace any whole-number digits dropped. (p. 62)

2. **Significant digits:**
   a. The following digits in a number are *significant*:
      - All nonzero digits.
      - All zeros between significant digits.
      - All zeros at the end of a decimal number.
   b. The following digits in a number are *not significant*:
      - All zeros at the beginning of a decimal number less than 1.
      - All zeros at the end of a whole number. (p. 63)

3. **Rounding a number to a given number of significant digits:** To round a number to a given number of significant digits:
   a. Count the given number of significant digits from left to right, starting with the first nonzero digit.
   b. If the next digit to the right is less than 5, drop that digit and all other following digits. Use zeros to replace any whole-number places dropped.
   c. If the next digit to the right is 5 or greater, add 1 to the digit in the place to which you are rounding. Drop all other following digits. Use zeros to replace any whole-number places dropped. (p. 63)

1.12 Multiplication and Division of Decimal Fractions

1. **Multiplying two decimal fractions:** To multiply two decimal fractions:
   a. Multiply the numbers as you would whole numbers.
   b. Count the total number of digits to the right of the decimal points in the two numbers being multiplied. Then place the decimal in the product so that it has that same total number of digits to the right of the decimal point. (p. 64)

2. **Dividing two decimal fractions:** To divide two decimal fractions:
   a. Use the same form as in dividing two whole numbers.
   b. Multiply both the divisor and the dividend (denominator and numerator) by a power of 10 that makes the divisor a whole number.
   c. Divide as you would whole numbers, and place the decimal point in the quotient directly above the decimal point in the dividend. (p. 65)

1.13 Percent

1. **Changing a percent to a decimal:** To change a percent to a decimal, move the decimal point two places to the *left* (divide by 100). Then remove the percent sign (%). (p. 72)

2. **Changing a decimal to a percent:** To change a decimal to a percent, move the decimal point two places to the *right* (multiply by 100). Write the percent sign (%) after the number. (p. 73)

3. **Changing a fraction to a percent:** To change a fraction to a percent:
   a. First, change the fraction to a decimal.
   b. Then change this decimal to a percent. (p. 73)

4. **Changing a percent to a fraction:** To change a percent to a fraction:
   a. Change the percent to a decimal.
   b. Then change the decimal to a fraction in lowest terms. (p. 74)

5. **Changing a percent that contains a mixed number to a fraction:** To change a percent that contains a mixed number to a fraction:
   a. Change the mixed number to an improper fraction.
   b. Then multiply this result by $\frac{1}{100}$ and remove the percent sign (%). (p. 75)

1.14 Rate, Base, and Part

1. **Percent problems:** Any percent problem calls for finding one of three things:
   a. The rate, R, usually has either a percent sign (%) or the word *percent* with it.
   b. The base, B, is usually the whole (or entire) amount. The base is often the number that follows the word *of*.
   c. The part, P, is usually some fractional part of the base, B. If you identify R and B first, then P will be the number that is not R or B. (p. 76)

2. **Formulas for finding part, base, and rate:**
   a. $P = BR$ Use to find the part.
   b. $B = \dfrac{P}{R}$ Use to find the base.
   c. $R = \dfrac{P}{B}$ Use to find the rate or percent. (p. 77)

3. **Percent increase (or percent decrease):** The process for finding the percent increase or percent decrease may be summarized by the following formula:

$$\text{percent increase (or percent decrease)} = \frac{\text{the change}}{\text{the original value}} \times 100\%$$

(p. 79)

1.16 Applications Involving Percent: Business and Personal Finance

1. **Simple interest:**

$$i = prt$$

where i = the amount of interest paid

p = the principal, which is the original amount of money borrowed or deposited

r = the interest rate written as a decimal

t = the time in years the money is being used (p. 87)

2. **Payment amount:**

$$\text{payment amount} = \frac{\text{principal} + \text{interest}}{\text{loan period in months or weeks}}$$

(p. 87)

3. **Compound interest:**

$$A = P\left(1 + \frac{r}{n}\right)^{nt}$$

where A = the amount after time t

P = the principal or initial amount

r = the annual interest rate written as a decimal

n = the number of times the interest is compounded each year

t = the number of years (p. 88)

4. **Amortization formula:**

$$A = P\left(\frac{i(1+i)^n}{(1+i)^n - 1}\right)$$

where A = the periodic payment amount

P = the amount of the principal

i = the interest rate as an annual percentage rate (APR) as a decimal divided by the number of annual interest payments (divide by 12 if monthly or by 52 if weekly)

n = the total number of payments (p. 89)

REVIEW | CHAPTER 1

1. Add: $435 + 2600 + 18 + 5184 + 6$
2. Subtract: $60{,}000 - 4{,}803$
3. Multiply: 7060×1300
4. Divide: $68{,}040 \div 300$
5. Evaluate: $12 - 3(5 - 2)$
6. Evaluate: $(6 + 4)8 \div 2 + 3$
7. Evaluate: $18 \div 2 \times 5 \div 3 - 6 + 4 \times 7$
8. Evaluate: $\dfrac{2 \cdot 4^2 + 3(4+5)^2}{10^2 - 3 \cdot 5^2}$
9. A homeowner has a lawn with the dimensions shown in Illustration 1. Fertilizer recommendations for lawns are given in pounds per thousand square feet (tsf). What is the area of the lawn in tsf?

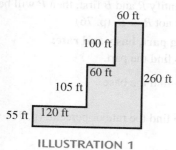

ILLUSTRATION 1

10. Find the volume of the figure in Illustration 2.

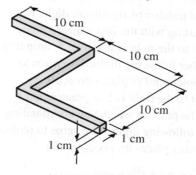

ILLUSTRATION 2

11. Given the formula $C = \dfrac{5}{9}(F - 32)$ and $F = 50$, find C.

12. Given the formula $P = \dfrac{Fs}{t}$, $F = 600$, $s = 50$, and $t = 10$, find P.

13. Is 460 divisible by 3?

14. Find the prime factorization of 54.

15. Find the prime factorization of 330.

Review

Simplify:

16. $\dfrac{36}{56}$ 17. $\dfrac{180}{216}$

Change each to a mixed number in simplest form:

18. $\dfrac{25}{6}$ 19. $3\dfrac{18}{5}$

Change each mixed number to an improper fraction:

20. $2\dfrac{5}{8}$ 21. $3\dfrac{7}{16}$

Perform the indicated operations and simplify:

22. $\dfrac{3}{8} + \dfrac{7}{8} + \dfrac{6}{8}$ 23. $\dfrac{1}{4} + \dfrac{5}{12} + \dfrac{5}{6}$

24. $\dfrac{29}{36} - \dfrac{7}{30}$ 25. $5\dfrac{3}{16} + 9\dfrac{5}{12}$

26. $6\dfrac{3}{8} - 4\dfrac{7}{12}$ 27. $18 - 6\dfrac{2}{5}$

28. $16\dfrac{2}{3} + 1\dfrac{1}{4} - 12\dfrac{11}{12}$ 29. $\dfrac{5}{6} \times \dfrac{3}{10}$

30. $3\dfrac{6}{7} \times 4\dfrac{2}{3}$ 31. $\dfrac{3}{8} \div 6$

32. $\dfrac{2}{3} \div 1\dfrac{7}{9}$ 33. $1\dfrac{4}{5} \div 1\dfrac{9}{16} \times 11\dfrac{2}{3}$

34. Find dimensions *A* and *B* in the figure in Illustration 3.

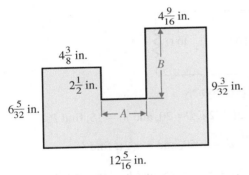

ILLUSTRATION 3

Fill in each blank:

35. 6 lb 9 oz = _____ oz 36. 168 ft = _____ in.
37. 72 ft = _____ yd 38. 36 mi = _____ yd

Write each common fraction as a decimal:

39. $\dfrac{9}{16}$ 40. $\dfrac{5}{12}$

Change each decimal to a common fraction or a mixed number and simplify:

41. 0.45 42. 19.625

Perform the indicated operations:

43. 8.6 + 140 + 0.048 + 19.63
44. 25 + 16.3 − 18 + 0.05 − 6.1
45. 86.7 − 18.035 46. 34 − 0.28
47. 0.605 × 5300 48. 18.05 × 0.106
49. 74.73 ÷ 23.5 50. 9.27 ÷ 0.45

51. Round 248.1563 to **a.** the nearest hundred, **b.** the nearest tenth, and **c.** the nearest ten.

52. Round 5.64908 to **a.** the nearest tenth, **b.** the nearest hundredth, and **c.** the nearest ten-thousandth.

Change each percent to a decimal:

53. 15% 54. $8\dfrac{1}{4}\%$

Change each decimal to a percent:

55. 0.065 56. 1.2

57. What is $8\dfrac{3}{4}\%$ of $12,000?

58. Complete the following table with the equivalents:

Fraction	Decimal	Percent
	0.25	
		$37\dfrac{1}{2}\%$
$\dfrac{5}{6}$		
$8\dfrac{3}{4}$		
	2.4	
		0.15%

59. In a small electronics business, the overhead is $32,000 and the gross income is $84,000. What percent is the overhead?

60. A new tire has a tread depth of $\dfrac{13}{32}$ in. At 16,000 miles, the tread depth is $\dfrac{11}{64}$ in. What percent of the tread is left?

61. A farmer bales 60 tons of hay, which contain 20% moisture. How many tons of dry matter does he harvest?

62. Charlotte wants to paint a room that measures 16 ft wide, 20 ft long, and 9 ft high. One gallon of the paint she selects covers 400 ft² and costs $16.99. The sales tax is 7.75%. She plans to paint the ceiling and the four walls with the same paint. How many gallons of paint should she buy? How much will it cost? She ignores the area of the two windows and the door into the room when she estimates the amount of paint she needs.

98　CHAPTER 1 ♦ Basic Concepts

63. A $\frac{7}{8}$-in.-thick board is being used as a base. If holes are drilled $\frac{9}{16}$ in. to insert posts, what is the thickness of the board left below each hole as shown in Illustration 4?

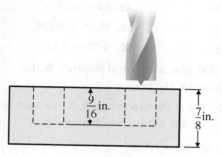

ILLUSTRATION 4

64. A dish manufacturing plant needs a shipping box 20 in. deep with a 10-in. square base. The box company is drawing out the die to cut the cardboard for this box. Find the dimensions of a sheet of cardboard needed to make one box that allows 1 in. for a glue edge as shown in Illustration 5.

ILLUSTRATION 5

Find each power or root rounded to three significant digits:

65. 15.9^3
66. $\sqrt{19200}$

67. Find the amount of money owed at the end of 5 years if $12,500 is borrowed at 4.5% per year compounded semiannually and no payments are made on the loan.

68. Maria purchased a new car for a price of $26,500. She received a rebate of $2500 and paid a sales tax of 7.25% after the rebate. She paid a down payment of 20% and obtained a loan for the balance at 6.5% APR over a 5-year period. How much is her monthly payment?

TEST | CHAPTER 1

1. Add: $47 + 4969 + 7 + 256$
2. Subtract: $4000 - 484$
3. Multiply: 4070×635
4. Divide: $96{,}000 \div 60$

Evaluate each expression:

5. $8 + 2(5 \times 6 + 8)$
6. $15 - 9 \div 3 + 3 \times 4$

7. Find the area of the figure in Illustration 1.

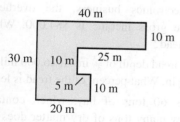

ILLUSTRATION 1

8. Find the volume of the figure in Illustration 2.

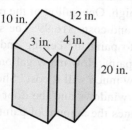

ILLUSTRATION 2

9. Ohm's law states that current (in A) equals voltage (in V) divided by resistance (in Ω). Find the current in the circuit in Illustration 3.

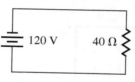

ILLUSTRATION 3

10. If $P = 2l + 2w$, $l = 20$, and $w = 15$, find P.
11. If $t = \dfrac{d}{r}$, $d = 1050$, and $r = 21$, find t.
12. If $P = 2a + b$, $a = 36$, and $b = 15$, find P.

Find the prime factorization of each number:

13. 90
14. 220

Simplify:

15. $\dfrac{30}{64}$
16. $\dfrac{28}{42}$

17. Change $\dfrac{23}{6}$ to a mixed number.

18. Change $3\dfrac{1}{4}$ to an improper fraction.

Perform the indicated operations and simplify:

19. $\dfrac{3}{8} + \dfrac{1}{4}$ 20. $\dfrac{5}{16} - \dfrac{5}{32}$

21. $3\dfrac{1}{8} + 2\dfrac{1}{2} + 4\dfrac{3}{4}$

22. $10\dfrac{1}{8} - 3\dfrac{5}{16}$

23. $3\dfrac{5}{8} + 2\dfrac{3}{16} - 1\dfrac{1}{4}$ 24. $\dfrac{3}{8} \times \dfrac{16}{27}$

25. $\dfrac{3}{8} \div 3\dfrac{5}{16}$ 26. $\dfrac{4}{3} \times \dfrac{1}{8} \times \dfrac{9}{20}$

27. $3\dfrac{5}{8} + 1\dfrac{3}{4} \times 6\dfrac{1}{5}$

28. Given the formula $P = 2l + 2w$, $l = 4\dfrac{3}{4}$, and $w = 2\dfrac{1}{2}$, find P.

29. Find the total current in the circuit in Illustration 4.

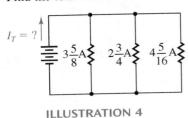

ILLUSTRATION 4

Fill in each blank:

30. 120 ft = _____ yd   31. 3 lb 5 oz = _____ oz

32. Express $\dfrac{5}{8}$ as a decimal.

33. Express 2.12 as a mixed number and simplify.

34. Add: 2.147 + 2.04 + 60 + 0.007 + 0.83

35. Subtract: 400 − 2.81

36. Round 27.2847 to the nearest **a.** tenth and **b.** hundredth.

37. Multiply: 6.12 × 1.32   38. Divide: $6.3 \overline{)0.315}$

39. 59.45 is 41% of what number?

40. 88 is what percent of 284? (to the nearest tenth)

41. Rachel receives a 6.7% increase in salary. If her salary was $612 per week, what is her new weekly salary?

Find each power or root rounded to three significant digits:

42. 0.235^2 43. $\sqrt[3]{304.8}$

44. Assume you invest $4000 at 4.5% compounded monthly for 3 years. How much money do you have at the end of 3 years?

CHAPTER 2 | Signed Numbers and Powers of 10

OBJECTIVES

- Find the absolute value of a signed number.
- Add, subtract, multiply, and divide signed numbers.
- Add, subtract, multiply, and divide signed numbers involving fractions.
- Use the rules for exponents for powers of 10 to multiply, divide, and raise a power to a power.
- Work with numbers in scientific notation.
- Work with numbers in engineering notation.

Mathematics at Work

Electronics technicians perform a variety of jobs. Electronic engineering technicians apply electrical and electronic theory and knowledge to design, build, test, repair, and modify experimental and production electrical equipment in industrial or commercial plants for use by engineering personnel in making engineering design and evaluation decisions.

Other electronics technicians repair electronic equipment such as industrial controls, telemetering and missile control systems, radar systems, and transmitters and antennas using testing instruments. Industrial controls automatically monitor and direct production processes on the factory floor. Transmitters and antennas provide communications links for many organizations. The federal government uses radar and missile control systems for national defense as well as other applications.

Electricians install, maintain, and repair wiring, equipment, and fixtures and ensure that work is in accordance with relevant codes. They also travel to locations to repair equipment and perform preventive maintenance on a regular basis. They use schematics and manufacturers' specifications that show connections and provide instructions on how to locate problems. They also use software programs and testing equipment to diagnose malfunctions. For more information, please visit **www.cengage.com** and access the Student Online Resources for this text.

Electronics Technician
Electronics technician checking a fuse box

2.1 Addition of Signed Numbers

Technicians use negative numbers in many ways. In an experiment using low temperatures, for example, you would record 10° below zero as −10°. Or consider sea level as zero altitude. If a submarine dives 75 m, you could consider its depth as −75 m (75 m below sea level). See Figure 2.1.

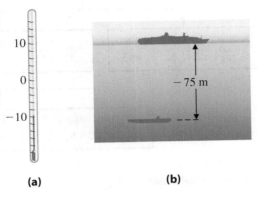

Figure 2.1

These measurements indicate a need for numbers other than positive numbers, which are the only numbers that we have used up to now. To illustrate the graphical relationship of these numbers, we draw a number line as in Figure 2.2 with a point representing zero and with evenly spaced points that represent the **positive integers** (1, 2, 3, . . .) to the right as shown. Then we mark off similarly evenly spaced points to the left of zero. These points correspond to the **negative integers** (−1, −2, −3, . . .) as shown. The negative integers are preceded by a negative (−) sign; −3 is read "negative 3," and −5 is read "negative 5." Each positive integer corresponds to a negative integer. For example, 3 and −3 are corresponding integers. Note that the distances from 0 to 3 and from 0 to −3 are equal.

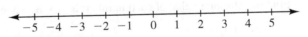

Figure 2.2
The real number line

The **rational numbers** are defined as those numbers that can be written as the ratio of two integers—that is, a/b, where $b \neq 0$. The **irrational numbers** are those numbers that cannot be written as the ratio of two integers, such as $\sqrt{2}$, $-\sqrt{30}$, or the square root of any nonperfect square; π; and several other kinds of numbers that you will study later. The **real numbers** consist of the rational and irrational numbers and are represented on the *real number line* as shown in Figure 2.2. The real number line is dense or full with real numbers; that is, each point on the number line represents a distinct real number, and each real number is represented by a distinct point on the number line. Examples of real numbers are illustrated in Figure 2.3.

The **absolute value of a number** is its distance from zero on the number line. Because distance is always considered positive, the absolute value of a number is never negative. We write the absolute value of a number x as $|x|$; it is read "the absolute value of x." Thus, $|x| \geq 0$.

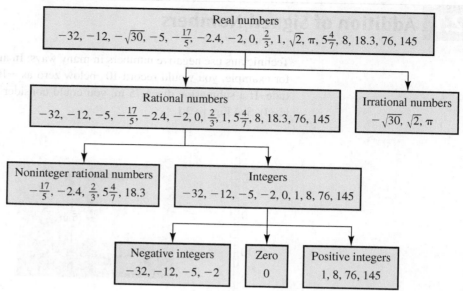

Figure 2.3
Examples of real numbers

("≥" means "is greater than or equal to.") For example, $|+6| = 6$, $|4| = 4$, and $|0| = 0$. However, if a number is less than 0 (negative), its absolute value is the corresponding positive number. For example, $|-6| = 6$ and $|-7| = 7$.

Remember:

> The absolute value of a number is never negative.

Example 1   Find the absolute value of each number: **a.** +3, **b.** −5, **c.** 0, **d.** −10, **e.** 15.

a. $|+3| = 3$ The distance between 0 and +3 on the number line is 3 units.
b. $|-5| = 5$ The distance between 0 and −5 on the number line is 5 units.
c. $|0| = 0$ The distance is 0 units.
d. $|-10| = 10$ The distance between 0 and −10 on the number line is 10 units.
e. $|15| = |+15| = 15$ The distance between 0 and +15 on the number line is 15 units.

♦

One number is *larger* than another number if the first number is to the *right* of the second on the number line in Figure 2.2. Thus, 5 is larger than 1, 0 is larger than −3, and 2 is larger than −4. Similarly, one number is *smaller* than another if the first number is to the *left* of the second on the number line in Figure 2.2. Thus, 0 is smaller than 3, −1 is smaller than 4, and −5 is smaller than −2.

The use of **signed numbers** (positive and negative numbers) is one of the most important operations that we will study. Signed numbers are used in work with exponents. Operations with signed numbers are also essential for success in the basic algebra that follows later.

Adding Two Numbers with Like Signs (the Same Signs)

1. To add two positive numbers, add their absolute values. The result is positive. A positive sign may or may not be used before the result. It is usually omitted.
2. To add two negative numbers, add their absolute values and place a negative sign before the result.

Example 2 Add:

a. $(+2) + (+3) = +5$
b. $(-4) + (-6) = -10$
c. $(+4) + (+5) = +9$
d. $(-8) + (-3) = -11$

Adding Two Numbers with Unlike Signs

To add a negative number and a positive number, find the difference of their absolute values. The sign of the number having the larger absolute value is placed before the result.

Example 3 Add:

a. $(+4) + (-7) = -3$
b. $(-3) + (+8) = +5$
c. $(+6) + (-1) = +5$
d. $(-8) + (+6) = -2$
e. $(-2) + (+5) = +3$
f. $(+3) + (-11) = -8$

Adding Three or More Signed Numbers

STEP 1 Add the positive numbers.
STEP 2 Add the negative numbers.
STEP 3 Add the sums from Steps 1 and 2 according to the rules for addition of two signed numbers.

Example 4 Add: $(-8) + (+12) + (-7) + (-10) + (+3)$.

STEP 1 $(+12) + (+3) = +15$ Add the positive numbers.
STEP 2 $(-8) + (-7) + (-10) = -25$ Add the negative numbers.
STEP 3 $(+15) + (-25) = -10$ Add the sums from Steps 1 and 2.

Therefore, $(-8) + (+12) + (-7) + (-10) + (+3) = -10$.

Example 5 Add: $(+4) + (-7) + (-2) + (+6) + (-3) + (-5)$.

STEP 1 $(+4) + (+6) = +10$
STEP 2 $(-7) + (-2) + (-3) + (-5) = -17$
STEP 3 $(+10) + (-17) = -7$

Therefore, $(+4) + (-7) + (-2) + (+6) + (-3) + (-5) = -7$.

SIGNED NUMBERS DRILL 1

Being able to work quickly and accurately with signed numbers is as important to success in algebra as being able to work quickly and accurately with the basic number facts in arithmetic. The following oral signed numbers drill is designed to be used in class by the instructor, who asks students in random order to add a given set of signed numbers and verbally give the answer. A friend may help you or a group of students may help each other by using this drill for extra practice. The answers are on the bottom of page 106 to make it easier for you to drill yourself.

Add the following signed numbers:

1. $(+4) + (+5)$
2. $(-6) + (-2)$
3. $(+7) + (-3)$
4. $(-8) + (+1)$
5. $(-3) + (+9)$
6. $(-7) + (-7)$
7. $(-4) + (0)$
8. $(0) + (+8)$
9. $(+2) + (+1)$
10. $(-5) + (-7)$
11. $(-6) + (+2)$
12. $(+6) + (-7)$
13. $(-10) + (+8)$
14. $(+3) + (-7)$
15. $(+6) + (+9)$
16. $(-8) + (-4)$
17. $(+4) + (-4)$
18. $(-7) + (+8)$
19. $(+1) + (-6)$
20. $(-9) + (-7)$
21. $(+3) + (-2)$
22. $(-6) + (+9)$
23. $(+2) + (-8)$
24. $(-7) + (-3)$
25. $(+7) + (-8)$
26. $(-3) + (-9)$
27. $(+1) + (+5)$
28. $(-2) + (-2)$
29. $(-3) + (+4)$
30. $(+9) + (-4)$

EXERCISES 2.1

Find the absolute value of each number:

1. 3
2. −4
3. −6
4. 0
5. +4
6. +8
7. 17
8. −37
9. −15
10. 49

Add:

11. $(+4) + (+6)$
12. $(-5) + (-9)$
13. $(+9) + (-2)$
14. $(-10) + (+4)$
15. $(+5) + (-7)$
16. $(-4) + (+6)$
17. $(-3) + (-9)$
18. $(+4) + (-9)$
19. $(12) + (-6)$
20. $(-12) + (6)$
21. $(-4) + (-5)$
22. $(+2) + (-11)$
23. $(-3) + (+7)$
24. $(+8) + (+2)$
25. $(-5) + (+2)$
26. $(-7) + (-6)$
27. $(+7) + (-8)$
28. $(8) + (-3)$
29. $(-10) + (6)$
30. $(+4) + (-11)$
31. $(-8) + (2)$
32. $(+3) + (+7)$
33. $(-2) + (0)$
34. $(0) + (+3)$
35. $(9) + (-5)$
36. $(+9) + (-9)$
37. $(16) + (-7)$
38. $(-19) + (-12)$
39. $(-6) + (+9)$
40. $(+20) + (-30)$
41. $(-1) + (-3) + (+8)$
42. $(+5) + (-3) + (+4)$
43. $(+1) + (+7) + (-1)$
44. $(-5) + (-9) + (-4)$
45. $(-9) + (+6) + (-4)$
46. $(+8) + (+7) + (-2)$
47. $(+8) + (-8) + (+7) + (-2)$
48. $(-6) + (+5) + (-8) + (+4)$
49. $(-4) + (-7) + (-7) + (-2)$
50. $(-3) + (-9) + (+5) + (+6)$
51. $(-1) + (-2) + (+9) + (-8)$
52. $(+6) + (+5) + (-7) + (-3)$
53. $(-6) + (+2) + (+7) + (-3)$
54. $(+8) + (-1) + (+9) + (+6)$
55. $(-5) + (+1) + (+3) + (-2) + (-2)$
56. $(+5) + (+2) + (-3) + (-9) + (-9)$
57. $(-5) + (+6) + (-9) + (-4) + (-7)$
58. $(-9) + (+7) + (-6) + (+5) + (-8)$
59. $(+1) + (-4) + (-2) + (+2) + (-9)$
60. $(-1) + (-2) + (-6) + (-3) + (-5)$
61. $(-2) + 8 + (-4) + 6 + (-1)$
62. $14 + (-5) + (-1) + 6 + (-3)$
63. $5 + 6 + (-2) + 9 + (-7)$
64. $(-5) + 4 + (-1) + 6 + (-7)$
65. $(-3) + 8 + (-4) + (-7) + 10$
66. $16 + (-7) + (-5) + 20 + (-5)$
67. $3 + (-6) + 7 + 4 + (-4)$
68. $(-8) + 6 + 9 + (-5) + (-4)$
69. $(-5) + 4 + (-7) + 2 + (-8)$
70. $7 + 9 + (-6) + (-4) + 9 + (-2)$

2.2 Subtraction of Signed Numbers

Subtracting Two Signed Numbers

To subtract two signed numbers, change the sign of the number being subtracted and add according to the rules for addition of signed numbers.

Example 1 Subtract:

a. $(+2) - (+5) = (+2) + (-5)$
$ = -3$ — To subtract, change the sign of the number being subtracted, $+5$, and add.

b. $(-7) - (-6) = (-7) + (+6)$
$ = -1$ — To subtract, change the sign of the number being subtracted, -6, and add.

c. $(+6) - (-4) = (+6) + (+4)$
$ = +10$ — To subtract, change the sign of the number being subtracted, -4, and add.

d. $(+1) - (+6) = (+1) + (-6)$
$ = -5$ — To subtract, change the sign of the number being subtracted, $+6$, and add.

e. $(-8) - (-10) = (-8) + (+10) = +2$

f. $(+9) - (-6) = (+9) + (+6) = +15$

g. $(-4) - (+7) = (-4) + (-7) = -11$

Subtracting More Than Two Signed Numbers

When more than two signed numbers are involved in subtraction, change the sign of *each* number being subtracted and add the resulting signed numbers.

Example 2 Subtract: $\quad (-4) - (-6) - (+2) - (-5) - (+7)$
$\ = (-4) + (+6) + (-2) + (+5) + (-7)$ — Change the sign of each number being subtracted and add the resulting signed numbers.

STEP 1 $(+6) + (+5) = +11$

STEP 2 $(-4) + (-2) + (-7) = -13$

STEP 3 $(+11) + (-13) = -2$

Therefore, $(-4) - (-6) - (+2) - (-5) - (+7) = -2$.

Adding and Subtracting Combinations of Signed Numbers

When combinations of additions and subtractions of signed numbers occur in the same problem, change *only* the sign of each number being subtracted. Then add the resulting signed numbers.

Example 3 Perform the indicated operations:

$(+4) - (-5) + (-6) - (+8) - (-2) + (+5) - (+1)$
$= (+4) + (+5) + (-6) + (-8) + (+2) + (+5) + (-1)$ — Change *only* the sign of each number being subtracted and add the resulting signed numbers.

STEP 1 $(+4) + (+5) + (+2) + (+5) = +16$

STEP 2 $(-6) + (-8) + (-1) = -15$

STEP 3 $(+16) + (-15) = +1$

Therefore, $(+4) - (-5) + (-6) - (+8) - (-2) + (+5) - (+1) = +1$.

Example 4 Perform the indicated operations:

$$(-12) + (-3) - (-5) - (+6) - (-1) + (+4) - (+3) - (-8)$$
$$= (-12) + (-3) + (+5) + (-6) + (+1) + (+4) + (-3) + (+8)$$

STEP 1 $\quad (+5) + (+1) + (+4) + (+8) = +18$
STEP 2 $\quad (-12) + (-3) + (-6) + (-3) = -24$
STEP 3 $\quad (+18) + (-24) = -6$

Therefore, $(-12) + (-3) - (-5) - (+6) - (-1) + (+4) - (+3) - (-8) = -6$.

SIGNED NUMBERS DRILL 2

The following oral signed numbers drill is designed to be used in class by the instructor, who will ask students in random order to add or subtract a given set of signed numbers by placing a plus or minus sign in the blank and verbally giving the answer. A friend may help you or a group of students may help each other by using this drill for extra practice. The answers are on the bottom of page 108 to make it easier for you to drill yourself.

Add or subtract the following signed numbers:

1. (+3) ___ (+5)   2. (−7) ___ (−2)   3. (+8) ___ (−3)
4. (−9) ___ (+1)   5. (−3) ___ (+7)   6. (−7) ___ (−7)
7. (−5) ___ (0)   8. (0) ___ (+2)   9. (+4) ___ (+1)
10. (−5) ___ (−8) 11. (−6) ___ (+3) 12. (+8) ___ (−7)
13. (−1) ___ (+7) 14. (+3) ___ (−9) 15. (+6) ___ (+4)
16. (−8) ___ (−3) 17. (+5) ___ (−5) 18. (−6) ___ (+8)
19. (+1) ___ (−7) 20. (−9) ___ (−5) 21. (+3) ___ (−2)
22. (−2) ___ (+4) 23. (+2) ___ (−7) 24. (−5) ___ (−3)
25. (+4) ___ (−8) 26. (−2) ___ (−9) 27. (+2) ___ (+5)
28. (−3) ___ (−3) 29. (+2) ___ (−4) 30. (+5) ___ (−4)

EXERCISES 2.2

Subtract:

1. $(+4) - (+6)$
2. $(-5) - (-9)$
3. $(+9) - (-2)$
4. $(-10) - (+4)$
5. $(+5) - (-7)$
6. $(-4) - (+6)$
7. $(-3) - (-9)$
8. $(+4) - (-9)$
9. $(12) - (-6)$
10. $(-12) - (6)$
11. $(3) - (9)$
12. $(-2) - (0)$
13. $(0) - (4)$
14. $(-5) - (-10)$
15. $(-7) - (-7)$
16. $(18) - (-18)$
17. $(-4) - (-5)$
18. $(+2) - (-11)$
19. $(-3) - (+7)$
20. $(+6) - (+8)$
21. $(-5) - (+2)$
22. $(-7) - (-6)$
23. $(+7) - (-8)$
24. $(8) - (-3)$
25. $(-10) - (6)$
26. $(+4) - (-11)$
27. $(-8) - (+2)$
28. $(+3) - (+7)$
29. $(-2) - (0)$
30. $(0) - (+3)$
31. $(9) - (-5)$
32. $(+9) - (-9)$
33. $(16) - (-7)$
34. $(-19) - (-12)$
35. $(-6) - (+9)$
36. $(+20) - (-30)$

ANSWERS TO SIGNED NUMBERS DRILL 1

1. +9 2. −8 3. +4 4. −7 5. +6 6. −14 7. −4 8. +8 9. +3 10. −12 11. −4 12. −1 13. −2
14. −4 15. +15 16. −12 17. 0 18. +1 19. −5 20. −16 21. +1 22. +3 23. −6 24. −10 25. −1
26. −12 27. +6 28. −4 29. +1 30. +5

Perform the indicated operations:

37. $(+6) - (-3) - (+1)$
38. $(+3) - (-7) - (+6)$
39. $(-3) - (-7) - (+8)$
40. $(+3) - (4) - (-9)$
41. $(+3) - (-6) - (+9) - (-8)$
42. $(+10) - (-4) - (6) - (-9)$
43. $(+5) - (-5) + (-8)$
44. $(+1) + (-7) - (-7)$
45. $(-3) + (-5) - (0) - (+7)$
46. $(+4) - (-3) + (+6) - (8)$
47. $(+4) - (-11) - (+12) + (-6)$
48. $(8) - (-6) - (+18) - (4)$
49. $(-7) - (+6) + (-3) - (-2) - (+9)$
50. $(-3) + (-4) + (+7) - (-2) - (+6)$
51. $-9 + 8 - 5 + 6 - 4$
52. $-12 + 2 + 30 - 6$
53. $-8 + 12 - 7 - 4 + 6$
54. $7 + 4 - 8 - 9 + 3$
55. $16 - 18 + 4 - 7 - 2 + 9$
56. $3 - 7 + 5 - 6 - 7 + 2$
57. $8 + 10 - 20 + 4 - 5 - 6 + 1$
58. $5 - 6 - 7 + 2 - 8 + 10$
59. $9 - 7 + 4 + 3 - 8 - 6 - 6 + 1$
60. $-4 + 6 - 7 - 5 + 6 - 7 - 1$

2.3 Multiplication and Division of Signed Numbers

Multiplying Two Signed Numbers

1. If the two numbers have the same sign, multiply their absolute values. This product is always positive.
2. If the two numbers have different signs, multiply their absolute values and place a negative sign before the product.

Example 1 Multiply:

a. $(+2)(+3) = +6$ — Multiply the absolute values of the signed numbers; the product is positive because the two numbers have the same sign.

b. $(-4)(-7) = +28$ — Multiply the absolute values of the signed numbers; the product is positive because the two numbers have the same sign.

c. $(-2)(+4) = -8$ — Multiply the absolute values of the signed numbers; the product is negative because the two numbers have different signs.

d. $(-6)(+5) = -30$ — Multiply the absolute values of the signed numbers; the product is negative because the two numbers have different signs.

e. $(+3)(+4) = +12$
f. $(-6)(-9) = +54$
g. $(-5)(+7) = -35$
h. $(+4)(-9) = -36$

♦

Multiplying More Than Two Signed Numbers

1. If the number of negative factors is even (divisible by 2), multiply the absolute values of the numbers. This product is positive.
2. If the number of negative factors is odd, multiply the absolute values of the numbers and place a negative sign before the product.

Example 2 Multiply: $(-11)(+3)(-6) = +198$ The number of negative factors is 2, which is even; therefore, the product is positive. ◆

Example 3 Multiply: $(-5)(-4)(+2)(-7) = -280$ The number of negative factors is 3, which is odd; therefore, the product is negative. ◆

> **Dividing Two Signed Numbers**
>
> 1. If the two numbers have the same sign, divide their absolute values. This quotient is always positive.
> 2. If the two numbers have different signs, divide their absolute values and place a negative sign before the quotient.

Since multiplication and division are related operations, the same rules for signed numbers apply to both operations.

Example 4 Divide:

a. $\dfrac{+12}{+2} = +6$ Divide the absolute values of the signed numbers; the quotient is positive because the two numbers have the same sign.

b. $\dfrac{-18}{-6} = +3$ Divide the absolute values of the signed numbers; the quotient is positive because the two numbers have the same sign.

c. $\dfrac{+20}{-4} = -5$ Divide the absolute values of the signed numbers; the quotient is negative because the two numbers have different signs.

d. $\dfrac{-24}{+6} = -4$ Divide the absolute values of the signed numbers; the quotient is negative because the two numbers have different signs.

e. $(+30) \div (+5) = +6$ g. $(+16) \div (-4) = -4$
f. $(-42) \div (-2) = +21$ h. $(-45) \div (+9) = -5$ ◆

ANSWERS TO SIGNED NUMBERS DRILL 2

The first answer is for addition and the second answer is for subtraction.

1. +8, −2 **2.** −9, −5 **3.** +5, +11 **4.** −8, −10 **5.** +4, −10 **6.** −14, 0 **7.** −5, −5 **8.** +2, −2 **9.** +5, +3
10. −13, +3 **11.** −3, −9 **12.** +1, +15 **13.** +6, −8 **14.** −6, +12 **15.** +10, +2 **16.** −11, −5 **17.** 0, +10
18. +2, −14 **19.** −6, +8 **20.** −14, −4 **21.** +1, +5 **22.** +2, −6 **23.** −5, +9 **24.** −8, −2 **25.** −4, +12
26. −11, +7 **27.** +7, −3 **28.** −6, 0 **29.** −2, +6 **30.** +1, +9

SIGNED NUMBERS DRILL 3

The following oral signed numbers drill is designed to be used in class by the instructor, who will ask students in random order to randomly add, subtract, or multiply a given set of signed numbers by placing a plus, minus, or times sign in the blank and verbally giving the answer. A friend may help you or a group of students may help each other by using this drill for extra practice. The answers are on the bottom of page 110 to make it easier for you to drill yourself.

Add, subtract, or multiply the following signed numbers:

1. (+4) __ (+8)   2. (−6) __ (−3)   3. (+8) __ (−3)
4. (−9) __ (+1)   5. (−3) __ (+5)   6. (−6) __ (−6)
7. (−7) __ (0)   8. (0) __ (−4)   9. (+3) __ (+1)
10. (−5) __ (−8) 11. (−6) __ (+3) 12. (+8) __ (−7)
13. (−1) __ (+7) 14. (+3) __ (−3) 15. (+5) __ (+9)
16. (−8) __ (−2) 17. (+5) __ (−5) 18. (−7) __ (+6)
19. (+1) __ (−4) 20. (−9) __ (−9) 21. (+3) __ (−1)
22. (−6) __ (+8) 23. (+2) __ (−7) 24. (−6) __ (−3)
25. (+7) __ (−4) 26. (−6) __ (−9) 27. (+7) __ (+3)
28. (−10) __ (−10) 29. (−3) __ (+5) 30. (+8) __ (−5)

EXERCISES 2.3

Multiply:

1. $(+4)(+6)$
2. $(-5)(-9)$
3. $(+9)(-2)$
4. $(-10)(+4)$
5. $(+5)(-7)$
6. $(-4)(+6)$
7. $(-3)(-9)$
8. $(+4)(-9)$
9. $(12)(-6)$
10. $(-12)(6)$
11. $(3)(9)$
12. $(-2)(0)$
13. $(0)(4)$
14. $(-5)(-10)$
15. $(-7)(-7)$
16. $(18)(-18)$
17. $(+15)(-20)$
18. $(-11)(-3)$
19. $(1)(-13)$
20. $(-22)(+14)$
21. $(+3)(-2)$
22. $(+5)(+7)$
23. $(-6)(-8)$
24. $(-9)(+2)$
25. $(-7)(-3)$
26. $(-4)(+4)$
27. $(-8)(+2)$
28. $(+5)(-3)$
29. $(-6)(-9)$
30. $(+4)(+7)$
31. $(-6)(+1)$
32. $(+4)(-2)$
33. $(-8)(-3)$
34. $(-2)(+6)$
35. $(-3)(-9)$
36. $(8)(-4)$
37. $(-9)(7)$
38. $(-8)(0)$
39. $(9)(-1)$
40. $(-4)(-10)$
41. $(+3)(-2)(+1)$
42. $(-5)(-9)(+2)$
43. $(-3)(+3)(-4)$
44. $(-2)(-8)(-3)$
45. $(+5)(+2)(+3)$
46. $(-4)(+3)(0)(+3)$
47. $(-3)(-2)(4)(-7)$
48. $(-3)(-1)(+1)(+2)$
49. $(-9)(-2)(+3)(+1)(-3)$
50. $(-6)(-2)(-4)(-1)(-2)(+2)$

Divide:

51. $\dfrac{+10}{+2}$
52. $\dfrac{-8}{-4}$
53. $\dfrac{+27}{-3}$
54. $\dfrac{-48}{+6}$
55. $\dfrac{-32}{-4}$
56. $\dfrac{+39}{-13}$
57. $\dfrac{+14}{+7}$
58. $\dfrac{45}{+15}$
59. $\dfrac{-54}{-6}$
60. $\dfrac{+72}{-9}$
61. $\dfrac{-100}{+25}$
62. $\dfrac{+84}{+12}$
63. $\dfrac{-75}{-25}$
64. $\dfrac{+36}{-6}$
65. $\dfrac{+85}{+5}$
66. $\dfrac{-270}{+9}$
67. $\dfrac{+480}{+12}$
68. $\dfrac{-350}{+70}$
69. $\dfrac{-900}{-60}$
70. $\dfrac{+4800}{-240}$
71. $(-49) \div (-7)$
72. $(+9) \div (-3)$
73. $(+80) \div (+20)$
74. $(-60) \div (+12)$
75. $(+45) \div (-15)$
76. $(+120) \div (-6)$
77. $(-110) \div (-11)$
78. $(+84) \div (+6)$
79. $(-96) \div (-12)$
80. $(-800) \div (+25)$

2.4 Signed Fractions

The rules for operations with signed integers also apply to fractions.

Example 1 Add: $\left(-\dfrac{1}{4}\right) + \left(-\dfrac{3}{16}\right) = \left(-\dfrac{4}{16}\right) + \left(-\dfrac{3}{16}\right)$ The LCD is 16.

$$= \dfrac{-4 - 3}{16}$$

$$= -\dfrac{7}{16} \quad \text{Combine the numerators.}$$

Example 2 Add: $\dfrac{3}{5} + \left(-\dfrac{2}{3}\right) = \dfrac{9}{15} + \left(-\dfrac{10}{15}\right)$ The LCD is 15.

$$= \dfrac{9 - 10}{15}$$

$$= -\dfrac{1}{15} \quad \text{Combine the numerators.}$$

Example 3 Add: $\left(-\dfrac{4}{9}\right) + \dfrac{2}{3} = \left(-\dfrac{4}{9}\right) + \dfrac{6}{9}$ The LCD is 9.

$$= \dfrac{-4 + 6}{9}$$

$$= \dfrac{2}{9} \quad \text{Combine the numerators.}$$

Example 4 Add: $\left(-2\dfrac{3}{4}\right) + \left(-1\dfrac{5}{6}\right) = \left(-2\dfrac{9}{12}\right) + \left(-1\dfrac{10}{12}\right)$ The LCD is 12.

$$= -3\dfrac{19}{12} \quad \text{Add the signed mixed numbers.}$$

$$= -\left(3 + \dfrac{12 + 7}{12}\right) \quad \text{Change the improper fraction to a mixed number.}$$

$$= -\left(3 + 1\dfrac{7}{12}\right)$$

$$= -4\dfrac{7}{12}$$

ANSWERS TO SIGNED NUMBERS DRILL 3

The first answer is for addition, the second answer is for subtraction, and the third answer is for multiplication.

1. +12, −4, +32 **2.** −9, −3, +18 **3.** +5, +11, −24 **4.** −8, −10, −9 **5.** +2, −8, −15 **6.** −12, 0, +36
7. −7, −7, 0 **8.** −4, +4, 0 **9.** +4, +2, +3 **10.** −13, −3, +40 **11.** −3, −9, −18 **12.** +1, +15, −56
13. +6, −8, −7 **14.** 0, +6, −9 **15.** +14, −4, +45 **16.** −10, −6, +16 **17.** 0, +10, −25 **18.** −1, −13, −42
19. −3, +5, −4 **20.** −18, 0, +81 **21.** +2, +4, −3 **22.** +2, −14, −48 **23.** −5, +9, −14 **24.** −9, −3, +18
25. +3, +11, −28 **26.** −15, +3, +54 **27.** +10, +4, +21 **28.** −20, 0, +100 **29.** +2, −8, −15 **30.** +3, +13, −40

Example 5   Subtract: $\left(-\dfrac{5}{9}\right) - \left(-\dfrac{5}{12}\right) = \left(-\dfrac{5}{9}\right) + \left(+\dfrac{5}{12}\right)$ Change the sign of the fraction being subtracted and add.

$$= \left(-\dfrac{20}{36}\right) + \left(+\dfrac{15}{36}\right)$$ The LCD is 36.

$$= \dfrac{-20 + 15}{36}$$

$$= -\dfrac{5}{36}$$ Combine the numerators. ◆

Example 6   Subtract: $\left(-1\dfrac{3}{8}\right) - \left(+2\dfrac{5}{6}\right) = \left(-1\dfrac{3}{8}\right) + \left(-2\dfrac{5}{6}\right)$ Change the sign of the mixed number being subtracted and add.
The LCD is 24.

$$= \left(-1\dfrac{9}{24}\right) + \left(-2\dfrac{20}{24}\right)$$

$$= -3\dfrac{29}{24}$$ Add the mixed numbers.

$$= -\left(3 + \dfrac{24 + 5}{24}\right)$$ Change the improper fraction to a mixed number.

$$= -\left(3 + 1\dfrac{5}{24}\right)$$

$$= -4\dfrac{5}{24}$$ ◆

Example 7   Subtract: $\left(-3\dfrac{5}{12}\right) - \left(-1\dfrac{2}{3}\right) = \left(-3\dfrac{5}{12}\right) + \left(+1\dfrac{2}{3}\right)$ Change the sign of the mixed number being subtracted and add.

$$= \left(-3\dfrac{5}{12}\right) + \left(1\dfrac{8}{12}\right)$$ The LCD is 12.

$$= \left(-2\dfrac{17}{12}\right) + \left(1\dfrac{8}{12}\right)$$ Borrow 1 or $\dfrac{12}{12}$.

$$= -1\dfrac{9}{12}$$ Add the mixed numbers.

$$= -1\dfrac{3}{4}$$ Reduce to lowest terms. ◆

Example 8   Multiply: $\left(-\dfrac{2}{5}\right)\left(-\dfrac{5}{8}\right) = \dfrac{10}{40}$ The product is positive.

$$= \dfrac{1}{4}$$ Reduce to lowest terms. ◆

Example 9   Multiply: $\left(\dfrac{4}{15}\right)\left(-\dfrac{5}{2}\right) = -\dfrac{20}{30}$ The product is negative.

$$= -\dfrac{2}{3}$$ Reduce to lowest terms. ◆

Example 10   Divide: $\left(-\dfrac{3}{7}\right) \div \left(-\dfrac{9}{14}\right) = \left(-\dfrac{\cancel{3}^{1}}{\cancel{7}_{1}}\right) \times \left(-\dfrac{\cancel{14}^{2}}{\cancel{9}_{3}}\right)$ Invert and multiply; cancels are shown.

$$= \dfrac{2}{3}$$ The product is positive. ◆

CHAPTER 2 ♦ Signed Numbers and Powers of 10

Example 11 Divide: $\left(-\dfrac{11}{15}\right) \div \dfrac{2}{3} = \left(-\dfrac{11}{\cancel{15}_{5}}\right) \times \dfrac{\cancel{3}^{1}}{2}$ Invert and multiply; cancels are shown.

$= -\dfrac{11}{10}$ or $-1\dfrac{1}{10}$ The product is negative. ♦

One more rule about fractions will help you.

> **Equivalent Signed Fractions**
>
> $\dfrac{a}{-b} = \dfrac{-a}{b} = -\dfrac{a}{b}$
>
> That is, a negative fraction may be written in three different but equivalent forms. However, the form $-\dfrac{a}{b}$ is the customary form.

For example, $\dfrac{3}{-4} = \dfrac{-3}{4} = -\dfrac{3}{4}$.

NOTE: $\dfrac{-a}{-b} = \dfrac{a}{b}$, using the rules for dividing signed numbers.

Example 12 Add: $\dfrac{3}{-4} + \dfrac{-2}{3} = \left(-\dfrac{3}{4}\right) + \left(-\dfrac{2}{3}\right)$ Change to customary form.

$= \left(-\dfrac{9}{12}\right) + \left(-\dfrac{8}{12}\right)$ The LCD is 12.

$= \dfrac{-9 + -8}{12}$

$= -\dfrac{17}{12}$ or $-1\dfrac{5}{12}$ Combine the numerators. ♦

Example 13 Add: $\dfrac{3}{-4} + \dfrac{2}{3} = \left(-\dfrac{3}{4}\right) + \dfrac{2}{3}$ Change to customary form.

$= \left(-\dfrac{9}{12}\right) + \left(\dfrac{8}{12}\right)$ The LCD is 12.

$= \dfrac{-9 + 8}{12}$

$= -\dfrac{1}{12}$ Combine the numerators. ♦

Example 14 Subtract: $\dfrac{3}{-4} - \dfrac{2}{3} = \left(-\dfrac{3}{4}\right) + \left(-\dfrac{2}{3}\right)$ Change to customary form, change the sign of the fraction being subtracted, and add.

$= \left(-\dfrac{9}{12}\right) + \left(-\dfrac{8}{12}\right)$ The LCD is 12.

$= \dfrac{-9 - 8}{12}$

$= -\dfrac{17}{12}$ or $-1\dfrac{5}{12}$ Combine the numerators. ♦

2.4 ♦ Signed Fractions

Example 15 Multiply: $\left(-\dfrac{1}{4}\right)\left(\dfrac{-3}{5}\right) = \left(-\dfrac{1}{4}\right)\left(-\dfrac{3}{5}\right)$ Change to customary form.

$= \dfrac{3}{20}$ The product is positive.

Example 16 Multiply: $\left(\dfrac{-1}{2}\right)\left(\dfrac{3}{4}\right) = \left(-\dfrac{1}{2}\right)\left(\dfrac{3}{4}\right)$ Change to customary form.

$= -\dfrac{3}{8}$ The product is negative.

Example 17 Divide: $\left(\dfrac{-2}{3}\right) \div 3 = \left(-\dfrac{2}{3}\right) \div 3$ Change to customary form.

$= \left(-\dfrac{2}{3}\right)\left(\dfrac{1}{3}\right)$ Invert and multiply.

$= -\dfrac{2}{9}$ The product is negative.

Example 18 Divide: $\left(\dfrac{-3}{7}\right) \div \dfrac{-5}{6} = \left(-\dfrac{3}{7}\right) \div \left(-\dfrac{5}{6}\right)$ Change to customary form.

$= \left(-\dfrac{3}{7}\right)\left(-\dfrac{6}{5}\right)$ Invert and multiply.

$= \dfrac{18}{35}$ The product is positive.

EXERCISES 2.4

Perform the indicated operations and simplify:

1. $\dfrac{1}{8} + \left(-\dfrac{5}{16}\right)$
2. $\left(-\dfrac{2}{3}\right) + \left(-\dfrac{2}{7}\right)$
3. $\dfrac{1}{2} + \left(-\dfrac{7}{16}\right)$
4. $\dfrac{2}{3} + \left(-\dfrac{7}{9}\right)$
5. $\left(-5\dfrac{3}{4}\right) + \left(-6\dfrac{2}{5}\right)$
6. $\left(-1\dfrac{3}{8}\right) + \left(5\dfrac{5}{12}\right)$
7. $\left(-3\dfrac{2}{3}\right) + \left(-\dfrac{4}{9}\right) + \left(4\dfrac{5}{6}\right)$
8. $\left(-\dfrac{3}{4}\right) + \left(-1\dfrac{1}{6}\right) + \left(-1\dfrac{1}{3}\right)$
9. $\left(-\dfrac{1}{4}\right) - \left(-\dfrac{1}{5}\right)$
10. $\left(-\dfrac{2}{9}\right) - \left(+\dfrac{1}{2}\right)$
11. $1\dfrac{3}{8} - \left(+\dfrac{5}{16}\right)$
12. $\left(-\dfrac{1}{3}\right) - \left(+3\dfrac{1}{2}\right)$
13. $\left(+1\dfrac{3}{4}\right) - (-4)$
14. $2\dfrac{3}{4} - \left(-3\dfrac{1}{4}\right)$
15. $\left(-\dfrac{2}{3}\right) + \left(-\dfrac{5}{6}\right) - \dfrac{1}{4}$
16. $\left(-\dfrac{3}{4}\right) - \left(-1\dfrac{2}{3}\right) + \left(-1\dfrac{5}{6}\right)$
17. $-\dfrac{1}{9} \times \dfrac{1}{7}$
18. $\left(-\dfrac{2}{3}\right)\left(-\dfrac{1}{2}\right)$
19. $\left(-3\dfrac{1}{3}\right)\left(-1\dfrac{4}{5}\right)$
20. $\dfrac{21}{8} \times 1\dfrac{7}{9}$
21. $\dfrac{4}{5} \div \left(-\dfrac{8}{9}\right)$
22. $\left(-1\dfrac{1}{4}\right) \div \dfrac{3}{5}$
23. $\left(-\dfrac{7}{9}\right) \div \left(-\dfrac{8}{3}\right)$
24. $2\dfrac{3}{4} \div \left(-3\dfrac{1}{6}\right)$
25. $\left(\dfrac{-1}{4}\right) + \left(\dfrac{1}{-5}\right)$
26. $\left(\dfrac{-4}{5}\right) + \left(-1\dfrac{1}{2}\right)$
27. $\dfrac{3}{4} + \left(\dfrac{-3}{8}\right)$
28. $\left(\dfrac{-3}{2}\right) + \left(\dfrac{-8}{3}\right)$
29. $\dfrac{5}{8} - \left(\dfrac{-5}{8}\right)$
30. $\left(\dfrac{-1}{4}\right) - \left(\dfrac{1}{-5}\right)$
31. $\left(\dfrac{-6}{8}\right) - (-4)$
32. $(-2) - \left(\dfrac{-1}{-4}\right)$

114   CHAPTER 2 ♦ Signed Numbers and Powers of 10

33. $\left(\dfrac{-1}{4}\right)\left(\dfrac{1}{-5}\right)$

34. $(-2)\left(\dfrac{-1}{-4}\right)$

42. $\left(\dfrac{-3}{4}\right)+\left(\dfrac{2}{-3}\right)-\left(\dfrac{-1}{-2}\right)-\left(\dfrac{-5}{6}\right)$

35. $\left(\dfrac{-5}{-8}\right)\left(-5\dfrac{1}{3}\right)$

36. $\left(\dfrac{-3}{-4}\right)(-12)$

43. $\left(\dfrac{-2}{5}\right)\left(\dfrac{3}{-4}\right)\left(\dfrac{-15}{-18}\right)$

37. $32 \div \left(\dfrac{-2}{3}\right)$

38. $\left(\dfrac{-4}{9}\right) \div (-2)$

44. $\left(-2\dfrac{3}{4}\right) \div \left(1\dfrac{3}{5}\right)\left(\dfrac{-2}{5}\right)$

39. $\left(\dfrac{-2}{-3}\right) \div \left(\dfrac{2}{-3}\right)$

40. $\left(-1\dfrac{3}{5}\right) \div \left(-3\dfrac{1}{5}\right)$

45. $\left(\dfrac{-2}{3}\right)+\left(\dfrac{-1}{2}\right)\left(\dfrac{5}{-6}\right)$

41. $\left(\dfrac{-2}{3}\right)+\left(-\dfrac{5}{6}\right)+\dfrac{1}{4}+\dfrac{1}{8}$

46. $\left(\dfrac{-4}{5}\right) \div \left(-1\dfrac{1}{2}\right)-\left(\dfrac{2}{-5}\right)$

2.5 Powers of 10

Multiplying Powers of 10

To multiply two powers of 10, add the exponents as follows:
$$10^a \times 10^b = 10^{a+b}$$

NOTE: The rules for working with powers of 10 shown in this section also apply to other bases, as shown in Section 5.4.

Example 1   Multiply: $(10^2)(10^3)$
Method 1: $(10^2)(10^3) = (10 \cdot 10)(10 \cdot 10 \cdot 10) = 10^5$
Method 2: $(10^2)(10^3) = 10^{2+3} = 10^5$ Add the exponents. ♦

Example 2   Multiply: $(10^3)(10^5)$
Method 1: $(10^3)(10^5) = (10 \cdot 10 \cdot 10)(10 \cdot 10 \cdot 10 \cdot 10 \cdot 10) = 10^8$
Method 2: $(10^3)(10^5) = 10^{3+5} = 10^8$ Add the exponents. ♦

Example 3   Multiply each of the following powers of 10:
a. $(10^9)(10^{12}) = 10^{9+12} = 10^{21}$ Add the exponents.
b. $(10^{-12})(10^{-7}) = 10^{(-12)+(-7)} = 10^{-19}$
c. $(10^{-9})(10^6) = 10^{(-9)+6} = 10^{-3}$
d. $(10^{10})(10^{-6}) = 10^{10+(-6)} = 10^4$
e. $10^5 \cdot 10^{-8} \cdot 10^4 \cdot 10^{-3} = 10^{5+(-8)+4+(-3)} = 10^{-2}$ ♦

Dividing Powers of 10

To divide two powers of 10, subtract the exponents as follows:
$$10^a \div 10^b = 10^{a-b}$$

2.5 ♦ Powers of 10 115

Example 4 Divide: $\dfrac{10^6}{10^2}$

Method 1: $\dfrac{10^6}{10^2} = \dfrac{\cancel{10} \cdot \cancel{10} \cdot 10 \cdot 10 \cdot 10 \cdot 10}{\cancel{10} \cdot \cancel{10}} = 10^4$

Method 2: $\dfrac{10^6}{10^2} = 10^{6-2} = 10^4$ Subtract the exponents. ♦

Example 5 Divide: $\dfrac{10^5}{10^3}$

Method 1: $\dfrac{10^5}{10^3} = \dfrac{\cancel{10} \cdot \cancel{10} \cdot \cancel{10} \cdot 10 \cdot 10}{\cancel{10} \cdot \cancel{10} \cdot \cancel{10}} = 10^2$

Method 2: $\dfrac{10^5}{10^3} = 10^{5-3} = 10^2$ Subtract the exponents. ♦

Example 6 Divide each of the following powers of 10:

a. $\dfrac{10^{12}}{10^4} = 10^{12-4} = 10^8$ Subtract the exponents.

b. $\dfrac{10^{-5}}{10^5} = 10^{(-5)-5} = 10^{-10}$

c. $\dfrac{10^6}{10^{-9}} = 10^{6-(-9)} = 10^{15}$

d. $10^{-8} \div 10^{-5} = 10^{-8-(-5)} = 10^{-3}$

e. $10^5 \div 10^9 = 10^{5-9} = 10^{-4}$ ♦

Raising a Power of 10 to a Power

To raise a power of 10 to a power, multiply the exponents as follows:
$$(10^a)^b = 10^{ab}$$

Example 7 Find the power $(10^2)^3$.

Method 1: $(10^2)^3 = 10^2 \cdot 10^2 \cdot 10^2$
$= 10^{2+2+2}$ Use the product of powers rule.
$= 10^6$

Method 2: $(10^2)^3 = 10^{(2)(3)} = 10^6$ Multiply the exponents. ♦

Example 8 Find each power of 10:

a. $(10^4)^3 = 10^{(4)(3)} = 10^{12}$ Multiply the exponents.

b. $(10^{-5})^2 = 10^{(-5)(2)} = 10^{-10}$

c. $(10^{-6})^{-3} = 10^{(-6)(-3)} = 10^{18}$

d. $(10^4)^{-4} = 10^{(4)(-4)} = 10^{-16}$

e. $(10^{10})^8 = 10^{(10)(8)} = 10^{80}$ ♦

CHAPTER 2 ♦ Signed Numbers and Powers of 10

In Section 1.10, we stated that $10^0 = 1$. Let's see why. To show this, we use the Substitution Principle, which states that if $a = b$ and $a = c$, then $b = c$.

$a = b$

$\dfrac{10^n}{10^n} = 10^{n-n}$ To divide powers, subtract the exponents.

$\quad\quad = 10^0$

$a = c$

$\dfrac{10^n}{10^n} = 1$ Any number other than zero divided by itself equals 1.

Therefore, $b = c$; that is, $10^0 = 1$.

Zero Power of 10

$10^0 = 1$

We also have used the fact that $10^{-a} = \dfrac{1}{10^a}$. To show this, we start with $\dfrac{1}{10^a}$:

$\dfrac{1}{10^a} = \dfrac{10^0}{10^a}$ $(1 = 10^0)$

$\quad\quad = 10^{0-a}$ To divide powers, subtract the exponents.

$\quad\quad = 10^{-a}$

Negative Power of 10

$10^{-a} = \dfrac{1}{10^a}$

For example, $10^{-3} = \dfrac{1}{10^3}$ and $10^{-8} = \dfrac{1}{10^8}$.

In a similar manner, we can also show that

$\dfrac{1}{10^{-a}} = 10^a$

For example, $\dfrac{1}{10^{-5}} = 10^5$ and $\dfrac{1}{10^{-2}} = 10^2$.

Combinations of multiplications and divisions of powers of 10 can also be done easily using the rules of exponents.

Example 9   Perform the indicated operations. Express the results using positive exponents.

a. $\dfrac{10^4 \cdot 10^0}{10^{-3}} = \dfrac{10^{4+0}}{10^{-3}}$ Add the exponents in the numerator.

$\quad\quad = \dfrac{10^4}{10^{-3}}$

$\quad\quad = 10^{4-(-3)}$ Subtract the exponents.

$\quad\quad = 10^7$

b. $\dfrac{10^{-2} \cdot 10^5}{10^2 \cdot 10^{-5} \cdot 10^8} = \dfrac{10^{-2+5}}{10^{2+(-5)+8}}$ Add the exponents in the numerator and in the denominator.

$= \dfrac{10^3}{10^5}$

$= 10^{3-5}$ Subtract the exponents.

$= 10^{-2}$

$= \dfrac{1}{10^2}$ Express the result using a positive exponent.

c. $\dfrac{10^2 \cdot 10^{-3}}{10^4 \cdot 10^{-7}} = \dfrac{10^{2+(-3)}}{10^{4+(-7)}}$ Add the exponents in the numerator and in the denominator.

$= \dfrac{10^{-1}}{10^{-3}}$

$= 10^{(-1)-(-3)}$ Subtract the exponents.

$= 10^2$

d. $\dfrac{10^{-5} \cdot 10^8 \cdot 10^{-6}}{10^3 \cdot 10^4 \cdot 10^{-1}} = \dfrac{10^{(-5)+8+(-6)}}{10^{3+4+(-1)}}$ Add the exponents in the numerator and in the denominator.

$= \dfrac{10^{-3}}{10^6}$

$= 10^{-3-6}$ Subtract the exponents.

$= 10^{-9}$

$= \dfrac{1}{10^9}$ Express the result using a positive exponent.

♦

EXERCISES 2.5

Perform the indicated operations using the laws of exponents. Express the results using positive exponents.

1. $10^4 \cdot 10^9$
2. $10^{-3} \cdot 10^5$
3. $\dfrac{10^4}{10^8}$
4. $10^4 \div 10^{-6}$
5. $\dfrac{1}{10^{-3}}$
6. $\dfrac{1}{10^{-5}}$
7. $(10^4)^3$
8. $(10^3)^{-2}$
9. $10^{-6} \cdot 10^{-4}$
10. $10^{-15} \cdot 10^{10}$
11. $\dfrac{10^{-3}}{10^{-6}}$
12. $10^{-2} \div 10^{-5}$
13. $(10^{-3})^4$
14. $(10^{-3})^{-5}$
15. $\left(\dfrac{10^6}{10^8}\right)^2$
16. $\left(\dfrac{10^{-2}}{10^{-5}}\right)^3$
17. $\dfrac{(10^0)^3}{10^{-2}}$
18. $\left(\dfrac{10^0}{10^{-3}}\right)^2$
19. $10^2 \cdot 10^{-5} \cdot 10^{-3}$
20. $10^{-6} \cdot 10^{-1} \cdot 10^4$
21. $10^3 \cdot 10^4 \cdot 10^{-5} \cdot 10^3$
22. $\dfrac{10^0 \cdot 10^{-3}}{10^{-6} \cdot 10^3}$
23. $\dfrac{10^3 \cdot 10^2 \cdot 10^{-7}}{10^5 \cdot 10^{-3}}$
24. $\dfrac{10^{-2} \cdot 10^{-3} \cdot 10^{-7}}{10^3 \cdot 10^4 \cdot 10^{-5}}$
25. $\dfrac{10^8 \cdot 10^{-6} \cdot 10^{10} \cdot 10^0}{10^4 \cdot 10^{-17} \cdot 10^8}$
26. $\dfrac{(10^{-4})^6}{10^4 \cdot 10^{-3}}$
27. $\dfrac{(10^{-9})^{-2}}{10^{16} \cdot 10^{-4}}$
28. $\left(\dfrac{10^4}{10^{-7}}\right)^3$
29. $\left(\dfrac{10^5 \cdot 10^{-2}}{10^{-4}}\right)^2$
30. $\left(\dfrac{10^{-7} \cdot 10^{-2}}{10^9}\right)^{-3}$

CHAPTER 2 ♦ Signed Numbers and Powers of 10

2.6 Scientific Notation

> **Scientific Notation**
>
> **Scientific notation** is a method that is especially useful for writing very large or very small numbers. To write a number in scientific notation, write it as a product of a number between 1 and 10 and a power of 10.

Example 1 Write 226 in scientific notation.

$$226 = 2.26 \times 10^2$$

Remember that 10^2 is a short way of writing $10 \times 10 = 100$. Note that multiplying 2.26 by 100 gives 226. ♦

Example 2 Write 52,800 in scientific notation.

$$52{,}800 = 5.28 \times 10{,}000 = 5.28 \times (10 \times 10 \times 10 \times 10)$$
$$= 5.28 \times 10^4$$

♦

> **Writing a Decimal Number in Scientific Notation**
>
> To write a decimal number in scientific notation,
>
> 1. Reading from left to right, place a decimal point after the first nonzero digit.
> 2. Place a caret (∧) at the position of the original decimal point.
> 3. If the decimal point is to the *left* of the caret, the exponent of the power of 10 is the same as the number of decimal places from the caret to the decimal point.
>
> $$26{,}638 = 2.6638. \times 10^{④} = 2.6638 \times 10^4$$
>
> 4. If the decimal point is to the *right* of the caret, the exponent of the power of 10 is the same as the negative of the number of places from the caret to the decimal point.
>
> $$0.00986 = 0.009.86 \times 10^{-③} = 9.86 \times 10^{-3}$$
>
> 5. If the decimal point is already after the first nonzero digit, the exponent of 10 is zero.
>
> $$2.15 = 2.15 \times 10^0$$

Example 3 Write 2738 in scientific notation.

$$2738 = 2.738 \times 10^{③} = 2.738 \times 10^3$$

♦

Example 4 Write 0.0000003842 in scientific notation.

$$0.0000003842 = 0.0000003.842 \times 10^{-⑦} = 3.842 \times 10^{-7}$$

♦

Writing a Number in Scientific Notation in Decimal Form

To change a number in scientific notation to decimal form,

1. Multiply the decimal part by the given *positive* power of 10 by moving the decimal point to the *right* the same number of decimal places as indicated by the exponent of 10. Supply zeros when needed.
2. Multiply the decimal part by the given *negative* power of 10 by moving the decimal point to the *left* the same number of decimal places as indicated by the exponent of 10. Supply zeros when needed.

Example 5 Write 2.67×10^2 as a decimal.

$2.67 \times 10^2 = 267$ Move the decimal point two places to the *right*, since the exponent of 10 is $+2$. ◆

Example 6 Write 8.76×10^4 as a decimal.

$8.76 \times 10^4 = 87{,}600$ Move the decimal point four places to the *right*, since the exponent of 10 is $+4$. It is necessary to write two zeros. ◆

Example 7 Write 5.13×10^{-4} as a decimal.

$5.13 \times 10^{-4} = 0.000513$ Move the decimal point four places to the *left*, since the exponent of 10 is -4. It is necessary to write three zeros. ◆

> You may find it useful to note that a number in scientific notation with
>
> (a) a positive exponent greater than 1 is *greater than* 10, and
> (b) a negative exponent is between 0 and 1.

That is, a number in scientific notation with a positive exponent represents a relatively large number. A number in scientific notation with a negative exponent represents a relatively small number.

Scientific notation may be used to compare two positive numbers expressed as decimals. First, write both numbers in scientific notation. The number having the greater power of 10 is the larger. If the powers of 10 are equal, compare the parts of the numbers that are between 1 and 10.

Example 8 Which is greater, 0.000876 or 0.0004721?

$0.000876 = 8.76 \times 10^{-4}$
$0.0004721 = 4.721 \times 10^{-4}$

Since the exponents are the same, compare 8.76 and 4.721. Since 8.76 is greater than 4.721, 0.000876 is greater than 0.0004721. ◆

Example 9 Which is greater, 0.0062 or 0.0382?

$$0.0062 = 6.2 \times 10^{-3}$$
$$0.0382 = 3.82 \times 10^{-2}$$

Since -2 is greater than -3, 0.0382 is greater than 0.0062.

Scientific notation is especially helpful for multiplying and dividing very large and very small numbers. To perform these operations, you must first know some rules for exponents. Many calculators perform multiplication, division, and powers of numbers entered in scientific notation and give the results, when very large or very small, in scientific notation.

Multiplying Numbers in Scientific Notation

To multiply numbers in scientific notation, multiply the decimals between 1 and 10. Then add the exponents of the powers of 10.

Example 10 Multiply $(4.5 \times 10^8)(5.2 \times 10^{-14})$. Write the result in scientific notation.

$$\begin{aligned}(4.5 \times 10^8)(5.2 \times 10^{-14}) &= (4.5)(5.2) \times (10^8)(10^{-14}) \\ &= 23.4 \times 10^{-6} \\ &= (2.34 \times 10^1) \times 10^{-6} \\ &= 2.34 \times 10^{-5}\end{aligned}$$

Note that 23.4×10^{-6} is not in scientific notation, because 23.4 is not between 1 and 10.

To find this product using a calculator that accepts numbers in scientific notation, use the following procedure.

4.5 [EE]* 8 [×] 5.2 [EE] [(−)] 14 [=]

[2.34 × 10⁻⁵]

NOTE: 1. You may need to set your calculator in scientific notation mode.

2. The [(−)] or [+/−] key is used to enter a negative number.

The product is 2.34×10^{-5}.

Dividing Numbers in Scientific Notation

To divide numbers in scientific notation, divide the decimals between 1 and 10. Then subtract the exponents of the powers of 10.

Example 11 Divide $\dfrac{4.8 \times 10^{-7}}{1.6 \times 10^{-11}}$. Write the result in scientific notation.

$$\begin{aligned}\frac{4.8 \times 10^{-7}}{1.6 \times 10^{-11}} &= \frac{4.8}{1.6} \times \frac{10^{-7}}{10^{-11}} \\ &= 3 \times 10^4\end{aligned}$$

---

*Key may also be [Exp]

2.6 ♦ Scientific Notation   121

Using a calculator, we have

4.8 [EE] [(−)] 7 [÷] 1.6 [EE] [(−)] 11 [=]

[3×10^4]

The quotient is 3×10^4.

Example 12 Evaluate $\dfrac{(6 \times 10^{-6})(3 \times 10^9)}{(2 \times 10^{-10})(4 \times 10^{-5})}$. Write the result in scientific notation.

$$\dfrac{(6 \times 10^{-6})(3 \times 10^9)}{(2 \times 10^{-10})(4 \times 10^{-5})} = \dfrac{(6)(3)}{(2)(4)} \times \dfrac{(10^{-6})(10^9)}{(10^{-10})(10^{-5})} = \dfrac{18}{8} \times \dfrac{10^3}{10^{-15}}$$
$$= 2.25 \times 10^{18}$$

Again, use a calculator.

6 [EE] [(−)] 6 [×] 3 [EE] 9 [÷] 2 [EE] [(−)] 10 [÷] 4 [EE] [(−)] 5 [=]

[2.25×10^{18}]

Note: Some calculators display this result as 2.25 E18.

The result is 2.25×10^{18}.

Powers of Numbers in Scientific Notation

To find the power of a number in scientific notation, find the power of the decimal between 1 and 10. Then multiply the exponent of the power of 10 by this same power.

Example 13 Find the power $(4.5 \times 10^6)^2$. Write the result in scientific notation.

$$(4.5 \times 10^6)^2 = (4.5)^2 \times (10^6)^2$$
$$= 20.25 \times 10^{12} \quad \text{Note that 20.25 is not between 1 and 10.}$$
$$= (2.025 \times 10^1) \times 10^{12}$$
$$= 2.025 \times 10^{13}$$

4.5 [EE] 6 [x^2] [=]

[2.025×10^{13}]

The result is 2.025×10^{13}.

Example 14 Find the power $(3 \times 10^{-8})^5$. Write the result in scientific notation.

$$(3 \times 10^{-8})^5 = 3^5 \times (10^{-8})^5$$
$$= 243 \times 10^{-40}$$
$$= (2.43 \times 10^2) \times 10^{-40}$$
$$= 2.43 \times 10^{-38}$$

The ∧ key is used to raise a number to a power.

3 EE (−) 8 ∧ or y^x 5 =

The result is 2.43×10^{-38}.

EXERCISES 2.6

Write each number in scientific notation:

1. 356
2. 15,600
3. 634.8
4. 24.85
5. 0.00825
6. 0.00063
7. 7.4
8. 377,000
9. 0.000072
10. 0.00335
11. 710,000
12. 1,200,000
13. 0.0000045
14. 0.0000007
15. 0.000000034
16. 4,500,000,000
17. 640,000
18. 85,000

Write each number in decimal form:

19. 7.55×10^4
20. 8.76×10^2
21. 5.31×10^3
22. 5.14×10^5
23. 7.8×10^{-2}
24. 9.44×10^{-3}
25. 5.55×10^{-4}
26. 4.91×10^{-6}
27. 6.4×10^1
28. 3.785×10^{-2}
29. 9.6×10^2
30. 7.3×10^3
31. 5.76×10^0
32. 6.8×10^{-5}
33. 6.4×10^{-6}
34. 7×10^8
35. 5×10^{10}
36. 5.05×10^0
37. 6.2×10^{-7}
38. 2.1×10^{-9}
39. 2.5×10^{12}
40. 1.5×10^{11}
41. 3.3×10^{-11}
42. 7.23×10^{-8}

Find the larger number:

43. 0.0037; 0.0048
44. 0.029; 0.0083
45. 0.000042; 0.00091
46. 148,000; 96,988
47. 0.00037; 0.000094
48. 0.8216; 0.792
49. 0.0613; 0.00812
50. 0.0000613; 0.01200

Find the smaller number:

51. 0.008; 0.0009
52. 295,682; 295,681
53. 1.003; 1.0009
54. 21.8; 30.2
55. 0.00000000998; 0.01
56. 0.10108; 0.10102
57. 0.000314; 0.000271
58. 0.00812; 0.0318

Perform the indicated operations (write each result in scientific notation with the decimal part rounded to three significant digits when necessary):

59. $(4 \times 10^{-6})(6 \times 10^{-10})$
60. $(3 \times 10^7)(3 \times 10^{-12})$
61. $\dfrac{4.5 \times 10^{16}}{1.5 \times 10^{-8}}$
62. $\dfrac{1.6 \times 10^6}{6.4 \times 10^{10}}$
63. $\dfrac{(4 \times 10^{-5})(6 \times 10^{-3})}{(3 \times 10^{-10})(8 \times 10^8)}$
64. $\dfrac{(5 \times 10^4)(3 \times 10^{-5})(4 \times 10^6)}{(1.5 \times 10^6)(2 \times 10^{-11})}$
65. $(1.2 \times 10^6)^3$
66. $(2 \times 10^{-9})^4$
67. $(6.2 \times 10^{-5})(5.2 \times 10^{-6})(3.5 \times 10^8)$
68. $\dfrac{(5 \times 10^{-6})^2}{4 \times 10^6}$
69. $\left(\dfrac{2.5 \times 10^{-4}}{7.5 \times 10^8}\right)^2$
70. $\left(\dfrac{2.5 \times 10^{-9}}{5 \times 10^{-7}}\right)^4$
71. $(18,000)(0.00005)$
72. $(4500)(69,000)(150,000)$
73. $\dfrac{2,400,000}{36,000}$
74. $\dfrac{(3500)(0.00164)}{2700}$
75. $\dfrac{84,000 \times 0.0004 \times 142,000}{0.002 \times 3200}$
76. $\dfrac{(0.0025)^2}{3500}$
77. $\left(\dfrac{48,000 \times 0.0144}{0.0064}\right)^2$
78. $\left(\dfrac{0.0027 \times 0.16}{12,000}\right)^3$
79. $\left(\dfrac{1.3 \times 10^4}{(2.6 \times 10^{-3})(5.1 \times 10^8)}\right)^5$
80. $\left(\dfrac{9.6 \times 10^{-3}}{(2.45 \times 10^{-4})(1.1 \times 10^5)}\right)^6$
81. $\left(\dfrac{18.4 \times 2100}{0.036 \times 950}\right)^8$
82. $\left(\dfrac{0.259 \times 6300}{866 \times 0.013}\right)^{10}$

2.7 Engineering Notation

Numbers may also be written in engineering notation, similar to scientific notation, as follows:

> **Engineering Notation**
>
> **Engineering notation** is used to write a number with its decimal part between 1 and 1000 and a power of 10 whose exponent is divisible by 3.

> **Writing a Decimal Number in Engineering Notation**
>
> To write a decimal number in engineering notation,
>
> 1. Move the decimal point *in groups of three digits* until the decimal point indicates a number between 1 and 1000.
> 2. If the decimal point has been moved to the *left*, the exponent of the power of 10 in engineering notation is the same as the number of places the decimal point was moved.
> 3. If the decimal point has been moved to the *right*, the exponent of the power of 10 in engineering notation is the same as the negative of the number of places the decimal point was moved.
>
> In any case, the exponent will be divisible by 3.

Example 1 Write 48,500 in engineering notation.

$$48{,}500 = 48.5 \times 10^3$$ Move the decimal point in groups of three decimal places until the decimal part is between 1 and 1000.

Check The exponent of the power of 10 must be divisible by 3. ♦

Example 2 Write 375,000,000,000 in engineering notation.

$$375{,}000{,}000{,}000 = 375 \times 10^9$$ Move the decimal point in groups of three decimal places until the decimal part is between 1 and 1000.

Check 9, the exponent of the power of 10, is divisible by 3. ♦

Example 3 Write 0.000000000002045 in engineering notation.

$$0.000000000002045 = 2.045 \times 10^{-12}$$ Move the decimal point in groups of three decimal places until the decimal part is between 1 and 1000.

Check -12, the exponent of the power of 10, is divisible by 3. In this case, this number is also written in scientific notation. ♦

Writing a number in engineering notation in decimal form is similar to writing a number in scientific notation in decimal form.

Example 4 Write 405×10^6 as a decimal.

$405 \times 10^6 = 405{,}000{,}000$ Move the decimal point six places to the *right*, since the exponent is $+6$.

Example 5 Write 87.035×10^{-6} as a decimal.

$87.035 \times 10^{-6} = 0.000087035$ Move the decimal point six places to the *left*, since the exponent is -6.

Operations with numbers in engineering notation using a calculator are very similar to operations with numbers in scientific notation. If your calculator has an engineering notation mode, set it in this mode. If not, use scientific notation and convert the result to engineering notation.

Example 6 Multiply $(26.4 \times 10^6)(722 \times 10^3)$. Write the result in engineering notation.

In this example, we will show an arithmetic step-by-step analysis and then show how to use a calculator to find this product.

$$(26.4 \times 10^6)(722 \times 10^3) = (26.4)(722) \times (10^6)(10^3)$$
$$= 19060.8 \times 10^9$$
$$= (19.0608 \times 10^3) \times 10^9$$
$$= 19.0608 \times 10^{12}$$

To find this product using a calculator that accepts numbers in engineering notation or scientific notation, use the following procedure:

26.4 EE 6 722 EE 3

19.0608×10^{12}

NOTE: If you use scientific notation, your result is 1.90608×10^{13}, which, when converted to engineering notation, is 19.0608×10^{12}.

Example 7 Divide $\dfrac{12.75 \times 10^{-15}}{236 \times 10^{-9}}$. Write the result in engineering notation rounded to three significant digits.

Find this quotient using a calculator as follows:

12.75 EE (−) 15 236 EE (−) 9

$54.02542373 \times 10^{-9}$

So, the result rounded to three significant digits is 54.0×10^{-9}.

NOTE: If you use scientific notation, your result is 5.40×10^{-8}, which, when converted to engineering notation, is 54.0×10^{-9}.

Example 8 Find the power $(15.4 \times 10^9)^2$ and write the result in engineering notation rounded to three significant digits.

15.4 EE 9

237.16×10^{18}

The result rounded to three significant digits is 237×10^{18}.

2.7 ♦ Engineering Notation

Example 9 — Find the square root $\sqrt{740.5 \times 10^{-18}}$ and write the result in engineering notation rounded to three significant digits.

√ 740.5 EE (−) 18 =

27.21212965 × 10⁻⁹

The result rounded to three significant digits is 27.2×10^{-9}. ♦

For comparison purposes, the following table shows six numbers written in both scientific notation and engineering notation:

Number	Scientific notation	Engineering notation
6,710,000	6.71×10^6	6.71×10^6
805,000	8.05×10^5	805×10^3
34,500,000	3.45×10^7	34.5×10^6
0.000096	9.6×10^{-5}	96×10^{-6}
0.000007711	7.711×10^{-6}	7.711×10^{-6}
0.000000444	4.44×10^{-7}	444×10^{-9}

EXERCISES 2.7

Write each number in engineering notation:

1. 28,000
2. 135,000
3. 3,450,000
4. 29,000,000
5. 220,000,000,000
6. 7,235,000,000,000,000
7. 0.0066
8. 0.00015
9. 0.0000000765
10. 0.0000000000044
11. 0.975
12. 0.0000000625

Write each number in decimal form:

13. 57.7×10^3
14. 135×10^6
15. 4.94×10^{12}
16. 46×10^9
17. 567×10^6
18. 3.24×10^{18}
19. 26×10^{-6}
20. 751×10^{-3}
21. 5.945×10^{-9}
22. 602.5×10^{-6}
23. 10.64×10^{-12}
24. 6.3×10^{-15}

Perform the indicated operations and write each result in engineering notation rounded to three significant digits:

25. $(35.5 \times 10^6)(420 \times 10^9)$
26. $(9.02 \times 10^{-6})(69.5 \times 10^{-24})$
27. $(2.7 \times 10^9)(27 \times 10^{-6})(270 \times 10^{-12})$
28. $(6 \times 10^{-12})(20 \times 10^{-9})(400 \times 10^{-6})$
29. $\dfrac{70.5 \times 10^6}{120 \times 10^{-9}}$
30. $\dfrac{450 \times 10^{-12}}{51 \times 10^6}$
31. $\dfrac{(5.15 \times 10^9)(65.3 \times 10^{-6})}{(27 \times 10^6)(800 \times 10^{12})}$
32. $\dfrac{(750 \times 10^{-12})(25 \times 10^{-6})(1.5 \times 10^{-3})}{(30 \times 10^{-9})(2 \times 10^{15})}$
33. $\dfrac{1}{2.95 \times 10^{-9}}$
34. $\dfrac{1}{55 \times 10^{12}}$
35. $(350 \times 10^9)^2$
36. $(92.5 \times 10^{-12})^2$
37. $\sqrt{80.5 \times 10^{12}}$
38. $\sqrt{750 \times 10^{-18}}$
39. $\dfrac{\sqrt{6.05 \times 10^9}}{(244 \times 10^{-6})^2}$
40. $\dfrac{1}{(24 \times 10^{-9})^2}$

SUMMARY | CHAPTER 2

Glossary of Basic Terms

Absolute value of a number. Its distance from zero on the number line. The absolute value of a number is never negative. (p. 101)

Engineering notation. A number written with its decimal part between 1 and 1000 and a power of 10 whose exponent is divisible by 3. (p. 123)

Irrational numbers. Those numbers that cannot be written as the ratio of two integers. (p. 101)

Negative integers. $-1, -2, -3, \ldots$, or those integers less than zero. (p. 101)

Positive integers. $1, 2, 3, \ldots$, or those integers greater than zero. (p. 101)

Rational numbers. Those numbers that can be written as the ratio of two integers; that is, a/b, where $b \neq 0$. (p. 101)

Real numbers. Those numbers that are either rational or irrational. (p. 101)

Scientific notation. A number written with its decimal part between 1 and 10 and a power of 10. (p. 118)

Signed numbers. Positive and negative numbers. (p. 102)

2.1 Addition of Signed Numbers

1. **Adding two numbers with like signs:**
   a. To add two positive numbers, add their absolute values. The result is positive.
   b. To add two negative numbers, add their absolute values and place a negative sign before the result. (p. 103)

2. **Adding two numbers with unlike signs:** To add a negative number and a positive number, find the difference of their absolute values. The sign of the number having the larger absolute value is placed before the result. (p. 103)

3. **Adding three or more signed numbers:** To add three or more signed numbers,
   a. Add the positive numbers.
   b. Add the negative numbers.
   c. Add the sums from steps (a) and (b) according to the rules for addition of two signed numbers. (p. 103)

2.2 Subtraction of Signed Numbers

1. **Subtracting two signed numbers:** To subtract two signed numbers, change the sign of the number being subtracted and add according to the rules for addition of signed numbers. (p. 105)

2. **Subtracting more than two signed numbers:** When more than two signed numbers are involved in subtraction, change the sign of *each* number being subtracted and add the resulting signed numbers. (p. 105)

3. **Adding and subtracting combinations of signed numbers:** When combinations of addition and subtraction of signed numbers occur in the same problem, change *only* the sign of each number being subtracted. Then add the resulting signed numbers. (p. 105)

2.3 Multiplication and Division of Signed Numbers

1. **Multiplying two signed numbers:**
   a. To multiply two numbers with the same sign, multiply their absolute values. This product is positive.
   b. To multiply two numbers with different signs, multiply their absolute values and place a negative sign before the result. (p. 107)

2. **Multiplying more than two signed numbers:** To multiply more than two signed numbers, multiply their absolute values; then
   a. place a positive sign before the result if the number of negative signed numbers is even, or
   b. place a negative sign before the result if the number of negative signed numbers is odd. (p. 107)

3. **Dividing two signed numbers:**
   a. To divide two numbers with the same sign, divide their absolute values. This product is positive.
   b. To divide two numbers with different signs, divide their absolute values and place a negative sign before the result. (p. 108)

2.4 Signed Fractions

1. **Equivalent signed fractions:** The following are equivalent signed fractions:
$$\frac{a}{-b} = \frac{-a}{b} = -\frac{a}{b}$$ (p. 112)

2.5 Powers of 10

1. **Multiplying powers of 10:** To multiply two powers of 10, add the exponents: $10^a \times 10^b = 10^{a+b}$ (p. 114)

2. **Dividing powers of 10:** To divide two powers of 10, subtract the exponents: $10^a \div 10^b = 10^{a-b}$ (p. 114)

3. **Raising a power of 10 to a power:** To raise a power of 10 to a power, multiply the exponents: $(10^a)^b = 10^{ab}$ (p. 115)
4. **Zero power of 10:** The zero power of 10 is 1: $10^0 = 1$ (p. 116)
5. **Negative power of 10:** Negative powers of 10 may be written $10^{-a} = \dfrac{1}{10^a}$ or $\dfrac{1}{10^{-a}} = 10^a$ (p. 116)

2.6 Scientific Notation

1. **Writing a decimal number in scientific notation:** To write a decimal number in scientific notation:
   a. Reading from left to right, place a decimal point after the first nonzero digit.
   b. Place a caret (∧) at the position of the original decimal point.
   c. If the decimal point is to the *left* of the caret, the exponent of the power of 10 is the same as the number of decimal places from the caret to the decimal point.
   d. If the decimal point is to the *right* of the caret, the exponent of the power of 10 is the same as the negative of the number of places from the caret to the decimal point.
   e. If the decimal point is already after the first nonzero digit, the exponent of 10 is zero. (p. 118)
2. **Writing a number in scientific notation in decimal form:** To change a number in scientific notation to decimal form:
   a. Multiply the decimal part by the given *positive* power of 10 by moving the decimal point to the *right* the same number of decimal places as indicated by the exponent of 10. Supply zeros when needed.
   b. Multiply the decimal part by the given *negative* power of 10 by moving the decimal point to the *left* the same number of decimal places as indicated by the exponent of 10. Supply zeros when needed. (p. 119)
3. **Useful note:** A number in scientific notation is greater than 10 if it has a positive exponent and between 0 and 1 if it has a negative exponent. (p. 119)

2.7 Engineering Notation

1. **Writing a decimal number in engineering notation:** To write a decimal number in engineering notation:
   a. Move the decimal point *in groups of three digits* until the decimal point indicates a number between 1 and 1000.
   b. If the decimal point has been moved to the *left*, the exponent of the power of 10 in engineering notation is the same as the number of places the decimal point was moved.
   c. If the decimal point has been moved to the *right*, the exponent of the power of 10 in engineering notation is the same as the negative of the number of places the decimal point was moved.

   In any case, the exponent will be divisible by 3. (p. 123)

REVIEW | CHAPTER 2

Find the absolute value of each number:

1. $+5$
2. -16
3. 13

Add:

4. $(-4) + (+7)$
5. $(-6) + (-2)$
6. $(+5) + (-8)$
7. $(-9) + (+2) + (-6)$

Subtract:

8. $3 - 6$
9. $(-7) - (+4)$
10. $(+9) - (-10)$
11. $(-6) - (+4) - (-8)$

Perform the indicated operations:

12. $(-2) - (+7) + (+4) + (-5) - (-10)$
13. $5 - 6 + 4 - 9 + 4 + 3 - 12 - 8$

Multiply:

14. $(-6)(+4)$
15. $(+4)(+9)$
16. $(-9)(-8)$
17. $(-2)(-7)(+1)(+3)(-2)$

Divide:

18. $\dfrac{-18}{-3}$
19. $(+30) \div (-5)$
20. $\dfrac{+45}{+9}$

Perform the indicated operations and simplify:

21. $\left(-\dfrac{6}{7}\right) - \left(\dfrac{5}{-6}\right)$
22. $\dfrac{-3}{16} \div \left(-2\dfrac{1}{4}\right)$
23. $\dfrac{-5}{8} + \left(-\dfrac{5}{6}\right) - \left(+1\dfrac{2}{3}\right)$
24. $\left(-\dfrac{9}{16}\right)\left(2\dfrac{2}{3}\right)$

Perform the indicated operations using the laws of exponents and express the results using positive exponents:

25. $10^9 \cdot 10^{-14}$
26. $10^6 \div 10^{-3}$
27. $(10^{-4})^3$
28. $\dfrac{(10^{-3} \cdot 10^5)^3}{10^6}$

Write each number in scientific notation:

29. 476,000
30. 0.0014

Write each number in decimal form:

31. 5.35×10^{-5}
32. 6.1×10^7

Find the larger number:

33. 0.00063; 0.00105
34. 0.056; 0.06

Find the smaller number:

35. 0.000075; 0.0006
36. 0.04; 0.00183

Perform the indicated operations and write each result in scientific notation:

37. $(9.5 \times 10^{10})(4.6 \times 10^{-13})$
38. $\dfrac{8.4 \times 10^8}{3 \times 10^{-6}}$
39. $\dfrac{(50{,}000)(640{,}000{,}000)}{(0.0004)^2}$
40. $(4.5 \times 10^{-8})^2$
41. $(2 \times 10^9)^4$
42. $\left(\dfrac{1.2 \times 10^{-2}}{3 \times 10^{-5}}\right)^3$

Write each number in engineering notation:

43. 275,000
44. 32,000,000
45. 0.00045

Write each number in decimal form:

46. 31.6×10^6
47. 746×10^{-3}

Perform the indicated operations and write each result in engineering notation rounded to three significant digits when necessary:

48. $(39.4 \times 10^6)(120 \times 10^{-3})$
49. $\dfrac{84.5 \times 10^{-9}}{3.48 \times 10^6}$
50. $\dfrac{1}{21.7 \times 10^{-6}}$

TEST | CHAPTER 2

Perform the indicated operations:

1. $-7 + (+9)$
2. $-28 + (-11)$
3. $18 + (-6)$
4. $-65 + 42$
5. $+112 + 241$
6. $-6 - (-10)$
7. $+16 - (+18)$
8. $(-4)(-7)$
9. $(+15)(-22)$
10. $(+17)(-8)$
11. $\dfrac{+20}{-4}$
12. $(-160) \div (-8)$
13. $(-2) + (-6) + (+5) + (-9) + (+10)$
14. $(-3)(-2)(+1)(+2)(-1)(+2)(+1)$
15. $(-5) + (-6) - (-7) - (+4) + (+3)$
16. $8 + (-1) + (-5) - (-3) + 10$
17. $-8 + 5 + 2 - 12 + 5 - 3$

Perform the indicated operations and simplify:

18. $\left(-\dfrac{2}{3}\right)(-6) + \left(-1\dfrac{1}{3}\right)$
19. $2\dfrac{1}{5} - \left(-1\dfrac{3}{10}\right) + 2\dfrac{3}{5}$
20. $\left(-\dfrac{5}{9}\right)\left(3\dfrac{2}{5}\right)$

21. Write 0.000182 in scientific notation.
22. Write 4.7×10^6 in decimal form.

Perform the indicated operations using the laws of exponents and express each result using positive exponents:

23. $(10^{-3})(10^6)$
24. $10^3 \div 10^{-5}$
25. $(10^2)^4$
26. $\dfrac{10^8 \cdot 10^{-6}}{(10^{-3})^{-2}}$
27. $\dfrac{(10^{-4})(10^{-8})^2}{(10^4)^{-6}}$

Perform the indicated operations. Write each result in scientific notation rounded to three significant digits when necessary:

28. $\dfrac{(7.6 \times 10^{13})(5.35 \times 10^{-6})}{4.64 \times 10^8}$
29. $\dfrac{(150{,}000)(18{,}000)(0.036)}{(0.0056)(48{,}000)}$
30. $\dfrac{(25{,}000)(0.125)}{(0.05)^3}$

Write each number in engineering notation:

31. 825,000
32. 0.0000751
33. Write 880×10^{-6} in decimal form.

Perform the indicated operations and write the result in engineering notation rounded to three significant digits:

34. $(39.4 \times 10^6)(120 \times 10^{-3})(45.0 \times 10^{12})$

35. $\dfrac{(3.03 \times 10^{12})^2}{\sqrt{615 \times 10^{-3}}}$

CUMULATIVE REVIEW | CHAPTERS 1–2

1. Evaluate: $16 \div 8 + 5 \times 2 - 3 + 7 \times 9$
2. Find the area of the figure in Illustration 1.

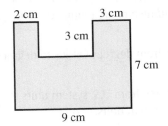

ILLUSTRATION 1

3. Find the total resistance in the series circuit in Illustration 2.

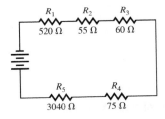

ILLUSTRATION 2

4. Is 2306 divisible by 6?
5. Find the prime factorization of 630.
6. Change $\dfrac{32}{9}$ to a mixed number in simplest form.
7. Find the area of a trapezoid with bases of 40 ft and 72 ft and height 80 ft.

Perform the indicated operations and simplify:

8. $\dfrac{5}{16} + \dfrac{1}{16} + \dfrac{7}{16}$
9. $\dfrac{3}{8} + \dfrac{1}{4} + \dfrac{7}{16}$
10. $6\dfrac{1}{2} - 4\dfrac{5}{8}$
11. $\dfrac{2}{5} \times \dfrac{1}{8}$

12. 5 lb 3 oz = _____ oz
13. Round 615.2875 to the nearest **a.** hundred, **b.** tenth, **c.** ten, and **d.** thousandth.
14. Change $7\dfrac{2}{5}\%$ to a decimal.
15. Find 28.5% of $14,000.
16. 212 is 32% of what number?
17. To the nearest hundredth, 58 is what percent of 615?
18. A used car is listed to sell at $6800. Joy bought it for $6375. What percent markdown did she receive?
19. Find the value of $-8 + (+9) + (-3) - (-12)$.
20. Multiply: $(-8)(-9)(+3)(-1)(+2)$

Perform the indicated operations and simplify:

21. $-\dfrac{3}{8} + \left(-\dfrac{1}{4}\right) - 2\dfrac{5}{16}$
22. $-\dfrac{5}{8} \times \dfrac{5}{8}$

23. Write 318,180 in scientific notation.
24. Find the larger number: 0.000618; 0.00213
25. Simplify. Express the result using positive exponents.

$$\dfrac{(10^{-4} \times 10^3)^{-2}}{10^6}$$

Write each number in engineering notation:

26. 4500
27. 0.00027

Write each number in decimal form:

28. 281×10^{-9}
29. 16.3×10^6

Perform the indicated operations and write each result in scientific notation rounded to three significant digits:

30. $(4.62 \times 10^4)(1.52 \times 10^6)$
31. $\dfrac{5.61 \times 10^7}{1.18 \times 10^{10}}$
32. $\dfrac{(5.62 \times 10^{-3})(6.28 \times 10^6)}{(5.1 \times 10^6)(2 \times 10^{12})}$
33. $\sqrt{4.28 \times 10^{-6}}$

CHAPTER 3 | The Metric System

OBJECTIVES

- Apply the basic concepts of the metric system, involving the SI prefixes and units of measure.

- Use conversion factors to change from one unit to another within the metric system of weights and measures using length, mass and weight, volume and area, time, current, and other units.

- Use the correct formula to change temperature measures from degrees Celsius to degrees Fahrenheit or vice versa.

- Use conversion factors to change units within the U.S. system, from U.S. system units to metric system units, and from metric system units to U.S. system units.

Mathematics at Work

Many health care professionals in the allied health areas provide essential, critical support in a variety of areas. A sampling of health care professional support areas includes registered nurses, licensed practical nurses, dental hygienists, dental assistants, occupational therapy assistants, physical therapy assistants, radiologic technologists, respiratory therapy assistants, surgical technologists and technicians, pharmacy technicians, and emergency medical technicians.

The health care professionals support our health care within prescribed duties as outlined by the various specific job descriptions in clinics and hospitals, and in public health, industrial, government, and private settings. Most health care degree and certificate programs are accredited by the corresponding national and/or state health accrediting agency. For more information, please visit **www.cengage.com** and access the Student Online Resources for this text.

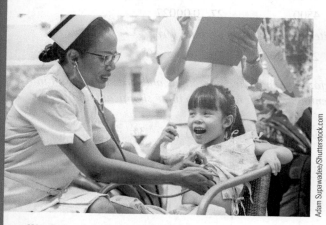

Allied Health Care Professionals
Nurse using stethoscope

3.1 Introduction to the Metric System

In early recorded history, parts of the human body were used for standards of measurement. However, these standards were neither uniform nor acceptable to all. So the next step was to define the various standards, such as the inch, the foot, and the rod. But each country introduced or defined its own standards, which were often not related to those in other countries. Then, in 1670, as the need for a single worldwide measurement system became recognized, Gabriel Mouton, a Frenchman, proposed a uniform decimal measurement system. By the 1800s, metric standards were first adopted worldwide. The U.S. Congress legalized the use of the metric system throughout the United States over 150 years ago on July 28, 1866. In 1960, the International System of Units was adopted for the modern metric system. The abbreviation for the International System of Units is **SI** (from the French *Système International d'Unités*) and is commonly called the **metric system**.

Throughout U.S. history, several attempts have been made to convert the nation to the metric system. By the 1970s, the United States was the only nonmetric industrialized nation left in the world, but the U.S. government did little to implement the system. Industry and business, however, found their foreign markets significantly limited because metric products were preferred. Now many segments of U.S. business and industry have independently gone metric because world trade requires it. Metric countries just naturally want metric products. And the inherent simplicity of the metric system of measurement and standardization of weights and measures have led to major cost savings in industries that have converted to it. Most major U.S. industries, such as the automotive, aviation, and farm implement industries, as well as the Department of Defense and other federal agencies, have effectively converted to the metric system. In some industries, you—the student and worker—will need to know and use both systems.

The SI metric system has seven base units, as shown in Table 3.1. Other commonly used metric units are shown in Table 3.2.

The metric system, a decimal or base 10 system, is very similar to our decimal number system. It is an easy system to use, because calculations are based on the number 10 and its multiples. The SI system has special prefixes that name multiples and submultiples; these can be used with almost all SI units. Table 3.3 shows the prefixes and the corresponding symbols.

Because the same prefixes are used with most all SI metric units, it is not necessary to memorize long lists or many tables.

Table 3.1 | Seven Base Metric Units

Basic unit	SI abbreviation	For measuring
metre*	m	length
kilogram	kg	mass
second	s	time
ampere	A	electric current
kelvin	K	temperature
candela	cd	light intensity
mole	mol	molecular substance

*At present, there is some difference of opinion on the spelling of metre and litre. We have chosen the "re" spelling because it is the internationally accepted spelling and because it distinguishes the unit from other meters, such as parking meters and electricity meters.

Table 3.2 | Other Commonly Used Metric Units

Unit	SI abbreviations	For measuring
litre*	L	volume
cubic metre	m^3	volume
square metre	m^2	area
newton	N	force
metre per second	m/s	speed
joule	J	energy
watt	W	power
radian	rad	plane angle

*See Table 3.1 footnote.

Table 3.3 | Prefixes for SI Units

Multiple or submultiple* decimal form	Power of 10	Prefix	Prefix symbol	Pronunciation	Meaning
1,000,000,000,000	10^{12}	tera	T	tĕr′ă	one trillion times
1,000,000,000	10^{9}	giga	G	jĭg′ă	one billion times
1,000,000	10^{6}	mega	M	mĕg′ă	one million times
1000	10^{3}	kilo**	k	kĭl′ō or kēl′ō	one thousand times
100	10^{2}	hecto	h	hĕk′tō	one hundred times
10	10^{1}	deka	da	dĕk′ă	ten times
0.1	10^{-1}	deci	d	dĕs′ĭ	one tenth of
0.01	10^{-2}	centi**	c	sĕnt′ĭ	one hundredth of
0.001	10^{-3}	milli**	m	mĭl′ĭ	one thousandth of
0.000001	10^{-6}	micro	μ	mī′krō	one millionth of
0.000000001	10^{-9}	nano	n	năn′ō	one billionth of
0.000000000001	10^{-12}	pico	p	pē′kō	one trillionth of

*Factor by which the unit is multiplied.
**Most commonly used prefixes.

NOTE: A common acronym used for remembering the order of the most common metric prefixes for all metric units is **king henry died by drinking chocolate milk**; that is,

king	kilo
henry	hecto
died	deca
by	base unit
drinking	deci
chocolate	centi
milk	milli

Example 1 Write the SI abbreviation for 45 kilometres.

The symbol for the prefix *kilo* is k.
The symbol for the unit *metre* is m.
The SI abbreviation for 45 kilometres is 45 km.

Example 2 Write the SI unit for the abbreviation 50 mg.

The prefix for m is *milli*.
The unit for g is *gram*.
The SI unit for 50 mg is 50 milligrams.

In summary, the U.S. or English system is an ancient one based on standards initially determined by parts of the human body, which is why there is no consistent relationship among units. In the metric system, however, standard units are subdivided into multiples of 10, similar to our number system, and the names associated with each subdivision have prefixes that indicate a multiple of 10.

EXERCISES 3.1

Give the metric prefix for each value:

1. 1000
2. 100
3. 0.01
4. 0.1
5. 0.001
6. 10
7. 1,000,000
8. 0.000001

Give the SI symbol for each prefix:

9. hecto
10. kilo
11. deci
12. milli
13. centi
14. deka
15. micro
16. mega

Write the abbreviation for each quantity:

17. 65 milligrams
18. 125 kilolitres
19. 82 centimetres
20. 205 millilitres
21. 36 microamperes
22. 75 kilograms
23. 19 hectolitres
24. 5 megawatts

Write the SI unit for each abbreviation:

25. 18 m
26. 15 L
27. 36 kg
28. 85 mm
29. 24 ps
30. 9 dam
31. 135 mL
32. 45 dL
33. 45 mA
34. 75 MW
35. The basic SI unit of length is _____.
36. The basic SI unit of mass is _____.
37. The basic SI unit of electric current is _____.
38. The basic SI unit of time is _____.
39. The common SI units of volume are _____ and _____.
40. The common SI unit of power is _____.

3.2 Length

The basic SI unit of length is the **metre (m)**. The height of a door knob is about 1 m. (See Figure 3.1.) One metre is a little more than 1 yd. (See Figure 3.2.)

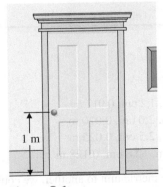

Figure 3.1
The height of a door knob is about 1 m.

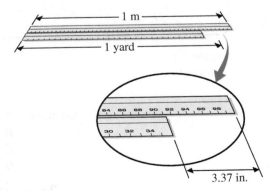

Figure 3.2
One metre is a little more than 1 yd.

Long distances are measured in kilometres (km) (1 km = 1000 m). The length of five city blocks is about 1 km. (See Figure 3.3.)

The centimetre (cm) is used to measure short distances, such as the width of this page (about 21 cm), or the width of a board. The width of your small fingernail is about 1 cm. (See Figure 3.4.)

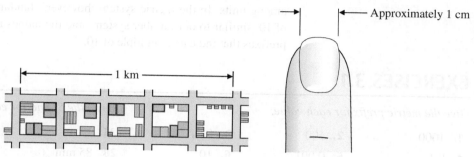

Figure 3.3
The length of five city blocks is about 1 km.

Figure 3.4
The width of your small fingernail is about 1 cm.

Figure 3.5
The thickness of a dime is about 1 mm.

The millimetre (mm) is used to measure very small lengths, such as the thickness of a sheet of metal or the depth of a tire tread. The thickness of a dime is about 1 mm. (See Figure 3.5.)

Each of the large numbered divisions on the metric ruler in Figure 3.6 marks one centimetre (cm). The smaller lines indicate halves of centimetres, and the smallest divisions show tenths of centimetres, or millimetres.

Example 1 Read A, B, C, and D on the metric ruler in Figure 3.6. Give each result in millimetres, centimetres, and metres.

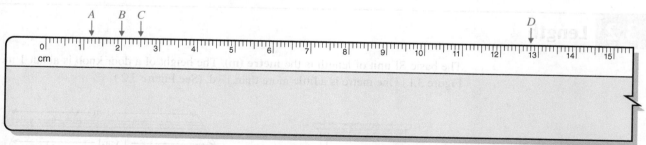

Figure 3.6
Metric ruler

Answers: $A = 12$ mm, 1.2 cm, 0.012 m
$B = 20$ mm, 2.0 cm, 0.020 m
$C = 25$ mm, 2.5 cm, 0.025 m
$D = 128$ mm, 12.8 cm, 0.128 m

To convert from one metric unit to another, we could use the same conversion factor procedure that we used in the U.S. or English system in Section 1.9.

> **Choosing Conversion Factors**
>
> The correct choice for a given conversion factor is the one in which the old units are in the numerator of the original expression and in the denominator of the conversion factor, or the old units are in the denominator of the original expression and in the numerator of the conversion factor. That is, set up the conversion factor so that the old units cancel each other.

Example 2 Change 3.6 km to metres.

Since *kilo* means 10^3 or 1000, 1 km = 1000 m. The two possible conversion factors are $\frac{1 \text{ km}}{1000 \text{ m}}$ and $\frac{1000 \text{ m}}{1 \text{ km}}$. Choose the one whose numerator is expressed in the new units (m) and whose denominator is expressed in the old units (km). This is $\frac{1000 \text{ m}}{1 \text{ km}}$.

$$3.6 \text{ km} \times \underbrace{\frac{1000 \text{ m}}{1 \text{ km}}}_{\text{conversion factor}} = 3600 \text{ m}$$

Example 3 Change 4 m to centimetres.

First, *centi* means 10^{-2} or 0.01; and 1 cm = 10^{-2} m. Choose the conversion factor with metres in the denominator and centimetres in the numerator.

$$4 \text{ m} \times \underbrace{\frac{1 \text{ cm}}{10^{-2} \text{ m}}}_{\text{conversion factor}} = 400 \text{ cm}$$

NOTE: Conversions within the metric system only involve moving the decimal point.

EXERCISES 3.2

Which is longer?

1. 1 metre or 1 millimetre
2. 1 metre or 1 centimetre
3. 1 metre or 1 kilometre
4. 1 millimetre or 1 kilometre
5. 1 centimetre or 1 millimetre
6. 1 kilometre or 1 centimetre

Fill in each blank:

7. 1 m = _____ mm
8. 1 km = _____ m
9. 1 cm = _____ m
10. 1 m = _____ cm
11. 1 m = _____ km
12. 1 hm = _____ m
13. 1 mm = _____ m
14. 1 m = _____ hm
15. 1 cm = _____ mm
16. 1 mm = _____ cm
17. 1 dam = _____ dm
18. 1 dm = _____ m

Which metric unit (km, m, cm, or mm) should you use to measure each item?

19. Diameter of an automobile tire
20. Thickness of sheet metal
21. Metric wrench sizes
22. Length of an automobile race
23. Length of a discus throw in track and field
24. Length and width of a table top
25. Distance between Chicago and St. Louis
26. Thickness of plywood

27. Thread size of a pipe
28. Length and width of a house lot

Fill in each blank with the most reasonable unit (km, m, cm, or mm):

29. A common metric wrench set varies in size from 6 to 19 _____.
30. The diameter of a wheel on a ten-speed bicycle is 56 _____.
31. A jet plane generally flies about 8–9 _____ high.
32. The width of a door in our house is 91 _____.
33. The length of the ridge on our roof is 24 _____.
34. Antonio's waist size is 95 _____.
35. The steering wheel on Brenda's automobile is 36 _____ in diameter.
36. Jan drives 12 _____ to school.
37. The standard metric size for plywood is 1200 _____ wide and 2400 _____ long.
38. The distance from home plate to the centerfield wall in a baseball park is 125 _____.
39. Read the measurements indicated by the letters on the metric ruler in Illustration 1 and give each result in millimetres and centimetres.

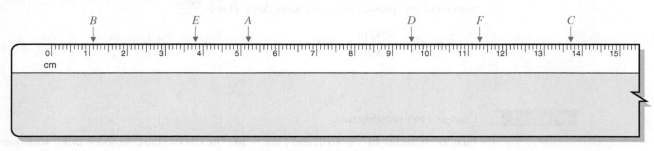

ILLUSTRATION 1

Use a metric ruler to measure each line segment. Give each result in millimetres and centimetres:

40.
41.
42.
43.
44.
45.
46.
47.
48.

49. Change 675 m to km.
50. Change 450 cm to m.
51. Change 1540 mm to m.
52. Change 3.2 km to m.
53. Change 65 cm to m.
54. Change 1.4 m to mm.
55. Change 7.3 m to cm.
56. Change 0.25 km to m.
57. Change 1250 m to km.
58. Change 4.5 m to cm.
59. Change 275 mm to cm.
60. Change 48 cm to mm.
61. Change 125 mm to cm.
62. Change 0.75 m to μm.
63. What is your height in metres and centimetres?

3.3 Mass and Weight

The **mass** of an object is the quantity of material making up the object. One unit of mass in the SI system is the *gram* (g). The gram is defined as the mass contained in 1 cubic centimetre (cm^3) of water, at its maximum density. A common paper clip has a mass of about 1 g. Three aspirin have a mass of about 1 g. (See Figure 3.7.)

3.3 ♦ Mass and Weight

(a) A common paper clip has a mass of about 1 g.

(b) Three aspirin have a mass of about 1 g.

Figure 3.7

Because the gram is so small, the **kilogram (kg)** is the basic unit of mass in the SI system. One kilogram is defined as the mass contained in 1 cubic decimetre (dm^3) of water at its maximum density.

For very large quantities, such as a trainload of coal or grain or a shipload of ore, the metric ton (1000 kg) is used. The milligram (mg) is used to measure very, very small masses such as medicine dosages. One grain of salt has a mass of about 1 mg.

The **weight** of an object is a measure of the earth's gravitational force—or pull—acting on the object. The SI unit of weight is the *newton* (N). As you are no doubt aware, the terms *mass* and *weight* are commonly used interchangeably by the general public. We have presented them here as technical terms, as they are used in the technical, engineering, and scientific professions. To further illustrate the difference, the mass of an astronaut remains relatively constant while his or her weight varies (the weight decreases as the distance from the earth increases). If the spaceship is in orbit or farther out in space, we say the crew is "weightless," because they seem to float freely in space. Their mass has not changed, although their weight is near zero. (See Figure 3.8.)

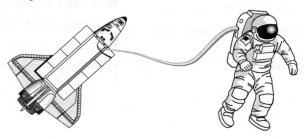

Figure 3.8
"Weightless" astronaut

Example 1

Change 12 kg to grams.

First, *kilo* means 10^3 or 1000; and 1 kg = 1000 g. Choose the conversion factor with kilograms in the denominator and grams in the numerator.

$$12 \text{ kg} \times \frac{1000 \text{ g}}{1 \text{ kg}} = 12{,}000 \text{ g}$$

↑ conversion factor ♦

Example 2

Change 250 mg to grams.

First, *milli* means 10^{-3} or 0.001; and 1 mg = 10^{-3} g. Choose the conversion factor with milligrams in the denominator and grams in the numerator.

$$250 \text{ mg} \times \frac{10^{-3} \text{ g}}{1 \text{ mg}} = 0.25 \text{ g}$$

↑ conversion factor ♦

EXERCISES 3.3

Which is larger?

1. 1 gram or 1 milligram
2. 1 gram or 1 kilogram
3. 1 milligram or 1 kilogram
4. 1 metric ton or 1 kilogram
5. 1 milligram or 1 microgram
6. 1 kilogram or 1 microgram

Fill in each blank:

7. 1 g = _____ mg
8. 1 kg = _____ g
9. 1 cg = _____ g
10. 1 mg = _____ g
11. 1 metric ton = _____ kg
12. 1 g = _____ cg
13. 1 mg = _____ μg
14. 1 μg = _____ mg

Which metric unit (kg, g, mg, or metric ton) should you use to measure the mass of each item?

15. A bar of handsoap
16. A vitamin capsule
17. A bag of flour
18. A four-wheel-drive tractor
19. A pencil
20. Your mass
21. A trainload of coal
22. A bag of potatoes
23. A contact lens
24. An apple

Fill in each blank with the most reasonable unit (kg, g, mg, or metric ton):

25. A slice of bread has a mass of about 25 _____.
26. Elevators in the college have a load limit of 2200 _____.
27. I take 1000 _____ of vitamin C every day.
28. My uncle's new truck can haul a load of 4 _____.
29. Postage rates for letters are based on the number of _____.
30. I take 1 _____ of vitamin C every day.
31. My best friend has a mass of 65 _____.
32. A jar of peanut butter contains 1200 _____.
33. The local grain elevator shipped 20,000 _____ of wheat last year.
34. One common size of aspirin tablets is 325 _____.
35. Change 875 g to kg.
36. Change 127 mg to g.
37. Change 85 g to mg.
38. Change 1.5 kg to g.
39. Change 3.6 kg to g.
40. Change 430 g to mg.
41. Change 270 mg to g.
42. Change 1350 g to kg.
43. Change 885 µg to mg.
44. Change 18 mg to µg.
45. Change 375 µg to mg.
46. Change 6.4 mg to µg.
47. Change 2.5 metric tons to kg.
48. Change 18,000 kg to metric tons.
49. Change 225,000 kg to metric tons.
50. Change 45 metric tons to kg.
51. What is your mass in kilograms?

3.4 Volume and Area

Volume

A common unit of volume in the metric system is the **litre (L)**. One litre of milk is a little more than 1 quart. The litre is commonly used for liquid volume. (See Figure 3.9.)

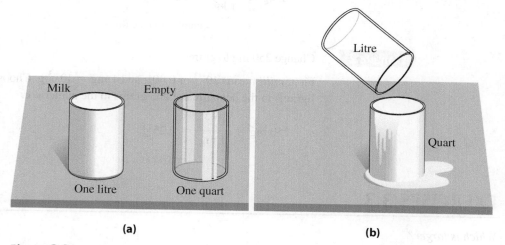

Figure 3.9
One litre is a little more than 1 quart.

The *cubic metre* (m^3) is used to measure large volumes. One cubic metre is the volume contained in a cube 1 m on an edge. The *cubic centimetre* (cm^3) is used to measure small volumes. It is the volume contained in a cube 1 cm on an edge.

NOTE: It is important to understand the relationship between the litre and the cubic centimetre. The litre is defined as the volume in 1 cubic decimetre (dm^3). That is, 1 L of liquid fills a cube 1 dm (10 cm) on an edge. (See Figure 3.10.)

The volume of the cube in Figure 3.10 can also be found by the formula

$V = lwh$

$V = (10 \text{ cm})(10 \text{ cm})(10 \text{ cm})$

$ = 1000 \text{ cm}^3$ Note: (cm)(cm)(cm) = cm^3

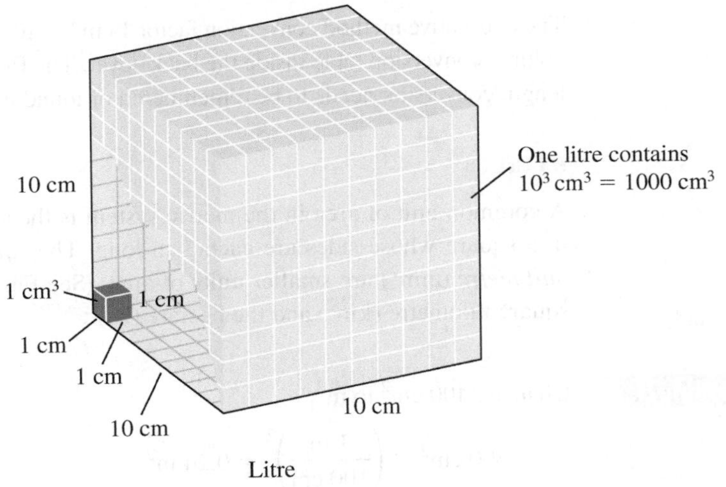

Figure 3.10
One litre contains 1000 cm^3.

Thus, 1 L = 1000 cm^3. Dividing each side by 1000, we have

$\dfrac{1}{1000} L = 1 \text{ cm}^3$

or   **1 mL = 1 cm^3**   1 mL = $\frac{1}{1000}$ L

Milk, soft drinks, and gasoline are sold by the litre. Liquid medicine and eye drops are sold by the millilitre. Large quantities of liquid are sold by the kilolitre (1000 L).

In Section 3.3, the kilogram was defined as the mass of 1 dm^3 of water. Since 1 dm^3 = 1 L, 1 litre of water has a mass of 1 kg.

Example 1 Change 0.5 L to millilitres.

$$0.5 \cancel{L} \times \dfrac{1000 \text{ mL}}{1 \cancel{L}} = 500 \text{ mL}$$

    ↑— conversion factor   ◆

Example 2 Change 4.5 cm^3 to mm^3.

$$4.5 \text{ cm}^3 \times \left(\dfrac{10 \text{ mm}}{1 \text{ cm}}\right)^3 = 4500 \text{ mm}^3$$

    ↑— conversion factor

Use the length conversion factor 1 cm = 10 mm and first form the conversion factor with cm in the denominator and mm in the numerator. Then raise the conversion factor to the third power to obtain cubic units in both numerator and denominator. Since the numerator equals the denominator, both the length conversion factor $\frac{10 \text{ mm}}{1 \text{ cm}}$ and its third power $\left(\frac{10 \text{ mm}}{1 \text{ cm}}\right)^3$ equal 1.

Alternative Method:

$$4.5 \text{ cm}^3 \times \frac{1000 \text{ mm}^3}{1 \text{ cm}^3} = 4500 \text{ mm}^3$$

conversion factor

The alternative method conversion factor $1 \text{ cm}^3 = 1000 \text{ mm}^3$ is taken directly from the metric volume conversion table inside the back cover. The first method is preferred, because only the length conversion needs to be remembered or found in a table.

Area

A common unit of area in the metric system is the **square metre (m^2)**, the area contained in a square whose sides are each 1 m long. The *square centimetre* (cm^2) and the *square millimetre* (mm^2) are smaller units of area. (See Figure 3.11.) The larger area units are the square kilometre (km^2) and the hectare (ha).

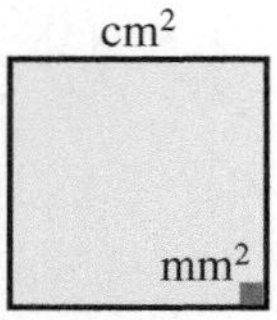

Figure 3.11
Relative sizes of 1 cm² and 1 mm²

Example 3

Change 2400 cm² to m².

$$2400 \text{ cm}^2 \times \left(\frac{1 \text{ m}}{100 \text{ cm}}\right)^2 = 0.24 \text{ m}^2$$

conversion factor

Use the length conversion factor 1 m = 100 cm and first form the conversion factor with cm in the denominator and m in the numerator. Then raise the conversion factor to the second power to obtain square units in both numerator and denominator. Since the numerator equals the denominator, both the length conversion factor $\frac{1 \text{ m}}{100 \text{ cm}}$ and its second power $\left(\frac{1 \text{ m}}{100 \text{ cm}}\right)^2$ equal 1.

Alternative Method:

$$2400 \text{ cm}^2 \times \frac{1 \text{ m}^2}{10{,}000 \text{ cm}^2} = 0.24 \text{ m}^2$$

conversion factor

The alternative method conversion factor $1 \text{ m}^2 = 10{,}000 \text{ cm}^2$ is taken directly from the metric area conversion table. The first method is again preferred, because only the length conversion needs to be remembered or found in a table.

Example 4

Change 1.2 km² to m².

$$1.2 \text{ km}^2 \times \left(\frac{1000 \text{ m}}{1 \text{ km}}\right)^2 = 1{,}200{,}000 \text{ m}^2$$

Alternative Method:

$$1.2 \text{ km}^2 \times \frac{10^6 \text{ m}^2}{1 \text{ km}^2} = 1{,}200{,}000 \text{ m}^2$$

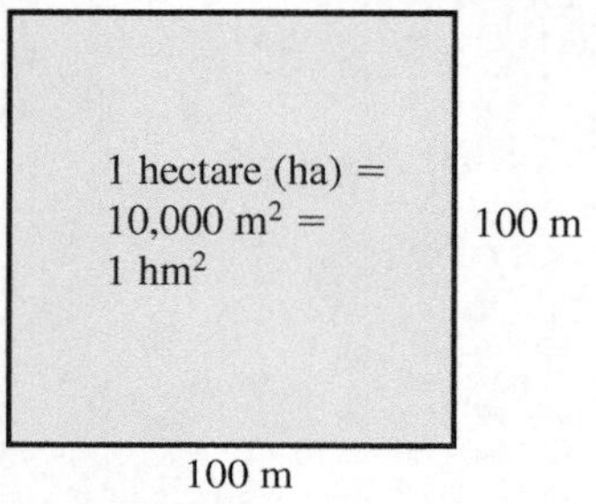

The **hectare (ha)** is the basic metric unit of land area. The area of 1 hectare equals the area of a square 100 m on a side, whose area is 10,000 m² or 1 square hectometre (hm²). (See Figure 3.12.)

Figure 3.12
Hectare

3.4 ♦ Volume and Area

The hectare is used because it is more convenient to say and use than "square hectometre." The metric prefixes are *not* used with the hectare unit. Instead of saying the prefix "kilo" with "hectare," we say "1000 hectares."

Example 5 How many hectares are contained in a rectangular field 360 m by 850 m?

The area in m² is

$$(360 \text{ m})(850 \text{ m}) = 306{,}000 \text{ m}^2 \text{ (Figure 3.13)}$$

$$306{,}000 \text{ m}^2 \times \frac{1 \text{ ha}}{10{,}000 \text{ m}^2} = 30.6 \text{ ha}$$

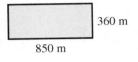

Figure 3.13

EXERCISES 3.4

Which is larger?

1. 1 litre or 1 millilitre
2. 1 millilitre or 1 kilolitre
3. 1 cubic millimetre or 1 cubic centimetre
4. 1 cubic metre or 1 litre
5. 1 square kilometre or 1 hectare
6. 1 square centimetre or 1 square millimetre

Fill in each blank:

7. 1 L = _____ mL
8. 1 mL = _____ L
9. 1 m³ = _____ cm³
10. 1 mm³ = _____ cm³
11. 1 cm² = _____ mm²
12. 1 km² = _____ ha
13. 1 m³ = _____ L
14. 1 cm³ = _____ mL

Which metric unit (m³, L, mL, m², cm², or ha) should you use to measure the following?

15. Oil in your automobile's crankcase
16. Cough syrup
17. Floor space in a warehouse
18. Size of a farm
19. Cross-sectional area of a piston
20. Piston displacement in an engine
21. Cargo space in a truck
22. Paint needed to paint a house
23. Eye drops
24. Page size of this book
25. Size of an industrial park
26. Gasoline in your automobile's fuel tank

Fill in each blank with the most reasonable unit (m³, L, mL, m², cm², or ha):

27. Lateesha ordered 12 _____ of concrete for her new driveway.
28. I drink 250 _____ of orange juice each morning.
29. Juan, a farmer, owns a 2500- _____ storage tank for diesel fuel.
30. Dwight planted 75 _____ of wheat this year.
31. Our house has 195 _____ of floor space.
32. We must heat 520 _____ of living space in our house.
33. When I was a kid, I mowed 6 _____ of lawns each week.
34. Our community's water tower holds 650 _____ of water.
35. The cross section of a log is 2500 _____.
36. Darnell bought a 25- _____ tarpaulin for his truck.
37. I need copper tubing with a cross section of 4 _____.
38. We should each drink 2 _____ of water each day.
39. Change 1500 mL to L.
40. Change 0.60 L to mL.
41. Change 1.5 m³ to cm³.
42. Change 450 mm³ to cm³.
43. Change 85 cm³ to mL.
44. Change 650 L to m³.
45. Change 85,000 m² to km².
46. Change 18 m² to cm².
47. Change 85,000 m² to ha.
48. Change 250 ha to km².
49. What is the mass of 500 mL of water?
50. What is the mass of 1 m³ of water?
51. How many hectares are contained in a rectangular field that measures 75 m by 90 m?
52. How many hectares are contained in a rectangular field that measures $\frac{1}{4}$ km by $\frac{1}{2}$ km?

3.5 Time, Current, and Other Units

The basic SI unit of time is the **second (s),** which is the same in all units of measurement. Time is also measured in minutes (min), hours (h), days, and years.

$$1 \text{ min} = 60 \text{ s}$$
$$1 \text{ h} = 60 \text{ min}$$
$$1 \text{ day} = 24 \text{ h}$$
$$1 \text{ year} = 365\tfrac{1}{4} \text{ days (approximately)}$$

Example 1 Change 4 h 15 min to seconds.

First, $4 \cancel{\text{h}} \times \dfrac{60 \text{ min}}{1 \cancel{\text{h}}} = 240 \text{ min}$

And $4 \text{ h } 15 \text{ min} = 240 \text{ min} + 15 \text{ min}$
$= 255 \text{ min}$

Then, $255 \cancel{\text{min}} \times \dfrac{60 \text{ s}}{1 \cancel{\text{min}}} = 15{,}300 \text{ s}$

Very short periods of time are commonly used in electronics. These are measured in parts of a second, given with the appropriate metric prefix.

Example 2 What is the meaning of each unit? **a.** 1 ms **b.** 1 μs **c.** 1 ns **d.** 1 ps

a. 1 ms = 1 millisecond = 10^{-3} s It means one-thousandth of a second.
b. 1 μs = 1 microsecond = 10^{-6} s It means one-millionth of a second.
c. 1 ns = 1 nanosecond = 10^{-9} s It means one-billionth of a second.
d. 1 ps = 1 picosecond = 10^{-12} s It means one-trillionth of a second.

Example 3 Change 25 ms to seconds.

First, *milli* means 10^{-3}, and 1 ms = 10^{-3} s. Then

$$25 \cancel{\text{ms}} \times \underbrace{\dfrac{10^{-3} \text{ s}}{1 \cancel{\text{ms}}}}_{\text{conversion factor}} = 0.025 \text{ s}$$

Example 4 Change 0.00000025 s to nanoseconds.

First, *nano* means 10^{-9}, and 1 ns = 10^{-9} s. Then

$$0.00000025 \cancel{\text{s}} \times \underbrace{\dfrac{1 \text{ ns}}{10^{-9} \cancel{\text{s}}}}_{\text{conversion factor}} = 250 \text{ ns}$$

The basic SI unit of electric current is the **ampere (A),** sometimes called the amp. This same unit is used in the U.S. system. The ampere is a fairly large amount of current, so smaller currents are measured in parts of an ampere and are given the appropriate SI prefix.

Example 5 What is the meaning of each unit? **a.** 1 mA **b.** 1 μA

a. 1 mA = 1 milliampere = 10^{-3} A It means one-thousandth of an ampere.
b. 1 μA = 1 microampere = 10^{-6} A It means one-millionth of an ampere.

Example 6
Change 275 μA to amperes.

First, *micro* means 10^{-6}, and $1\ \mu A = 10^{-6}$ A. Then

$$275\ \mu A \times \frac{10^{-6}\ A}{1\ \mu A} = 0.000275\ A$$

Example 7
Change 0.045 A to milliamps.

First, *milli* means 10^{-3}, and $1\ mA = 10^{-3}$ A. Then

$$0.045\ A \times \frac{1\ mA}{10^{-3}\ A} = 45\ mA$$

The common metric unit for both electrical and mechanical power is the *watt* (W).

Example 8
What is the meaning of each unit? **a.** 1 mW **b.** 1 kW **c.** 1 MW

a. $1\ mW = 1$ milliwatt $= 10^{-3}$ W   It means one-thousandth of a watt.
b. $1\ kW = 1$ kilowatt $= 10^{3}$ W   It means one thousand watts.
c. $1\ MW = 1$ megawatt $= 10^{6}$ W   It means one million watts.

Example 9
Change 0.025 W to milliwatts.

First, *milli* means 10^{-3}, and $1\ mW = 10^{-3}$ W. Then

$$0.025\ W \times \frac{1\ mW}{10^{-3}\ W} = 25\ mW$$

Example 10
Change 2.3 MW to watts.

First, *mega* means 10^{6}, and $1\ MW = 10^{6}$ W. Then

$$2.3\ MW \times \frac{10^{6}\ W}{1\ MW} = 2.3 \times 10^{6}\ W\ \text{or}\ 2{,}300{,}000\ W$$

A few other units that are commonly used in electronics are listed in Table 3.4. The metric prefixes are used with each of these units in the same way as with the other metric units we have studied.

Table 3.4	Units Commonly Used in Electronics	
Unit	**Symbol**	**Used to measure**
volt	V	voltage
ohm	Ω	resistance
hertz	Hz	frequency
farad	F	capacitance
henry	H	inductance
coulomb	C	charge

EXERCISES 3.5

Which is larger?

1. 1 amp or 1 milliamp
2. 1 microsecond or 1 picosecond
3. 1 second or 1 nanosecond
4. 1 megawatt or 1 milliwatt
5. 1 kilovolt or 1 megavolt
6. 1 volt or 1 millivolt

Write the abbreviation for each unit:

7. 43 kilowatts
8. 7 millivolts
9. 17 picoseconds
10. 1.2 amperes
11. 3.2 megawatts
12. 55 microfarads
13. 450 ohms
14. 70 nanoseconds

Fill in each blank:

15. 1 kW = _____ W
16. 1 mA = _____ A
17. 1 ns = _____ s
18. 1 day = _____ s
19. 1 A = _____ μA
20. 1 F = _____ μF
21. 1 V = _____ MV
22. 1 Hz = _____ kHz
23. Change 0.35 A to mA.
24. Change 18 kW to W.
25. Change 350 ms to s.
26. Change 1 h 25 min 16 s to s.
27. Change 13,950 s to h, min, and s.
28. Change 15 MV to kV.
29. Change 175 μF to mF.
30. Change 145 ps to ns.
31. Change 1500 kHz to MHz.
32. Change 5×10^{12} W to MW.

3.6 Temperature

The basic SI unit for temperature is **kelvin (K)**, which is used mostly in scientific and engineering work. Everyday temperatures are measured in *degrees Celsius* (°C). The United States also measures temperatures in degrees Fahrenheit (°F).

On the Celsius scale, water freezes at 0° and boils at 100°. Each degree Celsius is 1/100 of the difference between the boiling temperature and the freezing temperature of water. Figure 3.14 shows some approximate temperature readings in degrees Celsius and Fahrenheit and compares them with a related activity.

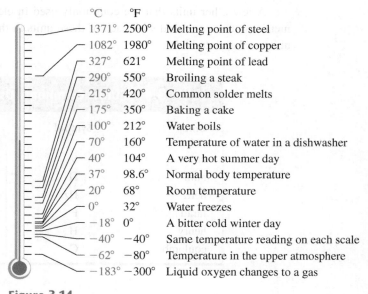

Figure 3.14
Related temperature readings in degrees Celsius and degrees Fahrenheit

3.6 ♦ Temperature

Digital infrared thermometers measure temperature by measuring the thermal radiation emitted by the object. These thermometers are commonly used for measuring the temperature of a person (see Figure 3.15); checking mechanical equipment or electrical circuit breaker boxes for hot spots; monitoring materials in the process of being heated or cooled, sometimes at very high and very low temperatures; and checking for hot spots in fighting fires, to name a few.

The formulas for changing between degrees Celsius and degrees Fahrenheit are

$$C = \frac{5}{9}(F - 32°)$$
$$F = \frac{9}{5}C + 32°$$

Figure 3.15
Digital thermometer for measuring the body temperature of a person

Example 1 Change 68°F to degrees Celsius.

$$C = \frac{5}{9}(F - 32°)$$
$$C = \frac{5}{9}(68° - 32°)$$
$$= \frac{5}{9}(36°) \quad \text{First, subtract within parentheses.}$$
$$= 20° \quad \text{Multiply.}$$

Thus, 68°F = 20°C.

Example 2 Change 35°C to degrees Fahrenheit.

$$F = \frac{9}{5}C + 32°$$
$$F = \frac{9}{5}(35°) + 32°$$
$$= 63° + 32° \quad \text{First, multiply.}$$
$$= 95° \quad \text{Add.}$$

That is, 35°C = 95°F.

Example 3 Change 10°F to degrees Celsius.

$$C = \frac{5}{9}(F - 32°)$$
$$C = \frac{5}{9}(10° - 32°)$$
$$= \frac{5}{9}(-22°)$$
$$= -12.2°$$

So 10°F = −12.2°C.

Example 4 Change $-60°C$ to degrees Fahrenheit.

$$F = \frac{9}{5}C + 32°$$
$$F = \frac{9}{5}(-60°) + 32°$$
$$= -108° + 32°$$
$$= -76°$$

So $-60°C = -76°F$.

EXERCISES 3.6

Use Figure 3.14 to choose the most reasonable answer for each statement:

1. The boiling temperature of water is **a.** 212°C, **b.** 100°C, **c.** 0°C, or **d.** 50°C.
2. The freezing temperature of water is **a.** 32°C, **b.** 100°C, **c.** 0°C, or **d.** $-32°C$.
3. Normal body temperature is **a.** 100°C, **b.** 50°C, **c.** 37°C, or **d.** 98.6°C.
4. The body temperature of a person who has a fever is **a.** 102°C, **b.** 52°C, **c.** 39°C, or **d.** 37°C.
5. The temperature on a hot summer day in the California desert is **a.** 108°C, **b.** 43°C, **c.** 60°C, or **d.** 120°C.
6. The temperature on a cold winter day in Chicago is **a.** 20°C, **b.** 10°C, **c.** 30°C, or **d.** $-10°C$.
7. The thermostat in your home should be set at **a.** 70°C, **b.** 50°C, **c.** 19°C, or **d.** 30°C.
8. Solder melts at **a.** 215°C, **b.** 420°C, **c.** 175°C, or **d.** 350°C.
9. Freezing rain is most likely to occur at **a.** 32°C, **b.** 25°C, **c.** $-18°C$, or **d.** 0°C.
10. The weather forecast calls for a low temperature of 3°C. What should you plan to do? **a.** Sleep with the windows open. **b.** Protect your plants from frost. **c.** Sleep with the air conditioner on. **d.** Sleep with an extra blanket.

Fill in each blank:

11. 77°F = _____ °C
12. 45°C = _____ °F
13. 325°C = _____ °F
14. 140°F = _____ °C
15. $-16°C$ = _____ °F
16. 5°F = _____ °C
17. $-16°F$ = _____ °C
18. $-40°C$ = _____ °F
19. $-78°C$ = _____ °F
20. $-10°F$ = _____ °C

3.7 Metric and U.S. Conversion

In technical work, you must sometimes change from one system of measurement to another. The approximate conversions between metric units and U.S. units are found in the Metric and U.S. Conversion Table inside the back cover. Most numbers are rounded to three or four significant digits. Due to this rounding and your choice of conversion factors, there may be a small difference in the last digit(s) of the answers involving conversion factors. This small difference is acceptable. In this section, round each result to three significant digits, when necessary. You may review significant digits in Section 1.11.

Figure 3.16 shows the relative sizes of each of four sets of common metric and U.S. units of area.

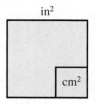 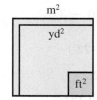 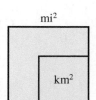 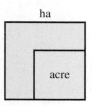

Figure 3.16

Relative sizes of some common metric and U.S. units of area

3.7 ♦ Metric and U.S. Conversion

Example 1 Change 17 in. to centimetres.

$$17 \text{ in.} \times \frac{2.54 \text{ cm}}{1 \text{ in.}} = 43.2 \text{ cm}$$

(conversion factor)

Example 2 Change 1950 g to pounds.

$$1950 \text{ g} \times \frac{1 \text{ lb}}{454 \text{ g}} = 4.30 \text{ lb}$$

(conversion factor)

NOTE: If you choose a different conversion factor, the result may vary slightly due to rounding. For example,

$$1950 \text{ g} \times \frac{0.00220 \text{ lb}}{1 \text{ g}} = 4.29 \text{ lb}$$

(conversion factor)

Example 3 Change 0.85 qt to millilitres.

$$0.85 \text{ qt} \times \frac{0.946 \text{ L}}{1 \text{ qt}} \times \frac{10^3 \text{ mL}}{1 \text{ L}} = 804 \text{ mL}$$

(conversion factors)

Example 4 Change 5 yd² to ft².

$$5 \text{ yd}^2 \times \left(\frac{3 \text{ ft}}{1 \text{ yd}}\right)^2 = 45 \text{ ft}^2$$

Use the length conversion factor 1 yd = 3 ft and first form the conversion factor with yd in the denominator and ft in the numerator. Then raise the conversion factor to the second power to obtain square units in both numerator and denominator. Since the numerator equals the denominator, both the length conversion factor $\frac{3 \text{ ft}}{1 \text{ yd}}$ and its second power $\left(\frac{3 \text{ ft}}{1 \text{ yd}}\right)^2$ equal 1.

Alternative Method:

$$5 \text{ yd}^2 \times \frac{9 \text{ ft}^2}{1 \text{ yd}^2} = 45 \text{ ft}^2$$

The alternative method conversion factor 1 yd² = 9 ft² is taken directly from the U.S. area conversion table inside the back cover. The first method is again preferred, because only the length conversion needs to be remembered or found in a table.

Example 5 How many square inches are in a metal plate 14 cm² in area?

$$14 \text{ cm}^2 \times \left(\frac{1 \text{ in.}}{2.54 \text{ cm}}\right)^2 = 2.17 \text{ in}^2$$

Alternative Method:

$$14 \text{ cm}^2 \times \frac{0.155 \text{ in}^2}{1 \text{ cm}^2} = 2.17 \text{ in}^2$$

Example 6 Change 147 ft³ to cubic yards.

$$147 \text{ ft}^3 \times \left(\frac{1 \text{ yd}}{3 \text{ ft}}\right)^3 = 5.44 \text{ yd}^3$$

Alternative Method:

$$147 \text{ ft}^3 \times \frac{1 \text{ yd}^3}{27 \text{ ft}^3} = 5.44 \text{ yd}^3$$

Example 7 How many cubic yards are in 12 m³?

$$12 \text{ m}^3 \times \left(\frac{1.09 \text{ yd}}{1 \text{ m}}\right)^3 = 15.5 \text{ yd}^3$$

Alternative Method:

$$12 \text{ m}^3 \times \frac{1.31 \text{ yd}^3}{1 \text{ m}^3} = 15.7 \text{ yd}^3$$

In the U.S. system, the *acre* is the basic unit of land area. Historically, the acre was the amount of ground that a yoke of oxen could plow in one day.

1 acre = 43,560 ft²
1 mi² = 640 acres = 1 section

Example 8 How many acres are in a rectangular field that measures 1350 ft by 2750 ft?

The area in ft² is

(1350 ft)(2750 ft) = 3,712,500 ft²

$$3{,}712{,}500 \text{ ft}^2 \times \frac{1 \text{ acre}}{43{,}560 \text{ ft}^2} = 85.2 \text{ acres}$$

Professional journals and publications in nearly all scientific areas, including agronomy and animal science, have been metric for several years, so that scientists around the world can better understand and benefit from U.S. research.

Land areas in the United States are still typically measured in the U.S. system. When converting between metric and U.S. land-area units, use the following relationship:

1 hectare = 2.47 acres

A good approximation is

1 hectare = 2.5 acres

Example 9 How many acres are in 30.6 ha?

$$30.6 \text{ ha} \times \frac{2.47 \text{ acres}}{1 \text{ ha}} = 75.6 \text{ acres}$$

Example 10 How many hectares are in the rectangular field in Example 8?

$$85.2 \text{ acres} \times \frac{1 \text{ ha}}{2.47 \text{ acres}} = 34.5 \text{ ha}$$

Considerable patience and education will be necessary before the hectare becomes the common unit of land area in the United States. The mammoth task of changing all property documents is only one of many obstacles.

Example 11 A recipe calls for 68 g of butter. How many ounces of butter is this?

$$68 \text{ g} \times \frac{0.0353 \text{ oz}}{1 \text{ g}} = 2.4 \text{ oz}$$

The following example shows how to use multiple conversion factors involving more complex units.

Example 12 Change 165 lb/in² to kg/cm².

This conversion requires a pair of conversion factors, as follows:

(a) from pounds to kilograms
(b) from in² to cm²

$$165 \, \frac{\text{lb}}{\text{in}^2} \times \frac{1 \text{ kg}}{2.20 \text{ lb}} \times \left(\frac{1 \text{ in.}}{2.54 \text{ cm}}\right)^2 = 11.6 \text{ kg/cm}^2$$

conversion factors for (a) (b)

EXERCISES 3.7

Fill in each blank, rounding each result to three significant digits when necessary. (Small differences in the last significant digit of answers are acceptable due to the choice of any conversion factor that has been rounded.)

1. 8 lb = _____ kg
2. 16 ft = _____ m
3. 38 cm = _____ in.
4. 81 m = _____ ft
5. 4 yd = _____ cm
6. 17 qt = _____ L
7. 30 kg = _____ lb
8. 15 mi = _____ km
9. 3.2 in. = _____ mm
10. 2 lb 4 oz = _____ g
11. A road sign reads "75 km to Chicago." What is this distance in miles?
12. A hole is 35 mm wide. How many inches wide is it?
13. The diameter of a bolt is 0.425 in. Express this diameter in mm.
14. Change $3\frac{13}{32}$ in. to cm.
15. A tank contains 8 gal of fuel. How many litres of fuel are in the tank?
16. How many pounds does a 150-kg satellite weigh?
17. An iron bar weighs 2 lb 6 oz. Express its weight in **a.** oz and **b.** kg.
18. A precision part is milled to 1.125 in. in width. What is the width in millimetres?
19. A football field is 100 yd long. What is its length **a.** in feet and **b.** in metres?
20. A micro wheel weighs 0.045 oz. What is its weight in mg?
21. A hole is to be drilled in a metal plate 5 in. in diameter. What is the diameter **a.** in cm and **b.** in mm?
22. A can contains 15 oz of tomato sauce. How many grams does the can contain?
23. Change 3 yd² to m².
24. Change 12 cm² to in².
25. How many ft² are in 140 yd²?
26. How many m² are in 15 yd²?
27. Change 18 in² to cm².
28. How many ft² are in a rectangle 12.6 yd long and 8.6 yd wide? (A = lw)
29. How many ft² are in a rectangle 12.6 m long and 8.6 m wide?
30. Find the area of the figure in Illustration 1 in in².

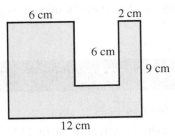

ILLUSTRATION 1

31. Change 15 yd³ to m³.
32. Change 5473 in³ to cm³.
33. How many mm³ are in 17 in³?
34. How many in³ are in 25 cm³?
35. Change 84 ft³ to cm³.
36. How many cm³ are in 98 in³?
37. A commercial space 60 ft wide and 90 ft deep rents for $48,600 annually. What is the price per square foot? What is the price per frontage foot?
38. A concrete sidewalk is to be built (as shown in Illustration 2) around the outside of a corner lot that measures 140 ft by 180 ft. The sidewalk is to be 5 ft wide. What is the surface area of the sidewalk? The sidewalk is to be 4 in. thick. How many yards (actually, cubic yards) of concrete are needed? Concrete costs $90/yd³ delivered. How much will the sidewalk cost?

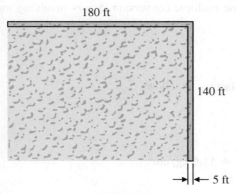

ILLUSTRATION 2

39. How many acres are in a rectangular field that measures 2400 ft by 625 ft?
40. How many acres are in a rectangular field that measures 820 yd by 440 yd?
41. How many hectares are in the field in Exercise 39?
42. How many hectares are in the field in Exercise 40?
43. A house lot measures 145 ft by 186 ft. What part of an acre is the lot?
44. How many acres are in $\frac{1}{4}$ mi^2?
45. How many acres are in $\frac{1}{8}$ section?
46. How many acres are in 520 square rods?
47. A corn yield of 10,550 kg/ha is equivalent to how many lb/acre? To how many bu/acre? (One bu of corn weighs 56 lb.)
48. A soybean yield of 45 bu/acre is equivalent to how many kg/ha? To how many metric tons/hectare? (One bu of soybeans weighs 60 lb.)
49. How many acres are in eight 30-in. rows 440 yards long?
50. a. How many rows 30 in. apart can be planted in a rectangular field 3300 ft long and 2600 ft wide? The rows run lengthwise.
    b. Suppose seed corn is planted at 20 lb/acre. How many bushels of seed corn will be needed to plant the field?
    c. Suppose 1 lb of seed corn contains 1200 kernels. How many bags, each containing 80,000 kernels, will be needed?
51. For Octoberfest, a vender purchased 15 kg of German brats. How many 6-oz brats can be made from the 15 kg of German brats?
52. Chef Dino has $2\frac{1}{2}$ gal of soup and plans to put the soup in 1-litre containers to sell. How many 1-litre containers are needed to hold all the soup?
53. Change 25.6 kg/cm^2 to lb/in^2.
54. Change 1.5 g/cm^2 to mg/mm^2.
55. Change 65 mi/h to m/s.
56. Change 415 lb/ft^3 to g/cm^3.

SUMMARY | CHAPTER 3

Glossary of Basic Terms

Ampere (A). The basic SI unit of electric current. (p. 142)
Hectare (ha). The basic metric unit for land area. (p. 140)
Kelvin (K). The basic SI unit for temperature; everyday metric temperatures are measured in degrees Celsius (°C). (p. 144)
Kilogram (kg). The basic SI unit of mass. (p. 137)
Litre (L). A common SI unit of volume. (p. 138)
Mass. The quantity of material making up an object. (p. 136)
Metre (m). The basic SI unit of length. (p. 133)
Second (s). The basic SI unit of time. (p. 142)
SI. Abbreviation for the International System of Units (from the French *Système International d'Unités*) and commonly called the **metric system.** (p. 131)
Square metre (m^2). A common SI unit of area. (p. 140)
Weight. A measure of the earth's gravitational force (pull) acting on an object. (p. 137)

3.1 Introduction to the Metric System

1. **SI base units:** Review the seven SI base units in Table 3.1 on page 131.
2. **Prefixes for SI units:** Review the prefixes for SI units in Table 3.3 on page 132.

3.2 Length

1. **Choosing conversion factors:** The correct choice for a given conversion factor is the one in which the old units are in the numerator of the original expression and in the denominator of the conversion factor or the old units are in the denominator of the original expression and in the numerator of the conversion factor. That is, set up the conversion factor so that the old units cancel each other. (p. 135)

3.6 Temperature

1. Formulas for changing between degrees Celsius and degrees Fahrenheit:

$$C = \frac{5}{9}(F - 32°)$$

$$F = \frac{9}{5}C + 32° \text{ (p. 145)}$$

REVIEW | CHAPTER 3

Give the metric prefix for each value:

1. 0.001
2. 1000

Give the SI abbreviation for each prefix:

3. mega
4. micro

Write the SI abbreviation for each quantity:

5. 42 millilitres
6. 8.3 nanoseconds

Write the SI unit for each abbreviation:

7. 18 km
8. 350 mA
9. 50 μs

Which is larger?

10. 1 L or 1 mL
11. 1 kW or 1 MW
12. 1 km² or 1 ha
13. 1 m³ or 1 L

Fill in each blank:

14. 650 m = _____ km
15. 750 mL = _____ L
16. 6.1 kg = _____ g
17. 4.2 A = _____ μA
18. 18 MW = _____ W
19. 25 μs = _____ ns
20. 250 cm² = _____ mm²
21. 25,000 m² = _____ ha
22. 0.6 m³ = _____ cm³
23. 250 cm³ = _____ mL
24. 72°F = _____ °C
25. −25°C = _____ °F
26. Water freezes at _____ °C.
27. Water boils at _____ °C.
28. 180 lb = _____ kg
29. 126 ft = _____ m
30. 360 cm = _____ in.
31. 275 in² = _____ cm²
32. 18 yd² = _____ ft²
33. 5 m³ = _____ ft³
34. 15.0 acres = _____ ha

Choose the most reasonable quantity:

35. Jorge and Maria drive **a.** 1600 cm, **b.** 470 m, **c.** 12 km, or **d.** 2400 mm to college each day.
36. Chuck's mass is **a.** 80 kg, **b.** 175 kg, **c.** 14 μg, or **d.** 160 Mg.
37. An automobile's fuel tank holds **a.** 18 L, **b.** 15 kL, **c.** 240 mL, or **d.** 60 L of gasoline.
38. Jamilla, being of average height, is **a.** 5.5 m, **b.** 325 mm, **c.** 55 cm, or **d.** 165 cm tall.
39. An automobile's average fuel consumption is **a.** 320 km/L, **b.** 15 km/L, **c.** 35 km/L, or **d.** 0.75 km/L.
40. On Illinois winter mornings, the temperature sometimes dips to **a.** −50°C, **b.** −30°C, **c.** 30°C, or **d.** −80°C.
41. Abdul drives **a.** 85 km/h, **b.** 50 km/h, **c.** 150 km/h, or **d.** 25 km/h on the interstate highway.
42. Complete the following table of metric system prefixes using the given sample metric unit:

Prefix	Symbol	Power of 10	Sample unit	How many?	How many?
tera	T	10^{12}	m	10^{12} m = 1 Tm	1 m = 10^{-12} Tm
giga	G	10^{9}	W	10^{9} W = 1 GW	1 W = 10^{-9} GW
mega	M	10^{6}	Hz		
kilo	k	10^{3}	g		
hecto	h	10^{2}	Ω		
deka	da	10^{1}	L		
deci	d	10^{-1}	g	10^{-1} g = 1 dg	1 g = 10 dg
centi	c	10^{-2}	m		
milli	m	10^{-3}	A	10^{-3} A = 1 mA	1 A = 10^{3} A
micro	μ	10^{-6}	W		
nano	n	10^{-9}	s		
pico	p	10^{-12}	s		

TEST | CHAPTER 3

1. Give the metric prefix for 1000.
2. Give the metric prefix for 0.01.
3. Which is larger, 200 mg or 1 g?
4. Write the SI unit for the abbreviation 240 μL.
5. Write the abbreviation for 30 hectograms.
6. Which is longer, 1 km or 25 cm?

Fill in each blank:

7. 4.25 km = _____ m
8. 7.28 mm = _____ μm
9. 72 m = _____ mm
10. 256 hm = _____ cm
11. 12 dg = _____ mg
12. 16.2 g = _____ mg
13. 7.236 metric tons = _____ kg
14. 310 g = _____ cg
15. 72 hg = _____ mg
16. 1.52 dL = _____ L
17. 175 L = _____ m^3
18. 2.7 m^3 = _____ cm^3
19. 400 ha = _____ km^2
20. 0.2 L = _____ mL

21. What is the basic SI unit of time?
22. Write the abbreviation for 25 kilowatts.

Fill in each blank:

23. 280 W = _____ kW
24. 13.9 mA = _____ A
25. 720 ps = _____ ns
26. What is the basic SI unit for temperature?
27. What is the freezing temperature of water on the Celsius scale?

Fill in each blank, rounding each result to three significant digits when necessary:

28. 25°C = _____ °F
29. 28°F = _____ °C
30. 98.6°F = _____ °C
31. 100 km = _____ mi
32. 200 cm = _____ in.
33. 1.8 ft^3 = _____ in^3
34. 37.8 ha = _____ acres
35. 80.2 kg = _____ lb

Measurement | CHAPTER 4

OBJECTIVES

- Distinguish the difference between an exact number and an approximate number (measurement).
- Find the number of significant digits (accuracy) of a measurement.
- Find the precision and greatest possible error of a measurement.
- Distinguish the difference between the accuracy and the precision of a measurement.
- Read metric and U.S. measurements on a vernier caliper and a micrometer caliper.
- Use the rules for measurement to add, subtract, multiply, and divide measurements.
- Find the relative error and percent of error of a measurement.
- Find the upper limit, the lower limit, and the tolerance interval when given a measurement and its tolerance.
- Find the value of a given resistor as well as its tolerance using the color code of electrical resistors.
- Read uniform and nonuniform scales.

Mathematics at Work

Automotive collision repair technicians repair, repaint, and refinish automotive vehicle bodies; straighten vehicle frames; and replace damaged glass and other automotive parts that cannot be economically repaired. Using modern techniques including diagnostics, electronic equipment, computer support equipment, and other specialized equipment, the technician's primary task is to restore damaged vehicles to their original condition. Training and education for this work are available at many community colleges and trade schools. Various automobile manufacturers and their participating dealers also sponsor programs at postsecondary schools across the United States. Good reading, mathematics, computer, and communications skills are needed.

Voluntary certification is available through the National Institute for Automotive Service Excellence (ASE) and is the recognized standard of achievement for automotive collision repair technicians. For more information, please visit **www.cengage.com** and access the Student Online Resources of this text.

Collision Repair Technician
Collision repair technician repairing an automobile

4.1 Approximate Numbers and Accuracy

Approximate Numbers (Measurements) versus Exact Numbers

As noted in Section 1.9, **measurement** is the comparison of an *observed* quantity with a *standard unit* quantity. Consider the measurement of the length of a metal block like the one in Figure 4.1.

1. First, measure the block with ruler A, graduated only in inches. This means measurements will be to the nearest inch. The measured length is 2 in.

2. Measure the same block with ruler B, graduated in half-inches. Measurements now are to the nearest half-inch. The measured length is $2\frac{1}{2}$ in. to the nearest half-inch.

3. Measure the block again with ruler C, graduated in fourths of an inch. To the nearest $\frac{1}{4}$ in., the measurement is $2\frac{1}{4}$ in.

4. Measure the block again with ruler D, graduated in eighths of an inch. To the nearest $\frac{1}{8}$ in., the measurement is $2\frac{3}{8}$ in.

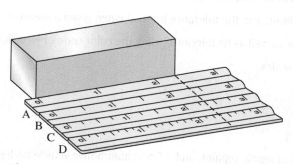

Figure 4.1
Measuring the length of a metal block with rulers of different precision

If you continue this process, by using finer and finer graduations on the ruler, will you ever find the "exact" length? No—since all measurements are only approximations, the "exact" length cannot be found. A measurement is only as good as the measuring instrument you use. It would be rather difficult for you to measure the diameter of a pinhead with a ruler.

A *tachometer* is used to measure the number of revolutions an object makes with respect to some unit of time. Since the unit of time is usually minutes, a tachometer usually measures revolutions per minute (rpm). This measurement is usually given in integral units, such as 10 rpm or 255 rpm. Tachometers are used in industry to test motors to see whether or not they turn at a specified rate. In the shop, both wood and metal lathes have specified rpm rates. Tachometers are also commonly used in sports cars to help drivers shift gears at the appropriate engine rpm. A tachometer normally measures the spindle speed or the rpm of a shaft, not the surface speed.

Consider the diagram of the tachometer in Figure 4.2. Each of the printed integral values on the dial indicates hundreds of rpm. That is, if the dial indicator is at 10, then the reading is 10 hundred rpm, or 1000 rpm. If the dial indicator is at 70, then the reading is 70 hundred rpm, or 7000 rpm. Each of the subdivisions between 0 and 10, 10 and 20, 20 and 30, and so forth, represents an additional 100 rpm.

Figure 4.2
Tachometer

4.1 ♦ Approximate Numbers and Accuracy

Tachometer readings are also approximate. Tachometers are calibrated in tens, hundreds, or thousands of revolutions per minute, and it is impossible to read the exact number of rpm.

Example 1 Read the tachometer in Figure 4.3.

The indicator is at the sixth division past 30; so the reading is 36 thousand, or 36,000 rpm. ♦

Figure 4.3

Up to this time in your study of mathematics, all measurements have probably been treated as exact numbers. An **exact number** is a number that has been determined as a result of counting (such as 24 students enrolled in a class) or by some definition (such as 1 hour (h) = 60 minutes (min), or 1 in. = 2.54 cm—these are conversion definitions accepted by all international government bureaus of standards). Addition, subtraction, multiplication, and division of exact numbers usually make up the main content of elementary-school mathematics.

However, nearly all data of a technical nature involve **approximate numbers**; that is, numbers that have been determined by some measurement instrument or process. This process may be direct, as with a ruler, or indirect, as with a surveying transit. Before studying how to perform calculations with approximate numbers (measurements), we must determine the "correctness" or certainty of an approximate number. First, we must realize that no measurement can be found exactly. The length of the cover of this book can be found using many instruments. The better the measuring device used, the better the certainty of the measurement.

In summary:

Exact versus Approximate Numbers:

1. Only counting numbers are exact.
2. All measurements are approximations.

Accuracy and Significant Digits

The **accuracy of a measurement** means the number of digits, called **significant digits**, that it contains. These indicate the number of units we are reasonably sure of having counted and of being able to rely on in a measurement. The greater the number of significant digits given in a measurement, the better the accuracy, and vice versa.

Why is the accuracy of a measurement important? First, you must realize that each measurement is an approximation. Then you need to know how "good" each measurement is (its certainty) before you use it to do calculations with other measurements. As noted earlier, your study of mathematics has likely involved calculations with only exact numbers. However, the results of calculations with measurements must be analyzed and adjusted so that the certainty of the results best matches the certainty of the individual measurements. No result of a calculation can be better than the worst measurement! As we shall see in Section 4.6, accuracy is important in determining the results of multiplying and dividing measurements.

Example 2 The average distance between the moon and the earth is 239,000 mi. This measurement indicates measuring 239 thousands of miles. Its accuracy is indicated by 3 significant digits. ♦

Example 3 A measurement of 10,900 m indicates measuring 109 hundreds of metres. Its accuracy is 3 significant digits. ♦

Example 4 A measurement of 0.042 cm indicates measuring 42 thousandths of a centimetre. Its accuracy is 2 significant digits.

Example 5 A measurement of 12.000 m indicates measuring 12,000 thousandths of metres. Its accuracy is 5 significant digits.

Notice that sometimes a zero is significant and sometimes it is not. Apply the following rules to determine whether a digit is significant or not.

Significant Digits

1. All nonzero digits are significant.

   For example, the measurement 1765 kg has 4 significant digits. (This measurement indicates measuring 1765 units of kilograms.)

2. All zeros between significant digits are significant.

   For example, the measurement 30,060 m has 4 significant digits. (This measurement indicates measuring 3006 tens of metres.)

3. A zero in a whole-number measurement that is specially tagged, such as by a bar above it, is significant.

   For example, the measurement 3$\overline{0}$,000 ft has 2 significant digits. (This measurement indicates measuring 3$\overline{0}$ thousands of feet.)

4. All zeros to the right of a significant digit *and* a decimal point are significant.

   For example, the measurement 6.100 L has 4 significant digits. (This measurement indicates measuring 610$\overline{0}$ thousandths of litres.)

5. Zeros to the right in a whole-number measurement that are not tagged are *not* significant.

   For example, the measurement 4600 V has 2 significant digits. (This measurement indicates measuring 46 hundreds of volts.)

6. Zeros to the left in a decimal measurement that is less than 1 are *not* significant.

   For example, the measurement 0.00960 s has 3 significant digits. (This measurement indicates measuring 96$\overline{0}$ hundred-thousandths of a second.)

Example 6 Determine the accuracy of each measurement.

Measurement	Accuracy (significant digits)
a. 109.006 m	6
b. 0.000589 kg	3
c. 75 V	2
d. 239,000 mi	3
e. 239,00$\overline{0}$ mi	6
f. 239,0$\overline{0}$0 mi	5
g. 0.03200 mg	4
h. 1.20 cm	3
i. 9.020 μA	4
j. 100.050 km	6

EXERCISES 4.1

Determine the accuracy of each measurement; that is, give the number of significant digits for each measurement:

1. 115 V
2. 47,000 lb
3. 7009 ft
4. 420 m
5. 6972 m
6. 320,070 ft
7. 440̄0 ft
8. 4400 Ω
9. 44̄00 m
10. 0.0040 g
11. 173.4 m
12. 2070 ft
13. 41,0̄00 mi
14. 0.025 A
15. 0.0350 in.
16. 6700 g
17. 173 m
18. 8060 ft
19. 240,0̄00 V
20. 2500 g
21. 72,0̄00 mi
22. 137 V
23. 0.047000 A
24. 7.009 g
25. 0.20 mi
26. 69.72 m
27. 32.0070 g
28. 61̄0 L
29. 15,0̄00 mi
30. 0.07050 mL
31. 100.020 in.
32. 250.0100 m
33. 900,200 ft
34. 15̄0 cm
35. 16,000 W
36. 0.001005 m

4.2 Precision and Greatest Possible Error

> **Precision**
>
> The **precision of a measurement** is the smallest unit with which the measurement is made, that is, the position of the last significant digit or the smallest unit or calibration on the measuring instrument.

Example 1 The precision of the measurement 239,000 mi is 1000 mi. (The position of the last significant digit is in the thousands place.) ◆

Example 2 The precision of the measurement 10,900 m is 100 m. (The position of the last significant digit is in the hundreds place.) ◆

Example 3 The precision of the measurement 6.90 L is 0.01 L. (The position of the last significant digit is in the hundredths place.) ◆

Example 4 The precision of the measurement 0.0016 A is 0.0001 A. (The position of the last significant digit is in the ten-thousandths place.) ◆

Example 5 Determine the precision of each measurement (see Example 6 in Section 4.1).

Measurement	Accuracy (significant digits)	Precision
a. 109.006 m	6	0.001 m
b. 0.000589 kg	3	0.000001 kg
c. 75 V	2	1 V
d. 239,000 mi	3	1000 mi
e. 239,00̄0 mi	6	1 mi
f. 239,0̄00 mi	5	10 mi
g. 0.03200 mg	4	0.00001 mg
h. 1.20 cm	3	0.01 cm
i. 9.020 μA	4	0.001 μA
j. 100.050 km	6	0.001 km

◆

158 CHAPTER 4 ♦ Measurement

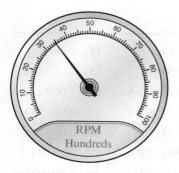

(a) Precision: 100 rpm

(b) Precision: 1000 rpm

Figure 4.4

The precision of a measuring instrument is determined by its smallest calibration.

Why is the precision of a measurement important? Precision is another distinctive element of the certainty of a measurement. It not only indicates the relative certainty of a measuring instrument but also is important in determining the results of adding and subtracting measurements, as we shall see in Section 4.6.

The precision of a measuring instrument is determined by the smallest unit or calibration on the instrument. The precision of the tachometer in Figure 4.4(a) is 100 rpm. The precision of the tachometer in Figure 4.4(b) is 1000 rpm.

The precision of a ruler graduated in eighths of an inch is $\frac{1}{8}$ in. The precision of a ruler graduated in fourths of an inch is $\frac{1}{4}$ in. However, if a measurement is given as $4\frac{5}{8}$ in., you have no way of knowing what ruler was used. Therefore, you cannot tell whether the precision is $\frac{1}{8}$ in., $\frac{1}{16}$ in., $\frac{1}{32}$ in., or what. The measurement could have been $4\frac{5}{8}$ in., $4\frac{10}{16}$ in., $4\frac{20}{32}$ in., or a similar measurement of some other precision. *Unless stated otherwise, you should assume that the smallest unit used is the one that is recorded.* In this case, the precision is assumed to be $\frac{1}{8}$ in.

Now study closely the enlarged portions of the tachometer (tach) readings in Figure 4.5, given in hundreds of rpm. Note that in each case the measurement is 4100 rpm, although the locations of the pointer are slightly different. Any actual speed between 4050 and 4150 is read 4100 rpm on the scale of this tachometer.

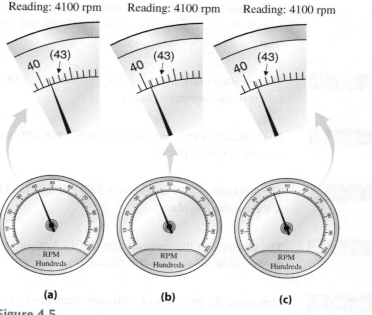

Figure 4.5

Any actual speed between 4050 rpm and 4150 rpm is read 4100 rpm on the scale of this tachometer.

The *greatest possible error* is one-half of the smallest unit on the scale on which the measurement is read. We see this in the tach readings in Figure 4.5, where any reading within 50 rpm of 4100 rpm is read as 4100 rpm. Therefore, the greatest possible error is 50 rpm.

If you have a tach reading of 5300 rpm, the greatest possible error is $\frac{1}{2}$ of 100 rpm, or 50 rpm. This means that the actual rpm is between $5300 - 50$ and $5300 + 50$—that is, between 5250 rpm and 5350 rpm.

Next consider the measurements of the three metal rods shown in Figure 4.6(a). Note that in each case, the measurement of the length is $4\frac{5}{8}$ in., although it is obvious that the rods are of different lengths. Any rod with actual length between $4\frac{9}{16}$ in. and $4\frac{11}{16}$ in. will measure $4\frac{5}{8}$ in. on this enlarged scale as shown in Figure 4.6(b). The greatest possible error is one-half of the smallest unit on the scale with which the measurement is made.

If the length of a metal rod is given as $3\frac{3}{16}$ in., the greatest possible error is $\frac{1}{2}$ of $\frac{1}{16}$ in., or $\frac{1}{32}$ in. This means that the actual length of the rod is between $3\frac{3}{16}$ in. $-\frac{1}{32}$ in. and $3\frac{3}{16}$ in. $+\frac{1}{32}$ in.; that is, between $3\frac{5}{32}$ in. and $3\frac{7}{32}$ in.

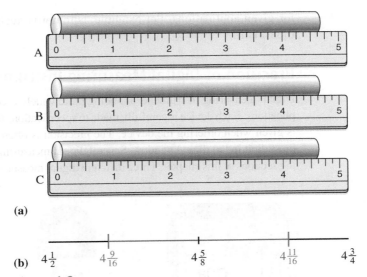

Figure 4.6
Any rod with actual length between $4\frac{9}{16}$ in. and $4\frac{11}{16}$ in. is read $4\frac{5}{8}$ in. on this scale.

> ### Greatest Possible Error
> The **greatest possible error of a measurement** is equal to one-half its precision.

Example 6 Find the precision and greatest possible error of the measurement 8.00 kg.

The position of the last significant digit is in the hundredths place; therefore, the precision is 0.01 kg.

The greatest possible error is one-half the precision.

$(0.5)(0.01 \text{ kg}) = 0.005 \text{ kg}$ ◆

Example 7 Find the precision and greatest possible error of the measurement 26,000 gal.

The position of the last significant digit is in the thousands place; therefore, the precision is 1000 gal.

The greatest possible error is one-half the precision.

$\frac{1}{2} \times 1000 \text{ gal} = 500 \text{ gal}$ ◆

Example 8 Find the precision and greatest possible error of the measurement 0.0460 mg.

Precision: 0.0001 mg
Greatest possible error: $(0.5)(0.0001 \text{ mg}) = 0.00005 \text{ mg}$ ◆

Example 9 Find the precision and the greatest possible error of the measurement $7\frac{5}{8}$ in.

The smallest unit is $\frac{1}{8}$ in., which is the precision.
The greatest possible error is $\frac{1}{2} \times \frac{1}{8}$ in. $= \frac{1}{16}$ in., which means that the actual length is within $\frac{1}{16}$ in. of $7\frac{5}{8}$ in. ◆

Many calculations with measurements are performed by people who do not make the actual measurements. Therefore, it is often necessary to agree on a method of recording measurements to indicate the precision of the instrument used. Some industries have agreed to set their own specific rounding, precision, and accuracy standards for given applications. For example, automotive technicians round down to the nearest thousandth.

Precision of Digital Measuring Instruments

Digital measuring instruments are being more widely used in many technical and industrial applications. Each device is carefully made to specifications that indicate its precision. The precision is given when ordering the device. The precision is often indicated by the position of the rightmost digit in the digital window. Several digital measuring instruments are shown in the following sections of this chapter. Three different digital measuring instruments are shown in Figure 4.7.

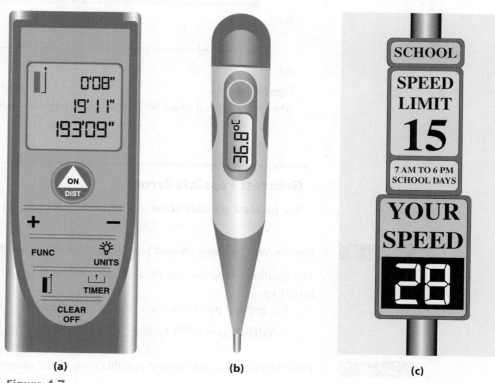

Figure 4.7
Digital measuring instruments. (a) Laser distance meter measures distance with given precision. This model has a precision of $\frac{1}{16}$ in. over medium distances. Length, width, and volume are common measurement modes. (b) This digital thermometer precisely measures body temperatures to nearest 0.1°C. (c) This digital speed zone traffic warning sign measures speeds precisely to nearest 1 mph, assuming it is calibrated correctly.

EXERCISES 4.2

Find **a.** *the precision and* **b.** *the greatest possible error of each measurement:*

1. 2.70 A
2. 13.0 ft
3. 14.00 cm
4. 1.000 in.
5. 15 km
6. 1.010 cm
7. 17.50 mi
8. 6.100 m
9. 0.040 A
10. 0.0001 in.
11. 0.0805 W
12. 10,$\overline{0}$00 W
13. 14$\overline{0}$0 Ω
14. 301,000 Hz
15. 3$\overline{0}$,000 L
16. 7,0$\overline{0}$0,000 g
17. 428.0 cm
18. 60.0 cm
19. 120 V
20. 30$\overline{0}$ km
21. 67.500 m
22. $1\frac{7}{8}$ in.
23. $3\frac{2}{3}$ yd
24. $3\frac{3}{4}$ yd
25. $9\frac{7}{32}$ in.
26. $4\frac{5}{8}$ mi
27. $9\frac{5}{16}$ mi
28. $5\frac{13}{64}$ in.
29. $9\frac{4}{9}$ in²
30. $18\frac{4}{5}$ in³

4.3 The Vernier Caliper

In your use of U.S. and metric rulers for making measurements, you have seen that very precise results are difficult to obtain. When more precise measurements are required, you must use a more precise instrument. One such instrument is the **vernier caliper**, which is used by technicians in machine shops, plant assembly lines, and many other workplaces. This instrument is a slide-type measuring instrument used to take precise inside, outside, and depth measurements. It has two metric scales and two U.S. scales. A vernier caliper is shown in Figure 4.8.

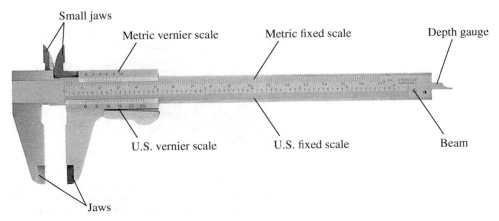

Figure 4.8
Vernier caliper. The metric scales on this vernier caliper are located above the U.S. scales.

To make an outside measurement, the jaws are closed snugly around the outside of an object, as in Figure 4.9(a). For an inside measurement, the smaller jaws are placed inside the object to be measured, as in Figure 4.9(b). For a depth measurement, the depth gauge is inserted into the opening to be measured, as in Figure 4.9(c).

  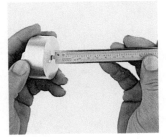

(a) Outside measurement   **(b)** Inside measurement   **(c)** Depth measurement

Figure 4.9
Measurements with a vernier caliper

Here are some tips for using a vernier caliper (and the micrometer in Section 4.4):

1. Check that the instrument is held perpendicular (that is, at 90°) to the surface of the part being measured.

2. When measuring the diameter of a round piece, check that the *full* diameter is being measured.
3. On rounds, take two readings at approximately 90° to each other. Then average the two readings.

Let's first consider the metric scales (Figure 4.10). One of them is fixed and located on the upper part of the beam. This fixed scale is divided into centimetres and subdivided into millimetres, so record all readings in millimetres (mm). The other metric scale, called the vernier scale, is the upper scale on the slide. The vernier scale is divided into tenths of millimetres (0.10 mm) and subdivided into halves of tenths of millimetres ($\frac{1}{20}$ mm or 0.05 mm). The precision of this vernier scale is therefore 0.05 mm.

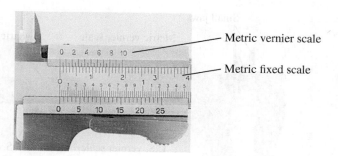

Figure 4.10
The fixed metric scale on the beam of this vernier caliper is divided into centimetres and further subdivided into millimetres (mm). Its movable metric vernier scale is divided into tenths of millimetres and subdivided into increments of 0.05 mm, which is the precision of this vernier caliper.

The figures in the examples and exercises have been computer generated for easier reading.

Reading a Vernier Caliper in Millimetres

STEP 1 Determine the number of whole millimetres in a measurement by counting—on the fixed scale—the number of millimetre graduations that are to the left of the zero graduation on the vernier scale. (Remember that each numbered graduation on the fixed scale represents 10 mm.) The zero graduation on the vernier scale may be directly in line with a graduation on the fixed scale. If so, read the total measurement directly from the fixed scale. Write it in millimetres, followed by a decimal point and two zeros.

STEP 2 The zero graduation on the vernier scale may not be directly in line with a graduation on the fixed scale. In that case, find the graduation on the vernier scale that is most nearly in line with any graduation on the fixed scale.

   a. If the vernier graduation is a long graduation, it represents the number of tenths of millimetres between the last graduation on the fixed scale and the zero graduation on the vernier scale. Then insert a zero in the hundredths place.

   b. If the vernier graduation is a short graduation, add 0.05 mm to the vernier graduation that is on the immediate left of the short graduation.

STEP 3 Add the numbers from Steps 1 and 2 to determine the total measurement.

4.3 ♦ The Vernier Caliper   163

Example 1 Read the measurement in millimetres on the vernier caliper in Figure 4.11.

STEP 1 The first mark to the left of the zero mark is   45.00 mm
STEP 2 The mark on the vernier scale that most nearly lines up with a mark on the fixed scale is   0.20 mm
STEP 3 The total measurement is   45.20 mm

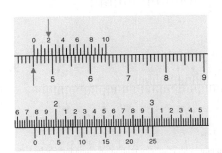

Figure 4.11   Figure 4.12

Example 2 Read the measurement in millimetres on the vernier caliper in Figure 4.12.

The total measurement is 21.00 mm, because the zero graduation on the vernier scale most nearly lines up with a mark on the fixed scale.

Example 3 Read the measurement in millimetres on the vernier caliper in Figure 4.13.

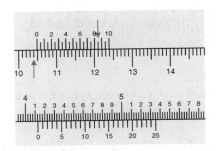

Figure 4.13

STEP 1 The first mark to the left of the zero mark is   104.00 mm
STEP 2 The mark on the vernier scale that most nearly lines up with a mark on the fixed scale is   0.85 mm
STEP 3 The total measurement is   104.85 mm

EXERCISES 4.3A

Read the measurement in millimetres shown on each vernier caliper:

1.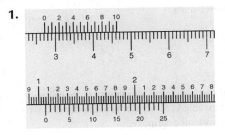

2.

3.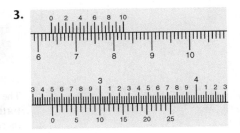

164 CHAPTER 4 ♦ Measurement

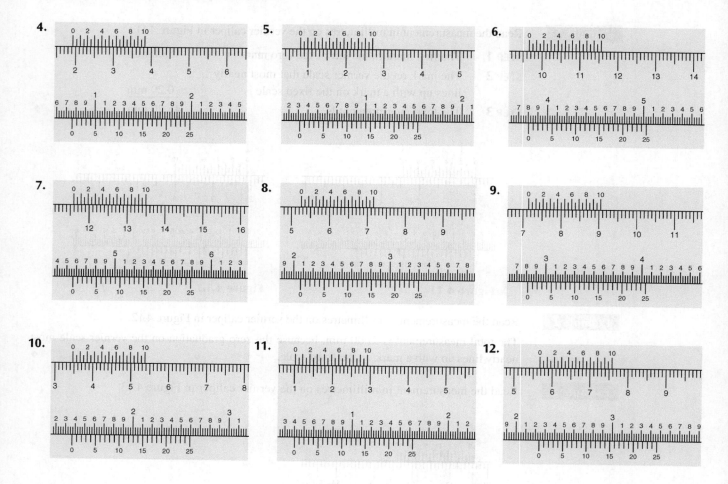

13–24. *Read the measurement in millimetres shown on each vernier caliper in Exercises 4.3B (pages 166–167).*

Now consider the two U.S. scales on the vernier caliper in Figure 4.14. One is fixed, the other movable. The fixed scale is located on the lower part of the beam, where each inch is divided into tenths (0.100 in., 0.200 in., 0.300 in., and so on). Each tenth is subdivided into four parts, each of which represents 0.025 in. The vernier scale is divided into 25 parts, each of which represents thousandths (0.001 in.). This means that the precision of this scale is 0.001 in.

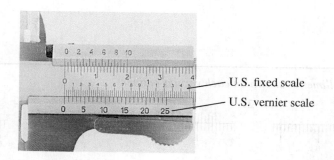

Figure 4.14
The fixed U.S. scale on the beam of this vernier caliper is divided into tenths of inches and further subdivided into increments of 0.025 in. Its movable U.S. vernier scale is divided into increments of 0.001 in., which is the precision of this vernier caliper.

Reading a Vernier Caliper in Thousandths of an Inch

STEP 1 Determine the number of inches and tenths of inches by reading the first numbered division that is to the left of the zero graduation on the vernier scale.

STEP 2 Add 0.025 in. to the number from Step 1 for each graduation between the last numbered division on the fixed scale and the zero graduation on the vernier scale. (If this zero graduation is directly in line with a graduation on the fixed scale, read the total measurement directly from the fixed scale.)

STEP 3 If the zero graduation on the vernier scale is not directly in line with a graduation on the fixed scale, find the graduation on the vernier scale that is most nearly in line with any graduation on the fixed scale. This graduation determines the number of thousandths of inches in the measurement.

STEP 4 Add the numbers from Steps 1, 2, and 3 to determine the total measurement.

Example 4 Read the measurement in inches shown on the vernier caliper in Figure 4.15.

STEP 1	The first numbered mark to the left of the zero mark is	1.600 in.
STEP 2	The number of 0.025-in. graduations is 3; 3 × 0.025 in. =	0.075 in.
STEP 3	The mark on the vernier scale that most nearly lines up with a mark on the fixed scale is	0.014 in.
STEP 4	The total measurement is	1.689 in.

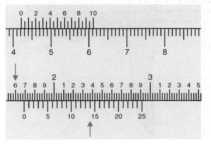
Figure 4.15

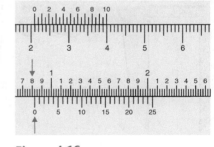
Figure 4.16

Example 5 Read the measurement in inches shown on the vernier caliper in Figure 4.16.

STEP 1	The first numbered mark to the left of the zero mark is	0.800 in.
STEP 2	The number of 0.025-in. graduations is 1; 1 × 0.025 in. =	0.025 in.
STEP 3	The total measurement is	0.825 in.

NOTE: The zero graduation is directly in line with a graduation on the fixed scale.

Example 6 Read the measurement in inches shown on the vernier caliper in Figure 4.17.

STEP 1	The first numbered mark to the left of the zero mark is	1.200 in.
STEP 2	The number of 0.025-in. graduations is 2; 2 × 0.025 in. =	0.050 in.
STEP 3	The mark on the vernier scale that most nearly lines up with a mark on the fixed scale is	0.008 in.
STEP 4	The total measurement is	1.258 in.

166 CHAPTER 4 ♦ Measurement

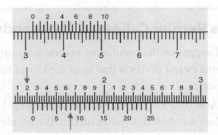

Figure 4.17

Digital Vernier Calipers

Digital vernier calipers that provide metric and U.S. readings in millimetres and decimal inches by means of a one-button process are also in common use—see Figure 4.18(a) and (b). This display reads to the nearest 0.01 mm or 0.0005 in., which indicates its precision. Before each measurement, carefully close the caliper and check that it reads 0.000 or press the zero button. Then make your outside, inside, or depth measurement as before. Just as we must learn to read clocks with hands as well as digital clocks because both types are in common use, we must learn to use both regular and digital vernier calipers. Some digital vernier calipers have a three-mode display with millimetre, decimal inch, and fractional inch readings. The fractional display in Figure 4.18(c) reads to the nearest 1/128 in., which indicates its precision.

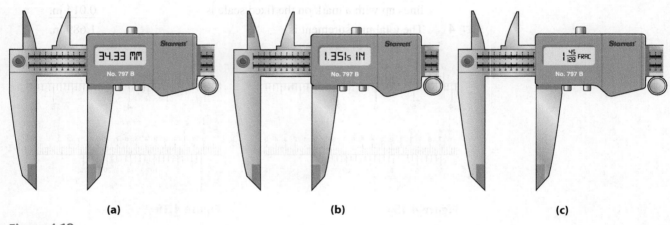

Figure 4.18
Digital vernier caliper with displays in (a) millimetre mode; (b) decimal inch mode; and (c) fractional inch mode

EXERCISES 4.3B

Read the measurement in inches shown on each vernier caliper:

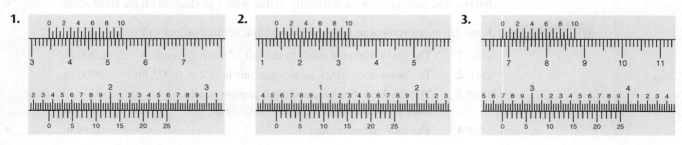

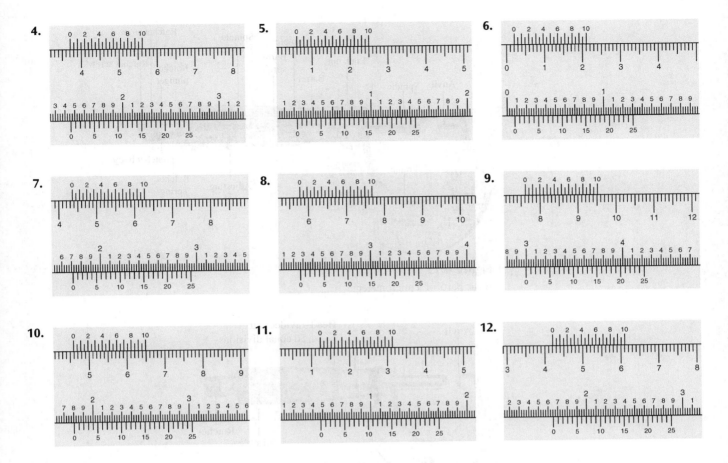

13–24. *Read the measurement in inches shown on each vernier caliper in Exercises 4.3A (pages 163–164).*

4.4 The Micrometer Caliper

The **micrometer caliper** (micrometer or "mike") is an instrument for measuring very small lengths using the movement of a finely threaded rotating screw, which gives it better precision than a vernier caliper. It is used in technical fields in which fine precision is required. Micrometers are available in metric units and U.S. units. The metric "mike" is graduated and read in hundredths of a millimetre (0.01 mm); the U.S. "mike" is graduated and read in thousandths of an inch (0.001 in.). The parts of a micrometer are labeled in Figure 4.19.

To use a micrometer properly, place the object to be measured between the anvil and spindle and turn the thimble until the object fits snugly. *Do not force the turning of the thimble, because this may damage the very delicate threads on the spindle that are located inside the thimble.* Some calipers have a ratchet to protect the instrument; the ratchet prevents the thimble from being turned with too much force. A metric micrometer is shown in Figure 4.20, and the basic parts are labeled. The barrels of most metric micrometers are graduated in millimetres. The micrometer in Figure 4.20 also has graduations of halves of millimetres, which are indicated by the lower set of graduations on the barrel. The threads on the spindle are made so that it takes two complete turns of the thimble for the spindle to move precisely one millimetre. The head is divided into 50 equal divisions—each division indicating 0.01 mm, which is the precision.

168 CHAPTER 4 ♦ Measurement

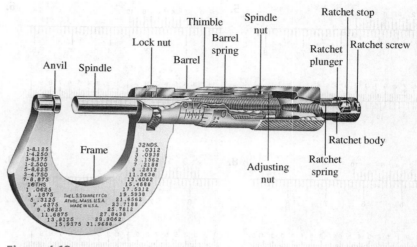

Figure 4.19
Basic parts of a micrometer

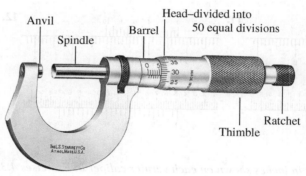

Figure 4.20
Metric micrometer

Reading a Metric Micrometer in Millimetres

STEP 1 Find the *whole number* of mm in the measurement by counting the number of mm graduations on the barrel to the left of the head.

STEP 2 Find the *decimal part* of the measurement by reading the graduation on the head (see Figure 4.20) that is most nearly in line with the center line on the barrel. Then multiply this reading by 0.01. If the head is at, or immediately to the right of, the half-mm graduation, then add 0.50 mm to the reading on the head.

STEP 3 Add the numbers found in Step 1 and Step 2.

Example 1   Read the measurement shown on the metric micrometer in Figure 4.21.

STEP 1	The barrel reading is	6.00 mm
STEP 2	The head reading is	0.24 mm
STEP 3	The total measurement is	6.24 mm

♦

Figure 4.21

Example 2
Read the measurement shown on the metric micrometer in Figure 4.22.

STEP 1	The barrel reading is	14.00 mm
STEP 2	The head reading is	0.12 mm
STEP 3	The total measurement is	14.12 mm

Figure 4.22

Example 3
Read the measurement shown on the metric micrometer in Figure 4.23.

STEP 1	The barrel reading is (Note that the head is past the half-mm mark.)	8.50 mm
STEP 2	The head reading is	0.15 mm
STEP 3	The total measurement is	8.65 mm

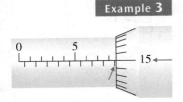

Figure 4.23

EXERCISES 4.4A

Read the measurement shown on each metric micrometer:

1.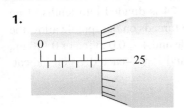

2.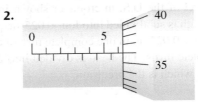

3.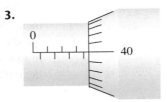

4.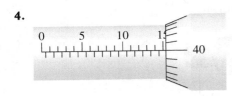

5.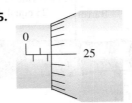

6.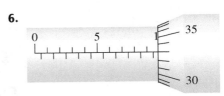

7.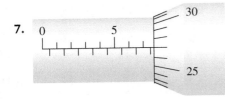

8.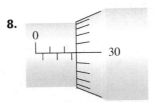

9.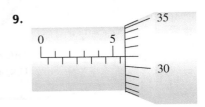

10.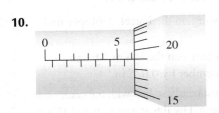

11.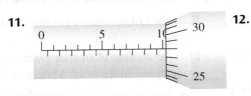

12.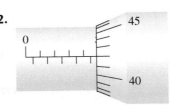

170   CHAPTER 4 ♦ Measurement

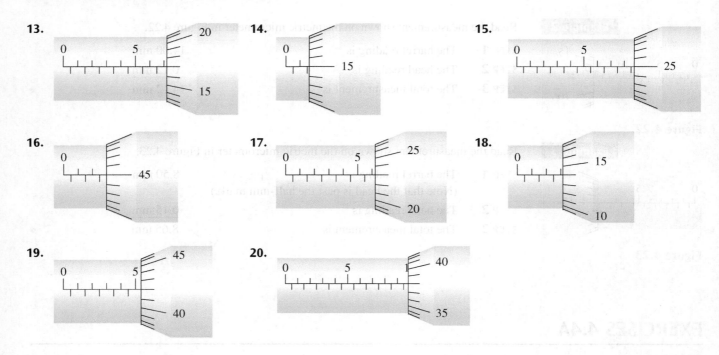

The barrel of the U.S. micrometer shown in Figure 4.24 is divided into tenths of an inch. Each tenth is subdivided into four 0.025-in. parts. The threads on the spindle allow the spindle to move 0.025 in. in one complete turn of the thimble and 4 × 0.025 in., or 0.100 in., in four complete turns. The head is divided into 25 equal divisions—each division indicating 0.001 in., which is the precision.

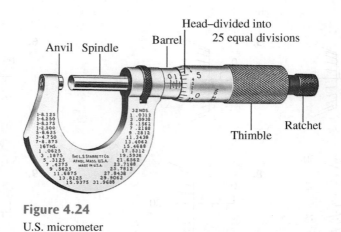

Figure 4.24
U.S. micrometer

Reading a U.S. Micrometer in Thousandths of an Inch

STEP 1 Read the last numbered graduation showing on the barrel. Multiply this number by 0.100 in.

STEP 2 Find the number of smaller graduations between the last numbered graduation and the head. Multiply this number by 0.025 in.

STEP 3 Find the graduation on the head that is most nearly in line with the center line on the barrel. Multiply the number represented by this graduation by 0.001 in.

STEP 4 Add the numbers found in Steps 1, 2, and 3.

4.4 ♦ The Micrometer Caliper

Example 4 Read the measurement shown on the U.S. micrometer in Figure 4.25.

STEP 1	3 numbered divisions on the barrel; 3×0.100 in. =	0.300 in.
STEP 2	1 small division on the barrel; 1×0.025 in. =	0.025 in.
STEP 3	The head reading is 17; 17×0.001 in. =	0.017 in.
STEP 4	The total measurement is	0.342 in.

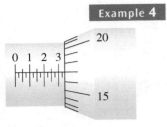

Figure 4.25

Example 5 Read the measurement shown on the U.S. micrometer in Figure 4.26.

STEP 1	4 numbered divisions on the barrel; 4×0.100 in. =	0.400 in.
STEP 2	2 small divisions on the barrel; 2×0.025 in. =	0.050 in.
STEP 3	The head reading is 21; 21×0.001 in. =	0.021 in.
STEP 4	The total measurement is	0.471 in.

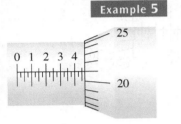

Figure 4.26

EXERCISES 4.4B

Read the measurement shown on each U.S. micrometer:

1.
2.
3.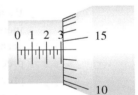
4.
5.
6.
7.
8.
9.
10.
11.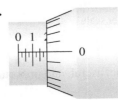
12.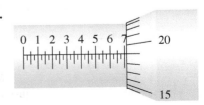

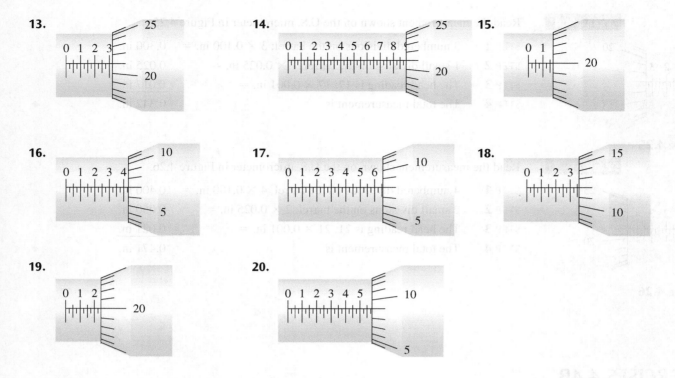

Other Micrometers

By adding a vernier scale on the barrel of a micrometer (as shown in Figure 4.27), we can increase the precision by one more decimal place. That is, the metric micrometer with vernier scale has a precision of 0.001 mm. The U.S. micrometer with vernier scale has a precision of 0.0001 in. Of course, these micrometers cost more because they require more precise threading than the ones previously discussed. Nevertheless, many jobs require this precision.

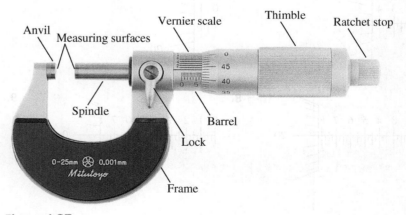

Figure 4.27
Micrometer with vernier scale

Micrometers are basic, useful, and important tools of the technician. Figure 4.28 shows just a few of their uses.

4.4 ♦ The Micrometer Caliper

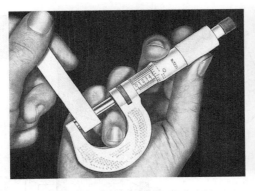

(a) Measuring a piece of die steel

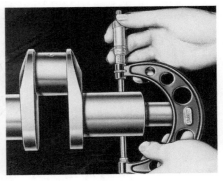

(b) Measuring the diameter of a crankshaft bearing

(c) Measuring tubing wall thickness with a round anvil micrometer

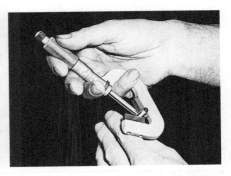

(d) Checking out-of-roundness on centerless grinding work

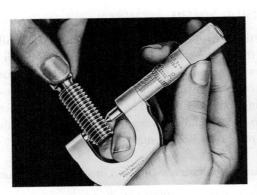

(e) Measuring the pitch diameter of a screw thread

(f) Measuring the depth of a shoulder with a micrometer depth gauge

Figure 4.28
Examples of how micrometers are used

Digital Micrometers

Digital micrometers that provide metric and U.S. (mm/inch) readings by means of a one-button process are in common use (see Figure 4.29). The display in this figure reads to the nearest 0.001 mm or 0.0001 in., which indicates its precision. Before each measurement, carefully close the anvil and spindle and check that it reads 0.000 or press the zero button. Then make your measurement as before.

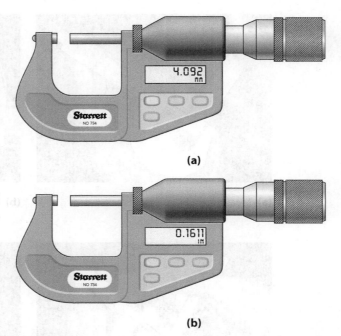

Figure 4.29
Digital micrometer with displays in (a) millimetre mode and (b) decimal inch mode

4.5 Addition and Subtraction of Measurements

Precision versus Accuracy

Recall that the *precision* of a measurement is the smallest unit with which a measurement is made; that is, the *position* of the last significant digit or the smallest unit or calibration on the measuring instrument. Recall also that the *accuracy* of a measurement is the *number* of digits, called significant digits, which indicate the number of units we are reasonably sure of having counted when making a measurement. Unfortunately, some people tend to use the terms *precision* and *accuracy* interchangeably even though each term expresses a different aspect of a given measurement.

Example 1 Compare the precision and the accuracy of the measurement 0.0007 mm.

Since the precision is 0.0001 mm, its precision is relatively good. However, since the accuracy is only one significant digit, its accuracy is relatively poor. ◆

Example 2 Given the measurements 13.00 m, 0.140 m, 3400 m, and 0.006 m, find the measurement that is **a.** the least precise, **b.** the most precise, **c.** the least accurate, and **d.** the most accurate.

First, let's find the precision and the accuracy of each measurement.

Measurement	Precision	Accuracy (significant digits)
13.00 m	0.01 m	4
0.140 m	0.001 m	3
3400 m	100 m	2
0.006 m	0.001 m	1

4.5 ♦ Addition and Subtraction of Measurements

From the table, we find the following:

a. The least precise measurement is 3400 m.
b. The most precise measurements are 0.140 m and 0.006 m.
c. The least accurate measurement is 0.006 m.
d. The most accurate measurement is 13.00 m. ♦

Consider finding the distance between points A and C in Figure 4.30 as follows:

One person measures the distance between A and B on level ground using a steel tape measure as 54 ft $7\frac{11}{16}$ in. A second person measures the distance between B and C over uneven ground by "stepping it off," assuming each step measures 3 ft, as 25 steps × 3 ft = 75 ft. What is the distance between points A and C?

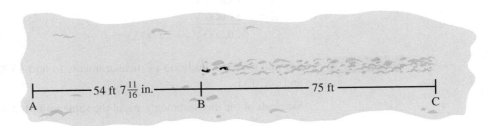

Figure 4.30
What is the sum of these two measurements of different precision?

How confident or certain are you that the total distance between A and C is

54 ft $7\frac{11}{16}$ in. + 75 ft = 129 ft $7\frac{11}{16}$ in.?

How confident or certain are you that the precision of the total measurement is to the nearest $\frac{1}{16}$ in., the nearest $\frac{1}{8}$ in., the nearest $\frac{1}{4}$ in., the nearest $\frac{1}{2}$ in., the nearest 1 in., or even the nearest 1 ft? Discuss with a classmate what an appropriate precision of the sum should be.

Before someone suggests to simply measure the distance from B to C using the steel tape, recall that this terrain is over uneven ground and has obstacles that make using the steel tape impractical. And assume that they did not have access to the laser distance meter in Figure 4.7(a). The point of this discussion is to learn how to add measurements with different precisions and obtain a sum that has a reasonable precision. Just simply treating and adding these measurements as exact numbers to obtain a sum with a good precision is not feasible.

In a series circuit, the electromagnetic force (emf) of the source equals the sum of the separate voltage drops across each resistor in the circuit. Suppose that someone measures the voltage across the first resistor R_1 in Figure 4.31. He uses a voltmeter calibrated in hundreds of volts and measures 15,800 V. Across the second resistor R_2, he uses a voltmeter calibrated in thousands of volts and measures 11,000 V. Does the total emf equal 26,800 V? Note that the first voltmeter and its reading indicate a precision of 100 V and a greatest possible error of 50 V. This means that the actual reading lies between 15,750 V and 15,850 V. The second voltmeter and its reading indicate a precision of 1000 V and a greatest possible error of 500 V. The actual reading, therefore, lies between 10,500 V and 11,500 V. This means that we are not very certain of the digit in the hundreds place in the sum 26,800 V.

To be consistent when adding or subtracting measurements of different precision, the sum or difference can be no more *precise* than the least precise measurement. Or, the result can be no more precise than its least precise measuring tool.

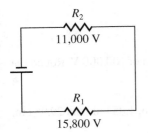

Figure 4.31

Adding or Subtracting Measurements of Different Precision

1. Make certain that all measurements are expressed in the same unit. If they are not, change them all to any common unit.
2. Add or subtract.
3. Then round the result to the same precision as the *least precise* measurement.

The total emf in the circuit shown in Figure 4.31 is therefore calculated as follows:

Measurement		Rounded
R_1:	15,800 V	
R_2:	11,000 V	
	26,800 V	→ 27,000 V

Example 3 Use the rules for addition of measurements to add 13,800 ft, 14,020 ft, 19,864 ft, 2490 ft, and 14,700 ft.

Since all of the measurements are in the same unit (that is, ft), add them together:

```
13,800 ft
14,020 ft
19,864 ft
 2,490 ft
14,700 ft
64,874 ft → 64,900 ft
```

Round this sum to the same precision as the least precise measurement. Since the precision of both 13,800 ft and 14,700 ft is 100 ft, round the sum to the nearest 100 ft. Thus, the sum is 64,900 ft. ♦

Example 4 Use the rules for addition of measurements to add 735,000 V, 490,000 V, 86,000 V, 1,300,000 V, and 20$\overline{0}$,000 V.

Since all of the measurements are in the same unit, add:

```
  735,000 V
  490,000 V
   86,000 V
1,300,000 V
  20̄0,000 V
2,811,000 V → 2,800,000 V
```

The least precise measurement is 1,300,000 V, which has a precision of 100,000 V. Round the sum to the nearest hundred thousand volts: 2,800,000 V. ♦

Example 5 Use the rules for addition of measurements to add 13.8 m, 140.2 cm, 1.853 m, and 29.95 cm.

First, change each measurement to a common unit (say, m) and add:

13.8	m	→	13.8	m
140.2	cm	→	1.402	m
1.853	m	→	1.853	m
29.95	cm	→	0.2995	m
			17.3545 m	→ 17.4 m

4.5 ♦ Addition and Subtraction of Measurements

The least precise measurement is 13.8 m, which is precise to the nearest tenth of a metre. So round the sum to the nearest tenth of a metre: 17.4 m. ♦

Example 6 Use the rules for subtraction of measurements to subtract 19.352 cm from 41.7 cm.
Since both measurements have the same unit, subtract:

$$\begin{array}{r} 41.7 \text{ cm} \\ \underline{19.352 \text{ cm}} \\ 22.348 \text{ cm} \rightarrow 22.3 \text{ cm} \end{array}$$

The least precise measurement is 41.7 cm, which is precise to the nearest tenth of a cm. Round the difference to the nearest tenth of a cm: 22.3 cm. ♦

EXERCISES 4.5

In each set of measurements, find the measurement that is **a.** *the most accurate and* **b.** *the most precise:*

1. 14.7 in.; 0.017 in.; 0.09 in.
2. 459 ft; 600 ft; 190 ft
3. 0.737 mm; 0.94 mm; 16.01 mm
4. 4.5 cm; 9.3 cm; 7.1 cm
5. 0.0350 A; 0.025 A; 0.00050 A; 0.041 A
6. 134.00 g; 5.07 g; 9.000 g; 0.04 g
7. 145 cm; 73.2 cm; 2560 cm; 0.391 cm
8. 15.2 km; 631.3 km; 20.0 km; 37.7 km
9. 205,000 Ω; 45,000 Ω; 5$\overline{0}$0,000 Ω; 90,000 Ω
10. 1,500,000 V; 65,000 V; 30,$\overline{0}$00 V; 20,000 V

In each set of measurements, find the measurement that is **a.** *the least accurate and* **b.** *the least precise:*

11. 15.5 in.; 0.053 in.; 0.04 in.
12. 635 ft; 400 ft; 240 ft
13. 43.4 cm; 0.48 cm; 14.05 cm
14. 4.9 kg; 670 kg; 0.043 kg
15. 0.0730 A; 0.043 A; 0.00008 A; 0.91 A
16. 197.0 m; 5.43 m; 4.000 m; 0.07 m
17. 2.1 m; 31.3 m; 461.5 m; 0.6 m
18. 295 m; 91.3 m; 1920 m; 0.360 m
19. 405,000 Ω; 35,000 Ω; 8$\overline{0}$0,000 Ω; 500,000 Ω
20. 1,600,000 V; 36,000 V; 40,$\overline{0}$00 V; 60,000 V

Use the rules for addition of measurements to find the sum of each set of measurements:

21. 14.7 m; 3.4 m
22. 168 in.; 34.7 in.; 61 in.
23. 42.6 cm; 16.41 cm; 1.417 cm; 34.4 cm
24. 407 g; 1648.5 g; 32.74 g; 98.1 g
25. 26,000 W; 19,600 W; 8450 W; 42,500 W
26. 5420 km; 1926 km; 850 km; 2$\overline{0}$00 km
27. 140,000 V; 76,200 V; 4700 V; 254,000 V; 370,000 V
28. 19,200 m; 8930 m; 50,040 m; 137 m
29. 14 V; 1.005 V; 0.017 V; 3.6 V
30. 120.5 cm; 16.4 cm; 1.417 m
31. 10.555 cm; 9.55 mm; 13.75 cm; 206 mm
32. 1350 cm; 1476 mm; 2.876 m; 4.82 m

Use the rules for subtraction of measurements to subtract the second measurement from the first:

33. 140.2 cm
    13.8 cm
34. 14.02 mm
    13.8 mm
35. 9200 mi
    627 mi
36. 1,900,000 V
    645,000 V
37. 167 mm
    13.2 cm
38. 16.41 oz
    11.372 oz
39. 98.1 g
    32.743 g
40. 4.000 in.
    2.006 in.
41. 0.54361 in.
    0.214 in.

42. ⌧ If you bolt four pieces of metal with thicknesses 0.136 in., 0.408 in., 1.023 in., and 0.88 in. together, what is the total thickness?

43. ⌧ If you clamp five pieces of metal with thicknesses 2.38 mm, 10.9 mm, 3.50 mm, 1.455 mm, and 8.2 mm together, what is the total thickness?

44. ⌧ What is the current going through R_5 in the circuit in Illustration 1? (*Hint:* In a parallel circuit, the current is divided among its branches. That is, $I_T = I_1 + I_2 + I_3 + \cdots$.) In this circuit containing two parallel circuits connected in series, the current through the top branch $I_1 + I_2 + I_3$ must equal the current through the bottom branch $I_4 + I_5$.

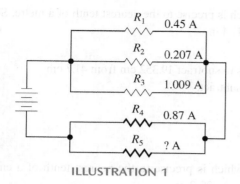

ILLUSTRATION 1

45. A welder cuts several pieces of steel angle of lengths 3.32 ft, 2.15 ft, 3.2 ft, and 4.0 ft. What is the total length of the pieces?

46. A welder weighed some bins of scrap metal. The bins weighed 266 lb, 620 lb, and 1200 lb, respectively. What was the total weight of scrap metal in all three bins?

47. A pilot loads baggage in the baggage compartment of a small plane. The baggage weighs 23.25 lb, 18.6 lb, and 25 lb. What is the total weight of the baggage?

48. To compensate for too much cargo on a plane with 38.35 gal of fuel, fuel is drained. First 8.2 gal are drained, then another 6.33 gal. After this, how much fuel is left?

49. An automobile cooling system is leaking. One day it lost 0.52 gal, a second day it lost 0.4 gal, and a third day it lost 0.34 gal. What was the total coolant lost over the 3-day period?

50. On a long trip an automobile is driven 340 mi the first day and 400 mi the second day, and the trip is finished with 253 mi the last day. What are the total miles driven?

51. A furnace burned 23.52 gal of gas in September, 25.8 gal in October, 33.24 gal in November, and 41 gal in December. What was the total gas burned over the 4-month period?

52. A 6-room building has the following supply air requirements. $A = 120$ ft³/min, $B = 265$ ft³/min, $C = 61$ ft³/min, and $D = 670$ ft³/min. What is the required HVAC unit supply air flow?

53. In making a specific CAD drawing, the pictorial representation must be precise to the nearest thousandth, and the dimensions must be identified with a precision of five decimal places. The shaft overall length must be shown on the drawing. Calculate the total length and show the dimension that you would put on the drawing in Illustration 2.

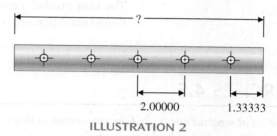

ILLUSTRATION 2

54. As part of an environmental science class, four families are selected to weigh their trash and recycling for a week using their bathroom scales, which have different precisions. The results are shown below in pounds:

	Trash	Recycle	Can/bottles for refund	Precision of scale
Family 1	35.3	21.5	4.9	0.1 lb
Family 2	14.4	28.6	3.8	0.2 lb
Family 3	18.5	36.0	2.5	0.5 lb
Family 4	46	12	4	1 lb

Find the total amount of each and the percentage of material that was recycled and returned for refund.

55. A fisherman brought home a cooler of fish and ice that weighed a total of 17.4 lb on his bathroom scale. He knows that the empty cooler weighs 3.6 lb. He added a 5-lb bag of ice and nothing else except the fish. How much did the fish weigh?

4.6 Multiplication and Division of Measurements

Suppose you want to find the area of a rectangular plot of ground with the measurements as described in Section 4.5 as the length and width. That is, one person measures the length of one side over level ground using a steel tape measure as 54 ft $7\frac{11}{16}$ in. A second person measures the length of the other side over uneven ground by "stepping it off," assuming each step measures 3 ft, as 25 steps × 3 ft = 75 ft as shown in Figure 4.32. The area can be found by multiplying these two measurements

$$(75 \text{ ft})(54 \text{ ft } 7\tfrac{11}{16} \text{ in.}) = 4098.046875 \text{ ft}^2$$

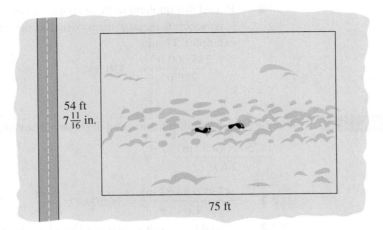

Figure 4.32
What is the area of the rectangular plot of ground formed by these two measurements?

How confident or certain are you about the accuracy (number of significant digits) of the product? Discuss with a classmate what an appropriate accuracy of the product should be.

The point of this discussion is to learn how to multiply such measurements and obtain a product that has a reasonable accuracy. Just simply treating and multiplying these measurements as exact numbers to obtain a product with a good accuracy is not feasible.

To be consistent when multiplying or dividing measurements, the product or quotient can be no more *accurate* than the least accurate measurement. Or, the result can be no more accurate than the results of the least accurate measuring tool.

Multiplying or Dividing Measurements

1. First, multiply and/or divide the measurements.
2. Then round the result to the same number of significant digits as the measurement that has the least number of significant digits. That is, round the result to the same accuracy as the *least accurate* measurement.

Example 1   Use the rules for multiplication of measurements: (20.41 g)(3.5 cm).

STEP 1   (20.41 g)(3.5 cm) = 71.435 g cm

STEP 2   Round this product to two significant digits, which is the accuracy of the least accurate measurement, 3.5 cm. That is,

(20.41 g)(3.5 cm) = 71 g cm   ♦

Example 2   Use the rules for multiplication of measurements: (125 m)(345 m)(204 m).

STEP 1   (125 m)(345 m)(204 m) = 8,797,500 m³

STEP 2   Round this product to three significant digits, which is the accuracy of the least accurate measurement (which is the accuracy of each measurement in this example). That is,

(125 m)(345 m)(204 m) = 8,8$\overline{0}$0,000 m³   ♦

Example 3 Use the rules for division of measurements to divide 288,000 ft³ by 216 ft.

STEP 1 $\dfrac{288{,}000 \text{ ft}^3}{216 \text{ ft}} = 1333.333\ldots \text{ ft}^2$

STEP 2 Round this quotient to three significant digits, which is the accuracy of the least accurate measurement (which is the accuracy of each measurement in this example). That is,

$$\dfrac{288{,}000 \text{ ft}^3}{216 \text{ ft}} = 1330 \text{ ft}^2$$

Example 4 Use the rules for multiplication and division of measurements to evaluate

$$\dfrac{(4750 \text{ N})(4.82 \text{ m})}{1.6 \text{ s}}$$

STEP 1 $\dfrac{(4750 \text{ N})(4.82 \text{ m})}{1.6 \text{ s}} = 14{,}309.375 \dfrac{\text{N m}}{\text{s}}$

STEP 2 Round this result to two significant digits, which is the accuracy of the least accurate measurement, 1.6 s. That is,

$$\dfrac{(4750 \text{ N})(4.82 \text{ m})}{1.6 \text{ s}} = 14{,}000 \dfrac{\text{N m}}{\text{s}} \text{ or } 14{,}000 \text{ N m/s}$$

There are even more sophisticated methods for dealing with the calculations of measurements. The method that one uses (and indeed, whether one should even follow any given procedure) depends on the number of measurements and the sophistication needed for a particular situation.

The procedures for addition, subtraction, multiplication, and division of measurements are based on methods followed and presented by the American Society for Testing and Materials.

NOTE: To multiply or divide measurements, the units do *not* need to be the same. (They *must* be the same in addition and subtraction of measurements.) Also note that the units are multiplied and/or divided in the same manner as the corresponding numbers.

EXERCISES 4.6

Use the rules for multiplication and/or division of measurements to evaluate:

1. (126 m)(35 m)
2. (470 mi)(1200 mi)
3. (1463 cm)(838 cm)
4. (2.4 A)(3600 Ω)
5. (18.7 m)(48.2 m)
6. (560 cm)(28.0 cm)
7. (4.7 Ω)(0.0281 A)
8. (5.2 km)(6.71 km)
9. (24.2 cm)(16.1 cm)(18.9 cm)
10. (0.045 m)(0.0292 m)(0.0365 m)
11. (2460 m)(960 m)(1970 m)
12. (46$\overline{0}$ in.)(235 in.)(368 in.)
13. (0.480 A)²(150 Ω)
14. 360 ft² ÷ 12 ft
15. 62,500 in³ ÷ 25 in.
16. 9180 yd³ ÷ 36 yd²
17. 1520 m² ÷ 40 m
18. 18.4 m³ ÷ 9.2 m²
19. 4800 V ÷ 14.2 A
20. $\dfrac{4800 \text{ V}}{6.72 \text{ Ω}}$
21. $\dfrac{5.63 \text{ km}}{2.7 \text{ s}}$
22. $\dfrac{0.497 \text{ N}}{(1.4 \text{ m})(8.0 \text{ m})}$
23. $\dfrac{(120 \text{ V})^2}{47.6 \text{ Ω}}$
24. $\dfrac{(19 \text{ kg})(3.0 \text{ m/s})^2}{2.46 \text{ m}}$

25. $\dfrac{140 \text{ g}}{(3.2 \text{ cm})(1.7 \text{ cm})(6.4 \text{ cm})}$

26. Find the area of a rectangle measured as 6.5 cm by 28.3 cm. ($A = lw$)

27. $V = lwh$ is the formula for the volume of a rectangular solid, where l = length, w = width, and h = height. Find the volume of a rectangular solid when $l = 16.4$ ft, $w = 8.6$ ft, and $h = 6.4$ ft.

28. Find the volume of a cube measuring 8.10 cm on each edge. ($V = e^3$, where e is the length of each edge.)

29. The formula $s = 4.90t^2$ gives the distance, s, in metres, that an object falls in a given time, t. Find the distance a ball falls in 2.4 seconds.

30. Given K.E. $= \frac{1}{2}mv^2$, $m = 2.87 \times 10^6$ kg, and $v = 13.4$ m/s. Find K.E.

31. A formula for determining the theoretical horsepower of an engine is $p = \dfrac{d^2 n}{2.50}$, where d is the diameter of each cylinder in inches and n is the number of cylinders. What is the horsepower of an 8-cylinder engine if each cylinder has a diameter of 3.00 in.? (*Note:* Eight is an exact number. Ignore the number of significant digits in an exact number when determining the number of significant digits in a product or quotient.)

32. Six pieces of metal, each 2.48 mm in thickness, are fitted together. What is the total thickness of the 6 pieces?

33. Find the volume of a storage bin cylinder with radius 6.2 m and height 8.5 m. The formula for the volume of a cylinder is $V = \pi r^2 h$.

34. In 2010, the United States harvested 12.4 billion bu of corn from 88.2 million acres. In 2004, 11.8 billion bu were harvested from 73.6 million acres. What was the yield in bu/acre for each year? What was the difference in yield between 2010 and 2004?

35. A room 24 ft long and 14 ft wide, with a ceiling height of 8.0 ft, has its air changed six times per hour. What are its ventilation requirements in CFM (ft³/min)?

36. A welder welds two pieces of pipe together and uses 2.25 rods. If this same weld is done 6 times, how many rods are used?

37. A rectangular metal storage bin has been welded together. Its dimensions are 13.5 in., 17.25 in., and $2\overline{0}$ in. What is the volume of such a storage bin?

38. A plane flew 1.8 h for each of 4 lessons. How many hours has it flown?

39. A plane flew 3.4 h and used 32.65 gal of gas. How many gallons per hour did it use?

40. A plane flies 60.45 mi due north, then flies 102.3 mi due east. Find the area of a rectangle formed by these dimensions.

41. What is the area of a windshield of an automobile if it measures 55.3 in. by 28.25 in.?

42. A vehicle traveled 620 mi and used 24.2 gal of fuel. How many miles per gallon did the vehicle get?

43. The trunk space of a small automobile measures 3.0 ft in width, 4.2 ft in length, and 1.5 ft in depth. Find the volume of the trunk.

44. An old furnace measures 26.5 in. wide, 35 in. long, and $7\overline{0}$ in. high. How much space does the furnace occupy?

45. 52.6 ft of duct is needed to put a furnace in a house. If the duct comes only in 6-ft sections, how many sections should be ordered?

46. The weather forecast for tonight calls for about 1 inch of rain per hour. The local shopping center has a paved parking lot that measures 2.50 acres and has two storm sewers to handle the runoff. **a.** Assuming 100% runoff, how many gallons of water per hour must the storm sewers handle tonight to avoid flooding if 1.00 in. of rain per hour falls? (*Note:* The volume of 1 acre of water 1 inch deep is 27,150 gallons.) **b.** If each storm sewer is rated at 25,000 gal/h, can we expect flooding from the parking lot?

47. According to climatologists, the carbon dioxide (CO_2) levels in the atmosphere in October 2016 were the highest levels in 650,000 years, standing at 405 parts per million (ppm). The safe upper limit is 350 ppm. The current rate of increase is 2.1 ppm per year. If that rate of increase remains constant from 2016 until 2100, what would be the expected CO_2 level in our atmosphere by the end of the year 2100?

48. According to the Environmental Protection Agency, a total of 258 million tons of solid waste was generated in the United States in 2014. Of that, 18% was food waste. If all of that food waste had been composted instead of put into landfills, how many pounds of food waste would have been kept out of our landfills?

49. There is so much water in the world that we normally measure large bodies of water in cubic miles or cubic kilometres instead of gallons. Cayuga Lake in upstate New York contains an estimated 9.4 km³. How many cubic miles of water does Cayuga Lake contain?

4.7 Relative Error and Percent of Error

Technicians must determine the importance of measurement error, which may be expressed in terms of relative error. The **relative error of a measurement** is found by comparing the greatest possible error with the measurement itself:

$$\text{relative error} = \frac{\text{greatest possible error}}{\text{measurement}}$$

Example 1 Find the relative error of the measurement 0.08 cm.

The precision is 0.01 cm. The greatest possible error is one-half the precision, which is 0.005 cm.

$$\text{relative error} = \frac{\text{greatest possible error}}{\text{measurement}} = \frac{0.005 \text{ cm}}{0.08 \text{ cm}} = 0.0625$$

Note that the units will always cancel, which means that the relative error is expressed as a unitless decimal. When this decimal is expressed as a percent, we have the percent of error. ◆

The **percent of error of a measurement** is the relative error expressed as a percent.

Percent of error can be used to compare different measurements because, being a percent, it compares each error in terms of 100. (The percent of error in Example 1 is 6.25%.)

Example 2 Find the relative error and percent of error of the measurement 13.8 m.

The precision is 0.1 m and the greatest possible error is then 0.05 m. Therefore,

$$\text{relative error} = \frac{\text{greatest possible error}}{\text{measurement}} = \frac{0.05 \text{ m}}{13.8 \text{ m}} = 0.00362$$

$$\text{percent of error} = 0.362\%$$ ◆

Example 3 Compare the measurements $3\frac{3}{4}$ in. and 16 mm. Which one is better? (Which one has the smaller percent of error?)

Measurement	$3\frac{3}{4}$ in.	16 mm
Precision	$\frac{1}{4}$ in.	1 mm
Greatest possible error	$\frac{1}{2} \times \frac{1}{4}$ in. $= \frac{1}{8}$ in.	$\frac{1}{2} \times 1$ mm $= 0.5$ mm
Relative error	$\dfrac{\frac{1}{8}\text{ in.}}{3\frac{3}{4}\text{ in.}} = \frac{1}{8} \div 3\frac{3}{4}$ $= \frac{1}{8} \div \frac{15}{4}$ $= \frac{1}{8} \times \frac{4}{15}$ $= \frac{1}{30}$ $= 0.0333$	$\dfrac{0.5 \text{ mm}}{16 \text{ mm}} = 0.03125$
Percent of error	3.33%	3.125%

Therefore, 16 mm is the better measurement, because its percent of error is smaller. ◆

Tolerance

In industry, the **tolerance** of a part or component is the acceptable amount that the part or component may vary from a given size. For example, a steel rod may be specified as $14\frac{3}{8}$ in. $\pm \frac{1}{32}$ in. The symbol "$\pm$" is read "plus or minus." This means that the rod may be as long as $14\frac{3}{8}$ in. $+ \frac{1}{32}$ in., that is, $14\frac{13}{32}$ in. This is called the *upper limit*. Or it may be as short as $14\frac{3}{8}$ in. $- \frac{1}{32}$ in., that is, $14\frac{11}{32}$ in. This is called the *lower limit*. Therefore, the specification means that any rod between $14\frac{11}{32}$ in. and $14\frac{13}{32}$ in. would be acceptable. We say that the tolerance is $\pm \frac{1}{32}$ in. The **tolerance interval**—the difference between the upper limit and the lower limit—is $\frac{2}{32}$ in., or $\frac{1}{16}$ in.

A simple way to check the tolerance of the length of a metal rod would be to carefully mark off lengths that represent the lower limit and upper limit, as shown in Figure 4.33. To check the acceptability of a rod, place one end of the rod flush against the metal barrier on the left. If the other end is between the upper and lower limit marks, the part is acceptable. If the rod is longer than the upper limit, it can then be cut to the acceptable limits. If the rod is shorter than the lower limit, it must be rejected. (It can be melted down for another try.)

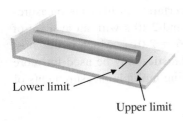

Figure 4.33
The tolerance interval is the difference between the upper limit and the lower limit.

Example 4

The specifications for a stainless steel cylindrical piston are given as follows:

Diameter: 10.200 cm ± 0.001 cm
Height: 14.800 cm ± 0.005 cm

Find the upper limit, the lower limit, and the tolerance interval for each dimension.

	Given length	Tolerance	Upper limit	Lower limit	Tolerance interval
Diameter:	10.200 cm	±0.001 cm	10.201 cm	10.199 cm	0.002 cm
Height:	14.800 cm	±0.005 cm	14.805 cm	14.795 cm	0.010 cm

◆

Tolerance may also be expressed as a percent. For example, resistors are color coded to indicate the tolerance of a given resistor. If the fourth band is silver, this indicates that the acceptable tolerance is ±10% of the given resistance. If the fourth band is gold, this indicates that the acceptable tolerance is ±5% of the given resistance. This is fully discussed in the next section.

Many times, bids may be accepted under certain conditions. They may be accepted, for example, when they are less than 10% over the architect's estimate.

Example 5

If the architect's estimate for a given project is $356,200 and bids may be accepted if they are less than 10% over the estimate, what is the maximum acceptable bid?

10% of $356,200 = (0.10)($356,200) = $35,620

The upper limit or maximum acceptable bid is $356,200 + $35,620 = $391,820. ◆

Single versus Multiple Measurements

Up to this point, we have been considering only single measurements of an event. When using experimental data, one can increase the certainty of the measurement of an event by repeating the same measurement multiple times. For example, for measuring a timed event, one may simply and carefully measure the time with a stopwatch multiple times and take the average. The more measurements you take, the better the certainty of the measurement of the event. For example, Zachary timed how long it takes for a heavy ball to drop from a height of 20.0 m to the ground with the same stopwatch. He obtained the following times: 1.94 s, 2.03 s, 2.10 s, 1.98 s, 1.96 s, and 2.05 s.

From these measurements, he has decreased the uncertainty of his time measurement. He knows the best time is likely between 1.94 s and 2.10 s with an average of (1.94 s + 2.03 s + 2.10 s + 1.98 s + 1.96 s + 2.05 s)/6 = 2.01 s, which is the most likely time. This assumes the stopwatch is not faulty. Statistics may be used for more sophisticated ways to determine the best results from using multiple measurements of the same event.

Random and Systematic Errors

When working with multiple measurements of an event, you need to understand that random and systematic errors are introduced. **Random errors** are variations that occur in the measurements of an event that do not have a predictable pattern. They can be reduced by careful attention to detail, but never eliminated. **Systematic errors** are variations that occur in the measurements of an event that have a predictable pattern. They often cause the value of the measurement to be consistently too large or too small. They are often caused by measuring instruments that are not calibrated correctly or have other defects. They can often be eliminated if you can determine they exist.

EXERCISES 4.7

For each measurement, find the precision, the greatest possible error, the relative error, and the percent of error (to the nearest hundredth percent):

1. 1400 lb
2. 240,000 Ω
3. 875 rpm
4. 12,500 V
5. 0.085 g
6. 0.188 cm
7. 2 g
8. 2.2 g
9. 2.22 g
10. 18,0$\overline{0}$0 W
11. 1.00 kg
12. 1.0 kg
13. 0.041 A
14. 0.08 ha
15. $11\frac{7}{8}$ in.
16. $1\frac{3}{4}$ in.
17. 12 ft 8 in.
18. 4 lb 13 oz

Compare each set of measurements by indicating which measurement is better or best:

19. 13.5 cm; $8\frac{3}{4}$ in.
20. 364 m; 36.4 cm
21. 16 mg; 19.7 g; $12\frac{3}{16}$ oz
22. 68,000 V; 3450 Ω; 3.2 A

Complete the table:

	Given measurement	Tolerance	Upper limit	Lower limit	Tolerance interval
23.	$3\frac{1}{2}$ in.	$\pm\frac{1}{8}$ in.	$3\frac{5}{8}$ in.	$3\frac{3}{8}$ in.	$\frac{1}{4}$ in.
24.	$5\frac{3}{4}$ in.	$\pm\frac{1}{16}$ in.			

(continued)

4.8 ♦ Color Code of Electrical Resistors

	Given measurement	Tolerance	Upper limit	Lower limit	Tolerance interval
25.	$6\frac{5}{8}$ in.	$\pm\frac{1}{32}$ in.			
26.	$7\frac{7}{16}$ in.	$\pm\frac{1}{32}$ in.			
27.	$3\frac{7}{16}$ in.	$\pm\frac{1}{64}$ in.			
28.	$\frac{9}{64}$ in.	$\pm\frac{1}{128}$ in.			
29.	$3\frac{3}{16}$ in.	$\pm\frac{1}{128}$ in.			
30.	$9\frac{3}{16}$ mi	$\pm\frac{1}{32}$ mi			
31.	1.19 cm	±0.05 cm			
32.	1.78 m	±0.05 m			
33.	0.0180 A	±0.0005 A			
34.	9.437 L	±0.001 L			
35.	24,000 V	±2000 V			
36.	375,000 W	±10,000 W			
37.	10.31 km	±0.05 km			
38.	21.30 kg	±0.01 kg			

Complete the table:

	Architect's estimate	Maximum rate above estimate	Maximum acceptable bid
39.	$48,250	10%	
40.	$259,675	7%	
41.	$1,450,945	8%	
42.	$8,275,625	5%	

4.8 Color Code of Electrical Resistors

The resistance of an electrical resistor is often given in a color code. A series of four colored bands is painted on the resistor. Each color on any of the first three bands stands for a digit or number as given in the following table:

Color on any of the first three bands	Digit or number
Black	0
Brown	1
Red	2
Orange	3
Yellow	4
Green	5
Blue	6
Violet	7
Gray	8
White	9
Gold on the third band	Multiply the value by 0.1
Silver on the third band	Multiply the value by 0.01

The fourth band indicates the tolerance of the resistor as given in the following table:

Color of the fourth band	Tolerance
Gold	±5%
Silver	±10%
Black or no fourth band	±20%

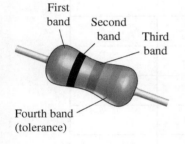

Electrical resistor

Figure 4.34

The value of each resistor is in ohms, Ω, and is given in two significant digits. The color bands are read from left to right when the resistor is in the position shown in Figure 4.34.

Finding the Value of a Resistor

STEP 1 The digit corresponding to the color of the first band is the first digit of the resistance.

STEP 2 The digit corresponding to the color of the second band is the second digit of the resistance.

STEP 3 a. The third band indicates the number of zeros to be written after the first two digits from Steps 1 and 2.

b. If the third band is gold, multiply the number corresponding to the digits from Steps 1 and 2 by 0.1. That is, place the decimal point *between* the two digits.

c. If the third band is silver, multiply the number corresponding to the digits from Steps 1 and 2 by 0.01. That is, place the decimal point *before* the two digits.

STEP 4 The fourth band indicates the tolerance written as a percent. The tolerance is

a. ±5% if the fourth band is gold,

b. ±10% if the fourth band is silver, or

c. ±20% if the fourth band is black or if there is no fourth band.

4.8 ♦ Color Code of Electrical Resistors   187

Example 1

Find the resistance of the resistor shown in Figure 4.35.

STEP 1   The first digit is 4—the digit that corresponds to *yellow*.
STEP 2   The second digit is 5—the digit that corresponds to *green*.
STEP 3A *Orange* on the third band means that three zeros should be written after the digits from Steps 1 and 2.

So the resistance is 45,000 Ω.   ◆

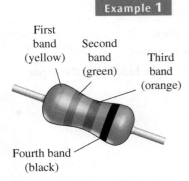

Figure 4.35

Example 2

Find the tolerance, the upper limit, the lower limit, and the tolerance interval for the resistor shown in Figure 4.35.

The black fourth band indicates a tolerance of ±20%.
20% of 45,000 Ω = (0.20)(45,000 Ω) = 9000 Ω.
That is, the tolerance is ±9000 Ω.

The upper limit is 45,000 Ω + 9000 Ω = 54,000 Ω.
The lower limit is 45,000 Ω − 9000 Ω = 36,000 Ω.
The tolerance interval is then 18,000 Ω.   ◆

Example 3

Find the resistance of the resistor shown in Figure 4.36.

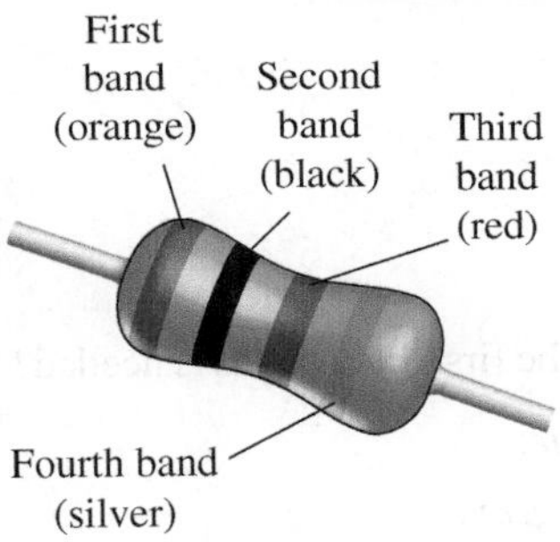

STEP 1   The first digit is 3—the digit that corresponds to *orange*.
STEP 2   The second digit is 0—the digit that corresponds to *black*.
STEP 3A *Red* on the third band means that two zeros should be written after the digits from Steps 1 and 2.

So the resistance is 3000 Ω.   ◆

Figure 4.36

Example 4

Find the tolerance, the upper limit, the lower limit, and the tolerance interval for the resistor shown in Figure 4.36.

The silver fourth band indicates a tolerance of ±10%.
10% of 3000 Ω = (0.10)(3000 Ω) = 300 Ω.
That is, the tolerance is ±300 Ω.

The upper limit is 3000 Ω + 300 Ω = 3300 Ω.
The lower limit is 3000 Ω − 300 Ω = 2700 Ω.
The tolerance interval is then 600 Ω.   ◆

Example 5

Find the resistance of the resistor shown in Figure 4.37.

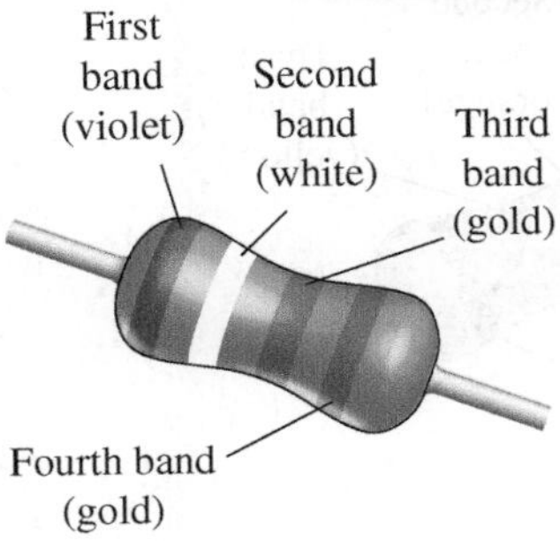

STEP 1   The first digit is 7—the digit that corresponds to *violet*.
STEP 2   The second digit is 9—the digit that corresponds to *white*.
STEP 3B *Gold* on the third band means to place the decimal point *between* the digits from Steps 1 and 2.

So the resistance is 7.9 Ω.   ◆

Figure 4.37

188 CHAPTER 4 ♦ Measurement

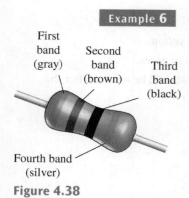

Figure 4.38

Example 6 Find the resistance of the resistor shown in Figure 4.38.

- **STEP 1** The first digit is 8—the digit that corresponds to *gray*.
- **STEP 2** The second digit is 1—the digit that corresponds to *brown*.
- **STEP 3A** *Black* on the third band means that no zeros are to be included after the digits from Steps 1 and 2.

So the resistance is 81 Ω.

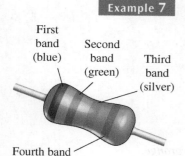

Figure 4.39

Example 7 Find the resistance of the resistor shown in Figure 4.39.

- **STEP 1** The first digit is 6—the digit that corresponds to *blue*.
- **STEP 2** The second digit is 5—the digit that corresponds to *green*.
- **STEP 3C** *Silver* on the third band means to place the decimal point *before* the digits from Steps 1 and 2.

So the resistance is 0.65 Ω.

Example 8 A serviceperson needs a 680,000-Ω resistor. What color code on the first three bands is needed?

- **STEP 1** The color that corresponds to the first digit, 6, is *blue*.
- **STEP 2** The color that corresponds to the second digit, 8, is *gray*.
- **STEP 3A** The color that corresponds to four zeros is *yellow*.

So the colors that the serviceperson is looking for are blue, gray, and yellow.

EXERCISES 4.8

For each resistor shown, find the resistance and the tolerance, written as a percent:

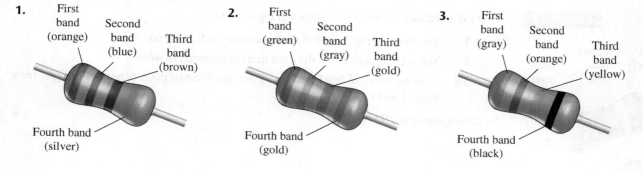

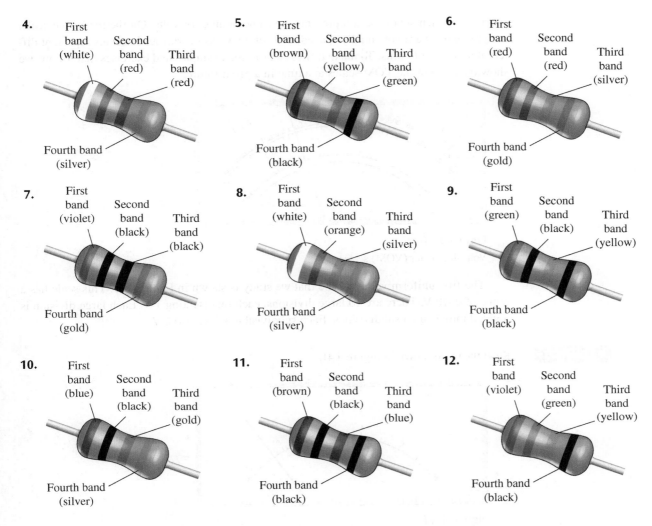

What color code on the first three bands is needed for each resistance?

*Find **a.** the tolerance in ohms, Ω, **b.** the upper limit, **c.** the lower limit, and **d.** the tolerance interval for each resistor:*

13. 4800 Ω
14. 95 Ω
15. 72,000 Ω
16. 3.1 Ω
17. 650,000 Ω
18. 100 Ω
19. 0.25 Ω
20. 9000 Ω
21. 4,500,000 Ω
22. 40 Ω
23. 7.6 Ω
24. 0.34 Ω

25. Exercise 1
26. Exercise 2
27. Exercise 3
28. Exercise 5
29. Exercise 7
30. Exercise 12

4.9 Reading Scales

Uniform Scales

A **uniform scale** has graduations equally spaced, and each subdivision represents the same quantity. Figure 4.40 shows some of the various scales that may be found on a *volt-ohm meter* (VOM). This instrument is used to measure voltage (measured in volts, V) and resistance (measured in ohms, Ω) in electrical circuits. Note that the voltage scales are uniform, while the resistance scale is nonuniform. On a given voltage scale, the graduations are equally spaced and each subdivision represents the same number of volts. On the resistance scale, the graduations are not equally spaced, and subdivisions on various intervals represent different numbers of ohms. To make things clear in the examples and exercises that follow, we show only one of the VOM scales at a time in a given figure.

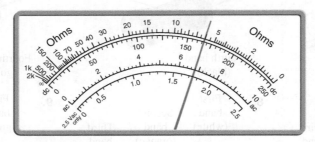

Figure 4.40
Volt-ohm meter (VOM) scales

The first uniform voltage scale that we study is shown in Figure 4.41. This scale has a range of 0–10 V. There are 10 large divisions, each representing 1 V. Each large division is divided into 5 equal subdivisions. Each subdivision is $\frac{1}{5}$ V, or 0.2 V.

Example 3   Read the scale shown in Figure 4.41.

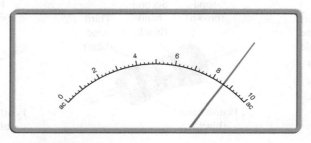

Figure 4.41

The needle is on the third graduation to the right of 8. Each subdivision is 0.2 V. Therefore, the reading is 8.6 V. ◆

Figure 4.42 shows a voltage scale that has a range of 0–2.5 V. There are 5 large divisions, each representing 0.5 V. Each division is divided into 5 subdivisions. Each subdivision is $\frac{1}{5} \times 0.5 \text{ V} = 0.1 \text{ V}$.

Example 4   Read the scale shown in Figure 4.42.

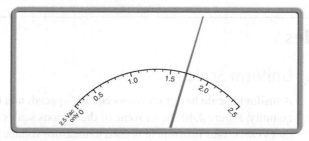

Figure 4.42

The needle is on the second graduation to the right of 1.5. Each subdivision is 0.1 V. Therefore, the reading is 1.7 V. ◆

Figure 4.43 shows a voltage scale that has a range of 0–250 V. There are 10 large divisions, each representing 25 V. Each division is divided into 5 subdivisions. Each subdivision is $\frac{1}{5} \times 25$ V = 5 V.

Example 5 Read the scale shown in Figure 4.43.

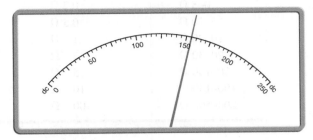

Figure 4.43

The needle is on the first graduation to the right of 150. Each subdivision is 5 V. Therefore, the reading is 155 V. ◆

Nonuniform Scales

A **nonuniform scale** has graduations not equally spaced, and each subdivision may represent different quantities. Figure 4.44 shows a nonuniform ohm scale usually found on a VOM. First, consider that part of the scale between 0 and 5. Each large division represents 1 ohm (Ω). Each large division is divided into 5 subdivisions. Therefore, each subdivision represents $\frac{1}{5} \times 1$ Ω, or 0.2 Ω.

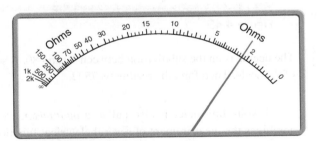

Figure 4.44

- Between 5 and 10, each subdivision is divided into 2 sub-subdivisions. Each sub-subdivision represents $\frac{1}{2} \times 1$ Ω, or 0.5 Ω.
- Between 10 and 20, each division represents 1 Ω.
- Between 20 and 100, each large division represents 10 Ω. Between 20 and 30, there are 5 subdivisions. Therefore, each subdivision represents $\frac{1}{5} \times 10$ Ω, or 2 Ω.
- Between 30 and 100, each large division has 2 subdivisions. Each subdivision represents $\frac{1}{2} \times 10$ Ω, or 5 Ω.
- Between 100 and 200, each large division represents 50 Ω.
- Between 100 and 150, there are 5 subdivisions. Each subdivision represents $\frac{1}{5} \times 50$ Ω, or 10 Ω.
- Between 200 and 500, there are 3 subdivisions. Each subdivision represents $\frac{1}{3} \times 300$ Ω, or 100 Ω.

The subdivisions on each part of the ohm scale may be summarized as follows:

Range	Each subdivision represents:
0–5 Ω	0.2 Ω
5–10 Ω	0.5 Ω
10–20 Ω	1 Ω
20–30 Ω	2 Ω
30–100 Ω	5 Ω
100–150 Ω	10 Ω
200–500 Ω	100 Ω

Example 6 Read the scale shown in Figure 4.44.

The needle is on the second subdivision to the left of 2, where each subdivision represents 0.2 Ω. Therefore, the reading is 2.4 Ω.

Example 7 Read the scale shown in Figure 4.45.

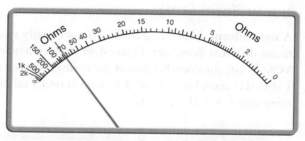

Figure 4.45

The needle is on the subdivision between 70 and 80. A subdivision represents 5 Ω on this part of the scale. Therefore, the reading is 75 Ω.

A volt-ohm meter is also called a *multimeter*, which is an electronic instrument that combines the measurement of several electrical functions in one unit. A typical multimeter can measure voltage, current, and resistance on a scale calibrated for all the different measurements that can be made with it. Multimeters are used to troubleshoot a wide variety of industrial and household devices. A digital multimeter is shown in Figure 4.46.

Figure 4.46
Digital multimeter

EXERCISES 4.9

Read each scale:

1.
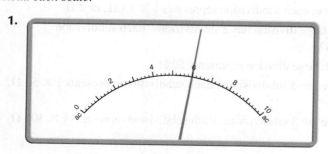

2.
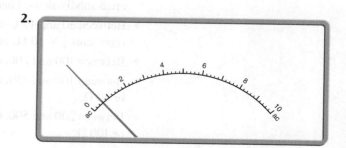

4.9 ◆ Reading Scales 193

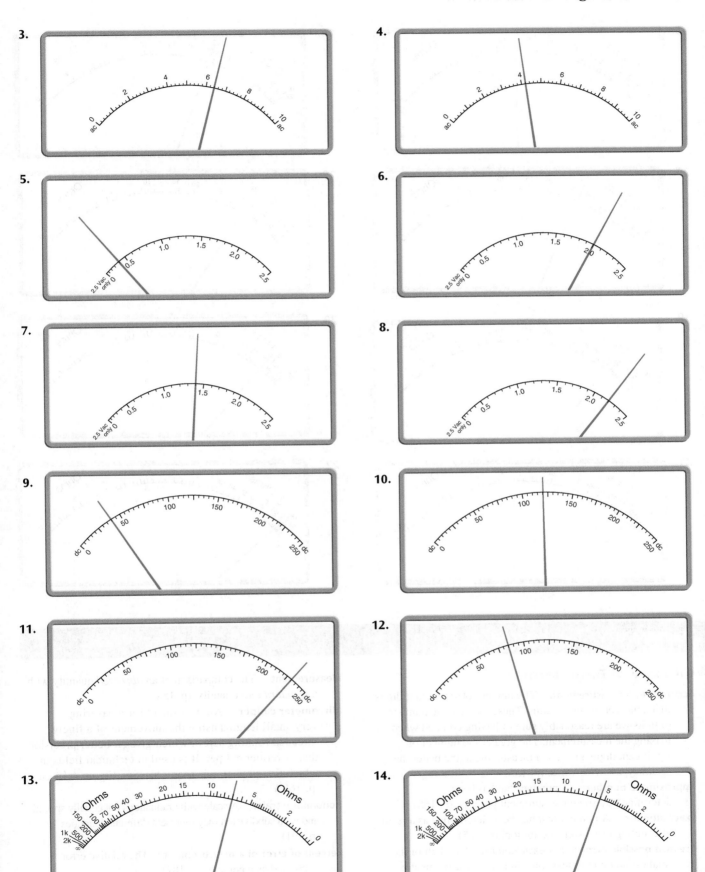

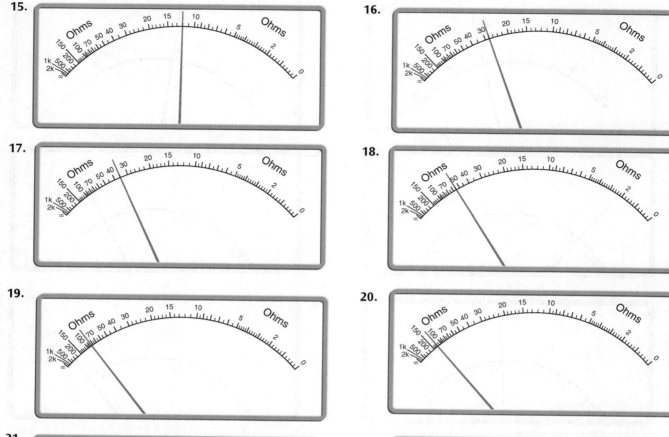

SUMMARY | CHAPTER 4

Glossary of Basic Terms

Accuracy of a measurement. The number of significant digits that a measurement contains. These indicate the number of units we are reasonably sure of having counted when making the measurement. The greater the number of significant digits given in a measurement, the better the accuracy, and vice versa. (p. 155)

Approximate number. A number that has been determined by some measurement process. (p. 155)

Exact number. A number that has been determined as a result of counting or by some definition. (p. 155)

Greatest possible error of a measurement. One-half of the smallest unit on the scale on which the measurement is read and equal to one-half of the measurement's precision. (p. 159)

Measurement. The comparison of an *observed* quantity with a *standard unit* quantity. (p. 154)

Micrometer caliper. An instrument for measuring very small lengths using the movement of a finely threaded rotating screw, which gives it better precision than a vernier caliper. It is used in technical fields in which fine precision is required. (See Figure 4.19 on p. 168.)

Nonuniform scale. A scale with graduations not equally spaced and each subdivision may represent different quantities. (p. 191)

Percent of error of a measurement. The relative error expressed as a percent. (p. 182)

Precision of a measurement. The smallest unit with which the measurement is made—that is, the position of the last significant digit or the smallest unit or calibration on the measuring instrument. (p. 157)

Random errors. Errors that result from variations that occur in the measurements of an event that do not have a predictable pattern. (p. 184)

Relative error of a measurement. The greatest possible error divided by the measurement itself. (p. 182)

Significant digits. Those digits in a number that we are reasonably sure of having counted and of being able to rely on in a measurement. (p. 155)

Systematic errors. Errors that result from variations that occur in the measurement of an event that have a predictable pattern. (p. 184)

Tolerance. The acceptable amount that a given part or component may vary from a given size. (p. 183)

Tolerance interval. The difference between the upper limit and the lower limit. (p. 183)

Uniform scale. A scale with graduations equally spaced and each subdivision represents the same quantity. (p. 189)

Vernier caliper. A slide-type measuring instrument used to take precise inside, outside, and depth measurements. (See Figure 4.8 on p. 161.)

4.1 Approximate Numbers and Accuracy

1. **Exact versus approximate numbers:**
   a. Only counting numbers are exact.
   b. All measurements are approximations. (p. 155)
2. **Significant digits:**
   a. The following digits are *significant:*
      - All nonzero digits
      - All zeros between significant digits
      - A zero in a whole-number measurement that is specially tagged, such as by a bar above it
      - All zeros to the right of a significant digit *and* a decimal point
   b. The following digits are *not significant:*
      - Zeros to the right in a whole-number measurement that are not tagged
      - Zeros to the left in a decimal measurement that is less than 1 (p. 156)

4.3 The Vernier Caliper

1. Review this section to read the various vernier calipers presented. (p. 161)

4.4 The Micrometer Caliper

1. Review this section to read the various micrometer calipers presented. (p. 167)

4.5 Addition and Subtraction of Measurements

1. **Adding or subtracting measurements of different precision:** To add or subtract measurements of different precision:
   a. Make certain that all measurements are expressed in the same unit. If they are not, change them all to any common unit.
   b. Add or subtract.
   c. Then round the result to the same precision as the *least precise* measurement. (p. 176)

4.6 Multiplication and Division of Measurements

1. **Multiplying or dividing measurements:** To multiply or divide measurements:
   a. First, multiply and/or divide the measurements.
   b. Then round the result to the same number of significant digits as the measurement that has the least number of significant digits. That is, round the result to the same accuracy as the *least accurate* measurement. (p. 179)

4.7 Relative Error and Percent of Error

1. **Relative error:**

   $$\text{relative error} = \frac{\text{greatest possible error}}{\text{measurement}}$$ (p. 182)

2. **Percent of error:**

   percent of error = the relative error expressed as a percent (p. 182)

4.8 Color Code of Electrical Resistors

1. Review this section to read the color code of electrical resistors, which gives the size and tolerance of a given resistor. (p. 185)

4.9 Reading Scales

1. Review this section to read the uniform and nonuniform scales presented. (p. 189)

REVIEW | CHAPTER 4

Give the number of significant digits (the accuracy) of each measurement:

1. 4.06 kg
2. 24,000 mi
3. 36$\overline{0}$0 V
4. 5.60 cm
5. 0.0070 W
6. 0.0651 s
7. 20.00 m
8. 20.050 km

*Find **a.** the precision and **b.** the greatest possible error of each measurement:*

9. 6.05 m
10. 15.0 mi
11. 160,500 L
12. 2300 V
13. 17.00 cm
14. 13,0$\overline{0}$0,000 V
15. $1\frac{5}{8}$ in.
16. $10\frac{3}{16}$ mi

Read the measurement shown on the vernier caliper in Illustration 1 in the following units:

17. Metric units
18. U.S. units

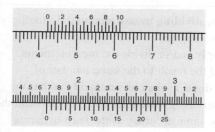

ILLUSTRATION 1

19. Read the measurement shown on the metric micrometer in Illustration 2.

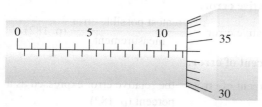

ILLUSTRATION 2

20. Read the measurement shown on the U.S. micrometer in Illustration 3.

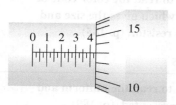

ILLUSTRATION 3

21. Find the measurement that is **a.** the most accurate and **b.** the most precise:

    2500 V; 36,500 V; 60,000 V; 9.6 V; 120 V

22. Find the measurement that is **a.** the least accurate and **b.** the least precise:

    0.0005 A; 0.0060 A; 0.425 A; 0.0105 A; 0.0055 A

Use the rules for addition of measurements to find the sum of each set of measurements:

23. 18,000 W; 260,000 W; 2300 W; 45,500 W; 398,000 W
24. 16.8 cm; 19.7 m; 0.14 km; 240 m
25. Use the rules for subtraction of measurements to subtract:

    1,500,000 V − 1,125,000 V

Use the rules for multiplication and/or division of measurements to evaluate:

26. 15.6 cm × 18.5 cm × 6.5 cm
27. $\dfrac{98.2 \text{ m}^3}{16.7 \text{ m}}$
28. $\dfrac{239 \text{ N}}{(24.8 \text{ m})(6.7 \text{ m})}$
29. $\dfrac{(220 \text{ V})^2}{365 \text{ }\Omega}$

*Find **a.** the relative error and **b.** the percent of error (to the nearest hundredth percent) for each measurement:*

30. $5\frac{7}{16}$ in.
31. 15.60 cm
32. Given a resistor of 2000 Ω with a tolerance of ±10%, find the upper and lower limits.

For each resistor find its resistance and its tolerance written as a percent:

33.
First band (brown) Second band (red) Third band (yellow) Fourth band (black)

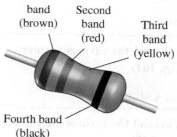

34.
First band (gray) Second band (green) Third band (silver) Fourth band (gold)

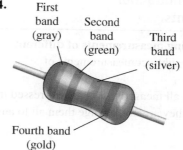

Read each scale:

35.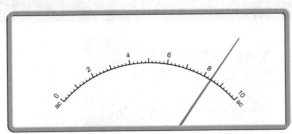

36.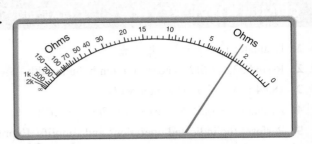

TEST | CHAPTER 4

Give the number of significant digits in each measurement:

1. 1.806 g   **2.** 7.00 L   **3.** 0.00015 A

Find **a.** *the precision and* **b.** *the greatest possible error of each measurement:*

4. 6.13 mm   **5.** 2400 Ω   **6.** $5\frac{3}{4}$ in.

Read the measurement shown on the vernier caliper in Illustration 1 in the following units:

7. Metric units   **8.** U.S. units

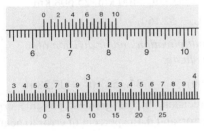

ILLUSTRATION 1

9. Read the measurement shown on the metric micrometer in Illustration 2.

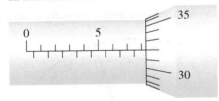

ILLUSTRATION 2

10. Read the measurement shown on the U.S. micrometer in Illustration 3.

ILLUSTRATION 3

11. Find the measurement that is **a.** the most accurate, **b.** the most precise, **c.** the least accurate, and **d.** the least precise:

208 m; 17,060 m; 25.9 m; 0.067 m

12. Find the measurement that is **a.** the most accurate, **b.** the most precise, **c.** the least accurate, and **d.** the least precise:

360 V; 0.5 V; 125,000 V; 600,000 V

13. Use the rules of measurement to multiply:

(4.0 m)(12 m)(0.60 m)

14. Use the rules of measurement to add:

12.9 L + 341 L + 2104 L

15. Use the rules of measurement to subtract:

108.07 g − 56.1 g

16. Use the rules of measurement to divide:

6.28 m² ÷ 25 m

17. Use the rules of measurement to evaluate:

$$\frac{(56.3 \text{ m})(25 \text{ m})(112.5 \text{ m})}{(21.275 \text{ m})^2}$$

18. Find **a.** the relative error and **b.** the percent of error (to the nearest hundredth percent) for the measurement 5.20 m.

Read each scale:

19.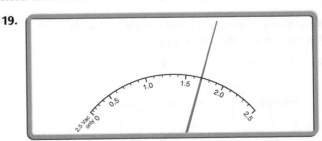

20.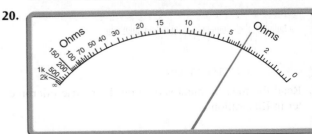

CUMULATIVE REVIEW | CHAPTERS 1–4

1. Evaluate: $1\frac{3}{8} - \frac{1}{2} \times \frac{3}{4} + 1\frac{5}{8} \div \frac{1}{16}$
2. Round 32,518.612 to nearest **a.** ten **b.** hundredth.
3. 18.84 is 31.4% of what number?
4. Evaluate: $(-4)(5) + (-6)(-4) - 7(-4) \div 2(7)$
5. Perform the indicated operations and simplify. Express the result using positive exponents. $\dfrac{(10^3 \cdot 10^{-2})^3}{10^3 \cdot 10^5}$
6. Perform the indicated operations and write the result in scientific notation. $\dfrac{(62.3 \times 10^3)(4.18 \times 10^{-5})}{(17.3 \times 10^{-4})^2}$
7. Give the SI abbreviation for milli.
8. Write the abbreviation for 25 kilograms.
9. Write the SI unit for 250 μs.
10. Which is larger: 1 amp or 1 mega-amp?

Round each result to three significant digits when necessary:

11. Change 120 km to m.
12. Change 250 cm to m.
13. Change 50 g to kg.
14. Change 4060 kg to metric tons.
15. Change 86°C to °F.
16. Change 50°F to °C.
17. Change 163 in^2 to cm^2.
18. Change 120 m to km.
19. Change 10 L to mL.

Give the number of significant digits in each measurement:

20. 0.25 Ω
21. 7.002 m

*Find **a.** the precision and **b.** the greatest possible error of each measurement:*

22. 14.28 mm
23. 62.3 lb

Read the measurement on the vernier caliper in Illustration 1 in the following units:

24. Metric units
25. U.S. units

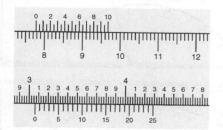

ILLUSTRATION 1

26. Read the measurement shown on the metric micrometer in Illustration 2.

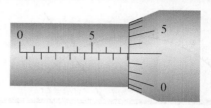

ILLUSTRATION 2

27. Read the measurement shown on the U.S. micrometer in Illustration 3.

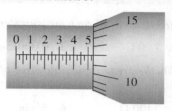

ILLUSTRATION 3

Use the rules for addition and subtraction of measurements to evaluate:

28. Add: 6120 km, 1743 km, 1400 km, 25,608 km
29. Subtract: 98.2 L − 52.16 L

Use the rules for multiplication and/or division of measurements to evaluate:

30. (283 cm)(150 cm)
31. 583 ft^2 ÷ 17.28 ft
32. Find **a.** the precision, **b.** the greatest possible error, **c.** the relative error, and **d.** the percent of error to nearest hundredth percent of the measurement 2.135 cm.

Read each scale:

33.

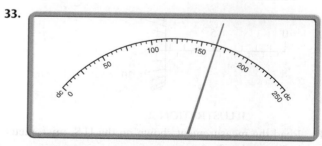

34.

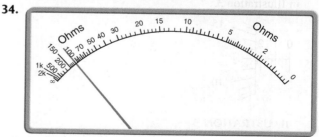

An Introduction to Algebra

CHAPTER 5

OBJECTIVES

- Apply the rules for order of operations to evaluate expressions with numbers and to evaluate algebraic expressions when the values of the letters are given.
- Simplify algebraic expressions by removing parentheses and combining like terms.
- Add and subtract polynomials.
- Multiply monomials.
- Multiply polynomials.
- Divide a monomial and a polynomial by a monomial.
- Divide a polynomial by a polynomial.

Mathematics at Work

The nation's construction industry depends on a technical and competent workforce. This workforce includes, but is not limited to, carpenters who cut, fit, and assemble wood and other materials in construction projects; plumbers, pipefitters, and steamfitters who install, maintain, and repair many different types of pipe systems that carry water, steam, air, and other liquids; painters who apply paint, stain, varnish, and other finishes to buildings and other structures; electricians who install, maintain, and repair electrical wiring, equipment, and fixtures; bricklayers and stonemasons who build walls and other structures with bricks, blocks, stones, and other masonry materials; and structural and reinforcing metal workers who use materials made from iron, steel, and other materials to construct highways, bridges, buildings, and towers.

Construction trade workers often learn their own trade through apprenticeship programs administered by local joint union–management committees or through community college or trade school programs, some of which are offered in partnership with the local joint union–management committees. For more information, please visit **www.cengage.com** and access the Student Online Resources for this text.

Construction Trades
Carpenter working in a building

5.1 Fundamental Operations

In arithmetic, we perform mathematical operations with specific numbers. In algebra, we perform these same basic mathematical operations with numbers and **variables**—letters that represent unknown quantities. Algebra allows us to express and solve general as well as specific problems that cannot be solved using only arithmetic. As a result, employers in technical and scientific areas require a certain level of skill and knowledge of algebra. Your problem-solving skills will increase significantly as your algebra skills increase.

To begin our study of algebra, some basic mathematical principles that you will apply are listed below. Most of them you probably already know; the rest will be discussed. Note that "$\neq$" means "is not equal to."

Basic Mathematical Principles

1. $a + b = b + a$ (Commutative Property for Addition)
2. $ab = ba$ (Commutative Property for Multiplication)
3. $(a + b) + c = a + (b + c)$ (Associative Property for Addition)
4. $(ab)c = a(bc)$ (Associative Property for Multiplication)
5. $a(b + c) = ab + ac$, or $(b + c)a = ba + ca$ (Distributive Property)
6. $a + 0 = a$
7. $a \cdot 0 = 0$
8. $a + (-a) = 0$ (Additive Inverse)
9. $a \cdot 1 = a$
10. $a \cdot \frac{1}{a} = 1$ ($a \neq 0$)   (Multiplicative Inverse)

In mathematics, letters are often used to represent numbers. Thus, it is necessary to know how to indicate arithmetic operations and carry them out using letters.

Addition: $x + y$ means add x and y.

Subtraction: $x - y$ means subtract y from x or add the negative of y to x; that is, $x + (-y)$.

Multiplication: xy or $x \cdot y$ or $(x)(y)$ or $(x)y$ or $x(y)$ means multiply x by y.

Division: $x \div y$ or $\frac{x}{y}$ means divide x by y, or find a number z such that $zy = x$.

Exponents: $xxxx$ means use x as a factor 4 times, which is abbreviated by writing x^4. In the expression x^4, x is called the *base*, and 4 is called the **exponent**. For example, 2^4 means $2 \cdot 2 \cdot 2 \cdot 2 = 16$.

Order of Operations

1. Perform all operations inside parentheses first. If the problem contains a fraction bar, treat the numerator and the denominator separately.
2. Evaluate all powers, if any. For example, $6 \cdot 2^3 = 6 \cdot 8 = 48$.
3. Perform any multiplications or divisions in order, from left to right.
4. Do any additions or subtractions in order, from left to right.

5.1 ♦ Fundamental Operations

Example 1   Evaluate: $4 - 9(6 + 3) \div (-3)$.

$$\begin{aligned}
&= 4 - 9(9) \div (-3) &&\text{Add within parentheses.}\\
&= 4 - 81 \div (-3) &&\text{Multiply.}\\
&= 4 - (-27) &&\text{Divide.}\\
&= 31 &&\text{Subtract.}
\end{aligned}$$

Example 2   Evaluate: $(-6) + 5(-2)^2(-9) - 7(3 - 5)^3$.

$$\begin{aligned}
&= (-6) + 5(-2)^2(-9) - 7(-2)^3 &&\text{Subtract within parentheses.}\\
&= (-6) + 5(4)(-9) - 7(-8) &&\text{Evaluate the powers.}\\
&= (-6) - 180 + 56 &&\text{Multiply.}\\
&= -130 &&\text{Add and subtract.}
\end{aligned}$$

To **evaluate an expression**, replace the letters with given numbers; then do the arithmetic using the order of operations. The result is the value of the expression.

Example 3   Evaluate: $\dfrac{x^2 - y + 5}{2x - 2}$, if $x = 4$ and $y = 3$.

$$\begin{aligned}
\frac{x^2 - y + 5}{2x - 2} &= \frac{4^2 - 3 + 5}{2(4) - 2} &&\text{Replace } x \text{ with 4 and } y \text{ with 3.}\\
&= \frac{16 - 3 + 5}{8 - 2} &&\text{Evaluate the power and multiply.}\\
&= \frac{18}{6} &&\text{Add and subtract.}\\
&= 3 &&\text{Divide.}
\end{aligned}$$

NOTE: In a fraction, the line between the numerator and denominator serves as parentheses for both. That is, do the operations in both numerator and denominator before doing the division.

Example 4   Evaluate: $\dfrac{ab}{3c} + c$, if $a = 6$, $b = 10$, and $c = -5$.

$$\begin{aligned}
\frac{ab}{3c} + c &= \frac{6 \cdot 10}{3(-5)} + (-5) &&\text{Replace } a \text{ with 6, } b \text{ with 10, and } c \text{ with } -5.\\
&= \frac{60}{-15} + (-5) &&\text{Multiply in the numerator and in the denominator.}\\
&= -4 + (-5) &&\text{Divide.}\\
&= -9 &&\text{Add.}
\end{aligned}$$

Example 5   Evaluate: $\dfrac{x^2 - y^2(z - 2)}{-7xz - 2(3 - x)^2}$, if $x = -9$, $y = -3$, and $z = 5$.

$$\begin{aligned}
\frac{x^2 - y^2(z - 2)}{-7xz - 2(3 - x)^2} &= \frac{(-9)^2 - (-3)^2(5 - 2)}{(-7)(-9)(5) - 2[3 - (-9)]^2} &&\text{Replace } x \text{ with } -9, y \text{ with } -3, \text{ and } z \text{ with 5.}\\
&= \frac{81 - (9)(3)}{315 - 2[12]^2} &&\text{Evaluate the powers and subtract within the parentheses in the numerator; then multiply and subtract within the brackets in the denominator.}\\
&= \frac{81 - 27}{315 - 288} &&\text{Multiply in the numerator; then evaluate the power and multiply in the denominator.}\\
&= \frac{54}{27} &&\text{Subtract in the numerator and the denominator.}\\
&= 2 &&\text{Divide.}
\end{aligned}$$

EXERCISES 5.1

Evaluate each expression:

1. $3(-5)^2 - 4(-2)$
2. $(-2)(-3)^2 + 3(-2) \div 6$
3. $4(-3) \div (-6) - (-18) \div 3$
4. $48 \div (-2)(-3) + (-2)^2$
5. $(-72) \div (-3) \div (-6) \div (-2) - (-4)(-2)(-5)$
6. $28 \div (-7)(2)^2 + 3(-4-2)^2 - (-3)^2$
7. $[(-2)(-3) + (-24) \div (-2)] \div [-10 + 7(-1)^2]$
8. $(-9)^2 \div 3^3(6) + [3(-2) - 5(-3)]$
9. $[(-2)(-8)^2 \div (-2)^3] - [-4 + (-2)^4]^2$
10. $[(-2)(3) + 5(-2)][5(-4) - 8(-3)]^2$

In Exercises 11–16, let $x = 2$ and $y = 3$, and evaluate each expression:

11. $2x - y$
12. $x - 2y$
13. $x^2 - y^2$
14. $5y^2 - x^2$
15. $\dfrac{3x + y}{3 + y}$
16. $\dfrac{2(x+y) - 2x}{2(y-x)}$

In Exercises 17–26, let $x = -1$ and $y = 5$, and evaluate each expression:

17. $xy^2 - x$
18. $4x^3 - y^2$
19. $\dfrac{2y}{x} - \dfrac{2x}{y}$
20. $3 + 4(x+y)$
21. $3 - 4(x+y)$
22. $1.7 - 5(2x - y)$
23. $\dfrac{1}{x} - \dfrac{1}{y} + \dfrac{2}{xy}$
24. $(2.4 - x)(x - xy)$
25. $\dfrac{y - 4x}{3x - 6xy}$
26. $\dfrac{(y-x)^2 - 4y}{4x^2 + 2}$

In Exercises 27–32, let $x = -3$, $y = 4$, and $z = 6$. Evaluate each expression:

27. $(2xy^2z)^2$
28. $(x^2 - y^2)z$
29. $(y^2 - 2x^2)z^2$
30. $\left(\dfrac{x + 3y}{z}\right)^2$
31. $\dfrac{(7 - x)^2}{z - y}$
32. $(2x + 3y)(y + z)$

In Exercises 33–40, let $x = -1$, $y = 2$, and $z = -3$. Evaluate each expression:

33. $(2x + 6)(3y - 4)$
34. $z^2 - 5yx^2$
35. $(3x + 5)(2y - 1)(5z + 2)$
36. $(3x - 4z)(2x + 3z)$
37. $(x - xy)^2(z - 2x)$
38. $3x^2(y - 3z)^2 - 6x$
39. $(x^2 + y^2)^2$
40. $(3x^2 - z^2)^2$
41. $\dfrac{x^2 + (z - y)^2}{4x^2 + z^2}$
42. $\dfrac{(3x^2 + 2)^2 - y^2}{6 - 3x^2 y^2}$

5.2 Simplifying Algebraic Expressions

Parentheses are often used to clarify the order of operations when the order of operations is complicated or may be ambiguous. Sometimes it is easier to simplify such an expression by first removing the parentheses—before doing the indicated operations. Two rules for removing parentheses are as follows:

Removing Parentheses

1. Parentheses preceded by a plus sign may be removed without changing the signs of the terms within. Think of using the Distributive Property, $a(b + c) = ab + ac$, from Section 5.1 and multiplying each term inside the parentheses by 1. That is,
$$3w + (4x + y) = 3w + 4x + y$$

2. Parentheses preceded by a minus sign may be removed if the signs of *all* the terms within the parentheses are changed; then the minus sign that preceded the parentheses is dropped. Think of using the Distributive Property, $a(b + c) = ab + ac$, from Section 5.1 and multiplying each term inside the parentheses by -1. That is,
$$3w - (4x - y) = 3w - 4x + y$$

(Notice that the sign of the term $4x$ inside the parentheses is not written. It is therefore understood to be plus.)

Example 1 Remove the parentheses from the expression $5x - (-3y + 2z)$.

$$5x - (-3y + 2z) = 5x + 3y - 2z$$

Change the signs of *all* of the terms within parentheses; then drop the minus sign that precedes the parentheses.

Example 2 Remove the parentheses from the expression $7x + (-y + 2z) - (w - 4)$.

$$7x + (-y + 2z) - (w - 4) = 7x - y + 2z - w + 4$$

Drop the plus sign before the first set of parentheses and do not change any of the signs within its parentheses. Change the signs of *all* of the terms within the second set of parentheses; then drop the minus sign that precedes its parentheses.

A **term** is a single number or a product of a number and one or more letters raised to powers. The following are examples of terms:

$$5x, \quad 8x^2, \quad 24y, \quad 15, \quad 3a^2b^3, \quad t$$

The **numerical coefficient** is the numerical factor of a term. The numerical factor of the term $16x^2$ is 16. The numerical coefficient of the term $-6a^2b$ is -6. The numerical coefficient of y is 1.

Terms are parts of an algebraic expression separated by plus and minus signs. For example, $3xy + 2y + 8x^2$ is an expression consisting of three terms:

$$3xy + 2y + 8x^2$$

1st term   2nd term   3rd term

Like Terms

Terms with the same variables with exactly the same exponents are called **like terms**. For example, $4x$ and $11x$ have the same variables and are like terms. The terms $-5x^2y^3$ and $8x^2y^3$ have the same variables with the same exponents and are like terms. The terms $8m$ and $5n$ have different variables, and the terms $7x^2$ and $4x^3$ have different exponents, so these are **unlike terms**.

Example 3 The following table gives examples of like terms and unlike terms:

Like terms	Unlike terms	
a. $2x$ and $3x$	**e.** $2x^2$ and $3x$	different exponents
b. $2ax$ and $5ax$	**f.** $2ax$ and $5bx$	different variables
c. $2x^3$ and $18x^3$	**g.** $2x^3$ and $18x^2$	different exponents
d. $2a^2x^4, a^2x^4,$ and $11a^2x^4$	**h.** $2a^2x^4, 3ax^4,$ and $11a^2x^3$	different exponents

Like terms that occur in a single expression can be combined into one term by combining coefficients (using the Distributive Property from Section 5.1). Thus, $ba + ca = (b + c)a$.

Example 4 Combine the like terms $2x + 3x$.

$$2x + 3x = (2 + 3)x$$
$$= 5x$$

Example 5 Combine the like terms $4ax + 6ax$.

$$4ax + 6ax = (4 + 6)ax$$
$$= 10ax$$

Example 6 Combine the like terms $2a^2x^4 + a^2x^4 + 11a^2x^4$.

$$2a^2x^4 + a^2x^4 + 11a^2x^4 = 2a^2x^4 + 1a^2x^4 + 11a^2x^4$$
$$= (2 + 1 + 11)a^2x^4$$
$$= 14a^2x^4$$

Example 7 Combine the like terms $9a^3b^4 + 2a^2b^3 + 7a^3b^4$.

$$9a^3b^4 + 2a^2b^3 + 7a^3b^4 = (9 + 7)a^3b^4 + 2a^2b^3$$
$$= 16a^3b^4 + 2a^2b^3$$

Some expressions contain parentheses that must be removed before combining like terms. Follow the order of operations.

Example 8 Simplify: $4x - (x - 2)$.

$4x - (x - 2) = 4x - x + 2$ Remove the parentheses by changing the signs of both terms within parentheses; then drop the minus sign that precedes the parentheses.

$ = 3x + 2$ Combine like terms.

Example 9 Simplify: $4x - (-2x - 3y) + 5y$.

$4x - (-2x - 3y) + 5y = 4x + 2x + 3y + 5y$

Remove the parentheses by changing the signs of both terms within parentheses; then drop the minus sign that precedes the parentheses.

$ = 6x + 8y$ Combine like terms.

Example 10 Simplify: $(7 - 2x) + (5x + 1)$.

$(7 - 2x) + (5x + 1) = 7 - 2x + 5x + 1$

Drop the implied plus sign before the first set of parentheses and do not change any of the signs within its parentheses. Drop the plus sign before the second set of parentheses and do not change any of the signs within its parentheses. Remove the parentheses.

$ = 3x + 8$ Combine like terms.

From Section 5.1, $a(b + c) = ab + ac$. The Distributive Property is applied to remove parentheses when a number, a letter, or some product precedes the parentheses.

Example 11 Remove the parentheses from each expression.

a. $3(6x + 5) = (3)(6x) + (3)(5)$ Apply the Distributive Property by multiplying each term within the parentheses by 3.

$ = 18x + 15$ Multiply.

5.2 ♦ Simplifying Algebraic Expressions

b. $-5(2a - 7) = (-5)(2a) - (-5)(7)$ Apply the Distributive Property by multiplying each term within the parentheses by -5.
$= -10a + 35$ Multiply.

c. $\frac{1}{2}(10x^2 + 28x) = \left(\frac{1}{2}\right)(10x^2) + \left(\frac{1}{2}\right)(28x)$ Apply the Distributive Property by multiplying each term within the parentheses by $\frac{1}{2}$.
$= 5x^2 + 14x$ Multiply.

Example 12   Simplify: $3x + 5(x - 3)$.

$3x + 5(x - 3) = 3x + 5x - 15$ Apply the Distributive Property by multiplying each term within the parentheses by 5.
$= 8x - 15$ Combine like terms.

Example 13   Simplify: $4y - 6(-y + 2)$.

$4y - 6(-y + 2) = 4y + 6y - 12$ Apply the Distributive Property by multiplying each term within the parentheses by -6.
$= 10y - 12$ Combine like terms.

EXERCISES 5.2

Remove the parentheses from each expression:

1. $a + (b + c)$
2. $a - (b + c)$
3. $a - (-b - c)$
4. $a - (-b + c)$
5. $a + (-b - c)$
6. $x + (y + z + 3)$
7. $x - (-y + z - 3)$
8. $x - (-y - z + 3)$
9. $x - (y + z + 3)$
10. $x + (-y - z - 3)$
11. $(2x + 4) + (3y + 4r)$
12. $(2x + 4) - (3y + 4r)$
13. $(3x - 5y + 8) + (6z - 2w + 3)$
14. $(4x + 6y - 9) + (-2z + 5w + 3)$
15. $(-5x - 3y - 2) - (6z - 3w - 5)$
16. $(-9x + 6) - (3z + 3w - 1)$
17. $(2x + 3y - 5) + (-z - w + 2) - (-3r + 2s + 7)$
18. $(5x - 11y - 2) - (7z + 3) + (3r + 7) - (4s - 2)$
19. $-(2x - 3y) - (z + 4w) - (4r - s)$
20. $-(3x + y) - (2z + 7w) - (3r - 5s + 2)$

Combine the like terms:

21. $b + b$
22. $4h + 6h$
23. $x^2 + 2x^2 + 3x + 7x$
24. $9k + 3k$
25. $5m - 2m$
26. $4x + 6x - 5x$
27. $3a + 5b - 2a + 7b$
28. $11 + 2m - 6 + m$
29. $6a^2 + a + 1 - 2a$
30. $5x^2 + 3x^2 - 8x^2$
31. $2x^2 + 16x + x^2 - 13x$
32. $13x^2 + 14xy + 6y^2 - 3y^2 + x^2$
33. $1.3x + 5.6x - 13.2x + 4.5x$
34. $2.3x^2 - 4.7x + 0.92x^2 - 2.13x$
35. $\frac{5}{9}x + \frac{1}{4}y + \frac{1}{3}x - \frac{3}{8}y$
36. $\frac{1}{2}x - \frac{2}{3}y - \frac{3}{4}x + \frac{5}{6}y$
37. $4x^2y - 2xy - y^2 - 3x^2 - 2x^2y + 3y^2$
38. $3x^2 - 5x - 2 + 4x^2 + x - 4 + 5x^2 - x + 2$
39. $2x^3 + 4x^2y - 4y^3 + 3x^3 - x^2y + y - y^3$
40. $4x^2 - 5x - 7x^2 - 3x - y^2 + 2x^2 + 3xy - 2y^2$

Simplify by first removing the parentheses and then combining the like terms:

41. $y - (y - 1)$
42. $x + (2x + 1)$
43. $4x + (4 - x)$
44. $5x - (2 - 3x)$
45. $10 - (5 + x)$
46. $x - (-x - y) + 2y$
47. $2y - (7 - y)$
48. $-y - (y + 3)$
49. $(5y + 7) - (y + 2)$
50. $(2x + 4) - (x - 7)$
51. $(4 - 3x) + (3x + 1)$
52. $10 - (y + 6) + (3y - 2)$
53. $-5y + 9 - (-5y + 3)$
54. $0.5x + (x - 1) - (0.2x + 8)$

55. $0.2x - (0.2x - 28)$
56. $(0.3x - 0.5) - (-2.3x + 1.4)$
57. $\left(\frac{1}{2}x - \frac{2}{3}\right) - \left(2 - \frac{3}{4}x\right)$
58. $\left(\frac{3}{4}x - 1\right) + \left(-\frac{1}{2}x - \frac{2}{3}\right)$
59. $4(3x + 9y)$
60. $6(-2a + 8b)$
61. $-12(3x^2 - 4y^2)$
62. $-3(-a^2 - 4a)$
63. $(5x + 13) - 3(x - 2)$
64. $(7x + 8) - 5(x - 6)$
65. $-9y - 0.5(8 - y)$
66. $12(x + 1) - 3(4x - 2)$
67. $2y - 2(y + 21)$
68. $3x - 3(6 - x)$
69. $6n - (2n - 8)$
70. $14x - 8(2x - 8)$
71. $0.8x - (-x + 7)$
72. $-(x - 3) - 3(4 + x)$
73. $4(2 - 3n) - 2(5 - 3n)$
74. $(x + 4) - 2(2x - 7)$
75. $\frac{2}{3}(6x - 9) - \frac{3}{4}(12x - 16)$
76. $13\left(7x - 2\frac{1}{2}\right) - 9\left(8x + 9\frac{2}{3}\right)$
77. $0.45(x + 3) - 0.75(2x + 13)$
78. $0.6(0.5x^2 - 0.9x) + 0.4(x^2 + 0.4x)$

5.3 Addition and Subtraction of Polynomials

A **monomial**, or *term*, is any algebraic expression that contains only products of numbers and variables, which have nonnegative integer exponents. The following expressions are examples of monomials:

$$2x, \quad 5, \quad -3b, \quad \frac{3}{4}a^2bw, \quad \sqrt{315}\,mn$$

A **polynomial** is either a monomial or the sum or difference of unlike monomials. We consider two special types of polynomials. A **binomial** is a polynomial that is the sum or difference of two unlike monomials. A **trinomial** is the sum or difference of three unlike monomials.

The following table shows examples of monomials, binomials, and trinomials:

Monomials	$3x$	$4ab^2$	$-15x^2y^3$	one term
Binomials	$a + b$	$5a^2 + 3$	$7xy^2 - 4x^2y$	two terms
Trinomials	$a + 3b - 2c$	$8x^2 - 3x + 12$	$2a^3b + 3a^2b^2 + ab^3$	three terms

Expressions that contain variables in the denominator are *not* polynomials. For example,

$$\frac{3}{4x}, \quad \frac{8x}{3x - 5}, \quad \text{and} \quad \frac{33}{4x^2} + \frac{8}{x - 1}$$

are *not* polynomials.

> The **degree of a monomial in one variable** is the same as the exponent of the variable.

Example 1 Find the degree of each monomial: **a.** $-7m$, **b.** $6x^2$, **c.** $5y^3$, **d.** 5.

a. $-7m$ has degree 1.   The exponent of m is 1.
b. $6x^2$ has degree 2.   The exponent of x is 2.
c. $5y^3$ has degree 3.   The exponent of y is 3.
d. 5 has degree 0.   5 may be written as $5x^0$.

♦

> The **degree of a polynomial in one variable** is the same as the degree of the highest-degree monomial contained in the polynomial.

5.3 ♦ Addition and Subtraction of Polynomials

Example 2 Find the degree of each polynomial: **a.** $5x^4 + x^2$ and **b.** $6y^3 + 4y^2 - y + 1$.

a. $5x^4 + x^2$ has degree 4, the degree of the highest-degree monomial.

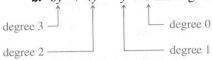

b. $6y^3 + 4y^2 - y + 1$ has degree 3, the degree of the highest-degree monomial.

A polynomial is in **decreasing order** if each term is of some degree less than the degree of the preceding term. The following polynomial is written in decreasing order:

$$4x^5 - 3x^4 - 4x^2 - x + 5$$

exponents decrease

A polynomial is in **increasing order** if each term is of some degree larger than the degree of the preceding term. The following polynomial is written in increasing order:

$$5 - x - 4x^2 - 3x^4 + 4x^5$$

exponents increase

Adding Polynomials

To add polynomials, add their like terms.

Example 3 Add: $(3x + 4) + (5x - 7)$.

$(3x + 4) + (5x - 7) = (3x + 5x) + [4 + (-7)]$ Add the like terms.

$= 8x - 3$

Example 4 Add: $(5x^2 + 6x - 8) + (4x^2 - 3)$.

$(5x^2 + 6x - 8) + (4x^2 - 3) = (5x^2 + 4x^2) + 6x + [(-8) + (-3)]$ Add the like terms.

$= 9x^2 + 6x - 11$

Example 5 Add: $(5a + 2b - 3c + 4) + (-2a - 4b + 7c - 3) + (6a + b - 4c)$.

$(5a + 2b - 3c + 4) + (-2a - 4b + 7c - 3) + (6a + b - 4c)$

$= [5a + (-2a) + 6a] + [2b + (-4b) + b] + [(-3c) + 7c + (-4c)] + [4 + (-3)]$ Add the like terms.

$= 9a - 1b + 0c + 1$

$= 9a - b + 1$

We sometimes find it easier to find the sum of polynomials by writing the like terms in columns and then adding the columns, as shown in the next example. Here, the polynomials are also written in decreasing order, which is also common as an organizational aid.

Example 6 Add: $(2x^2 - 5x) + (3x^2 + 2x - 4) + (-4x^2 + 5)$.

$$\begin{array}{r} 2x^2 - 5x \\ 3x^2 + 2x - 4 \\ -4x^2 + 5 \\ \hline x^2 - 3x + 1 \end{array}$$ Write the like terms in columns.

Add the like terms.

Subtracting Polynomials

To subtract two polynomials, change all the signs of the terms of the second polynomial and then add the two resulting polynomials.

Example 7 Subtract: $(5a - 9b) - (2a - 4b)$.

$$(5a - 9b) - (2a - 4b) = (5a - 9b) + (-2a + 4b)$$ Change all the signs of the terms of the second polynomial and add.
$$= [5a + (-2a)] + [(-9b) + 4b]$$ Add the like terms.
$$= 3a - 5b$$

Example 8 Subtract: $(5x^2 - 3x - 4) - (2x^2 - 5x + 6)$.

$$(5x^2 - 3x - 4) - (2x^2 - 5x + 6) = (5x^2 - 3x - 4) + (-2x^2 + 5x - 6)$$
Change all the signs of the terms of the second polynomial and add.
$$= [5x^2 + (-2x^2)] + [(-3x) + 5x] + [(-4) + (-6)]$$
Add the like terms.
$$= 3x^2 + 2x - 10$$

Subtraction can also be done in columns; the subtraction in long division of one polynomial by another polynomial is usually done using columns.

Example 9 Find the difference: $(5x^2 - 3x - 4) - (2x^2 - 7x + 5)$.

Subtract: $\begin{array}{r} 5x^2 - 3x - 4 \\ 2x^2 - 7x + 5 \end{array}$ → Add: $\begin{array}{r} 5x^2 - 3x - 4 \\ -2x^2 + 7x - 5 \\ \hline 3x^2 + 4x - 9 \end{array}$

The arrow indicates the change of the subtraction problem to an addition problem. Do this by changing the signs of each of the terms in the second polynomial.

EXERCISES 5.3

Classify each expression as a monomial, a binomial, or a trinomial:

1. $3m + 27$
2. $4a^2bc^3$
3. $-5x - 7y$
4. $2x^2 + 7y + 3z^2$
5. $-5xy$
6. $a + b + c$
7. $2x + 3y - 5z$
8. $2a - 3b^3$
9. $-42x^3 - y^4$
10. $15x^{14} - 3x^2 + 5x$

Rearrange each polynomial in decreasing order and state its degree:

11. $1 - x + x^2$
12. $2x^3 - 3x^4 + 2x$
13. $4x + 7x^2 - 1$
14. $y^3 - 1 + y^2$
15. $-4x^2 + 5x^3 - 2$
16. $3x^3 + 6 - 2x + 4x^5$
17. $7 - 3y + 4y^3 - 6y^2$
18. $1 - x^5$
19. $x^3 - 4x^4 + 2x^2 - 7x^5 + 5x - 3$
20. $360x^2 - 720x - 120x^3 + 30x^4 + 1 - 6x^5 + x^6$

Add the following polynomials:

21. $(5a^2 - 7a + 5) + (2a^2 - 3a - 4)$
22. $(2b - 5) + 3b + (-4b - 7)$
23. $(6x^2 - 7x + 5) + (3x^2 + 2x - 5)$
24. $(4x - 7y - z) + (2x - 5y - 3z) + (-3x - 6y - 4z)$
25. $(2a^3 - a) + (4a^2 + 7a) + (7a^3 - a - 5)$
26. $(5y - 7x + 4z) + (3z - 6y + 2x) + (13y + 7z - 6x)$
27. $(3x^2 + 4x - 5) + (-x^2 - 2x + 2) + (-2x^2 + 2x + 7)$
28. $(-x^2 + 6x - 8) + (10x^2 - 13x + 3) + (-12x^2 - 14x + 3)$

29. $(3x^2 + 7) + (6x - 7) + (2x^2 + 5x - 13) + (7x - 9)$
30. $(5x + 3y) + (-3x - 3y) + (-x - 6y) + (3x - 4y)$
31. $(5x^3 - 11x - 1) + (11x^2 + 3) + (3x + 7) + (2x^2 - 2)$
32. $(3x^4 - 5x^2 + 4) + (6x^4 - 6x^2 + 1) + (2x^4 - 7x^2)$

33. $\quad 4y^2 - 3y - 15$
$\quad7y^2 - 6y + 8$
$\quad\underline{-3y^2 + 4y + 13}$

34. $129a - 13b - 56c$
$-13a - 52b + 21c$
$\underline{44a + 11c}$

35. $3a^3 + 2a^2 + 5$
$a^3 - 7a - 2$
$-5a^2 + 4a$
$\underline{- 2a - 3}$

36. $4x^2 - 3xy + 5x - 6y$
$9xy - 4y^2 + 6y - 4$
$x^2 + y^2 - x + 3$
$\underline{-8x^2 + xy - 2y^2 + 3x - 10}$

Find each difference:

37. $(3x^2 + 4x + 7) - (x^2 - 2x + 5)$
38. $(2x^2 + 5x - 9) - (3x^2 - 4x + 7)$
39. $(3x^2 - 5x + 4) - (6x^2 - 7x + 2)$
40. $(1 - 3x - 2x^2) - (-1 - 5x + x^2)$
41. $(3a - 4b) - (2a - 7b)$
42. $(-13x^2 - 3y^2 - 4y) - (-5x - 4y + 5y^2)$
43. $(7a - 4b) - (3x - 4y)$
44. $(-16y^3 - 42y^2 - 3y - 5) - (12y^2 - 4y + 7)$
45. $(12x^2 - 3x - 2) - (11x^2 - 7)$
46. $(14z^3 - 6y^3) - (2y^2 + 4z^3)$
47. $(20w^2 - 17w - 6) - (13w^2 + 7w)$
48. $(y^2 - 2y + 1) - (2y^2 + 3y + 5)$
49. $(2x^2 - 5x - 2) - (x^2 - x + 8)$
50. $(3 - 5z + 3z^2) - (14 + z - 2z^2)$
51. Subtract $4x^2 + 2x - 7$ from $8x^2 - 2x + 5$.
52. Subtract $-6x^2 - 3x + 4$ from $2x^2 - 6x - 2$.
53. Subtract $9x^2 + 6$ from $3x^2 + 2x - 4$.
54. Subtract $-4x^2 - 6x + 2$ from $4x^2 + 6$.

Subtract the following polynomials:

55. $\quad 2x^3 + 4x - 1$
$\quad\underline{x^3 + x + 2}$

56. $\quad7x^4 + 3x^3 + 5x$
$\quad\underline{-2x^4 + x^3 - 6x + 6}$

57. $12x^5 - 13x^4 + 7x^2$
$\underline{4x^5 + 5x^4 + 2x^2 - 1}$

58. $8x^3 + 6x^2 - 15x + 7$
$\underline{14x^3 + 2x^2 + 9x - 1}$

5.4 Multiplication of Monomials

Earlier, we used exponents to write products of repeated number factors as follows:

$6^2 = 6 \cdot 6 = 36$
$5^3 = 5 \cdot 5 \cdot 5 = 125$
$2^4 = 2 \cdot 2 \cdot 2 \cdot 2 = 16$

A power with a variable base may also be used as follows:

$x^3 = x \cdot x \cdot x$
$c^4 = c \cdot c \cdot c \cdot c$

In the expression 2^4, the number 2 is called the *base*, and 4 is called the *exponent*. The expression may also be called *the fourth power of 2*. In x^3, the letter x is called the base and 3 is called the exponent.

Example 1 Multiply:

a. $x^3 \cdot x^2 = (x \cdot x \cdot x)(x \cdot x) = x^5$
b. $a^4 \cdot a^2 = (a \cdot a \cdot a \cdot a)(a \cdot a) = a^6$
c. $c^5 \cdot c^3 = (c \cdot c \cdot c \cdot c \cdot c)(c \cdot c \cdot c) = c^8$

Rule 1 for Exponents: Multiplying Powers

$$a^m \cdot a^n = a^{m+n}$$

That is, to multiply powers with the same base, add the exponents.

To multiply two monomials, multiply their numerical coefficients and combine their variable factors according to Rule 1 for exponents.

Example 2 Multiply: $(2x^3)(5x^4)$.

$(2x^3)(5x^4) = 2 \cdot 5 \cdot x^3 \cdot x^4 = 10x^{3+4}$ Add the exponents.
$= 10x^7$

Example 3 Multiply: $(3a)(-15a^2)(4a^4b^2)$.

$(3a)(-15a^2)(4a^4b^2) = (3)(-15)(4)(a)(a^2)(a^4)(b^2)$
$= -180a^7b^2$ Add the exponents.

A *special note* about the meaning of $-x^2$ is needed here. Note that x is squared, not $-x$. That is,

$-x^2 = -(x \cdot x)$

If $-x$ is squared, we have $(-x)^2 = (-x)(-x) = x^2$.

Rule 2 for Exponents: Raising a Power to a Power

$$(a^m)^n = a^{mn}$$

That is, to raise a power to a power, multiply the exponents.

Example 4 Find $(x^3)^5$.

By Rule 1:

$(x^3)^5 = x^3 \cdot x^3 \cdot x^3 \cdot x^3 \cdot x^3 = x^{15}$

By Rule 2:

$(x^3)^5 = x^{3 \cdot 5} = x^{15}$ Multiply the exponents.

Example 5 Find $(x^5)^9$.

$(x^5)^9 = x^{45}$ Multiply the exponents.

Rule 3 for Exponents: Raising a Product to a Power

$$(ab)^m = a^m b^m$$

That is, to raise a product to a power, raise each factor to that same power.

Example 6 Find $(xy)^3$.

$(xy)^3 = x^3 y^3$

Example 7 Find $(2x^3)^2$.

$(2x^3)^2 = 2^2(x^3)^2$
$= 4x^6$ Square each factor.

Example 8 Find $(-3x^4)^5$.

$(-3x^4)^5 = (-3)^5(x^4)^5$
$= -243x^{20}$ Raise each factor to the fifth power.

5.5 ♦ Multiplication of Polynomials

Example 9 Find $(ab^2c^3)^4$.

$$(ab^2c^3)^4 = a^4(b^2)^4(c^3)^4$$
$$= a^4b^8c^{12} \quad \text{Raise each factor to the fourth power.}$$

Example 10 Find $(2a^2bc^3)^2$.

$$(2a^2bc^3)^2 = 2^2(a^2)^2(b)^2(c^3)^2$$
$$= 4a^4b^2c^6$$

Example 11 Evaluate $(4a^2)(-5ab^2)$ when $a = 2$ and $b = 3$.

$$(4a^2)(-5ab^2) = 4(-5)(a^2)(a)(b^2) \quad \text{First, multiply.}$$
$$= -20a^3b^2$$
$$= -20(2)^3(3)^2 \quad \text{Substitute.}$$
$$= -20(8)(9)$$
$$= -1440$$

EXERCISES 5.4

Find each product:

1. $(3a)(-5)$
2. $(7x)(2x)$
3. $(4a^2)(7a)$
4. $(4x)(6x^2)$
5. $(-9m^2)(-6m^2)$
6. $(5x^2)(-8x^3)$
7. $(8a^6)(4a^2)$
8. $(-4y^4)(-9y^3)$
9. $(13p)(-2pq)$
10. $(4ab)(10a)$
11. $(6n)(5n^2m)$
12. $(-9ab^2)(6a^2b^3)$
13. $(-42a)\left(-\frac{1}{2}a^3b\right)$
14. $(28m^3)\left(\frac{1}{4}m^2\right)$
15. $\left(\frac{2}{3}x^2y^2\right)\left(\frac{9}{16}xy^2\right)$
16. $\left(-\frac{5}{6}a^2b^6\right)\left(\frac{9}{20}a^5b^4\right)$
17. $(8a^2bc)(3ab^3c^2)$
18. $(-4xy^2z^3)(4x^5z^3)$
19. $\left(\frac{2}{3}x^2y\right)\left(\frac{9}{32}xy^4z^3\right)$
20. $\left(\frac{3}{5}m^4n^7\right)\left(\frac{20}{9}m^2nq^3\right)$
21. $(32.6mnp^2)(-11.4m^2n)$
22. $(5.6a^2b^3c)(6.5a^4b^5)$
23. $(5a)(-17a^2)(3a^3b)$
24. $(-4a^2b)(-5ab^3)(-2a^4)$

Use the rules for exponents to simplify:

25. $(x^3)^2$
26. $(xy)^4$
27. $(x^4)^6$
28. $(2x^2)^5$
29. $(-3x^4)^2$
30. $(5x^2)^3$
31. $(-x^3)^3$
32. $(-x^2)^4$
33. $(x^2 \cdot x^3)^2$
34. $(3x)^4$
35. $(x^5)^6$
36. $(-3xy)^3$
37. $(-5x^3y^2)^2$
38. $(-x^2y^4)^5$
39. $(15m^2)^2$
40. $(-7w^2)^3$
41. $(25n^4)^3$
42. $(36a^5)^3$
43. $(3x^2 \cdot x^4)^2$
44. $(16x^4 \cdot x^5)^3$
45. $(2x^3y^4z)^3$
46. $(-4a^2b^3c^4)^4$
47. $(-2h^3k^6m^2)^5$
48. $(4p^5q^7r)^3$

Evaluate each expression when $a = 2$ and $b = -3$:

49. $(4a)(17b)$
50. $(3a)(-5b^2)$
51. $(9a^2)(-2a)$
52. $(a^2)(ab)$
53. $(41a^3)(-2b^3)$
54. $(ab)^2$
55. $(a^2)^2$
56. $(3a)^2$
57. $(4b)^3$
58. $(-2ab^2)^2$
59. $(5a^2b^3)^2$
60. $(-7ab)^2$
61. $(5ab)(a^2b^2)$
62. $(-3ab)^3$
63. $(9a)(ab^2)$
64. $(-2a^2)(6ab)$
65. $-a^2b^4$
66. $(-ab^2)^2$
67. $(-a)^4$
68. $-b^4$

5.5 Multiplication of Polynomials

To multiply a polynomial by a monomial, multiply each term of the polynomial by the monomial, and then add the products as shown in the following examples.

Example 1 Multiply: $3a(a^2 - 2a + 1)$.

$$3a(a^2 - 2a + 1) = 3a(a^2) + 3a(-2a) + 3a(1) \quad \text{Multiply each term of the}$$
$$= 3a^3 - 6a^2 + 3a \quad \text{polynomial by } 3a.$$

Example 2   Multiply: $(-5a^3b)(3a^2 - 4ab + 5b^3)$.

$(-5a^3b)(3a^2 - 4ab + 5b^3)$
$= (-5a^3b)(3a^2) + (-5a^3b)(-4ab) + (-5a^3b)(5b^3)$ Multiply each term of the polynomial by $-5a^3b$.
$= -15a^5b + 20a^4b^2 - 25a^3b^4$

To multiply a polynomial by another polynomial, multiply each term of the first polynomial by each term of the second polynomial. Then add the products. Arrange the work as shown in Example 3.

Example 3   Multiply: $(5x - 3)(2x + 4)$.

STEP 1 Write each polynomial in decreasing order, one under the other.

$5x - 3$
$2x + 4$

STEP 2 Multiply each term of the upper polynomial by the first term in the lower one.

$10x^2 - 6x$ ← $2x(5x - 3)$

STEP 3 Multiply each term of the upper polynomial by the second term in the lower one. Place like terms in the same columns.

$20x - 12$ ← $4(5x - 3)$

STEP 4 Add the like terms.

$10x^2 + 14x - 12$ Add.

To multiply two binomials, such as $(x + 3)(2x + 5)$, you may think of finding the area of a rectangle with sides $(x + 3)$ and $(2x + 5)$ as shown in Figure 5.1.

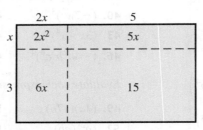

Figure 5.1

Note that the total area is $2x^2 + 6x + 5x + 15$ or $2x^2 + 11x + 15$. Using the four-step process above, we have

STEP 1 $2x + 5$
$x + 3$
STEP 2 $2x^2 + 5x$ ← $x(2x + 5)$
STEP 3 $6x + 15$ ← $3(2x + 5)$
STEP 4 $2x^2 + 11x + 15$ Add.

Example 4   Multiply: $(x + 3)(x^2 + 2x - 4)$.

STEP 1 $x^2 + 2x - 4$
$x + 3$
STEP 2 $x^3 + 2x^2 - 4x$ ← $x(x^2 + 2x - 4)$
STEP 3 $3x^2 + 6x - 12$ ← $3(x^2 + 2x - 4)$
STEP 4 $x^3 + 5x^2 + 2x - 12$ Add.

Example 5 Multiply: $(3a + b)(c + 2d)$.

STEP 1 $3a + b$
             $c + 2d$

STEP 2 $3ac + bc$ ← $c(3a + b)$

STEP 3 $6ad + 2bd$ ← $2d(3a + b)$

STEP 4 $3ac + bc + 6ad + 2bd$ Add.

Note that there are no like terms in Steps 2 and 3. ◆

EXERCISES 5.5

Find each product.

1. $4(a + 6)$
2. $3(a^2 - 5)$
3. $-6(3x^2 + 2y)$
4. $-5(8x - 4y^2)$
5. $a(4x^2 - 6y + 1)$
6. $c(2a + b + 3c)$
7. $x(3x^2 - 2x + 5)$
8. $y(3x + 2y^2 + 4y)$
9. $2a(3a^2 + 6a - 10)$
10. $5x(8x^2 - x + 5)$
11. $-3x(4x^2 - 7x - 2)$
12. $-6x(8x^2 + 5x - 9)$
13. $4x(-7x^2 - 3y + 2xy)$
14. $7a(2a + 3b - 4ab)$
15. $3xy(x^2y - xy^2 + 4xy)$
16. $-2ab(3a^2 + 4ab - 2b^2)$
17. $-6x^3(1 - 6x^2 + 9x^4)$
18. $5x^4(2x^3 + 8x^2 - 1)$
19. $5ab^2(a^3 - b^3 - ab)$
20. $7w^2y(w^2 - 4y^2 + 6w^2y^3)$
21. $\frac{2}{3}m(14n - 12m)$
22. $\frac{1}{2}a^2b(8ab^2 - 2a^2b)$
23. $\frac{4}{7}yz^3\left(28y - \frac{2}{5}z\right)$
24. $-\frac{1}{8}rs(3s - 16t)$
25. $-4a(1.3a^5 + 2.5a^2 + 1)$
26. $1.28m(2.3m^2 + 4.7n^2)$
27. $417a(3.2a^2 + 4a)$
28. $1.2m^2n^3(9.7m + 6.5mn - 13n^2)$
29. $4x^2y(6x^2 - 4xy + 5y^2)$
30. $x^2y^3z(x^4 - 3x^2y - 3yz + 4z^2)$
31. $\frac{2}{3}ab^3\left(\frac{3}{4}a^2 - \frac{1}{2}ab^2 + \frac{5}{6}b^3\right)$
32. $-\frac{5}{9}a^2b^4\left(\frac{3}{7}a^3b^2 - \frac{3}{5}ab - \frac{15}{16}b^4\right)$
33. $3x(x - 4) + 2x(1 - 5x) - 6x(2x - 3)$
34. $x(x - 2) - 3x(x + 8) - 2(x^2 + 3x - 5)$
35. $xy(3x + 2xy - y^2) - 2xy^2(2x - xy + 3y)$
36. $ab^2(2a - 3a^2b + b) - a^2b(1 + 2ab^2 - 4b)$
37. $(x + 1)(x + 6)$
38. $(x + 10)(x - 3)$
39. $(x + 7)(x - 2)$
40. $(x - 3)(x - 7)$
41. $(x - 5)(x - 8)$
42. $(x + 9)(x + 4)$
43. $(3a - 5)(a - 4)$
44. $(5x - 2)(3x - 4)$
45. $(6a + 4)(2a - 3)$
46. $(3x + 5)(6x - 7)$
47. $(4a + 8)(6a + 9)$
48. $(5x - 4)(5x - 4)$
49. $(3x - 2y)(5x + 2y)$
50. $(4x - 6y)(6x + 9y)$
51. $(2x - 3)(2x - 3)$
52. $(5m - 9)(5m + 9)$
53. $(2c - 5d)(2c + 5d)$
54. $(3a + 2b)(2a - 3b)$
55. $(-7m - 3)(-13m + 1)$
56. $(w - r)(w - s)$
57. $(x^5 - x^2)(x^3 - 1)$
58. $(7w^4 - 6r^2)(7w^4 + 5r^2)$
59. $(2y^2 - 4y - 8)(5y - 2)$
60. $(m^2 + 2m + 4)(m - 2)$
61. $(4x - 2y - 13)(6x + 3y)$
62. $(4y - 3z)(2y^2 - 5yz + 6z^2)$
63. $(g + h - 6)(g - h + 3)$
64. $(2x - 3y + 4)(4x - 5y - 2)$
65. $(8x - x^3 + 2x^4 - 1)(x^2 + 2 + 5x^3)$
66. $(y^5 - y^4 + y^3 - y^2 + y - 1)(y + 1)$

5.6 Division by a Monomial

To divide a monomial by a monomial, first write the quotient in fraction form. Then factor both numerator and denominator into prime factors. Reduce to lowest terms by dividing both the numerator and the denominator by their common factors. The remaining factors in the numerator and the denominator give the quotient.

Example 1   Divide: $6a^2 \div 2a$.

$$6a^2 \div 2a = \frac{6a^2}{2a}$$

$$= \frac{\overset{1}{\cancel{2}} \cdot 3 \cdot \overset{1}{\cancel{a}} \cdot a}{\underset{1}{\cancel{2}} \cdot \underset{1}{\cancel{a}}} = \frac{3a}{1} = 3a$$

Example 2   Divide: $4a \div 28a^3$.

$$4a \div 28a^3 = \frac{4a}{28a^3}$$

$$= \frac{\overset{1}{\cancel{2}} \cdot \overset{1}{\cancel{2}} \cdot \overset{1}{\cancel{a}}}{\underset{1}{\cancel{2}} \cdot \underset{1}{\cancel{2}} \cdot 7 \cdot \underset{1}{\cancel{a}} \cdot a \cdot a} = \frac{1}{7a^2}$$

Example 3   Divide: $\dfrac{6a^2bc^3}{10ab^2c}$.

$$\frac{6a^2bc^3}{10ab^2c} = \frac{\cancel{2} \cdot 3 \cdot \cancel{a} \cdot a \cdot \cancel{b} \cdot \cancel{c} \cdot c \cdot c}{\cancel{2} \cdot 5 \cdot \cancel{a} \cdot \cancel{b} \cdot b \cdot \cancel{c}} = \frac{3ac^2}{5b}$$

NOTE: Since division by zero is undefined, we will assume that there are no zero denominators here and for the remainder of this chapter.

To divide a polynomial by a monomial, divide each term of the polynomial by the monomial. Simplify each result when possible.

Example 4   Divide: $\dfrac{15x^3 - 3x^2 + 21x}{3x}$.

$$\frac{15x^3 - 3x^2 + 21x}{3x} = \frac{15x^3}{3x} - \frac{3x^2}{3x} + \frac{21x}{3x} \qquad \text{Divide each term of the polynomial by } 3x.$$

$$= \frac{\cancel{3} \cdot 5 \cdot \cancel{x} \cdot x \cdot x}{\cancel{3} \cdot \cancel{x}} - \frac{\cancel{3} \cdot \cancel{x} \cdot x}{\cancel{3} \cdot \cancel{x}} + \frac{\cancel{3} \cdot 7 \cdot \cancel{x}}{\cancel{3} \cdot \cancel{x}}$$

$$= 5x^2 - x + 7$$

Example 5   Divide: $(30d^4y - 28d^2y^2 + 12dy^2) \div (-6dy^2)$.

$$\frac{30d^4y - 28d^2y^2 + 12dy^2}{-6dy^2}$$

$$= \frac{30d^4y}{-6dy^2} + \frac{-28d^2y^2}{-6dy^2} + \frac{12dy^2}{-6dy^2} \qquad \text{Divide each term of the polynomial by } -6dy^2.$$

$$= \frac{\cancel{6} \cdot 5 \cdot \cancel{d} \cdot d \cdot d \cdot d \cdot \cancel{y}}{-\cancel{6} \cdot \cancel{d} \cdot \cancel{y} \cdot y} + \frac{-\cancel{2} \cdot 14 \cdot \cancel{d} \cdot d \cdot \cancel{y} \cdot \cancel{y}}{-\cancel{2} \cdot 3 \cdot \cancel{d} \cdot \cancel{y} \cdot \cancel{y}} + \frac{\cancel{6} \cdot 2 \cdot \cancel{d} \cdot \cancel{y} \cdot \cancel{y}}{-\cancel{6} \cdot \cancel{d} \cdot \cancel{y} \cdot \cancel{y}}$$

$$= \frac{5d^3}{-y} + \frac{14d}{3} + \frac{2}{-1}$$

$$= -\frac{5d^3}{y} + \frac{14d}{3} - 2$$

EXERCISES 5.6

Divide:

1. $\dfrac{9x^5}{3x^3}$
2. $\dfrac{15x^6}{5x^4}$
3. $\dfrac{18x^{12}}{12x^4}$
4. $\dfrac{20x^7}{4x^5}$
5. $\dfrac{18x^3}{3x^5}$
6. $\dfrac{4x^2}{12x^6}$
7. $\dfrac{8x^2}{12x}$
8. $\dfrac{-6x^3}{2x}$
9. $\dfrac{x^2y}{xy}$
10. $\dfrac{xy^2}{x^2y}$
11. $(15x) \div (6x)$
12. $(14x^2) \div (2x^3)$
13. $(15a^3b) \div (3ab^2)$
14. $(-13a^5) \div (7a)$
15. $(16m^2n) \div (2m^3n^2)$
16. $(-108m^4n) \div (27m^3n)$
17. $0 \div (113w^2r^3)$
18. $(-148wr^3) \div (148wr^3)$
19. $(207p^3) \div (9p)$
20. $(42x^2y^3) \div (-14x^2y^4)$
21. $\dfrac{92mn}{-46mn}$
22. $\dfrac{-132rs^3}{-33r^2s^2}$
23. $\dfrac{252}{7r^2}$
24. $\dfrac{118a^3}{-2a^4}$
25. $\dfrac{92x^3y}{-28xy^3}$
26. $\dfrac{45x^6}{-72x^3y^2}$
27. $\dfrac{-16a^5b^2}{-14a^8b^4}$
28. $\dfrac{35a^3b^4c^6}{63a^3b^2c^8}$
29. $(-72x^3yz^4) \div (-162xy^2)$
30. $(-144x^2z^3) \div (216x^5y^2z)$
31. $(4x^2 - 8x + 6) \div 2$
32. $(18y^3 + 12y^2 + 6y) \div 6$
33. $(x^4 + x^3 + x^2) \div x^2$
34. $(20r^2 - 16r - 12) \div (-4)$
35. $(ax - ay - az) \div a$
36. $(14c^3 - 28c^2 - 2c) \div (-2c)$
37. $\dfrac{24a^4 - 16a^2 - 8a}{8}$
38. $\dfrac{88x^5 - 110x^4 + 11x^3}{11x^3}$
39. $\dfrac{b^{12} - b^9 - b^6}{b^3}$
40. $\dfrac{27a^3 - 18a^2 + 36a}{-9a}$
41. $\dfrac{bx^4 - bx^3 + bx^2 - 4bx}{-bx}$
42. $\dfrac{4a^5 - 32a^4 + 8a^3 - 12a^2}{-4a^2}$
43. $\dfrac{24x^2y^3 + 12x^3y^4 - 6xy^3}{2xy^3}$
44. $\dfrac{3.5ax^2 - 0.42a^2x + 14a^2x^2}{0.07ax}$
45. $\dfrac{224x^4y^2z^3 - 168x^3y^3z^4 - 112xy^4z^2}{28xy^2z^2}$
46. $\dfrac{55w^2 - 11w - 33}{11w}$
47. $\dfrac{24y^5 - 18y^3 - 12y}{6y^2}$
48. $\dfrac{3a^2b + 4a^2b^2 - 6ab^2}{2ab^2}$
49. $\dfrac{1 - 6x^2 - 4x^4}{2x^2}$
50. $\dfrac{18w^4r^4 + 27w^3r^3 - 36w^2r^2}{9w^3r^3}$

5.7 Division by a Polynomial

Dividing a polynomial by a polynomial of more than one term is very similar to long division in arithmetic. We use the same names, as shown below.

$$
\begin{array}{r}
23 \\
25\overline{)593} \\
50 \\
\hline
93 \\
75 \\
\hline
18
\end{array}
\quad
\begin{array}{l}
\leftarrow \text{quotient} \\
\leftarrow \text{dividend} \\
\\
\\
\\
\leftarrow \text{remainder}
\end{array}
$$

divisor $\rightarrow 25\overline{)593}$

$$
\begin{array}{r}
x + 3 \\
x + 1\overline{)x^2 + 4x + 4} \\
x^2 + x \\
\hline
3x + 4 \\
3x + 3 \\
\hline
1
\end{array}
\quad
\begin{array}{l}
\leftarrow \text{quotient} \\
\leftarrow \text{dividend} \\
\\
\\
\\
\leftarrow \text{remainder}
\end{array}
$$

divisor $\rightarrow x + 1\overline{)x^2 + 4x + 4}$

As a check, you may use the relationship *dividend = divisor × quotient + remainder*. A similar procedure is followed in both cases. Compare the solutions of the two problems in the example that follows.

CHAPTER 5 ♦ An Introduction to Algebra

Example 1   Divide $337 \div 16$ (arithmetic) and $(2x^2 + x - 14) \div (x + 3)$ (algebra).

Arithmetic	Line	Algebra
$\phantom{16\overline{)}}21 \text{ r } 1$	**1.**	$\phantom{x+3\overline{)}2x^2+\ }2x - 5 \text{ r } 1$
$16\overline{)337}$	**2.**	$x + 3\overline{)2x^2 + x - 14}$
$\underline{32}$	**3.**	$\phantom{x+3\overline{)}}\underline{2x^2 + 6x}$
17	**4.**	$\phantom{x+3\overline{)2x^2+}}-5x - 14$
$\underline{16}$	**5.**	$\phantom{x+3\overline{)2x^2+}}\underline{-5x - 15}$
1	**6.**	$\phantom{x+3\overline{)2x^2+\ -5x-1}}1$

STEP

1. Divide 16 into 33. It will go at most 2 times, so write the 2 above the line over the dividend (337).

2. Multiply the divisor (16) by 2. Write the result (32) under the first two digits of 337.

3. Subtract 32 from 33, leaving 1. Bring down the 7, giving 17.

4. Divide 16 into 17. It will go at most 1 time, so write the 1 to the right of the 2 above the dividend.

5. Multiply the divisor (16) by 1. Write the result (16) under 17.

6. Subtract 16 from 17, leaving 1. The remainder (1) is less than 16, so the problem is finished. The quotient is 21 with remainder 1.

STEP

1. Divide the first term of the dividend, $2x^2$, by the first term of the divisor, x. Write the result, $2x$, above the dividend in line **1**.

2. Multiply the divisor, $x + 3$, by $2x$ and write the result, $2x^2 + 6x$, in line **3** as shown.

3. Subtract this result, $2x^2 + 6x$, from the first two terms of the dividend, leaving $-5x$. Bring down the last term of the dividend, -14, as shown in line **4**.

4. Divide the first term in line **4**, $-5x$, by the first term of the divisor, x. Write the result, -5, to the right of $2x$ above the dividend in line **1**.

5. Multiply the divisor, $x + 3$, by -5 and write the result, $-5x - 15$, in line **5** as shown.

6. Subtract line **5** from line **4**, leaving 1 as the remainder. Since the remainder, 1, is of a lower degree than the divisor, $x + 3$, the problem is finished. The quotient is $2x - 5$ with remainder 1. ♦

> The dividend and the divisor should always be arranged in decreasing order of degree. Any missing powers of x should be filled in by using zeros as coefficients. For example, $x^3 - 1 = x^3 + 0x^2 + 0x - 1$.

Example 2   Divide $8x^3 - 22x^2 + 27x - 18$ by $2x - 3$.

	Line
$\phantom{2x-3\overline{)}}4x^2 - 5x + 6$	**1.**
$2x - 3\overline{)8x^3 - 22x^2 + 27x - 18}$	**2.**
$\phantom{2x-3\overline{)}}\underline{8x^3 - 12x^2}$	**3.**
$\phantom{2x-3\overline{)8x^3}}-10x^2 + 27x$	**4.**
$\phantom{2x-3\overline{)8x^3}}\underline{-10x^2 + 15x}$	**5.**
$\phantom{2x-3\overline{)8x^3-10x^2+}}12x - 18$	**6.**
$\phantom{2x-3\overline{)8x^3-10x^2+}}\underline{12x - 18}$	**7.**
$\phantom{2x-3\overline{)8x^3-10x^2+12x-1}}0$	**8.**

STEP 1   Divide the first term in the dividend, $8x^3$, by the first term of the divisor, $2x$. Write $4x^2$ above the dividend in line **1**.

STEP 2   Multiply the divisor, $2x - 3$, by $4x^2$ and write the result in line **3**, as shown.

STEP 3   Subtract in line 4 and bring down the next term.
STEP 4   Divide $-10x^2$ by $2x$ and write $-5x$ above the dividend in line 1.
STEP 5   Multiply $2x - 3$ by $-5x$ and write the result in line 5.
STEP 6   Subtract in line 6 and bring down the next term.
STEP 7   Divide $12x$ by $2x$ and write 6 above the dividend in line 1.
STEP 8   Multiply $2x - 3$ by 6 and write the result in line 7.
STEP 9   Subtract in line 8; the 0 indicates that there is no remainder.

Check:
$$\begin{array}{r} 4x^2 - 5x + 6 \\ 2x - 3 \\ \hline 8x^3 - 10x^2 + 12x \\ -12x^2 + 15x - 18 \\ \hline 8x^3 - 22x^2 + 27x - 18 \end{array}$$
quotient
divisor

dividend

Example 3   Divide: $(x^3 - 1) \div (x - 1)$.

$$\begin{array}{r} x^2 + x + 1 \\ x - 1 \overline{\smash{)}x^3 + 0x^2 + 0x - 1} \\ \underline{x^3 - x^2} \\ x^2 + 0x \\ \underline{x^2 - x} \\ x - 1 \\ \underline{x - 1} \\ 0 \end{array}$$

Supply missing powers.

The remainder is 0, so the quotient is $x^2 + x + 1$.

Check:
$$\begin{array}{r} x^2 + x + 1 \\ x - 1 \\ \hline x^3 + x^2 + x \\ -x^2 - x - 1 \\ \hline x^3 \qquad\quad - 1 \end{array}$$
quotient
divisor

dividend

NOTE: When you subtract, be especially careful to change all of the signs of each term in the expression being subtracted. Then follow the rules for addition.

EXERCISES 5.7

Find each quotient and check:

1. $(x^2 + 3x + 2) \div (x + 1)$
2. $(y^2 - 5y + 6) \div (y - 2)$
3. $(6a^2 - 3a + 2) \div (2a - 3)$
4. $(21y^2 + 2y - 10) \div (3y + 2)$
5. $(12x^2 - x - 9) \div (3x + 2)$
6. $(20x^2 + 57x + 30) \div (4x + 9)$
7. $\dfrac{2y^2 + 3y - 5}{2y - 1}$
8. $\dfrac{3x^2 - 5x - 10}{3x - 8}$
9. $\dfrac{6b^2 + 13b - 28}{2b + 7}$
10. $\dfrac{8x^2 + 13x - 27}{x + 3}$
11. $(6x^3 + 13x^2 + x - 2) \div (x + 2)$
12. $(8x^3 - 18x^2 + 7x + 3) \div (x - 1)$
13. $(8x^3 - 14x^2 - 79x + 110) \div (2x - 7)$
14. $(3x^3 - 17x^2 + 18x + 10) \div (3x + 1)$
15. $\dfrac{2x^3 - 14x - 12}{x + 1}$
16. $\dfrac{x^3 + 7x^2 - 36}{x + 3}$
17. $\dfrac{4x^3 - 24x^2 + 128}{2x + 4}$
18. $\dfrac{72x^3 + 22x + 4}{6x - 1}$
19. $\dfrac{3x^3 + 4x^2 - 6}{x - 2}$
20. $\dfrac{2x^3 + 3x^2 - 9x + 5}{x + 3}$
21. $\dfrac{4x^3 + 2x^2 + 30x + 20}{2x - 5}$

22. $\dfrac{18x^3 + 6x^2 - 4x}{6x - 2}$

23. $\dfrac{8x^4 - 10x^3 + 16x^2 + 4x - 30}{4x - 5}$

24. $\dfrac{9x^4 + 12x^3 - 6x^2 + 10x + 24}{3x + 4}$

25. $\dfrac{8x^3 - 1}{2x + 1}$

26. $\dfrac{x^3 + 1}{x + 1}$

27. $\dfrac{x^4 - 16}{x + 2}$

28. $\dfrac{16x^4 + 1}{2x - 1}$

29. $\dfrac{3x^4 + 5x^3 - 17x^2 + 11x - 2}{x^2 + 3x - 2}$

30. $\dfrac{6x^4 + 5x^3 - 11x^2 + 9x - 5}{2x^2 - x + 1}$

SUMMARY | CHAPTER 5

Glossary of Basic Terms

Binomial. The sum or difference of two unlike monomials. (p. 206)

Decreasing order. A polynomial written with each term of some degree less than the preceding term. (p. 207)

Degree of a monomial in one variable. The same as the exponent of the variable. (p. 206)

Degree of a polynomial in one variable. The same as the degree of the highest-degree monomial contained in the polynomial. (p. 206)

Evaluate an expression. Replace the letters with given numbers; then do the arithmetic using the order of operations. (p. 201)

Exponent. In the expression, 2^4, the *base* is 2 and the *exponent* is 4, which indicates that the base 2 is multiplied as a factor 4 times, that is, $2^4 = 2 \cdot 2 \cdot 2 \cdot 2 = 16$. (p. 200)

Increasing order. A polynomial written with each term of some degree larger than the preceding term. (p. 207)

Like terms. Terms with the same variables with exactly the same exponents. (p. 203)

Monomial. An algebraic expression that contains only products of numbers and variables, which have nonnegative integer exponents. (p. 206)

Numerical coefficient. The numerical factor of a term. (p. 203)

Polynomial. Either a monomial or the sum or difference of unlike monomials. (p. 206)

Term. A single number or a product of a number and one or more letters raised to powers. (p. 203)

Trinomial. The sum or difference of three unlike monomials. (p. 206)

Unlike terms. Terms with different variables or different exponents. (p. 203)

Variables. Letters that represent unknown quantities. (p. 200)

5.1 Fundamental Operations

1. **Basic mathematical principles:**
   a. $a + b = b + a$ (Commutative Property for Addition)
   b. $ab = ba$ (Commutative Property for Multiplication)
   c. $(a + b) + c = a + (b + c)$ (Associative Property for Addition)
   d. $(ab)c = a(bc)$ (Associative Property for Multiplication)
   e. $a(b + c) = ab + ac$ or $(b + c)a = ba + ca$ (Distributive Property)
   f. $a + 0 = a$
   g. $a \cdot 0 = 0$
   h. $a + (-a) = 0$ (Additive Inverse)
   i. $a \cdot 1 = 0$
   j. $a \cdot \dfrac{1}{a} = 1$ ($a \neq 0$) (Multiplicative Inverse) (p. 200)

2. **Order of operations:**
   a. Perform all operations inside parentheses. If the problem contains a fraction bar, treat the numerator and the denominator separately.
   b. Evaluate all powers, if any.
   c. Perform any multiplications and divisions in order, from left to right.
   d. Do any additions and subtractions in order, from left to right. (p. 200)

5.2 Simplifying Algebraic Expressions

1. **Removing parentheses:**
   a. Parentheses preceded by a plus sign may be removed without changing the signs of the terms within.
   b. Parentheses preceded by a minus sign may be removed if the signs of *all* the terms within the parentheses are changed; then drop the minus sign that preceded the parentheses. (p. 202)

5.3 Addition and Subtraction of Polynomials

1. **Adding polynomials:** To add polynomials, add their like terms. (p. 207)

2. **Subtracting polynomials:** To subtract two polynomials, change all the signs of the terms of the second polynomial and then add the two resulting polynomials. (p. 208)

5.4 Multiplication of Monomials

1. **Multiplying powers:** To multiply powers with the same base, add the exponents: $a^m \cdot a^n = a^{m+n}$. (p. 209)
2. **Raising a power to a power:** To raise a power to a power, multiply the exponents: $(a^m)^n = a^{mn}$. (p. 210)
3. **Raising a product to a power:** To raise a product to a power, raise each factor to that same power: $(ab)^m = a^m b^m$. (p. 210)

5.5 Multiplication of Polynomials

1. **Multiplying a polynomial by a monomial:** To multiply a polynomial by a monomial, multiply each term of the polynomial by the monomial, and then add the products. (p. 211)
2. **Multiplying a polynomial by a polynomial:** To multiply a polynomial by a polynomial, multiply each term of the first polynomial by each term of the second polynomial, and then add the products. (p. 212)

5.6 Division by a Monomial

1. **Dividing a monomial by a monomial:** To divide a monomial by a monomial, first write the quotient in fraction form. Then factor both numerator and denominator into prime factors. Reduce to lowest terms by dividing both numerator and denominator by their common factors. The remaining factors in the numerator and the denominator give the quotient. (p. 213)
2. **Dividing a polynomial by a monomial:** To divide a polynomial by a monomial, divide each term of the polynomial by the monomial. (p. 214)

5.7 Division by a Polynomial

1. **Dividing a polynomial by a polynomial:** To divide a polynomial by a polynomial, use long division as shown in Section 5.7. (p. 215)

REVIEW | CHAPTER 5

1. For any number a, $a \cdot 1 = ?$
2. For any number a, $a \cdot 0 = ?$
3. For any number a except 0, $a \cdot \dfrac{1}{a} = ?$

Evaluate:

4. $10 - 4(3)$
5. $2 + 3 \cdot 4^2$
6. $(4)(12) \div 6 - 2^3 + 18 \div 3^2$

In Exercises 7–12, let $x = 3$ and $y = -2$. Evaluate each expression:

7. $x + y$
8. $x - 3y$
9. $5xy$
10. $\dfrac{x^2}{y}$
11. $y^3 - y^2$
12. $\dfrac{2x^3 - 3y}{xy^2}$

In Exercises 13–16, simplify by removing the parentheses and combining like terms:

13. $(5y - 3) - (2 - y)$
14. $(7 - 3x) - (5x + 1)$
15. $11(2x + 1) - 4(3x - 4)$
16. $(x^3 + 2x^2 y) - (3y^3 - 2x^3 + x^2 y + y)$
17. Is $1 - 8x^2$ a monomial, a binomial, or a trinomial?
18. Find the degree of the polynomial $x^4 + 2x^3 - 6$.

Perform the indicated operations:

19. $(3a^2 + 7a - 2) + (5a^2 - 2a + 4)$
20. $(6x^3 + 3x^2 + 1) - (-3x^3 - x^2 - x - 1)$
21. $(3x^2 + 5x + 2) + (9x^2 - 6x - 2) - (2x^2 + 6x - 4)$
22. $(6x^2)(4x^3)$
23. $(-7x^2 y)(8x^3 y^2)$
24. $(3x^2)^3$
25. $5a(3a + 4b)$
26. $-4x^2(8 - 2x + 3x^2)$
27. $(5x + 3)(3x - 4)$
28. $(3x^2 - 6x + 1)(2x - 4)$
29. $(49x^2) \div (7x^3)$
30. $\dfrac{15x^3 y}{3xy}$
31. $\dfrac{36a^3 - 27a^2 + 9a}{9a}$
32. $\dfrac{6x^2 + x - 12}{2x + 3}$
33. $\dfrac{3x^3 + 2x^2 - 6x + 4}{x + 2}$

TEST | CHAPTER 5

Evaluate:

1. $3 \cdot 5 - 2 \cdot 4^2$
2. $12 \div 2 \cdot 3 \div 2 + 3^3 - 16 \div 2^2$

Evaluate each expression when $x = 4$ and $y = -1$.

3. $4x^2 - 3y^3$
4. $\dfrac{3x^2y - 4x}{2y}$
5. Is $4 + 3x - 5x^2$ a monomial, a binomial, or a trinomial? Find its degree.
6. Rearrange the polynomial $3 + 4x^2 - 5x^3 - x$ in decreasing order.

Perform the indicated operations and simplify:

7. $(3a^2 - 17a) + (6a^2 + 4a)$
8. $5(2 + x) - 2(x + 4)$
9. Add: $\begin{array}{r} 3x^2 + 6x - 8 \\ -9x^2 + 6x \\ -3x^2 + 15 \\ - 7x + 4 \\ \hline \end{array}$
10. $(5a - 5b + 7) - (2a - 5b - 3)$
11. $(7a^2)(-4a^4)(a)$
12. $(6x^2)(-3x)^2$
13. $(-4x^3)^3$
14. $(6x^4y^2)^3$
15. $-5x(2x - 3)$
16. $8 - 2x(3x + 4)$
17. $\dfrac{85x^4y^2}{17x^2y^5}$
18. $(4a + 6)(a - 5)$
19. $(3x - 4)(6x - 5)$
20. $(x + y - 5)(x - y)$
21. $\dfrac{36a^4 - 20a^3 - 16a^2}{4a}$
22. $\dfrac{9x^4y^3 - 12x^2y + 18y^2}{3x^2y^3}$
23. $(5x^2y^3)(-7x^3y)$
24. $\dfrac{3x^2 - 13x - 10}{x - 5}$
25. $\dfrac{4x^5y^3}{-2x^3y^5}$
26. $\dfrac{6x^2 - 7x - 6}{3x - 5}$
27. $\dfrac{x^3 + 2x^2 + x + 12}{x + 3}$

Equations and Formulas

CHAPTER 6

OBJECTIVES

- Use the addition, subtraction, multiplication, and division properties of equality to solve simple equations.
- Solve equations with parentheses.
- Solve equations with fractions.
- Translate words into algebraic symbols.
- Solve application problems using equations.
- Solve a formula for a given letter.
- Substitute data into a formula and find the value of the indicated letter using the rules for working with measurements.
- Substitute data into a formula involving reciprocals and find the value of the indicated letter using a scientific calculator.

Mathematics at Work

Diesel technicians inspect, maintain, and repair diesel engines that power the agriculture, construction, trucking, locomotive, and heavy equipment industries. The diesel engine provides power for the nation's heavy vehicles and equipment because it delivers more power per unit of fuel and is more durable than the gasoline engine. Some diesel technicians specialize in farm tractors and farm equipment; large trucks; bulldozers, road graders, and construction equipment; and heavy industrial equipment, as well as automobiles and boats. These technicians work for companies or organizations that maintain their own equipment and spend their time doing preventive maintenance to keep their equipment operating dependably and safely. During routine maintenance, they inspect and repair the basic components to eliminate unnecessary wear and damage to avoid costly breakdowns.

The work of a diesel technician is becoming more complex as more electronic components are used to control engine operation. Diesel technicians use handheld computers and sophisticated equipment to diagnose problems and to adjust engine functions. They often need special equipment to handle large components. Many community colleges and trade and vocational schools offer associate degree and certificate diesel repair programs. Employers prefer graduates of formal training programs because of their basic understanding and their ability to more quickly advance to their journey technician level. Certifications are available within a variety of specialty areas, such as master heavy-duty truck repair and school bus repair in specific areas—for example, brakes, diesel engines, drive trains, electrical systems, and steering and suspension. For more information, please visit **www.cengage.com** and access the Student Online Resources for this text.

Diesel Technician
Diesel technician repairing a diesel engine

6.1 Equations

In technical work, the ability to use equations and formulas is essential. A **variable** is a symbol (usually a letter of the alphabet) used to represent an unknown number. An **algebraic expression** is a combination of numbers, variables, symbols for operations (plus, minus, times, divide), and symbols for grouping (parentheses or a fraction bar). Examples of algebraic expressions are

$$4x - 9, \qquad 3x^2 + 6x + 9, \qquad 5x(6x + 4), \qquad \frac{2x + 5}{-3x}$$

An **equation** is a statement that two quantities are equal. The symbol "=" is read "equals" and separates an equation into two parts: the left member and the right member. For example, in the equation

$$2x + 3 = 11$$

the left member is $2x + 3$ and the right member is 11. Other examples of equations are

$$x - 5 = 6, \qquad 3x = 12, \qquad 4m + 9 = 3m - 2, \qquad \frac{x - 2}{4} = 3(x + 1)$$

To **solve an equation** means to find what number or numbers can replace the variable to make the equation a true statement. In the equation $2x + 3 = 11$, the solution is 4. That is, when x is replaced by 4, the resulting equation is a true statement.

$$2x + 3 = 11$$
Let $x = 4$: $\quad 2(4) + 3 = 11 \quad$?
$$8 + 3 = 11 \quad \text{True}$$

A replacement number (or numbers) that produces a true statement in an equation is called a **solution** or a **root** of the equation. Note that replacing x in the equation above with any number other than 4, such as 5, results in a false statement.

$$2x + 3 = 11$$
Let $x = 5$: $\quad 2(5) + 3 = 11 \quad$?
$$10 + 3 = 11 \quad \text{False}$$

One method of solving equations involves changing the given equation to an equivalent equation by performing the same arithmetic operation on both sides of the equation. The basic arithmetic operations used are addition, subtraction, multiplication, and division.

Equivalent equations are equations that have the same solutions or roots. For example, $3x = 6$ and $x = 2$ are equivalent equations, since 2 is the root of each. In solving an equation by this method, continue to change the given equation to another equivalent equation until you find an equation whose root is obvious.

Four Basic Rules Used to Solve Equations

1. *Addition Property of Equality:* If the same quantity is added to both sides of an equation, the resulting equation is equivalent to the original equation.

    Example: Solve $x - 2 = 8$.
    $$x - 2 + 2 = 8 + 2 \qquad \text{Add 2 to both sides.}$$
    $$x = 10$$

♦

2. *Subtraction Property of Equality:* If the same quantity is subtracted from both sides of an equation, the resulting equation is equivalent to the original equation.

   Example: Solve $x + 5 = 2$.
   $$x + 5 - 5 = 2 - 5 \quad \text{Subtract 5 from both sides.}$$
   $$x = -3$$

3. *Multiplication Property of Equality:* If both sides of an equation are multiplied by the same (nonzero) quantity, the resulting equation is equivalent to the original equation.

   Example: Solve $\dfrac{x}{9} = 4$.
   $$9\left(\dfrac{x}{9}\right) = (4)9 \quad \text{Multiply both sides by 9.}$$
   $$x = 36$$

4. *Division Property of Equality:* If both sides of an equation are divided by the same (nonzero) quantity, the resulting equation is equivalent to the original equation.

   Example: Solve $4x = 20$.
   $$\dfrac{4x}{4} = \dfrac{20}{4} \quad \text{Divide both sides by 4.}$$
   $$x = 5$$

Basically, to solve a simple equation, use one of the rules and use a number that will *undo* what has been done to the variable.

Example 1

Solve: $x + 3 = 8$.

Since 3 has been added to the variable, use Rule 2 and subtract 3 from both sides of the equation.

$$x + 3 = 8$$
$$x + 3 - 3 = 8 - 3 \quad \text{Subtracting 3 was chosen to } undo \text{ adding 3.}$$
$$x + 0 = 5$$
$$x = 5$$

A check is recommended, since an error could have been made. To check, replace the variable in the original equation by 5, the apparent root, to make sure that the resulting statement is true.

Check:
$$x + 3 = 8$$
$$5 + 3 = 8 \quad \text{True}$$

Thus, the root is 5.

Example 2

Solve: $x - 4 = 7$.

Since 4 has been subtracted from the variable, use Rule 1 and add 4 to both sides of the equation.

$$x - 4 = 7$$
$$x - 4 + 4 = 7 + 4 \quad \text{Adding 4 was chosen to } undo \text{ subtracting 4.}$$
$$x + 0 = 11$$
$$x = 11$$

The apparent root is 11.

Check:
$$x - 4 = 7$$
$$11 - 4 = 7 \quad \text{True}$$

Thus, the root is 11.

Example 3   Solve: $2x = 9$.

Since the variable has been multiplied by 2, use Rule 4 and divide both sides of the equation by 2.

$$2x = 9$$
$$\frac{2x}{2} = \frac{9}{2} \qquad \text{Dividing by 2 was chosen to } undo \text{ multiplying by 2.}$$
$$x = \frac{9}{2}$$

NOTE: Each solution should be checked by substituting it into the original equation. When a check is not provided in this text, the check is left for you to do.

Example 4   Solve: $\frac{x}{3} = 9$.

Since the variable has been divided by 3, use Rule 3 and multiply both sides of the equation by 3.

$$\frac{x}{3} = 9$$
$$3\left(\frac{x}{3}\right) = (9)3 \qquad \text{Multiplying by 3 was chosen to } undo \text{ dividing by 3.}$$
$$x = 27$$

Example 5   Solve: $-4 = -6x$.

Since the variable has been multiplied by -6, use Rule 4 and divide both sides of the equation by -6.

$$-4 = -6x$$
$$\frac{-4}{-6} = \frac{-6x}{-6} \qquad \text{Dividing by } -6 \text{ was chosen to } undo \text{ multiplying by } -6.$$
$$\frac{2}{3} = x$$

The apparent root is $\frac{2}{3}$.

Check:
$$-4 = -6x$$
$$-4 = -6\left(\frac{2}{3}\right) \qquad ?$$
$$-4 = -4 \qquad \text{True}$$

Thus, the root is $\frac{2}{3}$.

Some equations have more than one operation indicated on the variable. For example, the equation $2x + 5 = 6$ has both addition of 5 and multiplication by 2 indicated on the variable. Use the following procedure to solve equations like this:

> When more than one operation is indicated on the variable, undo the additions and subtractions first, then undo the multiplications and divisions.

Example 6 Solve: $2x + 5 = 6$.

$$2x + 5 - 5 = 6 - 5 \quad \text{Subtract 5 from both sides.}$$
$$2x = 1$$
$$\frac{2x}{2} = \frac{1}{2} \quad \text{Divide both sides by 2.}$$
$$x = \frac{1}{2}$$

The apparent root is $\frac{1}{2}$.

Check:
$$2x + 5 = 6$$
$$2\left(\frac{1}{2}\right) + 5 = 6 \quad ?$$
$$1 + 5 = 6 \quad \text{True}$$

Thus, the root is $\frac{1}{2}$.

Example 7 Solve: $\frac{x}{3} - 6 = 9$.

$$\frac{x}{3} - 6 + 6 = 9 + 6 \quad \text{Add 6 to both sides.}$$
$$\frac{x}{3} = 15$$
$$3\left(\frac{x}{3}\right) = (15)3 \quad \text{Multiply both sides by 3.}$$
$$x = 45$$

Example 8 Solve: $118 - 22m = 30$.

$$118 - 22m - 118 = 30 - 118 \quad \text{Subtract 118 from both sides.}$$
$$-22m = -88$$
$$\frac{-22m}{-22} = \frac{-88}{-22} \quad \text{Divide both sides by } -22.$$
$$m = 4$$

The apparent root is 4.

Check:
$$118 - 22m = 30$$
$$118 - 22(4) = 30 \quad ?$$
$$118 - 88 = 30 \quad ?$$
$$30 = 30 \quad \text{True}$$

Thus, the root is 4.

Here is another approach to solving this equation:

$$118 - 22m = 30$$
$$118 - 22m + 22m = 30 + 22m \quad \text{Add } 22m \text{ to both sides.}$$
$$118 = 30 + 22m$$
$$118 - 30 = 30 + 22m - 30 \quad \text{Subtract 30 from both sides.}$$
$$88 = 22m$$
$$\frac{88}{22} = \frac{22m}{22} \quad \text{Divide both sides by 22.}$$
$$4 = m$$

EXERCISES 6.1

Solve each equation and check:

1. $x + 2 = 8$
2. $3a = 7$
3. $y - 5 = 12$
4. $\frac{2}{3}n = 6$
5. $w - 7\frac{1}{2} = 3$
6. $2m = 28.4$
7. $\frac{x}{13} = 1.5$
8. $n + 12 = -5$
9. $3b = 15.6$
10. $y - 17 = 25$
11. $17x = 5117$
12. $28 + m = 3$
13. $2 = x - 5$
14. $-29 = -4y$
15. $17 = -3 + w$
16. $49 = 32 + w$
17. $14b = 57$
18. $y + 28 = 13$
19. $5m = 0$
20. $28 + m = 28$
21. $x + 5 = 5$
22. $y + 7 = -7$
23. $4x = 64$
24. $5x - 125 = 0$
25. $\frac{x}{7} = 56$
26. $\frac{y}{5} = 35$
27. $-48 = 12y$
28. $13x = -78$
29. $-x = 2$
30. $-y = 7$
31. $5y + 3 = 13$
32. $4x - 2 = 18$
33. $10 - 3x = 16$
34. $8 - 2y = 4$
35. $\frac{x}{4} - 5 = 3$
36. $\frac{x}{5} + 4 = 9$
37. $2 - x = 6$
38. $8 - y = 3$
39. $\frac{2}{3}y - 4 = 8$
40. $5 - \frac{1}{4}x = 7$
41. $3x - 5 = 12$
42. $5y + 7 = 28$
43. $\frac{m}{3} - 6 = 8$
44. $\frac{w}{5} + 7 = 13$
45. $\frac{2x}{3} = 7$
46. $\frac{4b}{5} = 15$
47. $-3y - 7 = -6$
48. $28 = -7 - 3r$
49. $5 - x = 6$
50. $17 - 5w = -68$
51. $54y - 13 = 17.8$
52. $37a - 7 = 67$
53. $28w - 56 = -8$
54. $52 - 4x = -8$
55. $29r - 13 = 57$
56. $15x - 32 = 18$
57. $31 - 3y = 41$
58. $62 = 13y - 3$
59. $-83 = 17 - 4x$
60. $58 = 5m + 52$

6.2 Equations with Variables in Both Members

To solve equations with variables in both members (on both sides), such as

$$3x + 4 = 5x - 12$$

do the following:

First, add or subtract either variable term from both sides of the equation.

$$3x + 4 = 5x - 12$$
$$3x + 4 - 3x = 5x - 12 - 3x \quad \text{Subtract } 3x \text{ from both sides.}$$
$$4 = 2x - 12$$

Then, take the constant term (which now appears on the same side of the equation with the variable term) and add it to, or subtract it from, both sides. Solve the resulting equation.

$$4 + 12 = 2x - 12 + 12 \quad \text{Add 12 to both sides.}$$
$$16 = 2x$$
$$\frac{16}{2} = \frac{2x}{2} \quad \text{Divide both sides by 2.}$$
$$8 = x$$

6.2 ♦ Equations with Variables in Both Members

This equation could also have been solved as follows:

$$3x + 4 = 5x - 12$$
$$3x + 4 - 5x = 5x - 12 - 5x \quad \text{Subtract } 5x \text{ from both sides.}$$
$$-2x + 4 = -12$$
$$-2x + 4 - 4 = -12 - 4 \quad \text{Subtract 4 from both sides.}$$
$$-2x = -16$$
$$\frac{-2x}{-2} = \frac{-16}{-2} \quad \text{Divide both sides by } -2.$$
$$x = 8$$

Example 1 Solve: $5x - 4 = 8x - 13$.

$$5x - 4 - 8x = 8x - 13 - 8x \quad \text{Subtract } 8x \text{ from both sides.}$$
$$-3x - 4 = -13$$
$$-3x - 4 + 4 = -13 + 4 \quad \text{Add 4 to both sides.}$$
$$-3x = -9$$
$$\frac{-3x}{-3} = \frac{-9}{-3} \quad \text{Divide both sides by } -3.$$
$$x = 3$$

Check:
$$5x - 4 = 8x - 13$$
$$5(3) - 4 = 8(3) - 13 \quad ?$$
$$15 - 4 = 24 - 13 \quad ?$$
$$11 = 11 \quad \text{True}$$

Therefore, 3 is a root.

Example 2 Solve: $-2x + 5 = 6x - 11$.

$$-2x + 5 + 2x = 6x - 11 + 2x \quad \text{Add } 2x \text{ to both sides.}$$
$$5 = 8x - 11$$
$$5 + 11 = 8x - 11 + 11 \quad \text{Add 11 to both sides.}$$
$$16 = 8x$$
$$\frac{16}{8} = \frac{8x}{8} \quad \text{Divide both sides by 8.}$$
$$2 = x$$

Check:
$$-2x + 5 = 6x - 11$$
$$-2(2) + 5 = 6(2) - 11 \quad ?$$
$$-4 + 5 = 12 - 11 \quad ?$$
$$1 = 1 \quad \text{True}$$

Thus, 2 is a root.

Example 3 Solve: $5x + 7 = 2x - 14$.

$$5x + 7 - 2x = 2x - 14 - 2x \quad \text{Subtract } 2x \text{ from both sides.}$$
$$3x + 7 = -14$$
$$3x + 7 - 7 = -14 - 7 \quad \text{Subtract 7 from both sides.}$$
$$3x = -21$$
$$\frac{3x}{3} = \frac{-21}{3} \quad \text{Divide both sides by 3.}$$
$$x = -7$$

Check:
$$5x + 7 = 2x - 14$$
$$5(-7) + 7 = 2(-7) - 14 \quad ?$$
$$-35 + 7 = -14 - 14 \quad ?$$
$$-28 = -28 \quad \text{True}$$

So -7 is a root.

Example 4 Solve: $4 - 5x = 28 + x$.

$$4 - 5x + 5x = 28 + x + 5x \quad \text{Add } 5x \text{ to both sides.}$$
$$4 = 28 + 6x$$
$$4 - 28 = 28 + 6x - 28 \quad \text{Subtract 28 from both sides.}$$
$$-24 = 6x$$
$$\frac{-24}{6} = \frac{6x}{6} \quad \text{Divide both sides by 6.}$$
$$-4 = x$$

EXERCISES 6.2

Solve each equation and check:

1. $4y + 9 = 7y - 15$
2. $2y - 45 = -y$
3. $5x + 3 = 7x - 5$
4. $2x - 3 = 3x - 13$
5. $-2x + 7 = 5x - 21$
6. $5x + 3 = 2x - 15$
7. $3y + 5 = 5y - 1$
8. $3x - 4 = 7x - 32$
9. $-3x + 17 = 6x - 37$
10. $3x + 13 = 2x - 12$
11. $7x + 9 = 9x - 3$
12. $-5y + 12 = 12y - 5$
13. $3x - 2 = 5x + 8$
14. $13y + 2 = 20y - 5$
15. $-4x + 25 = 6x - 45$
16. $5x - 7 = 6x - 5$
17. $5x + 4 = 10x - 7$
18. $3x - 2 = 5x - 20$
19. $27 + 5x = 9 + 3x$
20. $2y + 8 = 5y - 1$
21. $-7x + 18 = 11x - 36$
22. $4x + 5 = 2x - 7$
23. $4y + 11 = 7y - 28$
24. $4x = 2x - 12$
25. $-4x + 2 = 8x - 7$
26. $6x - 1 = 9x - 9$
27. $13x + 6 = 6x - 1$
28. $6y + 7 = 18y - 1$
29. $3x + 1 = 17 - x$
30. $17 - 4y = 14 - y$

6.3 Equations with Parentheses

To solve an equation having parentheses in one or both members, always remove the parentheses first. Then combine like terms. Then use the previously explained methods to solve the resulting equation.

Example 1 Solve: $5 - (2x - 3) = 7$.

$$5 - 2x + 3 = 7 \quad \text{Remove parentheses.}$$
$$8 - 2x = 7 \quad \text{Combine like terms.}$$
$$8 - 2x - 8 = 7 - 8 \quad \text{Subtract 8 from both sides.}$$
$$-2x = -1$$
$$\frac{-2x}{-2} = \frac{-1}{-2} \quad \text{Divide both sides by } -2.$$
$$x = \frac{1}{2}$$

Check:
$$5 - (2x - 3) = 7$$
$$5 - \left[2\left(\frac{1}{2}\right) - 3\right] = 7 \quad ?$$

6.3 ♦ Equations with Parentheses

$$5 - (1 - 3) = 7 \quad ?$$
$$5 - (-2) = 7 \quad \text{True}$$

Therefore, $\frac{1}{2}$ is the root. ♦

Example 2   Solve: $7x - 6(5 - x) = 9$.

$7x - 30 + 6x = 9$	Remove parentheses.
$13x - 30 = 9$	Combine like terms.
$13x - 30 + 30 = 9 + 30$	Add 30 to both sides.
$13x = 39$	
$\dfrac{13x}{13} = \dfrac{39}{13}$	Divide both sides by 13.
$x = 3$	

♦

In the following examples, we have parentheses as well as the variable in both members.

Example 3   Solve: $3(x - 5) = 2(4 - x)$.

$3x - 15 = 8 - 2x$	Remove parentheses.
$3x - 15 + 2x = 8 - 2x + 2x$	Add $2x$ to both sides.
$5x - 15 = 8$	Combine like terms.
$5x - 15 + 15 = 8 + 15$	Add 15 to both sides.
$5x = 23$	
$\dfrac{5x}{5} = \dfrac{23}{5}$	Divide both sides by 5.
$x = \dfrac{23}{5}$	

Check: $3(x - 5) = 2(4 - x)$

$$3\left(\frac{23}{5} - 5\right) = 2\left(4 - \frac{23}{5}\right) \quad ?$$
$$3\left(-\frac{2}{5}\right) = 2\left(-\frac{3}{5}\right) \quad ?$$
$$-\frac{6}{5} = -\frac{6}{5} \quad \text{True}$$

Therefore, $\frac{23}{5}$ is the root. ♦

Example 4   Solve: $8x - 4(x + 2) = 12(x + 1) - 14$.

$8x - 4x - 8 = 12x + 12 - 14$	Remove parentheses.
$4x - 8 = 12x - 2$	Combine like terms.
$4x - 8 - 12x = 12x - 2 - 12x$	Subtract $12x$ from both sides.
$-8x - 8 = -2$	
$-8x - 8 + 8 = -2 + 8$	Add 8 to both sides.
$-8x = 6$	
$\dfrac{-8x}{-8} = \dfrac{6}{-8}$	Divide both sides by -8.
$x = -\dfrac{3}{4}$	

Check:
$$8x - 4(x + 2) = 12(x + 1) - 14$$
$$8\left(-\frac{3}{4}\right) - 4\left(-\frac{3}{4} + 2\right) = 12\left(-\frac{3}{4} + 1\right) - 14 \quad ?$$
$$-6 - 4\left(\frac{5}{4}\right) = 12\left(\frac{1}{4}\right) - 14 \quad ?$$
$$-6 - 5 = 3 - 14 \quad ?$$
$$-11 = -11 \quad \text{True}$$

Therefore, $-\dfrac{3}{4}$ is the root. ◆

EXERCISES 6.3

Solve each equation and check:

1. $2(x + 3) - 6 = 10$
2. $-3x + 5(x - 6) = 32$
3. $3n + (2n + 4) = 6$
4. $5m - (2m - 7) = -5$
5. $16 = -3(x - 4)$
6. $5y + 6(y - 3) = 15$
7. $5a - (3a + 4) = 8$
8. $2(b + 4) - 3 = 15$
9. $5a - 4(a - 3) = 7$
10. $29 = 4 + (2m + 1)$
11. $5(x - 3) = 21$
12. $27 - 8(2 - y) = -13$
13. $2a - (5a - 7) = 22$
14. $2(5m - 6) - 13 = -1$
15. $2(w - 3) + 6 = 0$
16. $6r - (2r - 3) + 5 = 0$
17. $3x - 7 + 17(1 - x) = -6$
18. $4y - 6(2 - y) = 8$
19. $6b = 27 + 3b$
20. $2a + 4 = a - 3$
21. $4(25 - x) = 3x + 2$
22. $4x - 2 = 3(25 - x)$
23. $x + 3 = 4(57 - x)$
24. $2(y + 1) = y - 7$
25. $6x + 2 = 2(17 - x)$
26. $6(17 - x) = 2 - 4x$
27. $5(x - 8) - 3x - 4 = 0$
28. $5(28 - 2x) - 7x = 4$
29. $3(x + 4) + 3x = 6$
30. $8x - 4(x + 2) + 11 = 0$
31. $y - 4 = 2(y - 7)$
32. $7(w - 4) = w + 2$
33. $9m - 3(m - 5) = 7m - 3$
34. $4(x + 18) = 2(4x + 18)$
35. $3(2x + 7) = 13 + 2(4x + 2)$
36. $5y - 3(y - 2) = 6(y + 1)$
37. $8(x - 5) = 13x$
38. $4(x + 2) = 30 - (x - 3)$
39. $5 + 3(x + 7) = 26 - 6(5x + 11)$
40. $2(y - 3) = 4 + (y - 14)$
41. $5(2y - 3) = 3(7y - 6) + 19(y + 1) + 14$
42. $2(7y - 6) - 11(y + 1) = 38 - 7(9y + 4)$
43. $16(x + 3) = 7(x - 5) - 9(x + 4) - 7$
44. $31 - 2(x - 5) = -3(x + 4)$
45. $4(y + 2) = 8(y - 4) + 7$
46. $12x - 13(x + 4) = 4x - 6$
47. $4(5y - 2) + 3(2y + 6) = 25(3y + 2) - 19y$
48. $6(3x + 1) = 5x - (2x + 2)$
49. $12 + 8(2y + 3) = (y + 7) - 16$
50. $-2x + 6(2 - x) - 4 = 3(x + 1) - 6$
51. $5x - 10(3x - 6) = 3(24 - 9x)$
52. $4y + 7 - 3(2y + 3) = 4(3y - 4) - 7y + 7$
53. $6(y - 4) - 4(5y + 1) = 3(y - 2) - 4(2y + 1)$
54. $2(5y + 1) + 16 = 4 + 3(y - 7)$
55. $-6(x - 5) + 3x = 6x - 10(-3 + x)$
56. $14x + 14(3 - 2x) + 7 = 4 - x + 5(2 - 3x)$
57. $2.3x - 4.7 + 0.6(3x + 5) = 0.7(3 - x)$
58. $5.2(x + 3) + 3.7(2 - x) = 3$
59. $0.089x - 0.32 + 0.001(5 - x) = 0.231$
60. $5x - 2.5(7 - 4x) = x - 7(4 + x)$

6.4 Equations with Fractions

To Solve an Equation with Fractions

1. Find the least common denominator (LCD) of all the fractional terms on both sides of the equation.

2. Multiply both sides of the equation by the LCD. (If this step has been done correctly, no fractions should now appear in the resulting equation.)
3. Solve the resulting equation from Step 2 using the methods introduced earlier in this chapter.

Example 1   Solve: $\dfrac{3x}{4} = \dfrac{45}{20}$.

The LCD of 4 and 20 is 20; therefore, multiply both sides of the equation by 20.

$$\dfrac{3x}{4} = \dfrac{45}{20}$$

$$20\left(\dfrac{3x}{4}\right) = \left(\dfrac{45}{20}\right)20$$

$$15x = 45$$

$$\dfrac{15x}{15} = \dfrac{45}{15} \qquad \text{Divide both sides by 15.}$$

$$x = 3$$

♦

Example 2   Solve: $\dfrac{3}{4} + \dfrac{x}{6} = \dfrac{13}{12}$.

The LCD of 4, 6, and 12 is 12; multiply both sides of the equation by 12.

$$\dfrac{3}{4} + \dfrac{x}{6} = \dfrac{13}{12}$$

$$12\left(\dfrac{3}{4} + \dfrac{x}{6}\right) = \left(\dfrac{13}{12}\right)12$$

$$12\left(\dfrac{3}{4}\right) + 12\left(\dfrac{x}{6}\right) = \left(\dfrac{13}{12}\right)12 \qquad \text{Apply the Distributive Property on the left side by multiplying each term within parentheses by 12.}$$

$$9 + 2x = 13$$

$$9 + 2x - 9 = 13 - 9 \qquad \text{Subtract 9 from both sides.}$$

$$2x = 4$$

$$\dfrac{2x}{2} = \dfrac{4}{2} \qquad \text{Divide both sides by 2.}$$

$$x = 2$$

Check:

$$\dfrac{3}{4} + \dfrac{x}{6} = \dfrac{13}{12}$$

$$\dfrac{3}{4} + \dfrac{2}{6} = \dfrac{13}{12} \qquad ?$$

$$\dfrac{9}{12} + \dfrac{4}{12} = \dfrac{13}{12} \qquad \text{True}$$

♦

Example 3 Solve: $\dfrac{2x}{9} - 4 = \dfrac{x}{6}$.

The LCD of 9 and 6 is 18; multiply both sides of the equation by 18.

$$\dfrac{2x}{9} - 4 = \dfrac{x}{6}$$

$$18\left(\dfrac{2x}{9} - 4\right) = \left(\dfrac{x}{6}\right)18$$

$$18\left(\dfrac{2x}{9}\right) - 18(4) = \left(\dfrac{x}{6}\right)18 \quad\quad \text{Apply the Distributive Property on the left side by multiplying each term within parentheses by 18.}$$

$$4x - 72 = 3x$$
$$4x - 72 - 4x = 3x - 4x \quad\quad \text{Subtract } 4x \text{ from both sides.}$$
$$-72 = -x$$
$$\dfrac{-72}{-1} = \dfrac{-x}{-1} \quad\quad \text{Divide both sides by } -1.$$
$$72 = x$$

Check:
$$\dfrac{2x}{9} - 4 = \dfrac{x}{6}$$
$$\dfrac{2(72)}{9} - 4 = \dfrac{72}{6} \quad ?$$
$$16 - 4 = 12 \quad \text{True}$$

♦

Example 4 Solve: $\dfrac{2}{3}x + \dfrac{3}{4}(36 - 2x) = 32$.

The LCD of 3 and 4 is 12; multiply both sides of the equation by 12.

$$\dfrac{2}{3}x + \dfrac{3}{4}(36 - 2x) = 32$$

$$12\left[\dfrac{2}{3}x + \dfrac{3}{4}(36 - 2x)\right] = (32)12$$

$$12\left(\dfrac{2}{3}x\right) + 12\left(\dfrac{3}{4}\right)(36 - 2x) = (32)12 \quad\quad \text{Apply the Distributive Property on the left to remove brackets.}$$

$$8x + 9(36 - 2x) = 384$$
$$8x + 324 - 18x = 384 \quad\quad \text{Apply the Distributive Property to remove parentheses.}$$
$$-10x + 324 = 384 \quad\quad \text{Combine like terms.}$$
$$-10x + 324 - 324 = 384 - 324 \quad\quad \text{Subtract 324 from both sides.}$$
$$-10x = 60$$
$$\dfrac{-10x}{-10} = \dfrac{60}{-10} \quad\quad \text{Divide both sides by } -10.$$
$$x = -6$$

Check:
$$\dfrac{2}{3}x + \dfrac{3}{4}(36 - 2x) = 32$$
$$\dfrac{2}{3}(-6) + \dfrac{3}{4}[36 - 2(-6)] = 32 \quad ?$$
$$-4 + \dfrac{3}{4}(36 + 12) = 32 \quad ?$$

6.4 ♦ Equations with Fractions

$$-4 + \frac{3}{4}(48) = 32 \quad ?$$

$$-4 + 36 = 32 \quad \text{True}$$

Example 5 Solve: $\dfrac{2x+1}{3} - \dfrac{x-6}{4} = \dfrac{2x+4}{8} + 2$.

The LCD of 3, 4, and 8 is 24; multiply both sides of the equation by 24.

$$\frac{2x+1}{3} - \frac{x-6}{4} = \frac{2x+4}{8} + 2$$

$$24\left(\frac{2x+1}{3} - \frac{x-6}{4}\right) = \left(\frac{2x+4}{8} + 2\right)24$$

$$24\left(\frac{2x+1}{3}\right) - 24\left(\frac{x-6}{4}\right) = \left(\frac{2x+4}{8}\right)24 + 2(24)$$

$$8(2x+1) - 6(x-6) = 3(2x+4) + 48$$

$16x + 8 - 6x + 36 = 6x + 12 + 48$ Remove parentheses.

$10x + 44 = 6x + 60$ Combine like terms.

$10x + 44 - 6x = 6x + 60 - 6x$ Subtract $6x$ from both sides.

$4x + 44 = 60$

$4x + 44 - 44 = 60 - 44$ Subtract 44 from both sides.

$4x = 16$

$\dfrac{4x}{4} = \dfrac{16}{4}$ Divide both sides by 4.

$x = 4$

Check:

$$\frac{2x+1}{3} - \frac{x-6}{4} = \frac{2x+4}{8} + 2$$

$$\frac{2(4)+1}{3} - \frac{4-6}{4} = \frac{2(4)+4}{8} + 2 \quad ?$$

$$\frac{9}{3} - \frac{-2}{4} = \frac{12}{8} + 2 \quad ?$$

$$3 + \frac{1}{2} = \frac{3}{2} + 2 \quad ?$$

$$\frac{7}{2} = \frac{7}{2} \quad \text{True}$$ ♦

When the variable appears in the denominator of a fraction in an equation, multiply both members by the LCD. Be careful that the replacement for the variable does not make the denominator zero.

Example 6 Solve: $\dfrac{12}{x} = 2$.

$x\left(\dfrac{12}{x}\right) = (2)x$ Multiply both sides by the LCD, x.

$12 = 2x$

$\dfrac{12}{2} = \dfrac{2x}{2}$ Divide both sides by 2.

$6 = x$

234 CHAPTER 6 ♦ Equations and Formulas

Check: $\dfrac{12}{x} = 2$

$\dfrac{12}{6} = 2$?

$2 = 2$ True

Thus, the root is 6.

Example 7 Solve: $\dfrac{5}{x} - 2 = 3$.

$\dfrac{5}{x} - 2 + 2 = 3 + 2$ Add 2 to both sides.

$\dfrac{5}{x} = 5$

$x\left(\dfrac{5}{x}\right) = (5)x$ Multiply both sides by x.

$5 = 5x$

$\dfrac{5}{5} = \dfrac{5x}{5}$ Divide both sides by 5.

$1 = x$

EXERCISES 6.4

Solve each equation and check:

1. $\dfrac{2x}{3} = \dfrac{32}{6}$
2. $\dfrac{5x}{7} = \dfrac{20}{14}$
3. $\dfrac{3}{8}y = 1\dfrac{14}{16}$
4. $\dfrac{5}{3}x = -13\dfrac{1}{3}$
5. $2\dfrac{1}{2}x = 7\dfrac{1}{2}$
6. $\dfrac{1}{2} + \dfrac{x}{3} = \dfrac{5}{2}$
7. $\dfrac{2}{3} + \dfrac{x}{4} = \dfrac{28}{6}$
8. $\dfrac{x}{7} - \dfrac{1}{14} = \dfrac{70}{28}$
9. $\dfrac{3}{4} - \dfrac{x}{3} = \dfrac{5}{12}$
10. $1\dfrac{1}{4} + \dfrac{x}{3} = \dfrac{7}{12}$
11. $\dfrac{3}{5}x - 25 = \dfrac{x}{10}$
12. $\dfrac{2x}{3} - 7 = -\dfrac{x}{2}$
13. $\dfrac{y}{3} - 1 = \dfrac{y}{6}$
14. $\dfrac{1}{2}x - 3 = \dfrac{x}{5}$
15. $\dfrac{3x}{4} - \dfrac{7}{20} = \dfrac{2}{5}x$
16. $\dfrac{5x}{6} + \dfrac{1}{3}(6 + x) = 37$
17. $\dfrac{x}{2} + \dfrac{2}{3}(2x + 3) = 46$
18. $\dfrac{1}{6}x - \dfrac{1}{9}(2x - 3) = 1$
19. $0.96 = 0.06(12 + x)$
20. $\dfrac{1}{2}(x + 2) + \dfrac{3}{8}(28 - x) = 11$
21. $\dfrac{3x - 24}{16} - \dfrac{3x - 12}{12} = 3$
22. $\dfrac{4x + 3}{15} - \dfrac{2x - 3}{9} = \dfrac{6x + 4}{6} - x$
23. $5x + \dfrac{6x - 8}{14} + \dfrac{10x + 6}{6} = 43$
24. $\dfrac{3x}{5} - \dfrac{9 - 3x}{10} = \dfrac{6}{10} - \dfrac{3x + 6}{10}$
25. $\dfrac{4x}{6} - \dfrac{x + 5}{2} = \dfrac{6x - 6}{8}$
26. $\dfrac{x}{3} + \dfrac{2x + 4}{4} = \dfrac{x - 1}{6} - \dfrac{3 - 2x}{2}$
27. $\dfrac{4}{x} = 6$
28. $\dfrac{2}{x} - 8 = -7$
29. $5 - \dfrac{1}{x} = 7$
30. $\dfrac{3}{x} - 6 = 8$
31. $\dfrac{5}{y} - 1 = 4$
32. $\dfrac{17}{x} = 8$
33. $\dfrac{3}{x} - 8 = 7$
34. $\dfrac{5}{2x} + 8 = 17$
35. $7 - \dfrac{6}{x} = 5$
36. $9 + \dfrac{3}{x} = 10\dfrac{1}{2}$
37. $\dfrac{6}{x} + 5 = 14$
38. $\dfrac{3}{x} - 3 = \dfrac{5}{2x} - 2$
39. $1 - \dfrac{2}{x} = \dfrac{14}{3x} - \dfrac{1}{3}$
40. $\dfrac{3}{x} + 2 = \dfrac{5}{x} - 4$
41. $\dfrac{7}{2x} + 14\dfrac{1}{2} = \dfrac{7}{x} - 10$
42. $\dfrac{8}{x} + \dfrac{1}{4} = \dfrac{5}{x} + \dfrac{1}{3}$

6.5 Translating Words into Algebraic Symbols

The ability to translate English words into algebra is very important for solving "applied" problems. To help you, we provide the following table of common English words for the common mathematical symbols:

+	−	×	÷	=
plus	minus	times	divide	equal or equals
increased by	decreased by	product	quotient	is or are
added to	subtract	multiply by	divided by	is equal to
more than	less than	double or twice		result is
sum of	difference	triple or thrice		
	subtract from			

Example 1 Translate into algebra: One number is four times another, and their sum is twenty.

Let x = first number
$4x$ = four times the number
$x + 4x$ = their sum

Sentence in algebra: $x + 4x = 20$

Example 2 Translate into algebra: The sum of a number and the number decreased by six is five.

Let x = the number
$x - 6$ = number decreased by six
$x + (x - 6)$ = sum

Sentence in algebra: $x + (x - 6) = 5$

Example 3 Translate into algebra: Fifteen more than twice a number is twenty-four.

Let x = the number
$2x$ = twice the number
$2x + 15$ = fifteen more than twice the number

Sentence in algebra: $2x + 15 = 24$

Example 4 Translate into algebra: Twice the sum of a number and five is eighty.

Let x = the number
$x + 5$ = sum of a number and five
$2(x + 5)$ = twice the sum of a number and five

Sentence in algebra: $2(x + 5) = 80$

EXERCISES 6.5

Translate each phrase or sentence into algebraic symbols:

1. A number decreased by twenty
2. A number increased by five
3. A number divided by six
4. A number times eighteen
5. The sum of a number and sixteen
6. Subtract twenty-six from a number
7. Subtract a number from twenty-six
8. Half a number
9. Twice a number
10. The difference between four and a number

11. The sum of six times a number and twenty-eight is forty.
12. The difference between twice a number and thirty is fifty.
13. The quotient of a number and six is five.
14. If seven is added to a number, the sum is 32.
15. If a number is increased by 28 and then the sum is multiplied by five, the result is 150.
16. The sum of a number and the number decreased by five is 25.
17. Seven less than the quotient of a number and six is two.
18. The product of five and five more than a number is 50.
19. The difference between thirty and twice a number is four.
20. Double the difference between a number and six is thirty.
21. The product of a number decreased by seven and the same number increased by five is thirteen.
22. Seven times a number decreased by eleven is 32.
23. The product of a number and six decreased by seventeen is seven.
24. If twelve is added to the product of a number and twelve, the sum is 72.
25. Seventeen less than four times a number is 63.

6.6 Applications Involving Equations*

An applied problem can often be expressed mathematically as a simple equation. The problem can then be solved by solving the equation. To solve such an application problem, we suggest the following steps:

Solving Application Problems

STEP 1   Read the problem carefully at least twice.
STEP 2   If possible, draw a diagram. This will often help you to visualize the mathematical relationship needed to write the equation.
STEP 3   Choose a letter to represent the unknown quantity in the problem, and write what it represents.
STEP 4   Write an equation that expresses the information given in the problem and that involves the unknown.
STEP 5   Solve the equation from Step 4.
STEP 6   Check your solution both in the equation from Step 4 and in the original problem itself.

Example 1   You need to tile the floor of a rectangular room with a wooden outer border of 6 in. The floor of the room is 10 ft by 8 ft 2 in. How many rows of 4-in.-by-4-in. tiles are needed to fit across the length of the room?

The sketch shown in Figure 6.1 is helpful in solving the problem.

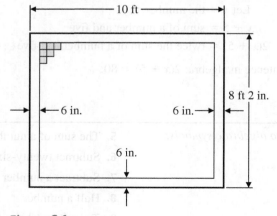

Figure 6.1

*Note: In this section, do not use the rules for calculating with measurements.

6.6 ♦ Applications Involving Equations

Let x = the number of tiles across the length of the room
$4x$ = the number of inches in x tiles

The total length of the rectangular room is then

$4x + 6 + 6 = 120$ 10 ft = 120 in.
$4x + 12 = 120$
$4x = 108$ Subtract 12 from both sides.
$x = 27$ Divide both sides by 4.

So there are 27 rows of tiles.

Example 2 An interior wall measures 30 ft 4 in. long. It is to be divided by 10 evenly spaced posts; each post is 4 in. by 4 in. (Posts are to be located in the corners.) What is the distance between posts? Note that there are 9 spaces between posts. (See Figure 6.2.)

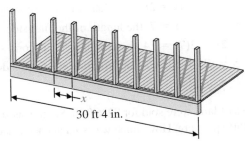

Figure 6.2

Let x = the distance between posts
$9x$ = the distance of 9 spaces
$(10)(4 \text{ in.})$ = the distance used up by ten 4-in. posts

The total length of the interior wall is then

$9x + (10)(4 \text{ in.}) = 364 \text{ in.}$ 30 ft 4 in. = 364 in.
$9x + 40 = 364$
$9x = 324$ Subtract 40 from both sides.
$x = 36$ in.   Divide both sides by 9.

Check:
$9x + 40 = 364$
$9(36) + 40 = 364$?
$324 + 40 = 364$?
$364 = 364$ True

Example 3 Two different automotive batteries cost a total of $270. One costs $12 more than twice the other. Find the cost of each battery.

Let x = the cost of one battery
$2x + 12$ = the cost of the other battery

The total cost of both batteries is then

$x + (2x + 12) = 270$
$3x + 12 = 270$ Combine like terms.
$3x = 258$ Subtract 12 from both sides.

$x = \$86$, the cost of the first battery
$2x + 12 = 2(86) + 12 = \$184$, the cost of the other battery

Example 4

One side of a triangle has a length twice another. The third side has a length 5 units more than the shortest side. The perimeter is 33. Find the length of each side.

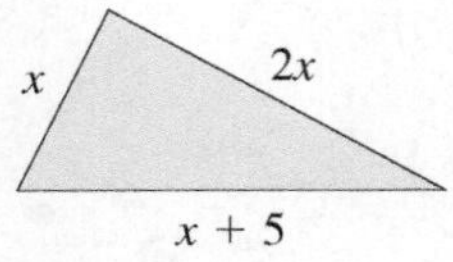

First, draw and label a triangle as in Figure 6.3.

Let $x =$ the length of the first side
$2x =$ the length of the second side
$x + 5 =$ the length of the third side

Figure 6.3

The sum of the three sides is then

$$x + 2x + (x + 5) = 33$$
$$4x + 5 = 33 \quad \text{Combine like terms.}$$
$$4x = 28 \quad \text{Subtract 5 from both sides.}$$
$$x = 7, \text{ the length of the first side}$$
$$2x = 2(7) = 14, \text{ the length of the second side}$$
$$x + 5 = (7) + 5 = 12, \text{ the length of the third side}$$

Example 5

Forty acres of land were sold for $216,000. Some was sold at $6400 per acre, and the rest was sold at $4800 per acre. How much was sold at each price?

Let $x =$ the amount of land sold at $6400/acre
$40 - x =$ the amount of land sold at $4800/acre

Then

$6400x =$ the value of the land sold at $6400/acre
$4800(40 - x) =$ the value of the land sold at $4800/acre
$216,000 =$ the total value of the land

Therefore, the equation for the total value of the land is

$$6400x + 4800(40 - x) = 216,000$$
$$6400x + 192,000 - 4800x = 216,000 \quad \text{Remove parentheses.}$$
$$1600x + 192,000 = 216,000 \quad \text{Combine like terms.}$$
$$1600x = 24,000 \quad \text{Subtract 192,000 from both sides.}$$
$$x = 15 \quad \text{Divide both sides by 1600.}$$
$$40 - x = 25$$

Thus, 15 acres were sold at $6400/acre and 25 acres were sold at $4800/acre.

NOTE: When you know the total of two parts, one possible equation-solving strategy is to

let $x =$ one part and
total $- x =$ the other part

Example 6

How much pure alcohol must be added to 200 mL of a solution that is 15% alcohol to make a solution that is 40% alcohol?

Let $x =$ the amount of pure alcohol (100%) added

You may find Figure 6.4 helpful.

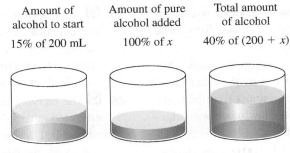

Figure 6.4

Write an equation in terms of the amount of pure alcohol; that is, the amount of pure alcohol in each separate solution equals the amount of pure alcohol in the final solution.

$$0.15(200) + 1.00x = 0.40(200 + x)$$
$$30 + x = 80 + 0.4x$$
$$x = 50 + 0.4x \qquad \text{Subtract 30 from both sides.}$$
$$0.6x = 50 \qquad \text{Subtract } 0.4x \text{ from both sides.}$$
$$x = 83.3 \qquad \text{Divide both sides by 0.6.}$$

Thus, 83.3 mL of pure alcohol must be added. ♦

EXERCISES 6.6

1. A built-in bookcase containing eight book shelves is to be constructed in a room. The floor-to-ceiling clearance is 8 ft 2 in. Each shelf is 1 in. thick. An equal space is to be left between the shelves, the top shelf and the ceiling, and the bottom shelf and the floor. What space should be between each shelf and the next? (There is no shelf against the ceiling and no shelf on the floor.)

2. Saw a board 8 ft 4 in. long into nine equal pieces. If the loss per cut is $\frac{1}{8}$ in., how long will each piece be?

3. Separate an order of 256 light fixtures so that the number of LED light fixtures will be 20 fewer than twice the number of incandescent light fixtures.

4. Distribute $1000 into three parts so that one part will be three times as large as the second and the third part will be as large as the sum of the other two.

5. Distribute $4950 among John, Maria, and Betsy so that Maria receives twice as much as John and Betsy receives three times as much as John.

6. Distribute $4950 among John, Maria, and Betsy so that Maria receives twice as much as John and Betsy receives three times as much as Maria.

7. A rectangle is twice as long as it is wide. Its perimeter (the sum of the lengths of its sides) is 60 cm. Find its length and width.

8. The length of a rectangle is 4 cm less than twice its width. Its perimeter is 40 cm. Find its length and width.

9. One side of a rectangular yard is bounded by the side of a house. The other three sides are to be fenced with 345 ft of fencing. The length of fence opposite the house is 15 ft less than either of the other two sides. Find the length and width of the yard.

10. A given type of concrete contains an amount of sand twice the amount of cement and an amount of gravel that is 50% more than the amount of sand. How many cubic yards of each must be used to make 9 yd³ of concrete? Assume no loss of volume in mixing.

11. The perimeter of a triangle is 122 ft. The lengths of two sides are the same. The length of the third side is 4 ft shorter than either of the other two sides. Find the lengths of the three sides.

12. Cut a board 20 ft long into three pieces so that the longest piece will be three times as long as each of the other two of equal lengths. Find the length of each piece.

13. Cut a 12-ft beam into two pieces so that one piece is 18 in. longer than the other. Find the length of the two pieces.

14. The total cost of three automobile batteries is $340. The most expensive battery is three times the cost of the least expensive. The third is $15 more than the least expensive. Find the cost of each battery.

15. The total cost of 20 boards is $166. One size costs $6.50, and the second size costs $9.50. How many boards are purchased at each price?

16. Amy and Kurt earned a total of $512 by working a total of 30 hours. If Amy earns $20/h and Kurt earns $16/h, how many hours did each work?

17. Joyce invests $7500 in two savings accounts. One account earns interest at 4% per year; the other earns 2.5% per year. The total interest earned from both accounts after one year is $232.50. How much was originally deposited in each account?

18. Chuck receives loans totaling $12,000 from two banks. One bank charges 7.5% annual interest, and the second bank charges 6% annual interest. He paid $840 in total interest in one year. How much was loaned at each bank?

19. Regular milk has 4% butterfat. How many litres of regular milk must be mixed with 40 L of milk with 1% butterfat to have milk with 2% butterfat?

20. How much pure alcohol must be added to 750 mL of a solution that is 40% alcohol to make a solution that is 60% alcohol?

21. Mix a solution that is 30% alcohol with a solution that is 80% alcohol to make 800 mL of a solution that is 60% alcohol. How much of each solution should you use?

22. Mix a solution that is 50% acid with a solution that is 100% water to make 4 L of a solution that is 10% acid. How much of each solution should you use?

23. A 12-quart cooling system is checked and found to be filled with a solution that is 40% antifreeze. The desired strength of the solution is 60% antifreeze. How many quarts of solution need to be drained and replaced with pure antifreeze to reach the desired strength?

24. In testing an engine, various mixtures of gasoline and methanol are being tried. How much of a 90% gasoline mixture and a 75% gasoline mixture are needed for 1200 L of an 85% gasoline mixture?

25. Assume you have sea water that weighs 64 lb/ft^3 and contains 35 parts per thousand (0.0035) dissolved salt by weight. How many cubic feet of the sea water would you need to place in an evaporation basin to collect 125 lb of sea salt?

26. The carrying capacity of any rangeland area can be estimated in terms of animal equivalent units (AEUs). A white-tailed deer requires 0.25 AEU of grazing capacity and an elk requires 0.67 AEU. A wildlife manager estimates that a state park will provide grazing to support about 150 AEUs on a sustained basis. If the park has 75 elk, what is the maximum number of deer that it will support?

27. Jeff invested $30,000 in a business whereas his partner Kris invested $40,000. How much more needs to be invested in order to generate a total that will yield $6250 annually with a 5% rate of return?

28. The dues for a country club membership are $725 per month plus a $700 allotment charged annually to be applied to food and beverage purchases. What amount must be invested in a bond at 8% to generate enough interest to cover the cost for each year's membership?

29. Cloth table coverings are to be cut for 8 rectangular tables. The length of the table is 3 times its width. The sum of the measurements of the four sides of a table is 240 in. Find the width and length of each table.

30. How many quarts of water must be added to 3 gal of soup that is 60% chicken broth to make the soup 40% chicken broth?

31. Ground beef that is 90% lean costs $1\frac{1}{4}$ times the cost of 80% lean ground beef. If the total cost of one pound of 90% lean ground beef plus one pound of 80% lean ground beef is $5.40, find the cost per pound of the 90% lean ground beef and the cost per pound of the 80% lean ground beef.

6.7 Formulas

A **formula** is a general rule written as an equation, usually expressed in letters, which shows the relationship between two or more quantities. For example, the formula

$$d = rt$$

states that the distance, d, that a body travels equals the product of its rate, r, of travel and the time, t, of travel. The formula

$$p = \frac{F}{A}$$

states that the pressure, p, equals the quotient of the force, F, and the area, A, over which the force is applied.

6.7 ♦ Formulas 241

Sometimes the letters in a formula do not match the first letter of the name of the quantity. For example, Ohm's law is often written

$$E = IR$$

where E is the voltage, I is the current, and R is the resistance.

Sometimes subscripts are used to distinguish between different readings of the same quantity. For example, the final velocity of an object equals the sum of the initial velocity and the product of its acceleration and the time of the acceleration. This is written

$$v_f = v_i + at$$

where v_f is the final velocity, v_i is the initial velocity, a is the acceleration, and t is the time.

Sometimes Greek letters are used. For example, the resistance of a wire is given by the formula

$$R = \frac{\rho L}{A}$$

where R is the resistance of the wire, ρ (rho) is the resistivity constant of the wire, L is the length of the wire, and A is the cross-sectional area of the wire.

Sometimes the formula is written with the letters and symbols used by the person who discovered the relationship. The letters may have no obvious relationship with the quantity.

Solving Formulas

To solve a formula for a given letter means to isolate the given letter on one side of the equation and express it in terms of all the remaining letters.

This means that the given letter appears on one side of the equation by itself; all the other letters appear on the opposite side of the equation. We solve a formula using the same methods that we use in solving an equation.

Example 1   Solve **a.** $d = rt$ for t and **b.** $15 = 5t$ for t.

a. $d = rt$

$\dfrac{d}{r} = \dfrac{rt}{r}$ Divide both sides by r.

$\dfrac{d}{r} = t$

b. $15 = 5t$

$\dfrac{15}{5} = \dfrac{5t}{5}$ Divide both sides by 5.

$3 = t$ ♦

Note that we use the same techniques for solving a formula as we learned earlier for solving equations.

Example 2   Solve $p = \dfrac{F}{A}$ for F and then for A.

First, solve for F:

$p = \dfrac{F}{A}$

$pA = \left(\dfrac{F}{A}\right)A$ Multiply both sides by A.

$pA = F$

Now solve for A:

$$p = \frac{F}{A}$$

$pA = F$ Multiply both sides by A.

$$\frac{pA}{p} = \frac{F}{p}$$ Divide both sides by p.

$$A = \frac{F}{p}$$

Example 3   Solve $V = E - Ir$ for I.

One way:

$$V = E - Ir$$

$V - E = E - Ir - E$ Subtract E from both sides.

$V - E = -Ir$

$$\frac{V - E}{-r} = \frac{-Ir}{-r}$$ Divide both sides by $-r$.

$$\frac{V - E}{-r} = I$$

Alternative way:

$$V = E - Ir$$

$V + Ir = E - Ir + Ir$ Add Ir to both sides.

$V + Ir = E$

$V + Ir - V = E - V$ Subtract V from both sides.

$Ir = E - V$

$$\frac{Ir}{r} = \frac{E - V}{r}$$ Divide both sides by r.

$$I = \frac{E - V}{r}$$

Note that the two results are equivalent. Take the first result,

$$\frac{V - E}{-r}$$

and multiply numerator and denominator by -1:

$$\left(\frac{V - E}{-r}\right)\left(\frac{-1}{-1}\right) = \frac{-V + E}{r} = \frac{E - V}{r}$$

Example 4   Solve $S = \dfrac{n(a + l)}{2}$ for n and then for l.

First, solve for n:

$$S = \frac{n(a + l)}{2}$$

$2S = n(a + l)$ Multiply both sides by 2.

$$\frac{2S}{a + l} = n$$ Divide both sides by $(a + l)$.

Now solve for l:

$$S = \frac{n(a+l)}{2}$$

$2S = n(a+l)$	Multiply both sides by 2.
$2S = na + nl$	Remove parentheses.
$2S - na = nl$	Subtract na from both sides.
$\dfrac{2S - na}{n} = l$	Divide both sides by n.

or $\quad \dfrac{2S}{n} - a = l$

Example 5 Solve $(\triangle L) = kL(T - T_0)$ for T.

$(\triangle L) = kL(T - T_0)$.	Note: Treat $(\triangle L)$ as one variable.
$(\triangle L) = kLT - kLT_0$	Remove parentheses.
$(\triangle L) + kLT_0 = kLT$	Add kLT_0 to both sides.
$\dfrac{(\triangle L) + kLT_0}{kL} = T$	Divide both sides by kL.

Can you show that $T = \dfrac{(\triangle L)}{kL} + T_0$ is an equivalent solution?

EXERCISES 6.7

Solve each formula for the given letter:

1. $E = Ir$ for r
2. $A = bh$ for b
3. $F = ma$ for a
4. $w = mg$ for m
5. $C = \pi d$ for d
6. $V = IR$ for R
7. $V = lwh$ for w
8. $X_L = 2\pi f L$ for f
9. $A = 2\pi rh$ for h
10. $C = 2\pi r$ for r
11. $v^2 = 2gh$ for h
12. $V = \pi r^2 h$ for h
13. $I = \dfrac{Q}{t}$ for t
14. $I = \dfrac{Q}{t}$ for Q
15. $v = \dfrac{s}{t}$ for s
16. $I = \dfrac{E}{Z}$ for Z
17. $I = \dfrac{V}{R}$ for R
18. $P = \dfrac{w}{t}$ for w
19. $E = \dfrac{I}{4\pi r^2}$ for I
20. $R = \dfrac{\pi}{2P}$ for P
21. $X_C = \dfrac{1}{2\pi fC}$ for f
22. $R = \dfrac{\rho L}{A}$ for L
23. $A = \dfrac{1}{2}bh$ for b
24. $V = \dfrac{1}{3}\pi r^2 h$ for h
25. $Q = \dfrac{I^2 Rt}{J}$ for R
26. $R = \dfrac{kl}{D^2}$ for l
27. $F = \dfrac{9}{5}C + 32$ for C
28. $C = \dfrac{5}{9}(F - 32)$ for F
29. $C_T = C_1 + C_2 + C_3 + C_4$ for C_2
30. $R_T = R_1 + R_2 + R_3 + R_4$ for R_4
31. $Ax + By + C = 0$ for x
32. $A = P + Prt$ for r
33. $Q_1 = P(Q_2 - Q_1)$ for Q_2
34. $v_f = v_i + at$ for v_i
35. $A = \left(\dfrac{a+b}{2}\right)h$ for h
36. $A = \left(\dfrac{a+b}{2}\right)h$ for b
37. $l = a + (n-1)d$ for d
38. $A = ab + \dfrac{d}{2}(a+c)$ for d
39. $Ft = m(V_2 - V_1)$ for m
40. $l = a + (n-1)d$ for n
41. $Q = wc(T_1 - T_2)$ for c
42. $Ft = m(V_2 - V_1)$ for V_2
43. $V = \dfrac{2\pi(3960 + h)}{P}$ for h
44. $Q = wc(T_1 - T_2)$ for T_2

6.8 Substituting Data into Formulas

Problem-solving skills are essential in all technical fields. Working with formulas is one of the most important tools that you can gain from this course.

> **Problem Solving**
>
> Necessary parts of problem solving include:
>
> 1. Analyzing the given data.
> 2. Finding an equation or formula that relates the given quantities with the unknown quantity.
> 3. Solving the formula for the unknown quantity.
> 4. Substituting the given data into this solved formula.

Actually, you may solve the formula for the unknown quantity and then substitute the data. Or you may substitute the data into the formula first and then solve for the unknown quantity. If you use a calculator, the first method is more helpful. Another good reason to use the first method is that some variables might cancel out so that any error introduced by a lower accuracy of this canceled variable would be eliminated entirely.

Example 1 Given the formula $V = IR$, $V = 120$, and $R = 2\overline{0}0$. Find I.

First, solve for I:

$$V = IR$$

$$\frac{V}{R} = \frac{IR}{R} \quad \text{Divide both sides by } R.$$

$$\frac{V}{R} = I$$

Then, substitute the data:

$$I = \frac{V}{R} = \frac{120}{2\overline{0}0} = 0.60$$

♦

Example 2 Given $v = v_0 + at$, $v = 60.0$, $v_0 = 20.0$, and $t = 5.00$. Find a.

First, solve for a:

$$v = v_0 + at$$

$$v - v_0 = at \quad \text{Subtract } v_0 \text{ from both sides.}$$

$$\frac{v - v_0}{t} = \frac{at}{t} \quad \text{Divide both sides by } t.$$

$$\frac{v - v_0}{t} = a$$

Then substitute the data:

$$a = \frac{v - v_0}{t} = \frac{60.0 - 20.0}{5.00} = \frac{40.0}{5.00} = 8.00$$

♦

Example 3

Given the formula $S = \dfrac{MC}{l}$, $S = 47.5$, $M = 190$, and $C = 8.0$. Find l.

First, solve for l:

$$S = \dfrac{MC}{l}$$

$Sl = MC$ Multiply both sides by l.

$l = \dfrac{MC}{S}$ Divide both sides by S.

Then, substitute the data:

$$l = \dfrac{MC}{S} = \dfrac{190(8.0)}{47.5} = 32$$ ♦

Example 4

Given the formula $Q = WC(T_1 - T_2)$, $Q = 15$, $W = 3.0$, $T_1 = 11\overline{0}$, and $T_2 = 6\overline{0}$. Find C.

First, solve for C:

$$Q = WC(T_1 - T_2)$$

$$\dfrac{Q}{W(T_1 - T_2)} = C \quad \text{Divide both sides by } W(T_1 - T_2).$$

Then, substitute the data:

$$\dfrac{15}{3.0(11\overline{0} - 6\overline{0})} = C$$

$$C = \dfrac{15}{150} = 0.10$$ ♦

Example 5

Given the formula $V = \dfrac{1}{2}lw(D + d)$, $V = 156.8$, $D = 2.00$, $l = 8.37$, and $w = 7.19$. Find d.

First, solve for d:

$$V = \dfrac{1}{2}lw(D + d)$$

$2V = lw(D + d)$ Multiply both sides by 2.

$2V = lwD + lwd$ Remove parentheses.

$2V - lwD = lwd$ Subtract lwD from both sides.

$\dfrac{2V - lwD}{lw} = d$ Divide both sides by lw.

$\dfrac{2V}{lw} - D = d$ Simplify the left side.

Then, substitute the data:

$$d = \dfrac{2V}{lw} - D = \dfrac{2(156.8)}{(8.37)(7.19)} - 2.00 = 3.21$$ ♦

Example 6

If you want to double the distance traveled and decrease the time by half, how would this affect your speed?

First, the relationship among these quantities is given by the formula $d = rt$.

Solve this formula for r.

$$d = rt$$

$$r = \dfrac{d}{t}$$

Substitute d by $2d$ (double the distance) and t by $\frac{1}{2}t$ (half the time).

$$r = \frac{2d}{\frac{1}{2}t} = 4\frac{d}{t}$$

This means the rate (speed) would be increased to 4 times its original speed. ◆

EXERCISES 6.8

a. *Solve for the indicated letter.* **b.** *Then substitute the given values to find the value of the indicated letter (use the rules for working with measurements):*

	Formula	Given	Find
1.	$A = lw$	$A = 414$, $w = 18.0$	l
2.	$V = IR$	$I = 9.20$, $V = 5.52$	R
3.	$V = \dfrac{\pi r^2 h}{3}$*	$V = 753.6$, $r = 6.00$	h
4.	$I = \dfrac{V}{R}$	$R = 44$, $I = 2.5$	V
5.	$E = \dfrac{mv^2}{2}$	$E = 484{,}000$; $v = 22.0$	m
6.	$v_f = v_i + at$	$v_f = 88$, $v_i = 10.0$, $t = 12$	a
7.	$v_f = v_i + at$	$v_f = 193.1$, $v_i = 14.9$, $a = 18.0$	t
8.	$y = mx + b$	$x = 3$, $y = 2$, $b = 9$	m
9.	$v_f^2 = v_i^2 + 2gh$	$v_f = 192$, $v_i = 0$, $g = 32.0$	h
10.	$A = P + Prt$	$r = 0.07$, $P = \$1500$, $A = \$2025$	t
11.	$L = \pi(r_1 + r_2) + 2d$	$L = 37.68$, $d = 6.28$, $r_2 = 5.00$	r_1
12.	$C = \dfrac{5}{9}(F - 32)$	$C = -20$	F
13.	$Fgr = Wv^2$	$F = 12{,}000$; $W = 24{,}000$; $v = 176$; $g = 32$	r
14.	$Q = WC(T_1 - T_2)$	$Q = 18.9$, $W = 3.0$, $C = 0.18$, $T_2 = 59$	T_1
15.	$A = \dfrac{1}{2}h(a + b)$	$A = 1160$, $h = 22.0$, $a = 56.5$	b
16.	$A = \dfrac{1}{2}h(a + b)$	$A = 5502$, $h = 28.0$, $b = 183$	a
17.	$V = \dfrac{1}{2}lw(D + d)$	$V = 226.8$, $l = 9.00$, $w = 6.30$, $D = 5.00$	d
18.	$S = \dfrac{n}{2}(a + l)$	$S = 575$, $n = 25$, $l = 15$	a
19.	$S = \dfrac{n}{2}(a + l)$	$S = 147.9$, $n = 14.5$, $l = 3.80$	a
20.	$S = \dfrac{n}{2}(a + l)$	$S = 96\dfrac{7}{8}$, $n = 15$, $a = 8\dfrac{2}{3}$	l

Note: Use the π key on your calculator.

21. A drill draws a current, I, of 4.50 A. The resistance, R, is 16.0 Ω. Find its power, P, in watts. $P = I^2R$.

22. The formula for piston displacement is $P = cd^2SN$, where c is a constant, d is the cylinder bore, S is the stroke, and N is the number of cylinders. For $c = 0.7854$, $d = 3$, $S = 4$, and $N = 4$, find the piston displacement.

23. The length of a cylinder is given by $l = \dfrac{V}{cd^2}$, where c is the constant 0.785, d is the diameter of the cylinder, and V is the volume of the cylinder. Find l if $V = 47.0$ in^3 and $d = 2.98$ in.

24. A flashlight bulb is connected to a 1.50-V source. Its current, I, is 0.250 A. What is its resistance, R, in ohms? $V = IR$.

25. The area of a rectangle is 84.0 ft^2. Its length is 12.5 ft. Find its width. $A = lw$.

26. The volume of a box is given by $V = lwh$. Find the width if the volume is 3780 ft^3, its length is 21.0 ft, and its height is 15.0 ft.

27. The volume of a cylinder is given by the formula $V = \pi r^2 h$, where r is the radius and h is the height. Find the height in m if the volume is 8550 m^3 and the radius is 15.0 m.

28. The pressure at the bottom of a lake is found by the formula $P = hD$, where h is the depth of the water and D is the density of the water. Find the pressure in lb/in^2 at 175 ft below the surface. $D = 62.4$ lb/ft^3.

29. The equivalent resistance R of two resistances connected in parallel is given by $\dfrac{1}{R} = \dfrac{1}{R_1} + \dfrac{1}{R_2}$. Find the equivalent resistance for two resistances of 20.0 Ω and 60.0 Ω connected in parallel.

30. The R value of insulation is given by the formula $R = \dfrac{L}{K}$, where L is the thickness of the insulating material and K is the thermal conductivity of the material. Find the R value of 8.0 in. of mineral wool insulation. $K = 0.026$. *Note:* L must be in feet.

31. A steel railroad rail expands and contracts according to the formula $\triangle l = \alpha l \triangle T$, where $\triangle l$ is the change in length, α is a constant called the *coefficient of linear expansion*, l is the original length, and $\triangle T$ is the change in temperature. If a 50.0-ft steel rail is installed at 0°F, how many inches will it expand when the temperature rises to 110°F? $\alpha = 6.5 \times 10^{-6}$/°F.

32. The inductive reactance, X_L, of a coil is given by $X_L = 2\pi f L$, where f is the frequency and L is the inductance. If the inductive reactance is 245 Ω and the frequency is 60.0 cycles/s, find the inductance L in henrys, H.

33. If you want to triple the distance traveled and double the time, how would this affect your speed?

34. If you want to halve the distance traveled and triple the time, how would this affect your speed?

35. The momentum, p, of an object equals the product of its mass, m, times its velocity, v. The mass of a given truck is 4 times the mass of an automotive. The truck is traveling $\frac{3}{4}$ the speed of the automobile. Compare the relative momentum of each.

36. A loaded truck is 8 times the mass of a small automobile. Its speed is $1\frac{1}{4}$ times the speed of the automobile. Compare the relative momentum of each.

6.9 Reciprocal Formulas Using a Calculator

The **reciprocal of a number** is 1 divided by that number. The product of a number (except 0) and its reciprocal is 1. Examples of numbers and their reciprocals are shown here:

Number	Reciprocal
4	$\frac{1}{4}$
-6	$-\frac{1}{6}$
$\frac{2}{3}$	$\frac{3}{2}$
$-\frac{12}{7}$	$-\frac{7}{12}$
0	None

The reciprocal of a number may be found by using the $1/x$ or x^{-1} key. This may require you to use the second function key on your calculator.

CHAPTER 6 ♦ Equations and Formulas

Example 1 Find the reciprocal of 12 rounded to three significant digits.

12 [x^{-1}] [=]

[0.083333333]

Thus, $\dfrac{1}{12} = 0.0833$ rounded to three significant digits.

Example 2 Find $\dfrac{1}{41.2}$ rounded to three significant digits.

41.2 [x^{-1}] [=]

[0.024271845]

The reciprocal of 41.2 is 0.0243 rounded to three significant digits.

Formulas involving reciprocals are often used in electronics and physics. We next consider an alternative method for substituting data into such formulas and solving for a specified letter using a calculator.

To use a calculator with formulas involving reciprocals,

1. Solve for the reciprocal of the specified letter.
2. Substitute the given data.
3. Follow the calculator steps shown in the following examples.

Example 3 Given the formula $\dfrac{1}{R} = \dfrac{1}{R_1} + \dfrac{1}{R_2}$, where $R_1 = 6.00\ \Omega$ and $R_2 = 12.0\ \Omega$, find R.

Since the formula is already solved for the reciprocal of R, substitute the data:

$$\dfrac{1}{R} = \dfrac{1}{R_1} + \dfrac{1}{R_2}$$

$$\dfrac{1}{R} = \dfrac{1}{6.00\ \Omega} + \dfrac{1}{12.0\ \Omega}$$

Then use your calculator as follows:

6 [x^{-1}] [+] 12 [x^{-1}] [=] [x^{-1}] [=]

[4]

So $R = 4.00\ \Omega$.

NOTE: This formula relates the electrical resistances in a parallel circuit.

Example 4 Given the formula $\dfrac{1}{f} = \dfrac{1}{s_0} + \dfrac{1}{s_i}$, where $f = 8.00$ cm and $s_0 = 12.0$ cm, find s_i.

First, solve the formula for the reciprocal of s_i:

$$\dfrac{1}{f} = \dfrac{1}{s_0} + \dfrac{1}{s_i}$$

$$\dfrac{1}{s_i} = \dfrac{1}{f} - \dfrac{1}{s_0} \qquad \text{Subtract } \dfrac{1}{s_0} \text{ from both sides.}$$

Next, substitute the data:

$$\frac{1}{s_i} = \frac{1}{8.00 \text{ cm}} - \frac{1}{12.0 \text{ cm}}$$

Then use your calculator as follows:

8 [x^{-1}] [−] 12 [x^{-1}] [=] [x^{-1}] [=]

24

So $s_i = 24.0$ cm.

Example 5 Given the formula $\frac{1}{C} = \frac{1}{C_1} + \frac{1}{C_2} + \frac{1}{C_3}$, where $C = 2.00 \ \mu\text{F}$, $C_1 = 3.00 \ \mu\text{F}$, and $C_3 = 18.0 \ \mu\text{F}$, find C_2.

First, solve the formula for the reciprocal of C_2:

$$\frac{1}{C} = \frac{1}{C_1} + \frac{1}{C_2} + \frac{1}{C_3}$$

$$\frac{1}{C_2} = \frac{1}{C} - \frac{1}{C_1} - \frac{1}{C_3} \qquad \text{Subtract } \frac{1}{C_1} \text{ and } \frac{1}{C_3} \text{ from both sides.}$$

Next, substitute the data:

$$\frac{1}{C_2} = \frac{1}{2.00 \ \mu\text{F}} - \frac{1}{3.00 \ \mu\text{F}} - \frac{1}{18.0 \ \mu\text{F}}$$

Then, use your calculator as follows:

2 [x^{-1}] [−] 3 [x^{-1}] [−] 18 [x^{-1}] [=] [x^{-1}] [=]

9

Therefore, $C_2 = 9.00 \ \mu\text{F}$.

NOTE: This formula relates the electrical capacitances of capacitors in a series circuit.

EXERCISES 6.9

Use the formula $\frac{1}{R} = \frac{1}{R_1} + \frac{1}{R_2}$ for Exercises 1–6:

1. Given $R_1 = 8.00 \ \Omega$ and $R_2 = 12.0 \ \Omega$, find R.
2. Given $R = 5.76 \ \Omega$ and $R_1 = 9.00 \ \Omega$, find R_2.
3. Given $R = 12.0 \ \Omega$ and $R_2 = 36.0 \ \Omega$, find R_1.
4. Given $R_1 = 24.0 \ \Omega$ and $R_2 = 18.0 \ \Omega$, find R.
5. Given $R = 15.0 \ \Omega$ and $R_2 = 24.0 \ \Omega$, find R_1.
6. Given $R = 90.0 \ \Omega$ and $R_1 = 125 \ \Omega$, find R_2.

Use the formula $\frac{1}{f} = \frac{1}{s_0} + \frac{1}{s_i}$ for Exercises 7–10:

7. Given $s_0 = 3.00$ cm and $s_i = 15.0$ cm, find f.
8. Given $f = 15.0$ cm and $s_i = 25.0$ cm, find s_0.
9. Given $f = 14.5$ cm and $s_0 = 21.5$ cm, find s_i.
10. Given $s_0 = 16.5$ cm and $s_i = 30.5$ cm, find f.

Use the formula $\frac{1}{R} = \frac{1}{R_1} + \frac{1}{R_2} + \frac{1}{R_3}$ for Exercises 11–16:

11. Given $R_1 = 30.0 \ \Omega$, $R_2 = 18.0 \ \Omega$, and $R_3 = 45.0 \ \Omega$, find R.
12. Given $R_1 = 75.0 \ \Omega$, $R_2 = 50.0 \ \Omega$, and $R_3 = 75.0 \ \Omega$, find R.
13. Given $R = 80.0 \ \Omega$, $R_1 = 175 \ \Omega$, and $R_2 = 275 \ \Omega$, find R_3.
14. Given $R = 145 \ \Omega$, $R_2 = 875 \ \Omega$, and $R_3 = 645 \ \Omega$, find R_1.
15. Given $R = 1250 \ \Omega$, $R_1 = 3750 \ \Omega$, and $R_3 = 4450 \ \Omega$, find R_2.
16. Given $R = 1830 \ \Omega$, $R_1 = 4560 \ \Omega$, and $R_2 = 9150 \ \Omega$, find R_3.

Use the formula $\dfrac{1}{C} = \dfrac{1}{C_1} + \dfrac{1}{C_2} + \dfrac{1}{C_3}$ for Exercises 17–22:

17. Given $C_1 = 12.0 \ \mu\text{F}$, $C_2 = 24.0 \ \mu\text{F}$, and $C_3 = 24.0 \ \mu\text{F}$, find C.
18. Given $C = 45.0 \ \mu\text{F}$, $C_1 = 85.0 \ \mu\text{F}$, and $C_3 = 115 \ \mu\text{F}$, find C_2.
19. Given $C = 1.25 \times 10^{-6}$ F, $C_1 = 8.75 \times 10^{-6}$ F, and $C_2 = 6.15 \times 10^{-6}$ F, find C_3.
20. Given $C = 1.75 \times 10^{-12}$ F, $C_2 = 7.25 \times 10^{-12}$ F, and $C_3 = 5.75 \times 10^{-12}$ F, find C_1.
21. Given $C_1 = 6.56 \times 10^{-7}$ F, $C_2 = 5.05 \times 10^{-6}$ F, and $C_3 = 1.79 \times 10^{-8}$ F, find C.
22. Given $C = 4.45 \times 10^{-9}$ F, $C_1 = 5.08 \times 10^{-8}$ F, and $C_3 = 7.79 \times 10^{-9}$ F, find C_2.

Use the formula $\dfrac{1}{R} = \dfrac{1}{R_1} + \dfrac{1}{R_2} + \dfrac{1}{R_3} + \dfrac{1}{R_4}$ for Exercises 23–24:

23. Given $R_1 = 655 \ \Omega$, $R_2 = 775 \ \Omega$, $R_3 = 1050 \ \Omega$, and $R_4 = 1250 \ \Omega$, find R.
24. Given $R = 155 \ \Omega$, $R_1 = 625 \ \Omega$, $R_3 = 775 \ \Omega$, and $R_4 = 1150 \ \Omega$, find R_2.

SUMMARY | CHAPTER 6

Glossary of Basic Terms

Algebraic expression. A combination of numbers, variables, symbols for operations, and symbols for grouping. (p. 222)
Equation. A statement that two quantities are equal. (p. 222)
Equivalent equations. Equations with the same solutions or roots. (p. 222)
Formula. A general rule written as an equation, usually expressed in letters, that shows the relationship between two or more quantities. (p. 240)
Reciprocal of a number. 1 divided by that number. The product of a number (except 0) and its reciprocal is 1. (p. 247)
Solution or **root.** A replacement number (or numbers) that produces a true statement in an equation. (p. 222)
Solve an equation. Find what number or numbers can replace the variable to make the equation a true statement. (p. 222)
Variable. A symbol (usually a letter) used to represent an unknown number. (p. 222)

6.1 Equations

1. **Four basic rules used to solve equations:**
   a. If the same quantity is added to both sides of an equation, the resulting equation is equivalent to the original equation.
   b. If the same quantity is subtracted from both sides of an equation, the resulting equation is equivalent to the original equation.
   c. If both sides of an equation are multiplied by the same (nonzero) quantity, the resulting equation is equivalent to the original equation.
   d. If both sides of an equation are divided by the same (nonzero) quantity, the resulting equation is equivalent to the original equation. (p. 222)
2. When more than one operation is indicated on the variable, undo the additions and subtractions first, then undo the multiplications and divisions. (p. 224)

6.3 Equations with Parentheses

1. **Solving an equation with parentheses:** To solve an equation having parentheses, remove the parentheses first. Then solve the resulting equation using other methods explained in this chapter. (p. 228)

6.4 Equations with Fractions

1. **Solving an equation with fractions:** To solve an equation with fractions:
   a. Find the least common denominator (LCD) of all the fractional terms on both sides of the equation.
   b. Multiply both sides of the equation by the LCD. (If this step has been done correctly, no fractions should now appear in the resulting equation.)
   c. Then solve the resulting equation using other methods explained in this chapter. (p. 230)

6.6 Applications Involving Equations

1. **Solving application problems:** To solve application problems:
   a. Read the problem carefully at least twice.
   b. If possible, draw a diagram. This will often help you to visualize the mathematical relationship needed to write the equation.
   c. Choose a letter to represent the unknown quantity in the problem, and write what it represents.
   d. Write an equation that expresses the information given in the problem and that involves the unknown.

e. Solve the equation.
f. Check your solution in the equation and in the original problem itself. (p. 236)

6.7 Formulas

1. **Solving formulas:** To solve a formula for a given letter, isolate the given letter on one side of the equation and express it in terms of all the remaining letters. (p. 241)

6.8 Substituting Data into Formulas

1. **Problem Solving:** To solve problems:
   a. Analyze the given data.
   b. Find an equation or formula that relates the given quantities with the unknown quantity.

c. Solve the formula for the unknown quantity.
d. Substitute the given data into this solved formula. (p. 244)

6.9 Reciprocal Formulas Using a Calculator

1. **To use a calculator with formulas involving reciprocals:**
   a. Solve for the reciprocal of the specified letter.
   b. Substitute the given data.
   c. Follow the calculator steps shown in Section 6.9. (p. 248)

REVIEW | CHAPTER 6

Solve each equation and check:

1. $2x + 4 = 7$
2. $11 - 3x = 23$
3. $\dfrac{x}{3} - 7 = 12$
4. $5 - \dfrac{x}{6} = 1$
5. $78 - 16y = 190$
6. $25 = 3x - 2$
7. $2x + 9 = 5x - 15$
8. $-6x + 5 = 2x - 19$
9. $3 - 2x = 9 - 3x$
10. $4x + 1 = 4 - x$
11. $7 - (x - 5) = 11$
12. $4x + 2(x + 3) = 42$
13. $3y - 5(2 - y) = 22$
14. $6(x + 7) - 5(x + 8) = 0$
15. $3x - 4(x - 3) = 3(x - 4)$
16. $4(x + 3) - 9(x - 2) = x + 27$
17. $\dfrac{2x}{3} = \dfrac{16}{9}$
18. $\dfrac{x}{3} - 2 = \dfrac{3x}{5}$
19. $\dfrac{3x}{4} - \dfrac{x-1}{5} = \dfrac{3+x}{2}$
20. $\dfrac{7}{x} - 3 = \dfrac{1}{x}$
21. $5 - \dfrac{7}{x} = 3\dfrac{3}{5}$

22. The length of a rectangle is 6 more than twice its width. Its perimeter is 48 in. Find its length and width.

23. Mix a solution that is 60% acid with a solution that is 100% acid to make 12 L of a solution that is 75% acid. How much of each solution should you use?

Solve each formula for the given letter:

24. $F = Wg$ for g
25. $P = \dfrac{W}{A}$ for A
26. $L = A + B + \dfrac{1}{2}t$ for t
27. $k = \dfrac{1}{2}mv^2$ for m
28. $P_2 = \dfrac{P_1 T_2}{T_1}$ for T_1
29. $v = \dfrac{v_f + v_0}{2}$ for v_0
30. $C = \dfrac{5}{9}(F - 32)$; find F if $C = 175$.
31. $P = 2(l + w)$; find w if $P = 112.8$ and $l = 36.9$.
32. $k = \dfrac{1}{2}mv^2$; find m if $k = 460$ and $v = 5.0$.
33. Given $\dfrac{1}{R} = \dfrac{1}{R_1} + \dfrac{1}{R_2}$, $R_1 = 50.0\ \Omega$, and $R_2 = 75.0\ \Omega$, find R.
34. Given $\dfrac{1}{C} = \dfrac{1}{C_1} + \dfrac{1}{C_2} + \dfrac{1}{C_3}$, $C = 25.0\ \mu F$, $C_1 = 75.0\ \mu F$, and $C_3 = 80.0\ \mu F$, find C_2.

TEST | CHAPTER 6

Solve each equation:

1. $x - 8 = -6$
2. $4x = 60$
3. $10 - 2x = 42$
4. $3x + 14 = 29$
5. $7x - 20 = 5x + 4$
6. $-2(x + 10) = 3(5 - 2x) + 5$
7. $\dfrac{1}{2}(3x - 6) = 3(x - 2)$
8. $\dfrac{8x}{9} = \dfrac{5}{6}$
9. $\dfrac{3x}{5} - 2 = \dfrac{x}{5} - \dfrac{x}{10}$
10. $\dfrac{8}{x} + 6 = 2$

11. $\dfrac{x}{2} - \dfrac{2}{5} = \dfrac{2x}{5} - \dfrac{3}{4}$

12. Distribute $2700 among Jose, Maria, and George so that Maria receives $200 more than Jose and George receives half of what Jose receives.

13. How much pure antifreeze must be added to 20 L of a solution that is 60% antifreeze to make a solution that is 80% antifreeze?

14. Solve $P = 2(l + w)$ for l.

15. Solve $C_T = C_1 + C_2 + C_3$ for C_2.

16. Solve $V = lwh$ for w.

17. Given $P = I^2R$, $P = 480$, and $I = 5.0$, find R.

18. Given $A = \dfrac{h}{2}(a + b)$, $A = 260$, $h = 13$, and $a = 15$, find b.

19. Given $\dfrac{1}{C} = \dfrac{1}{C_1} + \dfrac{1}{C_2}$, $C = 20.0\ \mu F$ and $C_2 = 30.0\ \mu F$, find C_1.

20. Given $\dfrac{1}{R} = \dfrac{1}{R_1} + \dfrac{1}{R_2} + \dfrac{1}{R_3}$, $R_1 = 225\ \Omega$, $R_2 = 475\ \Omega$, and $R_3 = 925\ \Omega$, find R.

CUMULATIVE REVIEW | CHAPTERS 1–6

1. Find the prime factorization of 696.
2. Change 0.081 to a percent.
3. Write 3.015×10^{-4} in decimal form.
4. Write 28,500 in scientific notation.
5. 5 ha = _____ m²
6. 101°F = _____ °C
7. 6250 in² = _____ ft²
8. Give the number of significant digits (accuracy) of each measurement:
   a. 110 cm   b. 6000 mi   c. 24.005 s
9. Read the measurement shown on the vernier caliper in Illustration 1 a. in metric units and b. in U.S. units.

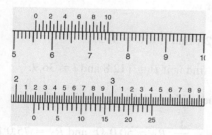

ILLUSTRATION 1

10. Read the measurement shown on the U.S. micrometer in Illustration 2.

ILLUSTRATION 2

11. Use the rules for addition of measurements to find the sum of 25,000 W; 17,900 W; 13,962 W; 8752 W; and 428,000 W.

Simplify:

12. $(2x - 5y) + (3y - 4x) - 2(3x - 5y)$
13. $(4y^3 + 3y - 5) - (2y^3 - 4y^2 - 2y + 6)$
14. $(3y^3)^3$
15. $-2x(x^2 - 3x + 4)$
16. $(6y^3 - 5y^2 - y + 2)(2y - 1)$
17. $(4x - 3y)(5x + 2y)$
18. $\dfrac{215\ x^2 y^3}{45 x^3 y^5}$
19. $(16x^2y^3)(-5x^4y^5)$
20. $\dfrac{x^3 + 2x^2 - 11x - 20}{x + 5}$
21. $3x^2 - 4xy + 5y^2 - (-3x^2) + (-7xy) + 10y^2$

Solve:

22. $4x - 2 = 12$
23. $\dfrac{x}{4} - 5 = 9$
24. $4x - 3 = 7x + 15$
25. $\dfrac{5x}{8} = \dfrac{3}{2}$
26. $5 - (x - 3) = (2 + x) - 5$
27. $C = \dfrac{1}{2}(a + b + c)$ for a
28. $A = lw$; find w if $l = 8.20$ m and $A = 91.3$ m².
29. Translate into algebraic symbols: The product of a number and 7 is 250.
30. The perimeter of a rectangle is 30 ft. The width is one-half of the length. What are the dimensions of the rectangle?

Ratio and Proportion

CHAPTER 7

OBJECTIVES

◆ Express a ratio and a rate in lowest terms.

◆ Solve a proportion.

◆ Solve application problems using ratios, rates, and proportions.

◆ Solve application problems involving direct variation.

◆ Solve application problems involving inverse variation.

Mathematics at Work

Heating, ventilation, air-conditioning, and refrigeration (HVAC/R) technicians install and repair such systems. Duties include installation and repair of oil burners, hot-air furnaces, heating stoves, and similar equipment in residential, commercial, and industrial buildings using hand and pipe threading. Heating and air-conditioning systems control the temperature, humidity, and total air quality in such locations. Refrigeration systems allow for the storing and transport of food, medicine, and other perishable items.

Heating, ventilation, air-conditioning, and refrigeration systems consist of many mechanical, electrical, and electronic components, such as motors, compressors, pumps, fans, ducts, pipes, thermostats, and switches. These technicians must be able to maintain, diagnose, and correct problems throughout the entire system by adjusting system controls to recommended settings and test the performance of the system using special tools and test equipment. Although trained to do installation or repair and maintenance, technicians often specialize in one or the other. Some specialize in one type of equipment, such as commercial refrigerators, oil burners, or solar panels. Technicians work for large or small contractors or directly for manufacturers or wholesalers. Employers prefer to employ those with technical school or apprenticeship training. Many community colleges and postsecondary and trade schools offer associate degree and certificate programs in which students study theory, design, equipment construction, and electronics as well as the basics of installation, maintenance, and repair. All technicians who work with refrigerants must be certified in their proper handling. North American Technician Excellence, Inc., offers one standard for certification of experienced technicians. For more information, please visit **www.cengage.com** and access the Student Online Resources for this text.

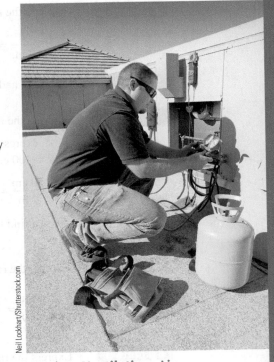

Heating, Ventilation, Air-Conditioning, and Refrigeration Technician
HVAC technician working on a large commercial unit

7.1 Ratio*

The comparison of two numbers is a very important concept, and one of the most important of all comparisons is the ratio. The **ratio of two numbers**, *a* and *b*, is the first number divided by the second number. Ratios may be written in several different ways. For example, the ratio of 3 to 4 may be written as $\frac{3}{4}$, 3/4, 3 : 4, or 3 ÷ 4. Each form is read "the ratio of 3 to 4."

If the quantities to be compared include units, the units should be the same whenever possible. To find the ratio of 1 ft to 15 in., first express both quantities in inches and then find the ratio:

$$\frac{1 \text{ ft}}{15 \text{ in.}} = \frac{12 \text{ in.}}{15 \text{ in.}} = \frac{12}{15} = \frac{4}{5}$$

Ratios are usually given in lowest terms.

Example 1 Express the ratio 18 : 45 in lowest terms.

$$18 : 45 = \frac{18}{45} = \frac{\cancel{9} \cdot 2}{\cancel{9} \cdot 5} = \frac{2}{5}$$

◆

Example 2 Express the ratio of 3 ft to 18 in. in lowest terms.

$$\frac{3 \text{ ft}}{18 \text{ in.}} = \frac{36 \text{ in.}}{18 \text{ in.}} = \frac{\cancel{18} \times 2 \text{ in.}}{\cancel{18} \times 1 \text{ in.}} = \frac{2}{1} \text{ or } 2$$

NOTE: $\frac{2}{1}$ and 2 indicate the same ratio, "the ratio of 2 to 1."

◆

Example 3 Express the ratio of 50 cm to 2 m in lowest terms.

First, express the measurements in the same units.
1 m = 100 cm, so 2 m = 200 cm.

$$\frac{50 \text{ cm}}{2 \text{ m}} = \frac{50 \cancel{\text{ cm}}}{200 \cancel{\text{ cm}}} = \frac{1}{4}$$

◆

To find the ratio of two fractions, use the technique for dividing fractions.

Example 4 Express the ratio $\frac{2}{3} : \frac{8}{9}$ in lowest terms.

$$\frac{2}{3} : \frac{8}{9} = \frac{2}{3} \div \frac{8}{9} = \frac{2}{3} \times \frac{9}{8} = \frac{18}{24} = \frac{\cancel{6} \cdot 3}{\cancel{6} \cdot 4} = \frac{3}{4}$$

◆

Example 5 Express the ratio of $2\frac{1}{2}$ to 10 in lowest terms.

$$2\frac{1}{2} \text{ to } 10 = \frac{5}{2} \div 10 = \frac{5}{2} \times \frac{1}{10} = \frac{5}{20} = \frac{\cancel{5} \cdot 1}{\cancel{5} \cdot 4} = \frac{1}{4}$$

◆

Example 6 Steel can be worked in a lathe at a cutting speed of 25 ft/min. Stainless steel can be worked in a lathe at a cutting speed of 15 ft/min. What is the ratio of the cutting speed of steel to the cutting speed of stainless steel?

$$\frac{\text{cutting speed of steel}}{\text{cutting speed of stainless steel}} = \frac{25 \text{ ft/min}}{15 \text{ ft/min}} = \frac{5}{3}$$

◆

---

*Note: In this chapter, do not use rules for calculating with measurements.

Example 7 A construction crew uses 4 buckets of cement and 12 buckets of sand to mix a supply of concrete. What is the ratio of cement to sand?

$$\frac{\text{amount of cement}}{\text{amount of sand}} = \frac{4 \text{ buckets}}{12 \text{ buckets}} = \frac{1}{3}$$

◆

In a ratio, we compare like or related quantities; for example,

$$\frac{18 \text{ ft}}{12 \text{ ft}} = \frac{3}{2} \quad \text{or} \quad \frac{50 \text{ cm}}{2 \text{ m}} = \frac{50 \text{ cm}}{200 \text{ cm}} = \frac{1}{4}$$

A ratio simplified into its lowest terms is a pair of unitless numbers.

Suppose you drive 75 miles and use 3 gallons of gasoline. Your mileage would be found as follows:

$$\frac{75 \text{ mi}}{3 \text{ gal}} = \frac{25 \text{ mi}}{1 \text{ gal}}$$

We say that your mileage is 25 miles per gallon. Note that each of these two fractions compares unlike quantities: miles and gallons. A **rate** is the comparison of two unlike quantities whose units do not cancel.

Example 8 Express the rate of $\frac{250 \text{ gal}}{50 \text{ acres}}$ in lowest terms.

$$\frac{250 \text{ gal}}{50 \text{ acres}} = \frac{5 \text{ gal}}{1 \text{ acre}} \text{ or } 5 \text{ gal/acre}$$

The symbol "/" is read "per." The rate is read "5 gallons per acre."

◆

A common medical practice is to give nourishment and/or medication to a patient by IV (intravenously). The number of drops per minute is related to the type of equipment being used. The number of drops per mL is called the *drop factor*. Common drop factors are 10 drops/mL, 12 drops/mL, and 15 drops/mL.

Example 9 A doctor orders 500 mL of glucose to be given to an adult patient by IV in 6 h. The drop factor of the equipment is 15 drops per millilitre. Determine the number of drops per minute in order to set up the IV.

First, change 6 h to minutes:

$$6 \text{ h} \times \frac{60 \text{ min}}{1 \text{ h}} = 360 \text{ min (time for IV)}$$

Thus, 500 mL of glucose is to be given during a 360-min time period, which gives us a rate of $\frac{500 \text{ mL}}{360 \text{ min}}$. Since the equipment has a drop factor of 15 drops/mL, the flow rate is

$$\frac{500 \text{ mL}}{360 \text{ min}} \times \frac{15 \text{ drops}}{\text{mL}} = 21 \text{ drops/min (rounded to the nearest whole number)}$$

◆

Sometimes the doctor orders an IV as a rate of flow, and the nurse must determine the time needed to administer the IV.

Example 10 Give 1500 mL of saline solution IV with a drop factor of 10 drops per millilitre at a rate of 50 drops/min to an adult patient. Find how long the IV should be administered.

First, determine the total number of drops to be administered:

$$1500 \text{ mL} \times \frac{10 \text{ drops}}{\text{mL}} = 15{,}000 \text{ drops}$$

Then, divide the total number of drops by the flow rate to find the time:

$$\frac{15{,}000 \text{ drops}}{50 \text{ drops/min}} = 300 \text{ min} \qquad \frac{\text{drops}}{\frac{\text{drops}}{\text{min}}} = \text{drops} \div \frac{\text{drops}}{\text{min}}$$

$$= 5 \text{ h}$$

$$= \cancel{\text{drops}} \times \frac{\text{min}}{\cancel{\text{drops}}} = \text{min}$$

♦

EXERCISES 7.1

Express each ratio in lowest terms:

1. 3 to 15
2. 6 : 12
3. 7 : 21
4. $\dfrac{4}{22}$
5. $\dfrac{80}{48}$
6. 28 to 20
7. 3 in. to 15 in.
8. 3 ft to 15 in.
9. 3 cm to 15 mm
10. 1 in. to 8 ft
11. 9 in^2 : 2 ft^2
12. 4 m : 30 cm
13. $\dfrac{3}{4}$ to $\dfrac{7}{6}$
14. $\dfrac{2}{3} : \dfrac{22}{9}$
15. $2\dfrac{3}{4} : 4$
16. 6 to $4\dfrac{2}{3}$
17. $\dfrac{5\dfrac{1}{3}}{2\dfrac{2}{3}}$
18. $\dfrac{18\dfrac{1}{2}}{2\dfrac{1}{4}}$
19. 10 to $2\dfrac{1}{2}$
20. $\dfrac{7}{8} : \dfrac{9}{16}$
21. $3\dfrac{1}{2}$ to $2\dfrac{1}{2}$
22. $2\dfrac{2}{3} : 3\dfrac{3}{4}$
23. $1\dfrac{3}{4}$ to 7
24. $4\dfrac{4}{5} : 12$

Express each rate in lowest terms:

25. $\dfrac{240 \text{ mi}}{8 \text{ gal}}$
26. $\dfrac{360 \text{ gal}}{18 \text{ acres}}$
27. $\dfrac{276 \text{ gal}}{6 \text{ h}}$
28. $\dfrac{\$36}{3 \text{ h}}$
29. $\dfrac{625 \text{ mi}}{12\dfrac{1}{2} \text{ h}}$
30. $\dfrac{150 \text{ mi}}{3\dfrac{3}{4} \text{ gal}}$
31. $\dfrac{2\dfrac{1}{4} \text{ lb}}{6 \text{ gal}}$
32. $\dfrac{\$64{,}800}{1800 \text{ ft}^2}$

33. A bearing bronze mix includes 96 lb of copper and 15 lb of lead. Find the ratio of copper to lead.

34. What is the alternator-to-engine drive ratio if the alternator turns at 1125 rpm when the engine is idling at 500 rpm?

35. A technician measured an oil flow rate of 90 gal in 5 min from the engine oil pump. Determine the oil flow rate in terms of gallons per minute.

36. A flywheel has 72 teeth, and a starter drive-gear has 15 teeth. Find the ratio of flywheel teeth to drive-gear teeth.

37. A transformer has a voltage of 18 V in the primary circuit and 4950 V in the secondary circuit. Find the ratio of the primary voltage to the secondary voltage.

38. The ratio of the voltage drops across two resistors wired in series equals the ratio of their resistances. Find the ratio of a 720-Ω resistor to a 400-Ω resistor.

39. A transformer has 45 turns in the primary coil and 540 turns in the secondary coil. Find the ratio of secondary turns to primary turns.

40. The resistance in ohms of a resistor is the ratio of the voltage drop across the resistor, in volts, to the current through the resistor, in amperes. A resistor has a voltage drop across it of 117 V and a current through it of 2.6 A. What is the resistance in ohms of the resistor?

41. A 150-bu wagon holds 2.7 tons of grain. Express the weight of grain in pounds per bushel.

42. The total yield from a 55-acre field is 7425 bu. Express the yield in bushels per acre.

43. A 350-gal spray tank covers 14 acres. Find the rate of application in gallons per acre.

44. Suppose 12 gal of herbicide concentrate are used for 28 acres. Find the rate of gallons of concentrate to acres.

45. A flower bed contains a mixture of 6 ft^3 of topsoil and 2 ft^3 of sand. What is the ratio of sand to the total mixture?

46. Yellow and red peppers were planted in the ratio of 2 yellow to 3 red. If 9 red peppers were planted, how many yellow peppers were planted?

47. Suppose 16 ft of copper tubing costs $27.04. Find its cost per foot.

48. A structure has 3290 ft^2 of wall area (excluding windows) and 1880 ft^2 of window area. Find the ratio of wall area to window area.

49. A 2150-ft² home sells for $268,750. Find the ratio of cost to area (price per ft²).

50. You need 15 ft³ of cement to make 80 ft³ of concrete. Find the ratio of volume of concrete to volume of cement.

51. A welder has 9 pieces of 4-ft steel angle and 12 pieces of 2-ft steel angle. What is the ratio of pieces of 4-ft steel angle to 2-ft steel angle?

52. A welder grabs a handful of 6011 welding rods and another handful of Super Strength 100 welding rods. When the welder sees how many of each she has, she has 32 of the 6011 welding rods and 60 of the Super Strength 100 welding rods. What is the ratio of the 6011 welding rods to the Super Strength 100 welding rods?

53. The total number of hours required for a private-pilot, single-engine land rating is 40 h of flight time. The total number of hours of flight time required for a commercial rating is 250 h. What is the ratio of the number of hours required for a private rating to those required for a commercial rating?

54. Two small window air conditioner units were purchased and put into opposite sides of a house. One air conditioner was 5000 Btu and the other, 7500 Btu. What is the ratio of the 5000-Btu to the 7500-Btu air conditioner?

55. A 1-litre (1000 mL) IV bag of dextrose solution contains 50 g of dextrose. Find the ratio of grams per millilitre of dextrose.

56. Over a period of 5 h, 1200 mL of an IV solution will be administered intravenously. How many mL/min is this?

57. A doctor prescribes 750 mL of normal saline to be given in 5 h. The drop factor of the equipment is 15 drops/mL. Determine the number of drops/min in order to set up the IV. (Round to the nearest whole number.)

58. A patient is given 500 mL of IV glucose using a drop factor of 15 drops/mL. The flow rate is 25 drops/min. How long will it take to infuse this IV?

59. Adult male cougars commonly require 50 to 150 mi² of territory to live. If 280 male cougars live in a forest area that covers 45,000 mi², of which 3000 mi² is covered by water, what is the ratio of male cougars per living area?

60. In tilapia fish farming, 5 tons of high-quality feed resulted in a weight gain of 5000 lb. What is the amount of feed-to-weight-gain ratio?

Find the flow rate for each given IV (assume a drop factor of 15 drops/mL):

61. 1200 mL in 6 h
62. 900 mL in 3 h
63. 1 L in 5.5 h
64. 2 L in 5 h

Find the length of time each IV should be administered (assume a drop factor of 10 drops/mL):

65. 1000 mL at a rate of 50 drops/min
66. 1600 mL at a rate of 40 drops/min
67. 2 L at a rate of 40 drops/min
68. 1.4 L at a rate of 35 drops/min

69. The siding replacement for a house of 2300 ft² costs $13,225. Find the cost per ft².

70. Ann worked 6 hours and her pay was $54.60. What was her rate per hour?

71. Twelve dozen pencils cost $5.88. What is the cost per pencil?

72. If 2 gal of paint covers 600 ft², how much area does 1 gal cover?

73. A recipe calls for 1 gal of tomato paste and 3 qt of peeled tomatoes. What is the ratio of tomato paste to peeled tomatoes?

74. The soup of the day has water as 35% of its volume and beef broth as 45% of its volume. What is the ratio of the volume of water to the volume of beef broth?

75. A chef pays $125 for a 25-lb pork loin. What is the ratio of cost to pound?

76. If 18 lb of potatoes are used in a recipe for 42 people, what is the ratio of lb/person?

7.2 Proportion

A **proportion** states that two ratios or two rates are equal. Thus,

$$\frac{3}{4} = \frac{9}{12}, \quad 2:3 = 4:6, \quad \text{and} \quad \frac{a}{b} = \frac{c}{d}$$

are proportions. A proportion has four terms. In the proportion $\frac{2}{5} = \frac{4}{10}$, the first term is 2, the second term is 5, the third term is 4, and the fourth term is 10.

The first and fourth terms of a proportion are called the **extremes**, and the second and third terms are called the **means** of the proportion. This is more easily seen when the proportion $\frac{a}{b} = \frac{c}{d}$ is written in the form

$$a : b = c : d$$

with means being b, c and extremes being a, d.

Example 1   Given the proportion $\frac{2}{3} = \frac{4}{6}$.

    a. The first term is 2.
    b. The second term is 3.
    c. The third term is 4.
    d. The fourth term is 6.
    e. The means are 3 and 4.
    f. The extremes are 2 and 6.
    g. The product of the means = $3 \cdot 4 = 12$.
    h. The product of the extremes = $2 \cdot 6 = 12$.

We see in **g** and **h** that the product of the means (that is, 12) equals the product of the extremes (also 12). Let us look at another proportion and see if this is true again.

Example 2   Given the proportion $\frac{5}{13} = \frac{10}{26}$, find the product of the means and the product of the extremes. The extremes are 5 and 26, and the means are 13 and 10. The product of the extremes is 130, and the product of the means is 130. Here again, the product of the means equals the product of the extremes. As a matter of fact, this will always be the case.

> **Proportion**
>
> In any proportion, the *product of the means* equals the *product of the extremes*. That is, if $\frac{a}{b} = \frac{c}{d}$, then $bc = ad$.

To determine whether two ratios are equal, put the two ratios in the form of a proportion. If the product of the means equals the product of the extremes, the ratios are equal.

Example 3   Determine whether or not the ratios $\frac{13}{36}$ and $\frac{29}{84}$ are equal.

If the product of the means (36×29) = the product of the extremes (13×84), then $\frac{13}{36} = \frac{29}{84}$.

However, $36 \times 29 = 1044$ and $13 \times 84 = 1092$. Therefore, $\frac{13}{36} \neq \frac{29}{84}$.

To *solve* a proportion means to find the missing term. To do this, form an equation by setting the product of the means equal to the product of the extremes. Then solve the resulting equation.

Example 4   Solve the proportion $\frac{x}{3} = \frac{8}{12}$.

$$\frac{x}{3} = \frac{8}{12}$$
$$12x = 24 \quad \text{The product of the means equals the product of the extremes.}$$
$$x = 2$$

Example 5 Solve the proportion $\dfrac{5}{x} = \dfrac{10}{3}$.

$$\dfrac{5}{x} = \dfrac{10}{3}$$

$10x = 15$ The product of the means equals the product of the extremes.

$x = \dfrac{3}{2}$ or 1.5

A calculator is helpful in solving a proportion with decimal fractions.

Example 6 Solve $\dfrac{32.3}{x} = \dfrac{17.9}{25.1}$.

$17.9x = (32.3)(25.1)$ The product of the means equals the product of the extremes.

$x = \dfrac{(32.3)(25.1)}{17.9} = 45.3$, rounded to three significant digits

32.3 × 25.1 ÷ 17.9 =

```
45.292179
```

Example 7 If 125 bolts cost $16.50, how much do 75 bolts cost?

First, let's find the rate of dollars/bolts in each case.

$\dfrac{\$16.50}{125 \text{ bolts}}$ and $\dfrac{x}{75 \text{ bolts}}$ where $x =$ the cost of 75 bolts

Since these two rates are equal, we have the proportion

$$\dfrac{16.5}{125} = \dfrac{x}{75}$$

$125x = (16.5)(75)$ The product of the means equals the product of the extremes.

$x = \dfrac{(16.5)(75)}{125}$ Divide both sides by 125.

$x = 9.9$

That is, the cost of 75 bolts is $9.90.

NOTE: A key to solving proportions like the one in Example 7 is to set up the proportion with the same units in each ratio—in this case,

$$\dfrac{\$}{\text{bolts}} = \dfrac{\$}{\text{bolts}}$$

Example 8 The pitch of a roof is the ratio of the rise to the run of a rafter. (See Figure 7.1.) The pitch of the roof shown is 2 : 7. Find the rise if the run is 21 ft.

Figure 7.1

$$\text{pitch} = \frac{\text{rise}}{\text{run}}$$

$$\frac{2}{7} = \frac{x}{21 \text{ ft}}$$

$$7x = (2)(21 \text{ ft})$$

$$x = \frac{(2)(21 \text{ ft})}{7}$$

$$x = 6 \text{ ft, which is the rise}$$

In Section 1.14, you studied percent, using the formula $P = BR$, where R is the rate written as a decimal. Knowing this formula and knowing the fact that *percent* means "per hundred," we can write the proportion

$$\boxed{\frac{P}{B} = \frac{R}{100}}$$

where R is the rate written as a percent. We can use this proportion to solve percent problems.

NOTE: You may find it helpful to review the meanings of P (part), B (base), and R (rate) in Section 1.14.

Example 9

A student answered 27 out of 30 questions correctly. What percent of the answers were correct?

P (part) = 27
B (base) = 30
R (rate) = x

$$\frac{27}{30} = \frac{x}{100}$$

$30x = 2700$ The product of the means equals the product of the extremes.

$x = 90$

Therefore, the student answered 90% of the questions correctly.

Example 10

A factory produces bearings used in automobiles. After inspecting 4500 bearings, the inspectors find that 127 are defective. What percent are defective?

P (part) = 127
B (base) = 4500
R (rate) = x

$$\frac{127}{4500} = \frac{x}{100}$$

$4500x = 12{,}700$

$x = 2.8$ (rounded to the nearest tenth)

Therefore, 2.8% of the bearings are defective.

Example 11

A 5% dextrose solution (D5W) contains 5 g of pure dextrose per 100 mL of solution. A doctor orders 500 mL of D5W IV for a patient. How much pure dextrose does the patient receive from that IV?

$$B \text{ (base)} = 500 \text{ mL}$$
$$R \text{ (rate)} = 5\%$$
$$P \text{ (part)} = x$$
$$\frac{x}{500} = \frac{5}{100}$$
$$100x = 2500$$
$$x = 25 \text{ g}$$

◆

Example 12 Prepare 2000 mL of a Lysol solution containing 1 part pure Lysol and 19 parts water. How much pure Lysol is needed?

$$R = 1 : (1 + 19) = 1 : 20 = 1/20 = 0.05 = 5\%$$
$$B = 2000 \text{ mL}$$
$$P = x$$
$$\frac{x}{2000} = \frac{5}{100}$$
$$100x = 10{,}000$$
$$x = 100 \text{ mL}$$

◆

Example 13 If it costs $24 to make 2 gal of soup, how much will it cost to make 10 qt of the same soup?

First, let's find the cost per quart in each case.

$$\frac{\$24}{8 \text{ qt}} \quad \text{and} \quad \frac{\$x}{10 \text{ qt}} \quad \text{where } x \text{ is the cost of 10 qt}$$

Since these two rates are equal, we have the proportion

$$\frac{24}{8} = \frac{x}{10}$$

$8x = 240$ The product of the means equals the product of the extremes.

$x = 30$ Divide both sides by 8.

Thus, the cost to make 10 qt is $30.

◆

Kitchen Ratios

The **kitchen ratio** is a very common term in the culinary field. It is used to express the relationship as a ratio among all the ingredients that are used in a recipe. This allows one to take any recipe and make more or less of the given recipe. Some recipes are given in terms of the number of parts of each ingredient.

Example 14 Cheese balls are formed using the kitchen ratio of 1, 1, 2 for blue cheese, cream cheese, and cheddar cheese. How much of each cheese is needed to make forty 2-oz cheese balls?

First, 40×2 oz = 80 oz of cheese is needed in the ratio of $1 : 1 : 2$.

Let x = the number of ounces of blue cheese

x = the number of ounces of cream cheese

$2x$ = the number of ounces of cheddar cheese

The total amount of cheese needed is

$$x + x + 2x = 80$$
$$4x = 80$$
$$x = 20$$

Thus, we need 20 oz of blue cheese, 20 oz of cream cheese, and 40 oz of cheddar cheese. ♦

Example 15 A recipe for tomato basil soup that serves four follows:

2 tbs onion

2 tbs butter

3 cups tomatoes

1 tbs basil

2 cups milk

What are the amounts needed for 10 servings?

The ratio between the number of servings needed and the number of servings in the recipe is 10 : 4, or 2.5. So, we multiply each ingredient in the given recipe by 2.5 to obtain the amount of each ingredient in the recipe that serves 10. So, we have

2 tbs onion × 2.5 = 5 tbs onion

2 tbs butter × 2.5 = 5 tbs butter

3 cups tomatoes × 2.5 = 7.5 cups tomatoes

1 tbs basil × 2.5 = 2.5 tbs basil

2 cups milk × 2.5 = 5 cups milk ♦

EXERCISES 7.2

In each proportion, find **a.** *the means,* **b.** *the extremes,* **c.** *the product of the means, and* **d.** *the product of the extremes:*

1. $\dfrac{1}{2} = \dfrac{3}{6}$
2. $\dfrac{3}{4} = \dfrac{6}{8}$
3. $\dfrac{7}{9} = \dfrac{28}{36}$
4. $\dfrac{x}{3} = \dfrac{6}{9}$
5. $\dfrac{x}{7} = \dfrac{w}{z}$
6. $\dfrac{a}{b} = \dfrac{4}{5}$

Determine whether or not each pair of ratios is equal:

7. $\dfrac{2}{3}, \dfrac{10}{15}$
8. $\dfrac{2}{3}, \dfrac{9}{6}$
9. $\dfrac{3}{5}, \dfrac{18}{20}$
10. $\dfrac{3}{7}, \dfrac{9}{21}$
11. $\dfrac{1}{3}, \dfrac{4}{12}$
12. $\dfrac{125}{45}, \dfrac{25}{9}$

Solve each proportion (round each result to three significant digits when necessary):

13. $\dfrac{x}{4} = \dfrac{9}{12}$
14. $\dfrac{1}{a} = \dfrac{4}{16}$
15. $\dfrac{5}{7} = \dfrac{4}{y}$
16. $\dfrac{12}{5} = \dfrac{x}{10}$
17. $\dfrac{2}{x} = \dfrac{4}{28}$
18. $\dfrac{10}{15} = \dfrac{y}{75}$
19. $\dfrac{5}{7} = \dfrac{3x}{14}$
20. $\dfrac{x}{18} = \dfrac{7}{9}$
21. $\dfrac{5}{7} = \dfrac{25}{y}$
22. $\dfrac{1.1}{6} = \dfrac{x}{12}$
23. $\dfrac{-5}{x} = \dfrac{2}{3}$
24. $\dfrac{4x}{9} = \dfrac{12}{7}$
25. $\dfrac{8}{0.04} = \dfrac{700}{x}$
26. $\dfrac{x}{9} = \dfrac{2}{0.6}$
27. $\dfrac{3x}{27} = \dfrac{0.5}{9}$
28. $\dfrac{0.25}{2x} = \dfrac{8}{48}$
29. $\dfrac{17}{28} = \dfrac{153}{2x}$
30. $\dfrac{3x}{10} = \dfrac{7}{50}$
31. $\dfrac{12}{y} = \dfrac{84}{144}$
32. $\dfrac{13}{169} = \dfrac{27}{x}$
33. $\dfrac{x}{48} = \dfrac{56}{72}$
34. $\dfrac{124}{67} = \dfrac{149}{x}$
35. $\dfrac{472}{x} = \dfrac{793}{64.2}$
36. $\dfrac{94.7}{6.72} = \dfrac{x}{19.3}$
37. $\dfrac{30.1}{442} = \dfrac{55.7}{x}$
38. $\dfrac{9.4}{291} = \dfrac{44.1}{x}$
39. $\dfrac{36.9}{104} = \dfrac{3210}{x}$
40. $\dfrac{0.0417}{0.355} = \dfrac{26.9}{x}$

41. $\dfrac{x}{4.2} = \dfrac{19.6}{3.87}$ **42.** $\dfrac{0.120}{3x} = \dfrac{0.575}{277}$

43. You need $2\tfrac{3}{4}$ ft^3 of sand to make 8 ft^3 of concrete. How much sand would you need to make 128 ft^3 of concrete?

44. The pitch of a roof is $\tfrac{1}{3}$. If the run is 15 ft, find the rise. (See Example 8.)

45. A builder sells an 1800-ft^2 home for $171,000. What would be the price of a 2400-ft^2 home of similar quality? Assume that the price per square foot remains constant.

46. Suppose 826 bricks are used in constructing a wall 14 ft long. How many bricks will be needed for a similar wall 35 ft long?

47. A buyer purchases 75 yd of material for $120. Then an additional 90 yd are ordered at the same unit cost. What is the additional cost?

48. A salesperson is paid a commission of $75 for selling $300 worth of goods. What is the commission on $760 of sales at the same rate of commission?

49. A plane flies for 3 h and uses 25 gal of 100*LL* aviation fuel. How much will be used if the plane flies for only 1.2 h?

50. Metal duct that is 6 in. in diameter costs $7.50 for 5 ft. If 16.5 ft are needed for an order, what is the cost?

51. Suppose 20 gal of water and 3 lb of pesticide are applied per acre. How much pesticide should you put in a 350-gal spray tank? Assume that the pesticide dissolves in the water and has no volume.

52. A farmer uses 150 lb of a chemical on a 40-acre field. How many pounds will he need for a 220-acre field? Assume the same rate of application.

53. Suppose a yield of 100 bu of corn per acre removes 90 lb of nitrogen, phosphorus, and potash (or potassium) (N, P, and K). How many pounds of N, P, and K would be removed by a yield of 120 bu per acre?

54. A farmer has a total yield of 42,000 bu of corn from a 350-acre farm. What total yield should he expect from a similar 560-acre farm?

55. An orchard has 3 rows with 5 apple trees each. The apple trees yielded 180 bu of apples. **a.** What was the average yield per tree? **b.** At $13 per half bushel, what is the income from the sale of apples?

56. A 25-lb bag of fertilizer covers 1200 ft^2 of lawn. **a.** At this rate how many pounds of fertilizer are needed for 3000 ft^2 of lawn? **b.** How many bags must be purchased?

57. A copper wire 750 ft long has a resistance of 1.563 Ω. How long is a copper wire of the same size whose resistance is 2.605 Ω? (The resistance of these wires is proportional to their length.)

58. The voltage drop across a 28-Ω resistor is 52 V. What is the voltage drop across a 63-Ω resistor that is in series with the first one? (Resistors in series have voltage drops proportional to their resistances.)

59. The ratio of secondary turns to primary turns in a transformer is 35 to 4. How many secondary turns are there if the primary coil has 68 turns?

60. If welding rods cost $75 per 50 lb, how much would 75 lb cost?

61. An engine with displacement of 2.0 L develops 270 hp. How many horsepower would be developed by a 1.6-L engine of the same design?

62. An 8-V automotive coil has 250 turns of wire in the primary circuit. The secondary voltage is 15,000 V. How many secondary turns are in the coil? (The ratio of secondary voltage to primary voltage equals the ratio of secondary turns to primary turns.)

63. An automobile uses 18 gal of fuel to go 560 mi. How many gallons are required to travel 820 mi?

64. A fuel pump delivers 35 mL of fuel in 420 strokes. How many strokes are needed to pump 50 mL of fuel?

65. A label reads: "2.5 mL of solution for injection contains 1000 mg of streptomycin sulfate." How many millilitres are needed to give 800 mg of streptomycin?

66. A multiple-dose vial has been mixed and labeled "200,000 units in 1 mL." How many millilitres are needed to give 900,000 units?

67. You are to administer 150 mg of aminophylline from a bottle marked 250 mg/10 mL. How many millilitres should you draw?

68. A label reads: "Gantrisin, 1.5 g in 20 mL." How many millilitres are needed to give 10.5 g of Gantrisin?

When finding the percent, round to the nearest tenth of a percent when necessary:

69. An automobile is listed to sell at $25,400. The salesperson offers to sell it to you for $23,900. What percent of the list price is the reduction?

70. A live hog weighs 254 lb, and its carcass weighs 198 lb. What percent of the live hog is carcass? What percent is waste?

71. In a 100-g sample of beef, there are 18 g of fat.
   a. What is the percent of fat in the beef?
   b. How many pounds of fat would there be in a 650-lb beef carcass? Assume the same percent of fat.

72. At the beginning of a trip, a tire on an automobile has a pressure of 32 psi (lb/in^2). At the end of the trip, the pressure is 38 psi. What is the percent increase in pressure?

73. A gasoline tank contains 5.7 hectolitres (hL) when it is 30% full. What is the capacity of the tank?

74. Jamal had a pay raise from $2650 to $2756 per month. Find the percent increase.

75. A concrete mix by volume is composed of 1 part cement, 2.5 parts sand, and 4 parts gravel. What is the percent by volume in the dry mix of **a.** cement, **b.** sand, and **c.** gravel?

76. You are to put 4 qt of pure antifreeze in a tractor radiator and then fill the radiator with 5 gal of water. What percent of antifreeze will be in the radiator?

77. In developing a paint line process for an appliance manufacturer, a given component on a conveyor belt passes a given point at a rate of 25 ft in 3 min into a drying booth. If an object must be in the drying booth for 10 min, how long should this booth be?

78. A barn with dimensions 32 ft × 14 ft × 8 ft (wall height) with a roof peak of 19 ft is being reproduced at the scale 1 in. = 4 ft to show the board of directors what it will look like when finished. Find the overall dimensions of the model.

79. A pound of sea water contains 35 parts per thousand of dissolved salt by weight. How much salt is contained in 1 ton of sea water?

80. The content of common fertilizer is listed using three numbers that represent the percentages of nitrogen, phosphorus, and potassium, commonly abbreviated as N-P-K. For example, a bag of 5-10-15 fertilizer contains 5% nitrogen, 10% phosphorus, and 15% potassium. If you apply enough 5-10-15 fertilizer to your lawn to apply a total of 24 lb of nitrogen, how much phosphorus and how much potassium are applied?

81. The gear ratio of a fishing reel refers to the number of times the spool rotates for each turn of the handle. If a reel requires 20 turns of the handle to retrieve 100 ft of line, how many turns would it take to retrieve 175 ft of line?

82. An acre inch is the amount of water needed to provide 1 inch of water on an acre of surface. There are approximately 27,150 gal of water in an acre inch. According to the USDA, only about $\frac{1}{4}$ of the initial water that is used in irrigation ever reaches the crop for which it is intended, with the rest evaporating or soaking into the ground before it reaches the crop. How many gallons of water would be required to apply an actual inch of water in a given field of one acre?

Find the number of millilitres of pure ingredient needed to prepare each solution as indicated:

83. 150 mL of 3% alcohol solution from pure alcohol

84. 1000 mL of 5% Lysol solution from pure Lysol

Find the number of grams of pure ingredient needed to prepare each solution as indicated.

85. 500 mL of 1% sodium bicarbonate solution from pure powdered sodium bicarbonate

86. 600 mL of 10% glucose solution from pure crystalline glucose

87. 1.5 L of 1:1000 epinephrine solution from pure epinephrine

88. 20 mL of 1:200 silver nitrate solution from pure silver nitrate

89. 300 mL of 1:10 glucose solution from pure glucose

90. 400 mL of 1:50 sodium bicarbonate solution from pure sodium bicarbonate

91. The ratio of beef to pork for meat loaf is 5:2. How much pork is needed if 6 lb of beef is used?

92. There are 6 bone-in prime rib cuts in each 109 beef loin and 7 bone-in prime rib cuts in each export beef loin. If the chef gets 42 bone-in prime rib cuts from the total export loins, how many bone-in prime rib cuts would come from the same number of 109 beef loins?

93. The following recipe is for sherry vinegar marinade: 125 mL vegetable oil, 50 mL sherry vinegar, and 25 mL salt. **a.** What is the kitchen ratio for these ingredients? **b.** How much of each ingredient is needed if we use 150 mL of sherry vinegar?

94. The following recipe is for popover batter: $1\frac{1}{2}$ qt whole milk, $4\frac{1}{2}$ cups all purpose flour, 1 qt large eggs, and $1\frac{1}{2}$ cups butter. **a.** What is the kitchen ratio for these ingredients? **b.** How much of each ingredient is needed for 3 gal of popover batter?

95. The following recipe is for 16 crêpes: $\frac{1}{4}$ cup margarine, 3 tbs sugar, $\frac{1}{2}$ cup orange juice, 2 tsp lemon juice, 1 tbs grated orange rind, $\frac{1}{2}$ tsp grated lemon rind, and $\frac{1}{2}$ cup toasted slivered almonds. You need to serve 5 crêpes per person for a group of 8. How much of each ingredient is needed?

96. The following recipe for la crème au chocolat serves 4: $\frac{1}{4}$ lb German sweet chocolate, 1 tbs sugar, $\frac{1}{2}$ cup cream, 2 egg yolks, and $\frac{1}{2}$ tsp vanilla. How much of each ingredient is needed to serve 25?

97. ✗ The following recipe for cheese soufflé serves 4: $\frac{1}{4}$ cup butter, $\frac{1}{4}$ cup all-purpose flour, 1 cup milk, $\frac{1}{2}$ tsp salt, 8 oz sharp processed American cheese, 4 egg yolks, and 4 stiff beaten egg whites. How much of each ingredient is needed for 18 servings?

98. ✗ The kitchen ratio of ingredients for cranberry salad is 8 : 4 : 4 : 2 : 2 : 1 of fresh cranberries, tokay grapes, sugar, pineapple, whipped cream, and walnuts. Each serving is 1.5 cups. How many cups of each ingredient are needed for 35 servings?

99. ✗ The kitchen ratio of ingredients for tenderloin of beef Wellington is 3 : 3 : 2 : 1 : 1 : 1 : $\frac{1}{2}$ for the following ingredients: lb beef tenderloin, sticks of pie crust, tbs soft butter, eggs, tsp salt, tbs water, and tsp pepper. If 12 lb of beef tenderloin is used, how much of each ingredient is needed?

100. ✗ The following recipe for a chicken or turkey casserole serves 4: 3 cups chicken or turkey, 1 cup mushrooms, 1 cup cooked rice, and 1 cup blanched almonds. How much of each ingredient is needed to serve 10?

7.3 Direct Variation

When two quantities, x and y, change so that their *ratios are constant*—that is,

$$\frac{y_1}{x_1} = \frac{y_2}{x_2}$$

they are said to *vary directly*. This relationship between the two quantities is called **direct variation**. If one quantity increases, the other increases by the same factor. Likewise, if one decreases, the other decreases by the same factor.

Consider the following data:

y	6	24	15	18	9	30
x	2	8	5	6	3	10

Note that y varies directly with x because the ratio $\frac{y}{x}$ is always 3, a constant. This x relationship may also be written $y = 3x$.

Direct Variation

$$\frac{y_1}{x_1} = \frac{y_2}{x_2}$$

Examples of direct variation are scale drawings such as maps and blueprints where

$$\frac{\text{scale measurement 1}}{\text{scale measurement 2}} = \frac{\text{actual measurement 1}}{\text{actual measurement 2}}$$

or

$$\frac{\text{scale measurement 1}}{\text{actual measurement 1}} = \frac{\text{scale measurement 2}}{\text{actual measurement 2}}$$

A *scale drawing* of an object has the same shape as the actual object, but the scale drawing may be smaller than, equal to, or larger than the actual object. The scale used in a drawing indicates what the ratio is between the size of the scale drawing and the size of the object drawn.

A portion of a map of the state of Illinois is shown in Figure 7.2. The scale is 1 in. = 32 mi.

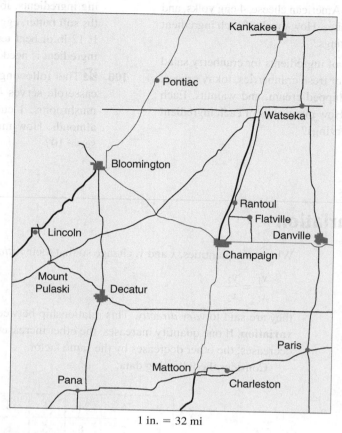

1 in. = 32 mi

Figure 7.2

Example 1   Find the approximate distance between Champaign and Kankakee using the map in Figure 7.2.

The distance on the map measures $2\frac{3}{8}$ in. Set up a proportion that has as its first ratio the scale drawing ratio and as its second ratio the length measured on the map to the actual distance.

$$\frac{1}{32} = \frac{2\frac{3}{8}}{d}$$

$$1d = 32\left(2\frac{3}{8}\right) \quad \text{The product of the means equals the product of the extremes.}$$

$$d = 32\left(\frac{19}{8}\right) = 76 \text{ mi}$$

♦

Square-ruled paper may also be used to represent scale drawings. Each square represents a unit of length according to some scale.

Example 2   The scale drawing in Figure 7.3 represents a metal plate a machinist is to make.

   a. How long is the plate?

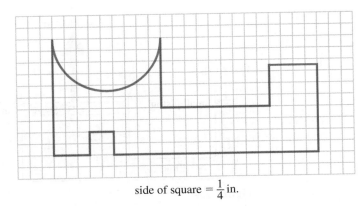

side of square = $\frac{1}{4}$ in.

Figure 7.3

Count the number of spaces, then set up the proportion:

$$\frac{\text{scale measurement 1}}{\text{actual measurement 1}} = \frac{\text{scale measurement 2}}{\text{actual measurement 2}}$$

$$\frac{1 \text{ space}}{\frac{1}{4} \text{ in.}} = \frac{22 \text{ spaces}}{x \text{ in.}}$$

$$x = 22\left(\frac{1}{4}\right) = 5\frac{1}{2} \text{ in.}$$

The product of the means equals the product of the extremes.

b. What is the width of the plate at its right end?

It is $7\frac{1}{2}$ spaces, so the proportion is

$$\frac{1 \text{ space}}{\frac{1}{4} \text{ in.}} = \frac{7\frac{1}{2} \text{ spaces}}{x \text{ in.}}$$

$$x = \left(7\frac{1}{2}\right)\left(\frac{1}{4}\right) = 1\frac{7}{8} \text{ in.}$$

c. What is the diameter of the semicircle?

$$\frac{1 \text{ space}}{\frac{1}{4} \text{ in.}} = \frac{9 \text{ spaces}}{x \text{ in.}}$$

$$x = 9\left(\frac{1}{4}\right) = 2\frac{1}{4} \text{ in.} \qquad \blacklozenge$$

Another example of direct variation is the hydraulic press or hydraulic pump, which allows one to exert a small force to move or raise a large object, such as an automobile. Other uses of hydraulics include compressing junk automobiles, stamping metal sheets to form automobile parts, and lifting truck beds.

A hydraulic press is shown in Figure 7.4. When someone presses a force of 50 lb on the small piston, a force of 5000 lb is exerted by the large piston. The *mechanical advantage* (MA) *of a hydraulic press* is the ratio of the force from the large piston (F_l) to the force on the small piston (F_s). The formula is

$$MA = \frac{F_l}{F_s}$$

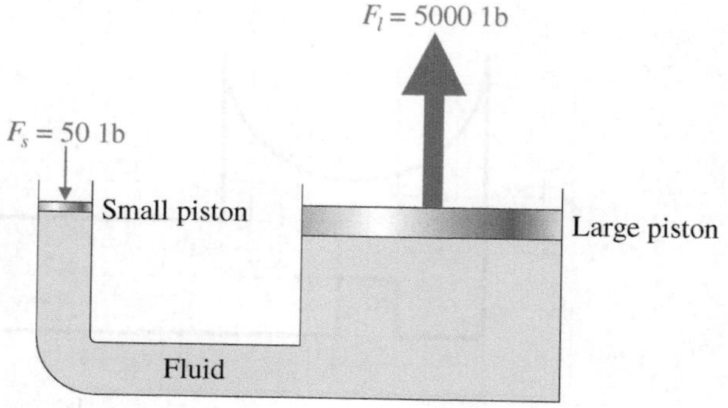

CHAPTER 7 ♦ Ratio and Proportion

Figure 7.4
Hydraulic press

Example 3 Find the mechanical advantage of the press shown in Figure 7.4.

$$MA = \frac{F_l}{F_s}$$

$$= \frac{5000 \text{ lb}}{50 \text{ lb}} = \frac{100}{1}$$

Thus, for every pound exerted on the small piston, 100 lb is exerted by the large piston.

♦

The *mechanical advantage of a hydraulic press* can also be calculated when the radii of the pistons are known.

$$MA = \frac{r_l^2}{r_s^2}$$

Example 4 The radius of the large piston of a hydraulic press is 12 in. The radius of the small piston is 2 in. Find the MA.

$$MA = \frac{r_l^2}{r_s^2}$$

$$= \frac{(12 \text{ in.})^2}{(2 \text{ in.})^2}$$

$$= \frac{144 \text{ in}^2}{4 \text{ in}^2}$$

$$= \frac{36}{1}$$

That is, for every pound exerted on the small piston, 36 lb is exerted by the large piston.

♦

You now have two ways of finding mechanical advantage: when F_l and F_s are known and when r_l and r_s are known. From this knowledge, you can find a relationship among F_l, F_s, r_l, and r_s.

Since $MA = \dfrac{F_l}{F_s}$ and $MA = \dfrac{r_l^2}{r_s^2}$, then $\dfrac{F_l}{F_s} = \dfrac{r_l^2}{r_s^2}$

Example 5 Given $F_s = 240$ lb, $r_l = 16$ in., and $r_s = 2$ in., find F_l.

$$\frac{F_l}{F_s} = \frac{r_l^2}{r_s^2}$$

$$\frac{F_l}{240 \text{ lb}} = \frac{(16 \text{ in.})^2}{(2 \text{ in.})^2}$$

$$F_l = \frac{(240 \text{ lb})(16 \text{ in.})^2}{(2 \text{ in.})^2}$$

$$= \frac{(240 \text{ lb})(256 \text{ in}^2)}{4 \text{ in}^2}$$

$$= 15{,}360 \text{ lb}$$

EXERCISES 7.3

Use the map in Figure 7.2 (page 266) to find the approximate distance between each pair of cities (find straight-line [air] distances only):

1. Champaign and Bloomington
2. Bloomington and Decatur
3. Rantoul and Kankakee
4. Rantoul and Bloomington
5. Pana and Rantoul
6. Champaign and Mattoon
7. Charleston and Pontiac
8. Paris and Bloomington
9. Lincoln and Danville
10. Flatville and Mt. Pulaski

Use the map in Illustration 1 to find the approximate air distance between each pair of cities:

11. St. Louis and Kansas City
12. Memphis and St. Louis
13. Memphis and Little Rock
14. Sedalia, MO, and Tulsa, OK
15. Fort Smith, AR, and Springfield, MO
16. Pine Bluff, AR, and Jefferson City, MO

Use the scale drawing of a metal plate cover in Illustration 2 in Exercises 17–26:

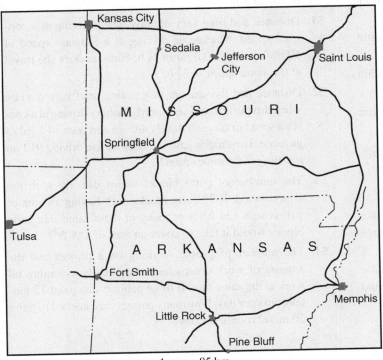

1 cm = 85 km

ILLUSTRATION 1

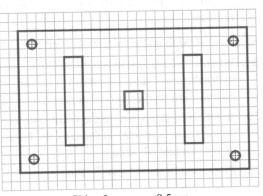

Side of square = 0.5 cm

ILLUSTRATION 2

17. What is the length of the plate cover?
18. What is the width of the plate cover?
19. What is the area of the plate cover?
20. What is the diameter of the circular holes?
21. What are the dimensions of the square hole?
22. What are the dimensions of the rectangular holes?
23. What is the distance between the rectangular holes, center to center?
24. What is the distance between the centers of the upper pair of circular holes?
25. What is the distance between the centers of the right pair of circular holes?
26. Answer each of the questions in Exercises 17–25 if the scale were changed so that the side of the square = $2\frac{1}{16}$ in.

With pencil and ruler make line drawings on square-ruled paper to fit each description in Exercises 27–30:

27. A rectangle 8 ft by 6 ft. Use this scale: Side of a square = 1 ft.
28. A square 16 cm on a side. Use this scale: Side of a square = 2 cm.
29. A circle 36 mm in diameter. Use this scale: Side of a square = 3 mm.
30. A rectangle 12 in. by 8 in. with a circle in its center 4 in. in diameter. Use this scale: Side of a square = 2 in.
31. a. Can the actual circle in Exercise 29 be placed within the actual rectangle in Exercise 27?
    b. Can the scale drawing of the circle be placed within the scale drawing of the rectangle?
32. a. Can the actual circle in Exercise 29 be placed within the actual square in Exercise 28?
    b. Can the scale drawing of the circle be placed within the scale drawing of the square?

Use the formulas for the hydraulic press to find each value in Exercises 33–50:

33. F_l = 4000 lb and F_s = 200 lb. Find MA.
34. When a force of 160 lb is applied to the small piston of a hydraulic press, a force of 4800 lb is exerted by the large piston. Find its mechanical advantage.
35. A 400-lb force applied to the small piston of a hydraulic press produces a 3600-lb force by the large piston. Find its mechanical advantage.
36. F_l = 2400 lb and MA = $\frac{50}{1}$. Find F_s.
37. F_l = 5100 lb and MA = $\frac{75}{1}$. Find F_s.
38. A hydraulic press has a mechanical advantage of 36 : 1. If a force of 2750 lb is applied to the small piston, what force is produced by the large piston?
39. A hydraulic press with an MA of 90 : 1 has a force of 2650 lb applied to its small piston. What force is produced by its large piston?
40. A hydraulic system with an MA of 125 : 1 has a force of 2450 lb exerted by its large piston. What force is applied to its small piston?
41. r_l = 27 in. and r_s = 3 in. Find MA.
42. r_l = 36 in. and r_s = 4 in. Find MA.
43. The radii of the pistons of a hydraulic press are 3 in. and 15 in. Find its mechanical advantage.
44. The radius of the small piston of a hydraulic press is 2 in., and the radius of its large piston is 18 in. What is its mechanical advantage?
45. F_s = 25 lb, r_l = 8 in., and r_s = 2 in. Find F_l.
46. F_s = 81 lb, r_l = 9 in., and r_s = 1 in. Find F_l.
47. F_l = 6400 lb, r_l = 16 in., and r_s = 4 in. Find F_s.
48. F_l = 7500 lb, r_l = 15 in., and r_s = 3 in. Find F_s.
49. A force of 40 lb is applied to a piston of radius 7 in. of a hydraulic press. The large piston has a radius of 28 in. What force is exerted by the large piston?
50. A force of 8100 lb is exerted by a piston of radius 30 in. of a hydraulic press. What force was applied to its piston of radius 3 in.?
51. Distance and time vary directly when driving at a constant speed. Mackenzie driving at a constant speed of 60 mi/h travels 90.0 mi in $1\frac{1}{2}$ h. How far does she travel at the same speed in $3\frac{3}{4}$ h?
52. Distance and the amount of gasoline used vary directly when driving at a constant speed. Zachary driving at a constant speed of 65 mi/h travels 495 mi and uses 14.1 gal of gasoline. How much gasoline does he use driving 912 mi traveling at the same speed?
53. The number of patio blocks varies directly with the area covered. If 93 patio blocks each having an area of $\frac{3}{4}$ ft^2 covers 124 ft^2, how many of these same size patio blocks would it take to cover an area of 188 ft^2?
54. The number of people working on a project and the amount of work completed vary directly assuming all work at the same rate. If three painters can paint 12 motel rooms per day, how many painters are needed to paint 20 motel rooms per day?

7.4 Inverse Variation

If two quantities, y and x, change so that their *product is constant* (that is, $y_1 x_1 = y_2 x_2$), they are said to *vary inversely*. This relationship between the two quantities is called **inverse variation**. This means that if one quantity increases, the other decreases and vice versa so that their product is always the same. Compare this with direct variation, where the ratio of the two quantities is always the same.

Consider the following data:

y	8	24	12	3	6	48
x	6	2	4	16	8	1

Note that y varies inversely with x because the product is always 48, a constant. This relationship may also be written $xy = 48$ or $y = \frac{48}{x}$.

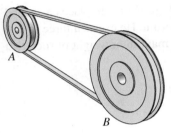

Figure 7.5 Pulley system

Inverse Variation

$$x_1 y_1 = x_2 y_2$$

or

$$\frac{y_1}{y_2} = \frac{x_2}{x_1}$$

One example of inverse variation is the relationship between two rotating pulleys connected by a belt (see Figure 7.5). This relationship is given by the following formula:

Pulley System Relationship

(diameter of A)(rpm of A) = (diameter of B)(rpm of B)

Example 1

A small pulley is 11 in. in diameter, and a larger one is 20 in. in diameter. How many rpm does the smaller pulley make if the larger one rotates at 44 rpm (revolutions per minute)?

(diameter of A)(rpm of A) = (diameter of B)(rpm of B)
(11)(x) = (20)(44)

$$x = \frac{(20)(44)}{11} = 80 \text{ rpm}$$

Another example of inverse variation is the relationship between the number of teeth and the number of rpm of two rotating gears, as shown in Figure 7.6.

Gear System Relationship

(no. of teeth in A)(rpm of A) = (no. of teeth in B)(rpm of B)

Figure 7.6 Gears

Example 2

A large gear with 14 teeth rotates at 40 rpm. It turns a small gear with 8 teeth. How fast does the small gear rotate?

(no. of teeth in A)(rpm of A) = (no. of teeth in B)(rpm of B)
(14)(40 rpm) = (8)(x)

$$\frac{(14)(40 \text{ rpm})}{8} = x$$

70 rpm = x

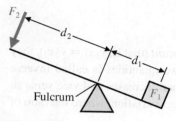

Figure 7.7 Lever

Figure 7.7 shows a lever, which is a rigid bar, pivoted to turn on a point (or edge) called a *fulcrum*. The parts of the lever on either side of the fulcrum are called *lever arms*.

To lift the box requires a force F_1. This is produced by pushing down on the other end of the lever with a force F_2. The distance from F_2 to the fulcrum is d_2. The distance from F_1 to the fulcrum is d_1. The principle of the lever is another example of inverse variation. It can be expressed by the following formula:

Lever Principle Relationship

$$F_1 d_1 = F_2 d_2$$

When the products are equal, the lever is balanced.

Example 3 A man places one end of a lever under a large rock, as in Figure 7.8. He places a second rock under the lever, 2 ft from the first rock, to act as a fulcrum. He exerts a force of 180 pounds at a distance of 6 ft from the fulcrum. Find F_1, the maximum weight of rock that could be lifted.

$$F_1 d_1 = F_2 d_2$$
$$(F_1)(2 \text{ ft}) = (180 \text{ lb})(6 \text{ ft})$$
$$F_1 = \frac{(180 \text{ lb})(\overset{3}{\cancel{6 \text{ ft}}})}{\cancel{2 \text{ ft}}}$$
$$F_1 = 540 \text{ lb}$$

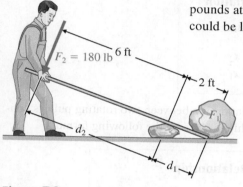

Figure 7.8

EXERCISES 7.4

Fill in the blanks:

	Pulley A		Pulley B	
	Diameter	rpm	Diameter	rpm
1.	25 cm	72	50 cm	
2.	18 cm		12 cm	96
3.	10 cm	120	15 cm	
4.		84	8 in.	48
5.	34 cm	440		680
6.	25 cm	600	48 cm	
7.		225	15 in.	465
8.	98 cm	240		360

9. A small pulley is 13 in. in diameter, and a larger one is 18 in. in diameter. How many rpm does the larger pulley make if the smaller one rotates at 720 rpm?

10. A 21-in. pulley, rotating at 65 rpm, turns a smaller pulley at 210 rpm. What is the diameter of the smaller pulley?

11. A large pulley turns at 48 rpm. A smaller pulley 8 in. in diameter turns at 300 rpm. What is the diameter of the larger pulley?

12. A pulley 32 in. in diameter turns at 825 rpm. At how many rpm will a pulley 25 in. in diameter turn?

13. A motor turning at 1870 rpm has a 4.0-in.-diameter pulley driving a fan that must turn at 680 rpm. What diameter pulley must be put on the fan?

14. A hydraulic pump is driven with an electric motor (see Illustration 1). The pump must rotate at 1200 rpm. The pump is equipped with a 6.0-in.-diameter belt pulley. The motor runs at 1800 rpm. What diameter pulley is required on the motor?

7.4 ♦ Inverse Variation

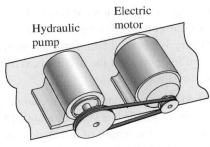

ILLUSTRATION 1

15. One pulley is 7 cm larger in diameter than a second pulley. The larger pulley turns at 80 rpm, and the smaller pulley turns at 136 rpm. What is the diameter of each pulley?

16. One pulley is twice as large in diameter as a second pulley. If the larger pulley turns at 256 rpm, what is the rpm of the smaller?

Use Illustration 2 for Exercises 17–18.

17. Find the rpm of pulley B if pulley A turns at $15\overline{0}0$ rpm.
18. Find the rpm of pulley C if pulley A turns at $15\overline{0}0$ rpm.

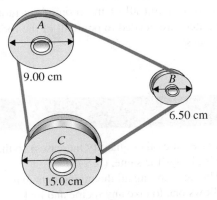

ILLUSTRATION 2

Use Illustration 3 for Exercises 19–20. Note: Pulleys B and C are fixed and rotate together on the same shaft.

19. The diameter of pulley A is 18.0 cm, pulley B is 5.00 cm, pulley C is 15.0 cm, and pulley D is 5.50 cm. Find the rpm of pulley D if pulley A turns at $12\overline{0}0$ rpm.

20. The diameter of pulley A is 12.0 in., pulley B is 4.50 in., pulley C is 15.0 in., and pulley D is 3.75 in. Find the rpm of pulley D if pulley A turns at $15\overline{0}0$ rpm.

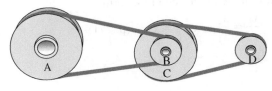

ILLUSTRATION 3

Fill in the blanks:

	Gear A		Gear B	
	Number of teeth	rpm	Number of teeth	rpm
21.	50	400		125
22.	220		45	440
23.	42	600	25	
24.	50	64		80
25.		$6\frac{1}{4}$	120	30
26.	80	$1\frac{1}{4}$		$3\frac{1}{3}$

27. A small gear with 25 teeth turns a large gear with 75 teeth at 32 rpm. How many rpm does the small gear make?

28. A large gear with 180 teeth running at 600 rpm turns a small gear at 900 rpm. How many teeth does the small gear have?

29. A large gear with 60 teeth turning at 72 rpm turns a small gear with 30 teeth. At how many rpm does the small gear turn?

30. A large gear with 80 teeth turning at 150 rpm turns a small gear with 12 teeth. At how many rpm does the small gear turn?

31. A large gear with 120 teeth turning at 30 rpm turns a small gear at 90 rpm. How many teeth does the small gear have?

32. A large gear with 200 teeth turning at 17 rpm turns a small gear at 100 rpm. How many teeth does the small gear have?

Complete the table:

	F_1	d_1	F_2	d_2
33.	18 lb	5 in.	9 lb	
34.	30 lb		70 lb	8 in.
35.	40 lb	9 in.		3 in.
36.		6.3 ft	458.2 lb	8.7 ft

In Exercises 33–37, draw a sketch for each and solve:

37. An object is 6 ft from the fulcrum and balances a second object 8 ft from the fulcrum. The first object weighs 180 lb. How much does the second object weigh?

38. A block of steel weighing 1800 lb is to be raised by a lever extending under the block 9 in. from the fulcrum. How far from the fulcrum must a 150-lb man apply his weight to the bar to balance the steel?
39. A rocker arm raises oil from an oil well. On each stroke, it lifts a weight of 1 ton on a weight arm 4 ft long. What force is needed to lift the oil on a force arm 8 ft long?
40. A carpenter needs to raise one side of a building with a lever 3.65 m in length. The lever, with one end under the building, is placed on a fulcrum 0.45 m from the building. A mass of 90 kg pulls down on the other end. What mass is being lifted when the building begins to rise?
41. A 1200-g mass is placed 72 cm from the fulcrum of a lever. How far from the fulcrum is a 1350-g mass that balances it?
42. A lever is in balance when a mass of 2000 g is placed 28 cm from a fulcrum. An unknown mass is placed 20 cm from the fulcrum on the other side. What is the amount of the unknown mass?
43. A 210-lb object is placed on a lever. It balances a 190-lb weight that is 28 in. from the fulcrum. How far from the fulcrum should the 210-lb weight be placed?
44. A piece of machinery weighs 3 tons. It is to be balanced by two men whose combined weight is 330 lb. The piece of machinery is placed 11 in. from the fulcrum. How far from the fulcrum must the two men exert their weight in order to balance it?
45. The length and width vary inversely in a rectangle of a given area. A given rectangle has sides of length 18 cm and width 12 cm. Find the length of a rectangle of the same area whose width is 8 cm.
46. The speed and time of travel vary inversely when driving a given distance. Juan drives 45 mi/h for 6 h to visit his grandmother on a snowy day. He drives at 55 mi/h for his return trip home. How long did the return trip take?
47. The current and resistance vary inversely in a circuit with a given voltage. In a 240-volt circuit, the current is 8 amperes and the resistance is 30 ohms. Find the current if the resistance is 80 ohms in a circuit with the same voltage.
48. The number of people working on a project and the time it takes to complete a project vary inversely, assuming everyone works at the same rate. If it takes three painters 24 days to paint all of the rooms in a motel, how many painters are needed to paint all of the rooms in 8 working days?

SUMMARY | CHAPTER 7

Glossary of Basic Terms

Direct variation. When two quantities change so that their ratios are constant. When one quantity increases, the other quantity increases (or when one quantity decreases, the other quantity decreases) so that their ratio is always the same. (p. 265)
Extremes. The first and fourth terms of a proportion. (p. 258)
Inverse variation. When two quantities change so that their products are constant. When one quantity increases, the other quantity decreases (or when one quantity decreases, the other quantity increases) so that their product is always the same. (p. 271)
Kitchen ratio. The ratio among all the ingredients in a recipe; this allows one to take any recipe and make more or less of the given recipe. (p. 261)
Means. The second and third terms of a proportion. (p. 258)
Proportion. An equation with two equal ratios. (p. 257)
Rate. The comparison of two unlike quantities whose units do not cancel. (p. 255)
Ratio of two numbers. The first number divided by the second number. (p. 254)

7.2 Proportion

1. In any proportion, the product of the means equals the product of the extremes. That is, if

$\dfrac{a}{b} = \dfrac{c}{d}$, then $bc = ad$. (p. 258)

7.3 Direct Variation

1. **Direct variation:** When two quantities vary directly, their ratios are constant in the form $\dfrac{y_1}{x_1} = \dfrac{y_2}{x_2}$. (p. 265)

7.4 Inverse Variation

1. **Inverse variation:** When two quantities vary inversely, their ratios are constant in the form

$\dfrac{y_1}{y_2} = \dfrac{x_2}{x_1}$ or $x_1 y_1 = x_2 y_2$. (p. 271)

REVIEW | CHAPTER 7

Write each ratio in lowest terms:

1. 7 to 28
2. 60 : 40
3. 1 g to 500 mg
4. $\dfrac{5 \text{ ft } 6 \text{ in.}}{9 \text{ ft}}$

Determine whether or not each pair of ratios is equal:

5. $\dfrac{7}{2}, \dfrac{35}{10}$
6. $\dfrac{5}{18}, \dfrac{30}{115}$

Solve each proportion (round each to three significant digits when necessary):

7. $\dfrac{x}{4} = \dfrac{5}{20}$
8. $\dfrac{10}{25} = \dfrac{x}{75}$
9. $\dfrac{3}{x} = \dfrac{8}{64}$
10. $\dfrac{72}{96} = \dfrac{30}{x}$
11. $\dfrac{73.4}{x} = \dfrac{25.9}{37.4}$
12. $\dfrac{x}{19.7} = \dfrac{144}{68.7}$
13. $\dfrac{61.1}{81.3} = \dfrac{592}{x}$
14. $\dfrac{243}{58.3} = \dfrac{x}{127}$

15. A piece of cable 180 ft long costs $67.50. How much will 500 ft cost at the same unit price?

16. A copper wire 750 ft long has a resistance of 1.89 Ω. How long is a copper wire of the same size whose resistance is 3.15 Ω?

17. A crew of electricians can wire 6 houses in 144 h. How many hours will it take them to wire 9 houses?

18. An automobile braking system has a 12-to-1 lever advantage on the master cylinder. A 25-lb force is applied to the pedal. What force is applied to the master cylinder?

19. Jones invests $6380 and Hernandez invests $4620 in a partnership business. What percent of the total investment does each have?

20. One gallon of a pesticide mixture weighs 7 lb 13 oz. It contains 11 oz of pesticide. What percent of the mixture is pesticide?

21. Indicate what kind of variation is shown by each equation.

   (a) $\dfrac{y_1}{x_1} = \dfrac{y_2}{x_2}$ (b) $y_1 x_1 = y_2 x_2$

22. What kind of variation is indicated when one quantity increases while the other increases so that their ratio is always the same?

23. Suppose $\tfrac{1}{4}$ in. on a map represents 25 mi. What distance is represented by $3\tfrac{5}{8}$ in.?

24. The scale on a map is 1 in. = 600 ft. Two places are known to be 2 mi apart. What distance will show between them on the map?

25. Two pulleys are connected by a belt. The numbers of rpm of the two pulleys vary inversely as their diameters. A pulley having a diameter of 25 cm is turning at 900 rpm. What is the number of rpm of the second pulley, which has a diameter of 40 cm?

26. A large gear with 42 teeth rotates at 25 rpm. It turns a small gear with 14 teeth. How fast does the small gear rotate?

27. In hydraulics, the formula relating the forces and the radii of the pistons is

$$\dfrac{F_l}{F_s} = \dfrac{r_l^2}{r_s^2}$$

Given $F_l = 6050$ lb, $r_l = 22$ in., and $r_s = 2$ in., find F_s.

28. An object 9 ft from the fulcrum of a lever balances a second object 12 ft from the fulcrum. The first object weighs 240 lb. How much does the second object weigh?

29. The current I varies directly as the voltage E. Suppose $I = 0.6$ A when $E = 30$ V. Find the value of I when $E = 100$ V.

30. The number of workers needed to complete a particular job is inversely proportional to the number of hours that they work. If 12 electricians can complete a job in 72 h, how long will it take 8 electricians to complete the same job? Assume that each person works at the same rate, no matter how many people are assigned to the job.

TEST | CHAPTER 7

Write each ratio in lowest terms:

1. 16 m to 64 m
2. 3 ft to 6 in.
3. 400 mL to 5 L

Solve each proportion: (Round to three significant digits when necessary.)

4. $\dfrac{x}{8} = \dfrac{18}{48}$
5. $\dfrac{8}{x} = \dfrac{24}{5}$

6. $\dfrac{7200}{84} = \dfrac{x}{252}$ 7. $\dfrac{72.6}{28.7} = \dfrac{x}{152}$

8. If 60 ft of fencing costs $138, how much does 80 ft cost?

9. Five quarts of pure antifreeze are added to 10 quarts of water to fill a radiator. What percent of antifreeze is in the mixture?

10. The scale on a map is 1 cm = 10 km. If two cities are 4.8 cm apart on the map, what is the actual distance between them?

11. What kind of variation is indicated when one quantity increases while the other decreases so that their product is always the same?

12. A used automobile sells for $7500. The down payment is $900. What percent of the selling price is the down payment?

13. A small gear with 36 teeth turns a large gear with 48 teeth at 150 rpm. What is the speed (in rpm) of the small gear?

14. A pulley is 20 cm in diameter and rotating at 150 rpm. Find the diameter of a smaller pulley that must rotate at 200 rpm.

15. Given the lever formula, $F_1 d_1 = F_2 d_2$, and $F_1 = 800$ lb, $d_1 = 9$ ft, $d_2 = 3.6$ ft, find F_2.

16. A man who weighs 200 lb is to be raised by a lever extending under the man 15 in. from the fulcrum. How much force must be applied at a distance of 24 in. from the fulcrum in order to lift him?

Graphing Linear Equations

CHAPTER 8

OBJECTIVES

- Find ordered pairs of numbers that are solutions to a linear equation with two variables.
- Plot points in the number plane.
- Graph a linear equation by plotting points.
- Find the slope of a line.
- Determine when two lines are parallel, perpendicular, or neither by finding the slope of the lines.
- Graph a linear equation given its slope and *y* intercept and through a given point with a given slope.
- Find the equation of a line with a given slope and *y* intercept, with a given slope through a given point, and through two given points.

Mathematics at Work

Drafters prepare working plans and detailed technical drawings used by construction and production workers to build a wide variety of products ranging from manufactured products to industrial machinery to buildings to oil and gas pipelines. Their drawings provide visual and technical details of the products and structures as well as specifying dimensions, materials to be used, and procedures and processes to be followed. Drafters also provide rough sketches, specifications, codes, and any calculations provided by engineers, architects, scientists, or surveyors. Most use computer-assisted drafting (CAD) equipment and software to prepare drawings. These systems use computer workstations to create a drawing on a video screen and store the drawings electronically so that revisions, copies, or variations can be made easily and quickly. While CAD is a useful tool, drafters need the basic drafting skills and standards as well as the CAD skills and knowledge.

Drafting work has many specializations due to special design and applications. Architectural drafters draw structures and buildings. Aeronautical drafters prepare engineering drawings detailing plans and specifications used for manufacturing aircraft, missiles, and related parts. Electrical drafters prepare wiring and layout diagrams used by workers to erect, install, and repair electrical equipment and wiring in a wide variety of settings. Civil drafters prepare drawings and topographical and relief maps used in construction projects such as highways, bridges, pipelines, and water and sewage systems. Mechanical drafters prepare detail and assembly drawings of a wide variety of machinery and mechanical devices showing dimensions, fastening methods, and other requirements.

Many community colleges and postsecondary and trade schools offer associate degree and certificate programs in which students develop drafting and mechanical skills; a basic knowledge of drafting standards, mathematics, science, and engineering technology; computer-aided drafting and design techniques; and communication and problem-solving skills. For more information, please visit **www.cengage.com** and access the Student Online Resources for this text.

Drafter/CAD Designer
Using CAD to design industrial components

8.1 Linear Equations with Two Variables

In Chapter 6, we studied linear equations with one variable, such as $2x + 4 = 10$ and $3x - 7 = 5$. We found that most linear equations in one variable have only one root. In this chapter, we study equations with *two* variables, such as

$$3x + 4y = 12 \quad \text{or} \quad x + y = 7$$

How many solutions does the equation $x + y = 7$ have? Any two numbers whose sum is 7 is a solution—for example, 1 for x and 6 for y, 2 for x and 5 for y, -2 for x and 9 for y, $5\frac{1}{2}$ for x and $1\frac{1}{2}$ for y, and so on. Most linear equations with *two variables* have *many possible solutions*.

Since it is very time consuming to write pairs of replacements in this manner, we use **ordered pairs** in the form (x, y) to write solutions of equations with two variables. Therefore, instead of writing the solutions of the equation $x + y = 7$ as above, we write them as $(1, 6), (2, 5), (-2, 9), \left(5\frac{1}{2}, 1\frac{1}{2}\right)$, and so on.

> **Linear Equation with Two Variables**
>
> A **linear equation with two variables** can be written in the form
>
> $ax + by = c$
>
> where the numbers a, b, and c are all real numbers such that a and b are both not 0.

Example 1 Determine whether the given ordered pair is a solution of the given equation.

a. $(5, 2)$; $3x + 4y = 23$

To determine whether the ordered pair $(5, 2)$ is a solution to $3x + 4y = 23$, substitute 5 for x and 2 for y as follows:

$$3x + 4y = 23$$
$$3(5) + 4(2) = 23 \qquad \text{Substitute } x = 5 \text{ and } y = 2.$$
$$15 + 8 = 23$$
$$23 = 23 \qquad \text{True}$$

The result is true, so $(5, 2)$ is a solution of $3x + 4y = 23$.

b. $(-5, 6)$; $2x - 4y = -32$

$$2(-5) - 4(6) = -32 \qquad \text{Substitute } x = -5 \text{ and } y = 6.$$
$$-10 - 24 = -32$$
$$-34 = -32 \qquad \text{False}$$

The result is false, so $(-5, 6)$ is not a solution of $2x - 4y = -32$. ◆

To find solutions of a linear equation with two variables, replace one variable with a number you have chosen and then solve the resulting linear equation for the remaining variable.

Example 2 Complete the three ordered-pair solutions of $2x + y = 5$.

a. $(4, \quad)$

Replace x with 4. Any number could be used, but for this example, we will use 4. The resulting equation is

$$2(4) + y = 5$$
$$8 + y = 5$$
$$8 + y - 8 = 5 - 8 \qquad \text{Subtract 8 from both sides.}$$
$$y = -3$$

Check: Replace x with 4 and y with -3.

$$2(4) + (-3) = 5 \quad ?$$
$$8 - 3 = 5 \quad \text{True}$$

Therefore, $(4, -3)$ is a solution.

b. $(-2,)$

Replace x with -2. The resulting equation is

$$2(-2) + y = 5$$
$$-4 + y = 5$$
$$-4 + y + 4 = 5 + 4 \quad \text{Add 4 to both sides.}$$
$$y = 9$$

Check: Replace x with -2 and y with 9.

$$2(-2) + 9 = 5 \quad ?$$
$$-4 + 9 = 5 \quad \text{True}$$

Thus, $(-2, 9)$ is a solution.

c. $(0,)$

Replace x with 0. The resulting equation is

$$2(0) + y = 5$$
$$0 + y = 5$$
$$y = 5$$

Check: Replace x with 0 and y with 5.

$$2(0) + (5) = 5 \quad ?$$
$$0 + 5 = 5 \quad \text{True}$$

Therefore, $(0, 5)$ is a solution. ♦

You may find it easier first to solve the equation for y and then make each replacement for x.

Example 3 Complete the three ordered-pair solutions of $3x - y = 4$ by first solving the equation for y.

a. $(5,)$ **b.** $(-2,)$ **c.** $(0,)$

$$3x - y = 4$$
$$3x - y - 3x = 4 - 3x \quad \text{Subtract } 3x \text{ from both sides.}$$
$$-y = 4 - 3x$$
$$y = -4 + 3x \quad \text{Divide both sides by } -1.$$

You may make a table to keep your work in order.

	x	$3x - 4$	y	Ordered pairs
a.	5	$3(5) - 4 = 15 - 4$	11	$(5, 11)$
b.	-2	$3(-2) - 4 = -6 - 4$	-10	$(-2, -10)$
c.	0	$3(0) - 4 = 0 - 4$	-4	$(0, -4)$

Therefore, the three solutions are (5, 11), (−2, −10), and (0, −4). Choosing a positive value, a negative value, and 0 for x is often a good approach.

Example 4 Complete the three ordered-pair solutions of $5x + 3y = 7$.

    a. (2,)   **b.** (0,)   **c.** (−1,)

We will first solve for y.

$$5x + 3y = 7$$
$$5x + 3y - 5x = 7 - 5x \quad \text{Subtract } 5x \text{ from both sides.}$$
$$3y = 7 - 5x$$
$$\frac{3y}{3} = \frac{7 - 5x}{3} \quad \text{Divide both sides by 3.}$$
$$y = \frac{7 - 5x}{3}$$

	x	$\dfrac{7 - 5x}{3}$	y	Ordered pairs
a.	2	$\dfrac{7 - 5(2)}{3} = \dfrac{-3}{3}$	-1	$(2, -1)$
b.	0	$\dfrac{7 - 5(0)}{3} = \dfrac{7}{3}$	$\dfrac{7}{3}$	$\left(0, \dfrac{7}{3}\right)$
c.	-1	$\dfrac{7 - 5(-1)}{3} = \dfrac{12}{3}$	4	$(-1, 4)$

The three solutions are $(2, -1)$, $\left(0, \dfrac{7}{3}\right)$, and $(-1, 4)$.

Solutions to linear equations with two variables may be shown visually by graphing them in a number plane. To construct a number plane, draw a horizontal number line, which is called the ***x* axis**, as in Figure 8.1. Then draw a second number line intersecting the first line at right angles so that both number lines have the same zero point, called the **origin**. The vertical number line is called the ***y* axis**.

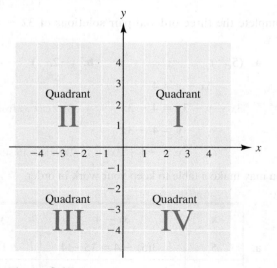

Figure 8.1
Rectangular coordinate system

Each number line, or axis, has a scale. The numbers on the *x* axis are *positive to the right* of the origin and *negative to the left* of the origin. Similarly, the numbers on the *y* axis are *positive above* the origin and *negative below* the origin.

All the points in the plane determined by these two intersecting axes make up the **number plane**. The axes divide the number plane into four regions, called **quadrants**. The quadrants are numbered as shown in Figure 8.1.

Points in the number plane are usually indicated by an *ordered pair* of numbers written in the form (x, y), where *x* is the first number in the ordered pair and *y* is the second number in the ordered pair. The numbers *x* and *y* are also called the **coordinates** of a point in the number plane. Figure 8.1 is often called the **rectangular coordinate system** or the *Cartesian coordinate system*.

Plotting Points in the Number Plane

To locate the point in the number plane which corresponds to an ordered pair (x, y):

STEP 1 Count right or left, from 0 (the origin) along the *x* axis, the number of spaces corresponding to the first number of the ordered pair (right if positive, left if negative).

STEP 2 Count up or down, from the point reached on the *x* axis in Step 1, the number of spaces corresponding to the second number of the ordered pair (up if positive, down if negative).

STEP 3 Mark the last point reached with a dot.

Example 5 Plot the point corresponding to the ordered pair $(3, -4)$ in Figure 8.2.

First, count three spaces to the right along the *x* axis. Then, count down four spaces from that point. Mark the final point with a dot.

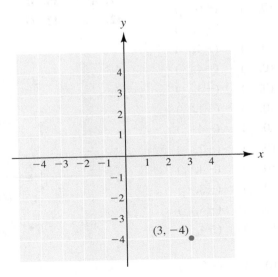

Figure 8.2

282 CHAPTER 8 ♦ Graphing Linear Equations

Example 6 Plot the points corresponding to the ordered pairs in the number plane in Figure 8.3:
$A(1, 2)$, $B(3, -2)$, $C(-4, 7)$, $D(5, 0)$, $E(-2, -3)$, $F(-5, -1)$, $G(-2, 4)$.

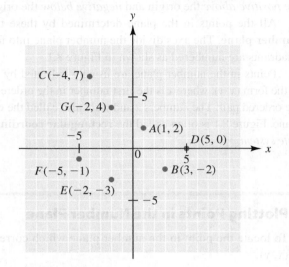

Figure 8.3

EXERCISES 8.1

Complete the three ordered-pair solutions of each equation:

	Equation		Ordered Pairs	
1.	$x + y = 5$	(3,)	(8,)	(−2,)
2.	$-2x + y = 8$	(2,)	(7,)	(−4,)
3.	$6x + 2y = 10$	(2,)	(0,)	(−2,)
4.	$6x - y = 0$	(3,)	(5,)	(−2,)
5.	$3x - 4y = 8$	(0,)	(2,)	(−4,)
6.	$5x - 3y = 8$	(1,)	(0,)	(−2,)
7.	$-2x + 5y = 10$	(5,)	(0,)	(−3,)
8.	$-4x - 7y = -3$	(−1,)	(0,)	(−8,)
9.	$9x - 2y = 10$	(2,)	(0,)	(−4,)
10.	$2x + 3y = 6$	(3,)	(0,)	(−6,)
11.	$y = 3x + 4$	(2,)	(0,)	(−3,)
12.	$y = 4x - 8$	(3,)	(0,)	(−4,)
13.	$5x + y = 7$	(2,)	(0,)	(−4,)
14.	$4x - y = 8$	(1,)	(0,)	(−3,)
15.	$2x = y - 4$	(3,)	(0,)	(−1,)
16.	$3y - x = 5$	(1,)	(0,)	(−4,)
17.	$5x - 2y = -8$	(4,)	(0,)	(−2,)
18.	$2x - 3y = 1$	(2,)	(0,)	(−4,)
19.	$9x - 2y = 5$	(1,)	(0,)	(−3,)
20.	$2x + 7y = -12$	(1,)	(0,)	(−8,)
21.	$y = 3$ (Think: $0x + 1y = 3$)	(2,)	(0,)	(−4,)
22.	$y + 4 = 0$	(3,)	(0,)	(−7,)
23.	$x = 5$ (Think: $1x + 0y = 5$)	(, 4)	(, 0)	(, −2)
24.	$x + 7 = 0$	(, 5)	(, 0)	(, −6)

Solve for y in terms of x:

25. $2x + 3y = 6$
26. $4x + 5y = 10$
27. $x + 2y = 7$
28. $2x + 2y = 5$
29. $x - 2y = 6$
30. $x - 3y = 9$
31. $2x - 3y = 9$
32. $4x - 5y = 10$
33. $-2x + 3y = 6$
34. $-3x + 5y = 25$
35. $-2x - 3y = -15$
36. $-3x - 4y = -8$

Write the ordered pair corresponding to each point in Illustration 1:

37. A 38. B 39. C 40. D 41. E
42. F 43. G 44. H 45. I 46. J

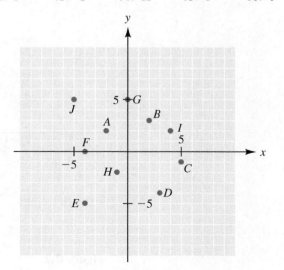

ILLUSTRATION 1

Plot each point in the number plane. Label each point by writing its ordered pair and letter:

47. $A\,(1, 3)$
48. $B\,(4, 0)$
49. $C\,(-6, -2)$
50. $D\,(-2, 4)$
51. $E\,(5, -4)$
52. $F\,(-4, -4)$
53. $G\,(0, 9)$
54. $H\,(3, 7)$
55. $I\,(5, -5)$
56. $J\,(-5, 5)$
57. $K\,(-6, -3)$
58. $L\,(3, -7)$
59. $M\,(-4, 5)$
60. $N\,(-2, -6)$
61. $O\,(1, -3)$
62. $P\,(5, 2)$
63. $Q\left(3, \frac{1}{2}\right)$
64. $R\left(4\frac{1}{2}, -3\frac{1}{2}\right)$
65. $S\left(-6, 2\frac{1}{2}\right)$
66. $T\left(-4\frac{1}{2}, 6\frac{1}{2}\right)$

8.2 Graphing Linear Equations

In Section 8.1, you learned that a linear equation with two variables has many solutions. In Example 2, you found that three of the solutions of $2x + y = 5$ were $(4, -3)$, $(-2, 9)$, and $(0, 5)$. Now plot the points corresponding to these ordered pairs and connect the points, as shown in Figure 8.4.

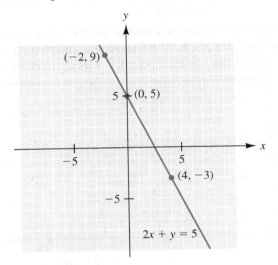

Figure 8.4

You can see from the figure that the three points lie on the same straight line. If you find another solution of $2x + y = 5$—say, $(1, 3)$—the point corresponding to this ordered pair also lies on the same straight line. The solutions of a linear equation with two variables always correspond to points lying on a straight line. Therefore, *the graph of the solutions of a linear equation with two variables is always a straight line*. Only part of this line can be shown on the graph; the line actually extends without limit in both directions.

Graphing Linear Equations

To draw the graph of a linear equation with two variables:

STEP 1 Find any three solutions of the equation.
(*Note:* Two solutions would be enough, since two points determine a straight line. However, a third solution gives a third point as a check. If the three points do not lie on the same straight line, you have made an error.)

STEP 2 Plot the three points corresponding to the three ordered pairs that you found in Step 1.

STEP 3 Draw a line through the three points. If the line is not straight, check your solutions.

We will show two methods for finding the three solutions of the equation. The first method involves solving the equation for y and then substituting three different values of x to find each corresponding y value. The second method involves substituting three different values of x and then solving each resulting equation for y. You may use either method. Depending on the equation, one method may be easier to use than the other.

Example 1 Draw the graph of $3x + 4y = 12$.

STEP 1 Find any three solutions of $3x + 4y = 12$. First, solve for y:

$$3x + 4y - 3x = 12 - 3x \quad \text{Subtract } 3x \text{ from both sides.}$$
$$4y = 12 - 3x$$
$$\frac{4y}{4} = \frac{12 - 3x}{4} \quad \text{Divide both sides by 4.}$$
$$y = \frac{12 - 3x}{4}$$

Choose any three values of x and solve for y. Here, we have chosen $x = 4$, $x = 0$, and $x = -2$.

NOTE: We often choose a positive number, 0, and a negative number for x to obtain a range of points in the graph. Although finding and plotting any two points will allow you to graph the straight line, the third point provides an excellent check.

x	$\frac{12 - 3x}{4}$	y	Ordered pairs
4	$\frac{12 - 3(4)}{4} = \frac{0}{4}$	0	$(4, 0)$
0	$\frac{12 - 3(0)}{4} = \frac{12}{4}$	3	$(0, 3)$
-2	$\frac{12 - 3(-2)}{4} = \frac{18}{4}$	$\frac{9}{2}$	$\left(-2, \frac{9}{2}\right)$

Three solutions are $(4, 0)$, $(0, 3)$, and $\left(-2, \frac{9}{2}\right)$.

STEP 2 Plot the points corresponding to $(4, 0)$, $(0, 3)$, and $\left(-2, \frac{9}{2}\right)$.

STEP 3 Draw a straight line through these three points. (See Figure 8.5.)

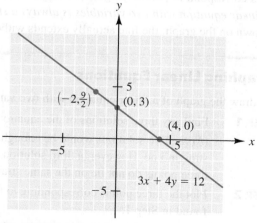

Figure 8.5

An alternative method is shown in Example 2.

8.2 ♦ Graphing Linear Equations

Example 2 Draw the graph of $2x - 3y = 6$.

STEP 1 Set up a table and write the values you choose for x—say, 3, 0, and -3.

	a.	b.	c.
x	3	0	-3
y			

STEP 2 Substitute the chosen values of x in the given equation and solve for y.

a. $2x - 3y = 6$
    $2(3) - 3y = 6$
    $6 - 3y = 6$
    $-3y = 0$
    $y = 0$

b. $2x - 3y = 6$
    $2(0) - 3y = 6$
    $0 - 3y = 6$
    $-3y = 6$
    $y = -2$

c. $2x - 3y = 6$
    $2(-3) - 3y = 6$
    $-6 - 3y = 6$
    $-3y = 12$
    $y = -4$

STEP 3 Write the values for y that correspond to the chosen values for x in the table, thus:

	a.	b.	c.
x	3	0	-3
y	0	-2	-4

That is, three solutions of $2x - 3y = 6$ are the ordered pairs $(3, 0)$, $(0, -2)$, and $(-3, -4)$.

STEP 4 Plot the points from Step 3 and draw a straight line through them, as in Figure 8.6.

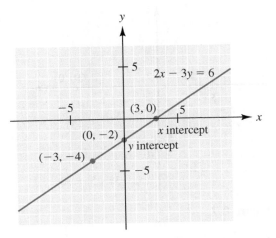

Figure 8.6

The x Intercept and the y Intercept of a Line

The line in Figure 8.6 crosses the x axis at the point $(3, 0)$. The number 3 is called the **x intercept**—the x coordinate of the point where the graph crosses the x axis.

To find the x intercept of a line, replace y in the equation by 0 and solve for x.

The line in Figure 8.6 crosses the y axis at the point $(0, -2)$. The number -2 is called the **y intercept**—the y coordinate of the point where the graph crosses the y axis.

To find the y intercept of a line, replace x in the equation by 0 and solve for y.

NOTE: The x intercept or the y intercept may also be given as an ordered pair.

CHAPTER 8 ♦ Graphing Linear Equations

Example 3 Find the x and y intercepts of the graph of $3x - 5y = 30$ and then graph the equation.

To find the x intercept, replace y in the equation by 0 and solve for x as follows:

$3x - 5y = 30$
$3x - 5(0) = 30$
$3x = 30$ Divide both sides by 3.
$x = 10$

So, the x intercept is $(10, 0)$.

To find the y intercept, replace x in the equation by 0 and solve for y as follows:

$3x - 5y = 30$
$3(0) - 5y = 30$
$-5y = 30$ Divide both sides by -5.
$y = -6$

So, the y intercept is $(0, -6)$.

Let's find a third point by letting $x = 5$ and solve for y as follows:

$3x - 5y = 30$
$3(5) - 5y = 30$
$15 - 5y = 30$
$-5y = 15$
$y = -3$

The third point is $(5, -3)$.

Plot these three points and draw a line through them. (See Figure 8.7.)

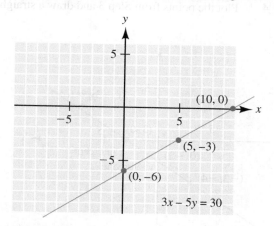

Figure 8.7

Finding the x and y intercepts is an excellent method for graphing a linear equation or for checking the graph of a linear equation.

Two special cases of linear equations have a graph of a horizontal line or a vertical line. The equation $y = 5$ is a linear equation with an x coefficient of 0. (This equation may also be written as $0x + 1y = 5$.) Similarly, $x = -7$ is a linear equation with a y coefficient of 0. (This equation may also be written as $1x + 0y = -7$.) These equations have graphs that are horizontal or vertical straight lines, as shown in the next two examples.

Example 4 Draw the graph of $y = 5$.

Set up a table and write the values you choose for x—say, 3, 0, and -4. As the equation states, y is always 5 for any value of x that you choose.

x	3	0	-4
y	5	5	5

Plot the points from the table: (3, 5), (0, 5), and (−4, 5). Then draw a straight line through them, as in Figure 8.8.

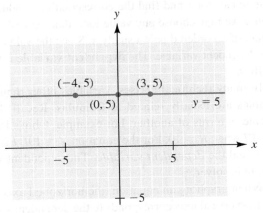

Figure 8.8

Horizontal Line

The graph of the linear equation $y = k$, where k is a constant, is the horizontal line through the point $(0, k)$. That is, $y = k$ is a horizontal line with a y intercept of k.

Example 5 Draw the graph of $x = -7$.

All ordered pairs that are solutions of $x = -7$ have an x value of -7. You can choose any number for y. Three ordered pairs that satisfy $x = -7$ are (−7, 3), (−7, 1), and (−7, −4). Plot these three points and draw a straight line through them, as in Figure 8.9.

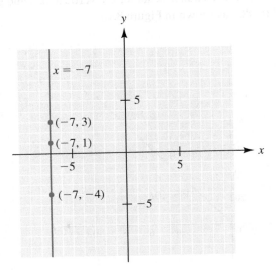

Figure 8.9

> **Vertical Line**
>
> The graph of the linear equation $x = k$, where k is a constant, is the vertical line through the point $(k, 0)$. That is, $x = k$ is a vertical line with an x intercept of k.

Solve the equation $2x - y = 8$ for y. The solution is $y = 2x - 8$. To graph this equation, assign values for x and find the corresponding y values. We call x the independent variable, because we may choose any value for x that we wish. The **independent variable** is the first element of an ordered pair, usually x. Since the value of y depends on the value of x, we call y the dependent variable. The **dependent variable** is the second element of an ordered pair, usually y.

In many technical classes, variables other than x and y are often used. These other variables are usually related to formulas. Recall that a formula can be solved for one variable in terms of another. For example, Ohm's law can be expressed as $V = IR$, or as $V = 10I$ when $R = 10$. For the equation $V = 10I$, I is called the *independent variable* and V is called the *dependent variable*. The dependent variable is the variable for which the formula is solved.

When graphing an equation, the horizontal axis corresponds to the independent variable; the vertical axis corresponds to the dependent variable. Think of graphing the ordered pairs

(independent variable, dependent variable)

Example 6

Draw the graph of $V = 10I$.

Since I is the independent variable, graph ordered pairs in the form (I, V). Set up a table and write the values you choose for I—say, 0, 5, and 8. For this example, limit your values of I to nonnegative numbers.

I	0	5	8
V	0	50	80

Next, choose a suitable scale for the vertical axis and graph the ordered pairs $(0, 0)$, $(5, 50)$, and $(8, 80)$, as shown in Figure 8.10.

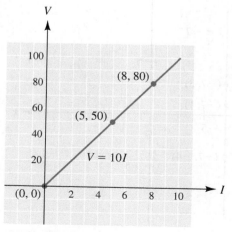

Figure 8.10

EXERCISES 8.2

Draw the graph of each equation:

1. $x + y = 7$
2. $x + 3y = 9$
3. $y = 2x + 3$
4. $y = 4x - 5$
5. $4y = x$
6. $2x + y = 6$
7. $6x - 2y = 10$
8. $2x + 3y = 9$
9. $3x - 4y = 12$
10. $3x - 5y = 15$
11. $5x + 4y = 20$
12. $2x - 3y = 18$
13. $2x + 7y = 14$
14. $2x - 5y = 20$
15. $y = 2x$
16. $y = -3x$
17. $3x + 5y = 11$
18. $4x - 3y = 15$
19. $y = -\frac{1}{2}x + 4$
20. $y = \frac{2}{3}x - 6$
21. $y = 3$
22. $y = -2$
23. $x = -4$
24. $x = 5$
25. $y - 6 = 0$
26. $y + 10 = 0$
27. $x + 3\frac{1}{2} = 0$
28. $x - 4 = 0$
29. $y = 0$
30. $x = 0$

Identify the independent and dependent variables for each equation:

31. $s = 4t + 7$
32. $V = 5t - 2$
33. $R = 0.5V$
34. $s = 65t$
35. $i = 30t - 10$
36. $E = 4V + 2$
37. $v = 50 - 6t$
38. $i = 18 - 3t$
39. $s = 3t^2 + 5t - 1$
40. $v = 2i^2 - 3i + 10$

41. The distance, s (in feet), that a body travels in t seconds is given by the equation $s = 5t + 10$. Graph the equation for nonnegative values of t.

42. The voltage, v (in mV), in an electrical circuit varies according to the equation $v = 10t - 5$, where t is in seconds. Graph the equation for nonnegative values of v.

43. The resistance, R, in an electrical circuit varies according to the equation $R = 1.5V$, where V is in volts. Graph the equation for nonnegative values of V.

44. The current, I (in amps), in an electrical circuit varies according to the equation $I = 0.05V$, where V is in volts. Choose a suitable scale and graph the equation for nonnegative values of V.

45. The voltage, v, in an electrical circuit is given by the equation $v = 60 - 5t$, where t is in μs. Graph the equation for nonnegative values of v and t.

46. The distance, s (in metres), that a point travels in t milliseconds is given by the equation $s = 24 - 2t$. Graph the equation for nonnegative values of s and t.

8.3 The Slope of a Line

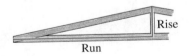

Figure 8.11
Slope

The slope of a line or the "steepness" of a roof (see Figure 8.11) can be measured by the following ratio:

$$\text{slope} = \frac{\text{vertical change}}{\text{horizontal change}} = \frac{\text{rise}}{\text{run}}$$

A straight line can also be graphed by using its slope and knowing one point on the line.

If two points on a line (x_1, y_1) (read "x-sub-one, y-sub-one") and (x_2, y_2) (read "x-sub-two, y-sub-two") are known (see Figure 8.12), the slope of the line is defined as follows:

Slope of a Line

$$\text{slope} = m = \frac{\text{vertical change}}{\text{horizontal change}} = \frac{\text{rise}}{\text{run}} = \frac{\text{difference in } y \text{ values}}{\text{difference in } x \text{ values}} = \frac{y_2 - y_1}{x_2 - x_1}$$

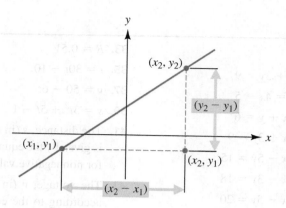

Figure 8.12
Slope of a line through two points

Example 1 Find the slope of the line passing through the points $(-2, 3)$ and $(4, 7)$. (See Figure 8.13.)

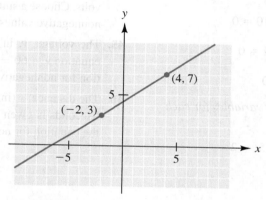

Figure 8.13

If we let $x_1 = -2$, $y_1 = 3$, $x_2 = 4$, and $y_2 = 7$, then

$$m = \frac{y_2 - y_1}{x_2 - x_1} = \frac{7 - 3}{4 - (-2)} = \frac{4}{6} = \frac{2}{3}$$

Note that if we reverse the order of taking the differences of the coordinates, the result is the same:

$$m = \frac{y_1 - y_2}{x_1 - x_2} = \frac{3 - 7}{-2 - 4} = \frac{-4}{-6} = \frac{2}{3}$$

Example 2 Find the slope of the line passing through $(-3, 2)$ and $(3, -6)$. (See Figure 8.14.)

If we let $x_1 = -3$, $y_1 = 2$, $x_2 = 3$, and $y_2 = -6$, then

$$m = \frac{y_2 - y_1}{x_2 - x_1} = \frac{-6 - 2}{3 - (-3)} = \frac{-8}{6} = -\frac{4}{3}$$

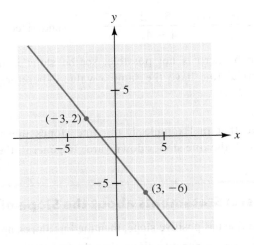

Figure 8.14

Example 3 Find the slope of the line through (−5, 2) and (3, 2). (See Figure 8.15.)

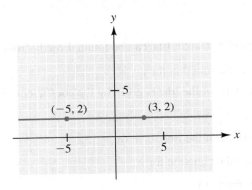

Figure 8.15

$$m = \frac{y_2 - y_1}{x_2 - x_1} = \frac{2 - 2}{3 - (-5)} = \frac{0}{8} = 0$$

Note that all points on any horizontal line have the same *y* value. Therefore, *the slope of any horizontal line is 0.*

Example 4 Find the slope of the line through (4, 2) and (4, −5). (See Figure 8.16.)

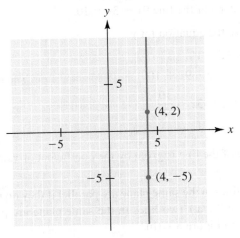

Figure 8.16

$$m = \frac{y_2 - y_1}{x_2 - x_1} = \frac{-5 - 2}{4 - 4} = \frac{-7}{0} \quad \text{(undefined)}$$

Division by zero is not possible, so the slope is undefined. Note that all points on any vertical line have the same x value. Therefore, *the slope of any vertical line is undefined.*

> **General Statements about the Slope of a Line**
>
> 1. If a line has positive slope, then the line slopes upward from left to right.
> 2. If a line has negative slope, then the line slopes downward from left to right.
> 3. If the slope of a line is zero, then the line is *horizontal.*
> 4. If the slope of a line is undefined, then the line is *vertical.*

The slope of a straight line can be found directly from its equation as follows:

1. Solve the equation for y.
2. The slope of the line is given by the coefficient of x.

Example 5   Find the slope of the line $4x + 6y = 15$.

First, solve the equation for y.

$$4x + 6y = 15$$
$$6y = -4x + 15 \quad \text{Subtract } 4x \text{ from both sides.}$$
$$y = -\frac{2}{3}x + \frac{5}{2} \quad \text{Divide both sides by 6.}$$

↑—slope

The slope of the line is given by the coefficient of x, or $m = -\frac{2}{3}$.

Example 6   Find the slope of the line $9x - 3y = 10$.

First, solve the equation for y.

$$9x - 3y = 10$$
$$-3y = -9x + 10 \quad \text{Subtract } 9x \text{ from both sides.}$$
$$y = 3x - \frac{10}{3} \quad \text{Divide both sides by } -3.$$

↑—slope

The slope of the line is given by the coefficient of x, or $m = 3$.

Two lines in the same plane are parallel if they do not intersect even if they are extended. (See Figure 8.17a.) Two lines in the same plane are perpendicular if they intersect at right angles, as in Figure 8.17(b).

Since parallel lines have the same steepness, they have the same slope.

(a) Parallel lines

(b) Perpendicular lines

Figure 8.17

Parallel Lines

Two lines are parallel if either one of the following conditions holds:
1. Both lines are perpendicular to the *x* axis (Figure 8.18a).
2. Both lines have the same slope (Figure 8.18b)—that is, if the equations of the two lines are $L_1: y = m_1 x + b_1$ and $L_2: y = m_2 x + b_2$, then
$$m_1 = m_2$$

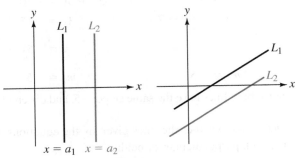

(a) Both lines are perpendicular to the *x* axis.

(b) $m_1 = m_2$

Figure 8.18
Parallel lines

Perpendicular Lines

Two lines are perpendicular if either one of the following conditions holds:
1. One line is vertical with equation $x = a$ and the other line is horizontal with equation $y = b$.
2. Neither is vertical and the product of the slopes of the two lines is -1; that is, if the equations of the lines are
$$L_1: y = m_1 x + b_1 \quad \text{and} \quad L_2: y = m_2 x + b_2$$
then
$$m_1 \cdot m_2 = -1$$

NOTE: The slopes of two perpendicular lines neither of which is vertical are negative reciprocals of each other; that is, $m_1 = -\dfrac{1}{m_2}$.

Example 7   Determine whether the lines given by the equations $2x + 3y = 6$ and $6x - 4y = 9$ are parallel, perpendicular, or neither.

First, find the slope of each line by solving its equation for *y*.

$$\begin{aligned} 2x + 3y &= 6 & 6x - 4y &= 9 \\ 3y &= -2x + 6 & -4y &= -6x + 9 \\ y &= -\frac{2}{3}x + 2 & y &= \frac{3}{2}x - \frac{9}{4} \\ m_1 &= -\frac{2}{3} & m_2 &= \frac{3}{2} \end{aligned}$$

Since the slopes are not equal, the lines are not parallel. Next, find the product of the slopes.

$$m_1 \cdot m_2 = \left(-\frac{2}{3}\right)\left(\frac{3}{2}\right) = -1$$

Thus, the lines are perpendicular.

Example 8 Determine whether the lines given by the equations $5x + y = 7$ and $15x + 3y = -10$ are parallel, perpendicular, or neither.

First, find the slope of each line by solving its equation for y.

$$5x + y = 7 \qquad\qquad 15x + 3y = -10$$
$$y = -5x + 7 \qquad\qquad 3y = -15x - 10$$
$$\qquad\qquad\qquad y = -5x - \frac{10}{3}$$
$$m_1 = -5 \qquad\qquad m_2 = -5$$

Since both lines have the same slope, -5, and different y intercepts, they are parallel.

Example 9 Determine whether the lines given by the equations $4x + 5y = 15$ and $-3x - 2y = 12$ are parallel, perpendicular, or neither.

First, find the slope of each line by solving its equation for y.

$$4x + 5y = 15 \qquad\qquad -3x - 2y = 12$$
$$5y = -4x + 15 \qquad\qquad -2y = 3x + 12$$
$$y = -\frac{4}{5}x + 3 \qquad\qquad y = -\frac{3}{2}x - 6$$
$$m_1 = -\frac{4}{5} \qquad\qquad m_2 = -\frac{3}{2}$$

Since the slopes are not equal and do not have a product of -1, the lines are neither parallel nor perpendicular. That is, the lines intersect but not at right angles.

EXERCISES 8.3

Find the slope of the line passing through each pair of points:

1. $(-3, 1), (2, 6)$
2. $(4, 7), (6, 2)$
3. $(-2, 1), (-3, -5)$
4. $(5, -3), (-4, -9)$
5. $(4, 0), (0, 5)$
6. $(1, -6), (-2, 0)$
7. $(-2, 4), (5, 4)$
8. $(-3, -7), (-2, -7)$
9. $(6, 1), (6, -4)$
10. $(-8, 5), (-8, -3)$
11. $(4, -2), (-6, -8)$
12. $(-9, -1), (-3, -5)$

Find the slope of each line:

13.
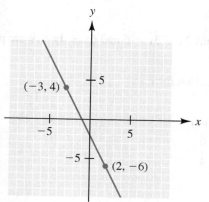

14.
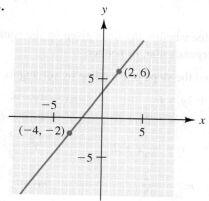

15.
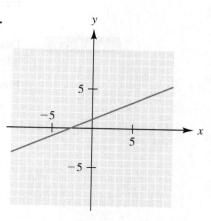

16. 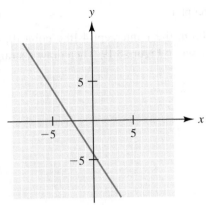 **17.** 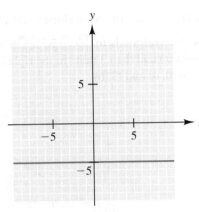 **18.**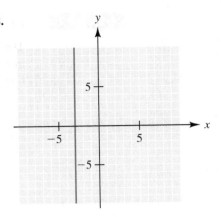

Find the slope of each line:

19. $y = 6x + 2$
20. $y = -4x + 3$
21. $y = -5x - 7$
22. $y = 9x - 13$
23. $3x + 5y = 6$
24. $9x + 12y = 8$
25. $-2x + 8y = 3$
26. $4x - 6y = 9$
27. $5x - 2y = 16$
28. $-4x - 2y = 7$
29. $x - 3 = 0$
30. $y + 15 = 0$

Determine whether the lines given by the equations are parallel, perpendicular, or neither:

31. $y = 4x - 5$
$y = 4x + 5$

32. $y = \dfrac{2}{3}x + 4$
$y = -\dfrac{3}{2}x - 5$

33. $y = \dfrac{3}{4}x - 2$
$y = -\dfrac{4}{3}x + \dfrac{5}{3}$

34. $y = \dfrac{4}{5}x - 2$
$y = -\dfrac{4}{5}x + \dfrac{2}{3}$

35. $x + 3y = 9$
$3x - y = 14$

36. $x + 2y = 11$
$2x + 4y = -5$

37. $x - 4y = 12$
$x + 4y = 16$

38. $2x + 7y = 6$
$14x - 4y = 18$

39. $y - 5x = 12$
$5x - y = -6$

40. $-3x + 9y = 20$
$x = 3y$

8.4 The Equation of a Line

We have learned to graph the equation of a straight line and to find the slope of a straight line given its equation or any two points on it. In this section, we will use the slope to graph the equation and to write its equation.

A fast and easy way to draw the graph of a straight line when the *slope* and *y intercept* are known is to first plot the *y*-intercept point on the graph and consider this to be the starting point. Then from this starting point and using the slope (*rise* over *run*), move up or down the number of units indicated by the rise and then move right or left the number of units indicated by the run to a second point. Mark this second point. Then draw a straight line through these two points.

SUGGESTION: If the slope is a fraction and negative, include the negative sign with the numerator (rise). If the slope is an integer, write the slope with 1 as its denominator (run). Also note that the *y* intercept does not have to be the starting point; any point of the line can be the starting point, if it is known.

Example 1 Draw the graph of the line with slope $\frac{2}{3}$ and y intercept 4.

The slope $\frac{2}{3}$ corresponds to $\frac{\text{difference in } y \text{ values}}{\text{difference in } x \text{ values}} = \frac{2}{3}$. From the y intercept 4 [the point (0, 4)], move 2 units up and then 3 units to the right, as shown in Figure 8.19. Then draw a straight line through (0, 4) and (3, 6).

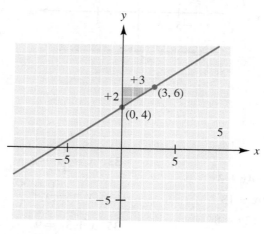

Figure 8.19

In Section 8.3, we learned how to find the slope of a line, given its equation, by solving for y. For example, for a line whose equation is $y = 3x + 5$ the slope is 3, the coefficient of x. What does the number 5 have to do with its graph? If we let $x = 0$, the equation is

$$y = 3x + 5$$
$$y = 3(0) + 5$$
$$= 0 + 5 = 5$$

The line crosses the y axis when $x = 0$. Therefore, the y intercept of the graph is 5.

Slope-Intercept Form

When the equation of a straight line is written in the form
$$y = mx + b$$
the slope of the line is m and the y intercept is b.

Example 2 Draw the graph of the equation $8x + 2y = -10$ using its slope and y intercept.

First, find the slope and y intercept by solving the equation for y as follows:

$$8x + 2y = -10$$
$$2y = -8x - 10 \quad \text{Subtract } 8x \text{ from both sides.}$$
$$y = -4x - 5 \quad \text{Divide both sides by 2.}$$

        ↗   ↖
      slope y intercept

8.4 ♦ The Equation of a Line   297

The slope is −4, and the y intercept is −5. The slope −4 corresponds to

$$\frac{\text{difference in } y \text{ values}}{\text{difference in } x \text{ values}} = \frac{-4}{1}$$

When the slope is an integer, write it as a ratio with 1 in the denominator. From the y intercept −5, move 4 units down and 1 unit to the right, as shown in Figure 8.20. Draw a straight line through the points (0, −5) and (1, −9).

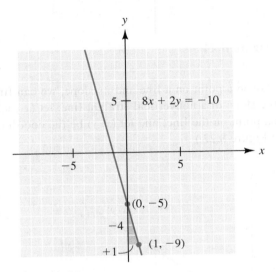

Figure 8.20   ♦

Example 3 Find the equation of the line with slope $\frac{3}{4}$ and y intercept −2.

Use the slope-intercept form with $m = \frac{3}{4}$ and $b = -2$.

$$y = mx + b$$
$$y = \frac{3}{4}x + (-2)$$
$$y = \frac{3}{4}x - 2$$
$$4y = 3x - 8 \quad \text{Multiply both sides by 4.}$$
$$0 = 3x - 4y - 8 \quad \text{Subtract } 4y \text{ from both sides.}$$
$$8 = 3x - 4y \quad \text{Add 8 to both sides.}$$
$$3x - 4y = 8$$

NOTE: Any of the last five equations in Example 3 is correct. The most common ways of writing equations are $y = mx + b$ and $cx + dy = f$.   ♦

Example 4 Draw the graph of the straight line through the point (−3, 6) with slope $-\frac{2}{5}$.

The slope $-\frac{2}{5}$ corresponds to $\frac{\text{difference in } y \text{ values}}{\text{difference in } x \text{ values}} = -\frac{2}{5}$. From the point (−3, 6), move 2 units down and 5 units to the right, as shown in Figure 8.21. Draw a straight line through the points (−3, 6) and (2, 4).

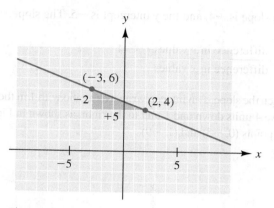

Figure 8.21

Given a point on a line and its slope, we can find its equation. To show this, let m be the slope of a nonvertical straight line; let (x_1, y_1) be the coordinates of a known or given point on the line; and let (x, y) be the coordinates of any other point on the line. (See Figure 8.22.)

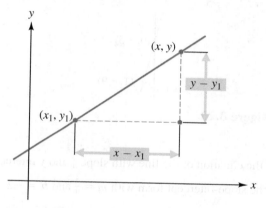

Figure 8.22

Then, by the definition of slope, we have

$$\frac{\text{difference in } y \text{ values}}{\text{difference in } x \text{ values}} = m$$

$$\frac{y - y_1}{x - x_1} = m$$

$$y - y_1 = m(x - x_1) \qquad \text{Multiply both sides by } (x - x_1).$$

The result is the point-slope form of the equation of a straight line.

Point-Slope Form

If m is the slope and (x_1, y_1) is any point on a nonvertical straight line, its equation is
$$y - y_1 = m(x - x_1)$$

Example 5 Find the equation of the line with slope -3 that passes through the point $(-1, 4)$.

Here, $m = -3$, $x_1 = -1$, and $y_1 = 4$. Using the point-slope form, we have

$$y - y_1 = m(x - x_1)$$
$$y - 4 = -3[x - (-1)]$$
$$y - 4 = -3(x + 1)$$
$$y - 4 = -3x - 3 \quad \text{Remove parentheses.}$$
$$y = -3x + 1 \quad \text{Add 4 to both sides.}$$

The point-slope form also can be used to find the equation of a line when two points on the line are known.

Example 6 Find the equation of the line through the points $(5, -4)$ and $(-1, 8)$.

First, find the slope.

$$m = \frac{y_2 - y_1}{x_2 - x_1} = \frac{8 - (-4)}{-1 - 5} = \frac{12}{-6} = -2$$

Substitute $m = -2$, $x_1 = 5$, and $y_1 = -4$ in the point-slope form.

$$y - y_1 = m(x - x_1)$$
$$y - (-4) = -2(x - 5)$$
$$y + 4 = -2x + 10 \quad \text{Remove parentheses.}$$
$$y = -2x + 6 \quad \text{Subtract 4 from both sides.}$$

We could have used the other point $(-1, 8)$, as follows:

$$y - y_1 = m(x - x_1)$$
$$y - 8 = -2[x - (-1)]$$
$$y - 8 = -2(x + 1)$$
$$y - 8 = -2x - 2 \quad \text{Remove parentheses.}$$
$$y = -2x + 6 \quad \text{Add 8 to both sides.}$$

EXERCISES 8.4

Draw the graph of each line with the given slope and y intercept:

1. $m = 2, b = 5$
2. $m = 4, b = -3$
3. $m = -5, b = 4$
4. $m = -1, b = 0$
5. $m = \frac{2}{3}, b = -4$
6. $m = \frac{3}{4}, b = 2$
7. $m = -\frac{2}{5}, b = -4$
8. $m = \frac{5}{6}, b = 3$
9. $m = \frac{4}{3}, b = -1$
10. $m = -\frac{9}{4}, b = 2$

Draw the graph of each equation using the slope and y intercept:

11. $2x + y = 6$
12. $4x + y = -3$
13. $3x + 5y = 10$
14. $4x + 3y = 9$
15. $3x - y = 7$
16. $5x - y = -2$
17. $3x - 2y = 12$
18. $2x - 5y = 20$
19. $2x - 6y = 0$
20. $4x + 7y = 0$

Find the equation of the line with the given slope and y intercept:

21. $m = 2, b = 5$
22. $m = 4, b = -3$
23. $m = -5, b = 4$
24. $m = -1, b = 0$
25. $m = \frac{2}{3}, b = -4$
26. $m = \frac{3}{4}, b = 2$
27. $m = -\frac{6}{5}, b = 3$
28. $m = -\frac{12}{5}, b = -1$
29. $m = -\frac{3}{5}, b = 0$
30. $m = 0, b = 0$

Draw the graph of the line through the given point with the given slope:

31. $(3, 5), m = 2$
32. $(-1, 4), m = -3$
33. $(-5, 0), m = \dfrac{3}{4}$
34. $(0, 0), m = -\dfrac{1}{3}$
35. $(6, -2), m = \dfrac{3}{2}$
36. $(-3, -3), m = \dfrac{1}{2}$
37. $(-1, -1), m = -1$
38. $(0, -4), m = 0$
39. $(2, -7), m = 1$
40. $(-8, 6), m = \dfrac{1}{8}$

Find the equation of the line through the given point with the given slope:

41. $(3, 5), m = 2$
42. $(-1, 4), m = -3$
43. $(-5, 0), m = \dfrac{3}{4}$
44. $(0, 0), m = -\dfrac{1}{3}$
45. $(6, -2), m = \dfrac{3}{2}$
46. $(-3, -3), m = \dfrac{1}{2}$
47. $(-3, -1), m = -\dfrac{10}{3}$
48. $(4, 5), m = -\dfrac{3}{7}$
49. $(12, -10), m = -1$
50. $(15, 20), m = 3$

Find the equation of the line through the given points:

51. $(2, 3), (5, 1)$
52. $(8, 5), (2, 1)$
53. $(-1, 4), (2, -2)$
54. $(-1, -2), (-4, 0)$
55. $(0, 1), (-3, 0)$
56. $(5, 5), (-2, -2)$
57. $(7, 4), (-1, 0)$
58. $(6, -3), (-3, 6)$
59. $(3, 3), (1, 5)$
60. $(16, 12), (4, 8)$

SUMMARY | CHAPTER 8

Glossary of Basic Terms

Coordinates. The numbers x and y written in the form (x, y). (p. 281)

Dependent variable. The second element of an ordered pair, usually y. (p. 288)

Independent variable. The first element of an ordered pair, usually x. (p. 288)

Linear equation with two variables. An equation that may be written in the form $ax + by = c$, where the numbers a, b, and c are all real numbers such that a and b are not both 0. (p. 278)

Number plane. All points in the plane determined by the intersecting, perpendicular x axis (horizontal axis) and y axis (vertical axis). (p. 281)

Ordered pair. Numbers in the form (x, y) that correspond to points in the number plane; also used to write solutions of systems of equations with two variables. (p. 278)

Origin. The zero point where the x and y axes intersect in the number plane. (p. 280)

Quadrant. Each of the four regions of the number plane formed by the intersection of the x and y axes. (p. 281)

Rectangular coordinate system. The number plane formed by the intersection of the x and y axes. Also called the *Cartesian coordinate system*. (p. 281)

x axis. The horizontal axis in the number plane. (p. 280)

x intercept. The x coordinate of the point where the graph crosses the x axis. (p. 285)

y axis. The vertical axis in the number plane. (p. 280)

y intercept. The y coordinate of the point where the graph crosses the y axis. (p. 285)

8.1 Linear Equations with Two Variables

1. **Plotting points in the number plane:** To locate or plot the point in the number plane which corresponds to an ordered pair (x, y):
   a. Count right or left, from 0 (the origin) along the x axis, the number of spaces corresponding to the first number of the ordered pair (right if positive, left if negative).
   b. Count up or down, from the point reached on the x axis in the step above, the number of spaces corresponding to the second number of the ordered pair (up if positive, down if negative).
   c. Mark the last point reached with a dot. (p. 281)

8.2 Graphing Linear Equations

1. **The graph of the solutions of a linear equation in two variables is always a straight line.** (p. 283)
2. **Graphing linear equations:** To graph a linear equation with two variables:
   a. Find any three solutions. Note: Two solutions would be enough, since two points determine a straight line. However, a third solution gives a third point as a check.
   b. Plot the three points corresponding to the three ordered pairs that you found above.
   c. Draw a line through the three points. If it is not a straight line, check your solutions. (p. 283)

3. To find the *x* intercept of a line, replace *y* in the equation by 0 and solve for *x*. (p. 285)
4. To find the *y* intercept of a line, replace *x* in the equation by 0 and solve for *y*. (p. 286)
5. **Horizontal line:** The graph of the linear equation $y = k$, where *k* is a constant, is the horizontal line through the point $(0, k)$. That is, $y = k$ is a horizontal line with a *y* intercept of *k*. (p. 287)
6. **Vertical line:** The graph of the linear equation $x = k$, where *k* is a constant, is the vertical line through the point $(k, 0)$. That is, $x = k$ is a vertical line with an *x* intercept of *k*. (p. 288)
7. **Independent/dependent variables and ordered pairs:** In graphing an equation, the horizontal axis corresponds to the independent variable; the vertical axis corresponds to the dependent variable. In general, think of graphing the ordered pairs:
   (independent variable, dependent variable) (p. 288)

8.3 The Slope of a Line

1. **Slope of a line:** If two points (x_1, y_1) and (x_2, y_2) on a line are known, the slope of the line is $m = \dfrac{y_2 - y_1}{x_2 - x_1}$. (p. 289)
2. **General statements about the slope of a line:**
   a. If a line has a positive slope, the line slopes upward from left to right.
   b. If a line has a negative slope, the line slopes downward from left to right.
   c. If the slope of a line is zero, the line is *horizontal*.
   d. If the slope of a line is undefined, the line is *vertical*. (p. 292)
3. **Parallel lines:** Two lines are parallel if either one of the following conditions holds:
   a. Both lines are perpendicular to the *x* axis.
   b. Both lines have the same slope; that is, $m_1 = m_2$. (p. 293)
4. **Perpendicular lines:** Two lines are perpendicular if either one of the following conditions holds:
   a. One is a vertical line with equation $x = a$ and the other line is horizontal with equation $y = b$.
   b. Neither is vertical and the product of the slopes of the two lines is -1; that is, $m_1 \cdot m_2 = -1$. (p. 293)

8.4 The Equation of a Line

1. **Slope-intercept form of a line:** When the equation of a straight line is written in the form $y = mx + b$, the slope of the line is *m* and the *y* intercept is *b*. (p. 296)
2. **Point-slope form of a line:** If *m* is the slope and (x_1, y_1) is any point on a nonvertical straight line, its equation is $y - y_1 = m(x - x_1)$. (p. 298)

REVIEW | CHAPTER 8

Complete the ordered-pair solutions of each equation:

1. $x + 2y = 8$ (3,) (0,) (−4,)
2. $2x - 3y = 12$ (3,) (0,) (−3,)
3. Solve for *y*: $6x + y = 15$.
4. Solve for *y*: $3x - 5y = -10$.

Write the ordered pair corresponding to each point in Illustration 1:

5. *A*
6. *B*
7. *C*
8. *D*

Plot each point in the number plane. Label each point by writing its ordered pair:

9. $E(3, -2)$
10. $F(-7, -4)$
11. $G(-1, 5)$
12. $H(0, -5)$

Draw the graph of each equation:

13. $x + y = 8$
14. $x - 2y = 5$
15. $3x + 6y = 12$
16. $4x - 5y = 15$
17. $4x = 9y$
18. $y = \dfrac{1}{3}x - 4$
19. $x = -6$
20. $y = 7$

Find the slope of the line passing through each pair of points:

21. $(3, -4), (10, 5)$
22. $(-4, 0), (2, 6)$

Find the slope of each line:

23. $y = 4x - 7$
24. $2x + 5y = 8$
25. $5x - 9y = -2$

ILLUSTRATION 1

Determine whether the lines given by the equations are parallel, perpendicular, or neither:

26. $y = 3x - 5$
$y = -\dfrac{1}{3}x - 5$

27. $3x - 4y = 12$
$8x - 6y = 15$

28. $2x + 5y = 8$
$4x + 10y = 25$

29. $x = 4$
$y = -6$

Draw the graph of each line with the given slope and y intercept:

30. $m = -2, b = 9$

31. $m = -\dfrac{2}{3}, b = -5$

Draw the graph of each equation using the slope and y intercept:

32. $3x + 5y = 20$

33. $5x - 8y = 32$

Find the equation of the line with the given slope and y intercept:

34. $m = -\dfrac{1}{2}, b = 3$

35. $m = \dfrac{8}{3}, b = 0$

36. $m = 0, b = 0$

Draw the graph of the line through the given point with the given slope:

37. $(6, -1), m = -3$

38. $(-5, -2), m = \dfrac{7}{2}$

Find the equation of the line through the given point with the given slope:

39. $(-2, 8), m = -1$

40. $(0, -5), m = -\dfrac{1}{4}$

Find the equation of the line through the given points:

41. $(2, -3), (10, 5)$

42. $(12, 0), (2, -5)$

TEST | CHAPTER 8

Given the equation $3x - 4y = 24$, complete each ordered pair:

1. $(-4,)$ **2.** $(0,)$ **3.** $(4,)$

4. Solve for y: $2x + 3y = 12$

5–6. Write the ordered pair corresponding to each point in Illustration 1.

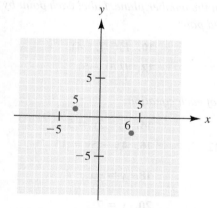

ILLUSTRATION 1

7. Draw the graph of $3x + y = 3$.

8. Draw the graph of $-2x + y = 4$.

9. Draw the graph of $s = 5 + 2t$ for nonnegative values of t.

10. Find the slope of the line containing the points $(-2, 4)$ and $(5, 6)$.

Find the slope of each line:

11. $y = 3x - 2$ **12.** $2x - 5y = 10$

13. Determine whether the graphs of the following pair of equations are parallel, perpendicular, or neither.
$2x - y = 10$
$y = 2x - 3$

14. Find the equation of the line having y intercept -3 and slope $\tfrac{1}{2}$.

15. Find the equation of the line containing the point $(-2, 3)$ and slope -2.

16. Find the equation of the line through $(4, 3)$ and $(-5, 10)$.

17. Draw the graph of the line containing the point $(3, 4)$ and having slope -3.

18. Draw the graph of the line $y = \tfrac{1}{2}x + 4$, using its slope and y intercept.

CUMULATIVE REVIEW | CHAPTERS 1–8

1. Evaluate: $2(6 - 5) + 3$
2. Subtract: $+6 - (-9)$
3. Find missing dimensions a and b in the figure in Illustration 1.

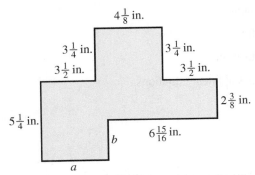

ILLUSTRATION 1

4. Change 250 cm to inches.
5. Multiply: $(6.2 \times 10^{-3})(1.8 \times 10^5)$
6. 61 mm = _____ m
7. Give the number of significant digits: 306,760 kg
8. Read the measurement shown on the metric micrometer in Illustration 2.

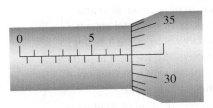

ILLUSTRATION 2

9. Use the rules for multiplication of measurements to evaluate: $1.8 \text{ m} \times 61.2 \text{ m} \times 3.2 \text{ m}$
10. Simplify: $(4x - 5) - (6 - 3x)$
11. Simplify: $(-5xy^2)(8x^3y^2)$
12. Simplify: $2x(4x - 3y)$
13. Solve: $3(x - 2) + 4(3 - 2x) = 9$
14. Solve: $\dfrac{2x}{3} + \dfrac{1}{5} = \dfrac{x}{4} - \dfrac{2}{3}$
15. Solve: $s = \dfrac{2V + t}{3}$ for V

Write each ratio in lowest terms:

16. 5 to 65
17. 32 in. : 3 yd

Solve each proportion (round each result to three significant digits when necessary):

18. $\dfrac{5}{13} = \dfrac{x}{156}$
19. $\dfrac{29.1}{73.8} = \dfrac{x}{104}$
20. $\dfrac{286}{x} = \dfrac{11.8}{59.7}$

21. If it costs $28.50 to repair 5 ft² of sidewalk, how much would it cost to repair 18 ft²?
22. A map shows a scale of 1 in. = 40 mi. What distance is represented by $4\frac{1}{4}$ in. on the map?
23. A large gear with 16 teeth rotates at 40 rpm. It turns a small gear at 64 rpm. How many teeth does the smaller gear have?
24. Complete the ordered-pair solutions of the equation: $2x + 3y = 12$ (3,), (0,), (−3,)
25. Solve for y: $4x + 2y = 7$
26. Draw the graph of $y = -2x - 5$.
27. Draw the graph of $3x - 2y = 12$.
28. Find the slope of the line containing the points $(-1, 3)$ and $(2, 6)$.
29. Find the equation of the line with slope $\frac{1}{2}$ and containing the point $(2, -4)$.
30. Determine whether the graphs of $2x - 3y = 6$ and $3x + 5y = 7$ are parallel, perpendicular, or neither.

CHAPTER 9 | Systems of Linear Equations

OBJECTIVES

- Solve a pair of linear equations by graphing.
- Solve a pair of linear equations by addition.
- Solve a pair of linear equations by substitution.
- Solve application problems involving pairs of linear equations.

Mathematics at Work

Welders join metal parts and fill holes or seams of metal products using handheld welding equipment. Related occupations include cutters, solderers, and braziers. Duties include reading blueprints, sketches, or specifications; calculating dimensions to be welded; inspecting structures or materials to be welded; igniting torches or starting power supplies; monitoring the welding process to avoid overheating; smoothing and polishing all surfaces; and maintaining equipment and machinery.

In the welding process, heat is applied to metal pieces to melt and fuse them to form a permanent bond. Because of its strength, welding is used to build ships, to manufacture and repair automobiles and farm equipment, to join beams in the construction of bridges and buildings, and to join pipes in pipelines, power plants, and refineries. Arc welding, the most common and simplest type of welding, uses electrical current to create heat and bond materials together. Cutters use heat to cut and trim metal objects to specific dimensions. Cutters also dismantle large objects, such as ships, railroad cars, and buildings. Solderers and braziers also use heat to join two or more metal items together. Soldering uses metals with a melting point below 840°F; brazing uses metals with a higher melting point.

Some welding positions require general certification in welding or certification in specific skills. Formal training is available in high school technical education courses and in postsecondary institutions such as community colleges, vocational–technical institutes, and private welding, soldering, and brazing schools. The U.S. Armed Forces also operate welding and soldering schools. Some highly skilled jobs require several years of combined school and on-the-job training.

Some employers are willing to hire inexperienced entry-level workers and train them on the job, but many prefer to hire workers who have been through formal training programs. Courses in blueprint reading, shop mathematics, mechanical drawing, physics, chemistry, and metallurgy are helpful.

For more information, please visit **www.cengage.com** and access the Student Online Resources for this text.

Welder
Welder working on a pipeline

9.1 Solving Pairs of Linear Equations by Graphing

Many problems can be solved by using two equations with two variables and solving them simultaneously. **To solve a pair of linear equations** with two variables simultaneously, you must find an ordered pair that will make both equations true at the same time.

As you know, the graph of a linear equation with two variables is a straight line. As shown in Figure 9.1, two straight lines (the graphs of *two* linear equations with two variables) in the same plane may be arranged as follows:

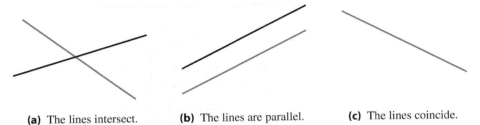

(a) The lines intersect. (b) The lines are parallel. (c) The lines coincide.

Figure 9.1
Possible relationships of two straight lines in the same plane

a. The lines may intersect. If so, they have one point in common. The equations have one common solution. The coordinates of the point of intersection define the common solution.

b. The lines may be parallel. If so, they have no point in common. The equations have no common solution.

c. The lines may coincide. That is, one line lies on top of the other. If so, any solution of one equation is also a solution of the other. With infinitely many points of intersection, there are infinitely many common solutions. The solution of the pair of equations consists of all points on the common line.

Example 1 Draw the graphs of $x - y = 2$ and $x + 3y = 6$ in the same number plane. Find the common solution of the equations.

STEP 1 Draw the graph of $x - y = 2$. First, solve for y.

$$x - y = 2$$
$$x - y - x = 2 - x \quad \text{Subtract } x \text{ from both sides.}$$
$$-y = 2 - x$$
$$\frac{-y}{-1} = \frac{2 - x}{-1} \quad \text{Divide both sides by } -1.$$
$$y = -2 + x$$

Then, find three solutions:

x	$-2 + x$	y
2	$-2 + 2$	0
0	$-2 + 0$	-2
-2	$-2 + (-2)$	-4

Three solutions are $(2, 0)$, $(0, -2)$, and $(-2, -4)$. Plot the three points that correspond to these three ordered pairs. Then draw a straight line through these three points, as in Figure 9.2.

STEP 2 Draw the graph of $x + 3y = 6$. First, solve for y.

$$x + 3y = 6$$
$$x + 3y - x = 6 - x \quad \text{Subtract } x \text{ from both sides.}$$
$$3y = 6 - x$$
$$\frac{3y}{3} = \frac{6-x}{3} \quad \text{Divide both sides by 3.}$$
$$y = \frac{6-x}{3}$$

x	$\frac{6-x}{3}$	y
6	$\frac{6-6}{3} = \frac{0}{3}$	0
0	$\frac{6-0}{3} = \frac{6}{3}$	2
−3	$\frac{6-(-3)}{3} = \frac{9}{3}$	3

Three solutions are (6, 0), (0, 2), and (−3, 3). Plot the three points that correspond to these three ordered pairs. Then draw a straight line through these three points, as in Figure 9.2.

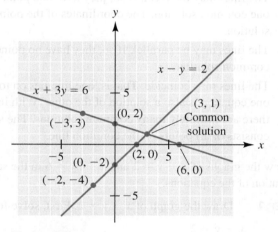

Figure 9.2

The point that corresponds to (3, 1) is the point of intersection. Therefore, (3, 1) is the common solution. That is, $x = 3$ and $y = 1$ are the only values that satisfy both equations. The solution (3, 1) should be checked in both equations. ♦

Example 2 Draw the graphs of $2x + 3y = 6$ and $4x + 6y = 30$ in the same number plane. Find the common solution of the equations.

STEP 1 Draw the graph of $2x + 3y = 6$. First, solve for y.

$$2x + 3y = 6$$
$$2x + 3y - 2x = 6 - 2x \quad \text{Subtract } 2x \text{ from both sides.}$$
$$3y = 6 - 2x$$
$$\frac{3y}{3} = \frac{6-2x}{3} \quad \text{Divide both sides by 3.}$$
$$y = \frac{6-2x}{3}$$

Then, find three solutions:

x	$\dfrac{6-2x}{3}$	y
3	$\dfrac{6-2(3)}{3} = \dfrac{0}{3}$	0
0	$\dfrac{6-2(0)}{3} = \dfrac{6}{3}$	2
-3	$\dfrac{6-2(-3)}{3} = \dfrac{12}{3}$	4

Three solutions are $(3, 0)$, $(0, 2)$, and $(-3, 4)$. Plot the three points that correspond to these three ordered pairs. Then draw a straight line through these three points, as in Figure 9.3.

STEP 2 Draw the graph of $4x + 6y = 30$. First, solve for y.

$$4x + 6y = 30$$
$$4x + 6y - 4x = 30 - 4x \qquad \text{Subtract } 4x \text{ from both sides.}$$
$$6y = 30 - 4x$$
$$\frac{6y}{6} = \frac{30 - 4x}{6} \qquad \text{Divide both sides by 6.}$$
$$y = \frac{30 - 4x}{6}$$

Then, find three solutions:

x	$\dfrac{30-4x}{6}$	y
3	$\dfrac{30-4(3)}{6} = \dfrac{18}{6}$	3
0	$\dfrac{30-4(0)}{6} = \dfrac{30}{6}$	5
-6	$\dfrac{30-4(-6)}{6} = \dfrac{54}{6}$	9

Three solutions are $(3, 3)$, $(0, 5)$, and $(-6, 9)$. Plot the three points that correspond to these ordered pairs in the same number plane as the points found in Step 1. Then draw a straight line through these three points. As you see in Figure 9.3, the lines are parallel. The lines have no points in common; therefore, there is no common solution.

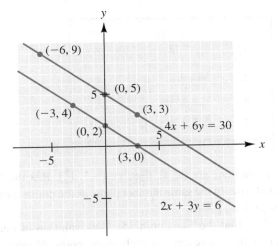

Figure 9.3

You may verify that the lines are parallel by showing that the slopes are equal.

CHAPTER 9 ♦ Systems of Linear Equations

Example 3 Draw the graphs of $-2x + 8y = 24$ and $-3x + 12y = 36$ in the same number plane. Find the common solution.

STEP 1 Draw the graph of $-2x + 8y = 24$. First, solve for y.

$$-2x + 8y = 24$$
$$-2x + 8y + 2x = 24 + 2x \quad \text{Add } 2x \text{ to both sides.}$$
$$8y = 24 + 2x$$
$$\frac{8y}{8} = \frac{24 + 2x}{8} \quad \text{Divide both sides by 8.}$$
$$y = \frac{24 + 2x}{8}$$

Then, find three solutions:

x	$\dfrac{24 + 2x}{8}$	y
-4	$\dfrac{24 + 2(-4)}{8} = \dfrac{16}{8}$	2
0	$\dfrac{24 + 2(0)}{8} = \dfrac{24}{8}$	3
8	$\dfrac{24 + 2(8)}{8} = \dfrac{40}{8}$	5

Three solutions are $(-4, 2)$, $(0, 3)$, and $(8, 5)$. Plot the three points that correspond to these three ordered pairs. Then draw a straight line through these three points.

STEP 2 Draw the graph of $-3x + 12y = 36$. First, solve for y.

$$-3x + 12y = 36$$
$$-3x + 12y + 3x = 36 + 3x \quad \text{Add } 3x \text{ to both sides.}$$
$$12y = 36 + 3x$$
$$\frac{12y}{12} = \frac{36 + 3x}{12} \quad \text{Divide both sides by 12.}$$
$$y = \frac{36 + 3x}{12}$$

Then, find three solutions:

x	$\dfrac{36 + 3x}{12}$	y
4	$\dfrac{36 + 3(4)}{12} = \dfrac{48}{12}$	4
-8	$\dfrac{36 + 3(-8)}{12} = \dfrac{12}{12}$	1
-5	$\dfrac{36 + 3(-5)}{12} = \dfrac{21}{12}$	$\dfrac{7}{4}$ or $1\dfrac{3}{4}$

Three solutions are $(4, 4)$, $(-8, 1)$, $\left(-5, 1\dfrac{3}{4}\right)$. Plot the three points that correspond to these ordered pairs in the same number plane as the points found in Step 1. Then draw a straight line through these points.

Note that the lines coincide. Any solution of one equation is also a solution of the other. Hence, there are infinitely many points of intersection and infinitely many common solutions (see Figure 9.4). The solutions are the coordinates of the points on either line.

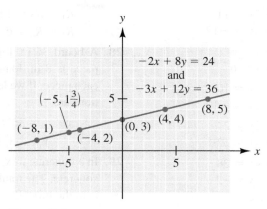

Figure 9.4

Example 4

The sum of two electric currents is 12 A. One current is three times the other. Find the two currents graphically.

Let x = first current
y = second current

The equations are then: $x + y = 12$
$y = 3x$

Draw the graph of each equation, as shown in Figure 9.5.

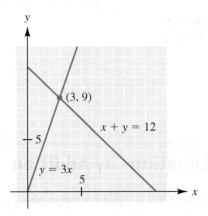

Figure 9.5

As you can see from Figure 9.5, the point of intersection is (3, 9). Thus, $x = 3$ and $y = 9$ is the common solution; the currents are 3 A and 9 A.

EXERCISES 9.1

Draw the graphs of each pair of linear equations. Find the point of intersection. If the lines do not intersect, tell whether the lines are parallel or coincide:

1. $y = 3x$
   $y = x + 4$

2. $x - y = 2$
   $x + 3y = 6$

3. $y = -x$
   $y - x = 2$

4. $x + y = 3$
   $2x + 2y = 6$

5. $x - 3y = 6$
   $2x - 6y = 18$

6. $x + y = 4$
   $2x + y = 5$

7. $2x - 4y = 8$
   $3x - 6y = 12$

8. $4x - 3y = 11$
   $6x + 5y = -12$

9. $3x - 6y = 12$
   $4x - 8y = 12$

10. $2x + y = 6$
    $2x - y = 6$

11. $3x + 2y = 10$
    $2x - 3y = 11$

12. $5x - y = 10$
    $x - 3y = -12$

13. $5x + 8y = -58$
    $2x + 2y = -18$

14. $6x + 2y = 24$
    $3x - 4y = 12$

15. $3x + 2y = 17$
    $x = 3$

16. $5x - 4y = 28$
    $y = -2$

17. $y = 2x$
    $y = -x + 2$

18. $y = -5$
    $y = x + 3$

19. $2x + y = 6$
    $y = -2x + 1$

20. $3x + y = -5$
    $2x + 5y = 1$

21. $4x + 3y = 2$
    $5x - y = 12$

22. $4x - 6y = 10$
    $2x - 3y = 5$

23. $2x - y = 9$
    $-2x + 3y = -11$

24. $x - y = 5$
    $2x - 3y = 5$

25. $8x - 3y = 0$
    $4x + 3y = 3$

26. $2x + 8y = 9$
    $4x + 4y = 3$

Solve Exercises 27–30 graphically:

27. The sum of two resistances is 14 Ω. Their difference is 6 Ω. Find the two resistances. If we let R_1 and R_2 be the two resistances, the equations are

    $R_1 + R_2 = 14$
    $R_1 - R_2 = 6$

28. A board 36 in. long is cut into two pieces so that one piece is 8 in. longer than the other. Find the length of each piece. If we let x and y be the two lengths, the equations are

    $x + y = 36$
    $y = x + 8$

29. In a concrete mix, there is four times as much gravel as concrete. The total volume is 20 ft³. How much of each is in the mix? If

    x = the amount of concrete
    y = the amount of gravel

    the equations are

    $y = 4x$
    $x + y = 20$

30. An electric circuit containing two currents may be expressed by the equations

    $3i_1 + 4i_2 = 15$
    $5i_1 - 2i_2 = -1$

    where i_1 and i_2 are the currents in microamperes (μA). Find the two currents.

9.2 Solving Pairs of Linear Equations by Addition

Often, solving a pair of linear equations by graphing results in only an approximate solution when an exact solution is needed.

Example 1 Solve the following pair of linear equations by graphing.

$$3x + 4y = -2$$
$$6x - 8y = 32$$

First, find any three solutions to the first equation by any method. Our solutions are $(2, -2)$, $(6, -5)$, and $\left(-3, 1\frac{3}{4}\right)$. Plot these three points and connect them with a straight line as shown in Figure 9.6.

Then, find any three solutions to the second equation by any method. Our solutions are $(0, -4)$, $(4, -1)$, and $\left(-2, -5\frac{1}{2}\right)$. Plot these three points and connect them with a straight line as shown in Figure 9.6.

9.2 ♦ Solving Pairs of Linear Equations by Addition

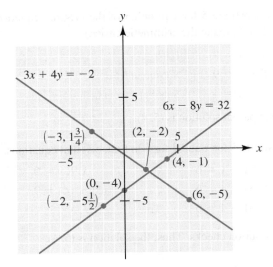

Figure 9.6

As you can see, only an approximate solution (point of intersection) can be read from this graph. The exact solution is found in Example 6. ♦

The *addition method* outlined below provides exact solutions to pairs of equations:

Solving a Pair of Linear Equations by the Addition Method

STEP 1 If necessary, multiply both sides of one or both equations by a number (or numbers) so that the numerical coefficients of one of the variables are negatives of each other.

STEP 2 Add the two equations from Step 1 to obtain an equation containing only one variable.

STEP 3 Solve the equation from Step 2 for the one remaining variable.

STEP 4 Solve for the second variable by substituting the solution from Step 3 in either of the original equations.

STEP 5 Check your solution by substituting the ordered pair in both original equations.

Example 2 Solve the following pair of linear equations by addition. Check your solution.

$$2x - y = 6$$
$$x + y = 9$$

Step 1 of the preceding rules is unnecessary, since you can eliminate the y variable by adding the two equations as they are.

$$\begin{aligned} 2x - y &= 6 \\ \underline{x + y} &= \underline{9} \\ 3x + 0 &= 15 \quad \text{Add the equations.} \\ 3x &= 15 \\ x &= 5 \quad \text{Divide both sides by 3.} \end{aligned}$$

312 CHAPTER 9 ♦ Systems of Linear Equations

Now substitute 5 for x in either of the original equations to solve for y. (Choose the simpler equation to make the arithmetic easier.)

$x + y = 9$
$5 + y = 9$ Substitute 5 for x.
$y = 4$ Subtract 5 from both sides.

The apparent solution is (5, 4).

Check: Substitute 5 for x and 4 for y in both original equations.

$2x - y = 6$ $x + y = 9$
$2(5) - 4 = 6$? $(5) + (4) = 9$ True
$10 - 4 = 6$ True

The solution checks. Thus, the solution is (5, 4). ♦

Example 3   Solve the following pair of linear equations by addition. Check your solution.

$2x + y = 5$
$x + y = 4$

First, multiply both sides of the second equation by -1 to eliminate y by addition.

$2x + y = 5$ $2x + y = 5$
$(-1)(x + y) = (-1)(4)$ or $-x - y = -4$
                                                $x = 1$

Now substitute 1 for x in the equation $x + y = 4$ to solve for y.

$x + y = 4$
$1 + y = 4$
$y = 3$

The solution is (1, 3). The check is left to the student. ♦

Example 4   Solve the following pair of linear equations by addition. Check your solution.

$4x + 2y = 2$
$3x - 4y = 18$

Multiply both sides of the first equation by 2 to eliminate y by addition.

$2(4x + 2y) = 2(2)$ or $8x + 4y = 4$
$3x - 4y = 18$ $3x - 4y = 18$
                                    $11x = 22$
                                    $x = 2$

Now substitute 2 for x in the equation $4x + 2y = 2$ to solve for y.

$4x + 2y = 2$
$4(2) + 2y = 2$
$8 + 2y = 2$
$2y = -6$
$y = -3$

The apparent solution is (2, −3).

9.2 ♦ Solving Pairs of Linear Equations by Addition

Check:
$$4x + 2y = 2$$
$$4(2) + 2(-3) = 2 \quad ?$$
$$8 - 6 = 2 \quad \text{True}$$

$$3x - 4y = 18$$
$$3(2) - 4(-3) = 18 \quad ?$$
$$6 + 12 = 18 \quad \text{True}$$

The solution checks. Thus, the solution is $(2, -3)$. ♦

Example 5

Solve the following pair of linear equations by addition. Check your solution.

$$3x - 4y = 11$$
$$4x - 5y = 14$$

Multiply both sides of the first equation by 4. Then multiply both sides of the second equation by -3 to eliminate x by addition.

$$4(3x - 4y) = 4(11)$$
$$-3(4x - 5y) = -3(14)$$
or
$$12x - 16y = 44$$
$$-12x + 15y = -42$$
$$0 - 1y = 2$$
$$y = -2$$

Now substitute -2 for y in the equation $3x - 4y = 11$ to solve for x.

$$3x - 4y = 11$$
$$3x - 4(-2) = 11$$
$$3x + 8 = 11$$
$$3x = 3$$
$$x = 1$$

The solution is $(1, -2)$.

Check: Left to the student. ♦

Example 6

Solve the following pair of linear equations from Example 1 by addition. Check your solution.

$$3x + 4y = -2$$
$$6x - 8y = 32$$

Multiply both sides of the first equation by 2 and eliminate y by addition.

$$2(3x + 4y) = (2)(-2)$$
$$6x - 8y = 32$$
or
$$6x + 8y = -4$$
$$6x - 8y = 32$$
$$12x \quad\quad = 28$$
$$x \quad = \tfrac{7}{3} \text{ or } 2\tfrac{1}{3} \quad\quad \text{Divide both sides by 12.}$$

Substitute $\tfrac{7}{3}$ for x in the equation $3x + 4y = -2$ to solve for y.

$$3x + 4y = -2$$
$$3\left(\tfrac{7}{3}\right) + 4y = -2$$
$$7 + 4y = -2$$
$$4y = -9 \quad\quad \text{Subtract 7 from both sides.}$$
$$y = -\tfrac{9}{4} \text{ or } -2\tfrac{1}{4} \quad\quad \text{Divide both sides by 4.}$$

The apparent solution is $\left(2\tfrac{1}{3}, -2\tfrac{1}{4}\right)$.

Check: Substitute $2\frac{1}{3}$ for x and $-2\frac{1}{4}$ for y in both equations.

$$3x + 4y = -2 \qquad\qquad 6x - 8y = 32$$
$$3\left(2\tfrac{1}{3}\right) + 4\left(-2\tfrac{1}{4}\right) = -2 \quad ? \qquad 6\left(2\tfrac{1}{3}\right) - 8\left(-2\tfrac{1}{4}\right) = 32 \quad ?$$
$$3\left(\tfrac{7}{3}\right) + 4\left(-\tfrac{9}{4}\right) = -2 \quad ? \qquad 6\left(\tfrac{7}{3}\right) - 8\left(-\tfrac{9}{4}\right) = 32 \quad ?$$
$$7 - 9 = -2 \quad \text{True} \qquad 14 + 18 = 32 \quad \text{True}$$

The solution checks. Compare the result in Figure 9.6 with this exact result. ◆

In the preceding examples, we considered only pairs of linear equations with one common solution. Thus, the graphs of these equations intersect at a point, and the ordered pair that names this point is the common solution for the pair of equations. Sometimes in solving a pair of linear equations by addition, the final statement is that two unequal numbers are equal, such as $0 = -2$. If so, the pair of equations does not have a common solution. The graphs of these equations are parallel lines.

Example 7 Solve the following pair of linear equations by addition.

$$2x + 3y = 7$$
$$4x + 6y = 12$$

Multiply both sides of the first equation by -2 to eliminate x by addition.

$$-2(2x + 3y) = -2(7) \qquad \text{or} \qquad -4x - 6y = -14$$
$$\underline{4x + 6y = 12} \qquad\qquad\qquad \underline{4x + 6y = 12}$$
$$\qquad\qquad\qquad\qquad\qquad\qquad\qquad 0 + 0 = -2$$
$$\qquad\qquad\qquad\qquad\qquad\qquad\qquad 0 = -2$$

Since $0 \neq -2$, there is no common solution, and the graphs of these two equations are parallel lines. ◆

If addition is used to solve a pair of linear equations and the resulting statement is $0 = 0$, then there are many common solutions. In fact, any solution of one equation is also a solution of the other. In this case, the graphs of the two equations coincide.

Example 8 Solve the following pair of linear equations by addition.

$$2x + 5y = 7$$
$$4x + 10y = 14$$

Multiply both sides of the first equation by -2 to eliminate x by addition.

$$-2(2x + 5y) = -2(7) \qquad \text{or} \qquad -4x - 10y = -14$$
$$\underline{4x + 10y = 14} \qquad\qquad \underline{4x + 10y = 14}$$
$$\qquad\qquad\qquad\qquad\qquad\qquad\qquad 0 + 0 = 0$$
$$\qquad\qquad\qquad\qquad\qquad\qquad\qquad 0 = 0$$

Since $0 = 0$, there are many common solutions, and the graphs of the two equations coincide.

NOTE: If you multiply both sides of the first equation by 2, you obtain the second equation. Thus, the two equations are equivalent. If two equations are equivalent, they should have the same graph. ◆

9.2 ♦ Solving Pairs of Linear Equations by Addition

In Section 9.1, we saw that the graphs of two straight lines in the same plane may (a) intersect, (b) be parallel, or (c) coincide (as shown in Figure 9.1). When using the addition method to solve a pair of linear equations, one of the same three possibilities occurs, as follows:

> **Addition Method Possible Cases**
>
> 1. The steps of the addition method result in exactly one ordered pair, such as $x = 2$ and $y = -5$. This ordered pair is the point at which the graphs of the two linear equations intersect.
> 2. The steps of the addition method result in a false statement, such as $0 = 7$ or $0 = -2$. This means that there is no common solution and that the graphs of the two linear equations are parallel.
> 3. The steps of the addition method result in a true statement, such as $0 = 0$. This means that there are many common solutions and that the graphs of the two linear equations coincide.

EXERCISES 9.2

Solve each pair of linear equations by addition. If there is one common solution, give the ordered pair of the point of intersection. If there is no one common solution, tell whether the lines are parallel or coincide.

1. $3x + y = 7$
   $x - y = 1$

2. $x + y = 8$
   $x - y = 4$

3. $2x + 5y = 18$
   $4x - 5y = 6$

4. $3x - y = 9$
   $2x + y = 6$

5. $-2x + 5y = 39$
   $2x - 3y = -25$

6. $-4x + 2y = 12$
   $-3x - 2y = 9$

7. $x + 3y = 6$
   $x - y = 2$

8. $3x - 2y = 10$
   $3x + 4y = 20$

9. $2x + 5y = 15$
   $7x + 5y = -10$

10. $4x + 5y = -17$
    $4x + y = 13$

11. $5x + 6y = 31$
    $2x + 6y = 16$

12. $6x + 7y = 0$
    $2x - 3y = 32$

13. $4x - 5y = 14$
    $2x + 3y = -4$

14. $6x - 4y = 10$
    $2x + y = 4$

15. $3x - 2y = -11$
    $7x - 10y = -47$

16. $3x + 2y = 10$
    $x + 5y = -27$

17. $x + 2y = -3$
    $2x + y = 9$

18. $5x - 2y = 6$
    $3x - 4y = 12$

19. $3x + 5y = 7$
    $2x - 7y = 15$

20. $12x + 5y = 21$
    $13x + 6y = 21$

21. $8x - 7y = -51$
    $12x + 13y = 41$

22. $5x - 7y = -20$
    $3x - 19y = -12$

23. $5x - 12y = -5$
    $9x - 16y = -2$

24. $2x + 3y = 2$
    $3x - 2y = 3$

25. $2x + 3y = 8$
    $x + y = 2$

26. $4x + 7y = 9$
    $12x + 21y = 12$

27. $3x - 5y = 7$
    $9x - 15y = 21$

28. $2x - 3y = 8$
    $4x - 3y = 0$

29. $2x + 5y = -1$
    $3x - 2y = 8$

30. $3x - 7y = -9$
    $2x + 14y = -6$

31. $16x - 36y = 70$
    $4x - 9y = 17$

32. $8x + 12y = 36$
    $16x + 15y = 45$

33. $4x + 3y = 17$
    $2x - y = -4$

34. $12x + 15y = 36$
    $7x - 12y = 187$

35. $2x - 5y = 8$
    $4x - 10y = 16$

36. $3x - 2y = 5$
    $7x + 3y = 4$

37. $5x - 8y = 10$
    $-10x + 16y = 8$

38. $-3x + 2y = 5$
    $-30x + y = 12$

39. $8x - 5y = 426$
    $7x - 2y = 444$

40. $3x - 10y = -21$
    $5x + 4y = 27$

41. $16x + 5y = 6$
    $7x + \dfrac{5}{8}y = 2$

42. $\dfrac{1}{4}x - \dfrac{2}{5}y = 1$
    $5x - 8y = 20$

43. $7x + 8y = 47$
    $5x - 3y = 51$

44. $2x - 5y = 13$
    $5x + 7y = 13$

9.3 Solving Pairs of Linear Equations by Substitution

For many problems, the *substitution* method is easier than the addition method for finding exact solutions. Use the substitution method when one equation has been solved or is easily solved for one of the variables:

> **Solving a Pair of Linear Equations by the Substitution Method**
>
> 1. From either of the two given equations, solve for one variable in terms of the other.
> 2. Substitute the result from Step 1 into the remaining equation. Note that this step should eliminate one variable.
> 3. Solve the equation from Step 2 for the remaining variable.
> 4. Solve for the second variable by substituting the solution from Step 3 into the equation resulting from Step 1.
> 5. Check by substituting the solution in both original equations.

Example 1 Solve the following pair of linear equations by substitution. Check your solution.

$$x + 3y = 15$$
$$x = 2y$$

First, substitute $2y$ for x in the first equation.

$$x + 3y = 15$$
$$2y + 3y = 15$$
$$5y = 15$$
$$y = 3$$

Now substitute 3 for y in the equation $x = 2y$ to solve for x.

$$x = 2y$$
$$x = 2(3)$$
$$x = 6$$

The apparent solution is (6, 3).

Check: Substitute 6 for x and 3 for y.

$x + 3y = 15$		$x = 2y$	
$6 + 3(3) = 15$	?	$6 = 2(3)$	True
$6 + 9 = 15$	True		

Thus, the solution is (6, 3). ◆

The addition method is often preferred if the pair of linear equations has no numerical coefficients equal to 1. For example,

$$3x + 4y = 7$$
$$5x + 7y = 12$$

Example 2 Solve the following pair of linear equations by substitution.

$$2x + 5y = 22$$
$$-3x + y = 18$$

First, solve the second equation for y and substitute in the first equation.

$$-3x + y = 18$$
$$y = 3x + 18 \quad \text{Add } 3x \text{ to both sides.}$$
$$2x + 5y = 22$$
$$2x + 5(3x + 18) = 22 \quad \text{Substitute } y = 3x + 18.$$
$$2x + 15x + 90 = 22 \quad \text{Remove parentheses.}$$
$$17x + 90 = 22 \quad \text{Combine like terms.}$$
$$17x = -68 \quad \text{Subtract 90 from both sides.}$$
$$x = -4 \quad \text{Divide both sides by 17.}$$

Now substitute -4 for x in the second equation to solve for y.

$$-3x + y = 18$$
$$(-3)(-4) + y = 18 \quad \text{Substitute } x = -4.$$
$$12 + y = 18$$
$$y = 6 \quad \text{Subtract 12 from both sides.}$$

The solution is $(-4, 6)$. The check is left to the student. ◆

Knowing both the addition and substitution methods, you may choose the one that seems easier to you for each problem.

EXERCISES 9.3

Solve, using the substitution method, and check:

1. $2x + y = 12$
   $y = 3x$
2. $3x + 4y = -8$
   $x = 2y$
3. $5x - 2y = 46$
   $x = 5y$
4. $2x - y = 4$
   $y = -x$
5. $3x + 2y = 30$
   $x = y$
6. $3x - 2y = 49$
   $y = -2x$
7. $5x - y = 18$
   $y = \frac{1}{2}x$
8. $15x + 3y = 9$
   $y = -2x$
9. $x - 6y = 3$
   $3y = x$
10. $4x + 5y = 10$
    $4x = -10y$
11. $3x + y = 7$
    $4x - y = 0$
12. $5x + 2y = 1$
    $y = -3x$
13. $4x + 3y = -2$
    $x + y = 0$
14. $7x + 8y + 93 = 0$
    $y = 3x$
15. $6x - 8y = 115$
    $x = -\dfrac{y}{5}$
16. $2x + 8y = 12$
    $x = -4y$
17. $3x + 8y = 27$
    $y = 2x + 1$
18. $4x - 5y = -40$
    $x = 3 - 2y$
19. $8y - 2x = -34$
    $x = 1 - 4y$
20. $2y + 7x = 48$
    $y = 3x - 2$
21. $3x + 4y = 25$
    $x - 5y = -17$
22. $5x - 4y = 29$
    $2x + y = 9$
23. $4x + y = 30$
    $-2x + 5y = 18$
24. $x + 6y = -20$
    $5x - 8y = 14$
25. $3x + 4y = 22$
    $-5x + 2y = -2$
26. $2x + 3y = 12$
    $5x - 6y = -51$

9.4 Applications Involving Pairs of Linear Equations*

Often, a technical application can be expressed mathematically as a system of linear equations. The procedure is similar to that outlined in Section 6.6, except that here you need to write two equations that express the information given in the problem and that involve both unknowns:

---

*In this chapter, do not use the rules for calculating with measurements.

318 CHAPTER 9 ♦ Systems of Linear Equations

> **Solving Applications Involving Equations with Two Variables**
>
> **1.** Choose a different variable for each of the two unknowns that you need to find. Write what each variable represents.
> **2.** Write the problem as two equations using both variables. To obtain these two equations, look for two different relationships that express the two unknown quantities in equation form.
> **3.** Solve this resulting system of equations using the methods given in this chapter.
> **4.** Answer the question or questions asked in the problem. Check your solution using the original problem to make certain that it makes sense.
> **5.** Check your solution in both original equations.

Example 1

The sum of two voltages is 120 V. The difference between them is 24 V. Find each voltage.

$$\text{Let } x = \text{large voltage}$$
$$y = \text{small voltage}$$

The sum of two voltages is 120 V; that is,

$$x + y = 120$$

The difference between them is 24 V; that is,

$$x - y = 24$$

The system of equations is

$$x + y = 120$$
$$\underline{x - y = 24}$$
$$2x = 144 \qquad \text{Add the equations.}$$
$$x = 72$$

Substitute $x = 72$ in the equation $x + y = 120$ and solve for y.

$$x + y = 120$$
$$72 + y = 120$$
$$y = 48 \qquad \text{Subtract 72 from both sides.}$$

Thus, the voltages are 72 V and 48 V.

Check: The sum of the voltages, 72 V + 48 V, is 120 V. The difference between them, 72 V − 48 V, is 24 V. ♦

Example 2

How many pounds of feed mix A that is 75% corn and how many pounds of feed mix B that is 50% corn will need to be mixed to make a 400-lb mixture that is 65% corn?

$$\text{Let } x = \text{number of pounds of mix A (75\% corn)}$$
$$y = \text{number of pounds of mix B (50\% corn)}$$

The sum of the two mixtures is 400 lb; that is,

$$x + y = 400$$

Thus, 75% of x is corn and 50% of y is corn. Adding these amounts together results in a 400-lb final mixture that is 65% corn. Write an equation in terms of the amount of corn—that is, the amount of corn in each mix equals the amount of corn in the final mixture.

$$0.75x + 0.50y = (0.65)(400)$$
$$\text{or} \quad 0.75x + 0.50y = 260$$

The system of equations is

$$x + y = 400$$
$$0.75x + 0.50y = 260$$

First, let's multiply both sides of the second equation by 100 to eliminate the decimals.

$$x + y = 400$$
$$75x + 50y = 26{,}000$$

Then, multiply both sides of the first equation by -50 to eliminate y by addition.

$$-50x - 50y = -20{,}000$$
$$\underline{75x + 50y = 26{,}000}$$
$$25x = 6000$$

$$\frac{25x}{25} = \frac{6000}{25} \qquad \text{Divide both sides by 25.}$$

$$x = 240$$

Now substitute 240 for x in the equation $x + y = 400$ to solve for y.

$$x + y = 400$$
$$240 + y = 400$$
$$y = 160$$

Therefore, we need 240 lb of mix A and 160 lb of mix B.

Check: Substitute 240 for x and 160 for y in both original equations.

$$x + y = 400 \qquad\qquad 0.75x + 0.50y = 260$$
$$240 + 160 = 400 \quad \text{True} \qquad 0.75(240) + 0.50(160) = 260 \quad ?$$
$$\qquad\qquad\qquad\qquad\qquad 180 + 80 = 260 \qquad \text{True} \quad \blacklozenge$$

Example 3 A company sells two grades of sand. One grade sells for 15¢/lb, and the other sells for 25¢/lb. How much of each grade needs to be mixed to obtain 1000 lb of a mixture worth 18¢/lb?

Let x = amount of sand selling at 15¢/lb

y = amount of sand selling at 25¢/lb

The total amount of sand is 1000 lb; that is,

$$x + y = 1000$$

One grade sells at 15¢/lb, and the other sells at 25¢/lb. The two grades are mixed to obtain 1000 lb of a mixture worth 18¢/lb. Here, we need to write an equation that relates the cost of the sand; that is, the cost of the sand separately equals the cost of the sand mixed.

$$15x + 25y = 18(1000)$$

That is, the cost of x pounds of sand at 15¢/lb is $15x$ cents. The cost of y pounds of sand at 25¢/lb is $25y$ cents. The cost of 1000 pounds of sand at 18¢/lb is 18(1000) cents. Therefore, the system of equations is

$$x + y = 1000$$
$$15x + 25y = 18{,}000$$

Multiply the first equation by -15 to eliminate x by addition.

$$-15x - 15y = -15{,}000$$
$$\underline{15x + 25y = 18{,}000}$$
$$10y = 3000$$
$$y = 300$$

Substitute $y = 300$ in the equation $x + y = 1000$ and solve for x.

$x + y = 1000$
$x + 300 = 1000$
$x = 700$

That is, 700 lb of sand selling at 15¢/lb and 300 lb of sand selling at 25¢/lb are needed to obtain 1000 lb of a mixture worth 18¢/lb.

Check: Left to the student. ◆

Example 4

Enclose a rectangular yard (Figure 9.7) with a fence so that the length is twice the width. The length of the 80-ft house is used to enclose part of one side of the yard. If 580 ft of fencing are used, what are the dimensions of the yard?

Let x = length of the yard

y = width of the yard

The amount of fencing used (two lengths plus two widths minus the length of the house) is 580 ft; that is,

$2x + 2y - 80 = 580$
$2x + 2y = 660$

The length of the yard is twice the width; that is,

$x = 2y$

The system of equations is

$2x + 2y = 660$
$x = 2y$

Substitute $2y$ for x in the equation $2x + 2y = 660$ and solve for y.

$2(2y) + 2y = 660$
$4y + 2y = 660$
$6y = 660$
$y = 110$

Now substitute 110 for y in the equation $x = 2y$ and solve for x.

$x = 2y$
$x = 2(110)$
$x = 220$

Therefore, the length is 220 ft and the width is 110 ft. (The check is left to the student.)

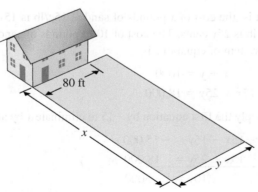

Figure 9.7 ◆

EXERCISES 9.4

1. A board 96 cm long is cut into two pieces so that one piece is 12 cm longer than the other. Find the length of each piece.

2. Find the capacity of two trucks if 6 trips of the smaller and 4 trips of the larger make a total haul of 36 tons, and 8 trips of the larger and 4 trips of the smaller make a total haul of 48 tons.

3. A plumbing contractor decides to field-test two new pumps. One is rated at 180 gal/h and the other at 250 gal/h. She tests one, then the other. Over a period of 6 h, she pumps a total of 1325 gal. Assume that both pumps operate as rated. How long is each in operation?

4. A bricklayer lays an average of 150 bricks per hour. During the job, he is called away and replaced by a less experienced man, who averages 120 bricks an hour. The two men laid 930 bricks in 7 h. **a.** How long did each work? **b.** How many bricks does each lay?

5. Two welders worked a total of 48 h on a project. One welder made $32/h, while the other made $41/h. If the gross earnings of the two welders was $1734 for the job, how many hours did each welder work?

6. A contractor finds a bill for $1110 for 720 ceiling tiles. She knows that there were two types of tiles used; one selling at $1.25 a tile and the other at $1.75 a tile. How many of each type were used?

7. A farmer has two types of feed. One has 5% digestible protein and the other 15% digestible protein. How much of each type must be mixed to have 100 lb of 12% digestible-protein feed?

8. A dairy farmer wants to make 125 lb of 12% butterfat cream. How many pounds of 40% butterfat cream and how many pounds of 2% butterfat milk must be mixed?

9. A farmer sells corn for $6.00/bu and soybeans for $15.00/bu. The entire 3150 bu sells for $22,950. How much of each is sold?

10. A farmer has a 1.4% solution and a 2.9% solution of a pesticide. How much of each must be mixed to have 2000 gal of 2% solution for his sprayer?

11. A farmer has a 6% solution and a 12% solution of pesticide. How much of each must be mixed to have 300 gal of an 8% solution for her sprayer?

12. The sum of two capacitors is 85 microfarads (μF). The difference between them is 25 μF. Find the size of each capacitor.

13. Nine batteries are hooked in series to provide a 33-V power source. Some of the batteries are 3 V and some are 4.5 V. How many of each type are used?

14. In a parallel circuit, the total current passing through two branches is 1.25 A. One branch has a resistance of 50 Ω, and the other has a resistance of 200 Ω. What current is flowing through each branch? *Note:* In a parallel circuit, the products of the current in amperes and the resistance in ohms are equal in all branches.

15. How much of an 8% solution and a 12% solution would you use to make 140 mL of a 9% electrolyte solution?

16. The total current in a parallel circuit with seven branches is 1.95 A. There are two levels of current within these seven branches. Some of the branches each total 0.25 A and each of the others total 0.35 A. How many of each type of branch are in the circuit? *Note:* The total current in a parallel circuit equals the sum of the currents in each branch.

17. A small single-cylinder engine was operated on a test stand for 14 min. It was run first at 850 rpm and was then increased to 1250 rpm. A total of 15,500 revolutions was counted during the test. How long was the engine operated at each speed?

18. In testing a hybrid engine, various solutions of gasoline and methanol are being tried. How much of a 95% gasoline solution and how much of an 80% gasoline solution would be needed to make 240 gal of a 90% gasoline solution?

19. An engine on a test stand was operated at two fixed settings, each with an appropriate load. At the first setting, fuel consumption was 1 gal every 12 min. At the second setting, it was 1 gal every 15 min. The test took 5 h, and 22 gal of fuel were used. How long did the engine run at each setting?

20. A technician stores a parts cleaner as a 65% solution, which is to be diluted to a 25% solution for use. Someone accidentally prepares a 15% solution. How much of the 65% solution and the 15% solution should be mixed to make 100 gal of the 25% solution?

21. Amy has a 3% solution and an 8% solution of a pesticide. How much of each must she mix to have 200 L of 4% solution?

22. A lawn seed mix containing 8% bluegrass is mixed with one that contains 15% bluegrass. How many pounds of each are needed to make 55 lb of a mixture that is 12% bluegrass?

23. When three identical compressors and five air-handling units are in operation, a total of 26.4 A are needed. When only two compressors and three air-handling units are being used, the current requirement is 17.2 A. How many amps are required

   a. by each compressor and
   b. by each air-handling unit?

322 CHAPTER 9 ♦ Systems of Linear Equations

24. A hospital has 35% saline solution on hand. How much water and how much of this solution should be used to prepare 140 mL of a 20% saline solution?

25. A nurse gives 1000 mL of an intravenous (IV) solution over a period of 8 h. It is given first at a rate of 140 mL/h, then at a reduced rate of 100 mL/h. How long should it be given at each rate?

26. A hospital has a 4% saline solution and an 8% saline solution on hand. How much of each should be used to prepare 1000 mL of 5% saline solution?

27. A medication is available in 2-mL vials and in 5-mL vials. In a certain month, 42 vials were used, totaling 117 mL of medication. How many of each type of vial were used?

28. A salesman turns in a ticket on two carpets for $3035. He sold a total of 75 yd^2 of carpet. One type was worth $36.50/yd^2, and the second type was worth $45/yd^2. He neglects to note, however, how much of each type he sold. How much did he sell of each type?

29. An apartment owner rents one-bedroom apartments for $750 and two-bedroom apartments for $975. A total of 13 apartments rent for $11,775 a month. How many of each type does she have?

30. A sporting goods store carries two types of snorkels. One sells for $14.95, and the other sells for $21.75. Records for July show that 23 snorkels were sold, for $357.45. How many of each type were sold?

31. The sum of two resistors is 550 Ω. One is 4.5 times the other. Find the size of each resistor.

32. The ratio of gravel to cement in a concrete mix is 4 to 1. The total volume is 15 yd^3. How much of each ingredient is used?

33. A wire 120 cm long is to be cut into two pieces so that one piece is three times as long as the other. Find the length of each piece.

34. The sum of two resistances is 1500 Ω. The larger is four times the smaller. Find the size of each resistance.

35. A rectangle is twice as long as it is wide. Its perimeter is 240 cm. Find the length and the width of the rectangle.

36. A duct from a furnace forks into two ducts. The air coming into the two ducts was 2300 ft^3/min. If the one duct took 800 ft^3/min more than the other duct, what was the flow in ft^3/min in each duct?

37. The sum of three currents is 210 mA. Two currents are the same. The third is five times either of the other two. Find the third current.

38. An experienced welder makes $\frac{5}{3}$ as many welds as a beginning welder. If an experienced welder and a beginning welder complete 240 welds, how many were from the experienced welder and how many from the beginning welder?

39. On a survey plat, the perimeter of a rectangular-shaped plot is noted as 2800 ft. If the length is 200 ft less than 3 times the width, find the dimensions of the plot.

40. A supply duct is to run the perimeter of a building having a length twice the width. The total duct length is 696 ft. What are the dimensions of the building?

41. In a drawing, the perimeter of a room is 40 ft. If the length is decreased by 6 ft and the width is doubled, the room would have the same perimeter.
   a. Find the original dimensions.
   b. Which room would have the greater square footage, or would they be equal?
   c. Find the percent increase or decrease, if any, in the square footage from the old room to the new room.

42. A rectangular walkway in front of an office building has a perimeter of 150 ft. If the length equals 45 ft more than twice the width, find the dimensions of the walkway.

43. If the length of a building is $2\frac{1}{2}$ times the width and each dimension is increased by 5 ft, then the perimeter is 230 ft. Find the dimensions of the original building. What is the percent increase or decrease in the perimeter and the area from the old building to the new building?

44. The center-to-center distance between a fan and a motor shaft is 30.0 in. See Illustration 1. Pulleys with a 4.5 : 1 ratio are installed. The distance between the pulleys is 19.0 in. Find the diameter of each pulley.

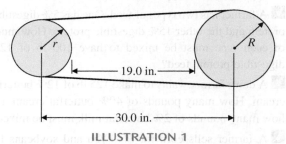

ILLUSTRATION 1

45. A carpenter starts with a board that is 12 ft long and cuts it into two pieces that differ in length by 8 in. Assuming no loss in length from the saw blade kerf, how long are the two pieces?

46. Taughannock Falls near Ithaca, New York, is about 48 ft taller than Niagara Falls. The two falls together would be 382 ft. How tall is each fall?

47. Ground corn and soybean meal are blended to produce 100 lb of animal feed that is 15% crude protein (CP). The ground corn is rated at 10% CP and supplies 8.5 lb of protein to the mix. How much of each grain is required for the mixture?

48. Two rental airplanes were flown a total of 54 h in one month. One plane rents for $105/h, and the other rents for $130/h. The total income from the two planes was

$6570. Find the number of hours each plane was flown for the month.

49. Jim bought both a municipal bond paying 3.5% and a corporate bond paying 8.5%. He paid $22,500 for both and receives a combined income of $1662.50 per year. How much did he invest in each bond?

50. Chef Ruth has 90% lean ground beef and also 75% lean ground beef. She needs 36 lb of 85% lean ground beef. How many pounds of 90% lean ground beef and how many pounds of 75% lean ground beef need to be mixed to get the 36 lb of 85% lean ground beef?

51. A coffee shop has tables that seat 6 guests and additional tables that seat 8 guests. If the total number of tables used is 21, how many 6-seat tables and how many 8-seat tables are used to seat a total of 148 guests?

52. The Open Late Restaurant sells chili in cups and bowls. A cup is priced at $3.79 and a bowl is priced at $4.89. If the total number of cups and bowls of chili sold is 29 and the total income from the chili sales is $130.81, how many cups and how many bowls of chili are sold?

SUMMARY | CHAPTER 9

Glossary of Basic Terms

Solving a pair of linear equations. To solve a pair of linear equations with two variables simultaneously, you must find an ordered pair that will make both equations true at the same time. (p. 305)

9.1 Solving Pairs of Linear Equations by Graphing

1. **Graphs of linear equations with two variables:** The graphs of a pair of linear equations with two variables consist of two straight lines and may be arranged as follows:
   a. The lines intersect. If so, they have one point in common whose coordinates define the point of intersection, which is the common solution.
   b. The lines are parallel. If so, they have no point in common, and the equations have no common solution.
   c. The lines coincide; that is, one line lies on top of the other. If so, any solution of one equation is also a solution of the other. With infinitely many points of intersection, there are infinitely many common solutions. The solution of the pair of equations consists of all points on the common line. (See Figure 9.1 on p. 305.)

9.2 Solving Pairs of Linear Equations by Addition

1. **Solving a pair of linear equations by the addition method:** The addition method is outlined as follows:
   a. If necessary, multiply both sides of one or both equations by a number (or numbers) so that the numerical coefficients of one of the variables are negatives of each other.
   b. Add the two equations from Step **a** to obtain an equation containing only one variable.
   c. Solve the equation from Step **b** for the one remaining variable.
   d. Solve for the second variable by substituting the solution from Step **c** in either of the original equations.
   e. Check your solutions by substituting the ordered pair in both original equations. (p. 311)

2. **Addition method possible cases:** When solving a pair of linear equations by graphing, we found three possible cases; that is, the lines may intersect, be parallel, or coincide. In using the addition method, one of these same three possibilities occurs as follows:
   a. The steps of the addition method result in exactly one ordered pair, which is the point of intersection, whose coordinates are the common solution.
   b. The steps of the addition method result in a false statement, such as $0 = 7$ or $0 = -2$. This means that there is no common solution and that the graphs of the two linear equations are parallel.
   c. The steps of the addition method result in a true statement, such as $0 = 0$. This means that there are many common solutions and that the graphs of the two linear equations coincide. (p. 315)

9.3 Solving Pairs of Linear Equations by Substitution

1. **Solving a pair of linear equations by the substitution method:** The substitution method is outlined as follows:
   a. From either of the two given equations, solve for one variable in terms of the other.
   b. Substitute the result from Step **a** into the remaining equation. Note that this step should eliminate one variable.
   c. Solve the equation from Step **b** for the remaining variable.

d. Solve for the second variable by substituting the solution from Step **c** into the equation resulting from Step **a**.

e. Check by substituting the solution in both original equations. (p. 316)

9.4 Applications Involving Pairs of Linear Equations

1. **Solving applications involving equations with two variables:** Follow these steps:
   a. Choose a different variable for each of the two unknowns you need to find. Write what each variable represents.
   b. Write the problem as two equations using both variables. To obtain these two equations, look for two different relationships that express the two unknown quantities in equation form.
   c. Solve this resulting system of equations using the methods given in this chapter.
   d. Answer the question or questions asked in the problem. Check your solution using the original problem to make certain that it makes sense.
   e. Check your solution in both original equations. (p. 318)

REVIEW | CHAPTER 9

Draw the graphs of each pair of linear equations on the same set of coordinate axes. Find the point of intersection. If the lines do not intersect, tell whether the lines are parallel or coincide:

1. $x + y = 6$
   $2x - y = 3$

2. $y = 2x + 5$
   $y = x + 2$

3. $4x + 6y = 12$
   $6x + 9y = 18$

4. $5x - 2y = 10$
   $10x - 4y = -20$

5. $3x + 4y = -1$
   $x = -3$

6. $y = 2x$
   $y = -5$

Solve each system of equations:

7. $x + y = 7$
   $2x - y = 2$

8. $3x + 2y = 11$
   $x + 2y = 5$

9. $3x - 5y = -3$
   $2x - 3y = -1$

10. $2x - 3y = 1$
    $4x - 6y = 5$

11. $3x + 5y = 8$
    $6x - 4y = 44$

12. $5x + 7y = 22$
    $4x + 8y = 20$

13. $x + 2y = 3$
    $3x + 6y = 9$

14. $3x + 5y = 52$
    $y = 2x$

15. $5y - 4x = -6$
    $x = \frac{1}{2}y$

16. $3x - 7y = -69$
    $y = 4x + 5$

17. You can buy twenty 20-amp electrical switches and eight 15-amp four-way electrical switches for $162 or sixty 20-amp electrical switches and forty 15-amp four-way electrical switches for $670. Find the price of each type of switch.

18. The sum of the length and width of a rectangular lot is 190 ft. The lot is 75 ft longer than it is wide. Find the length and width of the lot.

19. The sum of two inductors is 90 millihenrys (mH). The larger is 3.5 times the smaller. What is the size of each inductor?

20. The sum of two lengths is 90 ft, and their difference is 20 ft. Find the two lengths.

TEST | CHAPTER 9

Solve each system of equations by the method indicated.

1. $3x - y = 5$
   $2x - y = 0$ by graphing

2. $y = 3x - 5$
   $y = 2x - 1$ by graphing

3. $2x + 7y = -1$
   $x + 2y = 1$ by addition

4. $x - 3y = 8$
   $x + 4y = -6$ by addition

5. $y = -3x$
   $2x + 3y = 13$ by substitution

6. $x = 7y$
   $2x - 8y = 12$ by substitution

Solve each pair of linear equations. If there is one common solution, give the ordered pair of the point of intersection. If there is no one common solution, tell whether the lines are parallel or coincide:

7. $4x - 5y = 10$
   $-8x + 10y = 6$

8. $3x - y = 8$
   $12x - 4y = 32$

9. $x - 3y = -8$
   $2x + y = 5$

10. The perimeter of a rectangular lot is 600 m. The length is twice the width. Find its length and width.

11. The sum of two resistances is 550 Ω. The difference between them is 250 Ω. Find the size of each resistance.

Factoring Algebraic Expressions

CHAPTER 10

OBJECTIVES

- Find the greatest common monomial factor in an algebraic expression.
- Find the product of two binomials mentally.
- Factor trinomials.
- Find the square of a binomial.
- Identify and find the product of the sum and difference of two terms.
- Identify and factor perfect square trinomials.
- Identify and factor the difference of two squares.

Mathematics at Work

Agriculture requires a wide variety of support specialists. Those working in agricultural business management often specialize in agrimarketing, management, animal science, soils and fertilizers, grain merchandising, and crop production. Other areas include agricultural equipment management and marketing, equine management, and landscape design, construction, and management, as well as turf management. Precision farming requires specialist support in satellite-controlled soil sampling, fertilizer application, crop scouting, and yield mapping using geographic information systems (GIS) and precision farming technology. Farmers, soil testing labs, fertilizer and chemical companies, banks, and other agribusinesses need employees trained in these new techniques. Agricultural research colleges and companies need agricultural research technicians to assist in research development projects for seed, chemical production, and other agricultural products. Training and education for these careers are available at many community colleges and trade schools. For more information, please visit **www.cengage.com** and access the Student Online Resources for this text.

Farmer
Farmer harvesting corn

10.1 Finding Monomial Factors

Factoring an algebraic expression, like finding the prime factors of a number, means *writing the expression as a product of factors*. The prime factorization of 12 is $2 \cdot 2 \cdot 3$. Other factorizations of 12 are $2 \cdot 6$ and $4 \cdot 3$. Since factorization means writing a number or an algebraic expression as a product, then a number or an algebraic expression divided by one factor generates another factor. Thus, 12 divided by 2 gives 6, so 2 times 6 is a factorization of 12.

To factor the expression $2x + 2y$, notice that 2 is a factor common to both terms of the expression. In other words, 2 is a factor of $2x + 2y$. To find the other factor, divide by 2.

$$\frac{2x + 2y}{2} = \frac{2x}{2} + \frac{2y}{2} = x + y$$

Therefore, a factorization of $2x + 2y$ is $2(x + y)$.

A **monomial factor** is a one-term factor that divides each term of an algebraic expression. Here, 2 divides each term of the algebraic expression and is called a monomial factor. When factoring any algebraic expression, *always look first for monomial factors* that are common to all terms.

Example 1 Factor: $3a + 6b$.

First, look for a common monomial factor. Since 3 divides both $3a$ and $6b$, 3 is a common monomial factor of $3a + 6b$. Divide $3a + 6b$ by 3.

$$\frac{3a + 6b}{3} = \frac{3a}{3} + \frac{6b}{3}$$
$$= a + 2b$$

Thus, $3a + 6b = 3(a + 2b)$.

Check this result by multiplication: $3(a + 2b) = 3a + 6b$. ◆

Example 2 Factor: $4x^2 + 8x + 12$.

Since 4 divides each term of the expression, divide $4x^2 + 8x + 12$ by 4 to obtain the other factor.

$$\frac{4x^2 + 8x + 12}{4} = \frac{4x^2}{4} + \frac{8x}{4} + \frac{12}{4}$$
$$= x^2 + 2x + 3$$

Thus, $4x^2 + 8x + 12 = 4(x^2 + 2x + 3)$.

Check: $4(x^2 + 2x + 3) = 4x^2 + 8x + 12$

Your product should be the original expression.

NOTE: In this example, 2 is also a common factor of each term of the expression. However, 4 is the greatest common factor. The **greatest common factor of a polynomial** is the largest common factor that divides all terms in the expression. *When factoring, always choose the monomial factor that is the greatest common factor.* ◆

Example 3 Factor: $15ax - 6ay$.

Note that 3 divides both $15ax$ and $6ay$, so 3 is a common factor. However, a also divides $15ax$ and $6ay$, so a is also a common factor. We are looking for the greatest common factor (GCF), which in this case is $3a$. Then we divide $15ax - 6ay$ by $3a$ to obtain the other factor.

$$\frac{15ax - 6ay}{3a} = \frac{15ax}{3a} - \frac{6ay}{3a}$$
$$= 5x - 2y$$

Thus, $15ax - 6ay = 3a(5x - 2y)$.

Note that $3(5ax - 2ay)$ or $a(15x - 6y)$ are also factored forms of $15ax - 6ay$. However, we use the monomial factor that is the greatest common factor. ◆

Example 4   Factor: $15xy^2 - 25x^2y + 10xy$.

The greatest common factor is $5xy$. Dividing each term by $5xy$, we have $15xy^2 - 25x^2y + 10xy = 5xy(3y - 5x + 2)$. ◆

EXERCISES 10.1

Factor:

1. $4a + 4$
2. $3x - 6$
3. $bx + by$
4. $9 - 18y$
5. $15b - 20$
6. $12ab + 30ac$
7. $x^2 - 7x$
8. $3x^2 - 6x$
9. $a^2 - 4a$
10. $7xy - 21y$
11. $4n^2 - 8n$
12. $10x^2 + 5x$
13. $10x^2 + 25x$
14. $y^2 - 8y$
15. $3r^2 - 6r$
16. $x^3 + 13x^2 + 25x$
17. $4x^4 + 8x^3 + 12x^2$
18. $9x^4 - 15x^2 - 18x$
19. $9a^2 - 9ax^2$
20. $a - a^3$
21. $10x + 10y - 10z$
22. $2x^2 - 2x$
23. $3y - 6$
24. $y - 3y^2$
25. $14xy - 7x^2y^2$
26. $25a^2 - 25b^2$
27. $12x^2m - 7m$
28. $90r^2 - 10R^2$
29. $60ax - 12a$
30. $2x^2 - 100x^3$
31. $52m^2n^2 - 13mn$
32. $40x - 8x^3 + 4x^4$
33. $52m^2 - 14m + 2$
34. $27x^3 - 54x$
35. $36y^2 - 18y^3 + 54y^4$
36. $20y^3 - 10y^2 + 5y$
37. $6m^4 - 12m^2 + 3m$
38. $-16x^3 - 32x^2 - 16x$
39. $-4x^2y^3 - 6x^2y^4 - 10x^2y^5$
40. $18x^3y - 30x^4y + 48xy$
41. $3a^2b^2c^2 + 27a^3b^3c^2 - 81abc$
42. $15x^2yz^4 - 20x^3y^2z^2 + 25x^2y^3z^2$
43. $4x^3z^4 - 8x^2y^2z^3 + 12xyz^2$
44. $18a^2b^2c^2 + 24ab^2c^2 - 30a^2c^2$

10.2 Finding the Product of Two Binomials Mentally

In Section 5.5, you learned how to multiply two binomials such as $(2x + 3)(4x - 5)$ by the following method:

$$\begin{array}{r} 2x + 3 \\ 4x - 5 \\ \hline -10x - 15 \\ 8x^2 + 12x \\ \hline 8x^2 + 2x - 15 \end{array}$$

This process of multiplying two binomials can be shortened as follows:

Finding the Product of Two Binomials Mentally

1. The *first term* of the product is the product of the first terms of the binomials.
2. The *middle term* of the product is the sum of the outer product and the inner product of the binomials.
3. The *last term* of the product is the product of the last terms of the binomials.

CHAPTER 10 ♦ Factoring Algebraic Expressions

Let's use the above steps to find the product $(2x + 3)(4x - 5)$.

STEP 1 Product of the first terms: $(2x)(4x) = 8x^2$

STEP 2 Outer product $= (2x)(-5) = -10x$

$(2x + 3)(4x - 5)$

Inner product $= (3)(4x) = \underline{12x}$

Sum: $2x$

STEP 3 Product of the last terms: $(3)(-5) = \underline{-15}$

Therefore, $(2x + 3)(4x - 5) = 8x^2 + 2x - 15$

Note that in each method, we found the exact same terms. The second method is much quicker, especially when you become more familiar and successful with it. The second method is used to factor polynomials. Factoring polynomials is the content of the rest of this chapter and a necessary part of the next chapter. Therefore, it is very important that you learn to find the product of two binomials, mentally, before proceeding with the next section.

This method is often called the *FOIL method*, where *F* refers to the product of the *first* terms, *O* refers to the *outer* product, *I* refers to the *inner* product, and *L* refers to the product of the *last* terms.

Example 1 Find the product $(2x - 7)(3x - 4)$ mentally.

STEP 1 Product of the first terms: $(2x)(3x) = 6x^2$

STEP 2 Outer product $= (2x)(-4) = -8x$

$(2x - 7)(3x - 4)$

Inner product $= (-7)(3x) = \underline{-21x}$

Sum: $-29x$

STEP 3 Product of the last terms: $(-7)(-4) = \underline{28}$

Therefore, $(2x - 7)(3x - 4) = 6x^2 - 29x + 28$ ♦

Example 2 Find the product $(x + 4)(3x + 5)$ mentally.

STEP 1 Product of the first terms: $(x)(3x) = 3x^2$

STEP 2 Outer product $= (x)(5) = 5x$

$(x + 4)(3x + 5)$

Inner product $= (4)(3x) = \underline{12x}$

Sum: $17x$

STEP 3 Product of the last terms: $(4)(5) = \underline{20}$

Therefore, $(x + 4)(3x + 5) = 3x^2 + 17x + 20$ ♦

By now, you should be writing only the final result of each product. If you need some help, refer to the three steps at the beginning of this section and the outline shown in Examples 1 and 2.

Example 3 Find the product $(x + 8)(x + 5)$ mentally.
$$(x + 8)(x + 5) = x^2 + (5x + 8x) + 40 \quad \text{mental step}$$
$$= x^2 + 13x + 40 \quad \text{product}$$

Example 4 Find the product $(x - 6)(x - 9)$ mentally.
$$(x - 6)(x - 9) = x^2 + (-9x - 6x) + 54 \quad \text{mental step}$$
$$= x^2 - 15x + 54 \quad \text{product}$$

Example 5 Find the product $(x + 2)(x - 5)$ mentally.
$$(x + 2)(x - 5) = x^2 + (-5x + 2x) - 10$$
$$= x^2 - 3x - 10$$

Example 6 Find the product $(4x + 1)(5x + 8)$ mentally.
$$(4x + 1)(5x + 8) = 20x^2 + (32x + 5x) + 8$$
$$= 20x^2 + 37x + 8$$

Example 7 Find the product $(6x + 5)(2x - 3)$ mentally.
$$(6x + 5)(2x - 3) = 12x^2 + (-18x + 10x) - 15$$
$$= 12x^2 - 8x - 15$$

Example 8 Find the product $(4x - 5)(4x - 5)$ mentally.
$$(4x - 5)(4x - 5) = 16x^2 + (-20x - 20x) + 25$$
$$= 16x^2 - 40x + 25$$

EXERCISES 10.2

Find each product mentally:

1. $(x + 5)(x + 2)$
2. $(x + 3)(2x + 7)$
3. $(2x + 3)(3x + 4)$
4. $(x + 3)(x + 18)$
5. $(x - 5)(x - 6)$
6. $(x - 9)(x - 8)$
7. $(x - 12)(x - 2)$
8. $(x - 9)(x - 4)$
9. $(x + 8)(2x + 3)$
10. $(3x - 7)(2x - 5)$
11. $(x + 6)(x - 2)$
12. $(x - 7)(x - 3)$
13. $(x - 9)(x - 10)$
14. $(x - 9)(x + 10)$
15. $(x - 12)(x + 6)$
16. $(2x + 7)(4x - 5)$
17. $(2x - 7)(4x + 5)$
18. $(2x - 5)(4x + 7)$
19. $(2x + 5)(4x - 7)$
20. $(6x + 5)(5x - 1)$
21. $(7x + 3)(2x + 5)$
22. $(5x - 7)(2x + 1)$
23. $(x - 9)(3x + 8)$
24. $(x - 8)(2x + 9)$
25. $(6x + 5)(x + 7)$
26. $(16x + 3)(x - 1)$
27. $(13x - 4)(13x - 4)$
28. $(12x + 1)(12x + 5)$
29. $(10x + 7)(12x - 3)$
30. $(10x - 7)(12x + 3)$
31. $(10x - 7)(10x - 3)$
32. $(10x + 7)(10x + 3)$
33. $(2x - 3)(2x - 5)$
34. $(2x + 3)(2x + 5)$
35. $(2x - 3)(2x + 5)$
36. $(2x + 3)(2x - 5)$
37. $(3x - 8)(2x + 7)$
38. $(3x + 8)(2x - 7)$
39. $(3x + 8)(2x + 7)$
40. $(3x - 8)(2x - 7)$
41. $(8x - 5)(2x + 3)$
42. $(x - 7)(x + 5)$
43. $(y - 7)(2y + 3)$
44. $(m - 9)(m + 2)$
45. $(3n - 6y)(2n + 5y)$
46. $(6a - b)(2a + 3b)$
47. $(4x - y)(2x + 7y)$
48. $(8x - 12)(2x + 3)$
49. $\left(\dfrac{1}{2}x - 8\right)\left(\dfrac{1}{4}x - 6\right)$
50. $\left(\dfrac{2}{3}x - 6\right)\left(\dfrac{1}{3}x + 9\right)$

10.3 Finding Binomial Factors

A **binomial factor** is a two-term factor of an algebraic expression. The factors of a trinomial are often binomial factors. To find these binomial factors, you must "undo" the process of multiplication as presented in Section 10.2. The following steps will enable you to undo the multiplication in a trinomial such as $x^2 + 7x + 10$.

STEP 1 Factor any common monomial. In the expression $x^2 + 7x + 10$, there is no common factor.

STEP 2 If $x^2 + 7x + 10$ can be factored, the two factors will probably be binomials. Write parentheses for the binomials.
$$x^2 + 7x + 10 = (\quad)(\quad)$$

STEP 3 The product of the two first terms of the binomials is the first term of the trinomial. So the first term in each binomial must be x.
$$x^2 + 7x + 10 = (x\quad)(x\quad)$$

STEP 4 Here, all the signs in the trinomial are positive, so the signs in the binomials are also positive.
$$x^2 + 7x + 10 = (x+\quad)(x+\quad)$$

STEP 5 Find the last terms of the binomials by finding two numbers that have a product of $+10$ and a sum of $+7$. The only possible factorizations of 10 are $1 \cdot 10$ and $2 \cdot 5$. The sums of the pairs of factors are $1 + 10 = 11$ and $2 + 5 = 7$. Thus, the numbers that you want are 2 and 5.
$$x^2 + 7x + 10 = (x + 2)(x + 5)$$

STEP 6 Multiply the two binomials as a check, to see whether their product is the same as the original trinomial.

Example 1 Factor the trinomial $x^2 + 15x + 56$.

STEP 1 $x^2 + 15x + 56$ has no common monomial factor.
STEP 2 $x^2 + 15x + 56 = (\quad)(\quad)$
STEP 3 $x^2 + 15x + 56 = (x\quad)(x\quad)$
STEP 4 $x^2 + 15x + 56 = (x+\quad)(x+\quad)$ All signs in the trinomial are positive.

To determine which factors of 56 to use, list all possible pairs.

$$1 \cdot 56 = 56 \quad\quad 1 + 56 = 57$$
$$2 \cdot 28 = 56 \quad\quad 2 + 28 = 30$$
$$4 \cdot 14 = 56 \quad\quad 4 + 14 = 18$$
$$7 \cdot 8 = 56 \quad\quad 7 + 8 = 15$$

Since the coefficient of x in the trinomial is 15, choose 7 and 8 for the second terms of the binomial factors. There are no other pairs of positive whole numbers with a product of 56 and a sum of 15.

STEP 5 $x^2 + 15x + 56 = (x + 7)(x + 8)$

In actual work, all five steps are completed in one or two lines, depending on whether or not there is a common monomial.

STEP 6 Check: $(x + 7)(x + 8) = x^2 + 15x + 56$ ♦

Example 2 Factor the trinomial $x^2 - 13x + 36$.

Note that the only difference between this trinomial and the ones we have considered previously is the sign of the second term. Here, the sign of the second term is negative instead of positive. Thus, the steps for factoring will be the same except for Step 4.

STEP 1 $x^2 - 13x + 36$ has no common monomial factor.
STEP 2 $x^2 - 13x + 36 = x^2 + (-13)x + 36$
          $= (\quad)(\quad)$
STEP 3 $x^2 + (-13)x + 36 = (x\quad)(x\quad)$

Note that the sign of the third term (+36) is positive and the coefficient of the second term (−13x) is negative. Since 36 is positive, the two factors of +36 must have like signs, and since the coefficient of −13x is negative, the signs in the two factors must both be negative.

STEP 4 $x^2 + (-13)x + 36 = (x - \quad)(x - \quad)$

Find two integers whose product is 36 and whose sum is −13. Since $(-4)(-9) = 36$ and $(-4) + (-9) = -13$, these are the factors of 36 to be used.

STEP 5 $x^2 - 13x + 36 = (x - 4)(x - 9)$
STEP 6   Check: $(x - 4)(x - 9) = x^2 - 13x + 36$ ◆

Example 3   Factor the trinomial $3x^2 + 12x - 36$.

STEP 1 $3x^2 + 12x - 36 = 3[x^2 + 4x - 12]$ 3 is a common factor.
STEP 2 $3[x^2 + 4x + (-12)] = 3[(\quad)(\quad)]$
STEP 3 $3[x^2 + 4x + (-12)] = 3[(x\quad)(x\quad)]$

Note that the last term of the trinomial (−12) is negative. This means that the two factors of −12 must have unlike signs, since a positive number times a negative number gives a negative product.

STEP 4 $3[x^2 + 4x - 12] = 3[(x + \quad)(x - \quad)]$

Find two integers with a product of −12 and a sum of +4. All possible pairs of factors are shown below.

$(-12)(+1) = -12 \qquad (-12) + (+1) = -11$
$(+12)(-1) = -12 \qquad (+12) + (-1) = 11$
$(+6)(-2) = -12 \qquad (+6) + (-2) = 4$
$(-6)(+2) = -12 \qquad (-6) + (+2) = -4$
$(-4)(+3) = -12 \qquad (-4) + (+3) = -1$
$(+4)(-3) = -12 \qquad (+4) + (-3) = 1$

From these possibilities, you see that the two integers with a product of −12 and a sum of +4 are +6 and −2. Write these numbers as the last terms of the binomials.

STEP 5 $3[x^2 + 4x - 12] = 3(x + 6)(x - 2)$
STEP 6   Check: $3(x + 6)(x - 2) = 3[x^2 + 4x - 12]$
                              $= 3x^2 + 12x - 36$ ◆

Example 4   Factor the trinomial $x^2 - 11x - 12$.

The signs of the factors of −12 must be different. From the list in Example 3, choose the two factors with a sum of −11.

$x^2 - 11x - 12 = (x - 12)(x + 1)$

Check: $(x - 12)(x + 1) = x^2 - 11x - 12$ ◆

Factoring Trinomials

To factor a trinomial $x^2 + bx + c$, use the following steps. Assume that b and c are both positive numbers.

STEP 1 First, look for any common monomial factors.

STEP 2
a. For the trinomial $x^2 + bx + c$, use the form
$$x^2 + bx + c = (x +)(x +)$$
b. For the trinomial $x^2 - bx + c$, use the form
$$x^2 - bx + c = (x -)(x -)$$
c. For the trinomials $x^2 - bx - c$ and $x^2 + bx - c$, use the forms
$$x^2 - bx - c = (x +)(x -)$$
$$x^2 + bx - c = (x +)(x -)$$

EXERCISES 10.3

Factor each trinomial completely:

1. $x^2 + 6x + 8$
2. $x^2 + 8x + 15$
3. $y^2 + 9y + 20$
4. $2w^2 + 20w + 32$
5. $3r^2 + 30r + 75$
6. $a^2 + 14a + 24$
7. $b^2 + 11b + 30$
8. $c^2 + 21c + 54$
9. $x^2 + 17x + 72$
10. $y^2 + 18y + 81$
11. $5a^2 + 35a + 60$
12. $r^2 + 12r + 27$
13. $x^2 - 7x + 12$
14. $y^2 - 6y + 9$
15. $2a^2 - 18a + 28$
16. $c^2 - 9c + 18$
17. $3x^2 - 30x + 63$
18. $r^2 - 12r + 35$
19. $w^2 - 13w + 42$
20. $x^2 - 14x + 49$
21. $x^2 - 19x + 90$
22. $4x^2 - 84x + 80$
23. $t^2 - 12t + 20$
24. $b^2 - 15b + 54$
25. $x^2 + 2x - 8$
26. $x^2 - 2x - 15$
27. $y^2 + y - 20$
28. $2w^2 - 12w - 32$
29. $a^2 + 5a - 24$
30. $b^2 + b - 30$
31. $c^2 - 15c - 54$
32. $b^2 - 6b - 72$
33. $3x^2 - 3x - 36$
34. $a^2 + 5a - 14$
35. $c^2 + 3c - 18$
36. $x^2 - 4x - 21$
37. $y^2 + 17y + 42$
38. $m^2 - 18m + 72$
39. $r^2 - 2r - 35$
40. $x^2 + 11x - 42$
41. $m^2 - 22m + 40$
42. $y^2 + 17y + 70$
43. $x^2 - 9x - 90$
44. $x^2 - 8x + 15$
45. $a^2 + 27a + 92$
46. $x^2 + 17x - 110$
47. $2a^2 - 12a - 110$
48. $y^2 - 14y + 40$
49. $a^2 + 29a + 100$
50. $y^2 + 14y - 120$
51. $y^2 - 14y - 95$
52. $b^2 + 20b + 36$
53. $y^2 - 18y + 32$
54. $x^2 - 8x - 128$
55. $7x^2 + 7x - 14$
56. $2x^2 - 6x - 36$
57. $6x^2 + 12x - 6$
58. $4x^2 + 16x + 16$
59. $y^2 - 12y + 35$
60. $a^2 + 16a + 63$
61. $a^2 + 2a - 63$
62. $y^2 - y - 42$
63. $x^2 + 18x + 56$
64. $x^2 + 11x - 26$
65. $2y^2 - 36y + 90$
66. $ax^2 + 2ax + a$
67. $3xy^2 - 18xy + 27x$
68. $x^3 - x^2 - 156x$
69. $x^2 + 30x + 225$
70. $x^2 - 2x - 360$
71. $x^2 - 26x + 153$
72. $x^2 + 8x - 384$
73. $x^2 + 28x + 192$
74. $x^2 + 3x - 154$
75. $x^2 + 14x - 176$
76. $x^2 - 59x + 798$
77. $2a^2b + 4ab - 48b$
78. $ax^2 - 15ax + 44a$
79. $y^2 - y - 72$
80. $x^2 + 19x + 60$

10.4 Special Products

The square of a is $a \cdot a$ or a^2 (read "a squared"). The square of the binomial $x + y$ is $(x + y)(x + y)$ or $(x + y)^2$, which is read "the quantity $x + y$ squared." When the multiplication is performed, the product is

$$(x + y)^2 = (x + y)(x + y) = x^2 + 2xy + y^2$$

A **perfect square trinomial** is a trinomial with the same two binomial factors.

> ### The Square of a Binomial
>
> The square of the *sum* of two terms of a binomial equals the square of the first term *plus* twice the product of the two terms plus the square of the second term.
>
> $$(a + b)(a + b) = (a + b)^2 = a^2 + 2ab + b^2$$
>
> Similarly, the square of the *difference* of two terms of a binomial equals the square of the first term *minus* twice the product of the two terms plus the square of the second term.
>
> $$(a - b)(a - b) = (a - b)^2 = a^2 - 2ab + b^2$$

Example 1 Find $(x + 12)^2$.

The square of the first term is x^2. Twice the product of the terms is $2(x \cdot 12)$, or $24x$. The square of the second term is 144. Thus,

$$(x + 12)^2 = x^2 + 24x + 144$$

Example 2 Find $(5xy - 3)^2$.

The square of the first term is $25x^2y^2$. Twice the product of the terms is $2(5xy \cdot 3)$, or $30xy$. The square of the second term is 9. Thus,

$$(5xy - 3)^2 = 25x^2y^2 - 30xy + 9$$

Finding the *product of the sum and difference of two terms*, $(a + b)(a - b)$, is another special case in which the product is a binomial.

> ### The Product of the Sum and Difference of Two Terms
>
> This product is the difference of two squares: the square of the first term *minus* the square of the second term.
>
> $$(a + b)(a - b) = a^2 - b^2$$

Example 3 Find the product $(x + 3)(x - 3)$.

The square of the first term is x^2. The square of the second term is 9. Thus,

$$(x + 3)(x - 3) = x^2 - 9$$

Note that the sum of the outer and inner products is zero.

Example 4 Find the product $(4y + 7)(4y - 7)$.

The square of the first term is $16y^2$. The square of the second term is 49. Thus,

$$(4y + 7)(4y - 7) = 16y^2 - 49$$

Example 5 Find the product $(3x - 8y)(3x + 8y)$.

The square of the first term is $9x^2$. The square of the second term is $64y^2$. Thus,

$$(3x - 8y)(3x + 8y) = 9x^2 - 64y^2$$

EXERCISES 10.4

Find each product:

1. $(x + 3)(x - 3)$
2. $(x + 3)^2$
3. $(a + 5)(a - 5)$
4. $(y^2 + 9)(y^2 - 9)$
5. $(2b + 11)(2b - 11)$
6. $(x - 6)^2$
7. $(100 + 3)(100 - 3)$
8. $(90 + 2)(90 - 2)$
9. $(3y^2 + 14)(3y^2 - 14)$
10. $(y + 8)^2$
11. $(r - 12)^2$
12. $(t + 10)^2$
13. $(4y + 5)(4y - 5)$
14. $(200 + 5)(200 - 5)$
15. $(xy - 4)^2$
16. $(x^2 + y)(x^2 - y)$
17. $(ab + d)^2$
18. $(ab + c)(ab - c)$
19. $(z - 11)^2$
20. $(x^3 + 8)(x^3 - 8)$
21. $(st - 7)^2$
22. $(w + 14)(w - 14)$
23. $(x + y^2)(x - y^2)$
24. $(1 - x)^2$
25. $(x + 5)^2$
26. $(x - 6)^2$
27. $(x + 7)(x - 7)$
28. $(y - 12)(y + 12)$
29. $(x - 3)^2$
30. $(x + 4)^2$
31. $(ab + 2)(ab - 2)$
32. $(m - 3)(m + 3)$
33. $(x^2 + 2)(x^2 - 2)$
34. $(m + 15)(m - 15)$
35. $(r - 15)^2$
36. $(t + 7a)^2$
37. $(y^3 - 5)^2$
38. $(4 - x^2)^2$
39. $(10 - x)(10 + x)$
40. $(ay^2 - 3)(ay^2 + 3)$

10.5 Finding Factors of Special Products

To find the square root of a variable raised to a power, divide the exponent by 2 and use the result as the exponent of the given variable.

Example 1 Assuming that x and y are positive, find the square roots of **a.** x^2, **b.** x^4, **c.** x^6, and **d.** x^8y^{10}.

a. $\sqrt{x^2} = x$
b. $\sqrt{x^4} = x^2$
c. $\sqrt{x^6} = x^3$
d. $\sqrt{x^8y^{10}} = x^4y^5$ ◆

To factor a trinomial, first look for a common monomial factor. Then inspect the remaining trinomial to see if it is one of the special products. If it is not a perfect square trinomial and if it can be factored, use the methods shown in Section 10.3. If it is a perfect square trinomial, it may be factored using the reverse of the rule in Section 10.4.

Factoring Perfect Square Trinomials

Each of the two factors of a perfect square trinomial with a *positive* middle term is the square root of the first term *plus* the square root of the third term. That is,

$$a^2 + 2ab + b^2 = (a + b)(a + b)$$

Similarly, each of the two factors of the perfect square trinomial with a *negative* middle term is the square root of the first term *minus* the square root of the third term. That is,

$$a^2 - 2ab + b^2 = (a - b)(a - b)$$

Example 2 Factor: $9x^2 + 30x + 25$.

This perfect square trinomial has no common monomial factor. Since the middle term is positive, each of its two factors is the square root of the first term plus the square root of the third term. The square root of the first term is $3x$; the square root of the third term is 5. The sum is $3x + 5$. Therefore,

$$9x^2 + 30x + 25 = (3x + 5)(3x + 5)$$ ◆

Example 3 Factor: $x^2 - 12x + 36$.

This perfect square trinomial has no common monomial factor. Since the middle term is negative, each of its two factors is the square root of the first term minus the square root of the third term. The square root of the first term is x; the square root of the third term is 6. The difference is $x - 6$. Therefore,

$$x^2 - 12x + 36 = (x - 6)(x - 6)$$

◆

NOTE: If you do not recognize $x^2 - 12x + 36$ as a perfect square trinomial, you can factor it by trial and error as you would any trinomial (see Section 10.3). Your result should be the same.

Example 4 Factor: $4x^2 + 24xy + 36y^2$.

First, find the common monomial factor, 4.

$$4x^2 + 24xy + 36y^2 = 4(x^2 + 6xy + 9y^2)$$

This perfect square trinomial has a positive middle term. Each of its two factors is the square root of the first term plus the square root of the third term. The square root of the first term is x; the square root of the third term is $3y$. The sum is $x + 3y$. Therefore,

$$4x^2 + 24xy + 36y^2 = 4(x + 3y)(x + 3y)$$

◆

To factor a binomial that is the difference of two squares, use the reverse of the rule in Section 10.4. The factors of the difference of two squares are the square root of the first term *plus* the square root of the second term times the square root of the first term *minus* the square root of the second term, as shown below:

> **Factoring the Difference of Two Squares**
>
> $a^2 - b^2 = (a + b)(a - b)$

Note that a is the square root of a^2 and b is the square root of b^2.

Example 5 Factor: $x^2 - 4$.

First, find the square root of each term of the expression. The square root of x^2 is x, and the square root of 4 is 2. Thus, $x + 2$ is the sum of the square roots, and $x - 2$ is the difference of the square roots.

$$x^2 - 4 = (x + 2)(x - 2)$$

Check: $(x + 2)(x - 2) = x^2 - 4$

◆

Example 6 Factor: $1 - 36y^4$.

The square root of 1 is 1, and the square root of $36y^4$ is $6y^2$. Thus, the sum, $1 + 6y^2$, and the difference, $1 - 6y^2$, of the square roots are the factors.

$$1 - 36y^4 = (1 + 6y^2)(1 - 6y^2)$$

Check: $(1 + 6y^2)(1 - 6y^2) = 1 - 36y^4$

◆

Example 7 Factor: $81y^4 - 1$.

The square root of $81y^4$ is $9y^2$, and the square root of 1 is 1. The factors are the sum of the square roots, $9y^2 + 1$, and the difference of the square roots, $9y^2 - 1$.

$$81y^4 - 1 = (9y^2 + 1)(9y^2 - 1)$$

However, $9y^2 - 1$ is also the difference of two squares. Its factors are $9y^2 - 1 = (3y + 1)(3y - 1)$. Therefore,
$$81y^4 - 1 = (9y^2 + 1)(3y + 1)(3y - 1)$$

Example 8 Factor: $2x^2 - 18$.

First, find the common monomial factor, 2.
$$2x^2 - 18 = 2(x^2 - 9)$$
Then, $x^2 - 9$ is the difference of two squares whose factors are $x + 3$ and $x - 3$. Therefore,
$$2x^2 - 18 = 2(x + 3)(x - 3)$$

EXERCISES 10.5

Factor completely. Check by multiplying the factors:

1. $a^2 + 8a + 16$
2. $b^2 - 2b + 1$
3. $b^2 - c^2$
4. $m^2 - 1$
5. $x^2 - 4x + 4$
6. $2c^2 - 4c + 2$
7. $4 - x^2$
8. $4x^2 - 1$
9. $y^2 - 36$
10. $a^2 - 64$
11. $5a^2 + 10a + 5$
12. $9x^2 - 25$
13. $1 - 81y^2$
14. $16x^2 - 100$
15. $49 - a^4$
16. $m^2 - 2mn + n^2$
17. $49x^2 - 64y^2$
18. $x^2y^2 - 1$
19. $1 - x^2y^2$
20. $c^2d^2 - 16$
21. $4x^2 - 12x + 9$
22. $16x^2 - 1$
23. $R^2 - r^2$
24. $36x^2 - 12x + 1$
25. $49x^2 - 25$
26. $1 - 100y^2$
27. $y^2 - 10y + 25$
28. $x^2 + 6x + 9$
29. $b^2 - 9$
30. $16 - c^2d^2$
31. $m^2 + 22m + 121$
32. $n^2 - 30n + 225$
33. $4m^2 - 9$
34. $16b^2 - 81$
35. $4x^2 + 24x + 36$
36. $-2y^2 + 12y - 18$
37. $27x^2 - 3$
38. $225x^4 - 9x^2$
39. $am^2 - 14am + 49a$
40. $-bx^2 - 12bx - 36b$

10.6 Factoring General Trinomials

In previous examples, such as $x^2 + 7x + 10 = (x + 2)(x + 5)$, there was only one possible choice for the first terms of the binomials: x and x. When the coefficient of x^2 is greater than 1, however, there may be more than one possible choice. The idea is still the same: Find two binomial factors whose product equals the trinomial. Use the relationships outlined in Section 10.3 for finding the signs.

Example 1 Factor: $6x^2 - x - 2$.

The first terms of the binomial factors are either $6x$ and x or $3x$ and $2x$. The minus sign of the constant term of the trinomial tells you that one sign will be plus and the other will be minus in the last terms of the binomials. The last terms are either $+2$ and -1 or -2 and $+1$. The eight possibilities are

1. $(6x + 2)(x - 1) = 6x^2 - 4x - 2$
2. $(6x - 1)(x + 2) = 6x^2 + 11x - 2$
3. $(6x - 2)(x + 1) = 6x^2 + 4x - 2$
4. $(6x + 1)(x - 2) = 6x^2 - 11x - 2$

5. $(3x + 2)(2x - 1) = 6x^2 + x - 2$
6. $(3x - 1)(2x + 2) = 6x^2 + 4x - 2$
7. $(3x - 2)(2x + 1) = 6x^2 - x - 2$
8. $(3x + 1)(2x - 2) = 6x^2 - 4x - 2$

Only Equation 7 gives the desired middle term, $-x$. Therefore,

$$6x^2 - x - 2 = (3x - 2)(2x + 1)$$ ◆

When factoring trinomials of this type, sometimes you may have to make several guesses or look at several combinations until you find the correct one. It is a *trial-and-error process*. The rules for the signs, outlined in Section 10.3, simply reduce the number of possibilities you need to try.

Another way to reduce the possibilities is to eliminate any combination in which either binomial contains a common factor. In the above list, the factors $6x + 2$, $6x - 2$, $2x + 2$, and $2x - 2$ all have the common factor 2 and cannot be correct, so Equations 1, 3, 6, and 8 can be eliminated. It is important to look for common monomial factors as the first step in factoring any trinomial.

Example 2 Factor: $12x^2 - 23x + 10$.

First terms of the binomial: $12x$ and x, $6x$ and $2x$, or $4x$ and $3x$.
Signs of the last term of the binomial: both $(-)$.
Last terms of the binomial: -1 and -10 or -2 and -5.

You cannot use $12x$, $6x$, $4x$, or $2x$ with -10 or -2, since they contain the common factor 2, so the list of possible combinations is narrowed to

1. $(12x - 1)(x - 10) = 12x^2 - 121x + 10$
2. $(12x - 5)(x - 2) = 12x^2 - 29x + 10$
3. $(4x - 5)(3x - 2) = 12x^2 - 23x + 10$
4. $(4x - 1)(3x - 10) = 12x^2 - 43x + 10$

As you can see, Equation 3 is the correct one, since it gives the desired middle term, $-23x$. Therefore,

$$12x^2 - 23x + 10 = (4x - 5)(3x - 2)$$ ◆

Example 3 Factor: $12x^2 - 2x - 4$.

First, look for a common factor. In this case, it is 2, so we write

$$12x^2 - 2x - 4 = 2(6x^2 - x - 2)$$

Next, we try to factor the trinomial $6x^2 - x - 2$ into two binomial factors, as we did in Examples 1 and 2. Since the third term is negative (-2), we know that the signs of the second terms of the binomials are different. For the first terms of the binomials, we can try $6x$ and x or $3x$ and $2x$. The second terms can be $+2$ and -1 or -2 and $+1$. After eliminating all combinations with common factors, we have the following possibilities:

1. $(6x + 1)(x - 2) = 6x^2 - 11x - 2$
2. $(6x - 1)(x + 2) = 6x^2 + 11x - 2$
3. $(3x + 2)(2x - 1) = 6x^2 + x - 2$
4. $(3x - 2)(2x + 1) = 6x^2 - x - 2$

The correct one is Equation 4, since the middle term, $-x$, is the one we want. Therefore,

$$12x^2 - 2x - 4 = 2(3x - 2)(2x + 1)$$ ◆

Another method for factoring binomials in the form $ax^2 + bx + c$ is to list the pairs of the factors of ac and find the pair whose sum is b.

Example 4

Factor: $6x^2 - 19x + 10$.

First, list all factors of $ac = (6)(10) = 60$.

$1 \cdot 60$
$2 \cdot 30$
$3 \cdot 20$
$4 \cdot 15$
$5 \cdot 12$
$6 \cdot 10$

Note that only the factors $4 \cdot 15$ have a sum of 19. That is, the factors in the binomials must be some combination of 4 and 15.

Since the sign of the last term in both binomials is $(-)$, we have

$$6x^2 - 19x + 10 = (3x - 2)(2x - 5)$$

♦

This method is especially helpful if you are not successful in guessing by trial and error.

Example 5

Factor: $20x^2 + x - 12$.

First, list all the factors of $ac = (20)(12) = 240$:

$1 \cdot 240 \quad 6 \cdot 40$
$2 \cdot 120 \quad 8 \cdot 30$
$3 \cdot 80 \quad 10 \cdot 24$
$4 \cdot 60 \quad 12 \cdot 20$
$5 \cdot 48 \quad 15 \cdot 16$

Note that only the factors $15 \cdot 16$ can result in the middle term being 1. That is, the factors in the binomials must be some combination of 15 and 16. Since the sign of the last term of the trinomial is negative, we have

$$20x^2 + x - 12 = (5x + 4)(4x - 3)$$

♦

EXERCISES 10.6

Factor completely:

1. $5x^2 - 28x - 12$
2. $4x^2 - 4x - 3$
3. $10x^2 - 29x + 21$
4. $4x^2 + 4x + 1$
5. $12x^2 - 28x + 15$
6. $9x^2 - 36x + 32$
7. $8x^2 + 26x - 45$
8. $4x^2 + 15x - 4$
9. $16x^2 - 11x - 5$
10. $6x^2 + 3x - 3$
11. $12x^2 - 16x - 16$
12. $10x^2 - 35x + 15$
13. $15y^2 - y - 6$
14. $6y^2 + y - 2$
15. $8m^2 - 10m - 3$
16. $2m^2 - 7m - 30$
17. $35a^2 - 2a - 1$
18. $12a^2 - 28a + 15$
19. $16y^2 - 8y + 1$
20. $25y^2 + 20y + 4$
21. $3x^2 + 20x - 63$
22. $4x^2 + 7x - 15$
23. $12b^2 + 5b - 2$
24. $10b^2 - 7b - 12$
25. $15y^2 - 14y - 8$
26. $5y^2 + 11y + 2$
27. $90 + 17c - 3c^2$
28. $10x^2 - x - 2$
29. $6x^2 - 13x + 5$
30. $56x^2 - 29x + 3$
31. $2y^4 + 9y^2 - 35$
32. $2y^2 + 7y - 99$
33. $4b^2 + 52b + 169$
34. $6x^2 - 19x + 15$
35. $14x^2 - 51x + 40$
36. $42x^4 - 13x^2 - 40$
37. $28x^3 + 140x^2 + 175x$
38. $-24x^3 - 54x^2 - 21x$
39. $10ab^2 - 15ab - 175a$
40. $40bx^2 - 72bx - 70b$

SUMMARY | CHAPTER 10

Glossary of Basic Terms

Binomial factor. A two-term factor of an algebraic expression. (p. 330)
Factoring an algebraic expression. Writing the algebraic expression as a product of factors. (p. 326)
Greatest common factor of a polynomial. The largest common factor that divides all terms in the expression. (p. 326)
Monomial factor. A one-term factor that divides each term of an algebraic expression. (p. 326)
Perfect square trinomial. A trinomial with the same two binomial factors. (p. 333)

10.2 Finding the Product of Two Binomials Mentally

1. **Finding the product of two binomials mentally:**
   The mental process is outlined as follows:
   a. The *first term* of the product is the product of the first terms of the binomials.
   b. The *middle term* of the product is the sum of the outer product and the inner product of the binomials.
   c. The *last term* of the product is the product of the last terms of the binomials.
   This method is often called the *FOIL method,* where F refers to the product of the *first* terms, O refers to the *outer* product, I refers to the *inner* product, and L refers to the product of the *last* terms. (p. 327)

10.3 Finding Binomial Factors

1. **Factoring trinomials:** To factor a trinomial $x^2 + bx + c$, use the following steps:

   Assume that b and c are both positive numbers.

   First, look for any common monomial factors. Then:
   a. For the trinomial $x^2 + bx + c$, use the form
   $$x^2 + bx + c = (x +)(x +)$$
   b. For the trinomial $x^2 - bx + c$, use the form
   $$x^2 - bx + c = (x -)(x -)$$
   c. For the trinomials $x^2 - bx - c$ and $x^2 + bx - c$, use the forms
   $$x^2 - bx - c = (x +)(x -)$$
   $$x^2 + bx - c = (x +)(x -)$$ (p. 332)

10.4 Special Products

1. **The square of a binomial:** There are two forms.
   a. The square of the *sum* of two terms of a binomial equals the square of the first term *plus* twice the product of the two terms plus the square of the second term; that is, $(a + b)(a + b) = (a + b)^2 = a^2 + 2ab + b^2$.
   b. The square of the *difference* of two terms of a binomial equals the square of the first term *minus* twice the product of the two terms plus the square of the second term; that is, $(a - b)(a - b) = (a - b)^2 = a^2 - 2ab + b^2$. (p. 333)
2. **The product of the sum and difference of two terms:** This product is the difference of two squares—that is, the square of the first term *minus* the square of the second term, $(a + b)(a - b) = a^2 - b^2$. (p. 333)

10.5 Finding Factors of Special Products

1. **Factoring perfect square trinomials:** There are two forms.
   a. Each of the two factors of a perfect square trinomial with a *positive* middle term is the square root of the first term *plus* the square root of the third term—that is, $a^2 + 2ab + b^2 = (a + b)(a + b)$.
   b. Each of the two factors of a perfect square trinomial with a *negative* middle term is the square root of the first term *minus* the square root of the third term—that is, $a^2 - 2ab + b^2 = (a - b)(a - b)$. (p. 334)
2. **Factoring the difference of two squares:** The factors of the difference of two squares are the square root of the first term *plus* the square root of the second term times the square root of the first term *minus* the square root of the second term—that is, $a^2 - b^2 = (a + b)(a - b)$. (p. 335)

REVIEW | CHAPTER 10

Find each product mentally:

1. $(c + d)(c - d)$
2. $(x - 6)(x + 6)$
3. $(y + 7)(y - 4)$
4. $(2x + 5)(2x - 9)$
5. $(x + 8)(x - 3)$
6. $(x - 4)(x - 9)$
7. $(x - 3)^2$
8. $(2x - 6)^2$
9. $(1 - 5x^2)^2$

Factor each expression completely:

10. $6a + 6$
11. $5x - 15$
12. $xy + 2xz$
13. $y^4 + 17y^3 - 18y^2$
14. $y^2 - 6y - 7$
15. $z^2 + 18z + 81$
16. $x^2 + 10x + 16$
17. $4a^2 + 4x^2$
18. $x^2 - 17x + 72$
19. $x^2 - 18x + 81$
20. $x^2 + 19x + 60$
21. $y^2 - 2y + 1$
22. $x^2 - 3x - 28$
23. $x^2 - 4x - 96$
24. $x^2 + x - 110$
25. $x^2 - 49$
26. $16y^2 - 9x^2$
27. $x^2 - 144$
28. $25x^2 - 81y^2$
29. $4x^2 - 24x - 364$
30. $5x^2 - 5x - 780$
31. $2x^2 + 11x + 14$
32. $12x^2 - 19x + 4$
33. $30x^2 + 7x - 15$
34. $12x^2 + 143x - 12$
35. $4x^2 - 6x + 2$
36. $36x^2 - 49y^2$
37. $28x^2 + 82x + 30$
38. $30x^2 - 27x - 21$
39. $4x^3 - 4x$
40. $25y^2 - 100$

TEST | CHAPTER 10

Find each product mentally:

1. $(x + 8)(x - 3)$
2. $(2x - 8)(5x - 6)$
3. $(2x - 8)(2x + 8)$
4. $(3x - 5)^2$
5. $(4x - 7)(2x + 3)$
6. $(9x - 7)(5x + 4)$

Factor each expression completely:

7. $x^2 + 4x + 3$
8. $x^2 - 12x + 35$
9. $6x^2 - 7x - 90$
10. $9x^2 + 24x + 16$
11. $x^2 + 7x - 18$
12. $4x^2 - 25$
13. $6x^2 + 13x + 6$
14. $3x^2y^2 - 18x^2y + 27x^2$
15. $3x^2 - 11x - 4$
16. $15x^2 - 19x - 10$
17. $5x^2 + 7x - 6$
18. $3x^2 - 3x - 6$
19. $9x^2 - 121$
20. $9x^2 - 30x + 25$

CUMULATIVE REVIEW | CHAPTERS 1–10

1. Perform the indicated operations and simplify: $2 + 6^2 - 24 \div 3(4)$
2. Round 746.83 to the **a.** nearest tenth and **b.** nearest ten.
3. Do as indicated and simplify: $-\dfrac{2}{3} \div \dfrac{1}{5} + \dfrac{2}{3}$
4. Write 0.000318 in **a.** scientific notation and **b.** engineering notation.
5. Change 625 g to kg.
6. Change 6 m² to ft².
7. Read the voltmeter scale in Illustration 1.

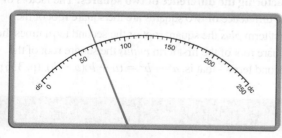

ILLUSTRATION 1

8. Use the rules of measurement to multiply: (5.0 cm)(148 cm)(0.128 cm)

Combine like terms and simplify:

9. $3(x - 2) - 4(2 - 3x)$
10. $(6a - 3b + 2c) - (-2a - 3b + c)$
11. Solve: $\dfrac{x}{3} - 4 = \dfrac{2x}{5}$
12. A rectangle is 5 m longer than it is wide. Its perimeter is 58 m. Find the length and the width.
13. Solve the proportion and round the result to three significant digits:
$$\dfrac{15.7}{8.2} = \dfrac{x}{10}$$
14. A pulley is 18 in. in diameter, is rotating at 125 rpm, and is connected to a smaller pulley rotating at 225 rpm. Find the diameter of the smaller pulley.
15. Complete the ordered-pair solutions of the equation: $2x + 3y = 12$
   (3,   ), (0,   ), (−3,   )
16. Solve for y: $3x - y = 5$
17. Draw the graph of $3x + 4y = 24$.

18. Draw the graphs of $2x - y = 4$ and $x + 3y = -5$. Find the point of intersection.

Solve each pair of linear equations:

19. $\dfrac{1}{2}x - \dfrac{2}{3}y = 1$

$6x - 8y = 12$

20. $y = 3x - 5$

$x + 3y = 8$

21. $x - y = 6$

$3x + y = 2$

22. $3x - 5y = 7$

$-6x + 10y = 5$

23. $135x + 40y = 29$

$60x - 45y = 38$

24. Two rental automobiles were leased for a total of 16 days. One automobile rents for $53.95 per day, and the other rents for $89.95 per day. The total cost for leasing both was $1223.20. Find the number of days each was rented.

Find each product mentally:

25. $(2x - 5)(3x + 8)$

26. $(5x - 7y)^2$

27. $(3x - 5)(5x - 7)$

Factor each expression completely:

28. $7x^3 - 63x$

29. $4x^3 + 12x^2$

30. $2x^2 - 7x - 4$

CHAPTER 11 | Quadratic Equations

OBJECTIVES

- Solve quadratic equations by factoring.
- Solve quadratic equations by using the quadratic formula.
- Solve application problems involving quadratic equations.
- Graph quadratic equations.
- Find the vertex of a parabola.
- Express the square root of a negative number as an imaginary number in terms of j.
- Simplify powers of j.
- Solve quadratic equations with imaginary roots.

Mathematics at Work

Aircraft mechanics and service technicians perform scheduled maintenance, make repairs, and make inspections to keep aircraft in peak operating condition. Many specialize in preventive maintenance by inspecting engines, landing gear, instruments, pressurized sections, and various accessories such as brakes, valves, pumps, and air-conditioning systems. Such inspections occur following a schedule based on the number of hours the aircraft has flown, calendar days, cycles of operation, or a combination of these factors. Powerplant mechanics are authorized to work on engines and do limited work on propellers. Airframe mechanics are authorized to work on any part of the aircraft except the instruments, power plants, and propellers. Combination airframe-and-powerplant mechanics (A&P mechanics) work on all parts of the plane except instruments. The majority of mechanics working on civilian aircraft are A&P mechanics. Avionics technicians repair and maintain components used for aircraft navigation and radio communications, weather radar systems, and other instruments and computers that control flight, engine, and primary functions. The Federal Aviation Administration (FAA) regulates certification of aircraft mechanics and service technicians as well as training programs. Mathematics, physics, chemistry, electronics, computer science, mechanical drawing, and communications skills are key to training programs and success on the job. For more information, please visit **www.cengage.com** and access the Student Online Resources for this text.

Aircraft Mechanics and Service Technicians
Airline mechanics checking maintenance records

11.1 Solving Quadratic Equations by Factoring

A **quadratic equation** in one variable is an equation in the form $ax^2 + bx + c = 0$, where $a \neq 0$.

Recall that linear equations, such as $2x + 3 = 0$, have at most *one* solution. Quadratic equations have at most *two* solutions. One way to solve quadratic equations is by factoring and using the following:

> If $ab = 0$, then either $a = 0$ or $b = 0$.

That is, if you multiply two factors and the product is 0, then one or both factors are 0.

Example 1 Solve: $4(x - 2) = 0$.

If $4(x - 2) = 0$, then $4 = 0$ or $x - 2 = 0$. However, the first statement, $4 = 0$, is false; thus, the solution is $x - 2 = 0$, or $x = 2$. ◆

Example 2 Solve: $(x - 2)(x + 3) = 0$.

If $(x - 2)(x + 3) = 0$, then either

$$x - 2 = 0 \quad \text{or} \quad x + 3 = 0$$

Therefore,

$$x = 2 \quad \text{or} \quad x = -3$$ ◆

Solving Quadratic Equations by Factoring

1. If necessary, write an equivalent equation in the form $ax^2 + bx + c = 0$.
2. Factor the polynomial.
3. Write equations by setting each factor containing a variable equal to zero.
4. Solve the two resulting first-degree equations.
5. Check.

Example 3 Solve $x^2 + 6x + 5 = 0$ for x.

STEP 1   Not needed.
STEP 2 $(x + 5)(x + 1) = 0$
STEP 3 $x + 5 = 0$ or $x + 1 = 0$
STEP 4 $x = -5$ or $x = -1$
STEP 5   Check:

Replace x with -5.

$$x^2 + 6x + 5 = 0$$
$$(-5)^2 + 6(-5) + 5 = 0 \quad ?$$
$$25 - 30 + 5 = 0$$
$$0 = 0 \quad \text{True}$$

Replace x with -1.

$$x^2 + 6x + 5 = 0$$
$$(-1)^2 + 6(-1) + 5 = 0 \quad ?$$
$$1 - 6 + 5 = 0$$
$$0 = 0 \quad \text{True}$$

Thus, the roots are -5 and -1. ◆

Example 4

Solve $x^2 + 5x = 36$ for x.

STEP 1 $x^2 + 5x - 36 = 0$

STEP 2 $(x + 9)(x - 4) = 0$

STEP 3 $x + 9 = 0$ or $x - 4 = 0$

STEP 4 $x = -9$ or $x = 4$

STEP 5 *Check:*

Replace x with -9.

$x^2 + 5x = 36$
$(-9)^2 + 5(-9) = 36$?
$81 - 45 = 36$
$36 = 36$ True

Replace x with 4.

$x^2 + 5x = 36$
$4^2 + 5(4) = 36$?
$16 + 20 = 36$
$36 = 36$ True

Thus, the roots are -9 and 4.

Example 5

Solve $3x^2 + 9x = 0$ for x.

STEP 1 Not needed.

STEP 2 $3x(x + 3) = 0$

STEP 3 $3x = 0$ or $x + 3 = 0$

STEP 4 $x = 0$ or $x = -3$

STEP 5 *Check:* Left to the student.

Example 6

Solve $x^2 = 4$ for x.

STEP 1 $x^2 - 4 = 0$

STEP 2 $(x + 2)(x - 2) = 0$

STEP 3 $x + 2 = 0$ or $x - 2 = 0$

STEP 4 $x = -2$ or $x = 2$

STEP 5 *Check:* Left to the student.

Example 7

Solve: $6x^2 = 7x + 20$.

STEP 1 $6x^2 - 7x - 20 = 0$

STEP 2 $(3x + 4)(2x - 5) = 0$

STEP 3 $3x + 4 = 0$ or $2x - 5 = 0$

STEP 4 $3x = -4$ $2x = 5$

$x = -\dfrac{4}{3}$ $x = \dfrac{5}{2}$

So the possible roots are $-\dfrac{4}{3}$ and $\dfrac{5}{2}$.

STEP 5 *Check:* Replace x with $-\dfrac{4}{3}$ and with $\dfrac{5}{2}$ in the original equation.

$6x^2 = 7x + 20$ $6x^2 = 7x + 20$

$6\left(-\dfrac{4}{3}\right)^2 = 7\left(-\dfrac{4}{3}\right) + 20$ $6\left(\dfrac{5}{2}\right)^2 = 7\left(\dfrac{5}{2}\right) + 20$

$6\left(\dfrac{16}{9}\right) = -\dfrac{28}{3} + 20$ $6\left(\dfrac{25}{4}\right) = \dfrac{35}{2} + 20$

$$\frac{32}{3} = -\frac{28}{3} + \frac{60}{3} \qquad \frac{75}{2} = \frac{35}{2} + \frac{40}{2}$$

$$\frac{32}{3} = \frac{32}{3} \quad \text{True} \qquad \frac{75}{2} = \frac{75}{2} \quad \text{True}$$

So the roots are $-\frac{4}{3}$ and $\frac{5}{2}$.

♦

EXERCISES 11.1

Solve each equation:

1. $x^2 + x = 12$
2. $x^2 - 3x + 2 = 0$
3. $x^2 + x - 20 = 0$
4. $d^2 + 2d - 15 = 0$
5. $x^2 - 2 = x$
6. $x^2 - 15x = -54$
7. $x^2 - 1 = 0$
8. $16n^2 = 49$
9. $x^2 - 49 = 0$
10. $4n^2 = 64$
11. $w^2 + 5w + 6 = 0$
12. $x^2 - 6x = 0$
13. $y^2 - 4y = 21$
14. $c^2 + 2 = 3c$
15. $n^2 - 6n - 40 = 0$
16. $x^2 - 17x + 16 = 0$
17. $9m = m^2$
18. $6n^2 - 15n = 0$
19. $x^2 = 108 + 3x$
20. $x^2 - x = 42$
21. $c^2 + 6c = 16$
22. $4x^2 + 4x - 3 = 0$
23. $10x^2 + 29x + 10 = 0$
24. $2x^2 = 17x - 8$
25. $4x^2 = 25$
26. $25x = x^2$
27. $9x^2 + 16 = 24x$
28. $24x^2 + 10 = 31x$
29. $3x^2 + 9x = 0$

30. A rectangle is 5 ft longer than it is wide. (See Illustration 1.) The area of the rectangle is 84 ft². Use a quadratic equation to find the dimensions of the rectangle.

31. The area of a triangle is 66 m², and its base is 1 m more than the height. (See Illustration 2.) Find the base and height of the triangle. (Use a quadratic equation.)

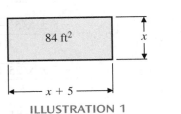
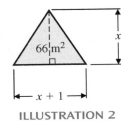

ILLUSTRATION 1   ILLUSTRATION 2

32. A rectangle is 9 ft longer than it is wide, and its area is 360 ft². Use a quadratic equation to find its length and width.

33. A heating duct has a rectangular cross section whose area is 40 in². If it is 3 in. longer than it is wide, find its length and width.

11.2 The Quadratic Formula

Many quadratic equations cannot be solved by factoring, so let's study a method by which any quadratic equation can be solved.

The roots of a quadratic equation in the form

$$ax^2 + bx + c = 0$$

may be found by using the following formula:

Quadratic Formula

$$x = \frac{-b \pm \sqrt{b^2 - 4ac}}{2a}$$

where a is the coefficient of the x^2 term,
b is the coefficient of the x term, and
c is the constant term.

CHAPTER 11 ♦ Quadratic Equations

The symbol (±) is used to combine two expressions into one. For example, "$a \pm 4$" means "$a + 4$ or $a - 4$." Similarly,

$$x = \frac{-b \pm \sqrt{b^2 - 4ac}}{2a} \quad \text{means}$$

$$x = \frac{-b + \sqrt{b^2 - 4ac}}{2a} \quad \text{or} \quad x = \frac{-b - \sqrt{b^2 - 4ac}}{2a}$$

Example 1 In the quadratic equation $3x^2 - x - 7 = 0$, find the values of a, b, and c.

$$a = 3, \quad b = -1, \quad \text{and} \quad c = -7$$

Example 2 Solve $x^2 + 5x - 14 = 0$ using the quadratic formula.

$$x = \frac{-b \pm \sqrt{b^2 - 4ac}}{2a}, \quad a = 1, \quad b = 5, \quad c = -14$$

So $x = \dfrac{-5 \pm \sqrt{5^2 - 4(1)(-14)}}{2(1)}$ Substitute.

$= \dfrac{-5 \pm \sqrt{25 + 56}}{2}$ Simplify.

$= \dfrac{-5 \pm \sqrt{81}}{2}$

$= \dfrac{-5 \pm 9}{2}$ $\sqrt{81} = 9$

$= \dfrac{-5 + 9}{2} \quad \text{or} \quad \dfrac{-5 - 9}{2}$

$= 2 \quad \text{or} \quad -7$

Check: Replace x with 2.   Replace x with -7.

$$x^2 + 5x - 14 = 0 \qquad\qquad x^2 + 5x - 14 = 0$$
$$2^2 + 5(2) - 14 = 0 \quad ? \qquad (-7)^2 + 5(-7) - 14 = 0 \quad ?$$
$$4 + 10 - 14 = 0 \qquad\qquad 49 - 35 - 14 = 0$$
$$0 = 0 \qquad\qquad\qquad\qquad 0 = 0$$

The roots are 2 and -7.

Before using the quadratic formula, make certain the equation is written in the form $ax^2 + bx + c = 0$, such that one member is zero.

Example 3 Solve $2x^2 = x + 21$ by using the quadratic formula.

First, write the equation in the form $2x^2 - x - 21 = 0$.

$$x = \frac{-b \pm \sqrt{b^2 - 4ac}}{2a}, \quad a = 2, \quad b = -1, \quad c = -21$$

So $x = \dfrac{-(-1) \pm \sqrt{(-1)^2 - 4(2)(-21)}}{2(2)}$ Substitute.

$= \dfrac{1 \pm \sqrt{1 + 168}}{4}$ Simplify.

$= \dfrac{1 \pm \sqrt{169}}{4}$

$= \dfrac{1 \pm 13}{4}$ $\sqrt{169} = 13$

$= \dfrac{1 + 13}{4}$ or $\dfrac{1 - 13}{4}$

$= \dfrac{7}{2}$ or -3

Check: Left to the student. ◆

The quantity under the radical sign, $b^2 - 4ac$, is called the **discriminant**. If the discriminant is not a perfect square, find the square root of the number by using a calculator and proceed as before. Round each final result to three significant digits.

Example 4 Solve $3x^2 + x - 5 = 0$ using the quadratic formula.

$x = \dfrac{-b \pm \sqrt{b^2 - 4ac}}{2a}$, $a = 3$, $b = 1$, $c = -5$

So $x = \dfrac{-1 \pm \sqrt{1^2 - 4(3)(-5)}}{2(3)}$ Substitute.

$= \dfrac{-1 \pm \sqrt{1 + 60}}{6}$ Simplify.

$= \dfrac{-1 \pm \sqrt{61}}{6}$

$= \dfrac{-1 \pm 7.81}{6}$ $\sqrt{61} = 7.81$

$= \dfrac{-1 + 7.81}{6}$ or $\dfrac{-1 - 7.81}{6}$

$= \dfrac{6.81}{6}$ or $\dfrac{-8.81}{6}$

$= 1.14$ or -1.47

The roots are 1.14 and -1.47. ◆

The check will not work out *exactly* when the number under the radical is not a perfect square.

EXERCISES 11.2

Find the values of a, b, and c in each equation:

1. $x^2 - 7x + 4 = 0$
2. $2x^2 + x - 3 = 0$
3. $3x^2 + 4x + 9 = 0$
4. $2x^2 - 14x + 37 = 0$
5. $-3x^2 + 4x + 7 = 0$
6. $17x^2 - x + 34 = 0$
7. $3x^2 - 14 = 0$
8. $2x^2 + 7x = 0$

Solve each equation using the quadratic formula. Check your solutions:

9. $x^2 + x - 6 = 0$
10. $x^2 - 4x - 21 = 0$
11. $x^2 + 8x - 9 = 0$
12. $2x^2 + 5x - 12 = 0$
13. $5x^2 + 2x = 0$
14. $3x^2 - 75 = 0$
15. $48x^2 - 32x - 35 = 0$
16. $13x^2 + 178x - 56 = 0$

Solve each equation using the quadratic formula (when necessary, round results to three significant digits):

17. $2x^2 + x - 5 = 0$
18. $-3x^2 + 2x + 5 = 0$
19. $3x^2 - 5x = 0$
20. $7x^2 + 9x + 2 = 0$
21. $-2x^2 + x + 3 = 0$
22. $5x^2 - 7x + 2 = 0$
23. $6x^2 + 9x + 1 = 0$
24. $16x^2 - 25 = 0$
25. $-4x^2 = 5x + 1$
26. $9x^2 = 21x - 10$
27. $3x^2 = 17$
28. $8x^2 = 11x - 3$
29. $x^2 = 15x + 7$
30. $x^2 + x = 1$
31. $3x^2 - 31 = 5x$
32. $-3x^2 - 5 = -7x^2$
33. $52.3x = -23.8x^2 + 11.8$
34. $18.9x^2 - 44.2x = 21.5$

11.3 Applications Involving Quadratic Equations

We now present some applications that involve quadratic equations. For consistency, all final results are rounded to three significant digits when necessary.

Example 1

A variable voltage in an electric circuit is given by the equation $V = 8t^2 - 28t + 20$, where t is in milliseconds (ms). Find the values of t when the voltage V equals **a.** 8 V and **b.** 15 V.

a. Substitute $V = 8$ into the equation.

$V = 8t^2 - 28t + 20$
$8 = 8t^2 - 28t + 20$
$0 = 8t^2 - 28t + 12$ Subtract 8 from both sides.
$0 = 2t^2 - 7t + 3$ Divide both sides by 4 to make the work easier.
$0 = (2t - 1)(t - 3)$ Factor.
$2t - 1 = 0$ or $t - 3 = 0$ Set each factor equal to 0 and solve for t.
$t = \frac{1}{2}$ ms $t = 3$ ms

b. Substitute $V = 15$ into the equation

$V = 8t^2 - 28t + 20$
$15 = 8t^2 - 28t + 20$
$0 = 8t^2 - 28t + 5$ Subtract 15 from both sides.

Note that the right side of the equation does not factor, so we use the quadratic formula with $a = 8$, $b = -28$, and $c = 5$.

$t = \frac{-b \pm \sqrt{b^2 - 4ac}}{2a}$

$t = \frac{-(-28) \pm \sqrt{(-28)^2 - 4(8)(5)}}{2(8)}$

$t = \frac{28 \pm \sqrt{784 - 160}}{16}$

$t = \frac{28 \pm \sqrt{624}}{16}$

$t = \frac{28 \pm 25.0}{16}$

$t = \frac{28 + 25.0}{16}$ or $t = \frac{28 - 25.0}{16}$

$t = 3.31$ ms $t = 0.188$ ms

◆

11.3 ♦ Applications Involving Quadratic Equations

Example 2 Design a rectangular metal plate so that its length is 6 cm more than twice its width and its area is 360 cm².

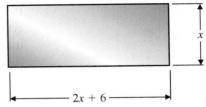

Figure 11.1

First, draw a diagram as in Figure 11.1 and let

$x =$ the width

$2x + 6 =$ the length

Then, use the formula for the area of a rectangle and substitute as follows:

$A = lw$

$360 = (2x + 6)x$

$360 = 2x^2 + 6x$ Remove parentheses.

$0 = 2x^2 + 6x - 360$ Subtract 360 from both sides.

$0 = x^2 + 3x - 180$ Divide both sides by 2 to make the work easier.

$0 = (x + 15)(x - 12)$ Factor.

$x + 15 = 0$ or $x - 12 = 0$ Set each factor equal to 0 and solve for x.

$x = -15$ $x = 12$

Note that the solution $x = -15$ is not meaningful, as x refers to a width measurement, which must be a positive quantity. Therefore,

$x =$ the width $= 12$ cm

$2x + 6 =$ the length $= 2(12) + 6 = 30$ cm

Check: $A = lw = (30 \text{ cm})(12 \text{ cm}) = 360 \text{ cm}^2$, which is the given area. ♦

Example 3 The perimeter of a rectangle is 20 cm, and its area is 16 cm². Find its dimensions (the length and the width).

First, note that the perimeter of a rectangle is the sum of the lengths of all four sides. Thus, if the perimeter is 20 cm, one width plus one length is 10 cm. So if the width is x, then the length must be $10 - x$.

Then, draw a diagram as in Figure 11.2 and let

$x =$ the width

$10 - x =$ the length

Then, use the formula for the area of a rectangle and substitute as follows:

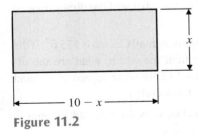

Figure 11.2

$A = lw$

$16 = (10 - x)x$

$16 = 10x - x^2$ Remove parentheses.

$x^2 - 10x + 16 = 0$ Set the equation equal to 0.

$(x - 2)(x - 8) = 0$ Factor.

$x - 2 = 0$ or $x - 8 = 0$ Set each factor equal to 0 and solve for x.

$x = 2$ $x = 8$

If $x = 2$,   If $x = 8$,

$x =$ the width $= 2$ cm $x =$ the width $= 8$ cm

$10 - x =$ the length $= 8$ cm $10 - x =$ the length $= 2$ cm

Note that the dimensions are the same 2 cm by 8 cm in both cases. Since the length is greater than the width, the length is 8 cm and the width is 2 cm. ♦

Example 4

A square is cut out of each corner of a rectangular sheet of metal 42 cm × 52 cm. The sides are then folded up to form a rectangular container with no top. What are the dimensions of the square if the area of the bottom of the container is 1200 cm²? Find the volume of the container.

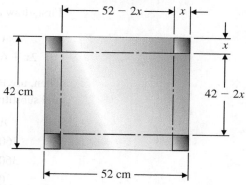

Figure 11.3

First, draw a diagram as in Figure 11.3 and let

x = the side of each square cutout
$42 - 2x$ = the width of the container
$52 - 2x$ = the length of the container

Use the formula for the area of the rectangular bottom of the container and substitute as follows:

$$A = lw$$
$$1200 = (52 - 2x)(42 - 2x)$$
$$1200 = 2184 - 188x + 4x^2 \quad \text{Remove parentheses.}$$
$$0 = 4x^2 - 188x + 984 \quad \text{Subtract 1200 from both sides.}$$
$$0 = x^2 - 47x + 246 \quad \text{Divide both sides by 4 to make the work easier.}$$
$$0 = (x - 41)(x - 6) \quad \text{Factor.}$$
$$x - 41 = 0 \quad \text{or} \quad x - 6 = 0 \quad \text{Set each factor equal to 0 and solve for } x.$$
$$x = 41 \qquad x = 6$$

Note that $x = 41$ cm is not physically possible. So the side of each square is 6 cm. The length of the container is $52 - 2x = 52 - 2(6) = 52 - 12 = 40$ cm. The width of the container is $42 - 2x = 42 - 2(6) = 42 - 12 = 30$ cm. The volume of the container is

$$V = lwh$$
$$V = (40 \text{ cm})(30 \text{ cm})(6 \text{ cm}) = 7200 \text{ cm}^3$$

♦

EXERCISES 11.3

1. A variable voltage in an electrical circuit is given by $V = t^2 - 12t + 40$, where t is in seconds. Find the values of t when the voltage V equals **a.** 8 V, **b.** 25 V, **c.** 104 V.

2. A variable electric current is given by $i = t^2 - 7t + 12$, where t is in seconds. At what times is the current i equal to **a.** 2 A? **b.** 0 A? **c.** 4 A?

3. A rectangular piece of sheet metal is 4 ft longer than it is wide. (See Illustration 1.) The area of the piece of sheet metal is 21 ft². Find its length and width.

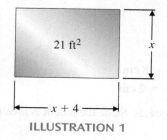

ILLUSTRATION 1

4. A hole in the side of a large metal tank needs to be repaired. A piece of rectangular sheet metal of area 16 ft² will patch the hole. If the length of the sheet metal must be 8 ft longer than its width, what will the dimensions of the sheet metal be?

5. The area of the wings of a small Cessna is 175 ft². If the length is 30 ft longer than the width, what are the dimensions of the wings? (This wing is one piece and goes along the top of the aircraft.)

6. The perimeter of a rectangle is 46 cm, and its area is 120 cm². Find its dimensions.

7. The perimeter of a rectangle is 160 m, and its area is 1200 m². Find its dimensions.

8. A rectangular field is fenced in by using a river as one side. If 1800 m of fencing are used for the 385,000-m² field, find its dimensions.

9. The dimensions of a doorway are 3 ft by 7 ft 6 in. If the same amount is added to each dimension, the area is increased by 18 ft². (See Illustration 2.) Find the dimensions of the new doorway.

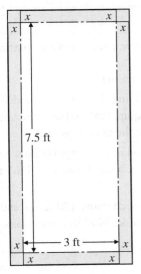

ILLUSTRATION 2

10. A square, 4 in. on a side, is cut out of each corner of a square sheet of aluminum. (See Illustration 3.) The sides are folded up to form a rectangular container with no top. The volume of the resulting container is 400 in³. What was the size of the original sheet of aluminum?

ILLUSTRATION 3

11. A square is cut out of each corner of a rectangular sheet of aluminum that is 40 cm by 60 cm. (See Illustration 4.) The sides are folded up to form a rectangular container with no top. The area of the bottom of the container is 1500 cm². **a.** What are the dimensions of each cut-out square? **b.** Find the volume of the container. ($V = lwh$)

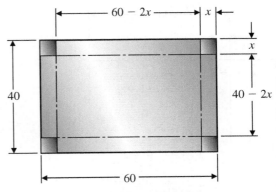

ILLUSTRATION 4

12. The area of a rectangular lot 80 m by 100 m is to be increased by 4000 m². (See Illustration 5.) The length and the width will be increased by the same amount. What are the dimensions of the larger lot? Find the percent increase for the length and width.

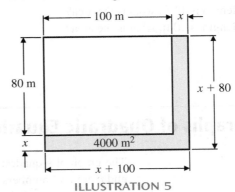

ILLUSTRATION 5

13. A border of uniform width is built around a rectangular garden that measures 16 ft by 20 ft. (See Illustration 6.) The area of the border is 160 ft². Find the width of the border.

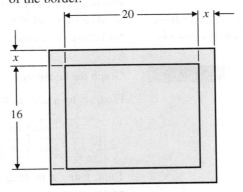

ILLUSTRATION 6

14. A border of uniform width is printed on a page measuring 11 in. by 14 in. (See Illustration 7.) The area of the border is 66 in². Find the width of the border.

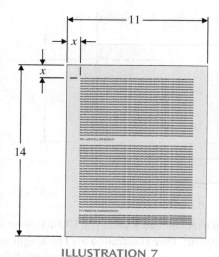

ILLUSTRATION 7

15. A company needs to build a warehouse with perimeter 300 ft. Find the dimensions to give maximum floor space.
   a. If the length is 10 ft, what is the area?
   b. If the length is 20 ft, what is the area?
   c. Write a formula (model) for the area in terms of the length.
   d. Complete the following table:

Length (ft)	30	40	50	60	70	80	90	100	110	120	130	140
Area (ft²)												

   e. Does one of these values give a maximum area? Explain.
   f. Graph the equation.
   g. Is there a different maximum?

16. A 2000-ft² storage building 9 ft high is needed to store yard maintenance equipment. What dimensions should be used to minimize the outside walls?

17. A landscaper is laying sod in a rectangular front lawn that is 76 ft longer than it is wide. Its area is 9165 ft². Find its dimensions.

18. A rectangular forest plot contains 120 acres and is three times as long as it is wide. Find its dimensions.

11.4 Graphs of Quadratic Equations

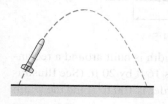

Figure 11.4
The trajectory of a rocket follows a path in the shape of a parabola.

The graph of a quadratic equation written in the form $y = ax^2 + bx + c = 0$, where a, b, and c are real numbers and $a \neq 0$, is called a **parabola**. The quadratic equation written in the form $x = ay^2 + by + c = 0$ also represents a parabola, but we will not work with this equation in this book. Many physical phenomena follow a curved path of a parabola. The trajectory of a rocket (Figure 11.4) or any projectile is one such example.

To draw the graph of a parabola, find points whose ordered pairs satisfy the equation by choosing various values of x and solving for y. Since this graph is not a straight line, you will need to find and plot many points to get an accurate graph of the curve. A table is helpful for listing these ordered pairs.

Example 1

Graph the equation $y = x^2$.

First, set up a table as follows:

x	-4	-3	-2	-1	0	1	2	3	4
$y = x^2$	16	9	4	1	0	1	4	9	16

Then, using a rectangular coordinate system, plot these points. Notice that with only the points shown in Figure 11.5(a), there isn't a definite outline of the curve. So let's look more closely at values of x between 0 and 1:

x	$\frac{1}{6}$	$\frac{1}{5}$	$\frac{1}{4}$	$\frac{1}{3}$	$\frac{2}{5}$	$\frac{1}{2}$	$\frac{3}{5}$	$\frac{3}{4}$	$\frac{2}{3}$	$\frac{4}{5}$	$\frac{5}{6}$
$y = x^2$	$\frac{1}{36}$	$\frac{1}{25}$	$\frac{1}{16}$	$\frac{1}{9}$	$\frac{4}{25}$	$\frac{1}{4}$	$\frac{9}{25}$	$\frac{9}{16}$	$\frac{4}{9}$	$\frac{16}{25}$	$\frac{25}{36}$

Plotting these additional ordered pairs, we get a better graph (Figure 11.5b). If we were to continue to choose more and more values, the graph would appear as a solid curve.

11.4 ♦ Graphs of Quadratic Equations 353

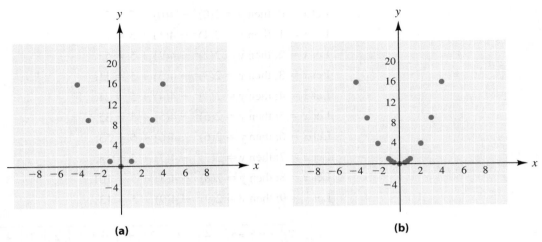

Figure 11.5
Plotting points that satisfy the equation $y = x^2$

Since it is impossible to find *all* ordered pairs that satisfy the equation, we will assume that all the points between any two of the ordered pairs already located could be found and that they do lie on the graph. Thus, assume that the graph of $y = x^2$ looks like the graph in Figure 11.6.

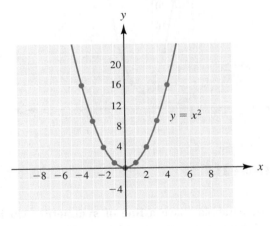

Figure 11.6

In summary, to draw the graph of a quadratic equation, form a table to find many ordered pairs that satisfy the equation by choosing various values of x and solving for y. Then plot these ordered pairs and connect them with a smooth curve.

Example 2   Graph the equation $y = 2x^2 - 4x + 5$.

Let $x = -7$; then $y = 2(-7)^2 - 4(-7) + 5 = 131$.
Let $x = -6$; then $y = 2(-6)^2 - 4(-6) + 5 = 101$.
Let $x = -5$; then $y = 2(-5)^2 - 4(-5) + 5 = 75$.
Let $x = -4$; then $y = 2(-4)^2 - 4(-4) + 5 = 53$.
Let $x = -3$; then $y = 2(-3)^2 - 4(-3) + 5 = 35$.
Let $x = -2$; then $y = 2(-2)^2 - 4(-2) + 5 = 21$.
Let $x = -1$; then $y = 2(-1)^2 - 4(-1) + 5 = 11$.

Let $x = 0$; then $y = 2(0)^2 - 4(0) + 5 = 5$.
Let $x = 1$; then $y = 2(1)^2 - 4(1) + 5 = 3$.
Let $x = 2$; then $y = 2(2)^2 - 4(2) + 5 = 5$.
Let $x = 3$; then $y = 2(3)^2 - 4(3) + 5 = 11$.
Let $x = 4$; then $y = 2(4)^2 - 4(4) + 5 = 21$.
Let $x = 5$; then $y = 2(5)^2 - 4(5) + 5 = 35$.
Let $x = 6$; then $y = 2(6)^2 - 4(6) + 5 = 53$.
Let $x = 7$; then $y = 2(7)^2 - 4(7) + 5 = 75$.
Let $x = 8$; then $y = 2(8)^2 - 4(8) + 5 = 101$.
Let $x = 9$; then $y = 2(9)^2 - 4(9) + 5 = 131$.

x	−7	−6	−5	−4	−3	−2	−1	0	1	2	3	4	5	6	7	8	9
y	131	101	75	53	35	21	11	5	3	5	11	21	35	53	75	101	131

Then plot the points from the table. Disregard those coordinates that cannot be plotted. Connect the points with a smooth curved line, as shown in Figure 11.7.

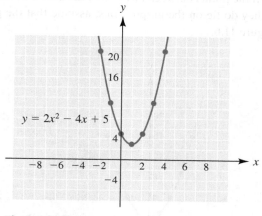

Figure 11.7

All parabolas have a **line of symmetry**, which means that a line can be drawn through a parabola dividing it into two parts that are mirror images of each other. (See Figure 11.8.)

The point of intersection of this line of symmetry and the graph of the parabola is called the **vertex**. In this section, the vertex is either the highest or lowest point of the parabola. Locating the vertex is most helpful in drawing the graph of a parabola.

The vertex of a parabola whose equation is in the form $y = ax^2 + bx + c$ may be found as follows:

Figure 11.8

The line of symmetry of a parabola divides the parabola into two parts that are mirror images of each other.

Vertex of a Parabola

1. The x coordinate is the value of $-\dfrac{b}{2a}$.

2. The y coordinate is found by substituting the x coordinate from Step 1 into the parabola equation and solving for y.

Example 3 Find the vertex of the parabola $y = 2x^2 - 4x + 5$ in Example 2.
Note that $a = 2$ and $b = -4$.

1. The x coordinate is $-\dfrac{b}{2a} = -\dfrac{-4}{2(2)} = 1$.

2. Substitute $x = 1$ into $y = 2x^2 - 4x + 5$ and solve for y.
$$y = 2(1)^2 - 4(1) + 5$$
$$= 2 - 4 + 5 = 3$$

Thus, the vertex is $(1, 3)$. ◆

Example 4 Graph the equation $y = -x^2 + 6x$.
First, find the vertex. The x coordinate is $-\dfrac{b}{2a} = -\dfrac{6}{2(-1)} = 3$. Then, substitute $x = 3$ into $y = -x^2 + 6x$.
$$y = -(3)^2 + 6(3)$$
$$y = -9 + 18 = 9$$
The vertex is $(3, 9)$. Graph the vertex in Figure 11.9.

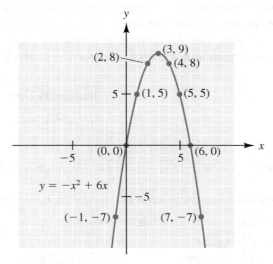

Figure 11.9

To find other points:

Let $x = 4$; then $y = -(4)^2 + 6(4) = -16 + 24 = 8$; graph $(4, 8)$.
Let $x = 5$; then $y = -(5)^2 + 6(5) = -25 + 30 = 5$; graph $(5, 5)$.
Let $x = 6$; then $y = -(6)^2 + 6(6) = -36 + 36 = 0$; graph $(6, 0)$.
Let $x = 7$; then $y = -(7)^2 + 6(7) = -49 + 42 = -7$; graph $(7, -7)$.

From symmetry, do you see that you can graph the points $(2, 8)$, $(1, 5)$, $(0, 0)$, and $(-1, -7)$ in Figure 11.9 without calculation? If not, let $x = 2$ and solve for y, etc. ◆

You may also note that the parabola in Figure 11.9 opens down. In general, the graph of $y = ax^2 + bx + c$

opens up with $a > 0$ and

opens down when $a < 0$.

EXERCISES 11.4

Draw the graph of each equation and label each vertex:

1. $y = 2x^2$
2. $y = -2x^2$
3. $y = \frac{1}{2}x^2$
4. $y = -\frac{1}{2}x^2$
5. $y = x^2 + 3$
6. $y = x^2 - 4$
7. $y = 2(x - 3)^2$
8. $y = -(x + 2)^2$
9. $y = x^2 - 2x + 1$
10. $y = 2(x + 1)^2 - 3$
11. $y = 2x^2 - 5$
12. $y = -3x^2 - 2x$
13. $y = x^2 - 2x - 5$
14. $y = -3x^2 + 6x + 15$
15. $y = x^2 - 2x - 15$
16. $y = 2x^2 - x - 15$
17. $y = -4x^2 - 5x + 9$
18. $y = 4x^2 - 12x + 9$
19. $y = \frac{1}{5}x^2 - \frac{2}{5}x + 4$
20. $y = -0.4x^2 + 2.4x + 0.7$

11.5 Imaginary Numbers

What is the meaning of $\sqrt{-4}$? What number squared is -4? Try to find its value on your calculator.

As you can see, this is a different kind of number. Up to now, we have considered only real numbers. The number $\sqrt{-4}$ is not a real number. The square root of a negative number is called an **imaginary number**. The imaginary unit is defined as $\sqrt{-1}$ and in many mathematics texts is given by the symbol *i*. However, in technical work, *i* is commonly used for current. To avoid confusion, many technical books use *j* for $\sqrt{-1}$, which is what we use in this book.

Imaginary Unit

$$\sqrt{-1} = j$$

Then $\sqrt{-4} = \sqrt{(-1)(4)} = (\sqrt{-1})(\sqrt{4}) = (j)(2)$ or $2j$.

Example 1 Express each number in terms of *j*: **a.** $\sqrt{-25}$, **b.** $\sqrt{-45}$, **c.** $\sqrt{-183}$.

a. $\sqrt{-25}$
$= \sqrt{(-1)(25)}$
$= (\sqrt{-1})(\sqrt{25})$
$= (j)(5)$ or $5j$

b. $\sqrt{-45}$
$= \sqrt{(-1)(45)}$
$= (\sqrt{-1})(\sqrt{45})$
$= (j)(6.71)$ or $6.71j$

c. $\sqrt{-183}$
$= \sqrt{(-1)(183)}$
$= (\sqrt{-1})(\sqrt{183})$
$= (j)(13.5)$ or $13.5j$

♦

Now let's consider powers of *j*, or $\sqrt{-1}$. Using the rules of exponents and the definition of *j*, carefully study the following powers of *j*:

$$j = j$$
$$j^2 = (\sqrt{-1})^2 = -1$$
$$j^3 = (j^2)(j) = (-1)(j) = -j$$
$$j^4 = (j^2)(j^2) = (-1)(-1) = 1$$
$$j^5 = (j^4)(j) = (1)(j) = j$$
$$j^6 = (j^4)(j^2) = (1)(-1) = -1$$
$$j^7 = (j^4)(j^3) = (1)(-j) = -j$$
$$j^8 = (j^4)^2 = 1^2 = 1$$

$$j^9 = (j^8)(j) = (1)(j) = j$$
$$j^{10} = (j^8)(j^2) = (1)(-1) = -1$$

As you can see, the values of j to a power repeat in the order of $j, -1, -j, 1, j, -1, -j, 1, \cdots$ Also, j to any power divisible by 4 equals 1.

Example 2 Simplify **a.** j^{15}, **b.** j^{21}, **c.** j^{72}.

a. $j^{15} = (j^{12})(j^3)$ **b.** $j^{21} = (j^{20})(j)$ **c.** $j^{72} = 1$
$\phantom{j^{15}} = (1)(-j) = -j$ $\phantom{j^{21}} = (1)(j) = j$

In general, we define an **imaginary number** as any number in the form bj, where b is a nonzero real number. We define a **complex number** as any number in the form $a + bj$, where a and b are real numbers. Note that in the general complex number $a + bj$, when $a = 0$ and b is nonzero, we have an imaginary number, and when $b = 0$, we have a real number. The following table contains examples of complex numbers, imaginary numbers, and real numbers:

Complex numbers	Imaginary numbers	Real numbers
$3 + 5j$	$8j$	9
$-2 - 6j$	$-12.5j$	$-5\frac{1}{2}$
$-1.6 + 4.44j$	$\frac{3}{4}j$	27.5
$\frac{1}{2} - \frac{1}{3}j$	$-0.322j$	-0.75

The solutions of the quadratic equation $ax^2 + bx + c = 0$ are given by the quadratic formula

$$x = \frac{-b \pm \sqrt{b^2 - 4ac}}{2a}$$

The part under the radical sign, $b^2 - 4ac$, is called the *discriminant*. The value of $b^2 - 4ac$ determines what kind of solutions (or roots) the quadratic equation has and how many solutions it has when a, b, and c are integers.

If $b^2 - 4ac$ is	Roots
positive and a perfect square,	both roots are rational.
positive and not a perfect square,	both roots are irrational.
zero,	there is only one rational root.
negative,	both roots are imaginary.

The relationship between the graph of $y = ax^2 + bx + c$ and the value of the discriminant is shown in Figure 11.10.

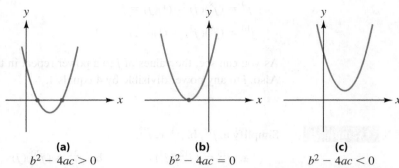

(a) $b^2 - 4ac > 0$
Two solutions, as indicated by the two points of intersection on the x axis

(b) $b^2 - 4ac = 0$
One solution, as indicated by the one point of intersection on the x axis

(c) $b^2 - 4ac < 0$
The graph does not cross the x axis; both roots are imaginary.

Figure 11.10

The value of the discriminant, $b^2 - 4ac$, determines the kinds and the number of solutions of the equation $y = ax^2 + bx + c$, where a, b, and c are integers.

Example 3 Determine the nature of the roots of $3x^2 + 5x - 2 = 0$ without solving the equation.

$a = 3, \quad b = 5, \quad c = -2$

The value of the discriminant is

$$b^2 - 4ac = (5)^2 - 4(3)(-2)$$
$$= 25 + 24 = 49$$

Since 49 is a perfect square, both roots are rational. ◆

Example 4 Determine the nature of the roots of $4x^2 - 12x + 9 = 0$ without solving the equation.

$a = 4, \quad b = -12, \quad c = 9$

The value of the discriminant is

$$b^2 - 4ac = (-12)^2 - 4(4)(9)$$
$$= 144 - 144 = 0$$

Therefore, there is only one rational root. ◆

Example 5 Determine the nature of the roots of $x^2 - 3x + 8 = 0$ without solving the equation.

$a = 1, \quad b = -3, \quad c = 8$

The value of the discriminant is

$$b^2 - 4ac = (-3)^2 - 4(1)(8)$$
$$= 9 - 32 = -23$$

Since -23 is negative, both roots are imaginary. ◆

Example 6 Solve $4x^2 - 6x + 5 = 0$ using the quadratic formula.

$$x = \frac{-b \pm \sqrt{b^2 - 4ac}}{2a}, \quad a = 4, \quad b = -6, \quad c = 5$$

$$x = \frac{-(-6) \pm \sqrt{(-6)^2 - 4(4)(5)}}{2(4)}$$

$$= \frac{6 \pm \sqrt{36 - 80}}{8}$$

$$= \frac{6 \pm \sqrt{-44}}{8}$$

$$= \frac{6 \pm 6.63j}{8} \qquad \sqrt{-44} = (\sqrt{-1})(\sqrt{44}) = 6.63j$$

$$= \frac{6 + 6.63j}{8} \quad \text{or} \quad \frac{6 - 6.63j}{8}$$

$$= 0.75 + 0.829j \quad \text{or} \quad 0.75 - 0.829j$$

The roots are $0.75 + 0.829j$ and $0.75 - 0.829j$. ◆

EXERCISES 11.5

Express each number in terms of j (when necessary, round the result to three significant digits):

1. $\sqrt{-49}$
2. $\sqrt{-64}$
3. $\sqrt{-14}$
4. $\sqrt{-5}$
5. $\sqrt{-2}$
6. $\sqrt{-3}$
7. $\sqrt{-56}$
8. $\sqrt{-121}$
9. $\sqrt{-169}$
10. $\sqrt{-60}$
11. $\sqrt{-27}$
12. $\sqrt{-40}$

Simplify:

13. j^3
14. j^6
15. j^{13}
16. j^{16}
17. j^{19}
18. j^{31}
19. j^{24}
20. j^{26}
21. j^{38}
22. j^{81}
23. $\dfrac{1}{j}$
24. $\dfrac{1}{j^6}$

Determine the nature of the roots of each quadratic equation without solving it:

25. $x^2 + 3x - 10 = 0$
26. $2x^2 - 7x + 3 = 0$
27. $5x^2 + 4x + 1 = 0$
28. $9x^2 + 12x + 4 = 0$
29. $3x + 1 = 2x^2$
30. $3x^2 = 4x - 8$
31. $2x^2 + 6 = x$
32. $2x^2 + 7x = 4$
33. $x^2 + 25 = 0$
34. $x^2 - 4 = 0$

Solve each quadratic equation using the quadratic formula (when necessary, round results to three significant digits):

35. $x^2 - 6x + 10 = 0$
36. $x^2 - x + 2 = 0$
37. $x^2 - 14x + 53 = 0$
38. $x^2 + 10x + 34 = 0$
39. $x^2 + 8x + 41 = 0$
40. $x^2 - 6x + 13 = 0$
41. $6x^2 + 5x + 8 = 0$
42. $4x^2 + 3x - 1 = 0$
43. $3x^2 = 6x - 7$
44. $5x^2 + 2x = -3$
45. $5x^2 + 8x + 4 = 0$
46. $2x^2 + x + 3 = 0$
47. $5x^2 + 14x = 3$
48. $2x^2 + 1 = x$
49. $x^2 + x + 1 = 0$
50. $12x^2 + 23x + 10 = 0$

SUMMARY | CHAPTER 11

Glossary of Basic Terms

Complex number. Any number in the form $a + bj$, where a and b are real numbers. (p. 357)

Discriminant. The quantity under the radical sign in the quadratic formula, $b^2 - 4ac$. (p. 347)

Imaginary number. Any number in the form bj, where b is a nonzero real number; the square root of a negative number is an imaginary number. The imaginary unit is defined as $\sqrt{-1} = j$. (p. 356)

Line of symmetry of a parabola. A *line of symmetry* can be drawn through a parabola, dividing it into two parts that are mirror images of each other. (p. 354)

Parabola. The graph of the quadratic equation $y = ax^2 + bx + c$, where a, b, and c are real numbers and $a \neq 0$. (p. 352)

Quadratic equation. A quadratic equation in one variable is an equation in the form $ax^2 + bx + c = 0$, where $a \neq 0$. A quadratic equation has at most two solutions. (p. 343)

Vertex of a parabola. The point of intersection of the line of symmetry and the graph of the parabola. (p. 354)

11.1 Solving Quadratic Equations by Factoring

1. **Solving quadratic equations by factoring:**
   a. If necessary, write an equivalent equation in the form $ax^2 + bx + c = 0$.
   b. Factor the polynomial.
   c. Write equations by setting each factor containing a variable equal to zero.
   d. Solve the two resulting first-degree equations.
   e. Check. (p. 343)

11.2 The Quadratic Formula

1. **Quadratic formula:** The roots of a quadratic equation in the form $ax^2 + bx + c = 0$ may be found by using the formula $x = \dfrac{-b \pm \sqrt{b^2 - 4ac}}{2a}$. (p. 345)

11.4 Graphs of Quadratic Equations

1. **Graphing quadratic equations:** To graph a quadratic equation with two variables, find and plot points whose ordered pairs satisfy the equation by choosing various values of x and solving for y. Then connect the points with a smooth curve. Finding additional points may be necessary.

2. **Vertex of a parabola:** The coordinates of the vertex of a parabola may be found as follows:
   a. The x coordinate is the value of $\dfrac{-b}{2a}$.
   b. The y coordinate is found by substituting the x coordinate from Step a into the parabola equation and solving for y. (p. 354)

11.5 Imaginary Numbers

1. The roots of a quadratic equation in the form $ax^2 + bx + c = 0$, where a, b, and c are integers, may be described by using the value of the discriminant as follows. (p. 356)

If $b^2 - 4ac$ is	Roots
positive and a perfect square,	both roots are rational.
positive and not a perfect square,	both roots are irrational.
zero,	there is only one rational root.
negative,	both roots are imaginary.

REVIEW | CHAPTER 11

1. If $ab = 0$, what is known about either a or b?
2. Solve for x: $3x(x - 2) = 0$

Solve each equation by factoring:

3. $x^2 - 4 = 0$
4. $x^2 - x = 6$
5. $5x^2 - 6x = 0$
6. $x^2 - 3x - 28 = 0$
7. $x^2 - 14x = -45$
8. $x^2 - 18 - 3x = 0$
9. $3x^2 + 20x + 32 = 0$

Solve each equation using the quadratic formula (when necessary, round results to three significant digits):

10. $3x^2 - 16x - 12 = 0$
11. $x^2 + 7x - 5 = 0$
12. $2x^2 + x = 15$
13. $x^2 - 4x = 2$
14. $3x^2 - 4x = 5$

15. The area of a piece of plywood is 36 ft². Its length is 5 ft more than its width. Find its length and width.

16. A variable electric current is given by the formula $i = t^2 - 12t + 36$, where t is in μs. At what times is the current i equal to **a.** 4 A? **b.** 0 A? **c.** 10 A?

Draw the graph of each equation and label each vertex:

17. $y = x^2 - x - 6$
18. $y = -3x^2 + 2$

Express each number in terms of j:

19. $\sqrt{-36}$
20. $\sqrt{-73}$

Simplify:

21. j^{12}
22. j^{27}

Determine the nature of the roots of each quadratic equation without solving it:

23. $9x^2 + 30x + 25 = 0$
24. $3x^2 - 2x + 4 = 0$

Solve each equation using the quadratic formula (when necessary, round results to three significant digits):

25. $x^2 - 4x + 5 = 0$
26. $5x^2 - 6x + 4 = 0$

27. A solar-heated house has a rectangular heat collector with a length 1 ft more than three times its width. The area of the collector is 21.25 ft². Find its length and width.

28. A rectangular opening is 15 in. wide and 26 in. long. (See Illustration 1.) A strip of constant width is to be removed from around the opening to increase the area to 672 in². How wide must the strip be?

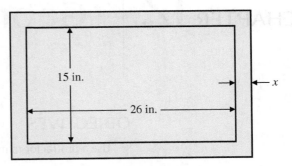

ILLUSTRATION 1

TEST | CHAPTER 11

Solve each equation:

1. $x^2 = 64$
2. $x^2 - 8x = 0$
3. $x^2 + 9x - 36 = 0$
4. $12x^2 + 4x = 1$

Solve each equation using the quadratic formula (when necessary, round results to three significant digits):

5. $5x^2 + 6x - 10 = 0$
6. $3x^2 = 4x + 9$

Solve each equation (when necessary, round results to three significant digits):

7. $21x^2 - 29x - 10 = 0$
8. $5x^2 - 7x = 2$
9. $3x^2 - 39x + 90 = 0$
10. $6x^2 = 8x + 5$
11. Draw the graph of $y = -x^2 - 8x - 15$ and label the vertex.
12. Draw the graph of $y = 2x^2 + 8x + 11$ and label the vertex.

Express each number in terms of j:

13. $\sqrt{-16}$
14. $\sqrt{-29}$

Simplify:

15. j^9
16. j^{28}
17. Determine the nature of the roots of $3x^2 - x + 4 = 0$ without solving it.
18. One side of a rectangle is 5 cm more than another. Its area is 204 cm². Find its length and width.

CHAPTER 12 | Geometry

OBJECTIVES

- Use a protractor to measure an angle.
- Apply the basic definitions and relationships for angles, lines, and geometric figures to solve application problems.
- Find the area and perimeter of quadrilaterals and triangles.
- Use the Pythagorean theorem to find the side of a right triangle when two sides are known.
- Use the relationships of similar polygons to solve application problems.
- Find the area and circumference of circles.
- Use the relationships of chords, secants, and tangent lines of a circle, arcs of a circle, and inscribed and central angles to solve application problems.
- Use radian measure to solve application problems.
- Find the volume, the lateral surface area, and the total surface area of prisms, cylinders, pyramids, cones, and spheres.

Mathematics at Work

Automotive service technicians inspect, maintain, and repair automobiles, light trucks, and vans. In the past, these workers were called mechanics. The increasing sophistication of automotive technology now requires workers to be able to use computerized shop equipment and work with electronic components in addition to the traditional hand tools. When a mechanical or electronic problem occurs, the technician uses a diagnostic approach to repair the problem based on information from the owner and the information obtained from the service equipment and computerized databases and service manuals.

The National Automotive Technicians Education Foundation (NATEF), an affiliate of the National Institute for Automotive Service Excellence (ASE), certifies automotive service technician, collision repair and refinish technician, engine specialist, and medium/heavy truck technician training programs offered by community colleges, postsecondary trade schools, technical institutes, and high schools. Although voluntary, NATEF certification signifies that the program meets uniform standards for instructional facilities, equipment, staff credentials, and curriculum. Various automobile manufacturers and their participating dealers also sponsor 2-year associate degree programs at postsecondary schools across the United States. For more information, please visit **www.cengage.com** and access the Student Online Resources for this text.

Automotive Service Technician
Automotive service technician working on an automobile

12.1 Angles and Polygons

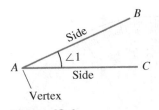

Figure 12.1
Basic parts of an angle

Some fundamentals of geometry must be understood in order to solve many technical applications. Geometry is also needed to follow some of the mathematical developments in technical mathematics courses, in technical support courses, and in on-the-job training programs. Here, we will cover the most basic and most often used geometric terms and relationships. **Plane geometry** is the study of the properties, measurement, and relationships of points, angles, lines, and curves in two dimensions: length and width.

An **angle** is formed by two lines that have a common point. The common point is called the **vertex** of the angle. The parts of the lines are called the **sides of the angle**. An angle is designated by a number, by a single letter, or by three letters. For example, the angle in Figure 12.1 is referred to as $\angle A$ or $\angle BAC$. The middle letter of the three letters is always the one at the vertex.

The measure of an angle is the amount of rotation needed to make one side coincide with the other side. The measure can be expressed in any one of many different units. The standard metric unit of plane angles is the radian (rad). While the radian is the metric unit of angle measurement, many ordinary measurements continue to be made in degrees (°). Although some trades subdivide the degree into the traditional minutes and seconds, most others use tenths and hundredths of degrees. Radian measure is developed in Section 12.6.

One degree is $\frac{1}{360}$ of one complete revolution; that is, $360° =$ one revolution. The protractor in Figure 12.2 is an instrument, marked in degrees, used to measure angles.

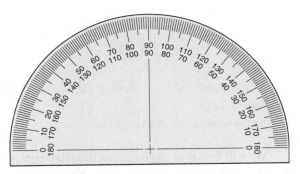

Figure 12.2
Protractor

Using a Protractor

STEP 1 Place the protractor so that the center mark on its base coincides with the vertex of the angle and so that the 0° mark is on one side of the angle.

STEP 2 Read the mark on the protractor that is on the other side of the angle (extended, if necessary).

   a. If the side of the angle under the 0° mark extends to the *right* from the vertex, read the inner scale to find the degree measure.

   b. If the side of the angle under the 0° mark extends to the *left* from the vertex, read the outer scale to find the degree measure.

Example 1 Find the degree measure of the angle in Figure 12.3.
The measure of the angle is 38°, using Step 2(a).

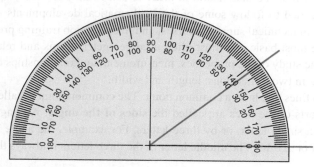

Figure 12.3

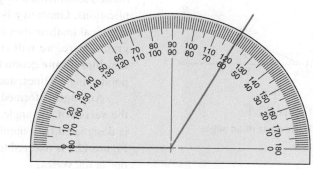

Figure 12.4

Example 2 Find the degree measure of the angle in Figure 12.4.
The measure of the angle is 120°, using Step 2(b).

Angles are often classified by degree measure. A **right angle** is an angle with a measure of 90°. In a sketch or diagram, a 90° angle is often noted by placing ⌐ or ¬ in the angle, as shown in Figure 12.5. An **acute angle** is an angle with a measure less than 90°. An **obtuse angle** is an angle with a measure greater than 90° but less than 180°.

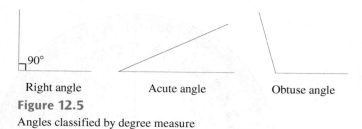

Right angle   Acute angle   Obtuse angle
Figure 12.5
Angles classified by degree measure

We will first study some geometric relationships of angles and lines in the same plane. Two lines **intersect** if they have only one point in common. (See Figure 12.6.)

Two lines in the same plane are **parallel** (∥) if they do not intersect even when extended. (See Figure 12.7.)

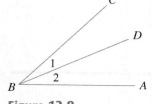

Figure 12.8
Adjacent angles

Figure 12.6
Lines l and m intersect at point A.

Figure 12.7
Lines l and m are parallel.

Two angles are **adjacent** if they have a common vertex and a common side lying between them. See Figure 12.8, where $\angle 1$ and $\angle 2$ are adjacent angles because they have a common vertex B and a common side BD between them. Angles CBA and CBD are not adjacent because although they have a common vertex, point B, their common side, BA, is not between the angles.

Two lines in the same plane are **perpendicular** (⊥) if they intersect and form equal adjacent angles. Each of these equal adjacent angles is a right angle. See Figure 12.9, where $l \perp m$ because $\angle 1 = \angle 2$. Angles 1 and 2 are right angles.

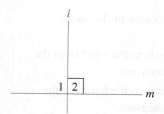

Figure 12.9
Perpendicular lines

Two angles are **complementary** if the sum of their measures is 90°. (See Figure 12.10.) Angles *A* and *B* in Figure 12.10(a) are complementary:

52° + 38° = 90°

Angles *LMN* and *NMP* in Figure 12.10(b) are complementary:

25° + 65° = 90°

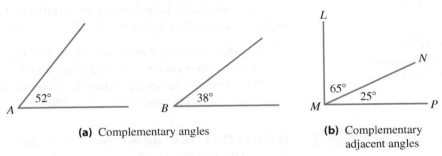

(a) Complementary angles (b) Complementary adjacent angles

Figure 12.10

Two angles are **supplementary** if the sum of their measures is 180°. (See Figure 12.11.) Angles *C* and *D* in Figure 12.11(a) are supplementary: 70° + 110° = 180°.

Two adjacent angles with their exterior sides in a straight line are *supplementary*. (See Figure 12.11(b).) Angles 1 and 2 have their exterior sides in a straight line, so they are supplementary: ∠1 + ∠2 = 180°.

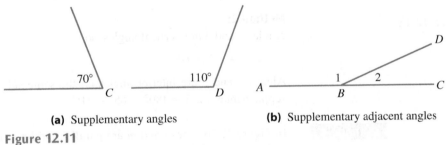

(a) Supplementary angles (b) Supplementary adjacent angles

Figure 12.11

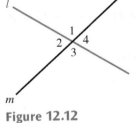

Figure 12.12
Vertical angles

When two lines intersect, the angles opposite each other are called **vertical angles**. (See Figure 12.12.) If two straight lines intersect, the vertical angles that are formed are equal. In Figure 12.12, angles 1 and 3 are vertical angles, so ∠1 = ∠3. Angles 2 and 4 are vertical angles, so ∠2 = ∠4.

A **transversal** is a line that intersects two or more lines in different points in the same plane (Figure 12.13). **Interior angles** are angles formed inside the lines by the transversal. Angles formed between lines *l* and *m* are called interior angles. **Exterior angles** are angles formed outside the lines by the transversal. Those angles outside lines *l* and *m* are called exterior angles.

Interior angles: ∠3, ∠4, ∠5, ∠6
Exterior angles: ∠1, ∠2, ∠7, ∠8

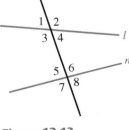

Figure 12.13
Line *t* is a transversal of lines *l* and *m*.

Exterior-interior angles with different vertices on the same side of the transversal are **corresponding angles**. For example, ∠3 and ∠7 in Figure 12.13 are corresponding angles. Angles 2 and 6 are also corresponding angles.

Pairs of angles with different vertices on opposite sides of the transversal but which are both interior or both exterior are **alternate angles**. Angles 3 and 6 in Figure 12.13 are alternate angles. Angles 1 and 8 are also alternate angles.

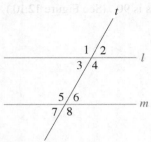

Figure 12.14
Line *t* is a transversal for parallel lines *l* and *m*.

If two *parallel* lines are cut by a transversal, then

- *corresponding* angles are *equal*.
- *alternate-interior* angles are *equal*.
- *alternate-exterior* angles are *equal*.
- *interior* angles on the same side of the transversal are *supplementary*.

In Figure 12.14, lines *l* and *m* are parallel and *t* is a transversal. The corresponding angles are equal. That is, $\angle 1 = \angle 5$, $\angle 2 = \angle 6$, $\angle 3 = \angle 7$, and $\angle 4 = \angle 8$.

The alternate-interior angles are equal. That is, $\angle 3 = \angle 6$ and $\angle 4 = \angle 5$.
The alternate-exterior angles are equal. That is, $\angle 1 = \angle 8$ and $\angle 2 = \angle 7$.
The interior angles on the same side of the transversal are supplementary. That is, $\angle 3 + \angle 5 = 180°$ and $\angle 4 + \angle 6 = 180°$.

Example 3

In Figure 12.15, lines *l* and *m* are parallel and line *t* is a transversal. The measure of $\angle 2$ is 65°. Find the measure of $\angle 5$.

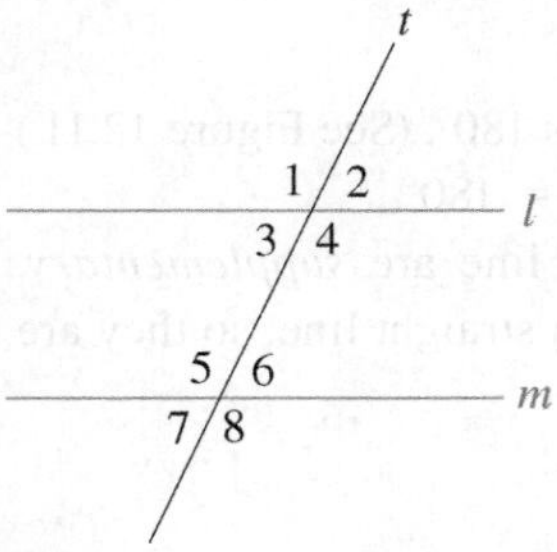

There are many ways of finding $\angle 5$. We show two ways.

Method 1:
Angles 2 and 4 are supplementary, so
$$\angle 4 = 180° - 65° = 115°$$
Angles 4 and 5 are alternate-interior angles, so
$$\angle 4 = \angle 5 = 115°$$

Method 2:
Angles 2 and 3 are vertical angles, so
$$\angle 2 = \angle 3 = 65°$$
Angles 3 and 5 are interior angles on the same side of the transversal; therefore, they are supplementary. $\angle 5 = 180° - 65° = 115°$

Figure 12.15

Example 4

In Figure 12.16, lines *l* and *m* are parallel and line *t* is a transversal. Given $\angle 2 = 2x + 10$ and $\angle 3 = 3x - 5$, find the measure of $\angle 1$.

Angles 2 and 3 are alternate-interior angles, so they are equal. That is,
$$2x + 10 = 3x - 5$$
$$15 = x$$

Then $\angle 2 = 2x + 10 = 2(15) + 10 = 40°$. Since $\angle 1$ and $\angle 2$ are supplementary, $\angle 1 = 180° - 40° = 140°$.

Figure 12.16

NOTE: We let $\overline{AB}$ be the line segment with endpoints at *A* and *B*:

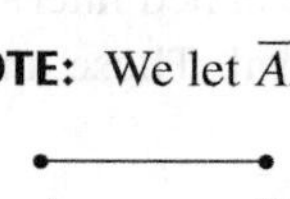

•——————•
A   B

Let $\overleftrightarrow{AB}$ be the line containing *A* and *B*:

←—•——————•—→
   A   B

And let AB be the length of $\overline{AB}$.

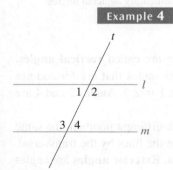

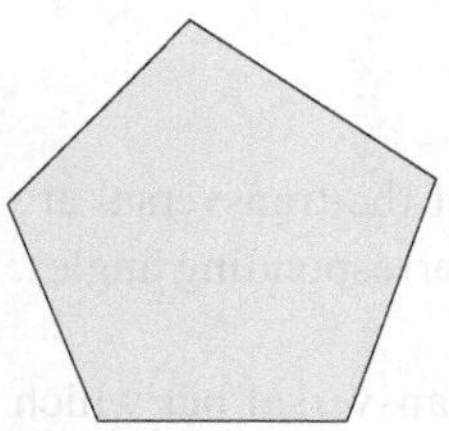

Polygon
Figure 12.17

A **polygon** is a closed figure whose sides are straight line segments. A polygon is shown in Figure 12.17. Polygons are named according to the number of sides they have (see Figure 12.18). A **triangle** is a polygon with three sides. A **quadrilateral** is a polygon with four sides. A **pentagon** is a polygon with five sides. A **regular polygon** has all its sides and interior angles equal.

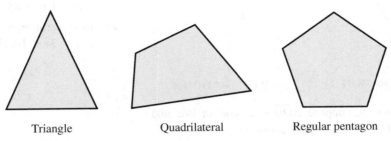

Triangle   Quadrilateral   Regular pentagon

Figure 12.18
A polygon is named according to the number of its sides.

Some polygons with more than five sides are named as follows.

Number of sides	Name of polygon
6	**Hexagon**
7	**Heptagon**
8	**Octagon**
9	**Nonagon**

EXERCISES 12.1

Classify each angle as right, acute, or obtuse:

1.
2.
3.
4.

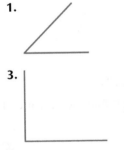

5.
6.

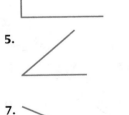
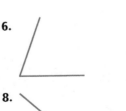

7.
8.

9. In Illustration 1, line l intersects line m and forms a right angle. Then $\angle 1$ is a(n) ___?___ angle. Lines l and m are ___?___.

10. Suppose $l \parallel m$ and $t \perp l$. Is $t \perp m$? Why or why not?

11. In Illustration 2, **a.** name the pairs of adjacent angles; **b.** name the pairs of vertical angles.

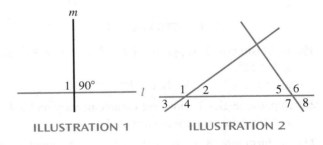

ILLUSTRATION 1   ILLUSTRATION 2

12. In Illustration 2, suppose $\angle 3 = 40°$ and $\angle 7 = 97°$. Find the measures of the other angles.

13. In Illustration 3, suppose $l \parallel m$ and $\angle 1 = 57°$. What are the measures of the other angles?

14. In Illustration 4, suppose $a \parallel b$, $a \perp c$, and $\angle 1 = 37°$. Find the measures of angles 2 and 3.

368 CHAPTER 12 ♦ Geometry

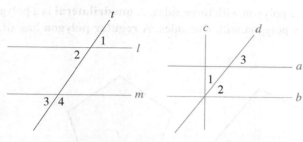

ILLUSTRATION 3   ILLUSTRATION 4

15. In Illustration 5, suppose $\overleftrightarrow{AOB}$ is a straight line and $\angle AOC = 119°$. What is the measure of $\angle COB$?

ILLUSTRATION 5

16. Suppose angles 1 and 2 are supplementary and $\angle 1 = 63°$. Then $\angle 2 = $?

17. Suppose angles 3 and 4 are complementary and $\angle 3 = 38°$. Then $\angle 4 = $?

18. In Illustration 6, suppose $l \parallel m$, $\overleftrightarrow{AOB}$ is a straight line, and $\angle 3 = \angle 6 = 68°$. Find the measure of each of the other angles.

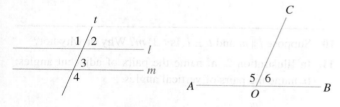

ILLUSTRATION 6

19. In Illustration 7, suppose $a \parallel b$, $t \parallel x$, $\angle 3 = 38°$, and $\angle 1 = 52°$.
  a. Is $x \perp a$?   **b.** Is $x \perp b$?

20. Suppose angles 1 and 2 are complementary and $\angle 1 = \angle 2$. Find the measure of each angle.

21. In Illustration 8, suppose $\angle 1$ and $\angle 3$ are supplementary. Find the measure of each angle.

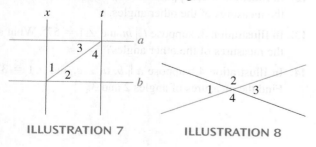

ILLUSTRATION 7   ILLUSTRATION 8

22. In Illustration 9, suppose $l \parallel m$, $\angle 1 = 3x - 50$, and $\angle 2 = x + 60$. Find the value of x.

23. In Illustration 9, suppose $l \parallel m$, $\angle 1 = 4x + 55$, and $\angle 3 = 10x - 85$. Find the value of x.

24. In Illustration 9, suppose $l \parallel m$, $\angle 1 = 8x + 60$, and $\angle 4 = 3x + 10$. Find the value of x.

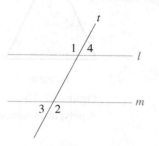

ILLUSTRATION 9

25. A plumber wishes to add a pipe parallel to an existing pipe as shown in Illustration 10. Find angle x.

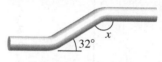

ILLUSTRATION 10

26. A machinist needs to weld a piece of iron parallel to an existing piece of iron as shown in Illustration 11. What is angle y?

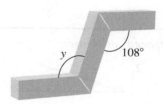

ILLUSTRATION 11

27. In Illustration 12, find angle z if $m \parallel n$.

28. Given $\overline{AB} \parallel \overline{CD}$ in Illustration 13, find the measure of
  a. angle 1, **b.** angle 2, **c.** angle 3.

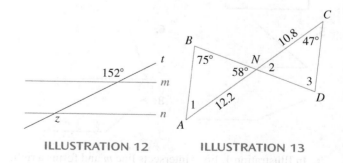

ILLUSTRATION 12   ILLUSTRATION 13

Name each polygon:

29.
30.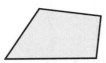
33.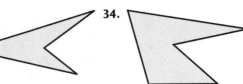
34.
31.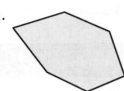
32.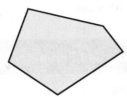
35.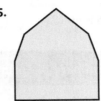
36.

12.2 Quadrilaterals

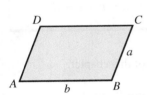

Figure 12.19
Parallelogram

A **parallelogram** is a quadrilateral with opposite sides parallel. In Figure 12.19, sides $\overline{AB}$ and $\overline{CD}$ are parallel, and sides $\overline{AD}$ and $\overline{BC}$ are parallel. Polygon $ABCD$ is therefore a parallelogram.

Figure 12.20(a) shows the same parallelogram with a perpendicular line segment drawn from point D to side $\overline{AB}$. This line segment is an **altitude**.

Figure 12.20(b) shows the result of removing the triangle at the left side of the parallelogram and placing it at the right side. You now have a rectangle with sides of lengths b and h. Note that the area of this rectangle, bh square units, is the same as the area of the parallelogram. So the area of a parallelogram is given by the formula $A = bh$, where b is the length of the base and h is the length of the altitude drawn to that base. The perimeter is $2a + 2b$ or $2(a + b)$.

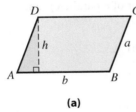 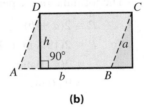

(a) (b)

Figure 12.20

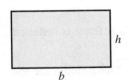

Figure 12.21
Rectangle

A **rectangle** is a parallelogram with four right angles. The area of the rectangle with sides of lengths b and h is given by the formula $A = bh$. (See Figure 12.21.)

Another way to find the area of a rectangle is to count the number of square units in it. In Figure 12.22, there are 15 squares in the rectangle, so the area is 15 square units.

The formula for the area of each of the following quadrilaterals follows from the formula for the area of a rectangle.

A **square** (Figure 12.23) is a rectangle with the lengths of all four sides equal. Its area is given by the formula $A = b \cdot b = b^2$. The perimeter is $b + b + b + b$, or $4b$. Note that the length of the altitude is also b.

A **rhombus** (Figure 12.24) is a parallelogram with the lengths of all four sides equal. Its area is given by the formula $A = bh$. The perimeter is $b + b + b + b$, or $4b$.

A **trapezoid** (Figure 12.25) is a quadrilateral with only two sides parallel. Its area is given by the formula $A = \left(\dfrac{a + b}{2}\right)h$. The perimeter is $a + b + c + d$.

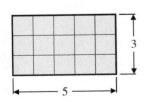

Figure 12.22

Figure 12.23
Square

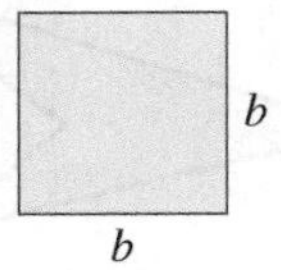

Figure 12.24
Rhombus

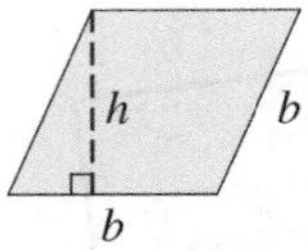

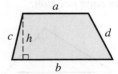
Figure 12.25
Trapezoid

Summary of Formulas for Area and Perimeter of Quadrilaterals

Quadrilateral	Area	Perimeter
Rectangle	$A = bh$	$P = 2(b + h)$
Square	$A = b^2$	$P = 4b$
Parallelogram	$A = bh$	$P = 2(a + b)$
Rhombus	$A = bh$	$P = 4b$
Trapezoid	$A = \left(\dfrac{a + b}{2}\right)h$	$P = a + b + c + d$

NOTE: Follow the rules for working with measurements in the rest of this chapter.

Example 1 Find the area and the perimeter of the parallelogram shown in Figure 12.26.

The formula for the area of a parallelogram is

$A = bh$

So $A = (27.2 \text{ m})(15.5 \text{ m})$

$\quad\ = 422 \text{ m}^2$

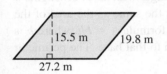
Figure 12.26

The formula for the perimeter of a parallelogram is

$P = 2(a + b)$

So $P = 2(19.8 \text{ m} + 27.2 \text{ m})$

$\quad\ = 2(47.0 \text{ m})$

$\quad\ = 94.0 \text{ m}$ ♦

Example 2 A rectangular lot 121.5 ft by 98.7 ft must be fenced. (See Figure 12.27.) A fence is installed for $7.50 per running foot. Find the cost of fencing the lot.

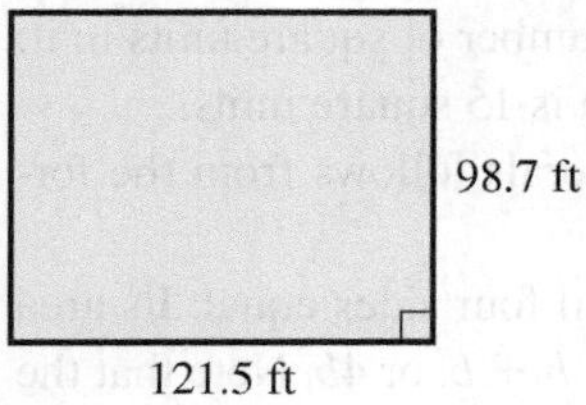

The length of fencing needed equals the perimeter of the rectangle. The formula for the perimeter is

$P = 2(b + h)$

So $P = 2(121.5 \text{ ft} + 98.7 \text{ ft})$

$\quad\ = 2(220.2 \text{ ft})$

$\quad\ = 440.4 \text{ ft}$

Figure 12.27

Cost $= \dfrac{\$7.50}{1 \text{ ft}} \times 440.4 \text{ ft} = \$3303$ ♦

Example 3 Find the cost of the fertilizer needed for the lawn in Example 2. One bag covers 2500 ft^2 and costs $18.95.

First, find the area of the rectangle. The formula for the area is

$A = bh$

So $A = (121.5 \text{ ft})(98.7 \text{ ft})$

$ = 12{,}0\overline{0}0 \text{ ft}^2$

The amount of fertilizer needed is found by dividing the total area by the area covered by one bag:

$$\frac{12{,}0\overline{0}0 \text{ ft}^2}{2500 \text{ ft}^2} = 4.8 \text{ bags, or } 5 \text{ bags.}$$

$$\text{Cost} = 5 \text{ bags} \times \frac{\$18.95}{1 \text{ bag}}$$

$$\phantom{\text{Cost}} = \$94.75$$

EXERCISES 12.2

Find the perimeter and the area of each quadrilateral:

1. 15.0 cm square (15.0 cm × 15.0 cm)

2. Rectangle 10.0 cm × 8.00 cm

3. Parallelogram, base 8.0 m, side 8.0 m, height 6.0 m

4. Parallelogram, base 10.0 dm, side 7.0 dm, height 5.0 dm

5. Trapezoid, top 10.0 m, bottom 24.5 m, left side 8.0 m, right side 11.0 m, height 6.0 m

6. Rectangle 18.6 m × 13.5 m

7. Parallelogram, base 23.9 in., side 20.8 in., height 17.2 in.

8. Parallelogram, base 13.7 m, side 13.7 m, height 11.9 m

9. Square 9.2 cm × 9.2 cm

10. Trapezoid, top 13.5 m, bottom 17.5 m, left side 8.01 m, height 6.91 m

In Exercises 11–12, use the formula $A = bh$.

11. $A = 24\overline{0} \text{ cm}^2$, $b = 10.0$ cm; find h.

12. $A = 792 \text{ m}^2$, $h = 25.0$ m; find b.

13. The area of a parallelogram is 486 ft². The length of its base is 36.2 ft. Find its height.

14. The area of a rectangle is 280 cm². Its width is 14 cm. Find its length.

15. ✴ A piece of 16-gauge steel has been cut into the shape of a trapezoid with height 16.0 in. and bases 21.0 in. and 23.0 in. What is the area of the trapezoidal piece of steel?

16. ✴ Looking at the side of a welded metal storage bin, the shape is a trapezoid. The lengths of the bases are 52.3 cm and 68.3 cm, and the height is 41.4 cm. Find its area.

17. ✈ On a sectional chart used for aviation navigation, a military operating zone has the shape of a trapezoid. One base is 20.0 mi in length; the other base is 14.0 mi. The lengths of the other two sides are 12.0 mi and 13.42 mi.
   a. What is the perimeter of the military operating zone?
   b. What is its area if the distance between the parallel sides is 11.6 mi?

18. ✈ A pilot flies from an airport to a VOR (Very high frequency Omnidirectional Range) site 82.0 mi away, then 55.0 mi on to another airport. After this, the pilot flies 82.0 mi to an NDB (Non-Directional Beacon) station and then back to the original airport. If the quadrilateral shape flown is a parallelogram, what is the distance from the NDB station back to the airport and what is the total distance of this trip?

19. ▦ A rectangular metal duct has a width of 8.0 in. If the area of a cross section is 128 in², what is the height? What is the perimeter of a cross section?

20. A rectangular X-ray film measures 15 cm by 32 cm. What is its area?

21. Each hospital bed and its accessories use 96 ft^2 of floor space. How many beds can be placed in a ward 24 ft by 36 ft?

22. A respirator unit needs a rectangular floor space of 0.79 m by 1.2 m. How many units could be placed in a storeroom having $2\overline{0}$ m^2 of floor space?

23. A 108-ft^2 roll of fiberglass is 36 in. wide. What is its length in feet?

24. The cost of the fiberglass in Exercise 23 is $9.16/yd^2. How much would this roll cost?

25. How many pieces of fiberglass, 36 in. wide and 72 in. long, can be cut from the roll in Exercise 23?

26. A business owner plans to build a storage garage for 85 automobiles. Each automobile needs a space of 15.0 ft by 10.0 ft. **a.** Find the floor area of the garage. **b.** At a cost of $14/ft^2, find the cost of the garage.

27. The rear view mirror of an automobile measures 2.9 in. high and has length 9.75 in. What is the area of the rear view mirror if the shape is a parallelogram?

28. A machinist plans to build a screen around his shop area. The area is rectangular and measures 16.2 ft by 20.7 ft. **a.** How many linear feet of screen will be needed? **b.** If the screen is to be 8.0 ft high, how many square feet of screen will be needed?

29. A piece of sheet metal in the shape of a parallelogram has a rectangular hole in it, as shown in Illustration 1. Find **a.** the area of the piece that was punched out, and **b.** the area of the metal that is left.

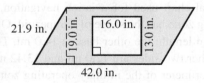

ILLUSTRATION 1

30. A rectangular piece of sheet metal has an area of 10,680 in^2. Its length is 72.0 in. Find its width.

31. What is the acreage of a ranch that is a square, 25.0 mi on a side?

32. A rectangular field of corn is averaging 115 bu/acre. The field measures 1020 yd by 928 yd. How many bushels of corn will there be?

33. Find the amount of sheathing needed for the roof in Illustration 2. How many squares of shingles must be purchased? (1 square = 100 ft^2.)

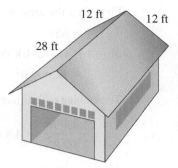

ILLUSTRATION 2

34. A ceiling is 12 ft by 15 ft. How many 1-ft by 3-ft suspension panels are needed to cover the ceiling?

35. The Smith family plan to paint the exterior of their home (shown in Illustration 3). The area of the openings not to be painted is 325 ft^2. The cost per square foot is $0.85. Find the cost of painting the house.

ILLUSTRATION 3

36. An 8-in.-thick wall uses 15 standard bricks (8 in. by $2\frac{1}{4}$ in. by $3\frac{3}{4}$ in.) for 1 square foot. (This includes mortar.) Find the number of bricks needed for a wall 8 in. thick, 18.0 ft long, and 8.5 ft high.

37. What is the display floor space of a parallelogram-shaped space that is 29.0 ft long and 8.7 ft deep?

38. Canvas that costs $\frac{3}{4}$¢/in^2 is used to make golf bags. Find the cost of 200 rectangular pieces of canvas, each 8 in. by 40 in.

39. By law, all businesses outside the Parkville city limits must fence their lots. How many feet of fence will be needed to fence the parallelogram-shaped lot shown in Illustration 4? If chain link fence costs $9.50 per foot installed, find the cost of the fence installed without gates.

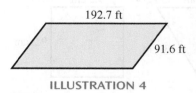

ILLUSTRATION 4

40. Find the area of the three trapezoid-shaped display floor spaces of the stores shown in Illustration 5.

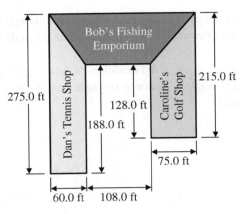

ILLUSTRATION 5

41. In a plant, an inside walkway is being laid out to go from one corner of the building along the perimeter to the opposite corner, where the offices are located. If the building is 80.0 ft × 100.0 ft and the walkway is 4.00 ft wide, how many square feet of the building are unavailable for manufacturing?

42. A rectangular lot is 155 ft × 175 ft. The house, driveway, and walks cover 7100 ft². What percent of the lot is lawn?

43. In the mid-20th century, engineers constructed a series of canals to move irrigation water from water sources to farming communities throughout the western United States. The volume of water a canal can move is partially dependent on its cross-sectional area. Suppose a canal is trapezoidal, 25.0 ft across the top and 15.0 ft across the bottom, with a planned depth of water of 10.0 ft as shown in Illustration 6. Find the cross-sectional area of the canal when it is full.

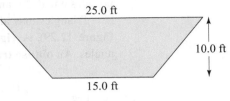

ILLUSTRATION 6

44. A game preserve manager fences off a pasture alongside a road with the dimensions shown in Illustration 7. Find the area of the pasture in acres.

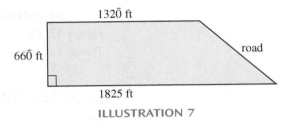

ILLUSTRATION 7

45. A plot of land in the shape of a parallelogram contains 27,800 ft². The frontage along the road is 265 ft as shown in Illustration 8. How deep is the lot?

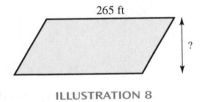

ILLUSTRATION 8

12.3 Triangles

Triangles are often classified in two ways:

1. by the number of equal sides
2. by the measures of the angles of the triangle

Triangles may be classified or named by the relative lengths of their sides. In each triangle in Figure 12.28, the lengths of the sides are represented by a, b, and c. An **equilateral**

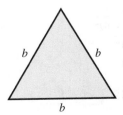 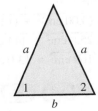 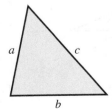

(a) Equilateral triangle   (b) Isosceles triangle   (c) Scalene triangle

Figure 12.28
Triangles named by sides

triangle has all three sides equal. All three angles are also equal. An **isosceles triangle** has two sides equal. The angles opposite these two sides are also equal. A **scalene triangle** has no sides equal. No angles are equal either.

Triangles may also be classified or named in terms of the measures of their angles (see Figure 12.29). A **right triangle** has one right angle. An **acute triangle** has three acute angles. An **obtuse triangle** has one obtuse angle.

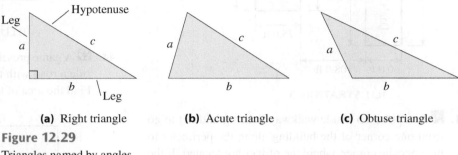

(a) Right triangle (b) Acute triangle (c) Obtuse triangle

Figure 12.29
Triangles named by angles

Pythagorean Theorem

In a right triangle, the side opposite the right angle is called the **hypotenuse**, which we label c. The other two sides, the sides opposite the acute angles, are called **legs**, which we label a and b. (See Figure 12.29a.) The **Pythagorean theorem** relates the lengths of the sides of any right triangle as follows:

> **Pythagorean Theorem**
>
> $$c^2 = a^2 + b^2 \quad \text{or} \quad c = \sqrt{a^2 + b^2}$$

The Pythagorean theorem states that *the square of the hypotenuse of a right triangle is equal to the sum of the squares of the lengths of the two legs.* Alternative forms of the Pythagorean theorem are

$$a^2 = c^2 - b^2 \quad \text{or} \quad a = \sqrt{c^2 - b^2}$$
and $$b^2 = c^2 - a^2 \quad \text{or} \quad b = \sqrt{c^2 - a^2}$$

Before we use the Pythagorean theorem, you may wish to review square roots in Section 1.15.

Example 1 Find the length of the hypotenuse of the triangle in Figure 12.30.

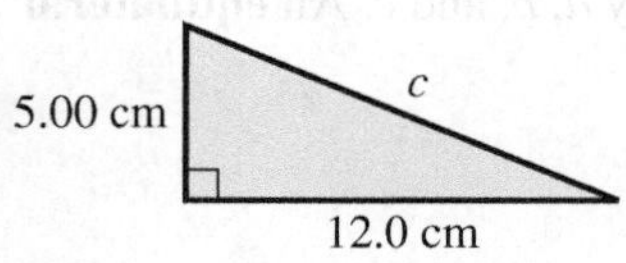

Substitute 5.00 cm for a and 12.0 cm for b in the formula:

$$c = \sqrt{a^2 + b^2}$$
$$c = \sqrt{(5.00 \text{ cm})^2 + (12.0 \text{ cm})^2}$$
$$= \sqrt{25.0 \text{ cm}^2 + 144 \text{ cm}^2}$$
$$= \sqrt{169 \text{ cm}^2} = 13.0 \text{ cm}$$

Figure 12.30

NOTE: You may need to use parentheses with some calculators.

Example 2 Find the length of the hypotenuse of the triangle in Figure 12.31.

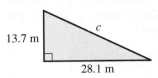

Figure 12.31

$$c = \sqrt{a^2 + b^2}$$
$$c = \sqrt{(13.7 \text{ m})^2 + (28.1 \text{ m})^2}$$
$$= 31.3 \text{ m}$$

◆

Example 3 Find the length of side b of the triangle in Figure 12.32.

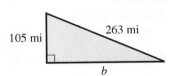

Figure 12.32

$$b = \sqrt{c^2 - a^2}$$
$$b = \sqrt{(263 \text{ mi})^2 - (105 \text{ mi})^2}$$
$$= 241 \text{ mi}$$

◆

Example 4 The right triangle in Figure 12.33 gives the relationship in a circuit among the applied voltage, the voltage across a resistance, and the voltage across a coil. The voltage across the resistance is 79 V. The voltage across the coil is 82 V. Find the applied voltage.

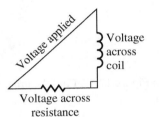

Figure 12.33

Using the Pythagorean theorem, we have

$$\text{voltage applied} = \sqrt{(\text{voltage across coil})^2 + (\text{voltage across resistance})^2}$$
$$= \sqrt{(82 \text{ V})^2 + (79 \text{ V})^2}$$
$$= 110 \text{ V}$$

◆

Perimeter and Area

To find the perimeter of a triangle, find the sum of the lengths of the three sides. The formula is $P = a + b + c$, where P is the perimeter and a, b, and c are the lengths of the sides.

An **altitude** of a triangle is a line segment drawn perpendicular from one vertex to the opposite side. Sometimes this opposite side must be extended. See Figure 12.34.

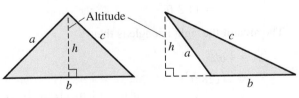

Figure 12.34

Look closely at a parallelogram (Figure 12.35(a)) to find the formula for the area of a triangle. Remember that the area of a parallelogram with sides of lengths a and b is given by $A = bh$. In this formula, b is the length of the base of the parallelogram, and h is the length of the altitude drawn to that base.

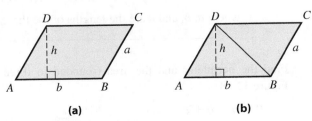

Figure 12.35

Next, draw a line segment from B to D in the parallelogram as in Figure 12.35(b). Two triangles are formed. We know from geometry that these two triangles have equal areas. Since the area of the parallelogram is bh square units, the area of one triangle is one-half the area of the parallelogram. So the formula for the area of a triangle is

Area of Triangle

$$A = \frac{1}{2}bh$$

where b is the length of the base of the triangle (the side to which the altitude is drawn) and h is the length of the altitude.

Example 5

The length of the base of a triangle is 10.0 cm. The length of the altitude to that base is 6.00 cm. Find the area of the triangle. (See Figure 12.36.)

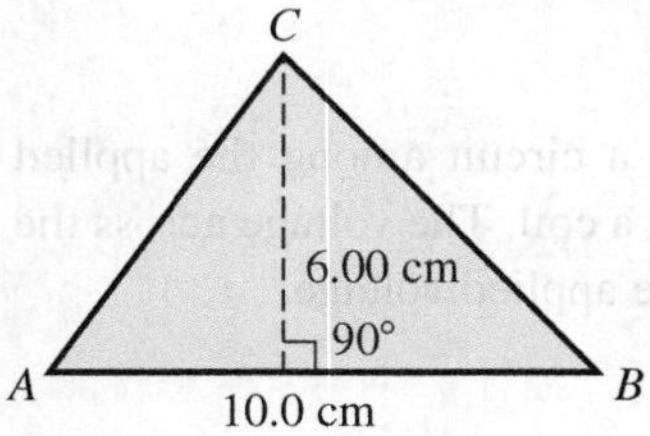

$$A = \frac{1}{2}bh$$

$$A = \frac{1}{2}(10.0 \text{ cm})(6.00 \text{ cm})$$

$$= 30.0 \text{ cm}^2$$

The area is 30.0 cm².

Figure 12.36

◆

Example 6

In the corner of an office, a 16-ft-long counter is built as shown in Figure 12.37. Find the area behind the counter if the two walls behind it are of equal length.

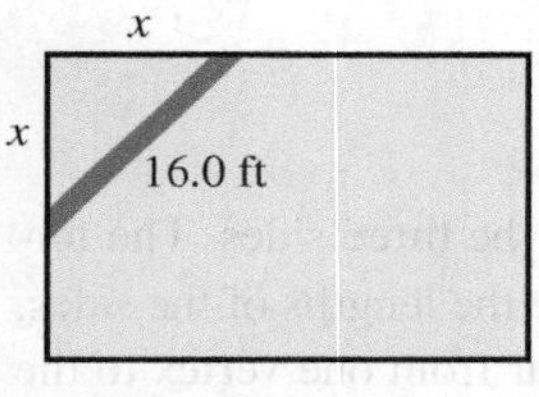

First, let x = the length of each wall behind the counter. Using the Pythagorean theorem, we have

$$a^2 + b^2 = c^2$$

$$x^2 + x^2 = (16.0 \text{ ft})^2$$

$$2x^2 = 256 \text{ ft}^2$$

$$x^2 = 128 \text{ ft}^2 \qquad \text{Divide both sides by 2.}$$

$$x = 11.3 \text{ ft} \qquad \text{Take the square root of both sides.}$$

Figure 12.37

The area of the right triangle is then

$$A = \tfrac{1}{2} bh$$

$$= \tfrac{1}{2}(128 \text{ ft}^2) \qquad \text{Since } b = h = x \text{ in this right triangle, we can replace } bh \text{ by } x^2 = 128 \text{ ft}^2 \text{ or replace each by 11.3 ft. The answers may differ due to rounding.}$$

$$= 64.0 \text{ ft}^2$$

◆

If only the lengths of the three sides are known, the area of a triangle is found by the following formula (called Heron's formula):

$$A = \sqrt{s(s-a)(s-b)(s-c)}$$

where a, b, and c are the lengths of the three sides and $s = \tfrac{1}{2}(a + b + c)$.

Example 7

Find the perimeter and the area (rounded to three significant digits) of the triangle in Figure 12.38.

$$P = a + b + c$$

$$P = 9 \text{ cm} + 15 \text{ cm} + 18 \text{ cm} = 42 \text{ cm}$$

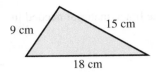

Figure 12.38

To find the area, first find s.

$$s = \frac{1}{2}(a + b + c)$$
$$s = \frac{1}{2}(9 + 15 + 18) = \frac{1}{2}(42) = 21$$
$$A = \sqrt{s(s-a)(s-b)(s-c)}$$
$$A = \sqrt{21(21-9)(21-15)(21-18)}$$
$$= \sqrt{21(12)(6)(3)}$$
$$= \sqrt{4536}$$
$$= 67.3 \text{ cm}^2$$

67.34983296

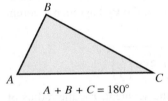

Figure 12.39

The following relationship is often used in geometry and trigonometry:

> The sum of the measures of the angles of any triangle is 180° (Figure 12.39).

Example 8

Two angles of a triangle have measures 80° and 40°. (See Figure 12.40.) Find the measure of the third angle of the triangle.

Since the sum of the measures of the angles of any triangle is 180°, we know that

$$40° + 80° + x = 180°$$
$$120° + x = 180°$$
$$x = 60°$$

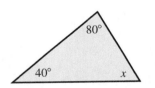

Figure 12.40

So the measure of the missing angle is 60°.

EXERCISES 12.3

Use the rules for working with measurements. Find the length of the hypotenuse in each triangle:

1.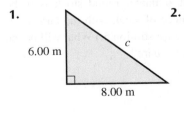

2.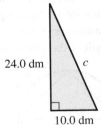

3., **4.**, **5.**, **6.**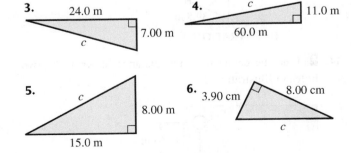

Note: You may need to insert this right parenthesis to clarify the order of operations. The square root key may also include the left parenthesis; if not, you need to key it in.

Find the length of the missing side in each triangle:

7.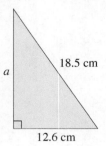

8.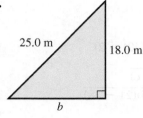

9.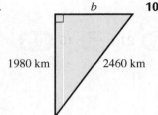

10.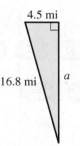

11.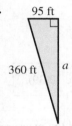

12.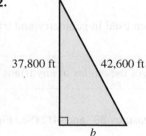

13. Find the length of the braces needed for the rectangular supports shown in Illustration 1.

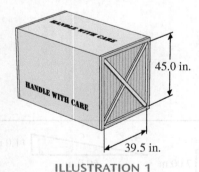

ILLUSTRATION 1

14. Find the center-to-center distance between the two holes in Illustration 2.

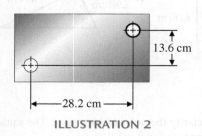

ILLUSTRATION 2

15. Find the total length of the brace material needed in Illustration 3.

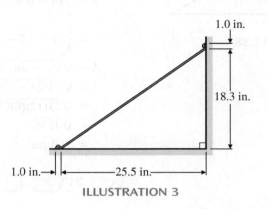

ILLUSTRATION 3

16. Often, a machinist must cut a keyway in a shaft. The total depth of cut equals the keyway depth plus the height of a circular segment. The height of a circular segment is found by applying the Pythagorean theorem or by using the formula

$$h = r - \sqrt{r^2 - \left(\frac{l}{2}\right)^2}$$

where h is the height of the segment, r is the radius of the shaft, and l is the length of the chord (or the width of the keyway). Find the total depth of cut shown in Illustration 4.

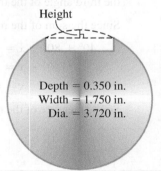

ILLUSTRATION 4

17. A piece of 4.00-in.-diameter round stock is to be milled into a square piece of stock with the largest dimensions possible. (See Illustration 5.) What will be the length of the side of the square?

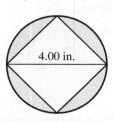

ILLUSTRATION 5

18. Find the length of the rafter in Illustration 6.

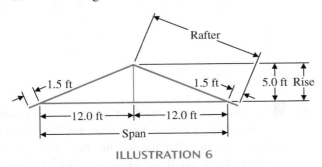

ILLUSTRATION 6

19. Find the offset distance x (rounded to nearest tenth of an inch) of the 6-ft length of pipe shown in Illustration 7.

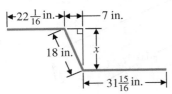

ILLUSTRATION 7

20. A conduit is run in a building (see Illustration 8).
    a. Find the length of the conduit from A_1 to A_6.
    b. Find the straight-line distance from A_1 to A_6.

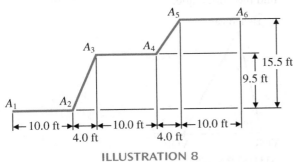

ILLUSTRATION 8

21. The voltage across a resistance is 85.2 V. The voltage across a coil is 78.4 V. Find the voltage applied in the circuit. (See Illustration 9.)

22. The voltage across a coil is 362 V. The voltage applied is 537 V. Find the voltage across the resistance. (See Illustration 9.)

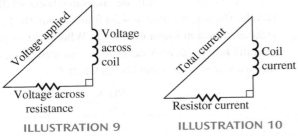

ILLUSTRATION 9   ILLUSTRATION 10

23. The resistor current is 24 A. The total current is 32 A. Find the coil current. (See Illustration 10.)

24. The resistor current is 50.2 A. The coil current is 65.3 A. Find the total current. (See Illustration 10.)

In Exercises 25–27, see Illustration 11.

25. Find the reactance of a circuit with impedance 165 Ω and resistance 105 Ω.

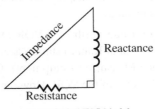

ILLUSTRATION 11

26. Find the impedance of a circuit with reactance 20.2 Ω and resistance 38.3 Ω.

27. Find the resistance of a circuit with impedance 4.5 Ω and reactance 3.7 Ω.

28. The base of a window is 7.2 m above the ground. The lower end of a ladder is 3.1 m from the side of the house. How long must a ladder be to reach the base of the window?

Find the area and perimeter of each isosceles triangle:

29.   30.

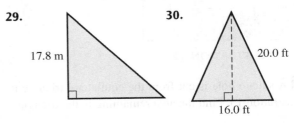

31. Find the area and perimeter of an equilateral triangle with one side 6.00 cm long.

32. Find the area and perimeter of an equilateral triangle with one side 18.0 m long.

Find the area and perimeter of each triangle:

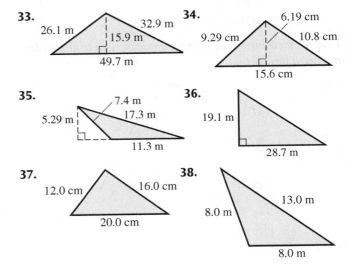

39. Two pieces of steel angle are welded to form right angles. The lengths of the two pieces are 6.0 ft and 9.0 ft, respectively. What is the distance between the two unwelded ends?

40. A triangular gusset (a triangular metal bracket to strengthen a joist) is 11.0 in. in height and 14.4 in. across the base. What is its area?

41. A helicopter is 62.0 mi due north of a VOR (Very high frequency Omnidirectional Range) station according to its DME (Distance Measuring Equipment). One hour later it is 41.0 mi due east of the VOR. How far has the helicopter flown?

42. An unusual architectural design requires triangular ducts that will be painted and exposed in the room. If the cross-sectional area is an equilateral triangle 3.6 ft², find the length of each side.

43. A steel plate is punched with a triangular hole as shown in Illustration 12. Find the area of the hole.

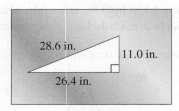

ILLUSTRATION 12

44. A square hole is cut from the equilateral triangle in Illustration 13. Find the area remaining in the triangle.

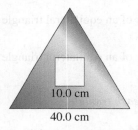

ILLUSTRATION 13

Find the measure of the missing angle in each triangle (do not use a protractor):

45.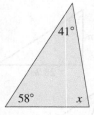

46.

47.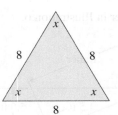

48.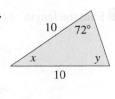

49. In a shaded corner outside of a manufacturing building, a decorative shrub garden as shown in Illustration 14 is to be planted. After the soil has been tilled, it will need to be fertilized. If one bag covers 75 ft², how many bags will be needed?

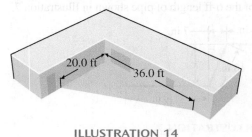

ILLUSTRATION 14

50. A helicopter is at a position from two VORs as in the diagram. Given the angles as shown in Illustration 15, find the third angle.

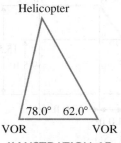

ILLUSTRATION 15

51. Slope may be defined as the steepness of an area and is sometimes given as a percent. For example, a hillside that falls 5 ft for each 100 ft of level length is said to have a 5% slope. The steeper an area of land, the more it is subject to erosion. On a hillside, a conservationist drove a stake into the ground at one point and a second stake 50.0 ft down the hillside, measured horizontally (level). The second stake is 4.50 ft lower than the first stake as shown in Illustration 16. **a.** What is the slope of the hill? **b.** If a sidewalk needs to be installed between the two stakes, what would be its length?

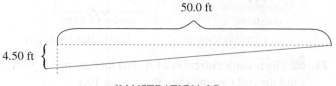

ILLUSTRATION 16

52. ✦ Starting at point A, a hiker walks due south for 7.00 mi to point B, then she walks due east for 5.00 mi to point C. How far must she walk to return directly from point C to point A?

53. ✦ When measuring land in odd-shaped lots, we often divide the lots into rectangles, triangles, and other geometric figures and calculate each area accordingly. For a lot with dimensions shown in Illustration 17, find its total area in acres.

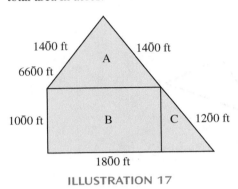

ILLUSTRATION 17

54. ✦ A local manufacturer keeps cows on site to eat the grass in given areas. A new water trough is needed to replace the old one. Find the amount of material needed to build a new trough as shown in Illustration 18.

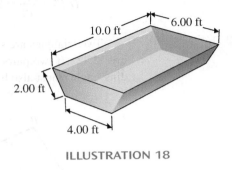

ILLUSTRATION 18

12.4 Similar Polygons

Polygons with the same shape are called **similar polygons**. Polygons are similar when the *corresponding angles are equal*. In Figure 12.41, polygon $ABCDE$ is similar to polygon $A'B'C'D'E'$ because the corresponding angles are equal. That is, $\angle A = \angle A'$, $\angle B = \angle B'$, $\angle C = \angle C'$, $\angle D = \angle D'$, and $\angle E = \angle E'$.

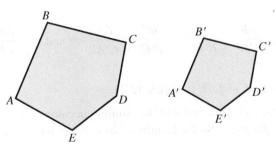

Figure 12.41
Similar polygons

When two polygons are similar, the lengths of the *corresponding sides are proportional*. That is,

$$\frac{AB}{A'B'} = \frac{BC}{B'C'} = \frac{CD}{C'D'} = \frac{DE}{D'E'} = \frac{EA}{E'A'}$$

Example 1 The polygons in Figure 12.41 are similar, and $AB = 12$, $DE = 3$, $A'B' = 8$. Find $D'E'$.

Since the polygons are similar,

$$\frac{AB}{A'B'} = \frac{DE}{D'E'}$$

$$\frac{12}{8} = \frac{3}{D'E'}$$

$12(D'E') = (8)(3)$ The product of the means equals the product of the extremes.

$$D'E' = \frac{24}{12} = 2$$

Two triangles are similar when two pairs of corresponding angles are equal, as in Figure 12.42. (If two pairs of corresponding angles are equal, then the third pair of corresponding angles must also be equal. Why?)

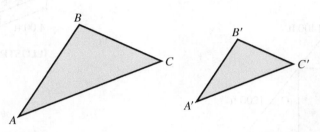

Figure 12.42
Similar triangles

Triangle ABC is similar to triangle $A'B'C'$ because $\angle A = \angle A'$, $\angle B = \angle B'$, and $\angle C = \angle C'$.

When two triangles are similar, the lengths of the corresponding sides are proportional. That is,

$$\frac{AB}{A'B'} = \frac{BC}{B'C'} = \frac{CA}{C'A'}$$

or

$$\frac{AB}{A'B'} = \frac{BC}{B'C'} \qquad \frac{BC}{B'C'} = \frac{CA}{C'A'} \quad \text{and} \quad \frac{AB}{A'B'} = \frac{CA}{C'A'}$$

Example 2 Find DE and AE in Figure 12.43.

Triangles ADE and ABC are similar, because $\angle A$ is common to both and each triangle has a right angle. So the lengths of the corresponding sides are proportional.

$$\frac{AB}{AD} = \frac{BC}{DE}$$

$$\frac{24}{20} = \frac{18}{DE}$$

$24(DE) = (20)(18)$

$$DE = \frac{360}{24} = 15$$

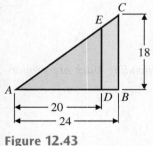

Figure 12.43

Use the Pythagorean theorem to find AE:

$$AE = \sqrt{(AD)^2 + (DE)^2}$$
$$AE = \sqrt{(20)^2 + (15)^2}$$
$$= \sqrt{400 + 225} = \sqrt{625} = 25$$

EXERCISES 12.4

Follow the rules for working with measurements beginning with Exercise 6.

1. In Illustration 1, suppose that $\overline{DE} \parallel \overline{BC}$. Find DE.

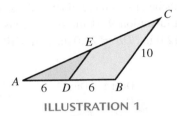

ILLUSTRATION 1

2. In Illustration 2, polygon $ABCD$ is similar to polygon $FGHI$. Find **a.** $\angle H$, **b.** FI, **c.** IH, **d.** BC.

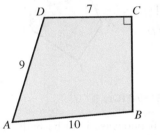 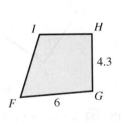

ILLUSTRATION 2

3. In Illustration 3, $\overline{AB} \parallel \overline{CD}$. Is triangle ABO similar to triangle DCO? Why or why not?

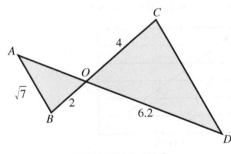

ILLUSTRATION 3

4. Find the lengths of $\overline{AO}$ and $\overline{CD}$ in Illustration 3.
5. In Illustration 4, quadrilaterals $ABCD$ and $XYZW$ are similar rectangles. $AB = 12$, $BC = 8$, and $XY = 8$. Find YZ.

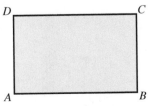 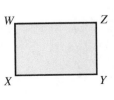

ILLUSTRATION 4

6. An inclined ramp is to be built so that it reaches a height of 6.00 ft over a 15.00-ft run. (See Illustration 5.) Braces are placed every 5.00 ft. Find the height of braces x and y.

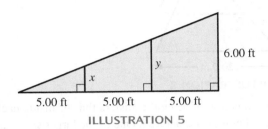

ILLUSTRATION 5

7. A tree casts a shadow $8\overline{0}$ ft long when a vertical rod 6.0 ft high casts a shadow 4.0 ft long. (See Illustration 6.) How tall is the tree?

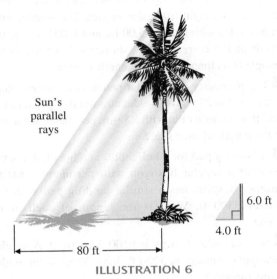

ILLUSTRATION 6

8. A machinist must follow a part drawing with scale 1 to 16. Find the dimensions of the finished stock shown in Illustration 7. That is, find lengths A, B, C, and D.

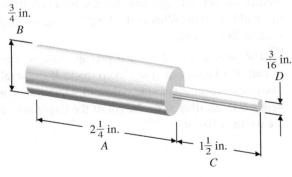

ILLUSTRATION 7

9. Find **a.** the length of *DE* and **b.** the length of *BC* in Illustration 8.

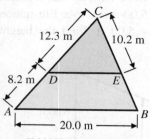

ILLUSTRATION 8

10. A collection of several canisters that fit inside each other is being manufactured. One of the larger sizes has a diameter of 6.00 in. and is 9.00 in. high. If one of the smaller sizes has a diameter of 4.00 in., what should its height be?

11. A right triangular support gusset is to be made similar to another right triangular gusset. The smaller gusset has sides with lengths 5.00 in. and 8.00 in. Find the length of the corresponding shorter side of the larger triangle if its longer side has length 17.0 in.

12. The perimeter of a regular pentagonal-shaped piece of flat steel is 25 in. If a welder cuts another piece of flat steel that is similar but with a 55-in. perimeter, what will be the length of each side?

13. A landing pad for a helicopter at a hospital has the shape of a regular hexagon with perimeter $300\overline{0}$ ft. Another hospital has a similar landing pad with perimeter $330\overline{0}$ ft. What is the length of each of its sides?

14. A rectangular runway is 6100 ft by 61 ft. A similar rectangular runway is 5200 ft. long. What is the width of this runway?

15. An older automobile has a fan belt that is the shape of an isosceles triangle. The two equal sides are 10.0 in. each, and the third side is 6.0 in. An older truck has a similar fan belt arrangement, but the isosceles triangle has width 12.0 in. What is the length of the two equal sides of this triangle?

16. The side mirror of a small pickup truck is similar to that of a larger full-size pickup truck. If the smaller truck has a rectangular side mirror of width 5.0 in. and height 8.0 in., what is the height of the larger mirror if the width is 10.0 in.?

17. A small heater has a rectangular filter that is 16 in. by 20 in. Another larger heater requires a similar filter that has a 48-in. width. What is the length of this larger filter?

18. A polygon cross-sectional duct is to be exposed and painted. It is to be attached to a smaller duct of the same shape as shown in Illustration 9. If the dimensions of the ducts are $AB = 12.0$ in., $DE = 20.0$ in., and $A'B' = 9.00$ in., find $D'E'$.

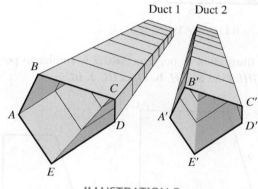

ILLUSTRATION 9

19. A 6.00-ft by 8.00-ft bookcase is to be built. It has horizontal shelves every foot. A support is to be notched in the shelves diagonally from one corner to the opposite corner. At what point should each of the shelves be notched? That is, find lengths *A*, *B*, *C*, *D*, and *E* in Illustration 10. How long is the crosspiece?

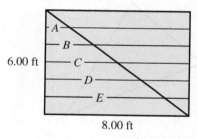

ILLUSTRATION 10

20. A vertical tower 132.0 ft high is anchored to the ground by guy wires as shown in Illustration 11. How long is each guy wire?

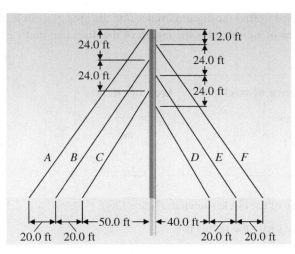

ILLUSTRATION 11

21. Illustration 12 shows the master mold for a symmetrical part for a model truck production. We need to reduce the dimensions to $\frac{1}{4}$ scale; that is, 1 in. = $\frac{1}{4}$ in. **a.** Find each given dimension to the new scale. **b.** At 2.00 oz/in^3, what is the weight of the finished model part?

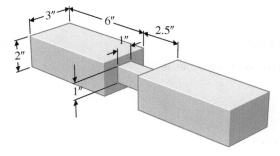

ILLUSTRATION 12

22. A team of rock climbers needs to estimate the height of a near-vertical cliff face. One holds a rod that is 36 in. tall vertically on the ground and measures its shadow as 25 in. long. Simultaneously, two other members of the group measure the length of the shadow cast by the cliff to be 117 ft from a point they estimate to be directly beneath the top of the cliff, as shown in Illustration 13. Approximately how tall is the cliff?

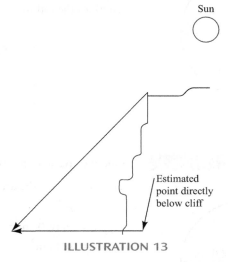

ILLUSTRATION 13

23. A cat owner has a scratching post that is shaped as a right triangle as shown in Illustration 14. It is 24 in. tall with a base of 26 in. She wants to build a larger scratching post in the same shape that will be 36 in. tall. Lumber is sold in even-foot lengths (6 ft, 8 ft, 10 ft, etc.). Find the lengths of the three sides of the larger scratching post and determine what length of lumber she must purchase to build it.

ILLUSTRATION 14

12.5 Circles

A **circle** is a plane curve consisting of all points at a given distance (called the **radius**, r) from a fixed point in the plane, called the **center**. (See Figure 12.44.) The **diameter**, d, of the circle is a line segment through the center of the circle with endpoints on the circle. Note that the length of the diameter equals the length of two radii—that is, $d = 2r$.

The **circumference** of a circle is the distance around the circle. The ratio of the circumference of a circle to the length of its diameter is a constant called π (pi). The number π cannot be written exactly as a decimal. Decimal approximations for π are 3.14 or 3.1416. When solving problems with π, use the π key on your calculator.

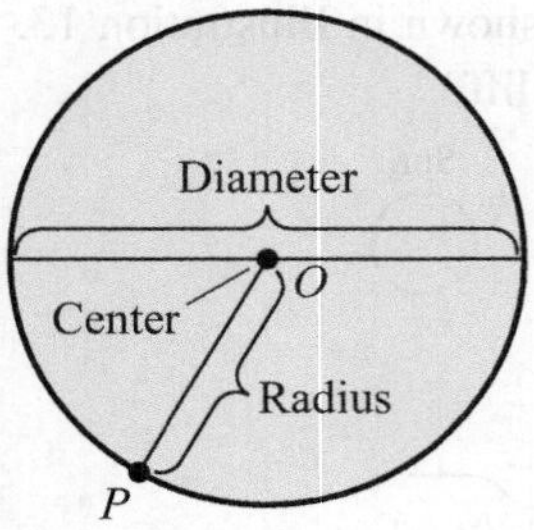

Figure 12.44
Circle

The following formulas are used to find the circumference and the area of a circle. C is the circumference and A is the area of a circle; d is the length of the diameter, and r is the length of the radius:

Circumference of circle:	Area of circle:
$C = 2\pi r$	$A = \pi r^2$
$C = \pi d$	$A = \dfrac{\pi d^2}{4}$

Example 1 Find the area and the circumference of the circle shown in Figure 12.45.

The formula for the area of a circle given the radius is

$$A = \pi r^2$$
$$A = \pi(16.0 \text{ cm})^2$$
$$= 804 \text{ cm}^2$$

The formula for the circumference of a circle given the radius is

$$C = 2\pi r$$
$$C = 2\pi(16.0 \text{ cm})$$
$$= 101 \text{ cm}$$

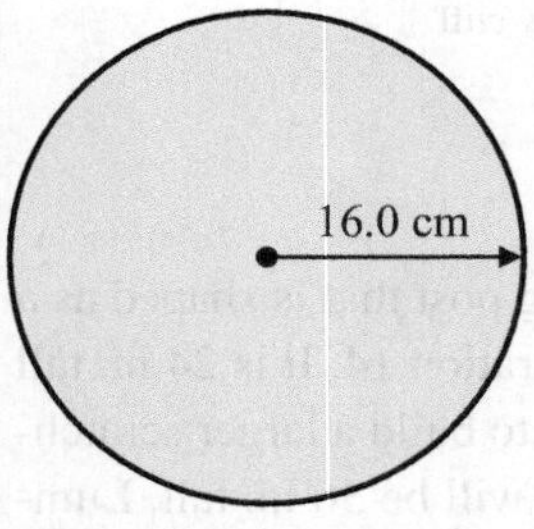

Figure 12.45

Example 2 The area of a circle is 576 m². Find the radius.

The formula for the area of a circle in terms of the radius is

$$A = \pi r^2$$
$$576 \text{ m}^2 = \pi r^2$$
$$\frac{576 \text{ m}^2}{\pi} = r^2 \quad \text{Divide both sides by } \pi.$$
$$\sqrt{\frac{576 \text{ m}^2}{\pi}} = r \quad \text{Take the square root of both sides.}$$
$$13.5 \text{ m} = r$$

Example 3 The circumference of a circle is 28.2 cm. Find the radius.

The formula for the circumference of a circle in terms of the radius is

$$C = 2\pi r$$
$$28.2 \text{ cm} = 2\pi r$$
$$\frac{28.2 \text{ cm}}{2\pi} = r \quad \text{Divide both sides by } 2\pi.$$
$$4.49 \text{ cm} = r$$

Example 4 The diameter of a circular table is 5.0 ft. How many people can be seated at the table if each person needs 22 in. for use?

The circumference is
$$C = \pi d$$
$$= \pi(5.0 \text{ ft})$$
$$= 15.7 \text{ ft}$$
$$= 188 \text{ in.}$$

Next, divide the circumference by the amount allotted per person; that is, 188 in. ÷ 22 in. = 8 people. ◆

An angle whose vertex is at the center of a circle is called a **central angle**. Angle A in Figure 12.46 is a central angle.

In general:

> The sum of the measures of all the central angles of any circle is 360°.

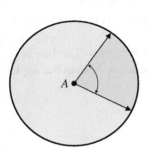

Figure 12.46
Central angle

Common Terms and Relationships of a Circle

A **chord** is a line segment that has its endpoints on the circle.

A **secant** is any line that intersects a circle at two points.

A **tangent** is a line that has only one point in common with a circle and lies totally outside the circle. A line segment along such a line with only one point in common with a circle is also considered tangent. In Figure 12.47, C is the center. $\overline{AB}$ is a chord. Line n is a secant. Line m is a tangent. $\overline{DE}$ is a diameter. $\overline{CF}$ is a radius.

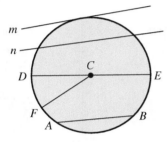

Figure 12.47

Arcs

An **inscribed angle** is an angle whose vertex is on the circle and whose sides are chords. The part of the circle between the two sides of an inscribed or central angle is called the **intercepted arc**. In Figure 12.48, C is the center and $\angle ACB$ is a central angle. $\angle DEF$ is an inscribed angle. $\overset{\frown}{AB}$ is the intercepted arc of $\angle ACB$. $\overset{\frown}{DF}$ is the intercepted arc of $\angle DEF$.

The following three relationships are often helpful to solve problems:

♦ The measure of a central angle in a circle is equal to the measure of its intercepted arc. (See Figure 12.49.)

♦ The measure of an inscribed angle in a circle is equal to one-half the measure of its intercepted arc. (See Figure 12.49.)

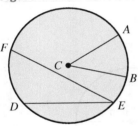

Figure 12.48
Arcs of a circle

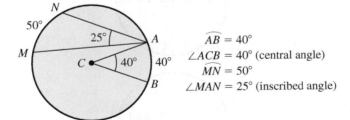

$\overset{\frown}{AB} = 40°$
$\angle ACB = 40°$ (central angle)
$\overset{\frown}{MN} = 50°$
$\angle MAN = 25°$ (inscribed angle)

Figure 12.49

♦ The measure of an angle formed by two intersecting chords in a circle is equal to one-half the sum of the measures of the intercepted arcs. (See Figure 12.50.)

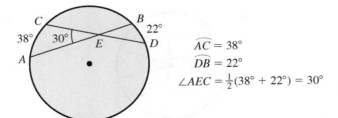

$\overset{\frown}{AC} = 38°$
$\overset{\frown}{DB} = 22°$
$\angle AEC = \frac{1}{2}(38° + 22°) = 30°$

Figure 12.50

388 CHAPTER 12 ♦ Geometry

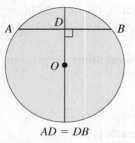

$AD = DB$

Figure 12.51

Other Chords and Tangents

A diameter that is perpendicular to a chord bisects the chord. (See Figure 12.51.)

A line segment from the center of a circle to the point of tangency is perpendicular to the tangent. (See Figure 12.52.)

Two tangents drawn from a point outside a circle to the circle are equal. The line segment drawn from the center of the circle to this point outside the circle bisects the angle formed by the tangents. (See Figure 12.53.)

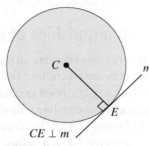

$CE \perp m$

Figure 12.52

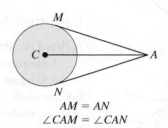

$AM = AN$
$\angle CAM = \angle CAN$

Figure 12.53

EXERCISES 12.5

Follow the rules for working with measurements.

*Find **a.** the circumference and **b.** the area of each circle:*

1.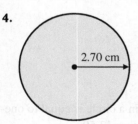
   5.00 in.

2.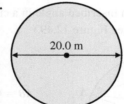
   20.0 m

3.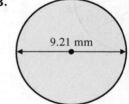
   9.21 mm

4.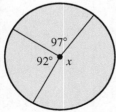
   2.70 cm

5.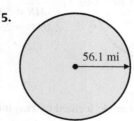
   56.1 mi

6.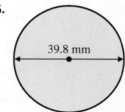
   39.8 mm

Find the measure of each unknown angle:

7.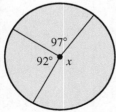
   97°, 92°, x

8.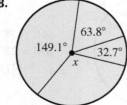
   149.1°, 63.8°, 32.7°, x

9.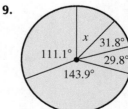
   111.1°, x, 31.8°, 29.8°, 143.9°

10. A round plate 12 in. in diameter is to have 10 holes drilled through it. To show this, draw half a circle and then "mirror" the other half as shown in Illustration 1. The $\frac{1}{2}$-in.-diameter holes are to be equally spaced on a 9-in. concentric circle (circles with the same center). What is angle A between the holes? Find angle B between the horizontal and each of the end holes.

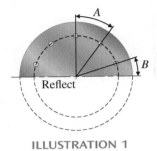

ILLUSTRATION 1

11. The area of a circle is 28.2 cm². Find its radius.
12. The area of a circle is 214 ft². Find its radius.
13. The circumference of a circle is 62.9 m. Find its radius.
14. The circumference of a circle is 17.2 in. Find its radius.
15. How many degrees are in a central angle whose arc is $\frac{1}{4}$ of a circle?
16. How many degrees are in a central angle whose arc is $\frac{2}{3}$ of a circle?
17. A welded circular metal tank has radius 24.0 in. A lid for this tank has the same radius. Find the area of the lid in square feet.
18. A circular hole is to be made in the side of a metal wall. If the area of the hole is to be 90.0 ft², what must the radius be?
19. The airspeed indicator of an airplane is circular with diameter 2.25 in. What are its area and circumference?
20. The side view of a tire resembles a doughnut. If the inner diameter is 15.0 in. and the outer diameter is 23.0 in., what is the area of the side of the tire?
21. If the rim diameter of a wheel of a vehicle is 16.0 in., what is its circumference?
22. Round metal duct has a cross-sectional area of 113 in². What is its diameter?
23. A wheel of radius 1.80 ft is used to measure a field. The wheel rotates 236 times while going the length of the field. How long is the field?
24. Find the length of the diameter of a circular silo with circumference 52.0 ft.

25. A rectangular piece of insulation is to be wrapped around a pipe 4.25 in. in diameter. (See Illustration 2.) How wide does the rectangular piece need to be?

ILLUSTRATION 2

26. How many 1.5-in.-diameter pipes will be needed to have approximately the same total cross-sectional area as one whose diameter is 5.0 in.?

27. A manifold is being designed to carry compressed gas from a tank to four processing stations where the gas is being used. (See Illustration 3.) The main line from the tank is 2.50 in. in diameter. The total cross-sectional area in the four outlet pipes must be the same as the cross-sectional area of the main line. For simplicity, we will not consider flow restriction due to friction, turbulence, or bends in the cylindrical lines. What diameter manifold discharge pipes are required?

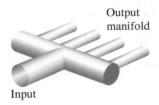

ILLUSTRATION 3

28. A pipe has a 3.50-in. outside diameter and a 3.25-in. inside diameter. (See Illustration 4.) Find the area of its cross section.

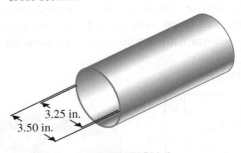

ILLUSTRATION 4

29. In Illustration 5, find the area of the rectangular piece of metal after the two circles are removed.

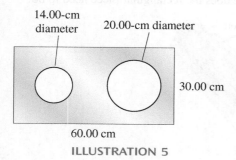

ILLUSTRATION 5

30. Find the area and perimeter of the figure in Illustration 6.

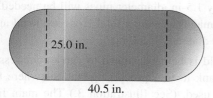

ILLUSTRATION 6

31. Find the length of strapping needed for the pipe in Illustration 7.

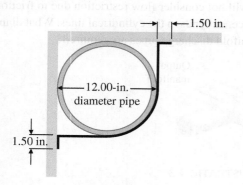

ILLUSTRATION 7

32. In a design for a workstation, in Illustration 8, each of the four circular sections is to be cut out and removed. **a.** Find the area of the workstation. **b.** How far would a worker have to reach to touch the center of the workstation?

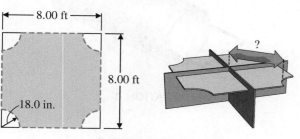

ILLUSTRATION 8

33. A boiler 5.00 ft in diameter is to be placed in a corner of a room shown in Illustration 9.
   a. How far from corner C are points A and B of the boiler?
   b. How long is a pipe from C to the center of the boiler M?

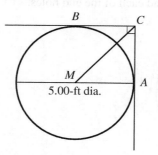

ILLUSTRATION 9

34. A pulley is connected to a spindle of a wheel by a belt. The distance from the spindle to the center of the pulley is 15.0 in. The diameter of the pulley is 15.0 in. What is the length of the belt? (See Illustration 10.)

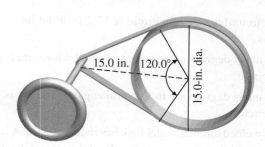

ILLUSTRATION 10

35. Mary needs to punch 5 equally spaced holes in a circular metal plate (see Illustration 11). Find the measure of each central angle.

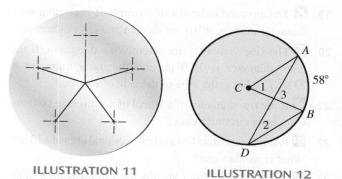

ILLUSTRATION 11   **ILLUSTRATION 12**

36. In Illustration 12, **a.** find the measure of $\angle 1$, where C is the center, **b.** find the measure of $\angle 2$, and **c.** find the measure of $\angle 3$, given that $\overline{AC} \parallel \overline{DB}$.

37. Find the measure of ∠1 in Illustration 13.

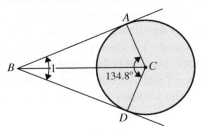

ILLUSTRATION 13

38. In Illustration 13, the length of $\overline{AC}$ is 5 and the distance between B and center C is 13. Find the length of $\overline{AB}$.

39. Illustration 14 shows a satellite at position P relative to a strange planet of radius $20\overline{0}0$ miles. The angle between the tangent lines is 11.14°. The distance from the satellite to Q is 20,500 miles. Find the altitude $\overline{SP}$ of the satellite above the planet.

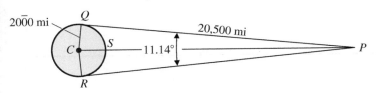

ILLUSTRATION 14

40. In Illustration 15, $\overline{CP} = 12.2$ m and $\overline{PB} = 10.8$ m. Find the radius of the circle, where C is the center.

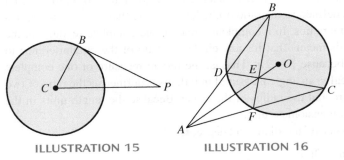

ILLUSTRATION 15   **ILLUSTRATION 16**

Exercises 41–44 refer to Illustration 16. $\overrightarrow{AB}$ *and* $\overrightarrow{AC}$ *are secants;* $\overline{CD}$ *and* $\overline{BF}$ *are chords.*

41. Suppose $\overset{\frown}{BD} = 85°$ and $\angle DEF = 52°$. Find the measure of $\overset{\frown}{CF}$.

42. Suppose $\overset{\frown}{BC} = 100°$ and $\overset{\frown}{DF} = 40°$. Find the measure of $\angle BAC$.

43. Suppose $\angle BEC = 78°$ and $\overset{\frown}{BC} = 142°$. Find the measure of $\overset{\frown}{DF}$.

44. Suppose $\angle DCF = 30°$, $\angle EFC = 52°$, and $\overset{\frown}{CF} = 110°$. Find the measure of $\overset{\frown}{BD}$.

45. Inscribe an equilateral triangle in a circle. **a.** How many degrees are contained in each arc? **b.** How many degrees are contained in each inscribed angle? **c.** Draw a central angle to each arc. How many degrees are contained in each central angle?

46. Inscribe a square in a circle. **a.** How many degrees are contained in each arc? **b.** How many degrees are contained in each inscribed angle? **c.** Draw a central angle to each arc. How many degrees are contained in each central angle?

47. Inscribe a regular hexagon in a circle. **a.** How many degrees are contained in each arc? **b.** How many degrees are contained in each inscribed angle? **c.** Draw a central angle to each arc. How many degrees are contained in each central angle?

48. An arc of a circle is doubled. Is its central angle doubled? Is its chord doubled?

49. In designing a bracket for use in a satellite, weight is of major importance. (See Illustration 17.) Find **a.** the area of the part in in², **b.** the overall length of the part, **c.** the overall height of the part, and **d.** its total weight.

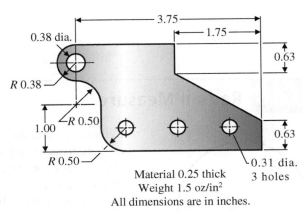

ILLUSTRATION 17

50. Windmills are used to generate electricity. The windmill in Illustration 18 in Madison County, New York, has a blade that is $13\overline{0}$ ft long from the center of its mounting. **a.** When the blade makes one complete rotation, how far (in ft) does the tip of the blade travel? **b.** What is the surface area of the rotating blades measured in acres?

ILLUSTRATION 18

51. A homeowner uses a water sprinkler that rotates in a circle and sends out a jet of water 60.0 ft. If she applies an average of 1 in. of water to the area being irrigated, how many gallons of water will she apply? (1 ft³ = 7.48 gal)

52. ⊠ A restaurant uses 9.00-in. (diameter) banquet plates and 12.0-in. dinner plates. How much more area is on the dinner plate than the banquet plate?

53. ⊠ The Highlington Farms Banquet Hall uses 8.00-ft × 30.0-in. rectangular tables and 6.00-ft diameter round tables. Each table has chairs around the entire perimeter. Chairs need to be placed 20 in. from the edge of the table. An additional 10 in. are needed behind each chair for waitstaff to walk between tables. Find the difference between the areas (including chairs and walk space) for each type of table.

54. ⊠ A bride has chosen the ballroom of a local hotel to have her wedding reception with a formal dinner. The ballroom is 78 ft × 92 ft. A space of 78 ft × 20 ft on one end is reserved for the head table, a small dance floor, gift table, and so on. For the dinner, she has chosen 6.00-ft round tables as described in Exercise 53 with 8 people seated at each table. She wants the tables arranged in a grid of parallel rows and columns. How many dinner guests (excluding the wedding party at the head table) can be seated?

55. ⊠ Ten dozen cookies are ordered for a special event. The cookies are 5 in. in diameter with a minimum of $\frac{1}{2}$-in. space between finished cookies, each requires $\frac{1}{3}$ cup of cookie dough, and they are placed on a cookie sheet with a baking area of 11 in. × 17 in. **a.** How many cups of cookie dough are needed? **b.** How many cookie sheets will be needed?

12.6 Radian Measure

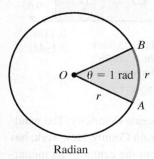

Figure 12.54 Radian

Radian measure, the metric unit of angle measure, is used in many applications, such as arc length and rotary motion. The **radian** (rad) unit is defined as the measure of an angle with its vertex at the center of a circle and with an intercepted arc on the circle equal in length to the radius. (In Figure 12.54, $\angle AOB$ forms the intercepted arc AB on the circle.)

In general, the radian is defined as the ratio of the length of arc that an angle intercepts on a circle to the length of its radius. In a complete circle or one complete revolution, the circumference $C = 2\pi r$. This means that for any circle the ratio of the circumference to the radius is constant (2π) because $\frac{C}{r} = 2\pi$. That is, the radian measure of one complete revolution is 2π rad. Technically, an angle measured in radians is defined as the ratio of two lengths: the lengths of the arc and the radius of the circle. Because the length units in the ratio cancel, the radian is a dimensionless unit.

What is the relationship between radians and degrees?

One complete revolution = 360°

One complete revolution = 2π rad

Therefore, 360° = 2π rad

180° = π rad

This gives us the conversion factors as follows:

$$\frac{\pi \text{ rad}}{180°} = 1 \quad \text{and} \quad \frac{180°}{\pi \text{ rad}} = 1$$

For comparison purposes,

$$1 \text{ rad} = \frac{180°}{\pi} = 57.3°$$

$$1° = \frac{\pi \text{ rad}}{180°} = 0.01745 \text{ rad}$$

12.6 ♦ Radian Measure

Example 1   How many degrees are in an angle that measures $\frac{\pi}{2}$ rad?

Use the conversion factor $\frac{180°}{\pi \text{ rad}}$.

$$\frac{\cancel{\pi}}{2} \cancel{\text{rad}} \times \frac{180°}{\cancel{\pi} \cancel{\text{rad}}} = \frac{180°}{2} = 90°$$

Example 2   How many radians are in an angle that measures 30°?

Use the conversion factor $\frac{\pi \text{ rad}}{180°}$.

$$\cancel{30}°\!_{1} \times \frac{\pi \text{ rad}}{\cancel{180}°\!_{6}} = \frac{\pi}{6} \text{ rad} \quad \text{or} \quad 0.524 \text{ rad}$$

In general, round lengths and angles in radians (when not expressed in terms of π) to three significant digits and angles in degrees to the nearest tenth of a degree.

As a wheel rolls along a surface, the distance s that a point on the wheel travels equals the product of the radius r and angle θ, measured in radians, through which the wheel turns. (See Figure 12.55.)

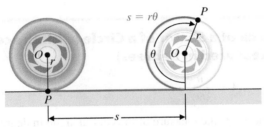

Figure 12.55

$$s = r\theta \quad (\theta \text{ in rad})$$

Example 3   Find the distance a point on the surface of a pulley travels if its radius is 10.0 cm and the angle of the turn is $\frac{5}{4}$ rad.

$s = r\theta$

$s = 10.0 \text{ cm} \times \frac{5}{4}$

$= 12.5 \text{ cm}$

Example 4   Find the distance a point on the surface of a gear travels if its radius is 15 cm and the angle of the turn is 420°.

$s = r\theta$

Since the angle is given in degrees, you must change 420° to radians. Use the conversion factor $\frac{\pi \text{ rad}}{180°}$.

$$\cancel{420}°\!^{7} \times \frac{\pi \text{ rad}}{\cancel{180}°\!_{3}} = \frac{7\pi}{3} \text{ rad}$$

$$s = \cancel{15}^{5} \text{ cm} \times \frac{7\pi}{\cancel{3}} = 35\pi \text{ cm or } 110 \text{ cm}$$

Example 5 A wheel with radius 5.40 cm travels a distance of 21.0 cm. Find angle θ **a.** in radians and **b.** in degrees that the wheel turns.

a. $$s = r\theta$$
Solve for θ:
$$\frac{s}{r} = \theta$$
$$\frac{21.0 \text{ cm}}{5.40 \text{ cm}} = \theta$$
$$3.89 \text{ rad} = \theta$$

b. $3.89 \text{ rad} \times \dfrac{180°}{\pi \text{ rad}} = 223.0°$ ◆

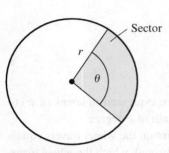

Figure 12.56
Sector of a circle

A **sector of a circle** is the region bounded by two radii of a circle and the arc intercepted by them. (See Figure 12.56.) The area of a given sector is proportional to the area of the circle itself; the area of the sector is a fraction of the area of the whole circle.

If the central angle of a sector is measured in degrees, the ratio of the measure of the central angle to 360° specifies the fraction of the area of the circle contained in the sector as follows:

> **Area of a Sector of a Circle (with the central angle measured in degrees)**
> $$A = \frac{\theta}{360°} \pi r^2$$
> where θ is the measure of the central angle in degrees and r is the radius.

If the central angle is measured in radians, the ratio of the measure of the central angle to 2π specifies the fraction of the area of the circle contained in the sector as follows:

$$A = \frac{\theta}{2\pi} \cdot \pi r^2 = \frac{1}{2} r^2 \theta$$

That is,

> **Area of a Sector of a Circle (with the central angle measured in radians)**
> $$A = \frac{1}{2} r^2 \theta$$
> where θ is the measure of the central angle in radians and r is the radius of the circle.

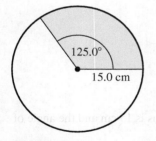

Figure 12.57

Example 6 Find the area of the sector of a circle of radius 15.0 cm with a central angle of 125.0°. (See Figure 12.57.)

$$A = \frac{\theta}{360°} \cdot \pi r^2$$
$$A = \frac{125.0°}{360°} \cdot \pi (15.0 \text{ cm})^2$$
$$= 245 \text{ cm}^2$$

◆

A segment of a circle is the region between a chord and an arc subtended by the chord. (See Figure 12.58.) Draw the radii to the ends of the chord as in Figure 12.59. The area of a segment equals the area of the sector minus the area of the isosceles triangle formed by the chord and two radii. Next, draw altitude h from the center, perpendicular to the chord. The area of the isosceles triangle is $A = \frac{1}{2}ch$, where c is the length of the chord. Using the Pythagorean theorem in $\triangle OMB$, we have

$(OM)^2 + (MB)^2 = (OB)^2$ Pythagorean theorem

$h^2 + \left(\dfrac{c}{2}\right)^2 = r^2$

$h^2 = r^2 - \dfrac{c^2}{4}$ Subtract $\frac{c^2}{4}$ from both sides.

$h^2 = \dfrac{4r^2 - c^2}{4}$ Write with LCD = 4.

$h = \dfrac{\sqrt{4r^2 - c^2}}{2}$ Take the square root of both sides.

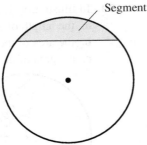

Figure 12.58
Segment of a circle

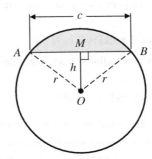

Figure 12.59

The area of the isosceles triangle is then

$A = \dfrac{1}{2}ch$

$ = \dfrac{1}{2}c\left(\dfrac{\sqrt{4r^2 - c^2}}{2}\right)$ Substitute for h from above.

$ = \dfrac{c\sqrt{4r^2 - c^2}}{4}$ Simplify.

The area of the segment is the area of the sector minus the area of the isosceles triangle:

$A = \dfrac{1}{2}r^2\theta - \dfrac{c\sqrt{4r^2 - c^2}}{4}$

where r is the radius, θ is the measure of the central angle, and c is the length of the chord.

If only the length of the chord and the radius of the circle are known, trigonometry is required. This application is treated in Chapter 13.

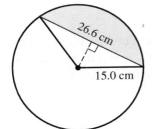

Figure 12.60

Example 7

The chord in Figure 12.60 has a length of 26.6 cm. The radius of the circle is 15.0 cm. The measure of the central angle is 125.0°, as in Example 6. Find the area of the segment.

The area of the isosceles triangle is

$A = \dfrac{c\sqrt{4r^2 - c^2}}{4}$

$A = \dfrac{(26.6 \text{ cm}) \sqrt{4(15.0 \text{ cm})^2 - (26.6 \text{ cm})^2}}{4}$

$ = 92.3 \text{ cm}^2$

Using the result from Example 6, the area of the segment is then
$$245 \text{ cm}^2 - 92.3 \text{ cm}^2 = 153 \text{ cm}^2$$

EXERCISES 12.6

In general, round lengths and angles in radians (when not expressed in terms of π) to three significant digits and angles in degrees to the nearest tenth of a degree.

1. π rad = _____ °
2. 1.7 rad = _____ °
3. 21.0° = _____ rad
4. 45.0° = _____ rad
5. Change $\frac{\pi}{3}$ rad to degrees.
6. Change 150.0° to radians.
7. Change 135.0° to radians.
8. Change $\frac{\pi}{12}$ rad to degrees.
9. How many radians are contained in a central angle that is $\frac{2}{3}$ of a circle?
10. What percent of 2π rad is $\frac{\pi}{2}$ rad?
11. Find the number of radians in a central angle whose arc is $\frac{2}{5}$ of a circle.
12. What percent of 2 rad is $\frac{\pi}{12}$ rad?

Complete the table using the formula $s = r\theta$ (θ in radians):

	Radius, r	Angle, θ	Distance, s
13.	25.0 cm	$\frac{2\pi}{5}$ rad	
14.	30.0 cm	$\frac{4\pi}{3}$ rad	
15.	6.00 cm	45.0°	
16.	172 mm	$\frac{\pi}{4}$ rad	
17.	18.0 cm	330.0°	
18.	3.00 m	250.0°	
19.	40.0 cm	rad	112 cm
20.	0.0081 mm	rad	0.011 mm
21.	0.500 m	°	0.860 m
22.	0.0270 m	°	0.0283 m
23.		$\frac{2\pi}{3}$ rad	18.5 cm
24.		315.0°	106 m

25. A pulley is turning at an angular velocity of 10.0 rad per second. How many revolutions is the pulley making each second? (*Hint:* One revolution equals 2π rad.)

26. The radius of a wheel is 20.0 in. It turns through an angle of 2.75 rad. What is the distance a point travels on the surface of the wheel?

27. The radius of a gear is 22.0 cm. It turns through an angle of 240.0°. What is the distance a point travels on the surface of the gear?

28. A wheel of diameter 6.00 m travels a distance of 31.6 m. Find the angle θ (in radians) that the wheel turns.

29. A wheel of diameter 15.2 cm turns through an angle of 3.40 rad. Find the distance a point travels on the surface of the wheel.

30. In Illustration 1, find
    a. the length of arc s
    b. the area of the sector
    c. the area of the segment

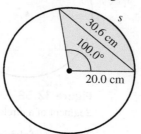

ILLUSTRATION 1   ILLUSTRATION 2

31. In Illustration 2, find
    a. the length of arc s
    b. the area of the sector
    c. the area of the segment

32. Given two concentric circles (circles with the same center) with central angle 45.0°, $r_1 = 4.00$ m, and $r_2 = 8.00$ m, find the shaded area in Illustration 3.

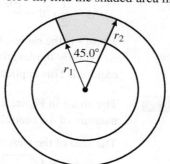

ILLUSTRATION 3

33. A round pizza, with a 12.0-in. diameter, is cut into 8 pie-shaped pieces. What is the area of each slice (sector) of pizza?

12.7 Prisms

Up to now, you have studied the geometry of two dimensions. **Solid geometry** is the geometry of three dimensions: length, width, and depth.

A **prism** is a solid whose sides are parallelograms and whose bases are one pair of parallel polygons that have the same size and shape. (See Figure 12.61.) (Recall that a closed figure made up of straight lines in two dimensions is called a *polygon*.) The two parallel polygons (which may be any type of polygon) are called the *bases* of the prism. The remaining polygons will be parallelograms and are called *lateral faces*. A *right prism* has lateral faces that are rectangles and are therefore perpendicular to the bases.

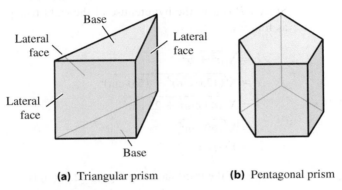

(a) Triangular prism   (b) Pentagonal prism

Figure 12.61
Basic parts of a prism. The name of the polygon used as the base names the type of prism.

The name of the polygon used as the base names the type of prism. For example, a prism with bases that are triangles is called a *triangular* prism (see Figure 12.61(a)), a prism with bases that are pentagons is called a *pentagonal* prism (see Figure 12.61(b)), and so on.

> The **lateral surface area** of a prism is the sum of the areas of the lateral faces of the prism.
>
> The **total surface area** of a prism is the sum of the areas of the lateral faces and the areas of the bases of the prism.

Example 1 Find the lateral surface area of the triangular right prism in Figure 12.62.

To find the lateral surface area, find the area of each lateral face. Then find the sum of these areas.

The area of the rectangular face located on the right front side (see Figure 12.63a) is

$$A = lw$$
$$A = (12.0 \text{ cm})(5.0 \text{ cm})$$
$$= 6\overline{0} \text{ cm}^2$$

The area of the square face on the left front side (see Figure 12.63b) is

$$A = s^2$$
$$A = (12.0 \text{ cm})^2$$
$$= 144 \text{ cm}^2$$

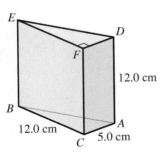

Figure 12.62

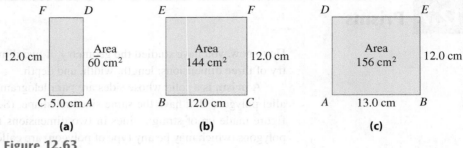

Figure 12.63

To find the area of the third face (the back side of the prism), first find length AB. Since AB is also the hypotenuse of the right triangle ABC, use the Pythagorean theorem as follows:

$$c = \sqrt{a^2 + b^2}$$
$$c = \sqrt{(12.0 \text{ cm})^2 + (5.0 \text{ cm})^2}$$
$$= \sqrt{144 \text{ cm}^2 + 25 \text{ cm}^2}$$
$$= \sqrt{169 \text{ cm}^2}$$
$$= 13.0 \text{ cm}$$

The area of the third face (see Figure 12.63(c)) is

$$A = lw$$
$$A = (13.0 \text{ cm})(12.0 \text{ cm})$$
$$= 156 \text{ cm}^2$$

Therefore, the lateral surface area is

$$6\overline{0} \text{ cm}^2 + 144 \text{ cm}^2 + 156 \text{ cm}^2 = 36\overline{0} \text{ cm}^2$$

◆

Example 2 Find the total surface area of the prism in Example 1.

To find the total surface area, first find the area of the bases. Then, add this result to the lateral surface area from Example 1. The bases have the same size and shape, so just find the area of one base and then double it.

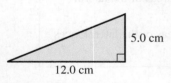

Figure 12.64

The area of one base as shown in Figure 12.64 is

$$B = A = \frac{1}{2}bh$$
$$B = \frac{1}{2}(12.0 \text{ cm})(5.0 \text{ cm})$$
$$= 3\overline{0} \text{ cm}^2$$

Double this to find the area of both bases.

$$2(3\overline{0} \text{ cm}^2) = 6\overline{0} \text{ cm}^2$$

Add this area to the lateral surface area to find the total area.

$$36\overline{0} \text{ cm}^2 + 6\overline{0} \text{ cm}^2 = 42\overline{0} \text{ cm}^2$$

So the total surface area is $42\overline{0} \text{ cm}^2$.

◆

The volume of a prism is found by the following formula:

> **Volume of Prism**
>
> $$V = Bh$$
>
> where B is the area of one of the bases and h is the altitude (perpendicular distance between the parallel bases).

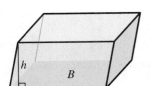

Figure 12.65

See Figure 12.65.

Example 3 Find the volume of the prism in Example 1.

To find the volume of the prism, use the formula $V = Bh$. B is the area of the base which we found to be $3\bar{0}$ cm². The altitude of a lateral face is 12.0 cm.

$$V = Bh$$
$$V = (3\bar{0} \text{ cm}^2)(12.0 \text{ cm})$$
$$= 360 \text{ cm}^3$$

Example 4 Find the volume of the prism in Figure 12.66.

Use the formula $V = Bh$. The base is a parallelogram with sides of length 10.0 cm and 4.0 cm. The altitude of the base is 3.0 cm. First, find B, the area of the base.

$$B = bh$$
$$B = (10.0 \text{ cm})(3.0 \text{ cm})$$
$$= 3\bar{0} \text{ cm}^2$$

The altitude of a lateral face of the prism is 7.0 cm. Therefore,

$$V = Bh$$
$$V = (3\bar{0} \text{ cm}^2)(7.0 \text{ cm})$$
$$= 210 \text{ cm}^3$$

Figure 12.66

Example 5 A rectangular piece of steel is 24.1 in. by 13.2 in. by 8.20 in. (Figure 12.67). Steel weighs 0.28 lb/in³. Find its weight, in pounds.

Find the volume using the formula for the volume of a prism, $V = Bh$. First, find B, the area of the base of the prism.

$$B = lw$$
$$B = (13.2 \text{ in.})(24.1 \text{ in.})$$

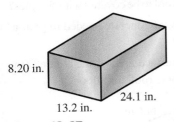

Figure 12.67

The volume is then

$$V = Bh$$
$$V = [(13.2 \text{ in.})(24.1 \text{ in.})](8.20 \text{ in.})$$
$$= 2610 \text{ in}^3$$

Since steel weighs 0.28 lb/in³, the total weight is

$$2610 \text{ in}^3 \times \frac{0.28 \text{ lb}}{1 \text{ in}^3} = 730 \text{ lb}$$

Example 6 A parking lot 256 ft by 124 ft will be repaved with asphalt $2\frac{1}{2}$ in. thick. **a.** How many cubic yards will be needed? **b.** If 1 cubic foot of asphalt weighs 137 lb, what is the total weight that will need to be trucked to the site?

a. The volume of asphalt is found by the formula

$$V = lwh$$

$$V = (256 \text{ ft})(124 \text{ ft})\left(2\frac{1}{2} \text{ in.} \times \frac{1 \text{ ft}}{12 \text{ in.}}\right) \quad \text{Note the conversion factor to obtain ft}^3.$$

$$= 6610 \text{ ft}^3 \times \frac{1 \text{ yd}^3}{27 \text{ ft}^3} \quad \text{Then convert this result to yd}^3.$$

$$= 245 \text{ yd}^3$$

b. $= 245 \text{ yd}^3 \times \frac{137 \text{ lb}}{\text{ft}^3} \times \frac{27 \text{ ft}^3}{1 \text{ yd}^3} \quad \text{Note the use of conversion factors.}$

$= 906{,}000 \text{ lb}$

EXERCISES 12.7

Follow the rules for working with measurements:

1. **a.** Find the lateral surface area of the prism shown in Illustration 1.
   b. Find the total surface area of the prism.
   c. Find the volume of the prism.

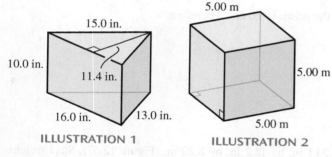

ILLUSTRATION 1   ILLUSTRATION 2

2. **a.** Find the lateral surface area of the rectangular prism shown in Illustration 2.
   b. Find the total surface area of the prism.
   c. Find the volume of the prism.
   d. What is the name given to this geometric solid?

3. A gusset is the shape of a right triangular prism. If the right triangular base has dimensions 3.0 in. by 4.0 in. by 5.0 in. and the height is 6.0 in., what is the total surface area of the prism? Find its volume.

4. In a drawing, a ceiling-to-floor bay window area is being added to a room. The shape of the added floor space area is an isosceles trapezoid with bases 12.0 ft and 6.00 ft and with slant sides 3 ft 6 in. **a.** Find the area of the floor space added to the room. **b.** If the room has 9 ft 6 in. ceilings, find the additional volume of the room. **c.** How many square yards of vinyl floor material would be purchased if it is available only in 6-ft widths and the owner wants no seams within the added bay window area?

5. The baggage compartment of a helicopter is a rectangular prism. The dimensions of the baggage compartment are 3.0 ft by 4.0 ft by 5.0 ft. What is the volume of the compartment?

6. A piece of 16.0-in. by 20.0-in. metal duct is a rectangular prism. If it is 4.00 ft long, what is the lateral surface area? What is its volume?

7. **a.** What is the area of the four sides of the building to be painted in Illustration 3? (Assume no windows.)
   b. What is the area of roof to be covered with shingles?
   c. What is the volume of concrete needed to pour a floor 16 cm deep?
   d. What is the total surface area of the figure? (Include painted surface, roof, and floor.)

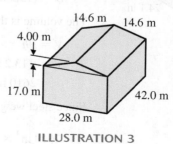

ILLUSTRATION 3

8. Find the volume of the wagon box in Illustration 4.

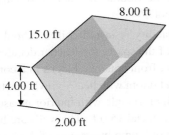

ILLUSTRATION 4

9. Find the volume of the gravity bin in Illustration 5.

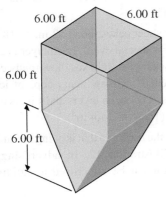

ILLUSTRATION 5

10. Steel weighs 0.28 lb/in^3. What is the weight of a rectangular piece of steel 0.3125 in. by 12.0 in. by 20.0 in.?

11. A steel rod of cross-sectional area 5.0 in^2 weighs 42.0 lb. Find its length. (Steel weighs 0.28 lb/in^3.)

12. The rectangular lead sleeve shown in Illustration 6 has a cored hole 2.0 in. by 3.0 in. How many cubic inches of lead are in this sleeve?

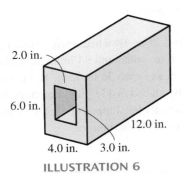

ILLUSTRATION 6

13. A triangular display pedestal is to be made of corrugated paper. Using Illustration 7, what size sheet needs to be designated on the drawing?

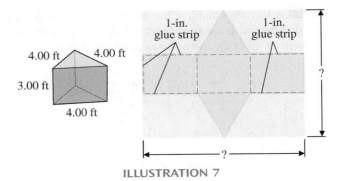

ILLUSTRATION 7

14. A concrete cube 3 ft on a side is being designed. To reduce the weight of the final product, it will be "voided" using a cardboard box. If 4 in. of concrete must be maintained and there is no concrete on the bottom, find the dimensions needed for a drawing of the cardboard box as shown in Illustration 8.

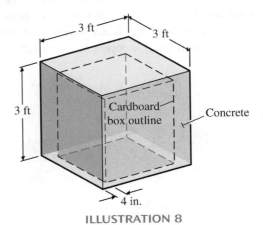

ILLUSTRATION 8

15. From what size sheet of cardboard can the cardboard box in Exercise 14 be cut? Assume a 1-in. glue strip as shown in Illustration 9.

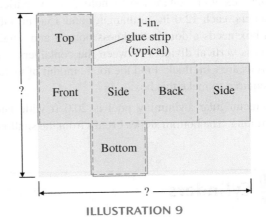

ILLUSTRATION 9

16. A Victorian building has one room in the shape of a rhombus with a ceiling height of 10.0 ft. Find the volume of air in the room shown in Illustration 10.

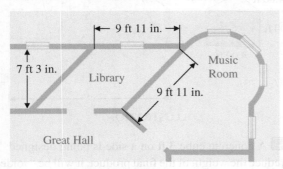

ILLUSTRATION 10

17. A swimming pool company is preparing drawings of a newly designed uniform 4.00-ft-deep lap pool as shown in Illustration 11. One overflow "scupper" valve is required for every 12,000 gal of water. The pool is 40.0 ft wide and 80.0 ft long. Water weighs 62.4 lb/ft³ and 8.34 lb/gal. How many scuppers are required?

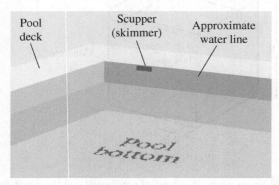

ILLUSTRATION 11

18. A cardboard box manufacturer must design a rectangular box with square top to hold four cylindrical containers, each 12.0 in. in diameter and 15.0 in. tall. Each box needs a double-thickness bottom and single thickness vertical dividers between the containers. All materials are $\frac{1}{8}$ in. thick. Find the total amount of material required to make this box.

19. A rectangular swimming pool is 20.0 ft wide and 40.0 ft long. The bottom slopes evenly from the shallow end, which is 3.00 ft deep, to the deep end, which is 6.00 ft deep. What is the volume of water (in ft³) that is required to fill the pool?

20. An aquaculture facility is installed in a school. To take best advantage of the space, the teacher decides to construct the fish tank from concrete blocks in a corner and line the sides and bottom with heavy vinyl. The tank is in the shape of a right triangle with interior measurements of 9.00 ft, 12.0 ft, and 15.0 ft. The walls are built 3.00 ft high. The rule of thumb in projects of this sort is to order 10% more than the actual required amount in order to account for waste and overlapping. At a cost of $0.26/ft², how much will the vinyl cost to line the interior walls and bottom?

21. Four double-layer sheet cakes, each 12 in. × 18 in., have been ordered for a special event. Each layer is 2 in. high. **a.** If the cake rises by $\frac{1}{3}$ as it bakes, how many cups of cake batter are needed? **b.** How many cups of icing are needed if $\frac{1}{4}$ in. is used between layers and $\frac{1}{2}$ in. is used on the top and sides? (1 cup = 14.4 in³)

22. Find the volume of the trash can that is being manufactured as shown in Illustration 12. In advertising it, how many quarts should it be advertised as holding? (1 gal = 231 in³)

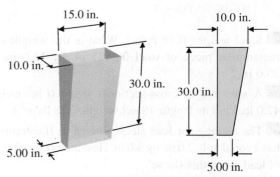

ILLUSTRATION 12

23. The inside dimensions of an 18-wheel tractor trailer are 52 ft 6 in. long, 8 ft 6 in. wide, and 94 in. high. A shipment of rectangular boxes each 36 in. long, 12 in. wide, and 6 in. high is to be placed inside. Find the maximum number of boxes that can be shipped if each box must be placed on its 12-in. × 36-in. bottom.

12.8 Cylinders

A **circular cylinder** is a geometric solid with a curved lateral surface and circles as parallel bases. The *axis* of a cylinder is the line segment between the centers of the bases. The altitude, h, is the shortest (perpendicular) distance between the bases. If the axis is perpendicular to the bases, the cylinder is called a *right circular cylinder* and the axis is the same length as the altitude. See Figure 12.68. You may also think of a right circular cylinder as the solid formed by rotating a rectangle about one of its sides.

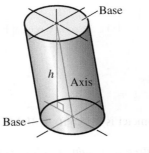
(a) Cylinder

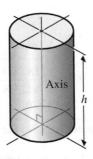
(b) Right circular cylinder

Figure 12.68

The volume of a right circular cylinder is found by the formula $V = Bh$, where B is the area of the base. The base is a circle with area $B = \pi r^2$. Therefore, the formula for the volume of a cylinder is written as follows:

Volume of Cylinder

$$V = \pi r^2 h$$

where r is the radius of the base and h is the altitude.

Example 1 Find the volume of the right circular cylinder in Figure 12.69.

The diameter is 24.0 m, so the radius is 12.0 m.

$$V = \pi r^2 h$$
$$V = \pi (12.0 \text{ m})^2 (40.0 \text{ m})$$
$$= 18{,}100 \text{ m}^3$$

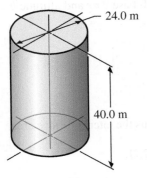

Figure 12.69

Example 2 Find the diameter of a cylindrical tank 23.8 ft high with a capacity of 136,000 gallons ($1 \text{ ft}^3 = 7.48 \text{ gal}$).

First, find the volume of the cylinder in ft³.

$$136{,}000 \text{ gal} \times \frac{1 \text{ ft}^3}{7.48 \text{ gal}} = 18{,}200 \text{ ft}^3$$

Since V and h are known, find r using the formula

$$V = \pi r^2 h$$

$$r^2 = \frac{V}{\pi h} \qquad \text{Divide both sides by } \pi h.$$

$$r = \sqrt{\frac{V}{\pi h}} \qquad \text{Take the square root of both sides.}$$

$$r = \sqrt{\frac{18{,}200 \text{ ft}^3}{\pi(23.8 \text{ ft})}}$$

$$= 15.6 \text{ ft}$$

Diameter is $2r$. So the diameter is $2(15.6 \text{ ft}) = 31.2 \text{ ft}$. ◆

The *lateral surface area of a right circular cylinder* can be visualized as a can without ends. Cut through the side of the can and then flatten it out, as shown in Figure 12.70.

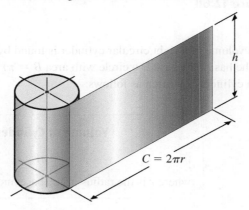

Figure 12.70
Lateral surface area of a cylinder

The lateral surface area of a right cylinder is a rectangle with base $2\pi r$ and altitude h. The formula for the lateral surface area is

> **Lateral Surface Area of Cylinder**
>
> $A = 2\pi rh$

The *total surface area of a cylinder* is the area of the bases plus the lateral surface area.

Example 3 Find the total surface area of the right circular cylinder in Figure 12.71.

The area of one base is

$$A = \pi r^2 = \pi(1.78 \text{ m})^2$$
$$= 9.95 \text{ m}^2$$

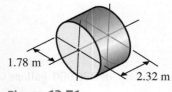

1.78 m
2.32 m

Figure 12.71

The area of both bases, then, is

$$2(9.95 \text{ m}^2) = 19.9 \text{ m}^2$$

The lateral surface area $= 2\pi rh$

$$= 2\pi(1.78 \text{ m})(2.32 \text{ m})$$
$$= 25.9 \text{ m}^2$$

The total surface area $= 19.9 \text{ m}^2 + 25.9 \text{ m}^2$

$$= 45.8 \text{ m}^2 \qquad ◆$$

12.8 ♦ Cylinders

Example 4 A manufacturer needs to plan for storage of 5-gallon oil drums. Find the height of a 5.00-gallon oil drum with a diameter of 12.00 in. Note: The volume of 1 gallon is 231 in^3.

The volume of the oil drum is found by using the formula

$V = \pi r^2 h$ Next, solve for h.

$h = \dfrac{V}{\pi r^2}$ Divide both sides by πr^2.

$h = \dfrac{5 \text{ gal} \times \dfrac{231 \text{ in}^3}{1 \text{ gal}}}{\pi (6.00 \text{ in.})^2}$ Substitute the data and simplify.

$= 10.2$ in.

EXERCISES 12.8

Follow the rules for working with measurements.

Find the volume of each cylinder:

1. 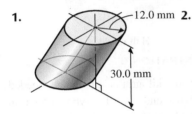 (12.0 mm, 30.0 mm)

2. (13.2 m, 17.9 m)

3. How many litres does a cylindrical tank of height 39.2 m with radius 8.20 m hold? (1 m^3 = 1000 L)

4. A steel cylindrical tank must hold 7110 gal of dyed water for a cloth process. Due to space constraints, the cylindrical tank is made 11.0 ft in diameter. How tall must the tank be? (Water weighs 8.34 lb/gal and 62.4 lb/ft^3.)

5. A technician draws plans for a 40$\overline{0}$,000-gallon cylindrical tank with radius 20.0 ft. What should the height be? (1 ft^3 = 7.48 gal)

6. An oil filter for a small automobile is cylindrical with radius 1.80 in. and height 3.60 in. What is the volume of the oil filter?

7. An air filter for an old automobile is cylindrical. If the inner radius is 4.00 in. and the outer radius is 5.00 in., what is the volume of the 2.00-in.-tall air filter?

8. An engine has 8 cylinders. Each cylinder has a bore of 4.70 in. in diameter and a length of 5.25 in. Find its total piston displacement.

9. A cylindrical tank is 25 ft 9 in. long and 7 ft 6 in. in diameter. How many cubic feet does it hold?

10. A 3.0-in.-diameter cylindrical rod is 16 in. long. Find its volume.

11. A cylindrical piece of steel is 10.0 in. long. Its volume is 25.3 in^3. Find its diameter.

12. If a metal cylindrical storage tank has a volume of 30$\overline{0}$0 ft^3 and a radius of 8.00 ft, what is its height? What is its total surface area?

13. Copper tubing $\frac{1}{2}$ in. in I.D. (inside diameter) is 12.0 ft long. What is the volume of the refrigerant contained in the tubing?

14. A rectangular steel plate 3.76 in. by 9.32 in. by 1.00 in. thick with a 2.00-in.-diameter hole in the center is being drawn. **a.** Find the volume of steel left after the hole is cut. **b.** If this steel weighs 30.0 oz/in^3, what weight should be given on the drawing?

15. Find **a.** the lateral surface area and **b.** the total surface area of the right circular cylinder shown in Illustration 1.

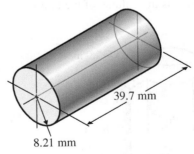

(39.7 mm, 8.21 mm)

ILLUSTRATION 1

16. Find **a.** the lateral surface area and **b.** the total surface area of the cylinder shown in Illustration 2.

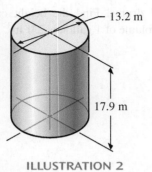

ILLUSTRATION 2

17. Find the total amount (area) of paper used for labels for 1000 cans like the one shown in Illustration 3.

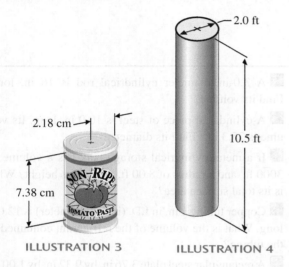

ILLUSTRATION 3   **ILLUSTRATION 4**

18. How many square feet of sheet metal are needed for the sides of the cylindrical tank shown in Illustration 4? (Allow 2.0 in. for seam overlap.)

19. A cylindrical piece of stock is turned on a lathe from 3.10 in. down to 2.24 in. in diameter. The cut is 5.00 in. long. What is the volume of the metal removed?

20. What is the volume of lead in the "pig" shown in Illustration 5? What is the volume of the mold?

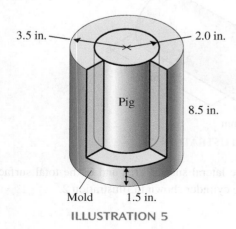

ILLUSTRATION 5

21. A cylinder bore is increased in diameter from 2.78 in. to 2.86 in. Its length is 5.50 in. How much has the surface area of the walls been increased?

22. Each cylinder bore of a 6-cylinder engine has a diameter of 2.50 in. and a length of 4.90 in. What is the lateral surface area of the six cylinder bores?

23. The sides of a cylindrical silo 15 ft in diameter and 26 ft high are to be painted. Each gallon of paint will cover 200 ft^2. How many gallons of paint will be needed?

24. How many square feet of sheet metal are needed to form the trough shown in Illustration 6?

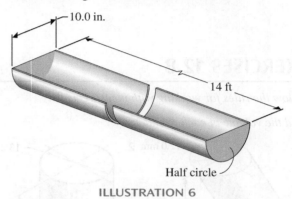

ILLUSTRATION 6

25. Find the number of kilograms of metal needed for 2,700,000 cans with ends of the type shown in Illustration 7. The metal has a density of 0.000147 g/cm^2.

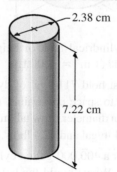

ILLUSTRATION 7

26. A cylindrical cooling tank has an outside diameter of 5.00 ft. The walls on all sides are 5.00 in. thick and the tank is 12.0 ft tall. How many gallons of water will this tank hold?

27. A concrete forming "paper" tube is used to void a column as shown in Illustration 8. The walls must be 3.00 in. thick and the outside column must be 37.0 in. in radius. The column is 20.0 ft tall. **a.** Find the diameter of the paper tube. **b.** If concrete weighs 148 lb/ft^3, what would you put on the drawing for the final column weight? **c.** At $186 \text{ ft}^2/\text{gal}$ per coat of paint, how much paint would you put in the bill of materials as required for one coat per each column exterior surface. **d.** If the column is not voided and poured solid, how much would it weigh?

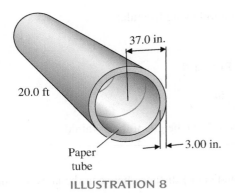

Paper tube

ILLUSTRATION 8

28. A machine shop needs a new parts washer. An ad in a shop journal shows the tank in Illustration 9, but someone forgot to state how many gallons the tank can hold. Find its capacity if it is filled to within 9.00 in. of the top. (1 gal = 231 in^3)

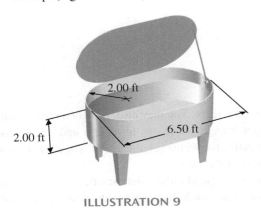

ILLUSTRATION 9

29. Two identical steel cylindrical tanks must hold a total of 22,020 gal of seal oil from a leather treatment plant. The tanks will be made in the shop and installed upstream of the sand filter. They can be only 11.0 ft in diameter. Ignoring the wall thickness, find the length of each tank. (1 gal = 231 in^3)

30. The Trans-Alaska oil pipeline, also known as the Alyeska pipeline, is approximately 800 mi long. Assuming that it has a uniform 48-in. inside diameter, how many ft^3 of oil would be required to fill the pipeline?

31. A silo is a cylinder-shaped structure that is used to store feeds and grain. If the inside dimensions of a silo are 40.0 ft in height and 20.0 ft in diameter, how many bushels of grain can be stored? (1 bu = 1.2445 ft^3)

32. A circular sediment tank at a wastewater treatment plant has a diameter of 110 ft. If the liquid inside the tank is 8.50 ft deep, how many ft^3 of waste will the tank hold?

33. A spaghetti sauce is made in a cylindrical stock pot with diameter 18.0 in. and height 20.0 in. If the sauce is 2.0 in. from the top of the pot, how many cups of sauce are in the pot? (1 cup = 14.4 in^3)

34. Chili is made in a cylindrical stock pot with diameter 14.0 in. and height 15.0 in. If the pot is $\frac{3}{4}$ full, how many $1\frac{1}{2}$-cup bowls can be served? (1 cup = 14.4 in^3)

35. A three-tiered, double-layered round wedding cake of diameters 8 in., 12 in., and 16 in. is ordered. Each layer is 3 in. high. If the cake rises by $\frac{1}{3}$ as it bakes, how many cups of cake batter are needed? (1 cup = 14.4 in^3)

12.9 Pyramids and Cones

A **pyramid** is a geometric solid whose base is a polygon and whose lateral faces are triangles with a common vertex. The common vertex is called the *apex* of the pyramid. If a pyramid has a base that is a triangle, then it is called a *triangular pyramid*. If a pyramid has a base that is a square, then it is called a *square pyramid*. In general, a pyramid is named by the shape of its base. Two types of pyramids are shown in Figure 12.72.

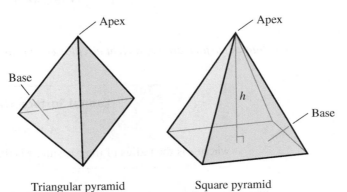

Figure 12.72
A pyramid is named by the shape of its base.

The volume of a pyramid is found using the following formula:

> **Volume of Pyramid**
>
> $$V = \frac{1}{3}Bh$$
>
> where B is the area of the base and h is the height of the pyramid.

The height of a pyramid is the shortest (perpendicular) distance between the apex and the base of the pyramid.

Example 1 Find the volume of the pyramid in Figure 12.73.

The base is a right triangle with legs 6.0 in. and 8.0 in. Therefore,

$$B = \frac{1}{2}(6.0 \text{ in.})(8.0 \text{ in.}) = 24 \text{ in}^2$$

The height is 7.0 in. Therefore,

$$V = \frac{1}{3}Bh = \frac{1}{3}(24 \text{ in}^2)(7.0 \text{ in.}) = 56 \text{ in}^3$$

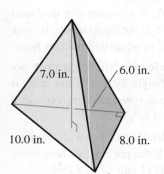

Figure 12.73

A **cone** is a geometric solid whose base is a circle. You may also think of a right circular cone as the solid formed by rotating a right triangle about one of its legs and which has a curved lateral surface that comes to a point called the *vertex* (Figure 12.74). The *axis* of a cone is a line segment from the vertex to the center of the base.

The height, h, of a cone is the shortest (perpendicular) distance between the vertex and the base. A *right circular cone* is a cone in which the height is the distance from the vertex to the center of the base. The *slant height* of a right circular cone is the length of a line segment that joins the vertex to any point on the circle that forms the base of the cone.

The *volume of a circular cone* is given by the formula $V = \frac{1}{3}Bh$. Since the base, B, is always a circle, its area is πr^2, where r is the radius of the base. Thus, the formula for the volume of a right circular cone is written as follows:

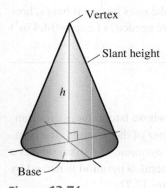

Figure 12.74
Cone

> **Volume of Cone**
>
> $$V = \frac{1}{3}\pi r^2 h$$

The *lateral surface area of a right circular cone* is found using the following formula:

> **Lateral Surface Area**
>
> $$A = \pi r s$$
>
> where r is the radius of the base and s is the slant height of the cone.

The *total surface area of a right circular cone* is the sum of the lateral surface area and the area of the base.

12.9 ♦ Pyramids and Cones

Example 2 Find the volume of the right circular cone in Figure 12.75.

$$V = \frac{1}{3}\pi r^2 h$$

$$V = \frac{1}{3}\pi (6.1 \text{ m})^2 (17.2 \text{ m})$$

$$= 670 \text{ m}^3$$

Example 3 Find the lateral surface area of the right circular cone in Figure 12.75.

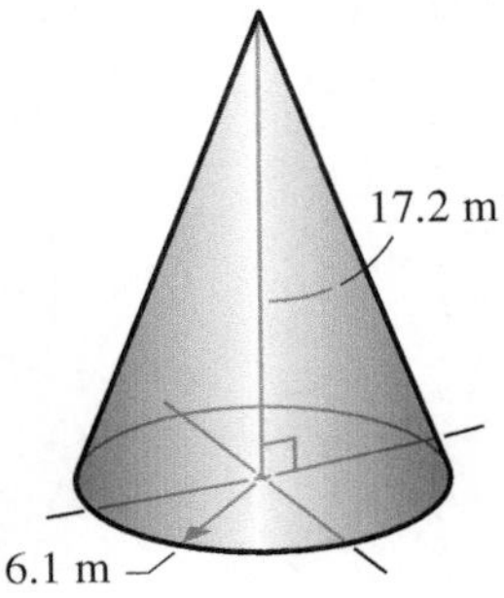

The formula for the lateral surface area is $A = \pi rs$. The slant height is not given. However, a right triangle is formed by the axis, the radius, and the slant height. Therefore, to find the slant height, s, use the formula

$$c = \sqrt{a^2 + b^2}$$

Then $s = \sqrt{(6.1 \text{ m})^2 + (17.2 \text{ m})^2}$

$$= 18.2 \text{ m}$$

The lateral surface area can then be found as follows:

$$A = \pi rs$$
$$= \pi (6.1 \text{ m})(18.2 \text{ m})$$
$$= 350 \text{ m}^2$$

Figure 12.75

The **frustum of a pyramid** is the section of a pyramid between the base and a plane parallel to the base, as shown in Figure 12.76.

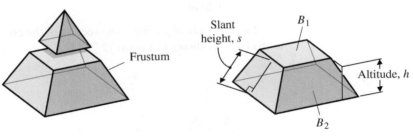

Figure 12.76
Frustum of pyramid

The altitude of the frustum is the perpendicular distance between the two bases. The *volume of the frustum of a pyramid* is

$$V = \frac{1}{3}h\left(B_1 + B_2 + \sqrt{B_1 B_2}\right)$$

where h is the altitude and B_1 and B_2 are the areas of the bases.

The *lateral surface area of the frustum of a pyramid* is

$$A = \frac{1}{2}s(P_1 + P_2)$$

where s is the slant height and P_1 and P_2 are the perimeters of the bases.

Example 4 Find the lateral surface area and the volume of the frustum in Figure 12.77.

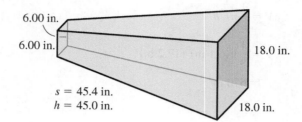

Figure 12.77
Air duct connection

$P_1 = 4(6.00 \text{ in.}) = 24.0 \text{ in.}$

$P_2 = 4(18.0 \text{ in.}) = 72.0 \text{ in.}$

$A = \frac{1}{2}s(P_1 + P_2)$

$A = \frac{1}{2}(45.4 \text{ in.})(24.0 \text{ in.} + 72.0 \text{ in.})$

$\quad = 2180 \text{ in}^2$

$B_1 = (6.00 \text{ in.})^2 = 36.0 \text{ in}^2$

$B_2 = (18.0 \text{ in.})^2 = 324 \text{ in}^2$

$V = \frac{1}{3}h\left(B_1 + B_2 + \sqrt{B_1 B_2}\right)$

$V = \frac{1}{3}(45.0 \text{ in.})\left(36.0 \text{ in}^2 + 324 \text{ in}^2 + \sqrt{(36.0 \text{ in}^2)(324 \text{ in}^2)}\right)$

$\quad = 7020 \text{ in}^3$ ♦

The **frustum of a cone** is the section of the cone between the base and a plane parallel to the base, as shown in Figure 12.78.

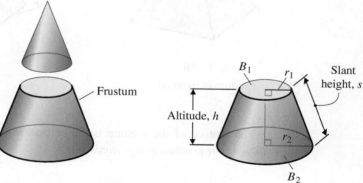

Figure 12.78
Frustum of cone

The altitude of the frustum is the perpendicular distance between the two bases. The *volume of the frustum of a cone* is

$$V = \frac{1}{3}h\left(B_1 + B_2 + \sqrt{B_1 B_2}\right)$$

where h is the altitude and B_1 and B_2 are the areas of the bases.

The *lateral surface area of a frustum of a right circular cone* (area of the curved surface) is

$$A = \pi s(r_1 + r_2)$$

where s is the slant height and r_1 and r_2 are the radii of the bases.

EXERCISES 12.9

Follow the rules for working with measurements.

Find the volume of each figure in Exercises 1–10:

1.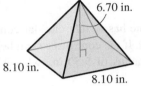

2.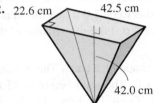

3.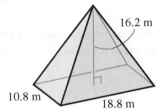

4.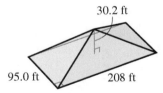

5.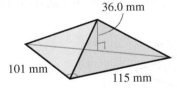

6.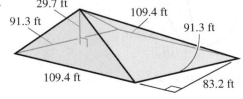

7.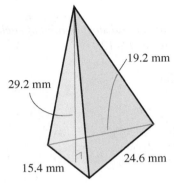

Hint: To find B, use $B = A = \sqrt{s(s-a)(s-b)(s-c)}$

8.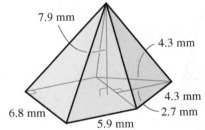

Hint: To find the area of the base of the pyramid, find the sum of the areas of the rectangle and the triangle.

9.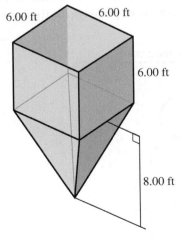

412 CHAPTER 12 ♦ Geometry

10.

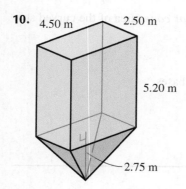

*Find **a.** the volume and **b.** the lateral surface area of each right circular cone in Exercises 11–12:*

11.

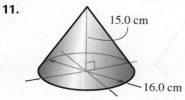

12.

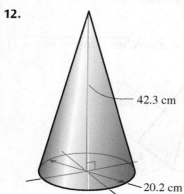

13. A loading chute in a flour mill goes directly into a feeding bin. The feeding bin is in the shape of an inverted right circular cone, as shown in Illustration 1. How many bushels of wheat can be placed in the feeding bin? (0.804 bu = 1 ft^3)

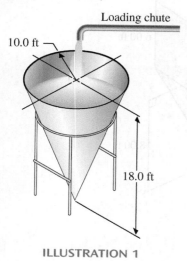

ILLUSTRATION 1

14. The circular tank in Illustration 2 is made of $\frac{1}{2}$-in. steel weighing 19.8 lb/ft^2. **a.** What is the total weight of the top? **b.** What is the total weight of the top, sides, and bottom of the tank?

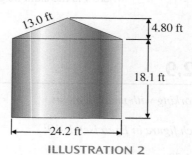

ILLUSTRATION 2

15. The nose of an airplane has a right circular cone in the center of the propeller. If the cone has a slant height of 13.0 in. with a base radius of 5.50 in., what is the lateral surface area?

16. A welder decides to make a pyramid out of flat steel. The height of the pyramid is to be 3.0 ft and the square base has edges of length 2.0 ft. Find the volume of the pyramid.

17. Gravel is piled in the shape of a cone. The circumference of the base is 224 ft. The slant height is 45 ft. Find the volume of gravel. If gravel weighs 3200 lb/yd^3, how many 22-ton truckloads are needed to transport the gravel?

18. Find the weight of the display model shown in Illustration 3. The model is made of pine. Pine weighs 31.2 lb/ft^3.

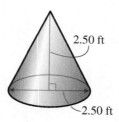

ILLUSTRATION 3

19. Find the volume of the frustum of the pyramid shown in Illustration 4.

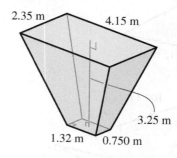

ILLUSTRATION 4

20. Find the volume and lateral surface area of the frustum of the cone shown in Illustration 5.

ILLUSTRATION 5

Find the volume and the lateral surface area of each storage bin:

21.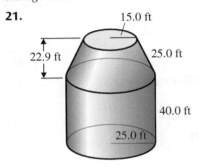

22.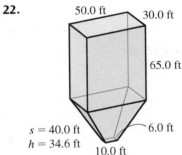

23. ![icon] A hopper must be designed to contain plastic resin pellets for an injection mold machine. (See Illustration 6.) The cylindrical portion of the tank is 18.0 in. in diameter. The spout is a conical frustum. For the hopper to hold 5.00 ft³ of resin pellets, find the length l of the cylindrical part.

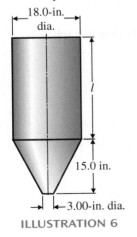

ILLUSTRATION 6

24. ![icon] A piece of 1-in. (diameter) round stock is tapered so that its tip is a cone. (See Illustration 7.) If the taper begins 3.00 in. from the end of the stock, find the volume of stock that was removed in order to produce the tapered end.

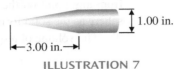

ILLUSTRATION 7

25. ![icon] A welder is assigned to fabricate luggage storage compartments to fit in the luggage storage area of an aircraft. What is the cubic foot displacement of the compartment shown in Illustration 8?

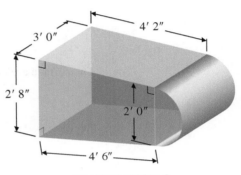

ILLUSTRATION 8

26. ![icon] Two different sizes of square metal duct are joined by a piece that is a frustum of a pyramid. Find the lateral surface area if the small end is square with one side of length 10.0 in. and the larger square end is of length 15.0 in. on one side with slant height 8.00 in.

27. ![icon] A lamp manufacturer decides to offset various rising costs by cutting the size of a lampshade, which in turn decreases the amount of material required. The original shade has a top diameter of 13.5 in., a bottom diameter of 15.0 in., and a slant height of 15.0 in. If 1.00 in. is cut from each of these three dimensions, will at least 10% of the material cost be saved?

28. ![icon] As part of a larger piece of work, an artist wishes to cast a solid plaster obelisk in the shape of a tall, slender square pyramid. She builds a form with a square base 26.0 in. on a side with a height of 128 in. into which she pours the plaster. How many gallons of plaster must she mix to fill the form? (1 gal = 231 in³)

29. ![icon] The artist in Exercise 28 wishes to pour a second obelisk in the shape of a right circular cone with the same dimensions (diameter 26.0 in. and height 128 in.). How many gallons of plaster must she mix to fill this form? (1 gal = 231 in³)

12.10 Spheres

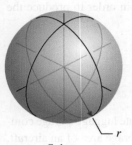

Sphere
Figure 12.79

A **sphere** (Figure 12.79) is a geometric solid formed by a closed curved surface, with all points on the surface the same distance from a given point (the center). The given distance from any point on the surface to the center is called the *radius*. You may also think of a sphere as the solid formed by rotating a circle about its diameter.

The volume of a sphere is found by using the following formula:

Volume of Sphere

$$V = \frac{4}{3}\pi r^3$$

where r is the radius of the sphere.

The surface area of a sphere is found by using the following formula:

Surface Area of Sphere

$$A = 4\pi r^2$$

Example 1 Find the surface area of a sphere of radius 2.80 cm. (See Figure 12.80.)

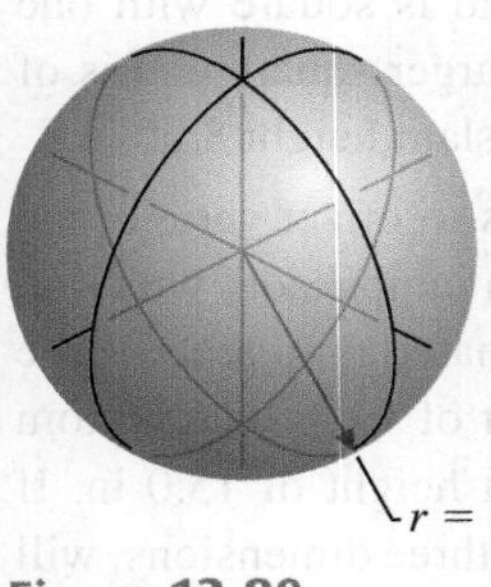

$$A = 4\pi r^2$$
$$A = 4\pi (2.80 \text{ cm})^2$$
$$= 4\pi (2.80)^2 \text{ cm}^2$$
$$= 98.5 \text{ cm}^2$$

$r = 2.80$ cm

Figure 12.80

Example 2 Find the volume of the sphere in Example 1. The formula for the volume of a sphere is

$$V = \frac{4}{3}\pi r^3$$
$$V = \frac{4}{3}\pi (2.80 \text{ cm})^3$$
$$= \frac{4}{3}\pi (2.80)^3 \text{ cm}^3$$
$$= 92.0 \text{ cm}^3$$

EXERCISES 12.10

Follow the rules for working with measurements.

*Find **a.** the surface area and **b.** the volume of each sphere:*

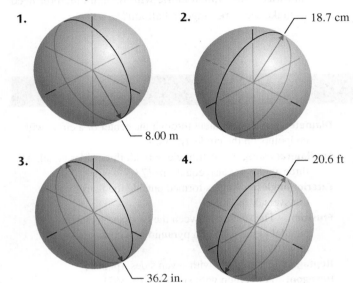

1. 8.00 m
2. 18.7 cm
3. 36.2 in.
4. 20.6 ft

5. A balloon 30.1 m in radius is to be filled with helium. How many m^3 of helium are needed to fill it?

6. An experimental balloon is to have a diameter of 5.72 m. How much material is needed for this balloon?

7. A welder has a large pan with the shape of a hemisphere that is used for scrap metal pieces. If the radius of this pan is 9.00 in., what is the volume?

8. An experimental aircraft has a Plexiglas covering over the cockpit that is hemispherical. If the radius of the hemisphere is 2.00 ft, what is the surface area?

9. A cooling water tower is in need of a larger ball float to shut off the inlet water when the tank is full. The present float is 6.00-in. in diameter and exerts 16.0 lb of force. Calculations show that a ball having twice the volume would provide the necessary force. What is the diameter of the new larger float?

10. A shift lever ball has diameter 2.125 in. What is its volume?

11. How many gallons of water can be stored in the spherical portion of the water tank shown in Illustration 1? ($7.48 \text{ gal} = 1 \text{ ft}^3$)

ILLUSTRATION 1

12. A city drains 150,000 gal of water from a full spherical tank with a radius of 26 ft. How many gallons of water are left in the tank? ($1 \text{ ft}^3 = 7.48 \text{ gal}$)

13. A spherical tank for liquefied petroleum is 16.0 in. in diameter. **a.** What is the ratio of surface area to the volume of the tank? **b.** Find the same ratio for a tank 24.0 in. in diameter. **c.** Find the same ratio for any tank of radius r.

14. Find the volume of the cylindrical silo with a hemispherical top shown in Illustration 2.

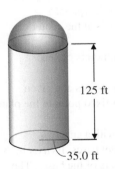

ILLUSTRATION 2

15. You are designing a dome house in the shape of a hemisphere with a 40.0-ft outside diameter. **a.** How many square feet of stucco should be listed on the bill of materials as needed to cover this home if 15% of it is windows and doors? **b.** Find the volume, which is needed to calculate the heat pump size. **c.** If the walls are 12 in. thick, what is the weight of this house used for the footing design? Concrete weighs 148 lb/ft³. Exclude the weight of the windows and doors. **d.** Two coats of sealer are required to cover and seal the exposed external concrete. If 1 gal covers 110 ft², how many gallons are required for the bill of materials?

16. A canister of coolant used to charge an air conditioner is cylindrical with a hemispherical top. What is the volume of the canister if the canister is 1.50 ft tall with radius 4.00 in.?

17. ✦ A meteorologist needs to launch a weather balloon that is roughly spherical. How many ft³ of helium will be required to inflate the balloon to a diameter of 9.0 ft?

18. ✦ A park ranger orders a hot air balloon to use for tourists in the park. She orders a balloon that is roughly spherical and has a diameter of 40.0 ft. **a.** Assuming it is spherical, what will be the volume of the balloon in ft³? **b.** Assuming the manufacturer of the balloon allows 10% more than the required amount of fabric for seams and waste, how much fabric will the manufacturer need to make the park ranger's balloon?

SUMMARY | CHAPTER 12

Glossary of Basic Terms

Acute angle. An angle with a measure less than 90°. (p. 364)

Acute triangle. A triangle with three acute angles. (p. 374)

Adjacent angles. Two angles with a common vertex and a common side lying between them. (p. 364)

Alternate angles. Pairs of angles with different vertices on opposite sides of the transversal but which are both interior or both exterior. (p. 365)

Altitude. A line segment drawn perpendicular to the base of a triangle or a quadrilateral. (p. 369)

Angle. Formed by two lines that have a common point. (p. 363)

Center of a circle. The fixed point that is the same given distance, r, from all points on a circle. (p. 385)

Central angle. An angle whose vertex is at the center of a circle. (p. 387)

Chord. A line segment that has its endpoints on the circle. (p. 387)

Circle. A plane curve consisting of all points at a given distance, called the *radius,* from a fixed point in the plane, called the *center.* (p. 385)

Circular cylinder. A geometric solid with a curved lateral surface and circles as parallel bases. Its *axis* is the line segment between the centers of the bases. The altitude, h, is the shortest (perpendicular) distance between the bases. If the axis is perpendicular to the bases, the cylinder is called a *right circular cylinder* and the axis is the same length as the altitude. (p. 402)

Circumference. The distance around a circle. (p. 385)

Complementary angles. Two angles for which the sum of their measures is 90°. (p. 365)

Cone. A geometric solid whose base is a circle and a curved lateral surface that comes to a point called the *vertex*. The *axis* of a cone is a line segment from the vertex to the center of the base. A *right circular cone* is a cone in which the height is the distance from the vertex to the center of the base. The *slant height* of a right circular cone is the length of a line segment that joins the vertex to any point on the circle that forms the base of the cone. (p. 408)

Corresponding angles. Exterior-interior angles with different vertices on the same side of the transversal. (p. 365)

Diameter. A line segment through the center of a circle with endpoints on the circle. (p. 385)

Equilateral triangle. A triangle with all three sides equal. All three angles are also equal. (p. 373)

Exterior angles. Angles formed outside the lines by the transversal. (p. 365)

Frustum. The section between the base and a plane parallel to the base of a pyramid or a cone. (p. 410)

Heptagon. A polygon with seven sides. (p. 367)

Hexagon. A polygon with six sides. (p. 367)

Hypotenuse. The side opposite the right angle in a right triangle. (p. 374)

Inscribed angle. An angle whose vertex is on the circle and whose sides are chords. (p. 387)

Intercepted arc. The part of the circle between two sides of an inscribed or central angle. (p. 387)

Interior angles. Angles formed inside the lines by the transversal. (p. 365)

Intersect. Two lines intersect if they have only one point in common. (p. 364)

Isosceles triangle. A triangle with two sides equal. The angles opposite the equal sides are also equal. (p. 374)

Lateral surface area. The sum of the areas of the lateral faces of a prism or a pyramid or the area of the curved side of a cylinder or a cone. (p. 397)

Legs. The sides of a right triangle opposite the acute angles. (p. 374)

Nonagon. A polygon with nine sides. (p. 367)

Obtuse angle. An angle with a measure greater than 90° but less than 180°. (p. 364)

Obtuse triangle. A triangle with one obtuse angle. (p. 374)

Octagon. A polygon with eight sides. (p. 367)

Parallel. Two lines are parallel (∥) if they do not intersect even when extended. (p. 364)

Parallelogram. A quadrilateral with opposite sides equal. (p. 369)

Pentagon. A polygon with five sides. (p. 367)

Perpendicular. Two lines are perpendicular (⊥) if they intersect and form equal adjacent angles. (p. 364)

Plane geometry. The study of the properties, measurement, and relationships of points, angles, lines, and curves in two dimensions: length and width. (p. 363)

Polygon. A closed figure whose sides are straight line segments. (p. 366)

Prism. A solid whose sides are parallelograms and whose bases are one pair of parallel polygons that have the same size and shape. The two parallel polygons (which may be any type of polygon) are called the *bases* of the prism. The remaining polygons will be parallelograms and are called *lateral faces*. A *right prism* has lateral faces that are rectangles and therefore are perpendicular to the bases. (p. 397)

Pyramid. A geometric solid whose base is a polygon and whose lateral surfaces are triangles with a common vertex, called the *apex*. (p. 407)

Pythagorean theorem. The square of the hypotenuse of a right triangle is equal to the sum of the squares of the lengths of the two legs, or $c^2 = a^2 + b^2$. (p. 374)

Quadrilateral. A polygon with four sides. (p. 367)

Radian. The measure of an angle with its vertex at the center of a circle and with an intercepted arc on the circle equal in length to the radius. (p. 392)

Radius. The distance, r, between the center of a circle and all points on the circle. (p. 385)

Rectangle. A parallelogram with four right angles. (p. 369)

Regular polygon. A polygon with all of its sides and interior angles equal. (p. 367)

Rhombus. A parallelogram with the lengths of all four sides equal. (p. 369)

Right angle. An angle with a measure of 90°. (p. 364)

Right triangle. A triangle with one right angle. (p. 374)

Scalene triangle. A triangle with no sides equal. No angles are equal either. (p. 374)

Secant. Any line that intersects a circle in two points. (p. 387)

Sector of a circle. The region of a circle bounded by two radii and the arc intercepted by them. (p. 394)

Sides of an angle. The parts of the lines that form an angle. (p. 363)

Similar polygons. Polygons with the same shape. (p. 381)

Solid geometry. The geometry of three dimensions: length, width, and depth. (p. 397)

Sphere. A geometric solid formed by a closed curved surface, with all points on the surface the same distance from a given point (the center). The given distance from any point on the surface to the center is called the *radius*. (p. 414)

Square. A rectangle with the lengths of all four sides equal. (p. 369)

Supplementary angles. Two angles for which the sum of their measures is 180°. (p. 365)

Tangent. A line that has only one point in common with a circle and lies totally outside the circle. A line segment along such a line with only one point in common with a circle is also considered tangent. (p. 387)

Total surface area. The area of the bases plus the lateral surface area, or all surface areas of the figure. (p. 397)

Transversal. A line that intersects two or more lines in different points in the same plane. (p. 365)

Trapezoid. A quadrilateral with only two sides parallel. (p. 369)

Triangle. A polygon with three sides. (p. 367)

Vertex. The common point of two lines that form an angle. (p. 363)

Vertical angles. The angles opposite each other when two lines intersect. (p. 365)

12.1 Angles and Polygons

1. Review this section on the use of a protractor.
2. If two parallel lines are cut by a transversal, then
   a. the *corresponding* angles are *equal*.
   b. the *alternate-interior* angles are *equal*.
   c. the *alternate-exterior* angles are *equal*.
   d. the *interior angles* on the same side of the transversal are *supplementary*. (p. 366)

12.2 Quadrilaterals

1. Summary of formulas for area and perimeter of quadrilaterals:

Quadrilateral	Area	Perimeter
Rectangle	$A = bh$	$P = 2(b + h)$
Square	$A = b^2$	$P = 4b$
Parallelogram	$A = bh$	$P = 2(a + b)$
Rhombus	$A = bh$	$P = 4b$
Trapezoid	$A = \left(\dfrac{a+b}{2}\right)h$	$P = a + b + c + d$

(p. 370)

12.3 Triangles

1. Pythagorean theorem: The square of the hypotenuse of a right triangle is equal to the sum of the squares of the lengths of the two legs. That is, $c^2 = a^2 + b^2$ or $c = \sqrt{a^2 + b^2}$. (p. 374)

2. Formulas for area and perimeter of a triangle: (p. 376)
$$A = \tfrac{1}{2}bh \quad P = a + b + c$$
$$A = \sqrt{s(s-a)(s-b)(s-c)},$$
where $s = \tfrac{1}{2}(a + b + c)$

3. The sum of the measures of the angles of any triangle is 180°. (p. 377)

12.4 Similar Polygons

1. Polygons are similar when the corresponding angles are equal. (p. 381)
2. When two polygons are similar, the lengths of their corresponding sides are proportional. (p. 381)

12.5 Circles

1. **Formulas for circumference and area of a circle:** (p. 386)

Circumference	Area
$C = 2\pi r$	$A = \pi r^2$
$C = \pi d$	$A = \dfrac{\pi d^2}{4}$

2. The sum of the measures of all the central angles of any circle is 360°. (p. 387)
3. **Arcs:**
   a. The measure of a central angle in a circle is equal to the measure of its intercepted arc.
   b. The measure of an inscribed angle in a circle is equal to one-half the measure of its intercepted arc.
   c. The measure of an angle formed by two intersecting chords in a circle is equal to one-half the sum of the measures of the intercepted arcs. (p. 387)
4. **Chords and tangents:**
   a. A diameter that is perpendicular to a chord bisects the chord.
   b. A line segment from the center of a circle to the point of tangency is perpendicular to the tangent.
   c. Two tangents drawn from a point outside a circle to the circle are equal. The line segment drawn from the center of the circle to this point outside the circle bisects the angle formed by the tangents. (p. 388)

12.6 Radian Measure

1. **Area of a sector of a circle with the central angle measured in degrees:**

   $A = \dfrac{\theta}{360°} \pi r^2$ (p. 394)

2. **Area of a sector of a circle with the central angle measured in radians:**

   $A = \dfrac{1}{2} r^2 \theta$ (p. 394)

12.7 Prisms

1. **Formulas for prisms:**
   a. The *lateral surface area of a prism* is the sum of the areas of the lateral faces of the prism.
   b. The *total surface area of a prism* is the sum of the areas of the lateral faces and the areas of the bases of the prism.
   c. The *volume of a prism* is $V = Bh$, where B is the area of one of the bases and h is the altitude of the prism. (pp. 397–399)

12.8 Cylinders

1. **Formulas for cylinders:**
   a. The *volume of a right circular cylinder* is $V = \pi r^2 h$, where r is the radius of the base and h is the altitude.
   b. The *lateral surface area of a right circular cylinder* is $A = 2\pi rh$, where r is the radius of the base and h is the altitude.
   c. The *total surface area* of a right circular cylinder is the area of the bases plus the lateral surface area. (pp. 403–404)

12.9 Pyramids and Cones

1. **Volume of a pyramid:** $V = \dfrac{1}{3} Bh$, where B is the area of the base and h is the height of the pyramid. The height of a pyramid is the shortest (perpendicular) distance between the apex and the base of the pyramid. (p. 408)
2. **Formulas for right circular cones:**
   a. The *volume of a right circular cone* is $V = \dfrac{1}{3}\pi r^2 h$, where r is the radius of the base and h is the altitude.
   b. The *lateral surface area of a right circular cone* is $A = \pi rs$, where r is the radius of the base and s is the slant height.
   c. The *total surface area of a right circular cone* is the sum of the lateral surface area and the area of the base. (p. 408)
3. **Formulas for the frustum of a pyramid:**
   a. Volume: $V = \dfrac{1}{3}h(B_1 + B_2 + \sqrt{B_1 B_2})$,

      where h is the altitude and B_1 and B_2 are the areas of the bases.
   b. Lateral surface area: $A = \dfrac{1}{2}s(P_1 + P_2)$,

      where s is the slant height and P_1 and P_2 are the perimeters of the bases. (p. 409)
4. **Formulas for the frustum of a cone:**
   a. Volume: $V = \dfrac{1}{3}h(B_1 + B_2 + \sqrt{B_1 B_2})$,

      where h is the altitude and B_1 and B_2 are the areas of the bases.
   b. Lateral surface area: $A = \pi s(r_1 + r_2)$, where s is the slant height and r_1 and r_2 are the radii of the bases. (p. 411)

12.10 Spheres

1. **Formulas for spheres:**
   a. Volume: $V = \dfrac{4}{3}\pi r^3$, where r is the radius of the sphere.
   b. Surface area: $A = 4\pi r^2$, where r is the radius of the sphere. (p. 414)

REVIEW | CHAPTER 12

Classify each angle as right, acute, or obtuse:

1.
2.

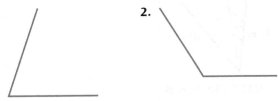

For Exercises 3–5, see Illustration 1:

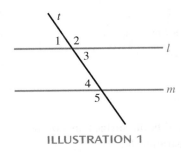

ILLUSTRATION 1

3. In Illustration 1, l is parallel to m and $\angle 5 = 121°$. Find the measure of each angle.
4. In Illustration 1, $\angle 4$ and $\angle 5$ are called ____?____ angles.
5. Suppose $\angle 1 = 4x + 5$ and $\angle 2 = 2x + 55$. Find the value of x.
6. Name the polygon that has **a.** 4 sides, **b.** 5 sides, **c.** 6 sides, **d.** 3 sides, and **e.** 8 sides.

Find the perimeter and the area of each quadrilateral:

7.

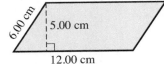

8.
9.

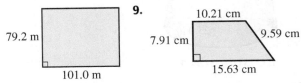

10. The area of a rectangle is 79.6 m². The length is 10.3 m. What is the width?
11. The area of a parallelogram is 2.53 cm². Find its height if the base is 1.76 cm.

Find the area and the perimeter of each triangle:

12.
13.

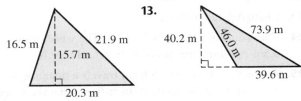

14.

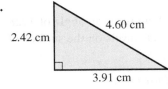

Find the length of the hypotenuse of each triangle:

15.
16.

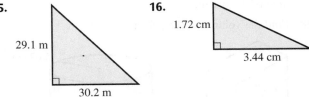

17. Find the measure of the missing angle in Illustration 2.

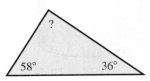

ILLUSTRATION 2

18. In Illustration 3, suppose $\overline{DE} \parallel \overline{BC}$. Find length BC.

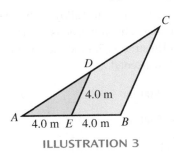

ILLUSTRATION 3

19. Find the area and circumference of the circle in Illustration 4.

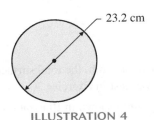

ILLUSTRATION 4

20. The area of a circle is 462 cm². Find its radius.
21. How many degrees are in a central angle whose arc is $\frac{3}{5}$ of a circle?
22. Change 24.0° to radians.
23. Change $\frac{\pi}{18}$ rad to degrees.
24. The radius of a wheel is 75.3 cm. The wheel turns 0.561 rad. Find the distance the wheel travels.
25. A wheel of diameter 25.8 cm travels a distance of 20.0 cm. Find the angle θ (in radians) that the wheel turns.
26. A wheel of radius 16.2 cm turns an angle of 1028°. Find the distance a point travels on the surface of the wheel.

For Exercises 27–28, see Illustration 5:

27. Find **a.** the lateral surface area and **b.** the total surface area of the prism.
28. Find the volume of the prism.

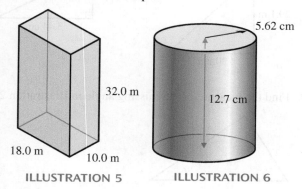

ILLUSTRATION 5   ILLUSTRATION 6

29. Find the volume of the right circular cylinder shown in Illustration 6.
30. A metallurgist needs to cast a molten alloy in the shape of a right circular cylinder. The dimensions of the mold are as shown in Illustration 7. Find the amount (volume) of molten alloy needed.

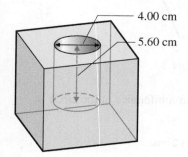

ILLUSTRATION 7

31. Find **a.** the lateral surface area and **b.** the total surface area of the cylinder that was cast in Exercise 30.
32. Find the volume of the pyramid shown in Illustration 8.

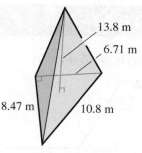

ILLUSTRATION 8

33. Find **a.** the volume and **b.** the lateral surface area of the right circular cone shown in Illustration 9.

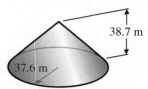

ILLUSTRATION 9

34. Find **a.** the volume and **b.** the surface area of the sphere shown in Illustration 10.

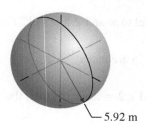

ILLUSTRATION 10

35. Find the volume and lateral surface area of the frustum of the cone shown in Illustration 11.

ILLUSTRATION 11

36. An open (no top) rectangular box with length 6.00 ft, width 4.00 ft, and height 2.25 ft is to be painted on all interior and exterior sides, including the bottom, with a waterproof material. Each gallon covers 11.0 ft² and costs $40. **a.** Find how many gallon containers must be purchased and the cost of painting this box. **b.** If the box is $\frac{4}{5}$ full of water, find the weight of the water. (Water weighs 62.4 lb/ft³.)

TEST | CHAPTER 12

1. Find the area of a rectangle 18.0 ft long and 6.00 ft wide.
2. Find the perimeter of a square lot 160 m on a side.

Given the trapezoid in Illustration 1, find

3. its area
4. its perimeter

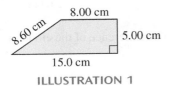

ILLUSTRATION 1

5. Find the length of side *a* in Illustration 2.

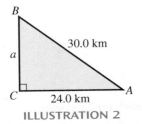

ILLUSTRATION 2

Given a circle of radius 20.0 cm, find

6. its area
7. its circumference
8. Change 240° to radians.
9. Change $\frac{7\pi}{4}$ rad to degrees.
10. Find the volume of a rectangular box 12.0 ft × 8.00 ft × 9.00 ft.
11. Find the total surface area of the box in Exercise 10.
12. Find the volume of a cylindrical tank 20.0 m in diameter and 30.0 m high.
13. Find **a.** the volume and **b.** the lateral surface area of the bin in Illustration 3.

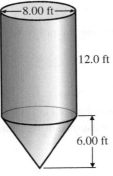

ILLUSTRATION 3

14. Find the volume of the frustum of the cone in Illustration 4.

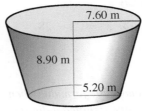

ILLUSTRATION 4

15. A welder is assigned to fabricate the sheet metal water trough shown in Illustration 5. **a.** Find its cubic foot capacity. **b.** What is the length of side *y*?

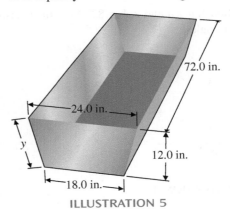

ILLUSTRATION 5

CUMULATIVE REVIEW | CHAPTERS 1–12

1. Add: $-8 + (+7) + (-3)$
2. Given $P = 2(l + w)$, where $l = 4\frac{1}{8}$ in. and $w = 2\frac{3}{4}$ in., find P.
3. The mass of a full-size automobile is **a.** 100 kg, **b.** 1500 kg, **c.** 10 kg, or **d.** 15,000 kg.
4. Find **a.** precision and **b.** greatest possible error: 20,400 L
5. Simplify: $\dfrac{14x^3 - 56x^2 - 28x}{7x}$
6. Find the product: $(-2x^2 + 7x - 3)(4x + 5)$
7. Solve: $4 - 2x = 18$
8. $E = mv^2$; find m if $E = 952$ and $v = 7.00$.
9. A 200-bu wagon holds 3.4 tons of grain. Express the weight of the grain in pounds per bushel.

10. A 160-lb object, 28.0 in. from the fulcrum of a lever, balances a second object 80.0 in. from the other end of the lever. What is the weight of the second object?
11. Solve for y: $5x - 8y = 10$
12. Find the slope of the line containing the points $(2, -1)$ and $(5, -8)$.
13. Solve: $5x - y = 12$
    $y = 2x$
14. The sum of the resistance of two resistors is 1300 Ω. The larger has three times the resistance of the smaller. Find the resistance of each.

Find each product mentally:

15. $(3x - 5)(2x + 7)$
16. $(4x - 3)^2$

Factor each expression completely:

17. $5x^3 - 15x$
18. $x^2 - 3x - 28$
19. Solve: $10x^2 - 5x = 105$
20. Solve: $2x^2 - x = 3$

Solve each equation using the quadratic formula (when necessary, round the results to three significant digits):

21. $5x^2 + 13x - 6 = 0$
22. $4x^2 - 10x - 29 = 0$
23. Draw the graph of $y = 2x^2 - 3x - 2$ and label the vertex.
24. In Illustration 1, $m \parallel n$ and $\angle 4 = 82°$. Find the measure of the other angles.

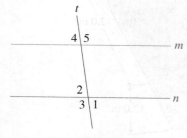

ILLUSTRATION 1

25. In Illustration 1, if $\angle 4 = 2x - 3$ and $\angle 5 = x + 6$, find the value of x.

Find the perimeter and area of each triangle:

26.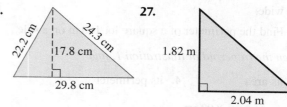

27.

28. Find the area and the circumference of the circle in Illustration 2.

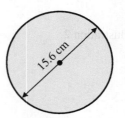

ILLUSTRATION 2

29. The area of a circle is 168 cm². Find its radius.
30. Find the volume and total surface area of the cylinder shown in Illustration 3.

ILLUSTRATION 3

Right Triangle Trigonometry

CHAPTER 13

OBJECTIVES

- Write the trigonometric ratios for the sine, cosine, and tangent of an angle using the basic terms of a right triangle.
- Find the value of a trigonometric ratio using a scientific calculator.
- Use a trigonometric ratio to find angles.
- Use a trigonometric ratio to find sides.
- Solve a right triangle.
- Solve application problems involving trigonometric ratios and right triangles.

Mathematics at Work

Computer support specialists provide technical assistance, support, and training to computer system users; investigate and resolve computer handware and software problems; and answer user questions and concerns in person or via a telephone help desk using automated diagnostic programs. Such concerns may include word processing, printing, e-mail, programming languages, and operating systems. Computer support specialists may work within an organization, work directly for a computer or software vendor, or work for help-desk or computer support service companies that provide customer support on an outsourced contract basis.

Other related computer specialists include database administrators, who work with database management systems software and determine ways to organize and store data. Network administrators design, install, and support an organization's local area network (LAN), wide area network (WAN), and Internet or intranet system. Computer security specialists plan, coordinate, and implement an organization's information security system. A computer scientist title applies widely to computer professionals who design computers and computer software, develop information technologies, and develop ways to use computers for new tasks. Many of these computer-related jobs require a 2-year associate degree; a bachelor's degree is usually required for the computer scientist. For more information, please visit **www.cengage.com** and access the Student Online Resources for this text.

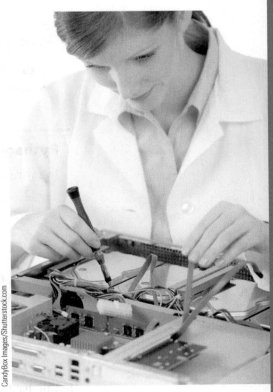

Computer Support Specialist
Computer technician reparing a computer

13.1 Trigonometric Ratios

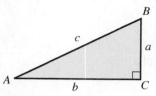

Figure 13.1
Right triangle

Many applications in science and technology require the use of triangles and trigonometry. Early applications of trigonometry, beginning in the second century B.C., were in astronomy, surveying, and navigation. Applications that you may study include electronics, the motion of projectiles, light refraction in optics, and sound.

In this chapter, we consider only right triangles. A right triangle has one right angle, two acute angles, a hypotenuse, and two legs. The right angle, as shown in Figure 13.1, is usually labeled with the capital letter *C*. The vertices of the two acute angles are usually labeled with the capital letters *A* and *B*. The hypotenuse is the side opposite the right angle, the longest side of a right triangle, and is usually labeled with the lowercase letter *c*. The legs are the sides opposite the acute angles. The leg (side) opposite angle *A* is labeled *a*, and the leg opposite angle *B* is labeled *b*. Note that each side of the triangle is labeled with the lowercase of the letter of the angle opposite that side.

The two legs are also named as the side *opposite* angle *A* and the side *adjacent to* (or next to) angle *A* or as the side opposite angle *B* and the side adjacent to angle *B*. See Figure 13.2.

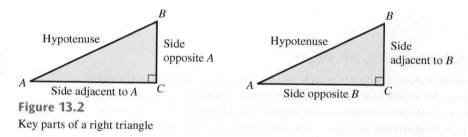

Figure 13.2
Key parts of a right triangle

> **Pythagorean Theorem**
>
> In any right triangle,
> $$c^2 = a^2 + b^2$$

That is, the square of the length of the hypotenuse is equal to the sum of the squares of the lengths of the legs. The following equivalent formulas are often more useful:

$c = \sqrt{a^2 + b^2}$ used to find the length of the hypotenuse
$a = \sqrt{c^2 - b^2}$ used to find the length of leg *a*
$b = \sqrt{c^2 - a^2}$ used to find the length of leg *b*

Recall that the Pythagorean theorem was developed in detail in Section 12.3.

Example 1   Find the length of side *b* in Figure 13.3.

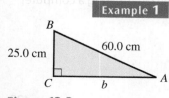

Figure 13.3

Using the formula to find the length of leg *b*, we have

$b = \sqrt{c^2 - a^2}$
$b = \sqrt{(60.0 \text{ cm})^2 - (25.0 \text{ cm})^2}$
$ = 54.5 \text{ cm}$

13.1 ♦ Trigonometric Ratios   425

Example 2 Find the length of side c in Figure 13.4.

Using the formula to find the hypotenuse c, we have

$$c = \sqrt{a^2 + b^2}$$
$$c = \sqrt{(29.7 \text{ m})^2 + (34.2 \text{ m})^2}$$
$$= 45.3 \text{ m}$$

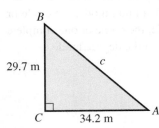

Figure 13.4

A **ratio** is the comparison of two quantities by division. The ratios of the sides of a right triangle can be used to find an unknown part—or parts—of that right triangle. Such a ratio is called a **trigonometric ratio** and expresses the relationship between an acute angle and the lengths of two of the sides of a right triangle.

The **sine of angle A**, abbreviated "sin A," equals the ratio of the length of the side opposite angle A, which is a, to the length of the hypotenuse, c.

The **cosine of angle A**, abbreviated "cos A," equals the ratio of the length of the side adjacent to angle A, which is b, to the length of the hypotenuse, c.

The **tangent of angle A**, abbreviated "tan A," equals the ratio of the length of the side opposite angle A, which is a, to the length of the side adjacent to angle A, which is b.

That is, in a right triangle (Figure 13.5), we have the following ratios:

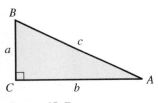

Figure 13.5

Trigonometric Ratios

$$\sin A = \frac{\text{length of side opposite angle } A}{\text{length of hypotenuse}} = \frac{a}{c}$$

$$\cos A = \frac{\text{length of side adjacent to angle } A}{\text{length of hypotenuse}} = \frac{b}{c}$$

$$\tan A = \frac{\text{length of side opposite angle } A}{\text{length of side adjacent to angle } A} = \frac{a}{b}$$

Similarly, the ratios can be defined for angle B.

$$\sin B = \frac{\text{length of side opposite angle } B}{\text{length of hypotenuse}} = \frac{b}{c}$$

$$\cos B = \frac{\text{length of side adjacent to angle } B}{\text{length of hypotenuse}} = \frac{a}{c}$$

$$\tan B = \frac{\text{length of side opposite angle } B}{\text{length of side adjacent to angle } B} = \frac{b}{a}$$

Example 3 Find the three trigonometric ratios rounded to four significant digits for angle A in the triangle in Figure 13.6.

$$\sin A = \frac{\text{length of side opposite angle } A}{\text{length of hypotenuse}} = \frac{a}{c} = \frac{144 \text{ m}}{156 \text{ m}} = 0.9231$$

$$\cos A = \frac{\text{length of side adjacent to angle } A}{\text{length of hypotenuse}} = \frac{b}{c} = \frac{60.0 \text{ m}}{156 \text{ m}} = 0.3846$$

$$\tan A = \frac{\text{length of side opposite angle } A}{\text{length of side adjacent to angle } A} = \frac{a}{b} = \frac{144 \text{ m}}{60.0 \text{ m}} = 2.400$$

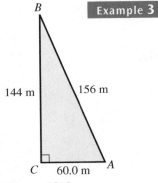

Figure 13.6

The values of the trigonometric ratios of various angles can be found with calculators. You will need a calculator that has sin, cos, and tan keys.

Very Important Note: When working with the trigonometric functions on your calculator, make certain that it is set in the degree mode. If your calculator has a DRG key, it is used to change angle measurement modes from degrees to radians to grads. In the degree mode, the circle or one complete revolution is divided into 360°. In the radian mode, the circle or one complete revolution is divided into 2π rad. In the grad mode, the circle or one complete revolution is divided into 400^g. We will be working exclusively in the degree mode.

Example 4 Find sin 37.5° rounded to four significant digits.

[SIN] 37.5 [=]

0.608761429

Thus, sin 37.5° = 0.6088 rounded to four significant digits.

Example 5 Find cos 18.63° rounded to four significant digits.

[COS] 18.63 [=]

0.947601273

Thus, cos 18.63° = 0.9476 rounded to four significant digits.

Example 6 Find tan 81.7° rounded to four significant digits.

[TAN] 81.7 [=]

6.854750833

Thus, tan 81.7° = 6.855 rounded to four significant digits.

A calculator may also be used to find the *angle* when the value of the trigonometric ratio is known. The procedure is shown in the following examples.

NOTE: The following keys on a calculator have the meanings that follow:

[SIN⁻¹] "the angle whose sine is"

[COS⁻¹] "the angle whose cosine is"

[TAN⁻¹] "the angle whose tangent is"

Example 7 Find angle A to the nearest tenth of a degree when sin A = 0.6372.

[SIN⁻¹] .6372 [=]

39.583346

NOTE: Make certain that your calculator is in the degree mode.

Thus, angle A = 39.6° rounded to the nearest tenth of a degree.

Example 8 Find angle B to the nearest tenth of a degree when $\tan B = 0.3106$.

  .3106 =

17.25479431

Thus, angle $B = 17.3°$ rounded to the nearest tenth of a degree.

Example 9 Find angle A to the nearest hundredth of a degree when $\cos A = 0.4165$.

65.3861858

Thus, angle $A = 65.39°$ rounded to the nearest hundredth of a degree.

EXERCISES 13.1

Refer to right triangle ABC in Illustration 1 for Exercises 1–10:

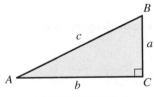

ILLUSTRATION 1

1. The side opposite angle A is ___?___.
2. The side opposite angle B is ___?___.
3. The hypotenuse is ___?___.
4. The side adjacent to angle A is ___?___.
5. The side adjacent to angle B is ___?___.
6. The angle opposite side a is ___?___.
7. The angle opposite side b is ___?___.
8. The angle opposite side c is ___?___.
9. The angle adjacent to side a is ___?___.
10. The angle adjacent to side b is ___?___.

Use right triangle ABC in Illustration 1 and the Pythagorean theorem to find each unknown side, rounded to three significant digits:

11. $c = 75.0$ m, $a = 45.0$ m
12. $a = 25.0$ cm, $b = 60.0$ cm
13. $a = 29.0$ mi, $b = 47.0$ mi
14. $a = 12.0$ km, $c = 61.0$ km
15. $c = 18.9$ cm, $a = 6.71$ cm
16. $a = 20.2$ mi, $b = 19.3$ mi
17. $a = 171$ ft, $b = 203$ ft
18. $c = 35.3$ m, $b = 25.0$ m
19. $a = 202$ m, $c = 404$ m
20. $a = 1.91$ km, $b = 3.32$ km
21. $b = 1520$ km, $c = 2160$ km
22. $a = 203,000$ ft, $c = 521,000$ ft
23. $a = 45,800$ m, $b = 38,600$ m
24. $c = 3960$ m, $b = 3540$ m

Use the triangle in Illustration 2 for Exercises 25–30.

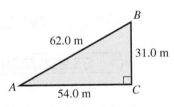

ILLUSTRATION 2

25. Find $\sin A$.
26. Find $\cos A$.
27. Find $\tan A$.
28. Find $\sin B$.
29. Find $\cos B$.
30. Find $\tan B$.

Find the value of each trigonometric ratio rounded to four significant digits:

31. $\sin 49.6°$
32. $\cos 55.2°$
33. $\tan 65.3°$
34. $\sin 69.7°$
35. $\cos 29.7°$
36. $\tan 14.6°$
37. $\sin 31.64°$
38. $\tan 13.25°$
39. $\cos 75.31°$
40. $\cos 84.83°$
41. $\tan 3.05°$
42. $\sin 6.74°$
43. $\sin 37.62°$
44. $\cos 18.94°$
45. $\tan 21.45°$
46. $\sin 11.31°$
47. $\cos 47.16°$
48. $\tan 81.85°$

428 CHAPTER 13 ♦ Right Triangle Trigonometry

Find each angle rounded to the nearest tenth of a degree:

49. $\sin A = 0.7941$
50. $\tan A = 0.2962$
51. $\cos B = 0.4602$
52. $\cos A = 0.1876$
53. $\tan B = 1.386$
54. $\sin B = 0.3040$
55. $\sin B = 0.1592$
56. $\tan B = 2.316$
57. $\cos A = 0.8592$
58. $\cos B = 0.3666$
59. $\tan A = 0.8644$
60. $\sin A = 0.5831$
63. $\cos A = 0.3572$
64. $\cos B = 0.2597$
65. $\sin A = 0.1506$
66. $\tan B = 2.500$
67. $\tan B = 3.806$
68. $\sin A = 0.4232$
69. $\cos B = 0.7311$
70. $\cos A = 0.6427$
71. $\sin B = 0.3441$
72. $\tan A = 0.5536$

73. In Exercises 25–30, there are two pairs of ratios that are equal. Name them.

Find each angle rounded to the nearest hundredth of a degree:

61. $\tan A = 0.1941$
62. $\sin B = 0.9324$

13.2 Using Trigonometric Ratios to Find Angles

A trigonometric ratio may be used to find an angle of a right triangle, given the lengths of any two sides.

Example 1

In Figure 13.7, find angle A using a calculator, as follows.

We know the sides opposite and adjacent to angle A. So we use the tangent ratio:

$$\tan A = \frac{\text{length of side opposite angle } A}{\text{length of side adjacent to angle } A}$$

$$\tan A = \frac{28.5 \text{ m}}{21.3 \text{ m}} = 1.338$$

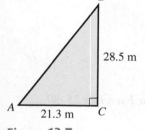

Figure 13.7

Next, find angle A to the nearest tenth of a degree when $\tan A = 1.338$. The complete set of operations on a calculator follows.

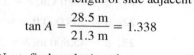

Thus, angle $A = 53.2°$ rounded to the nearest tenth of a degree. ♦

When calculations involve a trigonometric ratio, we shall use the following rule for significant digits:

Angles expressed to the nearest	The length of each side of the triangle contains
1°	Two significant digits
0.1°	Three significant digits
0.01°	Four significant digits

An example of each case is shown in Figure 13.8.

13.2 ♦ Using Trigonometric Ratios to Find Angles

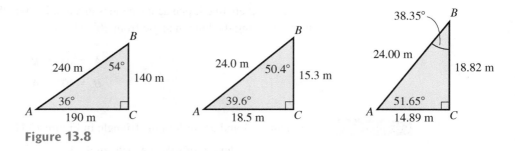

Figure 13.8

Example 2

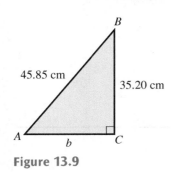

Figure 13.9

Find angle B in the triangle in Figure 13.9.

We know the hypotenuse and the side adjacent to angle B. So let's use the cosine ratio.

$$\cos B = \frac{\text{length of side adjacent to angle } B}{\text{length of hypotenuse}}$$

$$\cos B = \frac{35.20 \text{ cm}}{45.85 \text{ cm}}$$

Find angle B using a calculator as follows:

| 39.85033989 |

Thus, angle $B = 39.85°$ rounded to the nearest hundredth of a degree. ♦

The question is often raised, "Which of the three trig ratios do I use?" First, notice that each trigonometric ratio consists of two sides and one angle, or three quantities in all. To find the solution to such an equation, two of the quantities must be known. We will answer the question in two parts.

Which Trig Ratio to Use

1. If you are finding an angle, two sides must be known. Label these two known sides as *side opposite* the angle you are finding, *side adjacent* to the angle you are finding, or *hypotenuse*. Then choose the trig ratio that has these two sides.

2. If you are finding a side, one side and one angle must be known. Label the known side and the unknown side as *side opposite* the known angle, *side adjacent* to the known angle, or *hypotenuse*. Then choose the trig ratio that has these two sides.

A useful and time-saving fact about right triangles (Figure 13.10) is that the sum of the acute angles of any right triangle is 90°:

$$A + B = 90°$$

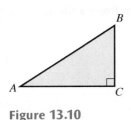

Figure 13.10

Why? We know that the sum of the interior angles of any triangle is 180°. A right triangle, by definition, contains a right angle, whose measure is 90°. That leaves 90° to be divided between the two acute angles.

Note, then, that if one acute angle is given or known, the other acute angle may be found by subtracting the known angle from 90°. That is,

$$A = 90° - B$$
$$B = 90° - A$$

Example 3 Find angle A and angle B in the triangle in Figure 13.11.

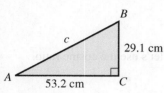

Figure 13.11

$$\tan A = \frac{\text{length of side opposite angle } A}{\text{length of side adjacent to angle } A}$$

$$\tan A = \frac{29.1 \text{ cm}}{53.2 \text{ cm}} = 0.5470$$

$$A = 28.7°$$

Angle $B = 90° - 28.7° = 61.3°$. ♦

EXERCISES 13.2

Using Illustration 1, find the measure of each acute angle for each right triangle:

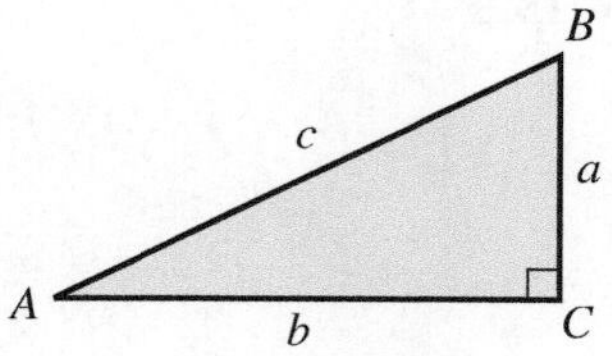

1. $a = 36.0$ m, $b = 50.9$ m
2. $a = 72.0$ cm, $c = 144$ cm
3. $b = 39.7$ cm, $c = 43.6$ cm
4. $a = 171$ km, $b = 695$ km
5. $b = 13.6$ m, $c = 18.7$ m
6. $b = 409$ km, $c = 612$ km
7. $a = 29.7$ m, $b = 29.7$ m, $c = 42.0$ m
8. $a = 36.2$ mm, $b = 62.7$ mm, $c = 72.4$ mm
9. $a = 2902$ km, $b = 1412$ km
10. $b = 21.34$ m, $c = 47.65$ m
11. $a = 0.6341$ cm, $c = 0.7982$ cm
12. $b = 4.372$ m, $c = 5.806$ m
13. $b = 1455$ ft, $c = 1895$ ft
14. $a = 25.45$ in., $c = 41.25$ in.
15. $a = 243.2$ km, $b = 271.5$ km
16. $a = 351.6$ m, $b = 493.0$ m
17. $a = 16.7$ m, $c = 81.4$ m
18. $a = 847$ m, $b = 105$ m
19. $b = 1185$ ft, $c = 1384$ ft
20. $a = 48.7$ cm, $c = 59.5$ cm
21. $a = 845$ km, $b = 2960$ km
22. $b = 2450$ km, $c = 3570$ km
23. $a = 8897$ m, $c = 9845$ m
24. $a = 58.44$ mi, $b = 98.86$ mi

ILLUSTRATION 1

13.3 Using Trigonometric Ratios to Find Sides

We also use a trigonometric ratio to find a side of a right triangle, given one side and the measure of one of the acute angles.

Example 1 Find side a in the triangle in Figure 13.12.

With respect to the known angle B, we know the hypotenuse and are finding the adjacent side. So we use the cosine ratio.

$$\cos B = \frac{\text{length of side adjacent to angle } B}{\text{length of hypotenuse}}$$

$$\cos 24.5° = \frac{a}{258 \text{ ft}}$$

$$a = (\cos 24.5°)(258 \text{ ft}) \quad \text{Multiply both sides by 258 ft.}$$

13.3 ♦ Using Trigonometric Ratios to Find Sides

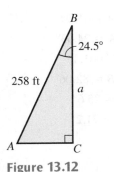

Figure 13.12

Side a can be found by using a calculator as follows:

[COS] 24.5 [ )]* [×] 258 [=]

[234.7700079]

Thus, side $a = 235$ ft rounded to three significant digits.

Example 2   Find sides b and c in the triangle in Figure 13.13.

If we find side b first, we are looking for the side adjacent to angle A, the known angle. We are given the side opposite angle A. Thus, we should use the tangent ratio.

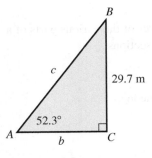

Figure 13.13

$$\tan A = \frac{\text{length of side opposite angle } A}{\text{length of side adjacent to angle } A}$$

$$\tan 52.3° = \frac{29.7 \text{ m}}{b}$$

$b(\tan 52.3°) = 29.7$ m   Multiply both sides by b.

$b = \dfrac{29.7 \text{ m}}{\tan 52.3°}$ Divide both sides by $\tan 52.3°$.

29.7 [÷] [TAN] 52.3 [=]

[22.95476858]

Thus, side $b = 23.0$ m rounded to three significant digits.

To find side c, we are looking for the hypotenuse, and we have the opposite side given. Thus, we should use the sine ratio.

$$\sin A = \frac{\text{length of side opposite angle } A}{\text{length of hypotenuse}}$$

$$\sin 52.3° = \frac{29.7 \text{ m}}{c}$$

$c(\sin 52.3°) = 29.7$ m   Multiply both sides by c.

$c = \dfrac{29.7 \text{ m}}{\sin 52.3°}$ Divide both sides by $\sin 52.3°$.

$ = 37.5$ m

The Pythagorean theorem may be used to check your work.

EXERCISES 13.3

Find the unknown sides of each right triangle (see Illustration 1):

1. $a = 36.7$ m, $A = 42.1°$
2. $b = 73.6$ cm, $B = 19.0°$
3. $a = 236$ km, $B = 49.7°$
4. $b = 28.9$ ft, $A = 65.2°$
5. $c = 49.1$ cm, $A = 36.7°$
6. $c = 236$ m, $A = 12.9°$

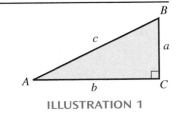

ILLUSTRATION 1

Note: You might need to insert a right parenthesis to clarify the order of operations. The trigonometry keys may also include the left parenthesis.

7. $b = 23.7$ cm, $A = 23.7°$
8. $a = 19.2$ km, $B = 63.2°$
9. $b = 29{,}200$ km, $A = 12.9°$
10. $c = 36.7$ mi, $B = 68.3°$
11. $a = 19.72$ m, $A = 19.75°$
12. $b = 125.3$ cm, $B = 23.34°$
13. $c = 255.6$ mi, $A = 39.25°$
14. $c = 7.363$ km, $B = 14.80°$
15. $b = 12{,}350$ m, $B = 69.72°$
16. $a = 3678$ m, $B = 10.04°$
17. $a = 1980$ m, $A = 18.4°$
18. $a = 9820$ ft, $B = 35.7°$
19. $b = 841.6$ km, $A = 18.91°$
20. $c = 289.5$ cm, $A = 24.63°$
21. $c = 185.6$ m, $B = 61.45°$
22. $b = 21.63$ km, $B = 82.06°$
23. $c = 256$ cm, $A = 25.6°$
24. $a = 18.3$ mi, $A = 71.2°$

13.4 Solving Right Triangles

The phrase **solving a right triangle** refers to finding the measures of the various parts of a triangle that are not given. We proceed as we did in the last two sections.

Example 1

Solve the right triangle in Figure 13.14.

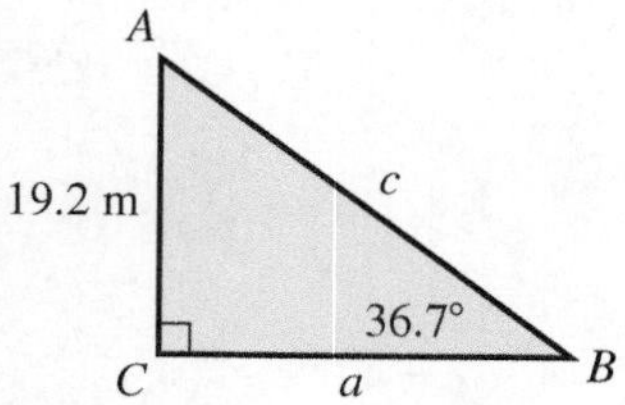

We are given the measure of one acute angle and the length of one leg.

$$A = 90° - B$$
$$A = 90° - 36.7° = 53.3°$$

We then can use either the sine or the cosine ratio to find side c.

$$\sin B = \frac{\text{length of side opposite angle } B}{\text{length of hypotenuse}}$$

$$\sin 36.7° = \frac{19.2 \text{ m}}{c}$$

$c(\sin 36.7°) = 19.2$ m   Multiply both sides by c.

$$c = \frac{19.2 \text{ m}}{\sin 36.7°}$$ Divide both sides by $\sin 36.7°$.

$\quad = 32.1$ m

Figure 13.14

Now we may use either a trigonometric ratio or the Pythagorean theorem to find side a.

Solution by a Trigonometric Ratio:

$$\tan B = \frac{\text{length of side opposite angle } B}{\text{length of side adjacent to angle } B}$$

$$\tan 36.7° = \frac{19.2 \text{ m}}{a}$$

$a(\tan 36.7°) = 19.2$ m   Multiply both sides by a.

$$a = \frac{19.2 \text{ m}}{\tan 36.7°}$$ Divide both sides by $\tan 36.7°$.

$\quad = 25.8$ m

Solution by the Pythagorean Theorem:

$$a = \sqrt{c^2 - b^2}$$
$$a = \sqrt{(32.1 \text{ m})^2 - (19.2 \text{ m})^2}$$
$\quad = 25.7$ m

Can you explain the difference in these two results?

Example 2

Solve the right triangle in Figure 13.15.

We are given the measure of one acute angle and the length of the hypotenuse.

$$A = 90° - B$$
$$A = 90° - 45.7° = 44.3°$$

To find side b, we must use the sine or the cosine ratio, since the hypotenuse is given.

$$\sin B = \frac{\text{length of side opposite angle } B}{\text{length of hypotenuse}}$$

$$\sin 45.7° = \frac{b}{397 \text{ km}}$$

$(\sin 45.7°)(397 \text{ km}) = b$ Multiply both sides by 397 km.

$$284 \text{ km} = b$$

Again, we can use either a trigonometric ratio or the Pythagorean theorem to find side a.

Solution by a Trigonometric Ratio:

$$\cos B = \frac{\text{length of side adjacent to angle } B}{\text{length of hypotenuse}}$$

$$\cos 45.7° = \frac{a}{397 \text{ km}}$$

$(\cos 45.7°)(397 \text{ km}) = a$ Multiply both sides by 397 km.

$$277 \text{ km} = a$$

Solution by the Pythagorean Theorem:

$$a = \sqrt{c^2 - b^2}$$
$$a = \sqrt{(397 \text{ km})^2 - (284 \text{ km})^2}$$
$$= 277 \text{ km}$$

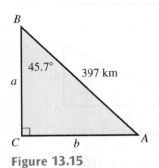

Figure 13.15

Example 3

Solve the right triangle in Figure 13.16.

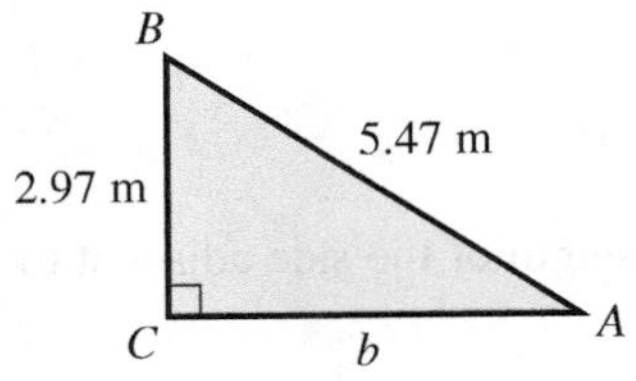

We are given two sides of the right triangle.

To find angle A or angle B, we could use either the sine or cosine, since the hypotenuse is given.

$$\sin A = \frac{\text{length of side opposite angle } A}{\text{length of hypotenuse}}$$

$$\sin A = \frac{2.97 \text{ m}}{5.47 \text{ m}} = 0.5430$$

$$A = 32.9°$$

Then

$$B = 90° - A$$
$$B = 90° - 32.9° = 57.1°$$

The unknown side b can be found by using the Pythagorean theorem:

$$b = \sqrt{c^2 - a^2}$$
$$b = \sqrt{(5.47 \text{ m})^2 - (2.97 \text{ m})^2}$$
$$= 4.59 \text{ m}$$

Figure 13.16

EXERCISES 13.4

Using Illustration 1, solve each right triangle:

1. $A = 50.6°, c = 49.0$ m
2. $a = 30.0$ cm, $b = 40.0$ cm
3. $B = 41.2°, a = 267$ ft
4. $A = 39.7°, b = 49.6$ km
5. $b = 72.0$ mi, $c = 78.0$ mi
6. $B = 22.4°, c = 46.0$ mi
7. $A = 52.1°, a = 72.0$ mm
8. $B = 42.3°, b = 637$ m
9. $A = 68.8°, c = 39.4$ m
10. $a = 13.6$ cm, $b = 13.6$ cm
11. $a = 12.00$ m, $b = 24.55$ m
12. $B = 38.52°, a = 4315$ m
13. $A = 29.19°, c = 2975$ ft
14. $B = 29.86°, a = 72.62$ m
15. $a = 46.72$ m, $b = 19.26$ m
16. $a = 2436$ ft, $c = 4195$ ft
17. $A = 41.1°, c = 485$ m
18. $a = 1250$ km, $b = 1650$ km
19. $B = 9.45°, a = 1585$ ft
20. $A = 14.60°, b = 135.7$ cm
21. $b = 269.5$ m, $c = 380.5$ m
22. $B = 75.65°, c = 92.75$ km
23. $B = 81.5°, b = 9370$ ft
24. $a = 14.6$ mi, $c = 31.2$ mi

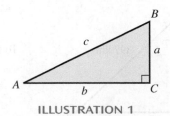

ILLUSTRATION 1

13.5 Applications Involving Trigonometric Ratios

Trigonometric ratios can be used to solve many applications similar to those problems we solved in the preceding sections. However, instead of having to find all the parts of a right triangle, we usually need to find only one.

Example 1 The roof in Figure 13.17 has a rise of 7.50 ft and a run of 18.0 ft. Find angle A.

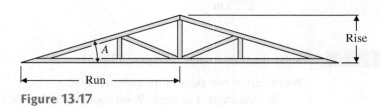

Figure 13.17

We know the length of the side opposite angle A and the length of the side adjacent to angle A. So we use the tangent ratio.

$$\tan A = \frac{\text{length of side opposite angle } A}{\text{length of side adjacent to angle } A}$$

$$\tan A = \frac{7.50 \text{ ft}}{18.0 \text{ ft}} = 0.4167$$

$$A = 22.6°$$

♦

The **angle of depression** is the angle between the horizontal and the line of sight to an object that is *below* the horizontal. The **angle of elevation** is the angle between the horizontal and the line of sight to an object that is *above* the horizontal.

In Figure 13.18, angle A is the angle of depression for an observer in the helicopter sighting down to the building on the ground, and angle B is the angle of elevation for an observer in the building sighting up to the helicopter.

13.5 ♦ Applications Involving Trigonometric Ratios

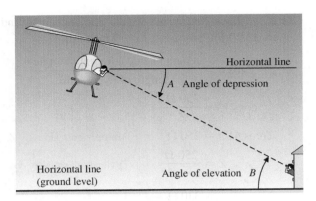

Figure 13.18

Example 2 A ship's navigator measures the angle of elevation to the beacon of a lighthouse to be 10.1°. He knows that this particular beacon is 225 m above sea level. How far is the ship from the lighthouse?

First, you should sketch the problem, as in Figure 13.19. Since this problem involves finding the length of the side adjacent to an angle when the opposite side is known, we use the tangent ratio.

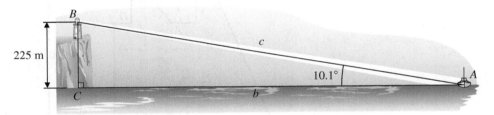

Figure 13.19

$$\tan A = \frac{\text{length of side opposite angle } A}{\text{length of side adjacent to angle } A}$$

$$\tan 10.1° = \frac{225 \text{ m}}{b}$$

$b(\tan 10.1°) = 225 \text{ m}$ Multiply both sides by b.

$b = \dfrac{225 \text{ m}}{\tan 10.1°}$ Divide both sides by $\tan 10.1°$.

$= 1260 \text{ m}$ ♦

Example 3 In ac (alternating current) circuits, the relationship between impedance Z (in ohms), the resistance R (in ohms), and the phase angle θ is shown by the right triangle in Figure 13.20(a). If the resistance is 250 Ω and the phase angle is 41°, find the impedance.

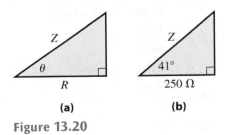

Figure 13.20

Here, we know the adjacent side and the angle and wish to find the hypotenuse (see Figure 13.20(b)). So we use the cosine ratio.

$$\cos\theta = \frac{\text{length of side adjacent to angle }\theta}{\text{length of hypotenuse}}$$

$$\cos 41° = \frac{250\ \Omega}{Z}$$

$$Z(\cos 41°) = 250\ \Omega \quad \text{Multiply both sides by } Z.$$

$$Z = \frac{250\ \Omega}{\cos 41°} \quad \text{Divide both sides by } \cos 41°.$$

$$= 330\ \Omega$$

Example 4   You are machining the part shown in Figure 13.21. Before you begin, you must find angle 1 and length AB.

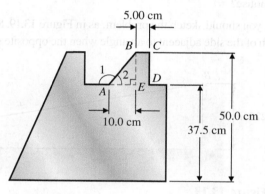

Figure 13.21

First, complete triangle ABE by drawing the dashed lines as shown.

Length BE = length CD = 50.0 cm − 37.5 cm = 12.5 cm

From right triangle ABE, we have

$$\tan \angle 2 = \frac{\text{length of side opposite } \angle 2}{\text{length of side adjacent to } \angle 2} = \frac{BE}{AE}$$

$$\tan \angle 2 = \frac{12.5\text{ cm}}{10.0\text{ cm}} = 1.25$$

$$\angle 2 = 51.3°$$

From right triangle ABE, we have

$$\sin \angle 2 = \frac{\text{length of side opposite } \angle 2}{\text{length of hypotenuse}}$$

$$\sin 51.3° = \frac{12.5\text{ cm}}{AB}$$

$$AB(\sin 51.3°) = 12.5\text{ cm} \quad \text{Multiply both sides by } AB.$$

$$AB = \frac{12.5\text{ cm}}{\sin 51.3°} \quad \text{Divide both sides by } \sin 51.3°.$$

$$AB = 16.0\text{ cm}$$

Since $\angle 1$ and $\angle 2$ are supplementary, $\angle 1 = 180° - \angle 2 = 180° - 51.3° = 128.7°$.

EXERCISES 13.5

1. Maria is to weld a support for a 23-m conveyor so that it will operate at a $2\overline{0}°$ angle. What is the length of the support? See Illustration 1.

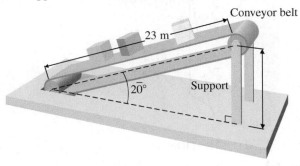

ILLUSTRATION 1

2. A conveyor is used to lift paper to a shredder. The most efficient operating angle of elevation for the conveyor is 35.8°. The paper is to be elevated 11.0 m. What length of conveyor is needed?

3. A bullet is found embedded in the wall of a room 2.3 m above the floor. The bullet entered the wall going upward at an angle of 12°. How far from the wall was the bullet fired if the gun was held 1.2 m above the floor?

4. The recommended safety angle of a ladder against a building is 78°. A $1\overline{0}$-m ladder will be used. How high up on the side of the building will the ladder safely reach? (See Illustration 2.)

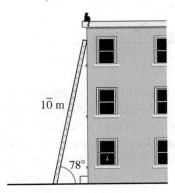

ILLUSTRATION 2

5. A piece of conduit 38.0 ft long is placed across the corner of a room, as shown in Illustration 3. Find length x and angle A.

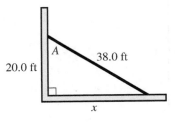

ILLUSTRATION 3

6. Find the width of the river in Illustration 4.

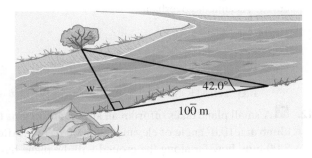

ILLUSTRATION 4

7. A roadway rises 220 ft for each 2300 ft of horizontal distance. (See Illustration 5.) Find the angle of inclination of the roadway. The percentage grade of a roadway is its slope written as a percent. Find the percentage grade of this roadway.

ILLUSTRATION 5

8. A smokestack is $18\overline{0}$ ft high. A guy wire must be fastened 20.0 ft below the top of the smokestack. The guy wire makes an angle of 40.0° with the ground. Find the length of the guy wire.

9. A railroad track has an angle of elevation of 1.0°. What is the difference in height (in feet) of two points on the track 1.00 mi apart?

10. A machinist needs to drill four holes 1.00 in. apart in a straight line in a metal plate, as shown in Illustration 6. If the first hole is placed at the origin and the line forms an angle of 32.0° with the vertical axis, find the coordinates of the other three holes (A, B, and C).

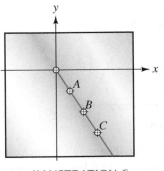

ILLUSTRATION 6

11. Enrico has to draft a triangular roof to a house. (See Illustration 7.) The roof is 30.0 ft wide. If the rafters are 17.0 ft long, at what angle will the rafters be laid at the eaves? Assume no overhang.

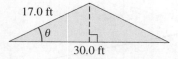

ILLUSTRATION 7

12. A small plane takes off from an airport and begins to climb at a 10.0° angle of elevation at $50\overline{0}0$ ft/min. After 3.00 min, how far along the ground will the plane have flown?

13. A gauge is used to check the diameter of a crankshaft journal. It is constructed to make measurements on the basis of a right triangle with a 60.0° angle. Distance AB in Illustration 8 is 11.4 cm. Find radius BC of the journal.

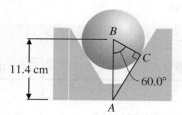

ILLUSTRATION 8

14. Round metal duct runs alongside some stairs from the floor to the ceiling. If the ceiling is 9.00 ft high and the angle of elevation between the floor and duct is 37.0°, how long is the duct?

15. The cables attached to a TV relay tower are $11\overline{0}$ m long. They meet level ground at an angle of 60.0°, as in Illustration 9. Find the height of the tower.

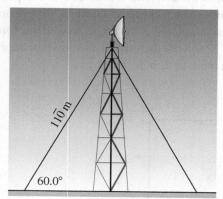

ILLUSTRATION 9

16. A lunar module is resting on the moon's surface directly below a spaceship in its orbit, 12.0 km above the moon. (See Illustration 10.) Two lunar explorers find that the angle from their position to that of the spaceship is 82.9°. What distance are they from the lunar module?

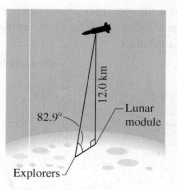

ILLUSTRATION 10

17. In ac (alternating current) circuits, the relationship between the impedance Z (in ohms), the resistance R (in ohms), the phase angle θ, and the reactance X (in ohms) is shown by the right triangle in Illustration 11.
   a. Find the impedance if the resistance is 350 Ω and the phase angle is 35°.
   b. Suppose the resistance is 550 Ω and the impedance is $7\overline{0}0$ Ω. What is the phase angle?
   c. Suppose the reactance is 182 Ω and the resistance is 240 Ω. Find the impedance and the phase angle.

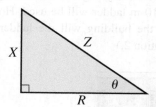

ILLUSTRATION 11

18. A right circular conical tank with its point down (Illustration 12) has a height of 4.00 m and a radius of 1.20 m. The tank is filled to a height of 3.70 m with liquid. How many litres of liquid are in the tank? (1000 litres = 1 m³.) How many litres would then need to be added to fill the tank to 100% capacity?

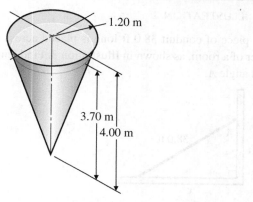

ILLUSTRATION 12

19. Use the right triangle in Illustration 13:
   a. Find the voltage applied if the voltage across the coil is 35.6 V and the voltage across the resistance is 40.2 V.
   b. Find the voltage across the resistance if the voltage applied is 378 V and the voltage across the coil is 268 V.
   c. Find the voltage across the coil if the voltage applied is 448 V and the voltage across the resistance is 381 V.

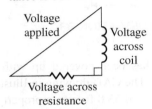

ILLUSTRATION 13

20. Machinists often use a coordinate system to drill holes by placing the origin at the most convenient location. A bolt circle is the circle formed by completing an arc through the centers of the bolt holes in a piece of metal. Find the coordinates of the centers of eight equally spaced $\frac{1}{4}$-in. holes on a bolt circle of radius 6.00 in., as shown in Illustration 14.

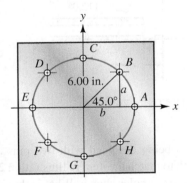

ILLUSTRATION 14

21. Twelve equally spaced holes must be drilled on a 14.500-in.-diameter bolt circle. (See Illustration 15.) What is the straight-line center-to-center distance between two adjacent holes?

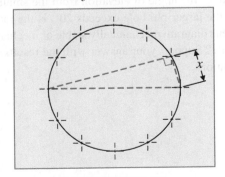

ILLUSTRATION 15

22. Dimension x in the dovetail shown in Illustration 16 is a check dimension. Precision steel balls of diameter 0.1875 in. are used in this procedure. What should this check dimension be?

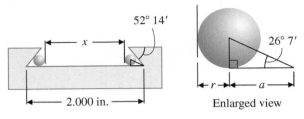

ILLUSTRATION 16

23. Find angle θ of the taper in Illustration 17.

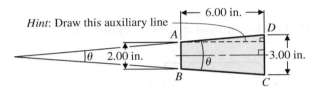

ILLUSTRATION 17

24. You need to use a metal screw with a head angle of angle A, which is not less than 65° and no larger than 70°. The team leader wants you to find angle A from the sketch shown in Illustration 18 and determine if the head angle will be satisfactory. Find the head angle A.

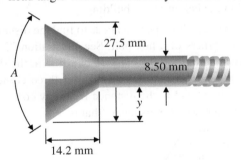

ILLUSTRATION 18

25. Find **a.** distance x and **b.** distance BD in Illustration 19. Length $BC = 5.50$ in.

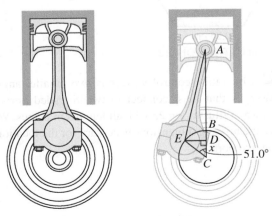

ILLUSTRATION 19

26. Find length AB along the roofline of the building in Illustration 20. The angle of elevation from point A to point B along the roof is 45.0°.

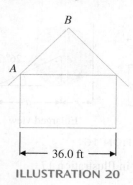

ILLUSTRATION 20

27. Find length x and angle A in Illustration 21.

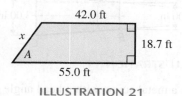

ILLUSTRATION 21

28. From the base of a building, measure out a horizontal distance of 215 ft along the ground. The angle of elevation to the top of the building from this point is 63.0°. Find the height of the building.

29. A mechanical draftsperson needs to find the distance across the corners of a hex-bolt. See Illustration 22. If the distance across the flats is 2.25 cm, find **a.** the distance from any corner to another corner three corners away (that is, x) and **b.** the distance from any corner to another corner two corners away (that is, y).

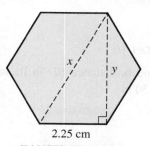

ILLUSTRATION 22

30. A hydraulic control valve has two parallel angular passages that must connect to two threaded ports, as shown in Illustration 23 with all lengths in inches. What are the missing dimensions necessary for the location of the two ports?

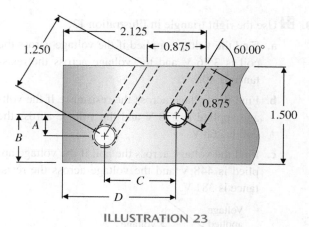

ILLUSTRATION 23

31. A benchmark has been covered up with dirt and needs to be found. The CAD drawing in Illustration 24 shows that it is located 33.0 ft from a property stake at an angle of 112.0°. Find distances a and b to help locate the benchmark.

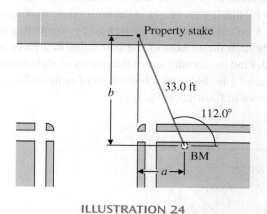

ILLUSTRATION 24

32. A mating part is being designed with two pins attached to a flat block to match the item in Illustration 25. The pins must fit into the holes shown. What is the distance from point C (center of small hole) to point D (center of larger hole)? Also find angle x. The placement of the pin holes produces a mechanically unstable predicament when the distance from pin hole to pin hole exceeds 8 in. or the angle of elevation from the smaller pin hole to the larger pin hole exceeds 20°. Is the mating part in this diagram mechanically stable or mechanically unstable? Support your answer with the results of your calculations.

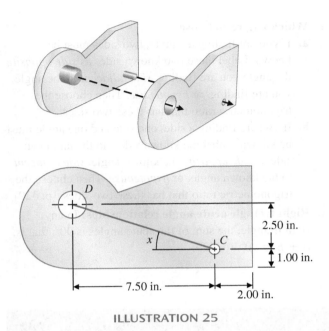

ILLUSTRATION 25

33. ⊕ Solar heating and electric panels should face the sun so that you gain as much sunlight as possible. Fixed panels are mounted at the best winter angle to the horizontal according to the following formula: $\theta = 0.90$ (your latitude in degrees) $+ 29°$. A homeowner in Ithaca, New York (latitude 42.44°N) wishes to mount a solar heating panel that is 4.0 ft wide and 8.0 ft long facing south with its 4.0-ft base mounted on a flat roof at the appropriate winter angle. **a.** Find the height above the roof of the top of the solar panel. **b.** Find the length of the base of the support if the top of the solar panel is supported by vertical posts.

34. ⊕ A lean-to is a simple shelter with three walls, a sloping roof, and an open front facing away from the prevailing winds. The back wall is short compared to the front opening. If the lean-to at a campsite has a front opening that is 6.0 ft tall, a back wall that is 2.0 ft tall, and a floor that is 8.0 ft deep, what angle does the roofline make with the ground?

SUMMARY | CHAPTER 13

Glossary of Basic Terms

Angle of depression. The angle between the horizontal and the line of sight to an object that is *below* the horizontal. (p. 434)

Angle of elevation. The angle between the horizontal and the line of sight to an object that is *above* the horizontal. (p. 434)

Cosine of angle A (cos A). Equals the ratio of the length of the side adjacent to angle A to the length of the hypotenuse. (p. 425)

Ratio. The comparison of two quantities by division. (p. 425)

Sine of angle A (sin A). Equals the ratio of the length of the side opposite angle A to the length of the hypotenuse. (p. 425)

Solving a right triangle. Find the measures of the various parts of a triangle that are not given. (p. 432)

Tangent of angle A (tan A). Equals the ratio of the length of the side opposite angle A to the length of the side adjacent to angle A. (p. 425)

Trigonometric ratio. Expresses the relationship between an acute angle and the lengths of two of the sides of a right triangle. (p. 425)

13.1 Trigonometric Ratios

1. **Pythagorean theorem**: In any right triangle, $c^2 = a^2 + b^2$. (See Illustration 1.) (p. 424)

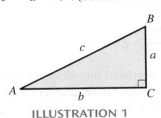

ILLUSTRATION 1

2. **Trigonometric ratios:**

$$\sin A = \frac{\text{length of side opposite angle } A}{\text{length of hypotenuse}}$$

$$\cos A = \frac{\text{length of side adjacent to angle } A}{\text{length of hypotenuse}}$$

$$\tan A = \frac{\text{length of side opposite angle } A}{\text{length of side adjacent to angle } A}$$

(p. 425)

13.2 Using Trigonometric Ratios to Find Angles

1. **Trigonometric ratios and significant digits:** When calculations involve a trigonometric ratio, use the following for significant digits. (p. 428)

Angles expressed to the nearest	The length of each side of the triangle contains
1°	Two significant digits
0.1°	Three significant digits
0.01°	Four significant digits

2. **Which trig ratio to use:**
   a. If you are finding an angle, two sides must be known. Label these two known sides as *side opposite* the angle you are finding, *side adjacent* to the angle you are finding, or *hypotenuse*. Then choose the trigonometric ratio that has these two sides.
   b. If you are finding a side, one side and one angle must be known. Label the known side and the unknown side as *side opposite* the known angle, *side adjacent* to the known angle, or *hypotenuse*. Then choose the trigonometric ratio that has these two sides. (p. 429)
3. **Right triangle acute angle relationships:** In any right triangle, the sum of the acute angles is 90°; that is, $A + B = 90°$. (p. 429)

REVIEW | CHAPTER 13

For Exercises 1–7, see Illustration 1.

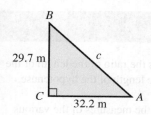

ILLUSTRATION 1

1. What is the length of the side opposite angle A in the right triangle?
2. What is the angle adjacent to the side whose length is 29.7 m?
3. Side c is known as the ___?___.
4. What is the length of side c?
5. $\dfrac{\text{length of side opposite angle } A}{\text{length of hypotenuse}}$ is what ratio?
6. $\cos A = \dfrac{?}{\text{length of hypotenuse}}$
7. $\tan B = \dfrac{?}{?}$

Find the value of each trigonometric ratio rounded to four significant digits:

8. $\cos 36.2°$
9. $\tan 48.7°$
10. $\sin 23.72°$

Find each angle rounded to the nearest tenth of a degree:

11. $\sin A = 0.7136$
12. $\tan B = 0.1835$
13. $\cos A = 0.4104$
14. Find angle A in Illustration 2.
15. Find angle B in Illustration 2.
16. Find side b in Illustration 3.
17. Find side c in Illustration 3.

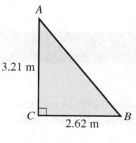

ILLUSTRATION 2

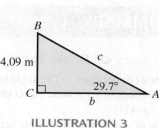

ILLUSTRATION 3

Solve each right triangle:

18.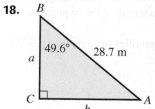

19.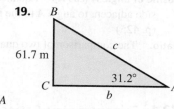

20.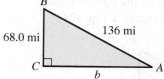

21. A satellite is directly overhead one observer station when it is at an angle of 68.0° from another observer station. (See Illustration 4.) The distance between the two stations is 20̄0̄0 m. What is the height of the satellite?

22. A ranger at the top of a fire tower observes the angle of depression to a fire on level ground to be 3.0°. If the tower is 275 ft tall, what is the ground distance from the base of the tower to the fire?

23. Find the angle of slope of the symmetrical roof in Illustration 5.

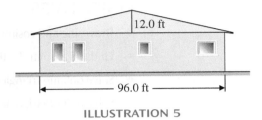

ILLUSTRATION 5

ILLUSTRATION 4

TEST | CHAPTER 13

Find the value of each trigonometric ratio rounded to four significant digits:

1. sin 35.5°
2. cos 16.9°
3. tan 57.1°

Find each angle rounded to the nearest tenth of a degree:

4. cos A = 0.5577
5. tan B = 0.8888
6. sin A = 0.4166
7. Find angle B in Illustration 1.
8. Find side a in Illustration 1.
9. Find side c in Illustration 1.
10. Find angle A in Illustration 2.
11. Find angle B in Illustration 2.
12. Find side b in Illustration 2.
13. A tower 50.0 ft high has a guy wire that is attached to its top and anchored in the ground 15.0 ft from its base. Find the length of the guy wire.
14. Find length x in the retaining wall in Illustration 3.
15. Find angle A in the retaining wall in Illustration 3.

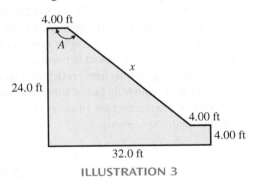

ILLUSTRATION 3

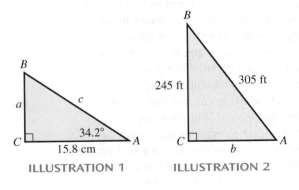

ILLUSTRATION 1   ILLUSTRATION 2

CHAPTER 14 | Trigonometry with Any Angle

OBJECTIVES

- Draw sine and cosine graphs by plotting points.
- Find the amplitude, the period, and the phase shift of sine and cosine graphs.
- Solve oblique triangles using the law of sines.
- Solve oblique triangles that have two possible solutions using the law of sines.
- Solve oblique triangles using the law of cosines.
- Solve application problems involving oblique triangles and the law of sines and the law of cosines.

Mathematics at Work

Land surveyors establish official land, airspace, and water boundaries. They write descriptions of land for deeds, leases, and other legal documents; define airspace for airports; and measure construction and mineral sites. Land surveyors also manage and plan the work of survey parties that measure distances, directions, and angles between points and elevations of points, lines, and contours on, above, and below the earth's surface. Other surveyors provide data relevant to the shape, contour, location, elevation, or dimension of land or land features. Surveying technicians assist land surveyors by operating surveying instruments, such as the theodolite (used to measure horizontal and vertical angles) and electronic distance-measuring equipment, and collecting information in the field and by performing computations and computer-aided drafting in offices. New technology, such as the satellite Global Positioning System (GPS) that locates points on the earth to a high degree of precision, is continually changing the nature of the work of surveyors and surveying technicians.

Land surveyors and surveying technicians often spend a lot of time outdoors and work longer hours during the summer, when weather and light conditions are related to the demand for specific surveying services. The work is often strenuous, requires long periods of walking carrying heavy equipment, and often requires traveling long distances from home.

In the past, many people with little formal training in surveying started as members of survey crews and worked their way up to become licensed surveyors. However, advancing technology and more stringent licensing standards are increasing formal education requirements. Specific requirements vary by state; many states have a continuing education requirement. Generally, the quickest path to licensure is a combination of 4 years of college, up to 4 years of experience under the supervision of an experienced surveyor, and passing the licensing examinations.

Surveying technicians often complete surveying technology programs in a community or technical college with emphasis placed on knowledge and hands-on skills needed for computer-aided drafting, construction layout, engineering surveys, and land surveying. Often, these associate degree programs may be transferred to universities that offer a 4-year degree program that also prepares the student to take the licensing examination. For more information, please visit **www.cengage.com** and access the Student Online Resources for this text.

Surveying Technician
Surveying technician working on a road project

14.1 Sine and Cosine Graphs

Up to this point, we have considered only the trigonometric ratios of angles between 0° and 90°, because we were working only with right triangles. For many applications, we need to consider values greater than 90°. In this section, we use a calculator to find the values of the sine and cosine ratios of angles greater than 90°. Then we use these values to construct various sine and cosine graphs.

The procedure for finding the value of the sine or cosine of an angle greater than 90° is the same as for angles between 0° and 90°, as shown in Section 13.1.

Example 1   Find sin 255° rounded to four significant digits.

SIN 255 =

-0.965925826

Thus, sin 255° = −0.9659 rounded to four significant digits.   ◆

Let's graph $y = \sin x$ for values of x between 0° and 360°. First, find a large number of values of x and y that satisfy the equation and plot them in the xy plane. For convenience, we will choose values of x in multiples of 30° and round the values of y to two significant digits.

x	0°	30°	60°	90°	120°	150°	180°	210°	240°	270°	300°	330°	360°
y	0	0.50	0.87	1.0	0.87	0.50	0	−0.50	−0.87	−1.0	−0.87	−0.50	0

Now choose a convenient scale for the x axis so that one unit equals 30°, and mark the x axis between 0° and 360°. Then choose a convenient scale for the y axis so that one unit equals 0.1, and mark the y axis between +1.0 and −1.0. Plot the points corresponding to the ordered pairs (x, y) from the table above. Then connect the points with a smooth, continuous curve. The graph is shown in Figure 14.1.

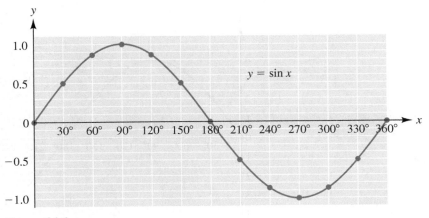

Figure 14.1

The graph of $y = \cos x$ for values of x between $0°$ and $360°$ can be found in a similar manner and is shown below.

x	0°	30°	60°	90°	120°	150°	180°	210°	240°	270°	300°	330°	360°
y	1.0	0.87	0.50	0	−0.50	−0.87	−1	−0.87	−0.50	0	0.50	0.87	1.0

Plot the points corresponding to these ordered pairs and connect them with a smooth curve, as shown in Figure 14.2.

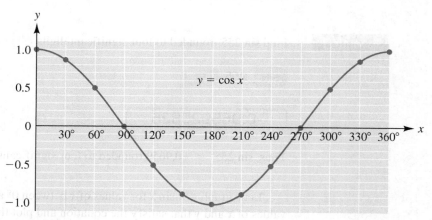

Figure 14.2

Example 2 Graph $y = 4 \sin x$ for values of x between $0°$ and $360°$.

Prepare a table listing values of x in multiples of $30°$. To find each value of y, find the sine of the angle, multiply this value by 4, and round to two significant digits.

x	0°	30°	60°	90°	120°	150°	180°	210°	240°	270°	300°	330°	360°
y	0	2.0	3.5	4	3.5	2.0	0	−2.0	−3.5	−4	−3.5	−2.0	0

Here, let's choose the scale for the y axis so that one unit equals 0.5 and mark the y axis between +4.0 and −4.0. Plot the points corresponding to these ordered pairs in the table and then connect them with a smooth curve, as shown in Figure 14.3.

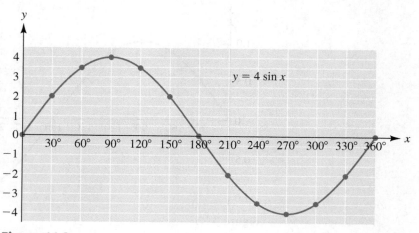

Figure 14.3

14.1 ♦ Sine and Cosine Graphs

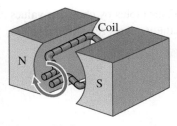

Figure 14.4
A coil of wire rotating in a magnetic field produces an alternating electric current.

Note that the graphs for $y = \sin x$ and $y = 4 \sin x$ are similar. That is, each starts at (0°, 0), reaches its maximum at $x = 90°$, crosses the x axis at (180°, 0), reaches its minimum at $x = 270°$, and meets the x axis at (360°, 0). In general, the graphs of equations in the form

$$y = A \sin x \quad \text{and} \quad y = A \cos x, \quad \text{where} \quad A > 0$$

reach a maximum value of A and a minimum value of $-A$. The value of A is usually called the **amplitude**.

One of the most common applications of waves is in alternating current. In a generator, a coil of wire is rotated in a magnetic field, which produces an alternating electric current. See Figure 14.4.

In a simple generator, the current i changes as the coil rotates according to the equation

$$i = I \sin x$$

where I is the maximum current and x is the angle through which the coil rotates.

Similarly, the voltage v also changes as the coil rotates according to the equation

$$v = V \sin x$$

where V is the maximum voltage and x is the angle through which the coil rotates.

Example 3

The maximum voltage V in a simple generator is 25 V. The changing voltage v as the coil rotates is given by

$$v = 25 \sin x$$

Graph the equation in multiples of 30° for one complete revolution of the coil.

First, prepare a table. To find each value of y, find the sine of the angle, multiply this value by 25, and round to two significant digits.

x	0°	30°	60°	90°	120°	150°	180°	210°	240°	270°	300°	330°	360°
y	0	13	22	25	22	13	0	−13	−22	−25	−22	−13	0

Let's choose the scale for the y axis so that one unit equals 5 V. Mark the y axis between +25 V and −25 V. Plot the points corresponding to the ordered pairs in the table. Then connect these with a smooth curve, as shown in Figure 14.5.

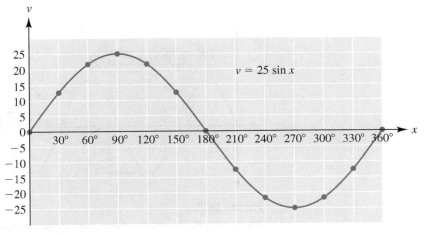

Figure 14.5

448 CHAPTER 14 ♦ Trigonometry with Any Angle

If you were to continue finding ordered pairs in the previous tables by choosing values of x greater than 360° and less than 0°, you would find that the y values repeat themselves and that the graphs form *waves*, as shown in Figure 14.6.

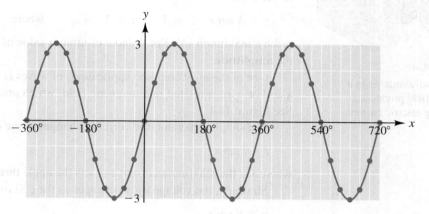

(a) $y = 3 \sin x$ for x between $-360°$ and $720°$

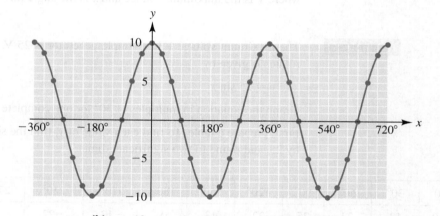

(b) $y = 10 \cos x$ for x between $-360°$ and $720°$

Figure 14.6

In general, the **period** of a sine or cosine graph is the x distance between any point on the graph and the corresponding point in the next cycle where the graph starts repeating itself. See Figure 14.7.

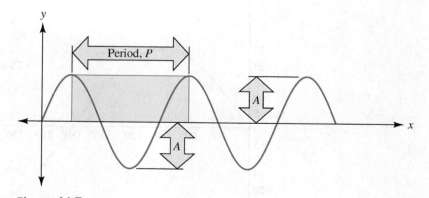

Figure 14.7

The period, P, of a sine or cosine graph is the x distance between any two successive corresponding points on the graph.

If the horizontal, or x axis, variable is *distance*, the length of one complete wave of the sine and cosine graphs along the x axis is called the **wavelength** and is given by the symbol λ, the Greek letter lambda. See Figure 14.8.

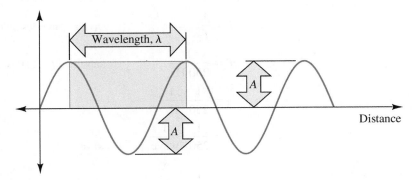

Figure 14.8
The wavelength, λ, is the length of one complete wave when the horizontal axis is distance.

If the horizontal, or x axis, variable is *time,* the time required for one complete wave of the sine and cosine graphs to pass a given point is called the **period**, T. See Figure 14.9.

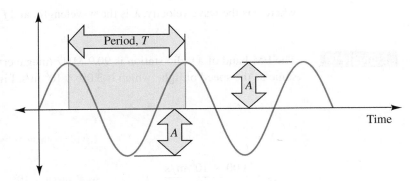

Figure 14.9
The period, T, is the length of time required for one complete wave to pass a given point when the horizontal axis is time.

The **frequency** f is the number of waves that pass a given point on the time axis each second. That is,

$$f = \frac{1}{T}$$

The unit of frequency is the hertz (Hz), where

1 Hz = 1 wave/s   or   1 cycle/s

Common multiples of the hertz are the kilohertz (kHz, 10^3 Hz) and the megahertz (MHz, 10^6 Hz).

Example 4 Find the period of a wave whose frequency is 250 Hz.

$$\text{Given } f = \frac{1}{T}$$

$$fT = 1 \qquad \text{Multiply both sides by } T.$$

$$T = \frac{1}{f} \qquad \text{Divide both sides by } f.$$

$$T = \frac{1}{250 \text{ Hz}} \qquad \text{Substitute } f = 250 \text{ Hz}.$$

$$= \frac{1}{250} \cdot \frac{1}{\text{Hz}}$$

$$= \frac{1}{250} \cdot \frac{1}{\frac{\text{wave}}{\text{s}}} \qquad \frac{1}{\frac{\text{wave}}{\text{s}}} = 1 \div \frac{\text{wave}}{\text{s}} = 1 \cdot \frac{\text{s}}{\text{wave}} = \frac{\text{s}}{\text{wave}}$$

$$= \frac{1}{250} \cdot \frac{\text{s}}{\text{wave}}$$

$$= 4.0 \times 10^{-3} \text{ s} \qquad \text{That is, one wave passes a given point each } 4.0 \times 10^{-3} \text{ s or } 4.0 \text{ ms.}$$

Frequency and wavelength are related to wave velocity by the formula

$$v = \lambda f$$

where v is the wave velocity, λ is the wavelength, and f is the frequency.

Example 5 The FM band of a radio station is 90.9 MHz (megahertz). The speed of a radio wave is the same as the speed of light, which is 3.00×10^8 m/s. Find its wavelength.

$$v = \lambda f$$

$$\lambda = \frac{v}{f} \qquad \text{Divide both sides by } f.$$

$$\lambda = \frac{3.00 \times 10^8 \text{ m/s}}{90.9 \text{ MHz}} \qquad \text{M = mega} = 10^6$$

$$= \frac{3.00 \times 10^8 \text{ m/s}}{90.9 \times 10^6 \text{ Hz}} \qquad 1 \text{ Hz} = 1 \text{ cycle/s} = 1 \text{ wavelength/s}$$

$$= \frac{3.00 \times 10^8 \text{ m/s}}{90.9 \times 10^6 \text{ wavelengths/s}} \qquad \frac{\text{m}}{\text{s}} \div \frac{\text{wavelengths}}{\text{s}} = \frac{\text{m}}{\text{s}} \cdot \frac{\text{s}}{\text{wavelengths}} = \frac{\text{m}}{\text{wavelength}}$$

$$= 3.30 \text{ m} \qquad \text{That is, each wavelength is 3.30 m.}$$

EXERCISES 14.1

Find each value rounded to four significant digits:

1. $\sin 137°$
2. $\sin 318°$
3. $\cos 246°$
4. $\cos 295°$
5. $\sin 205.8°$
6. $\sin 106.3°$
7. $\cos 166.5°$
8. $\cos 348.2°$
9. $\tan 217.6°$
10. $\tan 125.5°$
11. $\tan 156.3°$
12. $\tan 418.5°$

Graph each equation for values of x between 0° and 720° in multiples of 30°:

13. $y = 6 \sin x$
14. $y = 2 \sin x$
15. $y = 5 \cos x$
16. $y = 4 \cos x$

Graph each equation for values of x between 0° and 360° in multiples of 15°:

17. $y = \sin 2x$ **18.** $y = \cos 2x$

Graph each equation for values of x between 0° and 360° in multiples of 10°:

19. $y = 4 \cos 3x$ **20.** $y = 2 \sin 3x$

The maximum voltage in a simple generator is V. The changing voltage v as the coil rotates is given by

$$v = V \sin x$$

Graph this equation in multiples of 30° for two complete revolutions of the coil for each value of V:

21. $V = 36$ V   **22.** $V = 48$ V

The maximum current in a simple generator is I. The changing current i as the coil rotates is given by

$$i = I \sin x$$

Graph this equation in multiples of 30° for two complete revolutions of the coil for each value of I:

23. $I = 5.0$ A   **24.** $I = 7.5$ A

25. From the graph in Exercise 21, estimate the value of v at $x = 45°$ and $x = 295°$.

26. From the graph in Exercise 22, estimate the value of v at $x = 135°$ and $x = 225°$.

27. From the graph in Exercise 23, estimate the value of i at $x = 135°$ and $x = 225°$.

28. From the graph in Exercise 24, estimate the value of i at $x = 45°$ and $x = 190°$.

29. Find the period of a wave whose frequency is 5.0 kHz.

30. Find the period of a wave whose frequency is 1.1 MHz.

31. Find the frequency of a wave whose period is 0.56 s.

32. Find the frequency of a wave whose period is 25 μs.

33. A radar unit operates at a wavelength of 3.4 cm. Radar waves travel at the speed of light, which is 3.0×10^8 m/s. What is the frequency of the radar waves?

34. A local AM radio station broadcasts at 1400 kHz. What is the wavelength of its radio waves? (They travel at the speed of light, which is 3.0×10^8 m/s.)

35. Find the speed of a wave having frequency 4.50/s and wavelength 0.500 m.

36. Find the wavelength of water waves with frequency 0.55/s and speed 1.4 m/s.

37. Radio antennas are made so that the electromagnetic radio waves "fit" on the wires. Think of a radio antenna as the x axis of a sine or cosine graph, depending on the kind of antenna. The ribbon-like FM antenna that comes with the purchase of most stereo receivers is actually a "half-wave folded dipole antenna" so that its length is roughly one-half of the wavelength of a frequency in the middle of the FM band. Because all electromagnetic waves move at the speed of light (3.00×10^8 m/s), how long is the typical FM ribbon antenna for a frequency of 98.0 MHz? *Hint:* The length of a half-wave antenna is one-half of a wavelength. (See Example 5.)

14.2 Period and Phase Shift

As we saw in Figure 14.7, the period is the length of one complete cycle of the sine or cosine graph. In general, the period for $y = A \sin Bx$ and for $y = A \cos Bx$ may be found by the formula

$$P = \frac{360°}{B}$$

Example 1   Find the period and amplitude of $y = 2 \cos 3x$ and draw its graph.

$$P = \frac{360°}{3} = 120°$$

Draw the cosine graph with amplitude 2 and period 120°, as in Figure 14.10.

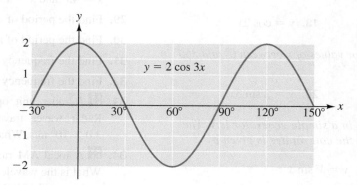

Figure 14.10

Example 2 Find the period and amplitude of $y = 5 \sin 4x$ and draw one period of its graph.

$$P = \frac{360°}{4} = 90°$$

Draw the sine graph with amplitude 5 and period 90°, as in Figure 14.11.

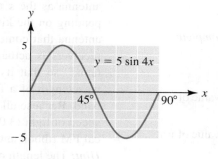

Figure 14.11

If the graph of a sine curve does not pass through the origin (0°, 0) or if the graph of a cosine curve does not pass through the point (0°, A), where A is the amplitude, the curve is **out of phase**. If the curve is out of phase, the **phase shift** is the horizontal distance between two successive corresponding points of the curve $y = A \sin Bx$ (or $y = A \cos Bx$) and the out-of-phase curve. (See Figure 14.12.)

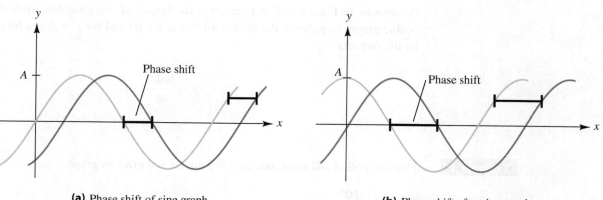

(a) Phase shift of sine graph **(b)** Phase shift of cosine graph

Figure 14.12

The phase shift is the horizontal distance between two successive corresponding points of either graph.

Example 3 Graph $y = 2 \sin x$ and $y = 2 \sin (x - 45°)$ on the same set of coordinate axes.
For $y = 2 \sin (x - 45°)$:

x	−45°	0°	45°	90°	135°	180°	225°	270°	315°	360°	405°	450°
y	−2	−1.4	0	1.4	2	1.4	0	−1.4	−2	−1.4	0	1.4

Graph both equations, as in Figure 14.13.

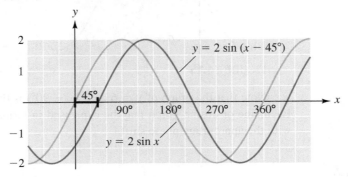

Figure 14.13

Graphing equations in the form $y = A \sin (Bx + C)$ or $y = A \cos (Bx + C)$ involves a phase shift:

Phase Shift

The effect of C in each equation is to shift the curve $y = A \sin Bx$ or $y = A \cos Bx$
1. to the *left* $\frac{C}{B}$ units if $\frac{C}{B}$ is positive.
2. to the *right* $\frac{C}{B}$ units if $\frac{C}{B}$ is negative.

Example 4 Graph $y = 4 \sin (3x + 60°)$.

The amplitude is 4. The period is $P = \frac{360°}{3} = 120°$. The phase shift is $\frac{C}{B} = \frac{60°}{3} = 20°$, or 20° to the *left*.

Graph $y = 4 \sin 3x$ and shift the curve 20° to the left, as shown in Figure 14.14.

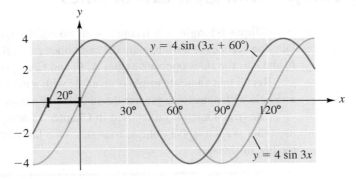

Figure 14.14

Example 5 Graph $y = 6 \cos (2x - 90°)$.

The amplitude is 6. The period is $\frac{360°}{2} = 180°$. The phase shift is $\frac{C}{B} = \frac{-90°}{2} = -45°$, or 45° to the *right*.

Graph $y = 6 \cos 2x$ and shift the curve 45° to the right, as shown in Figure 14.15.

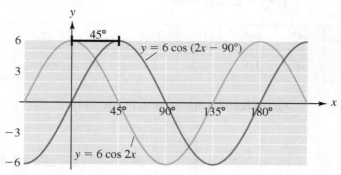

FIGURE 14.15

Note that the graph of $y = 6 \cos (2x - 90°)$ is the same as the graph of $y = 6 \sin 2x$. Each sine or cosine graph may be expressed in terms of the other trigonometric function with the appropriate phase shift.

EXERCISES 14.2

Find the period and amplitude, and graph at least two periods of each equation:

1. $y = 3 \sin 3x$
2. $y = 7 \cos 4x$
3. $y = 8 \cos 6x$
4. $y = 9 \sin 5x$
5. $y = 10 \sin 9x$
6. $y = 15 \cos 10x$
7. $y = 6 \cos \frac{1}{2}x$
8. $y = 4 \sin \frac{1}{3}x$
9. $y = 3.5 \sin \frac{2}{3}x$
10. $y = 1.8 \cos \frac{3}{4}x$
11. $y = 4 \sin \frac{5}{2}x$
12. $y = 6 \cos \frac{4}{3}x$

Find the period, amplitude, and phase shift, and graph at least two periods of each equation:

13. $y = \sin (x + 30°)$
14. $y = \cos (x + 45°)$
15. $y = 2 \cos (x - 60°)$
16. $y = 3 \sin (x - 120°)$
17. $y = 4 \sin (3x + 180°)$
18. $y = 5 \cos (2x + 60°)$
19. $y = 10 \sin (4x - 120°)$
20. $y = 12 \cos (4x + 180°)$
21. $y = 5 \sin \left(\frac{1}{2}x + 90°\right)$
22. $y = 6 \cos \left(\frac{3}{4}x - 240°\right)$
23. $y = 10 \cos \left(\frac{1}{4}x + 180°\right)$
24. $y = 15 \sin \left(\frac{2}{3}x - 120°\right)$

14.3 Solving Oblique Triangles: Law of Sines

An **oblique triangle** is a triangle with no right angle. We use the common notation of labeling vertices of a triangle by the capital letters A, B, and C, and using the small letters a, b, and c as the sides opposite angles A, B, and C, respectively.

Recall that an **acute angle** is an angle with a measure less than 90°. An **obtuse angle** is an angle with a measure greater than 90° but less than 180°.

The trigonometric ratios used in Chapter 13 apply *only* to right triangles. So we must use other ways to solve oblique triangles. One law that we use to solve oblique triangles is the law of sines:

Law of Sines

$$\frac{a}{\sin A} = \frac{b}{\sin B} = \frac{c}{\sin C}$$

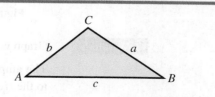

14.3 ♦ Solving Oblique Triangles: Law of Sines

That is, *for any triangle, the ratio of the length of any side to the sine of the opposite angle equals the ratio of the length of any other side to the sine of its opposite angle.*

When using this law, you must form a proportion by choosing two of the three ratios in which three of the four terms are known. In order to use the law of sines, you must know

a. two angles and a side opposite one of them (actually, knowing two angles and any side is enough, because in knowing two angles, the third is easily found) or

b. two sides and an angle opposite one of them.

Example 1

If $C = 28.0°$, $c = 46.8$ cm, and $B = 101.5°$, solve the triangle.*

First, find side b, in Figure 14.16.

$$\frac{c}{\sin C} = \frac{b}{\sin B}$$

$$\frac{46.8 \text{ cm}}{\sin 28.0°} = \frac{b}{\sin 101.5°}$$

$b(\sin 28.0°) = (\sin 101.5°)(46.8 \text{ cm})$ Multiply both sides by the LCD.

$$b = \frac{(\sin 101.5°)(46.8 \text{ cm})}{\sin 28.0°}$$ Divide both sides by sin 28.0°.

$= 97.7$ cm

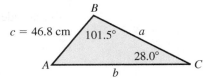

Figure 14.16

You may use a calculator to do this calculation as follows:

SIN 101.5) × 46.8 ÷ SIN 28 =

97.68531219

$A = 180° - B - C = 180° - 101.5° - 28.0° = 50.5°$

To find side a,

$$\frac{c}{\sin C} = \frac{a}{\sin A}$$

$$\frac{46.8 \text{ cm}}{\sin 28.0°} = \frac{a}{\sin 50.5°}$$

$a(\sin 28.0°) = (\sin 50.5°)(46.8 \text{ cm})$ Multiply both sides by the LCD.

$$a = \frac{(\sin 50.5°)(46.8 \text{ cm})}{\sin 28.0°}$$ Divide both sides by sin 28.0°.

$= 76.9$ cm

The solution is $a = 76.9$ cm, $b = 97.7$ cm, and $A = 50.5°$. ♦

A wide variety of applications may be solved using the law of sines.

Example 2

Find the lengths of rafters AC and BC for the roofline shown in Figure 14.17.

First, find angle C.

$C = 180° - A - B = 180° - 35.0° - 65.0° = 80.0°$

---

*As in previous sections, follow the rules for working with measurements; round sides to three significant digits and angles to the nearest tenth of a degree.

456 CHAPTER 14 ♦ Trigonometry with Any Angle

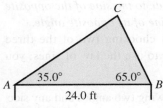

Figure 14.17

To find side AC,

$$\frac{AC}{\sin B} = \frac{AB}{\sin C}$$

$$\frac{AC}{\sin 65.0°} = \frac{24.0 \text{ ft}}{\sin 80.0°}$$

$AC(\sin 80.0°) = (\sin 65.0°)(24.0 \text{ ft})$ Multiply both sides by the LCD.

$$AC = \frac{(\sin 65.0°)(24.0 \text{ ft})}{\sin 80.0°}$$ Divide both sides by sin 80.0°.

$= 22.1$ ft

To find side BC,

$$\frac{BC}{\sin A} = \frac{AB}{\sin C}$$

$$\frac{BC}{\sin 35.0°} = \frac{24.0 \text{ ft}}{\sin 80.0°}$$

$BC(\sin 80.0°) = (\sin 35.0°)(24.0 \text{ ft})$ Multiply both sides by the LCD.

$$BC = \frac{(\sin 35.0°)(24.0 \text{ ft})}{\sin 80.0°}$$ Divide both sides by sin 80.0°.

$= 14.0$ ft

EXERCISES 14.3

Solve each triangle using the labels as shown in Illustration 1 (round lengths of sides to three significant digits and angles to the nearest tenth of a degree):

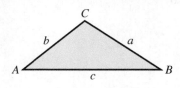

ILLUSTRATION 1

1. $A = 68.0°$, $a = 24.5$ m, $b = 17.5$ m
2. $C = 76.3°$, $b = 142$ cm, $c = 155$ cm
3. $A = 61.5°$, $B = 75.6°$, $b = 255$ ft
4. $B = 41.8°$, $C = 59.3°$, $c = 24.7$ km
5. $A = 14.6°$, $B = 35.1°$, $c = 43.7$ cm
6. $B = 24.7°$, $C = 136.1°$, $a = 342$ m
7. $A = 54.0°$, $C = 43.1°$, $a = 26.5$ m
8. $B = 64.3°$, $b = 135$ m, $c = 118$ m
9. $A = 20.1°$, $a = 47.5$ mi, $c = 35.6$ mi
10. $B = 75.2°$, $A = 65.1°$, $b = 305$ ft
11. $C = 48.7°$, $B = 56.4°$, $b = 5960$ m
12. $A = 118.0°$, $a = 5750$ m, $b = 4750$ m
13. $B = 105.5°$, $c = 11.3$ km, $b = 31.4$ km
14. $A = 58.2°$, $a = 39.7$ mi, $c = 27.5$ mi
15. $A = 16.5°$, $a = 206$ ft, $b = 189$ ft
16. $A = 35.0°$, $B = 49.3°$, $a = 48.7$ m
17. Find the distance AC across the river shown in Illustration 2.

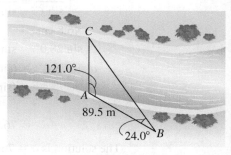

ILLUSTRATION 2

18. Find the lengths of rafters AC and BC of the roof shown in Illustration 3.

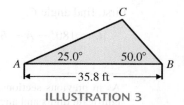

ILLUSTRATION 3

19. Find the distance AB between the ships shown in Illustration 4.

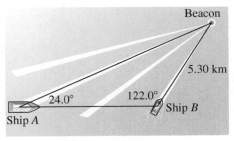

ILLUSTRATION 4

20. Find the height of the cliff shown in Illustration 5.

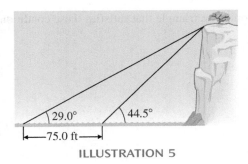

ILLUSTRATION 5

21. A contractor needs to grade the slope of a subdivision lot to place a house on level ground. (See Illustration 6.) The present slope of the lot is 12.5°. The contractor needs a level lot that is 105 ft deep. To control erosion, the back of the lot must be cut to a slope of 24.0°. How far from the street, measured along the present slope, will the excavation extend?

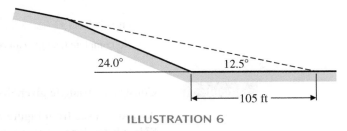

ILLUSTRATION 6

22. A weather balloon is sighted from points A and B, which are 4.00 km apart on level ground. The angle of elevation of the balloon from point A is 29.0°. Its angle of elevation from point B is 48.0°.

  a. Find the height (in m) of the balloon if it is between points A and B.

  b. Find its height (in m) if point B is between point A and the weather balloon.

14.4 Law of Sines: The Ambiguous Case

The solution of a triangle in which two sides and an angle opposite one of the sides are given needs special care. In this case, there may be one, two, or no triangles formed from the given information. Let's look at the possibilities.

Example 1   Construct a triangle given that $A = 32°$, $a = 18$ cm, and $b = 24$ cm.

As you can see from Figure 14.18, there are two triangles that satisfy these conditions: triangles ABC and $AB'C$. In one case, angle B is acute. In the other, angle B' is obtuse.

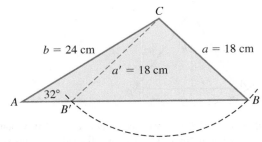

Figure 14.18
Two possible triangles can be drawn—one with an obtuse angle and two acute angles and one with all acute angles.   ◆

Example 2   Construct a triangle given that $A = 40°$, $a = 12$ cm, and $b = 24$ cm.

As you can see from Figure 14.19, there is no triangle that satisfies these conditions. Side a is just not long enough to reach AB.

458 CHAPTER 14 ♦ Trigonometry with Any Angle

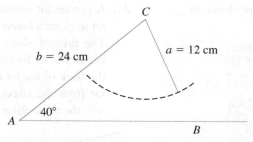

Figure 14.19

No complete triangle can be drawn.

Example 3   Construct a triangle given that $A = 50°$, $a = 12$ cm, and $b = 8$ cm.

As you can see from Figure 14.20, there is only one triangle that satisfies these conditions. Side a is too long for two solutions.

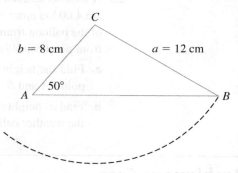

Figure 14.20

Only one possible triangle can be drawn.

Let's summarize the possible cases when two sides and an angle opposite one of the sides are given. Assume that angle A and adjacent side b are given. From these two parts, the altitude ($h = b \sin A$) is determined and fixed.

If angle A is *acute*, we have four possible cases as shown in Figure 14.21.

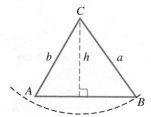 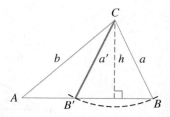

(a) When $h < b \leq a$, we have only one possible triangle. That is, when the side opposite the given acute angle is greater than the known adjacent side, there is only one possible triangle.

(b) When $h < a < b$, we have two possible triangles. That is, when the side opposite the given acute angle is less than the known adjacent side but greater than the altitude, there are two possible triangles.

Figure 14.21

The four possible cases when angle A is acute.

14.4 ♦ Law of Sines: The Ambiguous Case

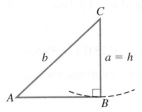 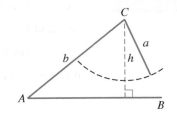

(c) When $a = h$, we have one possible triangle. That is, when the side opposite the given acute angle equals the altitude, there is only one possible triangle—a right triangle.

(d) When $a < h$, there is no possible triangle. That is, when the side opposite the given acute angle is less than the altitude, there is no possible triangle.

Figure 14.21
(continued)

If angle A is *obtuse*, we have two possible cases, as shown in Figure 14.22.

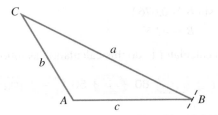 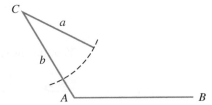

(a) When $a > b$, we have one possible triangle. That is, when the side opposite the given obtuse angle is greater than the known adjacent side, there is only one possible triangle.

(b) When $a \leq b$, there is no possible triangle. That is, when the side opposite the given obtuse angle is less than or equal to the known adjacent side, there is no possible triangle.

Figure 14.22
The two possible cases when angle A is obtuse.

NOTE: If the given parts are not angle A, side opposite a, and side adjacent b as in our preceding discussion, then you must substitute the given angle and sides accordingly. This is why it is so important to understand the general word description corresponding to each case.

The following table summarizes the possibilities for the ambiguous case of the law of sines.

If the given angle is acute:	We have:
a. altitude $<$ side adjacent $\leq$ side opposite	one triangle
b. altitude $<$ side opposite $<$ side adjacent	two triangles
c. side opposite $=$ altitude	one right triangle
d. side opposite $<$ altitude	no triangle
If the given angle is obtuse:	**We have:**
a. side opposite $>$ side adjacent	one triangle
b. side opposite $\leq$ side adjacent	no triangle

Example 4

If $A = 25.0°$, $a = 50.0$ m, and $b = 80.0$ m, solve the triangle.

First, find h.

$$h = b \sin A = (80.0 \text{ m})(\sin 25.0°) = 33.8 \text{ m}$$

Since $h < a < b$, we have two solutions. First, let's find *acute* angle B in Figure 14.23.

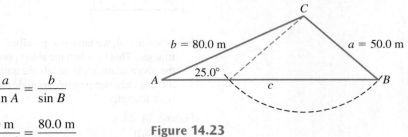

Figure 14.23

$$\frac{a}{\sin A} = \frac{b}{\sin B}$$

$$\frac{50.0 \text{ m}}{\sin 25.0°} = \frac{80.0 \text{ m}}{\sin B}$$

$(50.0 \text{ m})(\sin B) = (\sin 25.0°)(80.0 \text{ m})$ Multiply both sides by the LCD.

$$\sin B = \frac{(\sin 25.0°)(80.0 \text{ m})}{50.0 \text{ m}}$$ Divide both sides by 50.0 m.

$$\sin B = 0.6762$$

$$B = 42.5°$$

You may use a calculator to do this calculation as follows:

42.54656809

$$C = 180° - A - B = 180° - 25.0° - 42.5° = 112.5°$$

To find side c,

$$\frac{c}{\sin C} = \frac{a}{\sin A}$$

$$\frac{c}{\sin 112.5°} = \frac{50.0 \text{ m}}{\sin 25.0°}$$

$c(\sin 25.0°) = (\sin 112.5°)(50.0 \text{ m})$ Multiply both sides by the LCD.

$$c = \frac{(\sin 112.5°)(50.0 \text{ m})}{\sin 25.0°}$$ Divide both sides by $\sin 25.0°$.

$$= 109 \text{ m}$$

Next, let's find *obtuse* angle B in Figure 14.24.

$$\frac{a}{\sin A} = \frac{b}{\sin B}$$

$$\frac{50.0 \text{ m}}{\sin 25.0°} = \frac{80.0 \text{ m}}{\sin B}$$

$(50.0 \text{ m})(\sin B) = (\sin 25.0°)(80.0 \text{ m})$

$$\sin B = \frac{(\sin 25.0°)(80.0 \text{ m})}{50.0 \text{ m}}$$

$$\sin B = 0.6762$$

$$B = 180° - 42.5° = 137.5°$$

Figure 14.24

NOTE: If B is acute, B is SIN⁻¹ of 0.6762.

If B is obtuse, B is $180° -$ SIN^{-1} of 0.6762.

$$C = 180° - A - B = 180° - 25.0° - 137.5° = 17.5°$$

To find side c,

$$\frac{c}{\sin C} = \frac{a}{\sin A}$$

$$\frac{c}{\sin 17.5°} = \frac{50.0 \text{ m}}{\sin 25.0°}$$

$$c(\sin 25.0°) = (\sin 17.5°)(50.0 \text{ m})$$

$$c = \frac{(\sin 17.5°)(50.0 \text{ m})}{\sin 25.0°}$$

$$= 35.6 \text{ m}$$

The two solutions are $c = 109$ m, $B = 42.5°$, $C = 112.5°$ and $c = 35.6$ m, $B = 137.5°$, $C = 17.5°$. ◆

Example 5 If $A = 59.0°$, $a = 205$ m, and $b = 465$ m, solve the triangle.

First, find h.

$$h = b \sin A = (465 \text{ m})(\sin 59.0°) = 399 \text{ m}$$

Since $a < h$, there is no possible triangle.

What would happen if you tried to apply the law of sines anyway?

$$\frac{a}{\sin A} = \frac{b}{\sin B}$$

$$\frac{205 \text{ m}}{\sin 59.0°} = \frac{465 \text{ m}}{\sin B}$$

$$(205 \text{ m})(\sin B) = (\sin 59.0°)(465 \text{ m})$$

$$\sin B = \frac{(\sin 59.0°)(465 \text{ m})}{205 \text{ m}} = 1.944$$

NOTE: $\sin B = 1.944$ is *impossible*, because $-1 \leq \sin B \leq 1$. Recall that the graph of $y = \sin x$ has an amplitude of 1, which means that the values of $\sin x$ vary between 1 and -1. Your calculator will also indicate an error when you try to find angle B. ◆

As a final check to make certain that your solution is correct, verify that the following geometric triangle property is satisfied:

> In any triangle, the largest side is opposite the largest angle and the smallest side is opposite the smallest angle.

EXERCISES 14.4

For each general triangle, **a.** determine the number of solutions and **b.** solve the triangle, if possible, using the labels as shown in Illustration 1 (round lengths to three significant digits and angles to the nearest tenth of a degree):

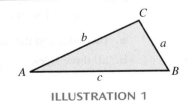

ILLUSTRATION 1

1. $A = 38.0°$, $a = 42.3$ m, $b = 32.5$ m
2. $C = 47.6°$, $a = 85.2$ cm, $c = 96.1$ cm
3. $A = 25.6°$, $b = 306$ m, $a = 275$ m
4. $B = 41.2°$, $c = 1860$ ft, $b = 1540$ ft
5. $A = 71.6°$, $b = 48.5$ m, $a = 15.7$ m
6. $B = 40.3°$, $b = 161$ cm, $c = 288$ cm
7. $C = 71.2°$, $a = 245$ cm, $c = 238$ cm
8. $A = 36.1°$, $b = 14.5$ m, $a = 12.5$ m
9. $B = 105.0°$, $b = 33.0$ mi, $a = 24.0$ mi
10. $A = 98.3°$, $a = 1420$ ft, $b = 1170$ ft
11. $A = 31.5°$, $a = 376$ m, $c = 406$ m
12. $B = 50.0°$, $b = 4130$ ft, $c = 4560$ ft
13. $C = 60.0°$, $c = 151$ m, $b = 181$ m
14. $A = 30.0°$, $a = 4850$ mi, $c = 3650$ mi
15. $B = 8.0°$, $b = 451$ m, $c = 855$ m
16. $C = 8.7°$, $c = 89.3$ mi, $b = 61.9$ mi
17. The owner of a triangular lot wishes to fence it along the lot lines. Lot markers at A and B have been located, but the lot marker at C cannot be found. The owner's attorney gives the following information by phone: $AB = 355$ ft, $BC = 295$ ft, and $A = 36.0°$. What is the length of AC?
18. The average distance from the sun to the earth is 1.5×10^8 km, and that from the sun to Venus is 1.1×10^8 km. Find the distance between the earth and Venus when the angle between the earth and the sun and the earth and Venus is $24.7°$. (Assume that the earth and Venus have circular orbits around the sun.)
19. A manufacturer has moved into a new building and wants to hang a sign outside the building based on the drawing in Illustration 2. How long is the lower support, and what angle does the upper support make with the building?

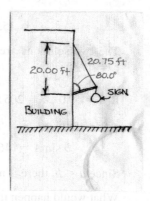

ILLUSTRATION 2

14.5 Solving Oblique Triangles: Law of Cosines

A second law used to solve oblique triangles is the law of cosines:

Law of Cosines

$$a^2 = b^2 + c^2 - 2bc \cos A$$
$$b^2 = a^2 + c^2 - 2ac \cos B$$
$$c^2 = a^2 + b^2 - 2ab \cos C$$

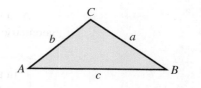

That is, *for any triangle, the square of any side equals the sum of the squares of the other two sides minus twice the product of these two sides and the cosine of their included angle.*

To use the law of cosines, you must know

a. two sides and the included angle or
b. all three sides.

Example 1

If $A = 115.2°$, $b = 18.5$ m, and $c = 21.7$ m, solve the triangle. (See Figure 14.25.)

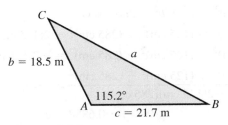

Figure 14.25

To find side a,

$$a^2 = b^2 + c^2 - 2bc \cos A$$
$$a^2 = (18.5 \text{ m})^2 + (21.7 \text{ m})^2 - 2(18.5 \text{ m})(21.7 \text{ m})(\cos 115.2°)$$
$$a = 34.0 \text{ m}$$

You may use a calculator to do this calculation as follows:

18.5 [x^2] [+] 21.7 [x^2] [−] 2 [×] 18.5 [×] 21.7 [×] [COS]

115.2 [=] [√] [ANS] [=]

> 33.98526435

To find angle B, use the law of sines, because it requires less computation.

$$\frac{a}{\sin A} = \frac{b}{\sin B}$$

$$\frac{34.0 \text{ m}}{\sin 115.2°} = \frac{18.5 \text{ m}}{\sin B}$$

$(34.0 \text{ m})(\sin B) = (\sin 115.2°)(18.5 \text{ m})$ Multiply both sides by the LCD.

$\sin B = \dfrac{(\sin 115.2°)(18.5 \text{ m})}{34.0 \text{ m}}$ Divide both sides by 34.0 m.

$$\sin B = 0.4923$$
$$B = 29.5°$$
$$C = 180° - A - B = 180° - 115.2° - 29.5° = 35.3°$$

The solution is $a = 34.0$ m, $B = 29.5°$, and $C = 35.3°$. ◆

Example 2

If $a = 125$ cm, $b = 285$ cm, and $c = 382$ cm, solve the triangle. (See Figure 14.26.)

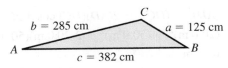

Figure 14.26

When three sides are given, you are advised to find the angle opposite the largest side first. Why?

CHAPTER 14 ♦ Trigonometry with Any Angle

To find angle C,

$$c^2 = a^2 + b^2 - 2ab \cos C$$
$$(382 \text{ cm})^2 = (125 \text{ cm})^2 + (285 \text{ cm})^2 - 2(125 \text{ cm})(285 \text{ cm}) \cos C$$
$$(382 \text{ cm})^2 - (125 \text{ cm})^2 - (285 \text{ cm})^2 = -2(125 \text{ cm})(285 \text{ cm}) \cos C$$
$$\frac{(382 \text{ cm})^2 - (125 \text{ cm})^2 - (285 \text{ cm})^2}{-2(125 \text{ cm})(285 \text{ cm})} = \cos C$$
$$-0.6888 = \cos C$$
$$133.5° = C$$

You may use a calculator to do this calculation as follows:

382 [x^2] [−] 125 [x^2] [−] 285 [x^2] [=] [ANS] [÷] [(−)] 2 ÷ 125 ÷ 285 [=] [COS⁻¹] [ANS] [=]

    133.5318657

To find angle A, let's use the law of sines.

$$\frac{c}{\sin C} = \frac{a}{\sin A}$$
$$\frac{382 \text{ cm}}{\sin 133.5°} = \frac{125 \text{ cm}}{\sin A}$$
$$(382 \text{ cm})(\sin A) = (\sin 133.5°)(125 \text{ cm})$$
$$\sin A = \frac{(\sin 133.5°)(125 \text{ cm})}{382 \text{ cm}}$$
$$\sin A = 0.2374$$
$$A = 13.7°$$
$$B = 180° - A - C = 180° - 13.7° - 133.5° = 32.8°$$

The solution is $A = 13.7°$, $B = 32.8°$, and $C = 133.5°$. ♦

Example 3 Find the lengths of guy wires AC and BC for a tower located on a hillside, as shown in Figure 14.27(a). The height of the tower is 50.0 m; $\angle ADC = 120.0°$; $AD = 20.0$ m; $BD = 15.0$ m.

First, let's use triangle ACD in Figure 14.27(b) to find length AC. Using the law of cosines,

$$(AC)^2 = (AD)^2 + (DC)^2 - 2(AD)(DC) \cos ADC$$
$$(AC)^2 = (20.0 \text{ m})^2 + (50.0 \text{ m})^2 - 2(20.0 \text{ m})(50.0 \text{ m}) \cos 120.0°$$
$$AC = 62.4 \text{ m}$$

Next, use triangle CDB and the law of cosines to find length BC. Note that $\angle CDB = 180° - 120.0° = 60.0°$.

$$(BC)^2 = (BD)^2 + (DC)^2 - 2(BD)(DC) \cos CDB$$
$$(BC)^2 = (15.0 \text{ m})^2 + (50.0 \text{ m})^2 - 2(15.0 \text{ m})(50.0 \text{ m}) \cos 60.0°$$
$$BC = 44.4 \text{ m}$$

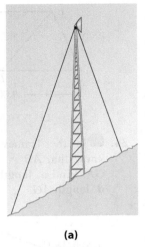
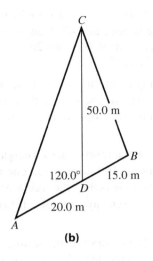

(a)   (b)

Figure 14.27

EXERCISES 14.5

Solve each triangle using the labels as shown in Illustration 1 (round lengths of sides to three significant digits and angles to the nearest tenth of a degree):

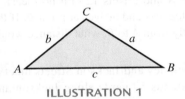

ILLUSTRATION 1

1. $A = 55.0°$, $b = 20.2$ m, $c = 25.0$ m
2. $B = 14.5°$, $a = 37.6$ cm, $c = 48.2$ cm
3. $C = 115.0°$, $a = 247$ ft, $b = 316$ ft
4. $A = 130.0°$, $b = 15.2$ km, $c = 9.50$ km
5. $a = 38,500$ mi, $b = 67,500$ mi, $c = 47,200$ mi
6. $a = 146$ cm, $b = 271$ cm, $c = 205$ cm
7. $B = 19.3°$, $a = 4820$ ft, $c = 1930$ ft
8. $C = 108.5°$, $a = 415$ m, $b = 325$ m
9. $a = 19.5$ m, $b = 36.5$ m, $c = 25.6$ m
10. $a = 207$ mi, $b = 106$ mi, $c = 142$ mi
11. Find the distance a across the pond in Illustration 2.

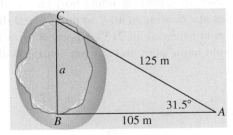

ILLUSTRATION 2

12. Find the length of rafter AC in Illustration 3.

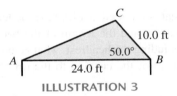

ILLUSTRATION 3

13. Find angles A and C in the roof in Illustration 4.

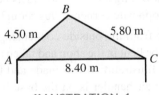

ILLUSTRATION 4

14. **a.** Find angles A and ACB in the roof in Illustration 5. $AC = BC$
    b. Find length CD.

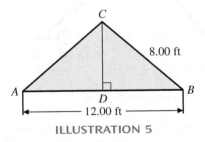

ILLUSTRATION 5

15. A piece of sheet metal is to be cut in the shape of a triangle with sides of 24.0 in., 12.0 in., and 21.0 in. Find the measures of the angles.

466 CHAPTER 14 ♦ Trigonometry with Any Angle

16. Three pieces of steel angle are welded to form a triangle. If two pieces are welded at a 42.0° angle and the lengths of these two pieces are 36.0 in. and 20.0 in., what is the length of the third piece?

17. A plane flies 70.0 mi due north from its base airport. Then it makes a 70.0° turn northeast and flies another 90.0 mi. How far will it be to go straight back to the base airport?

18. The taxiways for a small airport make a triangle with the runway. The runway is 61$\overline{0}$0 ft long, and one of the taxiways is 33$\overline{0}$0 ft. How long is the other taxiway if the angle between the 61$\overline{0}$0-ft runway and 33$\overline{0}$0-ft taxiway is 62.0°?

19. An automobile has been sideswiped, causing much damage. The distance from the driver's side front wheel to the driver's side rear wheel is 120.0 in., and the length between the rear wheels is 54.0 in. If the angle between these measurements is 110.0°, what is the distance from the driver's side front tire to the passenger's side rear tire?

20. An automobile seat is reclined at 140.0°. If the length of the cushion is 20.0 in. and the length of the back of the seat is 34.0 in. including the headrest, what is the distance from the front tip of the cushion to the tip of the headrest?

21. A room is shaped like a kite. The lengths of two adjacent walls are 20.0 ft and 28.0 ft. If the angle between these walls is 130.0°, how long must the two ducts be that stretch from corner to opposite corner of this room?

22. The refrigerant line from the outside condensing unit to the air handler must be bent in the shop and taken to the job site. The sketch in Illustration 6 was made and taken to the shop. Notice the air handler is on the base line of the triangle. Find the three angles.

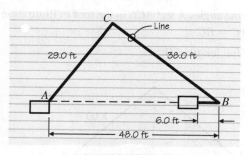

ILLUSTRATION 6

23. In the framework shown in Illustration 7, we know that $AE = CD$, $AB = BC$, $BD = BE$, and $\overline{AC} \parallel \overline{ED}$. Find **a.** $\angle BEA$, **b.** $\angle A$, **c.** length BE, and **d.** length DE.

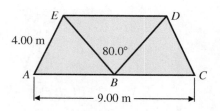

ILLUSTRATION 7

24. In the framework shown in Illustration 8, we know that $AB = DE$, $BC = CD$, $AH = FE$, $HG = GF$. Find **a.** length HB, **b.** $\angle AHB$, **c.** length GC, and **d.** length AG.

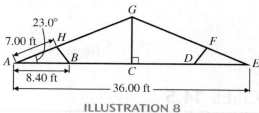

ILLUSTRATION 8

25. A triangular lot has sides 1580 ft, 2860 ft, and 1820 ft long. Find its largest angle.

26. A ship starts at point A and travels 125 mi northeast. It then travels 150 mi due east and arrives at point B. If the ship had sailed directly from A to B, what distance would it have traveled?

27. See Illustration 9. Deloney and Jackson Streets meet at a 45° angle. A lot extends 5$\overline{0}$ yards along Jackson and 4$\overline{0}$ yards along Deloney. Find the length of the back border.

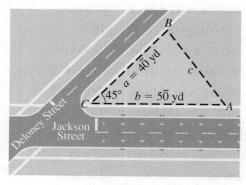

ILLUSTRATION 9

28. See Illustration 10. From its home port (H), a ship sails 52 miles at a bearing of 147° to point A and then sails 78 miles at a bearing of 213° to point B. How far is point B from home port, and at what bearing will it return?

Summary 467

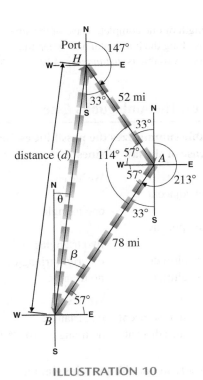

ILLUSTRATION 10

29. An 8.00-ft antenna must be mounted on the roof of a warehouse. Find the length of a guy wire to be attached 1.00 ft from the top of the antenna to the edge of the roof as shown in Illustration 11.

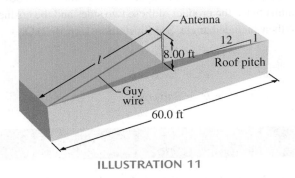

ILLUSTRATION 11

30. Snakes, an important part of the food chain, cannot chew their prey, so they generally swallow their prey whole. A special hinged jaw permits some snakes to open their mouth at an angle of as much as 150° so that they can swallow animals larger than the size of their own heads. If a particular snake's jaws are 3.0 in. long and it opens its mouth at 150°, how wide is the jaw opening? That is, what is the largest prey it can swallow?

31. A kite is flying from the top of a hill as shown in Illustration 12. How far is the kite from the person standing at point A?

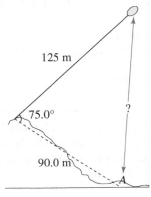

ILLUSTRATION 12

32. A game preserve manager is fencing the triangular plot of land in Illustration 13. Find the most acute angle.

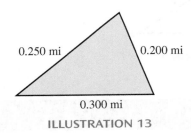

ILLUSTRATION 13

SUMMARY | CHAPTER 14

Glossary of Basic Terms

Acute angle. An angle with a measure less than 90°. (p. 454)

Amplitude. The maximum y value for the sine and cosine graphs. (p. 447)

Frequency. The number of waves that pass a given point on the time axis each second. (p. 449)

Oblique triangle. A triangle with no right angle. (p. 454)

Obtuse angle. An angle with a measure greater than 90° but less than 180°. (p. 454)

Out of phase. When the graph of the sine curve does not pass through the origin (0°, 0) or when the graph of the cosine curve does not pass through the point (0°, A), where A is the amplitude. (p. 452)

Period. The x distance between any point on a sine or cosine graph and the corresponding point in the next cycle where the graph starts repeating itself. Also, the time required for one complete wave to pass a given point on the horizontal axis when the horizontal, or x axis, variable is *time*. (p. 448)

Phase shift. The horizontal distance between two successive corresponding points of the curve $y = A \sin Bx$ (or $y = A \cos Bx$) and the out-of-phase curve. (p. 452)

Wavelength. The length of one complete wave of the sine and cosine graphs along the horizontal axis when the horizontal, or x axis, variable is *distance*. (p. 449)

14.1 Sine and Cosine Graphs

1. **Frequency and period:** $f = 1/T$. (p. 449)
2. **Frequency and wavelength:** $v = \lambda f$. (p. 450)

14.2 Period and Phase Shift

1. **Period:** the period for $y = A \sin Bx$ and for $y = \cos Bx$ is $P = 360°/B$. (p. 451)
2. **Phase shift:** the phase shift for $y = A \sin (Bx + C)$ and for $y = \cos (Bx + C)$ is
   a. to the *left* C/B units if C/B is positive.
   b. to the *right* C/B units if C/B is negative. (p. 453)

14.3 Solving Oblique Triangles: Law of Sines

1. **Law of Sines:** For any triangle, the ratio of the length of any side to the sine of the opposite angle equals the ratio of the length of any other side to the sine of its opposite angle. See Illustration 1. (p. 454) Or

$$\frac{a}{\sin A} = \frac{b}{\sin B} = \frac{c}{\sin C}$$

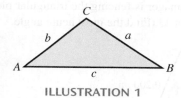

ILLUSTRATION 1

14.4 Law of Sines: The Ambiguous Case

1. **The following table summarizes the possibilities for the ambiguous case of the law of sines:**

If the given angle is acute:	We have:
a. altitude < side adjacent ≤ side opposite	one triangle
b. altitude < side opposite < side adjacent	two triangles
c. side opposite = altitude	one right triangle
d. side opposite < altitude	no triangle
If the given angle is obtuse:	We have:
a. side opposite > side adjacent	one triangle
b. side opposite ≤ side adjacent	no triangle (p. 459)

2. **One check when solving an oblique triangle:** the largest side is opposite the largest angle and the smallest side is opposite the smallest angle. (p. 461)

14.5 Solving Oblique Triangles: Law of Cosines

1. **Law of Cosines:** For any triangle, the square of any side equals the sum of the squares of the other two sides minus twice the product of these two sides and the cosine of their included angle. See Illustration 1. (p. 462) Or

$$a^2 = b^2 + c^2 - 2bc \cos A$$
$$b^2 = a^2 + c^2 - 2ac \cos B$$
$$c^2 = a^2 + b^2 - 2ab \cos C$$

REVIEW | CHAPTER 14

Find each value rounded to four significant digits:

1. tan 143°
2. sin 209.8°
3. cos 317.4°

Graph each equation for values of x between 0° and 360° in multiples of 15°:

4. $y = 6 \cos x$
5. $y = 3 \sin 2x$

Find the period and amplitude, and graph at least two periods of each equation:

6. $y = 5 \sin 3x$
7. $y = 3 \cos 4x$

Find the period, amplitude, and phase shift, and graph at least two periods of each equation:

8. $y = 4 \cos (x + 60°)$
9. $y = 6 \sin (2x - 180°)$

Solve each triangle using the labels as shown in Illustration 1 (round lengths of sides to three significant digits and angles to the nearest tenth of a degree):

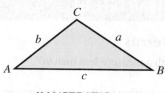

ILLUSTRATION 1

10. $B = 52.7°$, $b = 206$ m, $a = 175$ m
11. $A = 61.2°$, $C = 75.6°$, $c = 88.0$ cm
12. $B = 17.5°$, $a = 345$ m, $c = 405$ m
13. $a = 48.6$ cm, $b = 31.2$ cm, $c = 51.5$ cm

14. $A = 29.5°$, $b = 20.5$ m, $a = 18.5$ m
15. $B = 18.5°$, $a = 1680$ m, $b = 1520$ m
16. $a = 575$ ft, $b = 1080$ ft, $c = 1250$ ft
17. $C = 73.5°$, $c = 58.2$ ft, $b = 81.2$ ft
18. Find **a.** angle B and **b.** length x in Illustration 2.

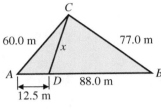

ILLUSTRATION 2

19. The centers of five holes are equally spaced around a circle of diameter 16.00 in. Find the distance between the centers of two successive holes.

20. In the roof truss in Illustration 3, $AB = DE$, $BC = CD$, and $AE = 36.0$ m. Find the lengths **a.** AF, **b.** BF, **c.** CF, and **d.** BC.

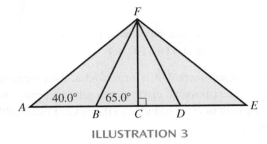

ILLUSTRATION 3

TEST | CHAPTER 14

Find each value rounded to four significant digits:

1. $\cos 182.9°$
2. $\tan 261°$
3. Find the period, amplitude, and phase shift, and draw at least two periods of $y = 2 \sin(3x + 45°)$.
4. Find angle B in Illustration 1.
5. Find angle C in Illustration 1.
6. Find side c in Illustration 1.

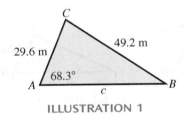

ILLUSTRATION 1

7. Find angle C in Illustration 2.
8. Find angle B in Illustration 2.
9. Find angle A in Illustration 2.

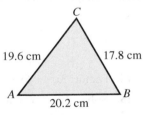

ILLUSTRATION 2

10. Find the length of the rafter shown in Illustration 3.
11. Find angle A in Illustration 3.

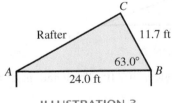

ILLUSTRATION 3

CUMULATIVE REVIEW | CHAPTERS 1–14

1. Given the formula $R = \dfrac{Vt}{I}$, where $V = 32$, $t = 5$, and $I = 20$, find R.
2. Simplify: $-\dfrac{5}{6} - \left(-\dfrac{3}{5}\right)$
3. Write 41,800 in scientific notation.
4. 90 kg = _____ lb

5. Read the scale in Illustration 1.

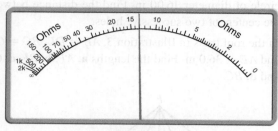

ILLUSTRATION 1

6. For the measurement 0.0018 mm, find **a.** the precision, **b.** the greatest possible error, **c.** the relative error, and **d.** the percent of error rounded to the nearest hundredth of a percent.

7. Simplify: $(2a^2 - 5a + 3) + (4a^2 + 3a - 1)$

8. Solve: $6 + 3(x - 2) = 24$

9. Solve: $\dfrac{x}{2} - \dfrac{2}{7} = \dfrac{1}{3}$

10. The area of a rectangle with constant width varies directly as its length. The area is 30.8 m² when the length is 12.8 m. Find the area when the length is 42.5 m.

11. Solve for y: $-5x - 3y = -8$

12. Solve: $7x - y = 4$
    $14x - 2y = 8$

13. Factor: $x^2 - 2x - 168$ 14. Factor: $3x^2 - 6x - 189$

15. Solve: $3x^2 - 13x = 10$ 16. Solve: $2x^2 - x - 8 = 0$

17. Find **a.** the area and **b.** the perimeter of the triangle in Illustration 2.

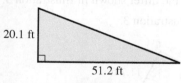

ILLUSTRATION 2

18. The area of a rectangle is 307 ft². The length is 22.4 ft. Find the width.

19. Find the volume of the frustum of the rectangular pyramid in Illustration 3.

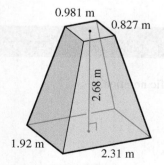

ILLUSTRATION 3

20. Find tan 67.2° rounded to four significant digits.

21. If cos $A = 0.6218$, find angle A in degrees.

From the triangle in Illustration 4, find:

22. $\angle B$ 23. side a 24. side c

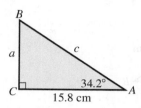

ILLUSTRATION 4

25. A roof has a rise of 8.00 ft and a span of 24.0 ft. Find its pitch and the distance rounded to the nearest inch, from the eave to its peak.

26. Find cos 191.13° rounded to four significant digits.

27. Draw the graph of $y = \dfrac{3}{2} \cos 2x$ for values of x between 0° and 360° in multiples of 15°.

28. Find the period and amplitude and draw at least two periods of the graph of the equation $y = 2 \sin \dfrac{1}{2}x$.

29. Given the triangle in Illustration 5, find angle C.

30. Find angle A in Illustration 6.

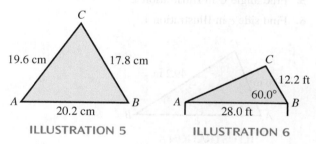

ILLUSTRATION 5   ILLUSTRATION 6

Basic Statistics | CHAPTER 15

OBJECTIVES

- Use bar graphs, circle graphs, line graphs, and frequency distributions to present data.
- Find and use the mean, the median, and the mode of a data set.
- Find and use percentiles to describe ranked data.
- Find the range and the sample standard deviation of a set of data.
- Find the sample standard deviation of a set of grouped data.
- Use statistical process control tools and techniques to determine whether a process is in control.
- Use histograms, run charts, and scattergrams to view data.
- Use a normal distribution to find the number or percent of data within a given interval.
- Find the sample space for a given event and the probability that a given event will happen.
- Determine when two events are independent and find the probability of two independent events.

Mathematics at Work

Careers in the culinary arts involve the art and skills of preparing and cooking food. Culinary artists use their skills to responsibly prepare foods that taste good, are nutritious, and are pleasing to the eye. They must have a working knowledge of food science and an understanding of diet and nutrition. They work in restaurants, hospitals, businesses, nursing homes, and other institutions where food is prepared and served.

A sampling of careers in the culinary arts includes chefs and head cooks who oversee the daily food preparation, direct kitchen staff, and handle any food-related concerns; food service managers who are responsible for the daily operations of restaurants and other establishments that prepare and serve food; and beverage managers who manage all food and beverage outlets in large hotels and other large establishments.

Increasingly, a college education with formal qualifications is required for success in this field. Public and private colleges offer a wide variety of classroom, hands-on skills, and on-the-job experiences. For more information, please visit **www.cengage.com** and access the Student Online Resources for this text.

Professional chefs
Cooking in a commercial kitchen

15.1 Bar Graphs

Statistics is the branch of mathematics that deals with the collection, analysis, interpretation, and presentation of masses of numerical data. In this chapter, we will first study the different ways in which data can be presented using graphs. Then, in the later sections, we will study some of the different ways to describe small sets of data. We will study only the most basic parts to help you read and better understand newspapers, magazines, and some of the technical reports in your field of interest. The chapter includes an examination of statistical process control, a technique that is widely used in manufacturing.

A graph is a picture that shows the relationship between several types of collected information. A graph is very useful when there are large quantities of information to analyze. There are many ways of graphing. A **bar graph** is a graph with parallel bars whose lengths are proportional to the frequency of the given quantities in a data set. Look closely at the bar graph in Figure 15.1.

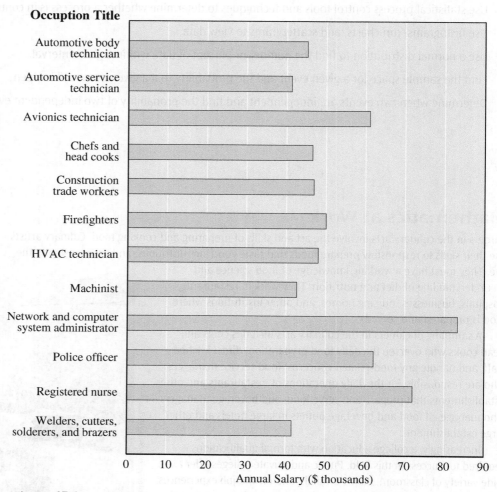

Figure 15.1
Bar graph of annual salaries of selected occupational titles, May 2015, Occupational Employment Statistics, Bureau of Labor Statistics, US Department of Labor

Example 1 What is the annual salary of an automotive service technician as shown in the bar graph in Figure 15.1?

Find the automotive service technician in the "Occupational Title" column. Read the right end of the bar down to the Annual Salary ($ thousands) scale: approximately $41,000. ◆

EXERCISES 15.1

Find the approximate annual earnings of each of the following occupations from the bar graph in Figure 15.1:

1. Machinist
2. Chefs and head cooks
3. HVAC technician
4. Police officer
5. Registered nurse
6. Welders, cutters, solderers, and brazers
7. Network and computer system administrator
8. Firefighters
9. Construction trade workers
10. Automotive body technician

Find the following information from Illustration 1:

11. How many barrels per day were used by France?
12. How many barrels per day were used by Japan?
13. What country used the most barrels per day?
14. What country used the fewest barrels per day?
15. How many barrels per day were used by Russia?
16. How many barrels per day were used by Italy?
17. How many barrels per day were used by the United States?
18. How many barrels per day were used by China?
19. How many barrels per day were used by Canada?
20. What was the total number of barrels per day used by all the countries listed?
21. A survey of 100 families was taken to find the number of times the families had gone out to eat in the past month. The data are given in Illustration 2. Draw a bar graph for this survey.

Times out in past month	Number of families
0	2
1	15
2	51
3	17
4	10
5 or more	5

ILLUSTRATION 2

22. Illustration 3 shows the average test scores on chapter tests given in a mathematics class. Draw a bar graph for these scores.
23. Illustration 4 gives the number of U.S. workers employed in the given industries in 2014. Draw a bar graph for these data.

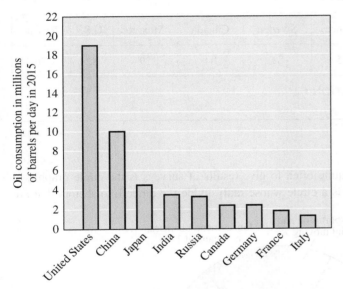

ILLUSTRATION 1

Chapter	1	2	3	4	5	6	7	8	9	10	11	12	13	14
Score	78	81	75	77	84	81	79	70	72	73	75	69	81	72

ILLUSTRATION 3

Industry	Mining	Manufacturing	State and local government	Construction	Transportation	Health	Retail trade	Financial
Number (in 1000s)	843	12,188	19,134	6138	4640	18,057	15,364	7980

ILLUSTRATION 4

24. **a.** Illustration 5 lists the 2015 male life expectancy for the given countries. Draw a bar graph for these data.

b. Illustration 6 gives the 2015 female life expectancy for the given countries. Draw a bar graph for these data.

25. The data in Illustration 7 shows the median usual weekly offerings and the unemployment rates by educational attainment in 2015 for persons age 25 and over. Earnings are for full-time wage and salary workers. Source: U.S. Bureau of Labor Statistics, Current Population Survey. Draw side-by-side bar graphs for comparison purposes.

	Median usual weekly earnings	Unemployment rate
Doctoral degree	$1623	1.7%
Professional degree	$1730	1.5%
Master's degree	$1341	2.5%
Bachelor's degree	$1137	2.8%
Associate's degree	$ 798	3.8%
Some college, no degree	$ 738	5.0%
High school diploma	$ 678	5.4%
Less than a high school diploma	$ 493	8.0%
All workers	**$ 860**	**4.3%**

ILLUSTRATION 7

Country	Egypt	Nigeria	Israel	Japan	France	Sweden	Canada	Mexico	U.S.	Russia
Years	69	53	81	80	79	84	80	76	77	65

ILLUSTRATION 5

Country	Egypt	Nigeria	Israel	Japan	France	Sweden	Canada	Mexico	U.S.	Russia
Years	73	56	84	87	85	81	84	79	82	76

ILLUSTRATION 6

15.2 Circle Graphs

Another type of graph used quite often to give results of surveys is the circle graph (see Figure 15.2). A **circle graph** is a circle whose radii divide the circle into sectors that are

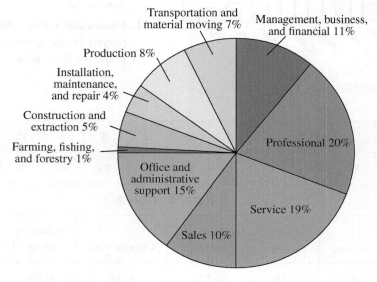

Figure 15.2

Circle graph of U.S. major occupational groups 2010 (U.S. Bureau of Labor Statistics)

proportional in angle and area in relative size to the quantities represented. The circle graph is used to show the relationship between the parts and the whole.

To make a circle graph with data given in percents, first draw the circle. Since there are 360° in a circle, multiply the percent of an item by 360 to find what part of the circle is used by that item. When a circle graph is to be drawn from data that are not in percent form, the data must be converted to percents. Once the data are in this form, the steps in drawing the graph are the same as those already given.

Example 1

You are asked to draw a circle graph illustrating the following June weather patterns: sunny or mostly sunny (16 days), cloudy or mostly cloudy with no measureable precipitation (8 days), rainy (6 days).

First, find the percent of the days of the month for each pattern:

16 days ÷ 30 days = $53\frac{1}{3}\%$
8 days ÷ 30 days = $26\frac{2}{3}\%$
6 days ÷ 30 days = 20%

Then, find the central angle of the circle that each pattern represents.

$53\frac{1}{3}\%$ of 360° = 0.533 × 360° = 192°
$26\frac{2}{3}\%$ of 360° = 0.267 × 360° = 96°
20% of 360° = 0.2 × 360° = 72°

With a protractor draw central angles of 192°, 96°, and 72°. Then label the sections as in Figure 15.3. ◆

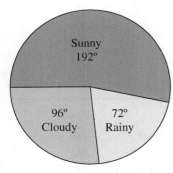

June Weather Patterns
Figure 15.3

Example 2

Draw a circle graph with the following data.

Suggested semester credit-hour requirements for a community college curriculum in engineering technology are as follows:

Course	Semester hours
Mathematics (technical)	10
Applied science	10
Technical courses in major	34
General education courses	12
	66

Write each area of study as a percent of the whole program.

Mathematics: $\dfrac{10}{66} = \dfrac{r}{100}$

$66r = 1000$ The product of the means equals the product of the extremes.

$r = 15.2\%$ Divide both sides by 66.

15.2% × 360° = 0.152 × 360°
= 55° (rounded to nearest whole degree)

Science: same as mathematics, 55°

Technical courses: $\dfrac{34}{66} = \dfrac{r}{100}$

$66r = 3400$

$r = 51.5\%$

51.5% × 360° = 0.515 × 360° = 185°

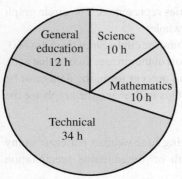

General education: $\dfrac{12}{66} = \dfrac{r}{100}$

$66r = 1200$

$r = 18.2\%$

$18.2\% \times 360° = 0.182 \times 360° = 66°$

Then draw the central angles and label the sections (see Figure 15.4).

Engineering Technology Semester Credit-Hour Requirements

Figure 15.4

EXERCISES 15.2

1. Find 26% of 360°.
2. Find 52% of 360°.
3. Find 15.2% of 360°.
4. Find 37.1% of 360°.
5. Find 75% of 360°.
6. Find 47.7% of 360°.
7. Of 744 students, 452 are taking mathematics. What angle of a circle would show the percent of students taking mathematics?
8. Of 2017 students, 189 are taking technical physics. What angle of a circle would show the percent of students taking technical physics?
9. Of 5020 cellphones, 208 are found to be defective. What angle of a circle would show the percent defective?
10. Candidate A was one of four candidates in an election. Of 29,106 votes cast, 4060 were for Candidate A. What angle of a circle would show the percent of votes *not* cast for Candidate A?
11. A department spends $16,192 of its $182,100 budget for supplies. What angle of a circle would show the percent of money the department spends on things other than supplies?
12. In June, the sales of an automobile dealership were as follows:

    Brand A: 29   Brand B: 52   Brand C: 15

    Brand D: 75   Brand E: 43

    What central angle of a circle graph would show Brand B's sales as a percent of the total sales for the month?

Draw a circle graph for Exercises 13–20:

13. United States work-related deaths by cause in 2014:

Transportation	41%
Violence with persons or animals	17%
Contact with object or equipment	16%
Falls	16%
Exposure to harmful substances	7%
Fire	3%

14. United States estimated population by age in 2016:

14 and under	18.7%
15–24	13.5%
25–54	39.6%
55–64	12.8%
65 and older	15.3%

15. The suggested semester credit-hour requirements for a community college curriculum in industrial technology are shown in Illustration 1.

Course	Semester hours
Mathematics	6
Applied science	8
Technical specialties	34
General education courses	12
	60

    ILLUSTRATION 1

16. A company interviewed its 473 employees to find the toughest day to work of a 5-day work week as shown in Illustration 2.

Day	Number
Monday	251
Tuesday	33
Wednesday	57
Thursday	43
Friday	89

    ILLUSTRATION 2

17. Total forest by continent in year 2015 in millions of hectares:

    Africa 624
    Asia 593
    Europe 1015
    North and Central America 751
    Oceania 174
    South America 842

18. The highest level of education for persons age 25 and older in the United States in 2015 can be found in the data shown in Illustration 3.

Less than high school	5.9%	Some college	21.3%
Some high school	8.0%	Bachelor's degree	18.0%
High school graduate	28.1%	Advanced degree	10.9%
Associate degree	7.8%		

ILLUSTRATION 3

19. The 2015 United Nations population data of the following regions of the world in millions were: Northern America, 358; Latin America and the Caribbean, 634; Europe, 738; Asia, 4393; Africa, 1186; Oceania, 39.

20. Some 2030 United Nations projected population data for the regions of the world are as follows in millions: Northern America, 396; Latin America and the Caribbean, 721; Africa, 1680; Europe, 734; Asia, 4900; Oceania, 47.

15.3 Line Graphs

A **line graph** is a graph formed by segments of straight lines that join the plotted points that represent given data. The line graph is used to show changing conditions, often over a certain time interval.

Example 1 An industrial technician must keep a chemical at a temperature below 60°F. He must also keep an hourly record of its temperature and record each day's temperatures on a line graph. The following table shows the data he collected:

Time	8:00	9:00	10:00	11:00	12:00	1:00	2:00	3:00	4:00	5:00
Temp. (°F)	60°	58°	54°	51°	52°	57°	54°	52°	54°	58°

When drawing line graphs, (a) use graph paper, because it is already subdivided both vertically and horizontally; (b) choose horizontal and vertical scales so that the line uses up most of the space allowed for the graph; (c) name and label each scale so that all marks on the scale are the same distance apart and show equal intervals; (d) plot the points from the given data; (e) connect each pair of points in order by a straight line. When you have taken all these steps, you will have a line graph (see Figure 15.5).

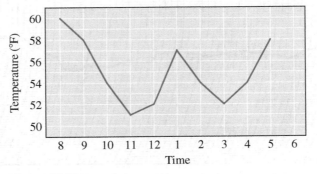

Figure 15.5

Line graph

EXERCISES 15.3

1. The data in Illustration 1 are from the records of the industrial technician in Example 1. These data were recorded on the following day. Draw a line graph for them.

Time	8:00	9:00	10:00	11:00	12:00	1:00	2:00	3:00	4:00	5:00
Temp. (°F)	59°	57°	55°	54°	53°	57°	55°	53°	56°	59°

ILLUSTRATION 1

2. An inspector recorded the number of faulty wireless routers and the hour in which they passed by his station, as shown in Illustration 2. Draw a line graph for these data.

Time	7–8	8–9	9–10	10–11	11–12	1–2	2–3	3–4	4–5	5–6
Number of faulty units	1	2	2	3	6	2	4	4	7	10

ILLUSTRATION 2

3. Illustration 3 lists the 6-months sales performance for Martha and George (in $). Draw a line graph for these data.

Month	Jan	Feb	Mar	Apr	May	June
Martha	84,262	72,596	24,680	65,141	81,270	72,490
George	61,180	74,296	93,240	61,724	48,244	75,419

ILLUSTRATION 3

A technician is often asked to read graphs drawn by a machine. The machine records measurements by plotting them on a graph. Any field in which quality control or continuous information is needed might use this way of recording measurements. Illustration 4 shows a microbarograph used by the weather service to record atmospheric pressure in inches. For example, the reading on Monday at 8:00 P.M. was 29.34 in.

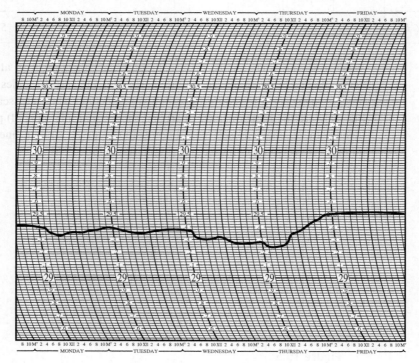

ILLUSTRATION 4
Microbarograph (atmospheric pressure in inches)

Use the microbarograph in Illustration 4 to answer Exercises 4–8.

4. What was the atmospheric pressure recorded for Tuesday at 2:00 P.M.?
5. What was the highest atmospheric pressure recorded? When was it recorded?
6. What was the lowest atmospheric pressure recorded?
7. What was the atmospheric pressure recorded for Thursday at 10:00 P.M.?
8. What was the atmospheric pressure recorded for Monday at noon?

A hygrothermograph is used by the weather services to record temperature and relative humidity (see Illustration 5). The lower part of the graph is used to measure relative humidity from 0% to 100%. The upper part of the graph is used to measure temperature from 10°F to 110°F. For example, at 8:00 P.M., the temperature was 86°F, and the relative humidity was 82%.

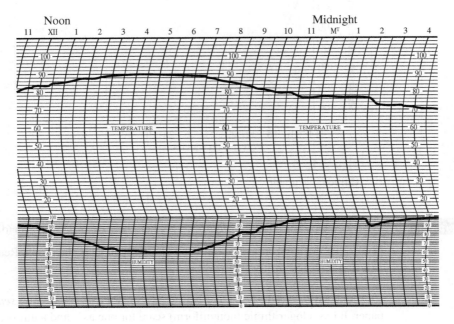

ILLUSTRATION 5
Hygrothermography (temperature in °F and relative humidity in %)

Use the hygrothermograph in Illustration 5 to answer Exercises 9–14.

9. What was the relative humidity at 12:00 midnight?
10. What was the temperature at 3:30 A.M.?
11. What was the highest temperature recorded?
12. What was the lowest temperature recorded?
13. What was the relative humidity at 2:00 A.M.?
14. What was the lowest relative humidity recorded?
15. According to National Oceanic and Atmospheric Administration (NOAA) records, the concentration of CO_2 gas in the atmosphere measured at the Mauna Loa Observatory in Hawaii has increased steadily for the past 55 years as shown in Illustration 6. **a.** What was the change in average annual concentrations in CO_2 (measured in parts per million by volume, ppmv) from 1960 to 2015? **b.** In what year did the average annual concentrations first exceed 360 ppmv?

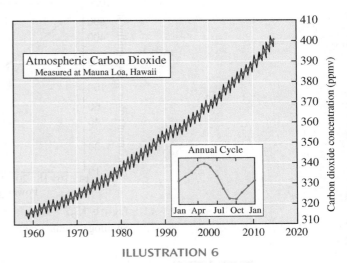

ILLUSTRATION 6
Data provided by the Scripps Institution of Oceanography
NOAA Earth System Research Laboratory.

15.4 Other Graphs

A graph can be a curved line, as shown in Figure 15.6. This graph shows typical power gain for class B push-pull amplifiers with 9-volt power supply.

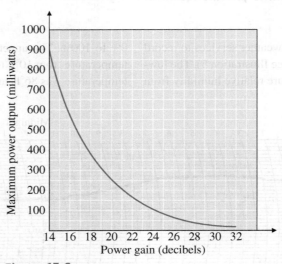

Figure 15.6
Curved line graph

Example 1   What is the power output in Figure 15.6 when the gain is 22 decibels (dB)?

Find 22 on the horizontal axis and read up until you meet the graph. Read left to the vertical axis and read 160 milliwatts (mW).   ♦

One way to avoid having to use a curved line for a graph is to use *semilogarithmic* graph paper. It has a logarithmic (nonuniform) scale for one axis and a uniform scale for the other axis. Figure 15.7 shows the data from Figure 15.6 plotted on semilogarithmic graph paper.

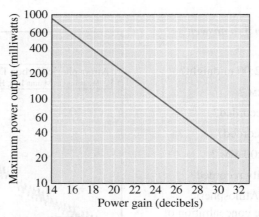

Figure 15.7
Semilogarithmic graph

Example 2   Find the power gain in Figure 15.7 when the power output is 60 mW.

Find 60 on the vertical axis. Read across until you meet the graph; read down to the horizontal axis and read 27 dB.   ♦

EXERCISES 15.4

Use Figure 15.6 to find the answers for Exercises 1–5.

1. What is the highest power output? What is the power gain at the highest power output?
2. What is the power gain when the power output is 600 mW?
3. What is the power output when the power gain is 25 dB?
4. Between what two decibel readings is the greatest change in power output found?
5. Between what two decibel readings is the least change in power output found?

Use Figure 15.7 to find the answers for Exercises 6–10.

6. What is the highest power output? What is the power gain at the highest power output?
7. What is the power gain when the power output is 600 mW?
8. What is the power output when the power gain is 25 dB?
9. Between what two decibel readings is the greatest change in power output found?
10. Between what two decibel readings is the least change in power output found?

15.5 Mean Measurement

We have already seen in other chapters that with each technical measurement, a certain amount of error is made. One way in which a technician can offset this error is to use what is called the mean of the measurements or the mean measurement. The **mean measurement** (or **mean**) is the *average* of a set of measurements. To find the mean measurement, the technician takes several measurements. The mean measurement is then found by dividing the sum of these measurements by the number of measurements taken:

$$\text{mean measurement} = \frac{\text{sum of measurements}}{\text{number of measurements}}$$

Example 1 A machinist measured the thickness of a metal disk with a micrometer at four different places. She found the following values:

2.147 in., 2.143 in., 2.151 in., 2.148 in.

Find the mean measurement.

STEP **1** Add the measurements.

2.147 in.
2.143 in.
2.151 in.
<u>2.148 in.</u>
8.589 in.

STEP 2   Divide the sum of the measurements by the number of measurements.

$$\text{mean measurement} = \frac{\text{sum of measurements}}{\text{number of measurements}}$$

$$= \frac{8.589 \text{ in.}}{4}$$

$$= 2.14725 \text{ in.}$$

So the mean measurement is 2.147 in.

Note that the mean measurement is written so that it has the same precision as each of the measurements. ♦

EXERCISES 15.5

Find the mean measurement for each set of measurements:

1. 609$\overline{1}$; 505$\overline{0}$; 710$\overline{2}$; 411$\overline{1}$; 606$\overline{0}$; 591$\overline{0}$; 711$\overline{2}$; 585$\overline{5}$; 628$\overline{0}$; 10,171; 902$\overline{0}$; 10,172

2. 2.7; 8.1; 9.3; 7.2; 10.6; 11.4; 12.9; 13.5; 16.1; 10.9; 12.7; 15.9; 20.7; 21.9; 30.6; 42.9

3. 205$\overline{0}$; 1951; 2132; 2232; 2147; 1867; 1996; 1785

4. 0.018; 0.115; 0.052; 0.198; 0.222; 0.189; 0.228; 0.346; 0.196; 0.258; 0.337; 0.532

5. 1.005; 1.102; 1.112; 1.058; 1.068; 1.115; 1.213

6. 248; 625; 324; 125; 762; 951; 843; 62; 853; 192; 346; 367; 484; 281; 628; 733; 801; 97; 218

7. 21; 53; 78; 42; 63; 28; 57; 83; 91; 32; 18

8. 0.82; 0.31; 1.63; 0.79; 1.08; 0.78; 1.14; 1.93; 0.068

9. 1.69; 2.38; 4.17; 7.13; 3.68; 2.83; 4.17; 8.29; 4.73; 3.68; 6.18; 1.86; 6.32; 4.17; 2.83; 1.08; 9.62; 7.71

10. 3182; 444$\overline{0}$; 2967; 7632; 1188; 6653; 2161; 8197; 5108; 9668; 5108; 6203; 1988; 4033; 1204; 3206; 4699; 3307; 7226

11. 47.61 cm; 48.23 cm; 47.92 cm; 47.81 cm

12. 9234 m; 9228 m; 9237 m; 9235 m; 9231 m

13. 0.2617 in.; 0.2614 in.; 0.2624 in.; 0.2620 in.; 0.2619 in.; 0.2617 in.

14. 6.643 mm; 6.644 mm; 6.647 mm; 6.645 mm; 6.650 mm

15. The mileage on six vehicles leased for 1 year was recorded as follows: 25,740 mi, 32,160 mi, 41,005 mi, 21,612 mi, 35,424 mi, 25,810 mi. What is the mean measurement?

16. A trucking company had hauls of 2018 km, 2101 km, 2005 km, 2025 km, 2035 km. What is the mean measurement of the hauls?

17. Over an 8-day period of time the high temperature of each day was recorded in degrees Fahrenheit as follows: 69, 81, 74, 83, 67, 71, 75, 63. What is the mean measurement?

18. A pharmacist weighed ten different capsules of the same compound and recorded these measurements: 3414 mg, 3433 mg, 3431 mg, 3419 mg, 3441 mg, 3417 mg, 3427 mg, 3434 mg, 3435 mg, 3432 mg. What is the mean measurement?

19. A technician measured the power usage of six appliances and recorded the following results: 108 kW, 209 kW, 176 kW, 162 kW, 188 kW, 121 kW. What is the mean measurement?

20. A trucking company had seven items with the following weights: 728 lb, 475 lb, 803 lb, 915 lb, 1002 lb, 256 lb, 781 lb. What is the mean measurement?

21. ✴ According to the U.S. Energy Information Administration, the total U.S. coal production (in millions of tons, where 1 U.S. short ton = 2000 lb) by year ending in 2015 was:

2006	1026.6	2011	1095.6
2007	1146.6	2012	1016.5
2008	1171.8	2013	984.8
2009	1174.9	2014	1000.0
2010	1084.4	2015	897.0

What was the mean annual U.S. coal production?

22. ✴ As trees grow in diameter, the wood produced in spring and early summer is less dense and lighter in color than the wood produced in late summer and fall. The result is rings of light and dark wood known as annual tree rings. The thickness of the annual ring is determined largely by the weather that the tree experienced during that particular year. Hot, dry years tend to produce thin rings, whereas cool, wet years tend to produce thick rings. In a cross section of the 100+-year-old oak tree in Illustration 1, the annual rings for the 10 years from 1920 to 1929 were measured as follows:

Annual tree ring thickness (in mm)

1920	6	1924	8	1927	6
1921	8	1925	7	1928	9
1922	8	1926	6	1929	11
1923	8				

What was the mean growth of the annual rings for the 1920s?

ILLUSTRATION 1

15.6 Other Average Measurements and Percentiles

There are other procedures to determine an average measurement besides finding the mean measurement. The **median measurement** is the measurement that falls in the middle of a group of measurements arranged in order of size. That is, one-half of the measurements are larger than or equal to the median, and one-half of the measurements are less than or equal to the median.

Example 1 Find the median of the following set of measurements.

2.151 mm, 2.148 mm, 2.146 mm, 2.143 mm, 2.149 mm

STEP 1 Arrange the measurements in order of size.

2.151 mm
2.149 mm
2.148 mm
2.146 mm
2.143 mm

STEP 2 Find the middle measurement.

Since there are five measurements, the third measurement, 2.148 mm, is the median. ♦

In Example 1, there was an odd number of measurements. When there is an even number of measurements, there is no one middle measurement. In this case, the median is found by taking the mean of the two middle measurements.

Example 2 Find the median measurement of the following set of measurements.

54°, 57°, 59°, 55°, 53°, 57°, 50°, 56°

STEP 1 Arrange the measurements in order of size.

59°
57°
57°
56°
55°
54°
53°
50°

STEP 2 Since there are eight measurements, find the mean of the two middle measurements.

$$\frac{55° + 56°}{2} = \frac{111°}{2} = 55.5°$$

So the median measurement is 55.5°.

Another kind of average often used is the mode. The **mode** is the measurement that appears most often in a set of measurements. In Example 2, 57° is the mode, because two of the measurements have this value. However, the mode can present problems. There can be more than one mode, and the mode may or may not be near the middle.

Example 3 Find the mode of the following set of measurements:

3.8 cm, 3.2 cm, 3.7 cm, 3.5 cm, 3.8 cm, 3.9 cm, 3.5 cm, 3.1 cm

The measurements 3.5 cm and 3.8 cm are both modes because each appears most often—twice.

Related to averages is the idea of measuring the position of a piece of data relative to the rest of the data. **Percentiles** are numbers that divide a given data set into 100 equal parts. The ***n*th percentile** is the number P_n such that n percent of the data (ranked from smallest to largest) is at or below P_n. For example, if you score in the 64th percentile on some standardized test, this means that you scored at the same level or higher than 64% of those who took the test and you scored lower than 36% of those who took the test.

Example 4 The following list gives 50 pieces of ranked data (ranked from smallest to largest).

Ranked Data				
16	49	82	121	147
19	50	88	125	148
23	51	89	126	150
27	52	99	129	155
31	57	101	130	156
32	64	103	131	161
32	71	104	138	163
39	72	107	142	169
43	78	118	143	172
47	79	120	145	179

a. Find the 98th percentile.
b. Find the 75th percentile.
c. Find the 26th percentile.

Solution

a. The 98th percentile is 172 (the 49th piece of data: $0.98 \times 50 = 49$); that is, 98% of the data is at or smaller in value than 172.

b. The 75th percentile is 142 (the 38th piece of data: $0.75 \times 50 = 37.5$ or 38); that is, 75% of the data is at or smaller in value than 142.

c. The 26th percentile is 51 (the 13th piece of data: $0.26 \times 50 = 13$); that is, 26% of the data is at or smaller in value than 51.

EXERCISES 15.6

1–20. *Find the median measurement for each set of measurements in Exercises 1–20 of Exercises 15.5.*

Find the following percentiles for the data listed in Example 4:

21. 94th percentile
22. 80th percentile
23. 55th percentile
24. 12th percentile
25. 5th percentile
26. 50th percentile

Find the mode for each set of measurements in Exercises 27–36.

27. 2.81 mm, 2.90 mm, 2.78 mm, 2.85 mm, 2.82 mm, 2.85 mm, 2.81 mm, 2.85 mm
28. 105, 110, 211, 313, 415, 475
29. 7291, 7288, 7285, 7287, 7285, 7291
30. 41,265; 42,051; 43,006; 42,051; 41,258; 42,051
31. 28, 30, 41, 30, 19, 25, 28, 19, 28, 42, 36
32. 5129 mi, 7025 mi, 6291 mi, 7025 mi, 8207 mi, 5129 mi, 6292 mi
33. 6.2 in., 3.8 in., 5.2 in., 7.1 in., 6.2 in., 3.9 in.
34. 25,607 km, 26,108 km, 27,203 km, 26,512 km, 27,203 km, 26,607 km
35. 0.018 in., 0.024 in., 0.022 in., 0.019 in., 0.024 in., 0.023 in.
36. 625 lb, 571 lb, 652 lb, 537 lb, 553 lb, 652 lb, 718 lb, 652 lb

37. The cross section of the oak tree mentioned in the previous section showed the following annual ring thicknesses (in mm) from 1940 to 1949:

1940	11	1944	7	1947	8
1941	9	1945	10	1948	9
1942	8	1946	9	1949	7
1943	11				

Find **a.** the median annual ring thickness for the 1940s, **b.** the mode, **c.** the mean, and **d.** the **midrange** (the average of the highest and lowest values in a set of data).

38. Total worldwide coal production (in millions of tons) by year for the decade ending in 2004 was:

1995	5096	1999	4941	2002	5265
1996	5106	2000	4935	2003	5648
1997	5132	2001	5233	2004	6079
1998	5046				

What was the median annual worldwide coal production?

15.7 Range and Standard Deviation

The mean measurement gives the technician the average value of a group of measurements, but the mean does not give any information about how the actual data vary in values. Some type of measurement that gives the amount of variation is often helpful in analyzing a set of data.

One way of describing the variation in the data is to find the range. The **range** is the difference between the highest value and the lowest value in a set of data.

Example 1 Find the range of the following measurements:

$$54°, 57°, 59°, 55°, 53°, 57°, 50°, 56°$$

The range is the difference between the highest value, 59°, and the lowest value, 50°. The range is $59° - 50° = 9°$. ◆

The range gives us an idea of how much the data are spread out, but another measure, the standard deviation, is often more helpful. The **standard deviation** tells how the data typically vary from the mean. Suppose we are given data sets $A = 4, 5, 5, 6$ and $B = 2, 3, 7, 8$. Both sets have a mean of 5, but the data in set A are "nearer" to the mean than are the data in set B. A mathematical way of describing this is to use standard

CHAPTER 15 ♦ Basic Statistics

deviation. There are two types of standard deviation: population standard deviation, denoted by σ, and sample standard deviation, denoted by s. We shall use sample standard deviation in this text; that is,

$$s = \sqrt{\frac{\text{sum of (measurement} - \text{mean)}^2}{\text{number of measurements} - 1}}$$

In set A, $s = \sqrt{\dfrac{(4-5)^2 + (5-5)^2 + (5-5)^2 + (6-5)^2}{3}} = \sqrt{\dfrac{2}{3}} = 0.82$

In set B, $s = \sqrt{\dfrac{(2-5)^2 + (3-5)^2 + (7-5)^2 + (8-5)^2}{3}} = \sqrt{\dfrac{26}{3}} = 2.9$

In set A, the data deviate from the mean by 0.82; in set B, the deviation is 2.9.

Example 2 Find the sample standard deviation for the data given in Example 1.

STEP 1 Find the mean.

$$\text{mean} = \frac{\text{sum of measurements}}{\text{number of measurements}}$$

$$\text{mean} = \frac{54° + 57° + 59° + 55° + 53° + 57° + 50° + 56°}{8}$$

$$= \frac{441°}{8} = 55.1°$$

STEP 2 Find the difference between each piece of data and the mean.

$54 - 55.1 = -1.1$
$57 - 55.1 = 1.9$
$59 - 55.1 = 3.9$
$55 - 55.1 = -0.1$
$53 - 55.1 = -2.1$
$57 - 55.1 = 1.9$
$50 - 55.1 = -5.1$
$56 - 55.1 = 0.9$

STEP 3 Square each difference and find the sum of the squared amounts.

$(-1.1)^2 = 1.21$
$(1.9)^2 = 3.61$
$(3.9)^2 = 15.21$
$(-0.1)^2 = 0.01$
$(-2.1)^2 = 4.41$
$(1.9)^2 = 3.61$
$(-5.1)^2 = 26.01$
$(0.9)^2 = 0.81$
54.88

STEP 4 $s = \sqrt{\dfrac{54.88}{7}} = 2.80°$

♦

15.7 ♦ Range and Standard Deviation

Example 3 Find the sample standard deviation for the following data:

2015 mi, 1926 mi, 3251 mi, 4007 mi, 1821 mi, 5238 mi, 9111 mi, 7212 mi, 5778 mi, 6661 mi

STEP 1 Find the mean.

$$\text{mean} = \frac{2015 \text{ mi} + 1926 \text{ mi} + 3251 \text{ mi} + 4007 \text{ mi} + 1821 \text{ mi} + 5238 \text{ mi} + 9111 \text{ mi} + 7212 \text{ mi} + 5778 \text{ mi} + 6661 \text{ mi}}{10}$$

$$= \frac{47{,}020}{10} = 4702$$

STEP 2 Find the difference between each piece of data and the mean.

$$2015 - 4702 = -2687$$
$$1926 - 4702 = -2776$$
$$3251 - 4702 = -1451$$
$$4007 - 4702 = -695$$
$$1821 - 4702 = -2881$$
$$5238 - 4702 = 536$$
$$9111 - 4702 = 4409$$
$$7212 - 4702 = 2510$$
$$5778 - 4702 = 1076$$
$$6661 - 4702 = 1959$$

STEP 3 Square each difference and find the sum of the squared amounts.

$$(-2687)^2 = 7{,}219{,}969$$
$$(-2776)^2 = 7{,}706{,}176$$
$$(-1451)^2 = 2{,}105{,}401$$
$$(-695)^2 = 483{,}025$$
$$(-2881)^2 = 8{,}300{,}161$$
$$(536)^2 = 287{,}296$$
$$(4409)^2 = 19{,}439{,}281$$
$$(2510)^2 = 6{,}300{,}100$$
$$(1076)^2 = 1{,}157{,}776$$
$$(1959)^2 = \underline{3{,}837{,}681}$$
$$56{,}836{,}866$$

STEP 4

$$s = \sqrt{\frac{\text{sum of (measurement} - \text{mean})^2}{\text{number of measurements} - 1}}$$

$$= \sqrt{\frac{56{,}836{,}866}{10 - 1}} = 2513 \quad \blacklozenge$$

EXERCISES 15.7

1–20. Find the range for each set of measurements in Exercises 1–20 in Exercises 15.5.

21–40. Find the sample standard deviation for each set of measurements in Exercises 1–20 in Exercises 15.5.

15.8 Grouped Data

Finding the mean of a large number of measurements can take much time and can be subject to mistakes. Grouping the measurements (the data) can make the work in finding the mean much easier.

Grouped data are data arranged in groups that are determined by setting up intervals. An **interval** contains all data between two given numbers a and b. We will show such an interval here by the form a–b. For example, 2–8 means all numbers between 2 and 8.

The number a is called the *lower limit* and b is called the *upper limit* of the interval. The number midway between a and b, $\frac{a+b}{2}$, is called the *midpoint* of the interval. In the above example, the lower limit is 2, the upper limit is 8, and the midpoint is $\frac{2+8}{2} = 5$.

While there are no given rules for choosing these intervals, the following general rules are helpful:

General Rules for Choosing Intervals for Grouped Data

1. The number of intervals chosen should be between 6 and 20.
2. The length of all intervals should be the same and should always be an odd number.
3. The midpoint of each interval should have the same number of digits as each of the measurements that falls within that interval. The lower limit and the upper limit of each interval will have one more digit than the measurements within the interval. In this way, no actual measurement will have exactly the same value as any of these limits. It will therefore be clear to which interval each measurement belongs.
4. The lower limit of the first interval should be lower than the lowest measurement value, and the upper limit of the last interval should be higher than the highest measurement value.

Once the intervals have been chosen, form a frequency distribution. A **frequency distribution** is a list of each interval, its midpoint, and the number of measurements (frequency) that lie in that interval.

Example 1 Make a frequency distribution for the recorded high temperatures for the days from November 1 to January 31 as given in Table 15.1.

First, choose the number and size of the group intervals to be used. We must have enough group intervals to cover the range of the data (the difference between the highest and the lowest values). Here, the range is $55° - 2° = 53°$. Since 53 is close to 54, let us choose the odd number 9 as the interval length. This means that we will need $54 \div 9 = 6$ group intervals. This satisfies our general rule for the number of intervals.

Our first interval is 1.5–10.5 with a midpoint of 6. Here, 1.5 is the lower limit and 10.5 is the upper limit of the interval. We then make the frequency distribution as shown in Table 15.2.

Table 15.1

November	High temperature (°F)	December	High temperature (°F)	January	High temperature (°F)
1	42	1	20	1	29
2	45	2	27	2	29
3	36	3	32	3	30
4	41	4	45	4	26
5	29	5	26	5	2
6	40	6	24	6	45
7	29	7	28	7	41
8	18	8	45	8	12
9	45	9	13	9	31
10	49	10	32	10	26
11	30	11	41	11	25
12	38	12	49	12	15
13	20	13	32	13	52
14	41	14	23	14	42
15	26	15	46	15	22
16	15	16	31	16	30
17	46	17	12	17	19
18	50	18	31	18	19
19	31	19	40	19	19
20	36	20	9	20	55
21	31	21	42	21	23
22	38	22	40	22	17
23	22	23	15	23	26
24	29	24	24	24	12
25	39	25	28	25	16
26	52	26	27	26	21
27	29	27	29	27	39
28	25	28	8	28	20
29	30	29	36	29	23
30	36	30	45	30	22
		31	12	31	9

Table 15.2

Temperature (°F)	Midpoint x	Tally	Frequency f
1.5–10.5	6	////	4
10.5–19.5	15	LH1 LH1 ////	14
19.5–28.5	24	LH1 LH1 LH1 LH1 ///	23
28.5–37.5	33	LH1 LH1 LH1 LH1 ///	23
37.5–46.5	42	LH1 LH1 LH1 LH1 //	22
46.5–55.5	51	LH1 /	6
			92

To find the mean from the frequency distribution, (a) multiply the frequency of each interval, f, by the midpoint of that interval, x, forming the product, xf, (b) add the products xf, and (c) divide by the number of data, sum of f.

$$\text{mean} = \frac{\text{sum of } xf}{\text{sum of } f}$$

Example 2 Find the mean of the data given in Example 1.

A frequency distribution table (Table 15.3) gives the information for finding the mean.

Table 15.3

Temperature (°F)	Midpoint x	Frequency f	Product xf
1.5–10.5	6	4	24
10.5–19.5	15	14	210
19.5–28.5	24	23	552
28.5–37.5	33	23	759
37.5–46.5	42	22	924
46.5–55.5	51	6	306
		92	2775

The mean temperature is found as follows:

$$\text{mean} = \frac{\text{sum of } xf}{\text{sum of } f} = \frac{2775}{92} = 30.2°F$$

NOTE: If the mean of the data in Example 2 were found by summing the actual temperatures and dividing by the number of temperatures, the mean would be 29.86°F, or 29.9°F. There may be a small difference between the two calculated means. This is because we are using the midpoints of the intervals rather than the actual data. However, since the mean is easier to find by this method, the small difference in values is acceptable.

15.8 ♦ Grouped Data

Example 3 Find the mean of the data given in Example 1, this time using an interval length of 5.

The range of the data is 53, which is close to 55, a number that is divisible by 5. Since $55 \div 5 = 11$, we will use 11 intervals, each of length 5. Now make a frequency distribution with 1.5–6.5 as the first interval, using 4 as the first midpoint. The frequency distribution then becomes as shown in Table 15.4.

Table 15.4

Temperature (°F)	Midpoint x	Frequency f	Product xf
1.5–6.5	4	1	4
6.5–11.5	9	3	27
11.5–16.5	14	9	126
16.5–21.5	19	9	171
21.5–26.5	24	15	360
26.5–31.5	29	20	580
31.5–36.5	34	7	238
36.5–41.5	39	11	429
41.5–46.5	44	11	484
46.5–51.5	49	3	147
51.5–56.5	54	3	162
		92	2728

Find the mean temperature:

$$\text{mean} = \frac{\text{sum of } xf}{\text{sum of } f} = \frac{2728}{92} = 29.7°F$$

♦

EXERCISES 15.8

1. From the following grouped data, find the mean.

Interval	Midpoint x	Frequency f	Product xf
41.5–48.5		12	
48.5–55.5		15	
55.5–62.5		20	
62.5–69.5		25	
69.5–76.5		4	
76.5–83.5		2	

2. Make a frequency distribution of the following scores from a mathematics test and use it to find the mean score.

85, 73, 74, 69, 87, 81, 68, 76, 78, 75, 88, 85, 67, 83, 82, 95, 63, 84, 94, 66, 84, 78, 96, 67, 63, 59, 100, 90, 100, 94, 79, 79, 74

3. A laboratory technician records the life span (in months) of rats treated at birth with a fertility hormone. From the following frequency distribution, find the mean life span.

Life span (months)	Midpoint x	Frequency f	Product xf
−0.5–2.5		12	
2.5–5.5		18	
5.5–8.5		22	
8.5–11.5		30	
11.5–14.5		18	

4. The life expectancy of a fluorescent light bulb is given by the number of hours that it will burn. From the following frequency distribution, find the mean life of this type of bulb.

CHAPTER 15 ♦ Basic Statistics

Life of bulb (hours)	Midpoint x	Frequency f	Product xf
−0.5–499.5		2	
499.5–999.5		12	
999.5–1499.5		14	
1499.5–1999.5		17	
1999.5–2499.5		28	
2499.5–2999.5		33	
2999.5–3499.5		14	
3499.5–3999.5		5	

5. The shipment times in hours for a load of goods from a factory to market are tabulated in the following frequency distribution. Find the mean shipment time.

Shipment time (hours)	Midpoint x	Frequency f	Product xf
22.5–27.5		2	
27.5–32.5		41	
32.5–37.5		79	
37.5–42.5		28	
42.5–47.5		15	
47.5–52.5		6	

6. The cost of goods stolen from a department store during the month of December has been tabulated by dollar amounts in the following frequency distribution. Find the mean cost of the thefts.

Cost ($)	Midpoint x	Frequency f	Product xf
−0.5–24.5		2	
24.5–49.5		17	
49.5–74.5		25	
74.5–99.5		51	
99.5–124.5		38	
124.5–149.5		32	

7. The number of passengers and their luggage weight in pounds on Flight 2102 have been tabulated in the following frequency distribution. Find the mean luggage weight.

Weight (lb)	Midpoint x	Frequency f	Product xf
0.5–9.5		1	
9.5–18.5		3	
18.5–27.5		22	
27.5–36.5		37	
36.5–45.5		56	
45.5–54.5		19	
54.5–63.5		17	
63.5–72.5		10	
72.5–81.5		5	
81.5–90.5		2	

8. The income of the residents in a neighborhood was tabulated. The results are shown in the following frequency distribution. Find the mean income.

Income ($)	Midpoint x	Frequency f	Product xf
2,500–12,500		1	
12,500–22,500		2	
22,500–32,500		15	
32,500–42,500		25	
42,500–52,500		8	

9. The number of defective parts per shipment has been tabulated in the following frequency distribution. Find the mean of defective parts per shipment.

Number of defective parts per shipment	Frequency f
0.5–3.5	1
3.5–6.5	7
6.5–9.5	20
9.5–12.5	9
12.5–15.5	32
15.5–18.5	3

10. The following dollar amounts are traffic fines collected in one day in a village. Make a frequency distribution and use it to find the mean amount of the fines.

$30, $28, $15, $14, $32, $67, $45, $30, $17, $25, $30, $19, $27, $32, $51, $45, $36, $42, $72, $50, $18, $41, $23, $32, $35, $46, $50, $61, $82, $78, $39, $42, $27, $20

11. The length of hospital stays for patients at a local hospital has been tabulated, and the results are shown in the following frequency distribution. Find the mean length for a hospital stay.

Length of stay (days)	Frequency
0.5–1.5	50
1.5–2.5	32
2.5–3.5	18
3.5–4.5	10
4.5–5.5	8
5.5–6.5	5
6.5–7.5	26
7.5–8.5	17
8.5–9.5	22

12. The frequency of repair for the trucks owned by a trucking firm over a 5-year period has been tabulated. The results are shown in the following frequency distribution. Find the mean number of repairs over the 5-year period.

Times repaired	Frequency
1.5–2.5	22
2.5–3.5	53
3.5–4.5	71
4.5–5.5	108
5.5–6.5	102
6.5–7.5	120
7.5–8.5	146
8.5–9.5	135
9.5–10.5	98
10.5–11.5	84
11.5–12.5	42
12.5–13.5	12
13.5–14.5	8

13. The scores that golfers shot on 18 holes at a local course were tabulated. The results are shown in the following frequency distribution. Find the mean score.

Score	Frequency
68.5–73.5	5
73.5–78.5	7
78.5–83.5	10
83.5–88.5	12
88.5–93.5	20
93.5–98.5	22
98.5–103.5	25
103.5–108.5	32
108.5–113.5	17
113.5–118.5	12
118.5–123.5	9

14. The corn yield in bushels per acre for a certain hybrid planted by farmers during the year was tabulated in the following frequency distribution. Find the mean yield.

Yield (bu/acre)	Frequency
45.5–54.5	2
54.5–63.5	1
63.5–72.5	3
72.5–81.5	6
81.5–90.5	27
90.5–99.5	43
99.5–108.5	201
108.5–117.5	197
117.5–126.5	483
126.5–135.5	332
135.5–144.5	962
144.5–153.5	481
153.5–162.5	512
162.5–171.5	193
171.5–180.5	185
180.5–189.5	92
189.5–198.5	87
198.5–207.5	53
207.5–216.5	38

15. The following are the squad sizes of the football teams in a regional area. Make a frequency distribution and use it to find the mean.

    108, 115, 97, 68, 72, 63, 19, 24, 202, 38, 43, 52, 83, 74, 39, 40, 51, 22, 37, 43, 48, 19, 23, 56, 72, 63, 23, 31, 43

16. The number of miles traveled by an experimental tire before it became unfit for use is recorded below. Make a frequency distribution and use it to find the mean.

8,457; 22,180; 15,036; 32,168; 9,168; 25,068; 32,192; 38,163; 18,132; 34,186; 36,192; 37,072; 14,183; 42,183; 19,182; 33,337; 38,162; 28,048; 20,208; 34,408; 35,108; 40,002; 29,208; 32,225; 33,207

15.9 Standard Deviation for Grouped Data

For grouped data, the sample standard deviation is found similarly to how the mean is found. A frequency table is used. Columns are inserted to show the difference, D, between the midpoints and the mean ($D = x -$ mean); the square of D, D^2; and the frequency times D^2, $D^2 f$. The following formula gives the sample standard deviation for grouped data:

Sample Standard Deviation for Grouped Data

$$s = \sqrt{\frac{\text{sum of } D^2 f}{n - 1}}$$

where n is the number of pieces of data.

Example 1 Given the grouped data in Table 15.5, find **a.** the mean and **b.** the sample standard deviation.

Table 15.5

Interval (cm)	Frequency f
3.5–12.5	3
12.5–21.5	3
21.5–30.5	7
30.5–39.5	6
39.5–48.5	4
48.5–57.5	1

First, add the x, xf, D, D^2, and $D^2 f$ columns in Table 15.5 as shown in Table 15.6 and complete the corresponding entries in the frequency distribution as shown in Table 15.6.

Table 15.6

Interval (cm)	Midpoint x	Frequency f	Product xf	$x -$ mean D	D^2	$D^2 f$
3.5–12.5	8	3	24	−21	441	1323
12.5–21.5	17	3	51	−12	144	432
21.5–30.5	26	7	182	−3	9	63
30.5–39.5	35	6	210	6	36	216
39.5–48.5	44	4	176	15	225	900
48.5–57.5	53	1	53	24	576	576
		$n = 24$	696			3510

15.9 ♦ Standard Deviation for Grouped Data

a. The mean = $\dfrac{\text{sum of } xf}{n} = \dfrac{696}{24} = 29.0$ cm

b. $s = \sqrt{\dfrac{\text{sum of } D^2 f}{n-1}} = \sqrt{\dfrac{3510}{24-1}} = 12.4$ cm

So the mean measurement is 29.0 cm, and the data typically tend to vary from the mean by 12.4 cm. ♦

Many scientific calculators have statistical functions. These can be used to find the mean (usually denoted by $\bar{x}$) and the sample standard deviation (denoted by s). If you have a calculator with these functions, you should read its manual.

Example 2 Given the grouped data in Table 15.7, find **a.** the mean and **b.** the sample standard deviation.

Table 15.7

Mass (nearest kg)	Frequency f
1.5–4.5	10
4.5–7.5	15
7.5–10.5	8
10.5–13.5	24
13.5–16.5	125
16.5–19.5	62
19.5–22.5	89
22.5–25.5	51
25.5–28.5	28
28.5–31.5	17

First, add the x, xf, D, D^2, and $D^2 f$ columns in Table 15.7 as shown in Table 15.8 and complete the corresponding entries in the frequency distribution as shown in Table 15.8.

Table 15.8

Mass (nearest kg)	Midpoint x	Frequency f	Product xf	$x -$ mean D	D^2	$D^2 f$
1.5–4.5	3	10	30	−15	225	2250
4.5–7.5	6	15	90	−12	144	2160
7.5–10.5	9	8	72	−9	81	648
10.5–13.5	12	24	288	−6	36	864
13.5–16.5	15	125	1875	−3	9	1125
16.5–19.5	18	62	1116	0	0	0
19.5–22.5	21	89	1869	3	9	801
22.5–25.5	24	51	1224	6	36	1836
25.5–28.5	27	28	756	9	81	2268
28.5–31.5	30	17	510	12	144	2448
		429	7830			14,400

a. The mean $x = \dfrac{\text{sum of } xf}{n} = \dfrac{7830}{429} = 18.3$ or 18 kg

b. $s = \sqrt{\dfrac{\text{sum of } D^2 f}{n-1}} = \sqrt{\dfrac{14{,}400}{429-1}} = 5.80$ or 6 kg

So, the mean is 18 kg, and the data typically tend to vary from the mean by 6 kg. ♦

EXERCISES 15.9

1–16. Find the sample standard deviation for each set of measurements in Exercises 1–16 of Exercises 15.8.

17. Measuring the annual tree rings for the 100+-year-old oak tree in Exercise 22 in Section 15.5 produces the following thicknesses (in mm) for 1930 to 1939:

Year	mm	Year	mm	Year	mm
1930	9	1934	7	1937	10
1931	9	1935	8	1938	11
1932	8	1936	10	1939	9
1933	7				

a. What is the range of the annual ring thicknesses for the 1930s?

b. What is the standard deviation?

15.10 Statistical Process Control

One of the many uses of statistics is in random sampling of processed goods to improve quality control. **Statistical process control** is a primary analysis tool for quality improvement that helps companies collect, organize, interpret, and track a wide variety of information during production of materials, delivery of services, and monitoring their normal work processes, business-related processes, and customer satisfaction. By watching the production process, technicians can make changes early rather than waiting until a large number of defective goods has been produced.

Control charts are used to help find the information to improve quality control. Different kinds of control charts give different information; three types of control charts are listed below:

Median Chart: Easy to use; shows the variation of the process. It is usually used to compare the output of several processes or various stages of the same process.

Individual Reading Chart: Used for expensive measurements or when the output at any point in time remains relatively constant. Such a chart does not isolate individual steps of the process, so it can be hard to find out why there is a variation.

Mean Control Chart: Shows the sample means plotted over time, to show whether the process is changing and whether it is in control. We will study this type of chart in this section. The chart has a center line at the *target value* of the process mean or at the process mean as determined by the data. Dashed lines represent *control lines,* which are located at the mean plus or minus three times the standard deviation divided by the square root of the number of items in each sample. There are two cases when the process is out of control: (i) when any point falls outside the central limits and (ii) when any run of nine or more consecutive points falls on the same side of the center line. Figure 15.8 shows examples of two processes that are out of control (parts a and b) and one that is in control (part c).

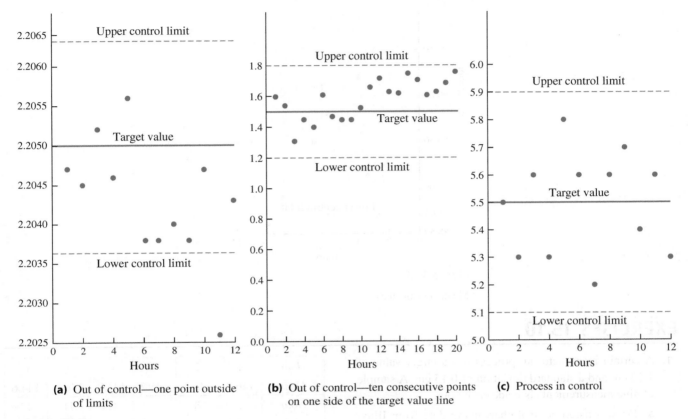

(a) Out of control—one point outside of limits

(b) Out of control—ten consecutive points on one side of the target value line

(c) Process in control

Figure 15.8
Three mean control charts

Example 1 A manufacturer of golf balls checks a sample of 100 balls every hour. The compression of the ball has a target value of 90. The standard deviation is 4.5. Construct a mean control chart using this and the information in the following table:

Hour	1:00	2:00	3:00	4:00	5:00	6:00	7:00	8:00	9:00	10:00	11:00	12:00
Mean	91.1	90.8	90.5	90.2	90.1	89.8	89.5	89.2	88.9	88.6	89.5	89.5

The lower and upper control limits are found by the mean ±3 standard deviations divided by the square root of the number of samples as follows:

$$90 \pm \frac{3(4.5)}{\sqrt{100}} = 90 \pm \frac{13.5}{10} = 90 \pm 1.35 = 91.35 \text{ or } 88.65$$

The mean control chart is shown in Figure 15.9. The process is out of control. One point is beyond the control limits; this happened at the tenth hour. At that time, some source caused the process to go out of control. It is then up to technicians to locate the trouble and fix it.

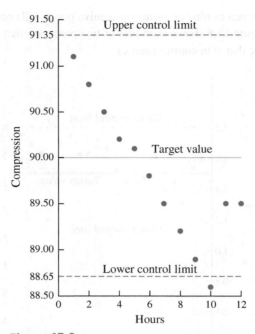

Figure 15.9

Mean control chart

EXERCISES 15.10

1. A certain manufacturing process has a target value of 1.20 cm and a standard deviation of 0.15 cm. A sample of nine measurements is made each hour.

    a. Draw a mean control chart using data from Illustration 1.

    b. Is the process out of control? If it is, at what time does the mean control chart signal lack of control?

Hour	1	2	3	4	5	6
Mean	1.30	1.15	1.10	1.25	1.08	1.11
Hour	7	8	9	10	11	12
Mean	1.18	1.15	1.07	1.12	1.08	1.16
Hour	13	14	15	16	17	18
Mean	1.12	1.21	1.25	1.50	1.30	1.26
Hour	19	20	21	22	23	24
Mean	1.29	1.19	1.26	1.31	1.17	1.15

ILLUSTRATION 1

2. A sporting goods manufacturer makes baseballs. The target mass of a baseball is 145.5 grams, with a standard deviation of 3.5 grams. A technician selects 100 balls at random per hour and records the mean mass of the samples. Illustration 2 lists the mass in grams for a 36-hour period.

    a. Make a mean control chart for the data shown.

    b. Is the process out of control? If it is, at what time does the mean control chart signal lack of control?

Hour	1	2	3	4	5	6
Mass	146.2	145.3	145.2	144.8	146.3	144.6
Hour	7	8	9	10	11	12
Mass	145.0	146.1	144.8	145.1	145.4	143.7
Hour	13	14	15	16	17	18
Mass	145.0	146.3	145.2	145.9	146.0	145.7
Hour	19	20	21	22	23	24
Mass	146.3	144.8	144.9	144.9	145.6	143.8
Hour	25	26	27	28	29	30
Mass	145.7	145.8	144.8	144.9	144.6	145.3
Hour	31	32	33	34	35	36
Mass	146.3	144.0	146.2	145.5	145.4	144.6

ILLUSTRATION 2

3. The depth of a silicon wafer is targeted at 1.015 mm. If properly functioning, the process produces items with mean 1.015 mm and has a standard deviation of 0.004 mm. A sample of 16 items is measured once each hour. The sample means for the past 12 h are given in Illustration 3. From the data, make a mean control chart and determine whether the process is in control.

Hour	1	2	3	4	5	6
Mean	1.016	1.013	1.015	1.017	1.013	1.014
Hour	7	8	9	10	11	12
Mean	1.017	1.016	1.014	1.013	1.016	1.017

ILLUSTRATION 3

4. The illumination of a light bulb is targeted at 1170 lumens. The standard deviation is 16.6. A technician randomly selects 15 bulbs per hour and records the mean illumination each hour. Use the data in Illustration 4 to make a mean control chart and determine if the process is in control.

Hour	1	2	3	4	5	6
Illumination	1156	1141	1145	1180	1183	1180
Hour	7	8	9	10	11	12
Illumination	1177	1191	1193	1171	1188	1172
Hour	13	14	15	16	17	18
Illumination	1179	1161	1159	1173	1190	1143
Hour	19	20	21	22	23	24
Illumination	1187	1191	1181	1144	1158	1181
Hour	25	26	27	28	29	30
Illumination	1192	1191	1165	1168	1195	1181

ILLUSTRATION 4

5. The target load weight for a ladder is 250 lb. A technician selects 5 ladders at random per day to test load limit and records the mean load for each day. Use the data in Illustration 5 to make a mean control chart and determine if the process is in control.

Day	1	2	3	4	5	6
Pounds	238	275	260	258	260	238
Day	7	8	9	10	11	12
Pounds	261	250	270	265	241	248
Day	13	14	15	16	17	18
Pounds	273	265	260	253	240	245

ILLUSTRATION 5

6. A capsule is targeted to contain 50 mg of garlic. A technician selects 25 capsules at random out of a batch of 2000 and records the mean amount of garlic in the samples. Use the data in Illustration 6 to make a mean control chart and determine if the process is in control.

Batch	1	2	3	4	5	6
Garlic	53	51	45	47	52	57
Batch	7	8	9	10	11	12
Garlic	42	45	48	53	32	52
Batch	13	14	15	16	17	18
Garlic	57	55	48	51	50	44
Batch	19	20	21	22	23	24
Garlic	47	57	50	47	52	55

ILLUSTRATION 6

15.11 Other Graphs for Statistical Data

In Sections 15.1 through 15.4, we saw how we can view statistical data using various types of graphs. In this section, you will be introduced to three more types.

A **histogram** is a bar graph that reflects the frequency of the number displayed in a frequency distribution.

Example 1 Use the grouped data in Table 15.6 on page 494 to draw a histogram.

For grouped data, use the intervals for the base and use the frequency for the height. The intervals are 3.5–12.5, 12.5–21.5, 21.5–30.5, 30.5–39.5, 39.5–48.5, and 48.5–57.5. The corresponding frequencies (heights) are 3, 3, 7, 6, 4, and 1. See Figure 15.10.

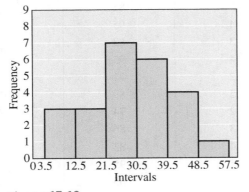

Figure 15.10

Histogram

500 CHAPTER 15 ♦ Basic Statistics

Example 2 A **run chart** is a line graph in which data are collected over a period of time.

In Table 15.1 on page 489, choose the December readings and create a run chart. Pair each day in December with the high temperature reading. Plot the points in order from left to right. See Figure 15.11.

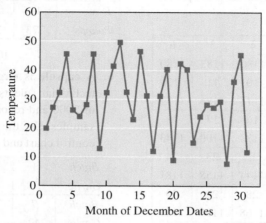

Figure 15.11
Run chart

A **scattergram** is a graph of two variables as distinct points that is useful in trying to determine whether a relationship between the two variables can be inferred.

Example 3 Use the information from Table 15.9 showing hours of study for a final exam and the grade received to construct a scattergram. See Figure 15.12.

Table 15.9	
Study time (h)	Grade received
3	71
4	90
6	93
5	98
3	70
4	82
3	87
5	93
3	82
4	95
3	82
4	95
2	71
5	89
2	74
3	86
4	83
5	95
2	65
3	82

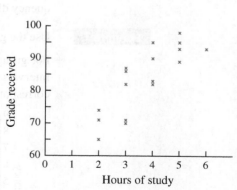

Figure 15.12
Scattergram

The scattergram shows a positive linear correlation, since the grade received increases as the number of study hours increases. ◆

A graph with the points descending, as in Figure 15.13a, shows a negative linear correlation. The points in the graph in Figure 15.13b show no linear correlation.

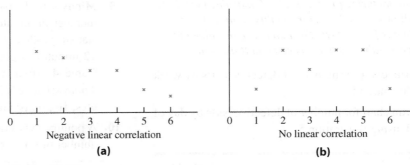

Figure 15.13

EXERCISES 15.11

In Exercises 1–4, draw a histogram for the data displayed in each frequency distribution.

1.

Interval	Frequency f
41.5–48.5	12
48.5–55.5	15
55.5–62.5	20
62.5–69.5	25
69.5–76.5	4
76.5–83.5	2

ILLUSTRATION 1

2. The number of passengers and their luggage weight in pounds on Flight 2102 have been tabulated in the frequency distribution in Illustration 2.

Weight (lb)	Frequency f
0.5–9.5	1
9.5–18.5	3
18.5–27.5	22
27.5–36.5	37
36.5–45.5	56
45.5–54.5	19
54.5–63.5	17
63.5–72.5	10
72.5–81.5	5
81.5–90.5	2

ILLUSTRATION 2

3. The life expectancy of a fluorescent light bulb is given by the number of hours that it will burn, tabulated in the frequency distribution in Illustration 3.

Life of bulb (hours)	Frequency f
−0.5–499.5	2
499.5–999.5	12
999.5–1499.5	14
1499.5–1999.5	17
1999.5–2499.5	28
2499.5–2999.5	33
2999.5–3499.5	14
3499.5–3999.5	5

ILLUSTRATION 3

4. The shipment times in hours for a load of goods from a factory to market are tabulated in the frequency distribution in Illustration 4.

Shipment time (hours)	Frequency f
22.5–27.5	2
27.5–32.5	41
32.5–37.5	79
37.5–42.5	28
42.5–47.5	15
47.5–52.5	6

ILLUSTRATION 4

Use the data displayed in Table 15.1 on page 489 to draw a run chart for the daily high temperature using the:

5. November data
6. January data

Precision Manufacturing produces 20,000 machine nuts daily. There is an allowance of 1% error. Quality control checks 250 nuts each day for 40 days. The number of machine nuts that are not acceptable each day is given in Illustration 5.

7. Draw a run chart for number of defective parts by week using Illustration 5.
8. Draw a run chart for number of defective parts by day of the week using Illustration 5.

Week	Mon	Tue	Wed	Thu	Fri	Weekly totals
1	18	65	42	22	67	214
2	8	70	35	42	51	206
3	8	40	50	30	20	148
4	8	42	44	68	5	167
5	81	0	53	13	67	214
6	57	70	22	39	40	228
7	45	72	22	62	8	209
8	20	35	5	41	6	107
Daily totals	245	394	273	317	264	1493

ILLUSTRATION 5

In Exercises 9–12,

a. Draw a scattergram for the data given.
b. Does the scattergram have a positive, a negative, or no linear correlation?

9. Mindy's basketball coach kept records on each team member in minutes played and points scored. Mindy's statistics follow: Game 1, 20 minutes, 5 points; Game 2, 12 minutes, 8 points; Game 3, 24 minutes, 13 points; Game 4, 16 minutes, 6 points; Game 5, 8 minutes, 4 points; Game 6, 22 minutes, 14 points; Game 7, 28 minutes, 16 points; Game 8, 30 minutes, 20 points.

10. A poultry-eviscerating line processed the following number of boxes of turkey for each hour of the day.

Hour	7–8	8–9	9–10	10–11	11–12	1–2	2–3	3–4
Boxes of turkey	8	9	9	8	6	8	7	5

11. x is paired with y by the following table:

x	4	8	9	11	13	15
y	2	14	6	18	2	20

12. x is paired with y by the following table:

x	−6	−5	−4	0	2	3	5
y	−3	−3	−2	−2	−1	−1	0

15.12 Normal Distribution

The **normal distribution** of large data sets tends to group data around the mean and/or the median in a way that the result resembles a bell-shaped curve, as shown in Figure 15.14. The shape of the normal distribution curve will depend on the size of the standard deviation. The empirical rule states that approximately 68% of the data will be within one standard deviation of the mean, 95% of the data will fall within two standard deviations, and 99.7% of the data will fall between three standard deviations (see Figure 15.15).

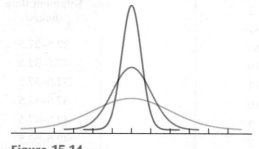

Figure 15.14
Different standard deviations

15.12 ♦ Normal Distribution 503

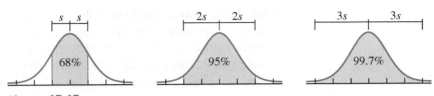

Figure 15.15

s represents one standard deviation.

Example 1 Given the data in Table 15.10, determine the mean, median, mode, and standard deviation. Draw the normal distribution curve.

Table 15.10

Data	Frequency	Data	Frequency	Data	Frequency
70	1	74	175	78	30
71	10	75	250	79	10
72	20	76	160	80	2
73	117	77	93		Total 868

The mean is 75, the median is 75, and the mode is 75, and the standard deviation is 1.5. In this example:

- One standard deviation below the mean is $75 - 1.5 = 73.5$, and one standard deviation above the mean is $75 + 1.5 = 76.5$. This includes $175 + 250 + 160 = 585$ of the 868 scores, or 67%.
- Two standard deviations below the mean is $75 - 2(1.5) = 72$, and two standard deviations above the mean is $75 + 2(1.5) = 78$. This includes $20 + 117 + 175 + 250 + 160 + 93 + 30 = 845$ of the 868 scores, or 97%.
- Three standard deviations below the mean is $75 - 3(1.5) = 70.5$, and three standard deviations above the mean is $75 + 3(1.5) = 79.5$. This includes $10 + 20 + 117 + 175 + 250 + 160 + 93 + 30 + 10 = 865$ of the 868 scores, or 99.7%. See Figure 15.16.

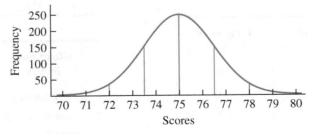

Figure 15.16

Normal distribution

Example 2 A clothing manufacturer is going to produce 100,000 women's blouses in a normal distribution of sizes with a mean of 10 and a standard deviation of 2.5. Given that the sizes are 2, 4, 6, 8, 10, 12, 14, 16, and 18, how many of each size should the manufacturer make?

- 68% of the blouses should be within one standard deviation of the mean 10. One standard deviation below this mean is 10 − 2.5 = 7.5, and one standard deviation above this mean is 10 + 2.5 = 12.5, which includes sizes 8, 10, and 12. Thus, he would manufacture 68,000 of the blouses in sizes 8, 10, and 12.
- 95% of the blouses should be within two standard deviations of the mean 10. Two standard deviations below this mean is 10 − 2(2.5) = 5, and two standard deviations above it is 10 + 2(2.5) = 15. This includes all of the blouses in sizes 8, 10, and 12 plus all in sizes 6 and 14. Thus, the manufacturer would produce 95,000 blouses in sizes 6 through 14, including the 68,000 in sizes 8, 10, and 12. 95,000 − 68,000 = 27,000 blouses in sizes 6 and 14.
- 99.7% of the blouses should be within three standard deviations of the mean 10. Three standard deviations below this mean is 10 − 3(2.5) = 2.5, and three standard deviations above it is 10 + 3(2.5) = 17.5. This includes all blouses in sizes 6, 8, 10, and 12, plus those in sizes 4 and 16. The manufacturer would make 99,700 blouses, which include the 95,000 in the sizes 6, 8, 10, 12, and 14, plus those in sizes 4 and 16. 99,700 − 95,000 = 4700 blouses in sizes 4 and 16.
- The remaining 300 blouses would be in sizes 2 and 18.

EXERCISES 15.12

1. Given a normal distribution with a mean of 85 and a sample standard deviation of 15, how much of the data should be in the interval between 55 and 115?

2. Given the 20 numbers 32, 34, 35, 41, 42, 43, 44, 45, 48, 49, 51, 53, 55, 55, 57, 58, 59, 63, 65, and 71, find **a.** the mean, **b.** the median, **c.** the mode, and **d.** the sample standard deviation. **e.** Construct a normal distribution.

*In Exercises 3–6, use the frequency distribution to find **a.** the mean and **b.** the sample standard deviation. **c.** Does the data form a normal distribution? Remember: If the data fall within 2% of the empirical rule of 68%, 95%, and 99.7% for one, two, or three standard deviations, respectively, they form a normal distribution.*

3.

Interval	Midpoint x	Frequency f
41.5–48.5		10
48.5–55.5		50
55.5–62.5		200
62.5–69.5		188
69.5–76.5		40
76.5–83.5		20

4.

Weight (lb)	Midpoint x	Frequency f
0.5–9.5		10
9.5–18.5		40
18.5–27.5		120
27.5–36.5		210
36.5–45.5		360
45.5–54.5		340
54.5–63.5		220
63.5–72.5		100
72.5–81.5		30
81.5–90.5		10

5.

Life of bulb (hours)	Frequency f
−0.5–999.5	105
999.5–1999.5	480
1999.5–2999.5	2050
2999.5–3999.5	4100
3999.5–4999.5	2450
4999.5–5999.5	420
5999.5–6999.5	155

6.

Shipment time (hours)	Frequency f
22.5–27.5	20
27.5–32.5	410
32.5–37.5	790
37.5–42.5	700
42.5–47.5	500
47.5–52.5	40

7. In Example 2, find the number of blouses to be made of each size if the mean were 12 and the sample standard deviation were 2.8.

8. A trouser manufacturer has an order for 80,000 men's trousers in a normal distribution of waist sizes 28, 30, 32, 34, 36, 38, 40, 42, and 44 with a mean of 36 and a sample standard deviation of 2.1. Find the number of trousers to be made of each waist size.

15.13 Probability

Probability is another useful mathematical tool that a technician can use to make decisions. A **sample space** is the set of all possible outcomes of an event.

Example 1 Find the sample space of rolling one die.

The die has 6 sides with dots on each of the six sides and this gives the sample space

[1, 2, 3, 4, 5, 6] ◆

Example 2 Find the sample space of rolling two dice.

This gives a sample space of the following 36 possible outcomes:

[(1, 1), (1, 2), (1, 3), (1, 4), (1, 5), (1, 6), (2, 1), (2, 2), (2, 3), (2, 4), (2, 5), (2, 6), (3, 1), (3, 2), (3, 3), (3, 4), (3, 5), (3, 6), (4, 1), (4, 2), (4, 3), (4, 4), (4, 5), (4, 6), (5, 1), (5, 2), (5, 3), (5, 4), (5, 5), (5, 6), (6, 1), (6, 2), (6, 3), (6, 4), (6, 5), (6, 6)] ◆

Example 3 Find the sample space when two balls are drawn out of a bag containing 4 balls: 3 red and 1 white.

Two balls drawn at a time will have a sample space of

[(red, red), (red, red), (red, red), (red, white), (white, red), (white, red)] ◆

The **probability** p is the likelihood that an event will happen and is given by the ratio of number of events happening n to the total number of possible events in the sample space s.

$$p = \frac{n}{s}$$

Example 4 Find the probability that a 3 will be rolled in Example 1.

Here, $n = 1$ and $s = 6$. So

$$p = \frac{n}{s}$$

$$p = \frac{1}{6}$$ ◆

Example 5 Find the probability that a total of 3 will be rolled in Example 2.

Here, $n = 2$ [the pairs (2, 1) and (1, 2)] and $s = 36$. So

$$p = \frac{n}{s}$$

$$p = \frac{2}{36} = \frac{1}{18}$$

Example 6 Find the probability in Example 3 that two red balls will be drawn.

Here, $n = 3$ [three pairs of (red, red)] and $s = 6$. Then

$$p = \frac{n}{s}$$

$$p = \frac{3}{6} = \frac{1}{2}$$

All probabilities are between 0 and 1 inclusive. That is, $0 \leq p \leq 1$. The probability of an event that must happen is 1. The probability of an event that is impossible is 0. The sum of all the probabilities in a sample space is 1.

Example 7 Find the probability of rolling a 7 when rolling one die.

A regular die has a sample space of [1, 2, 3, 4, 5, 6]. The probability that a 7 will happen is 0.

Example 8 Find the probability of a head when a two-headed coin is flipped.

The sample space is [1]. The probability that a head will occur when flipped is 1.

EXERCISES 15.13

In Exercises 1–6, find each sample space.

1. The hearts from a standard deck of 52 cards
2. The cards from a standard deck of cards that are less than 3
3. The red face cards from a standard deck of cards
4. The cards taken two at a time from a standard deck of cards that are less than 3
5. Marbles taken two at a time from a bag with 2 red marbles and 1 white marble
6. Pieces of paper taken two at a time with the numbers 1–7 written on them
7. From the sample space in Exercise 1, what is the probability that an ace will be drawn?
8. From the sample space in Exercise 2, what is the probability that one ace will be drawn?
9. From the sample space in Exercise 3, what is the probability that the queen of hearts will be drawn?
10. From the sample space in Exercise 1, what is the probability that the 7 of spades will be drawn?
11. From the sample space in Exercise 2, what is the probability that the card drawn will be a jack?
12. From the sample space in Exercise 3, what is the probability that the card drawn will be greater than 10?
13. From the sample space in Exercise 4, what is the probability that the sum of two cards drawn will be 4?
14. From the sample space in Exercise 5, what is the probability that two red marbles will be drawn?
15. From the sample space in Exercise 6, what is the probability that the sum of the two numbers drawn will be 7?
16. A bag contains 6 red and 4 white marbles. One marble is drawn. **a.** What is the probability that a white marble will be drawn? **b.** What is the probability that a red marble will be drawn? **c.** What is the sum of the answers in part **a** and part **b**?

17. During a manufacturing process, 52 defective items are found out of 10,000 produced. What is the probability that when an item is selected, it will be defective?

18. In a classroom, there are 18 female students and 7 male students. If a teacher picks a student at random, what is the probability that the student will be a female?

15.14 Independent Events

Events can happen in many different ways. One of these is when the events are independent. Two events are **independent** if the probability of one event does not change the probability of the second event.

Example 1 Find whether drawing 1 red marble then replacing it and drawing a second red marble from a bag containing 1 red marble and 1 white marble are independent events.

The same number of marbles exists the second time as the first, so the probability is the same and the events are independent. ♦

Example 2 Find whether rolling a die and getting a 3 and rolling a second die and getting a 5 are independent events.

In each case, the probability $p = \frac{1}{6}$, so the events are independent. ♦

Example 3 Find whether the events of drawing 2 marbles from a bag containing 3 red marbles and 1 white marble taken in order are independent events.

When a marble, either red or white, is removed from the bag, the number of marbles left is different from the first time. The probabilities are different and the events are not independent. ♦

The probability of two independent events occurring in a given order can be found by finding the product of the probabilities of each separate event. That is,

$$p(A \text{ and } B) = p(A) \cdot p(B)$$

Example 4 Find the probability in Example 1 that both marbles drawn will be red.

$$p(\text{red and red}) = p(\text{red}) \cdot p(\text{red}) = \frac{1}{2} \cdot \frac{1}{2} = \frac{1}{4}$$ ♦

Example 5 Find the probability in Example 2 that a 3 and 5 will be rolled.

$$p(3 \text{ and } 5) = p(3) \cdot p(5) = \frac{1}{6} \cdot \frac{1}{6} = \frac{1}{36}$$ ♦

EXERCISES 15.14

1. A bag contains 1 red marble, 1 blue marble, 1 green marble, and 1 white marble. What is the probability of drawing 1 red marble, replacing it, and then drawing a red marble again?

2. A card is chosen from a deck of 52 cards. It is put back in the deck and a second card is chosen. What is the probability of drawing an ace and a 10?

3. A bag contains 8 green marbles, 3 white marbles, and 5 red marbles. A marble is removed and then placed back in the bag, and a second marble is picked. What is the probability of drawing a green marble and a red marble?

4. A large box of vegetables contains 4 tomatoes, 3 heads of lettuce, and 7 onions. A vegetable is chosen at random and

then replaced in the box, and a second vegetable is picked. What is the probability of choosing a tomato and an onion?

5. Six out of ten motorcyclists wear safety helmets in states that do not require them. If two motorcyclists are chosen, what is the probability that both wear a safety helmet?

6. A card is chosen from a deck of cards and placed back in the deck, and a second card is chosen from the deck. What is the probability of drawing the ace of spades and the queen of hearts?

7. A coin is tossed, and a die is thrown. What is the probability of a head and a 5?

8. A card is drawn and replaced four times from a deck of 52 cards. What is the probability of drawing 4 clubs?

9. A card is drawn and replaced four times from a deck of 52 cards. What is the probability of drawing 4 aces?

10. A pair of dice is rolled. What is the probability of rolling 12?

11. Three dice are rolled. What is the probability of rolling three 3's?

12. A card is chosen from a deck of cards and then placed back in the deck, and a second card is chosen. What is the probability of drawing 2 kings?

13. A spinner has numbers 1–7 marked equally on the face. If the spinner is spun 2 times, what is the probability of having a 6 and a 4?

14. A spinner has numbers 1–7 marked equally on the face. If the spinner is spun 3 times, what is the probability of an even and an odd and a 4?

15. A bag of marbles contains 5 yellow marbles, 4 white marbles, 3 blue marbles, and 7 red marbles. A marble is drawn and replaced. What is the probability of having a red marble, a white marble, and a blue marble?

16. A card is drawn from a deck of 52 cards and then replaced. What is the probability of having a heart, the 10 of spades, and a jack?

SUMMARY | CHAPTER 15

Glossary of Basic Terms

Bar graph. A graph with parallel bars whose lengths are proportional to the frequency of the given quantities in a data set. (p. 472)

Circle graph. A circle whose radii divide the circle into sectors that are proportional in angle and area relative in size to the quantities represented. (p. 474)

Control charts. Charts used to help find information to improve quality control. (p. 496)

Frequency distribution. A list of each interval, its midpoint, and the number of measurements (frequency) that lie in that interval. (p. 488)

Grouped data. Data arranged in groups that are determined by setting up intervals. (p. 488)

Histogram. A bar graph that reflects the frequency of the number displayed in a frequency distribution. (p. 499)

Independent events. Two events for which the probability of one event does not change the probability of the second event. (p. 507)

Interval. All data between two given numbers such as a and b. The smaller number a is called the *lower limit* of the interval. The larger number b is called the *upper limit* of the interval. The *midpoint* of the interval is the number midway between a and b; that is, $\frac{a+b}{2}$. (p. 488)

Line graph. A graph formed by segments of straight lines that join the plotted points that represent given data. (p. 477)

Mean measurement (or **mean**). The average of a set of measurements. (p. 481)

Median measurement. The measurement that falls in the middle of a group of measurements arranged in order of size. (p. 483)

Midrange. The average of the highest and lowest values in a set of data. (p. 485)

Mode. The measurement that appears most often in a set of measurements. (p. 484)

Normal distribution. The distribution of large data sets tends to group around the mean and/or median in a way that resembles a bell-shaped curve. (p. 502)

nth percentile. The number P_n such that n percent of the data (ranked from smallest to largest) is at or below P_n. (p. 484)

Percentiles. Numbers that divide a given data set into 100 equal parts. (p. 484)

Probability. The likelihood that an event will happen; the ratio of number of events happening to the total number of possible events in the sample space. (p. 505)

Range. The difference between the highest value and the lowest value in a set of data. (p. 485)

Run chart. A line graph in which data are collected over a period of time. (p. 500)

Sample space. The set of all possible outcomes of an event. (p. 505)

Scattergram. A graph of two variables as distinct points that is useful in trying to determine whether a relationship between the two variables can be inferred. (p. 500)

Standard deviation. Describes how much the data typically vary from the mean. (p. 485)

Statistical process control. A primary analysis tool for quality improvement that helps companies collect, organize, interpret, and track a wide variety of information during production of materials, delivery of services, and monitoring their normal work processes, business-related processes, and customer satisfaction. (p. 496)

15.5 Mean Measurement

1. **Mean measurement:**

$$\text{mean measurement} = \frac{\text{sum of measurements}}{\text{number of measurements}}$$

(p. 481)

15.7 Range and Standard Deviation

1. **Sample standard deviation:**

$$s = \sqrt{\frac{\text{sum of (measurement} - \text{mean})^2}{\text{number of measurements} - 1}}$$

(p. 486)

15.8 Grouped Data

1. **General rules for choosing intervals for grouped data:**
   a. The number of intervals chosen should be between 6 and 20.
   b. The length of all intervals should be the same and should always be an odd number.
   c. The midpoint of each interval should have the same number of digits as each of the measurements that fall within that interval. The lower limit and the upper limit of each interval will have one more digit than the measurements within the interval. In this way, no actual measurement will have exactly the same value as any of these limits. It will therefore be clear to which interval each measurement belongs.
   d. The lower limit of the first interval should be lower than the lowest measurement value, and the upper limit of the last interval should be higher than the highest measurement value. (p. 488)

2. **To find the mean from a frequency distribution:**
   a. Multiply the frequency of each interval, f, by the midpoint of that interval, x, forming the product, xf.
   b. Add the products xf.
   c. Divide by the number of data, sum of f. That is,

$$\text{mean} = \frac{\text{sum of } xf}{\text{sum of } f}$$ (p. 490)

Statistics. The branch of mathematics that deals with the collection, analysis, interpretation, and presentation of masses of numerical data. (p. 472)

15.9 Standard Deviation for Grouped Data

1. **Sample standard deviation for grouped data:**

$$s = \sqrt{\frac{\text{sum of } D^2 f}{n - 1}}$$ (p. 494)

15.10 Statistical Process Control

1. *Control charts* are used to help find the information to improve quality control. Review the following three types of control charts discussed in this section: median charts, individual reading chart, and mean control chart. (p. 496)

15.12 Normal Distribution

1. The *normal distribution* of large data sets tends to group data around the mean and/or median in a way that resembles a bell-shaped curve. The empirical rule states that approximately 68% of the data will be within one standard deviation of the mean, 95% of the data will be within two standard deviations of the mean, and 99.7% of the data will be within three standard deviations of the mean. (p. 502)

15.13 Probability

1. The *probability p* is the likelihood that an event will happen and is given by the ratio of number of events happening n to the total number of possible events in the sample space s, or

$$p = \frac{n}{s}$$ (p. 505)

15.14 Independent Events

1. The *probability of two independent events* is the product of the probabilities of each separate event; that is, $p(A \text{ and } B) = p(A) \cdot p(B)$. (p. 507)

REVIEW | CHAPTER 15

1. Find 35% of 360°.
2. Find 56.1% of 360°.
3. Draw a circle graph using the following U.S. data. In fall 2016, 35,400,000 students attended public elementary schools; 15,000,000 attended grades 9 to 12; and 20,500,000 attended American colleges and universities.
4. Draw a line graph using the data in Exercise 3.

5. In Illustration 5 on page 479, what was the temperature at 10:00 P.M.?

For Exercises 6–8, use the following data.

A technician, using a very precise tool, measured a piece of metal to be used in a satellite. He recorded the following measurements: 7.0036 mm; 7.0035 mm; 7.0038 mm; 7.0035 mm; 7.0036 mm.

6. What was the mean measurement?
7. What was the median?
8. What was the sample standard deviation?
9. Given the frequency distribution in Illustration 1, find **a.** the mean and **b.** the sample standard deviation.

Interval	Frequency f
10.5–21.5	4
21.5–32.5	17
32.5–43.5	10
43.5–54.5	28
54.5–65.5	13
65.5–76.5	12
76.5–87.5	9

ILLUSTRATION 1

For Exercises 10–12, use the data below.

A student's test and quiz scores for a quarter were recorded as follows: 72, 83, 79, 85, 91, 93, 80, 95, 82.

10. What is the mean of the scores recorded?
11. What is the median?
12. What is the sample standard deviation?

For Exercises 13 and 14, use the frequency distribution shown in Illustration 2.

13. Find the mean of the data.
14. Find the sample standard deviation.

Interval	Midpoint	Frequency
6.5–9.5	8	3
9.5–12.5	11	10
12.5–15.5	14	4
15.5–18.5	17	9
18.5–21.5	20	15
21.5–24.5	23	28
24.5–27.5	26	3
27.5–30.5	29	2

ILLUSTRATION 2

15. A spinner with numbers 1–5 equally spaced on the face is spun.
    a. What is the sample space?
    b. What is the probability that an odd number will be spun?
16. A bag contains 4 white marbles, 3 red marbles, 1 green marble, and 10 black marbles. A marble is drawn and replaced in the bag each time. What is the probability of drawing a red marble and a red marble and a black marble?

TEST | CHAPTER 15

1. See Illustration 1. What country has the longest life expectancy?
2. See Illustration 1. What country has a life expectancy of 52 years?

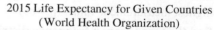
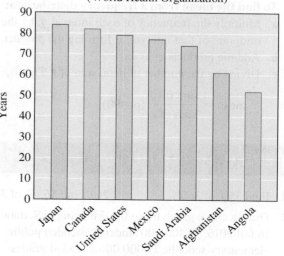

ILLUSTRATION 1

3. Find 38% of 360°.

4. Draw a circle graph using the following data on cargo traffic between cities in the United States in billions of ton-miles: rail, 975; road, 866; air, 8.7; inland water, 435; pipeline, 587.

5. Draw a line graph using the following decade data on population (in thousands) of the United States: 1940—132,594; 1950—152,271; 1960—180,671; 1970—204,879; 1980—227,757; 1990—249,246; 2000—281,422; 2010—308,746.

6. Using the data in Exercise 5, find the population of the United States in 1975.

7. See Illustration 2. What is the power output when the power gain is 28 dB?

8. See Illustration 2. What is the power gain when the power output is 250 mW?

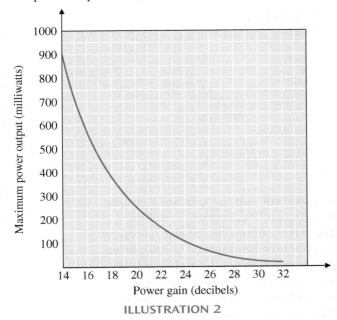

ILLUSTRATION 2

9. Draw a histogram using the data in the frequency distribution in Illustration 3.

Interval	Frequency
6.5–9.5	3
9.5–12.5	6
12.5–15.5	12
15.5–18.5	20
18.5–21.5	14
21.5–24.5	8
24.5–27.5	2

ILLUSTRATION 3

10. x is paired with y by the following table. **a.** Draw a scattergram using this data. **b.** Is a positive, a negative, or no linear correlation shown?

x	0	1	2	3	4	5
y	4	2	6	3	8	4

11. What is the sample space of cards lower than 4 in a deck of 52 playing cards?

For Exercises 12–14, use the data in Illustration 4.

12. Write a frequency distribution.
13. Find the mean using grouped data.
14. Find the sample standard deviation.

6.0, 3.1, 0.6, 1.8, 2.1, 1.5, 4.1, 3.7, 3.3, 3.5, 2.5, 5.2, 2.5, 1.1, 3.2, 3.7, 2.7, 1.7, 4.4, 4.6, 4.0, 3.9, 2.9, 2.0, 1.9, 5.9, 2.4, 3.5, 0.9, 2.4, 0.6, 3.4, 0.5, 3.0, 3.0, 3.9, 3.3, 1.1, 3.2, 3.3, 2.5, 3.0, 3.7, 3.5, 4.2, 3.5, 1.6, 5.6, 5.2, 3.0, 3.5, 2.0, 2.6, 3.4, 3.3, 3.0, 3.0, 1.4

ILLUSTRATION 4

15. A card is chosen at random from a deck of 52 playing cards. What is the probability that it is a spade?

16. A coin is tossed, and a die is rolled. What is the probability of a head and an even number?

CHAPTER 16 | Binary and Hexadecimal Numbers

OBJECTIVES

- Change a binary number to decimal form.
- Add, subtract, and multiply binary numbers.
- Change a decimal number to binary form.
- Change a hexadecimal number to decimal form.
- Change a decimal number to hexadecimal form.
- Add and subtract hexadecimal numbers.
- Change a binary number to hexadecimal form.
- Change a hexadecimal number to binary form.

Mathematics at Work

Telecommunications technicians install, troubleshoot, and maintain the connections used on copper and fiber optic communications cables as well as cell phone and wireless communications. This exciting career offers excellent opportunities. The telecommunications technician needs a solid background in electronics, telecommunications, and networking to employ hands-on troubleshooting and problem-solving skills in an office or field working environment. An understanding of business telephone systems, transmission line and multiplexing systems, structured cabling, and fiber optics is needed. Communications and computer information systems skills are also critical. For more information, please visit **www.cengage.com** and access the Student Online Resources for this text.

Chubykin Arkady/Shutterstock.com

Telecommunications Technician
Technician checking fiber optic communications lines

16.1 Introduction to Binary Numbers

The decimal system of numbers has ten symbols, or digits: 0, 1, 2, 3, 4, 5, 6, 7, 8, and 9. This system of numeration is based on the historical and natural way humans have used their fingers to count objects and maintain records of their possessions by groups of tens.

Computers use a **binary number system**, which has only two symbols or numerals: 0 and 1. These can represent the two positions in a transistor, "off" and "on." Off is assigned 0, and on is assigned 1.

Table 16.1 should help you to understand the relationship between these two systems by comparing place values in the decimal system with place values in the binary system.

Table 16.1

Decimal System						
Millions	Hundred thousands	Ten thousands	Thousands	Hundreds	Tens	Ones
10^6	10^5	10^4	10^3	10^2	10^1	10^0
$10 \times 10 \times 10 \times 10 \times 10 \times 10$	$10 \times 10 \times 10 \times 10 \times 10$	$10 \times 10 \times 10 \times 10$	$10 \times 10 \times 10$	10×10	10	0
1,000,000	100,000	10,000	1000	100	10	0

Binary System						
Sixty-fours	Thirty-twos	Sixteens	Eights	Fours	Twos	Ones
2^6	2^5	2^4	2^3	2^2	2^1	2^0
$2 \times 2 \times 2 \times 2 \times 2 \times 2$	$2 \times 2 \times 2 \times 2 \times 2$	$2 \times 2 \times 2 \times 2$	$2 \times 2 \times 2$	2×2	2	1
64	32	16	8	4	2	1
1000000_2	100000_2	10000_2	1000_2	100_2	10_2	1_2

Note: In base ten, there are powers of ten and ten numerals.
In base two, there are powers of two and two numerals.

The binary equivalents of the decimal numbers 0–17 are given in Table 16.2.

Table 16.2

Decimal form	Binary form	Decimal form	Binary form
0	0	9	1001
1	1	10	1010
2	10	11	1011
3	11	12	1100
4	100	13	1101
5	101	14	1110
6	110	15	1111
7	111	16	10000
8	1000	17	10001

514 CHAPTER 16 ♦ Binary and Hexadecimal Numbers

Some notation is needed to distinguish between decimal and binary numbers. The notation generally used is a subscript after the number to indicate the base of the system in which the number is written. For instance, 1101_{10} is a decimal number, and 1101_2 is a binary number.

Example 1 1101_{10} is a decimal number and means

1 thousand + 1 hundred + 0 tens + 1 one
$= 1 \times 10^3 + 1 \times 10^2 + 0 \times 10^1 + 1 \times 10^0$
$= 1000 + 100 + 0 + 1$
$= 1101_{10}$ ♦

Example 2 1101_2 is a binary number and means

1 eight + 1 four + 0 twos + 1 one
$= 1 \times 2^3 + 1 \times 2^2 + 0 \times 2^1 + 1 \times 2^0$
$= 8 + 4 + 0 + 1$
$= 13_{10}$ ♦

Example 3 What is 1100111_2 in base 10?

$1100111_2 = 1 \times 2^6 + 1 \times 2^5 + 0 \times 2^4 + 0 \times 2^3 + 1 \times 2^2 + 1 \times 2^1 + 1 \times 2^0$
$= 64 + 32 + 0 + 0 + 4 + 2 + 1$
$= 103_{10}$ ♦

EXERCISES 16.1

Change each binary number to decimal form:

1. 11
2. 101
3. 110
4. 1100
5. 1001
6. 11101
7. 110011
8. 10001
9. 101111
10. 11111
11. 1001110
12. 10010010
13. 111011
14. 1000001
15. 10011100
16. 1110001
17. 10001100
18. 1100111
19. 111111
20. 11100011

16.2 Addition of Binary Numbers

Addition of binary numbers is relatively easy because only two numerals are used. The addition facts for binary addition are as follows:

$$\begin{array}{cccc} 0 & 0 & 1 & 1 \\ +0 & +1 & +0 & +1 \\ \hline 0 & 1 & 1 & 10 \end{array}$$

In binary numbers, $1 + 1 = 10$ is read, "One plus one equals *one-zero*."

Example 1 Add:
```
  11
 101
  11
————
1000
```
For convenience, write the binary number to be carried at the top of the next column to the left.

Check the result by decimal addition:

$101 = 5$
$\underline{11 = 3}$
$1000 = 8$ ♦

16.2 ♦ Addition of Binary Numbers

Example 2   Add. Check by decimal addition:
$$\begin{array}{r} 11 \\ 10010 = 18 \\ \underline{111} = \underline{7} \\ 11001 = 25 \end{array}$$

Example 3   Add. Check by decimal addition:
$$\begin{array}{r} 111001 \\ \underline{11011} \\ 1010100 \end{array}$$

$$\begin{array}{r} 1111 \\ 111001 = 57 \\ \underline{11011} = \underline{27} \\ 1010100 = 84 \end{array}$$

NOTE: In the second column from the left, $1 + 1 + 1 = 10 + 1 = 11$. ♦

Example 4   Add. Check by decimal addition:

$$\begin{array}{r} 1111 \\ 11101 = 29 \\ \underline{1111} = \underline{15} \\ 101100 = 44 \end{array}$$
♦

Example 5   Add. Check by decimal addition:

$$\begin{array}{r} 101 \\ 1010 = 10 \\ 11 = 3 \\ 101 = 5 \\ \underline{1011} = \underline{11} \\ 11101 = 29 \end{array}$$

NOTE: In the second column from the right, $1 + 1 + 1 + 1 = 10 + 1 + 1 = 11 + 1 = 100$. Here, you must carry 10. Write 10 at the top of the next columns so that the 0 is above the next column to the left and the 1 is above the second column to the left. ♦

EXERCISES 16.2

Add the following binary numbers and check your result by decimal addition:

1. 110
     10
2. 101
   101
3. 111
   100
4. 110
     11
5. 101
   111
6. 1011
    101
7. 1001
    111
8. 11010
     111
9. 10101
    1100
10. 11100
      111
11. 11011
     1001
12. 111010
      1101
13. 101110
     11001
14. 101001
     11111
15. 101010
     11011
16. 1110111
     111001
17. 1011001
      11100
18. 1000111
     101011
19. 11001110
     1011001
20. 11101011
     1100111
21. 10001
    10101
22. 10001
    11001
23. 110110
     11011
24. 111111
     11111
25. 100101
     11011
26. 1011
     101
    1001
27. 11101
     1001
    11101
28. 11010
    10101
    11100
     1101
29. 10001
     1011
    11010
     1001
30. 1111
     111
      11
       1

16.3 Subtraction of Binary Numbers

Two methods for subtraction can be used. The first method is to use the following subtraction facts: $0 - 0 = 0$, $1 - 1 = 0$, $1 - 0 = 1$, and $0 - 1$ is found by "borrowing" and having $10 - 1 = 1$.

Example 1

Subtract:
$$\begin{array}{r} {}^{0\,1}\\ 1\not{1}011\\ -\ 110\\ \hline 10101 \end{array}$$

Example 2

Subtract:
$$\begin{array}{r} {}^{1}\\ {}^{0\,1\,\not{1}1}\\ \not{1}0\not{1}00\\ -\ 1001\\ \hline 1011 \end{array}$$

The second method for subtraction of binary numbers is using the 1's complement. $a - b$ can be written as $a + (-b)$. In binary form, the negative is called the complement, and we will use the 1's complement. To find the 1's complement of a binary number, reverse each digit of the number being subtracted. For example, the 1's complement of 11100101 is 00011010. Subtraction can then be done as addition.

The following summarizes how to subtract binary numbers using the 1's complement. Given $a - b$, find the 1's complement of b by reversing all the digits of b to form $-b$. Find $a + (-b)$. If the sum has more than n digits (where n is the number of digits in the first number a), add the extra digit to the result. If the sum does not have more than n digits, leave the result. However, if the result starts with a 1, then the result is a negative number and the complement is used.

Example 3

Find $100011 - 1111$.

$$100011 - 1111 = \begin{array}{r} 100011\\ -1111\\ \end{array} = \begin{array}{r} 100011\\ -001111\\ \end{array} = \begin{array}{r} 100011\\ +110000\\ \hline 1010011 \end{array}$$

Since there are more than six digits in the result, you must add 1. The final answer is $10011 + 1$ or 10100.

Example 4

Find $10 - 1000$.

$$\begin{array}{r} 10\\ -1000\\ \end{array} = \begin{array}{r} 0010\\ -1000\\ \end{array} = \begin{array}{r} 0010\\ +0111\\ \hline 1001 \end{array}$$

The result does not have more than four digits, but it does start with 1, which means that the answer is negative and can be found by the complement of the result. The answer is -0110 or -110.

EXERCISES 16.3

Subtract the following binary numbers and check in the binary system:

1. 110
     10
2. 111
     101
3. 1011
     101
4. 110
     11
5. 1001
     111
6. 11001
     10101
7. 100
     11
8. 1000
     101
9. 100101
     11011
10. 11100
      111
11. 110001
      10101
12. 10100
      1010
13. 11000
      101
14. 1101101
      1010101
15. 10001
      1111
16. 10100
      1001
17. 10010
      1001
18. 100010
      1101
19. 10100100
      1011101
20. 10100000
      1101001

Use the 1's complement method to subtract the following binary numbers.

21. 11
      01
22. 101
      100
23. 01
      11
24. 100
      101
25. 1011
      1001
26. 1110
      1010
27. 1000
      1011
28. 10000
      1111
29. 101101
      1001010
30. 10110011
      1001010
31. 11110001
      110111
32. 10011
      1000100
33. 1011101
      10001011
34. 1001011
      1110111
35. 111110001100
      111000001010
36. 100010011001
      100001011110

16.4 Multiplication of Binary Numbers

The multiplication facts for binary numbers are

$0 \times 0 = 0$

$1 \times 0 = 0$

$0 \times 1 = 0$

$1 \times 1 = 1$

Since these facts are the same as for decimal numbers, binary multiplication is very simple to perform.

Example 1

Multiply:
```
    110
  × 11
    110
   110
  10010
```

NOTE: The positioning of the binary numbers is the same as in multiplication of decimal numbers.

Example 2

Multiply:
```
   1011   →  11
  × 101   →   5
   1011      55
  0000
 1011
 110111  →  55
```

EXERCISES 16.4

Multiply the following binary numbers:

1. 101
   × 11

2. 101
   × 10

3. 110
   × 10

4. 111
   × 11

5. 101
   × 101

6. 100
   × 110

7. 110
   × 110

8. 110
   × 101

9. 1011
   × 11

10. 1101
    × 11

11. 1100
    × 110

12. 1001
    × 101

13. 1110
    × 110

14. 10110
    × 1101

15. 111001
    × 1011

16. 11111
    × 1111

17. 1110
    × 1001

18. 10001
    × 1100

19. 110101
    × 101

20. 1011010
    × 101110

16.5 Conversion from Decimal to Binary System

We stated earlier that the computer performs computations in the binary system. Does this mean that you must change all decimal numbers to binary numbers before you can use the computer? The answer is no. Most computers have converters that automatically change numbers in the decimal system to the binary system.

How does a computer perform a simple computation? Consider multiplying 5 × 7. Figure 16.1 traces the flow of the computation through the computer.

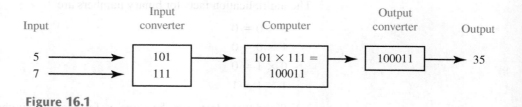

Figure 16.1

Notice that the input converter changes a number from decimal form to binary form; the computer performs the computations; and the output converter changes a number from binary form back to decimal form.

The work done by the input converter is a technique that the technician must understand. The process of conversion from decimal to binary form is not difficult. Read the following procedure carefully:

Changing a Number from Decimal Form to Binary Form

1. Write the given number in decimal form.
2. Divide it by 2.
3. Write the quotient below the decimal and write the remainder at the right, even if it is 0.
4. Divide the first quotient by 2 and write the remainder at the right; repeat the process until the final quotient is 0.
5. Obtain the binary form of the decimal number by using the remainders from each step in order from bottom to top.

Example 1 Write 13 in binary form.

```
2|13    remainders
2| 6       1
2| 3       0
2| 1       1
   0       1    Read digits up!
```

Thus, $13_{10} = 1101_2$.

Example 2 Write 84 in binary form.

```
2|84    remainders
2|42       0
2|21       0
2|10       1
2| 5       0
2| 2       1
2| 1       0
   0       1    Read digits up!
```

Thus, $84_{10} = 1010100_2$.

EXERCISES 16.5

Write each decimal number in binary form:

1. 14
2. 39
3. 63
4. 17
5. 72
6. 40
7. 20
8. 47
9. 24
10. 56
11. 32
12. 80
13. 37
14. 85
15. 100
16. 162
17. 111
18. 128
19. 113
20. 170

16.6 Conversion from Binary to Decimal System

Remember that in decimal numbers the powers of 10 and of $\frac{1}{10}$ are determined by the places of the digits in the number.

$$10^6 \quad 10^5 \quad 10^4 \quad 10^3 \quad 10^2 \quad 10^1 \quad 10^0 \quad . \quad \frac{1}{10^1} \quad \frac{1}{10^2} \quad \frac{1}{10^3} \quad \frac{1}{10^4} \quad \frac{1}{10^5} \quad \frac{1}{10^6}$$

↑ decimal point

Similarly, in the binary system, the powers of 2 and of $\frac{1}{2}$ are determined by the places of the digits in the number.

$$2^6 \quad 2^5 \quad 2^4 \quad 2^3 \quad 2^2 \quad 2^1 \quad 2^0 \quad . \quad \frac{1}{2^1} \quad \frac{1}{2^2} \quad \frac{1}{2^3} \quad \frac{1}{2^4} \quad \frac{1}{2^5} \quad \frac{1}{2^6}$$

↑ binary point

NOTE: $2^0 = 1$; remember, also, that the names of the places to the left of the binary point are names of powers of 2 and to the right of the binary point are names of powers of $\frac{1}{2}$.

CHAPTER 16 ♦ Binary and Hexadecimal Numbers

Read binary numbers by naming the digits in the order they occur from left to right and read the binary point as "point." For example,

101.01 is read, "one-zero-one-point-zero-one."

110.101 is read, "one-one-zero-point-one-zero-one."

> ### Changing a Number from Binary Form to Decimal Form
>
> Find the place value of each digit in the given binary number and multiply it by its corresponding power of 2 or $\frac{1}{2}$.

Example 1 Write 1010.11_2 in decimal form.

$$1010.11 = 1 \times 2^3 + 0 \times 2^2 + 1 \times 2^1 + 0 \times 2^0 + 1 \times \frac{1}{2} + 1 \times \frac{1}{4}$$

$$= 8 + 0 + 2 + 0 + \frac{1}{2} + \frac{1}{4}$$

$$= 10\frac{3}{4} = 10.75 \qquad ♦$$

Example 2 Write 11011.001_2 in decimal form.

$$11011.001 = 1 \times 2^4 + 1 \times 2^3 + 0 \times 2^2 + 1 \times 2^1 + 1 \times 2^0 + 0 \times \frac{1}{2} + 0 \times \frac{1}{4} + 1 \times \frac{1}{8}$$

$$= 16 + 8 + 0 + 2 + 1 + 0 + 0 + \frac{1}{8}$$

$$= 27\frac{1}{8} = 27.125 \qquad ♦$$

EXERCISES 16.6

Change each binary number to decimal form.

1. 10.1
2. 1.1
3. 10.01
4. 11.11
5. 101.11
6. 110.11
7. 100.001
8. 110.101
9. 100.101
10. 1101.1
11. 1100.11
12. 1001.101
13. 111.111
14. 1100.1011
15. 11010.1001
16. 101.1
17. 100.1101
18. 1.11111
19. 11.1101
20. 11010.1101

16.7 Hexadecimal System

The use of binary numbers when working with large decimal numbers is very cumbersome. The use of hexadecimal numers simplifies this problem. Hexadecimal numbers are those numbers with base 16 as compared with decimal (base 10) and binary (base 2). The primary use of a hexadecimal number system is as a more compact, friendly representation of binary coded values often used in digital electronics and computer programming.

55 in decimal form is $5(10^1) + 5(10^0) = 55$

55 in binary form is $110111_2 = 1(2^5) + 1(2^4) + 0(2^3) + 1(2^2) + 1(2^1) + 1(2^0)$

55 in hexadecimal form is $37_{16} = 3(16^1) + 7(16^0)$

The binary system uses 2 numerals 0 and 1. The decimal system uses 10 numerals 0, 1, 2, 3, 4, 5, 6, 7, 8, 9. The **hexadecimal number system** uses 16 numerals. We will use the numerals 0, 1, 2, 3, 4, 5, 6, 7, 8, 9, and the following:

$$A = 10 \quad B = 11 \quad C = 12 \quad D = 13 \quad E = 14 \quad F = 15$$

The hexadecimal number 2A is written in decimal form as

$$2(16^1) + A(16^0) = 2(16^1) + 10(16^0)$$
$$= 32 + 10$$
$$= 42$$

Changing a Number from Hexadecimal Form to Decimal Form

Find the place value of each digit of the given binary number and multiply it by its corresponding power of 16.

Example 1 Write the hexadecimal number DB2 in decimal form.

$$DB2 = D(16^2) + B(16^1) + 2(16^0)$$
$$= 13(16^2) + 11(16^1) + 2(16^0)$$
$$= 13(256) + 11(16) + 2(1)$$
$$= 3328 + 176 + 2$$
$$= 3506$$

Example 2 Write the hexadecimal number 5DF2 in decimal form.

$$5DF2 = 5(16^3) + D(16^2) + F(16^1) + 2(16^0)$$
$$= 5(16^3) + 13(16^2) + 15(16^1) + 2(16^0)$$
$$= 5(4096) + 13(256) + 15(16) + 2(1)$$
$$= 20{,}480 + 3328 + 240 + 2$$
$$= 24{,}050$$

To write a decimal number in hexadecimal form, you can use the same basic method as writing a decimal number in binary form. In the binary form, you used repeated divisions by 2 and used the remainders. In the hexadecimal form, you use repeated divisions by 16 and use the remainders, keeping in mind that you now get remainders 0 through 15. You use the hexadecimal letter representations for 10 through 15.

Changing a Number from Decimal Form to Hexadecimal Form

1. Write the given number in decimal form.
2. Divide it by 16.
3. Write the quotient below the decimal and write the remainder at the right, even if it is 0.
4. Divide the first quotient by 16 and write the remainder at the right; repeat the process until the final quotient is 0.
5. Obtain the hexadecimal form of the decimal number by using the remainders from each step in order from bottom to top.

Example 3

Change 258 to hexadecimal form.

```
16|258    remainders
16| 16    2
16|  1    0
    0     1   ↑ Read digits up.
```

Thus, $258_{10} = 102_{16}$.

Example 4

Write 3527 in hexadecimal form.

```
16|3527   remainders
16| 220    7
16|  13   12   C
     0    13   D  Read digits up.
```

Thus, $3527_{10} = DC7_{16}$.

Example 5

Write 42508 in hexadecimal form.

```
16|42508   remainders
16| 2656   12   C
16|  166    0
16|   10    6
      0    10   A  Read digits up.
```

Thus, $42508_{10} = A60C_{16}$.

EXERCISES 16.7

Change each hexadecimal number to decimal form:

1. 25
2. 37
3. 125
4. 208
5. 1E
6. A3
7. C5
8. 7B
9. 407
10. 579
11. A22
12. C51
13. BC2
14. AE8
15. B2B
16. DD4
17. CDE
18. ACA
19. 2A5B
20. 4D7A

Change each decimal number to hexadecimal form:

21. 235
22. 579
23. 58
24. 97
25. 3352
26. 7369
27. 89,504
28. 92,713
29. 33,558
30. 52,185

16.8 Addition and Subtraction of Hexadecimal Numbers

Use Table 16.3 to add and subtract hexadecimal numbers.

Addition of Hexadecimal Numbers

To add two hexadecimal numbers, find the first number in the left vertical column, find the second number in the top horizontal row, and find their sum at the intersection in Table 16.3. For example, to add 3 and D, locate 3 in the left column and D in the top row and find 10 at the intersection. Continue this process until all pairs of digits have been added.

16.8 ♦ Addition and Subtraction of Hexadecimal Numbers

Table 16.3 | Hexadecimal Table for Addition

+	0	1	2	3	4	5	6	7	8	9	A	B	C	D	E	F
0	0	1	2	3	4	5	6	7	8	9	A	B	C	D	E	F
1	1	2	3	4	5	6	7	8	9	A	B	C	D	E	F	10
2	2	3	4	5	6	7	8	9	A	B	C	D	E	F	10	11
3	3	4	5	6	7	8	9	A	B	C	D	E	F	10	11	12
4	4	5	6	7	8	9	A	B	C	D	E	F	10	11	12	13
5	5	6	7	8	9	A	B	C	D	E	F	10	11	12	13	14
6	6	7	8	9	A	B	C	D	E	F	10	11	12	13	14	15
7	7	8	9	A	B	C	D	E	F	10	11	12	13	14	15	16
8	8	9	A	B	C	D	E	F	10	11	12	13	14	15	16	17
9	9	A	B	C	D	E	F	10	11	12	13	14	15	16	17	18
A	A	B	C	D	E	F	10	11	12	13	14	15	16	17	18	19
B	B	C	D	E	F	10	11	12	13	14	15	16	17	18	19	1A
C	C	D	E	F	10	11	12	13	14	15	16	17	18	19	1A	1B
D	D	E	F	10	11	12	13	14	15	16	17	18	19	1A	1B	1C
E	E	F	10	11	12	13	14	15	16	17	18	19	1A	1B	1C	1D
F	F	10	11	12	13	14	15	16	17	18	19	1A	1B	1C	1D	1E

Example 1 Add the hexadecimal numbers B52 and 3A4 and check using decimals.

$$\begin{array}{ll} \text{B52} & 2 + 4 = 6 \\ \underline{\text{3A4}} & 5 + \text{A} = \text{F} \\ \text{EF6} & \text{B} + 3 = \text{E} \end{array}$$

Check using decimals:

$$\begin{array}{ll} \text{B52} & 11(16^2) + 5(16^1) + 2(16^0) = 2898 \\ \underline{\text{3A4}} & 3(16^2) + 10(16^1) + 4(16^0) = \underline{932} \\ \text{EF6} & 14(16^2) + 15(16^1) + 6(16^0) = 3830 \end{array}$$

♦

Example 2 Add the hexadecimal numbers and check using decimals.

$$\begin{array}{ll} \text{F2E7} & 7 + 2 = 9 \\ \underline{\text{E3B2}} & \text{E} + \text{B} = 19 \text{ (carry 1)} \\ \text{1D699} & 2 + 3 + 1 = 6 \\ & \text{F} + \text{E} = \text{1D} \end{array}$$

Check using decimals:

$$\begin{array}{ll} \text{F2E7} & 15(16^3) + 2(16^2) + 14(16^1) + 7(16^0) = 62{,}183 \\ \underline{\text{E3B2}} & 14(16^3) + 3(16^2) + 11(16^1) + 2(16^0) = \underline{58{,}290} \\ \text{1D699} & 1(16^4) + 13(16^3) + 6(16^2) + 9(16^1) + 9(16^0) = 120{,}473 \end{array}$$

♦

Subtraction of Hexadecimal Numbers

As with binary numbers, we again have two methods available for subtracting two hexadecimal numbers. The first method can be used only when the number you are subtracting is less than or equal to the number you are subtracting it from. Use the hexadecimal table (Table 16.3). Find the number being subtracted in the first column and the number it is being subtracted from in the table to the right; the answer will be above the second number in the top row. Note that you may need to borrow; use the table when borrowing (subtracting) 1 from A, B, C, D, E, and F.

Example 3

Subtract: 58A3 − 2E57

$$\begin{array}{r} 4\ 1\ 9 \\ \cancel{5}\cancel{8}\cancel{A}3 \\ -2E57 \\ \hline 2A4C \end{array}$$

3 − 7 means need to "borrow 1" from A; A − 1 = 9; 13 − 7 = C from Table 16.3.

9 − 5 = 4

8 − E means need to "borrow 1" from 5; 5 − 1 = 4; 18 − E = A from Table 16.3.

4 − 2 = 2 ◆

The second method has no conditions. It uses the idea that a subtraction can be written as an addition. The complement of a hexadecimal number is the negative of the number. The 15's complement will be used. The following are *complementary pairs* of digits: 0 and F, 1 and E, 2 and D, 3 and C, 4 and B, 5 and A, 6 and 9, and 7 and 8. Change each digit in the number being subtracted to its complement and add the resulting hexadecimal numbers. Using this method, you may find the result gives you more digits than when you started. When this happens, add the extra digit 1 to the result to get the answer.

Example 4

Subtract: 58A3 − 2E57

$$\begin{array}{rr} & 1 \\ 58A3 = & 58A3 \\ -2E57 & +D1A8 \\ \hline & 12A4B \\ & 1 \\ \hline & 2A4C \end{array}$$

3 + 8 = B

A + A = 14, carry 1

8 + 1 + 1 = A

5 + D = 12

There is an extra digit; therefore, add 1 as shown. ◆

Example 5

Subtract: 1E05 − 314C

$$\begin{array}{rr} 1E05 = & 1E05 \\ -314C & +CEB3 \\ \hline & ECB8 \end{array}$$

Since the number being subtracted is larger, the answer is negative, so we use the complement of each digit, −1347. ◆

Example 6

Subtract: 241A − 46C2

$$\begin{array}{rr} & 1 \\ 241A = & 241A \\ -46C2 & +B93D \\ \hline & DD57 \end{array}$$

A + D = 17, carry 1

1 + 1 + 3 = 5

4 + 9 = D

2 + B = D

Since the number being subtracted is larger, the answer is negative, so we use the complement of each digit, −22A8. ◆

Example 7

Subtract: 5CC1 − F59

$$\begin{array}{rr} & 1 \\ 5CC1 = & 5CC1 \\ -0F59 & F0A6 \\ \hline & 14D67 \\ & 1 \\ \hline & 4D68 \end{array}$$

1 + 6 = 7

C + A = 16, carry 1

1 + C + 0 = D

5 + F = 14

There is an extra digit; therefore, add 1 as shown. ◆

EXERCISES 16.8

Add the following hexadecimal numbers. Check using decimals.

1. 2 + 9
2. 4 + 8
3. 5 + F
4. E + 3
5. A + D
6. E + C
7. 78 + 31
8. 22 + 78
9. 5B + E2
10. C3 + 6A
11. AE + BB
12. DC + FA
13. 456 + 327
14. 288 + 705
15. A5D + 7EA

16. C3E + 8AB	**17.** AAA + BCE	**18.** DAF + CBD	**37.** F − 35	**38.** 7 − 1B	**39.** 1B − 24
19. 4527 + 8713	**20.** 2851 + 3277		**40.** 2A − 36	**41.** 12 − A3	**42.** 27 − B4
21. 59C2 + 708F	**22.** 3B25 + 52A1		**43.** 129 − 248	**44.** 259 − 743	**45.** 43B − 7A2
23. 5D1E + A2F3	**24.** C7B1 + 2B3D		**46.** 2C5 − B43	**47.** 2DE − 4CC	**48.** 3AF − 7DD
25. AC2B + 4C3	**26.** 2FCA + C25		**49.** AFA − DAC	**50.** BCB − EAF	
27. ABFE + 3ACF	**28.** FFEA + D4CE		**51.** 1990 − 2418	**52.** 3855 − 2763	
29. ABFF + CEED	**30.** FABE + ABED		**53.** 25A7 − 1992	**54.** 5882 − 7B93	

Subtract the following hexadecimal numbers. Check using decimals.

55. 32B2 − 4B5C   **56.** 7D3F − 51C6
57. D9BB − BE2C   **58.** ACCF − BA3C
59. CAAB − FEDA   **60.** BCCF − DDBA

31. 2 − E   **32.** 5 − F   **33.** A − D
34. 9 − F   **35.** 9 − 1C   **36.** B − 25

16.9 Binary to Hexadecimal Conversion

Changing a Number from Binary Form to Hexadecimal Form

If the binary number is larger than any of the numbers in Table 16.4,

1. insert enough zeros at the beginning of the binary number so that you have a total number of digits divisible by 4,
2. mark the binary number in groups of 4, and
3. use Table 16.4 to make the corresponding hexadecimal substitution for each group of 4 binary digits.

Table 16.4

Decimal	Binary	Hexadecimal
0	0	0
1	1	1
2	10	2
3	11	3
4	100	4
5	101	5
6	110	6
7	111	7
8	1000	8
9	1001	9
10	1010	A
11	1011	B
12	1100	C
13	1101	D
14	1110	E
15	1111	F

Example 1   Change binary 1011100 to hexadecimal form.

Since there are 7 digits, we must add one 0 at the beginning, forming 01011100. We group in 4's and get 0101 and 1100. By Table 16.4, 0101 in binary is 5 in hexadecimal and 1100 is C. The answer is 5C. ♦

Example 2   Change binary 11111 to hexadecimal form.

There are 5 digits, so we must add three zeros at the beginning, forming 00011111. We group in 4's and get 0001 and 1111. By Table 16.4, 0001 in binary is 1 in hexadecimal and 1111 is F. The answer is 1F. ♦

Example 3   Change binary 1110010101 to hexadecimal form.

There are 10 digits, so we must add two zeros at the beginning, forming 001110010101. We group in 4's and get 0011 1001 0101. Using Table 16.4, we get 3 9 5. The answer is 395. ♦

Example 4   Change binary 1110001111011111 to hexadecimal form.

There are 16 digits; therefore, we need not add any zeros. We group in 4's, get 1110 0011 1101 1111, and then get E 3 D F using Table 16.4. The answer is E3DF. ♦

To change hexadecimal numbers to binary numbers, we reverse the process by replacing each hexadecimal digit by its group of 4 binary equivalent. Note that if fewer than 4 digits are given in Table 16.4, you must add enough zeros at the beginning to make a group of 4.

> **Changing a Number from Hexadecimal Form to Binary Form**
>
> Replace each hexadecimal digit by its group of 4 binary equivalent from right to left.

Example 5   Change hexadecimal 7A2 to binary form.

From Table 16.4, 7 is 111 or 0111, A is 1010, and 2 is 10 or 0010. The answer is 011110100010 or 11110100010. ♦

Example 6   Change hexadecimal 8FA2 to binary form.

From Table 16.4, 8 is 1000, F is 1111, A is 1010, and 2 is 0010. The answer is 1000111110100010. ♦

Hexadecimal Code for Colors

One use of the hexadecimal system is for color codes. The basic computer colors are red, green, and blue (referred to as RGB). We can obtain other colors by using various amounts of these colors. The hexadecimal numbers 00 and FF represent the strongest and the weakest colors in a range.

Red is hexadecimal code FF0000, green is 00FF00, and blue is 0000FF. Various combinations of RGB result in other colors (e.g., white is FFFFFF, black is 000000, misty rose is FFE4E1, midnight blue is 191970, and lawn green is 7CFC00). Charts are available that show the colors and their hexadecimal codes.

EXERCISES 16.9

Change each binary number to hexadecimal form:

1. 110
2. 1010
3. 1011
4. 111
5. 10111
6. 11001
7. 100100
8. 110110
9. 1000100
10. 1101111
11. 11100110
12. 10011011
13. 100111000
14. 110011100
15. 11100111101
16. 1111110000
17. 110011100111
18. 11110111001
19. 110011001100
20. 110100101001
21. 110100010010001
22. 11001011101101
23. 1111000111001010
24. 1101110010010001

Change each hexadecimal number to binary form:

25. 6
26. E
27. 24
28. 79
29. 2A
30. F4
31. 251
32. 628
33. A32
34. C43
35. 7E4
36. 3F7
37. 4DD
38. 5FC
39. ACD
40. FBF
41. 4A3B
42. 2B7E
43. BCAF
44. CACE

SUMMARY | CHAPTER 16

Glossary of Basic Terms

Binary number system. A number system with only two symbols or numerals: 0 and 1. (p. 513)

Hexadecimal number system. A number system with 16 numerals: 0, 1, 2, 3, 4, 5, 6, 7, 8, 9, A = 10, B = 11, C = 12, D = 13, E = 14, and F = 15. (p. 521)

16.2 Addition of Binary Numbers

1. **Addition facts for binary addition:**
   a. $0 + 0 = 0$
   b. $0 + 1 = 1$
   c. $1 + 0 = 1$
   d. $1 + 1 = 10$ (p. 514)

16.3 Subtraction of Binary Numbers

1. **Subtraction facts for binary subtraction:**
   a. $0 - 0 = 0$
   b. $1 - 1 = 0$
   c. $1 - 0 = 1$
   d. $0 - 1$ is found by "borrowing" and having $10 - 1 = 1$. (p. 516)

16.4 Multiplication of Binary Numbers

1. **Multiplication facts for binary numbers:**
   a. $0 \times 0 = 0$
   b. $1 \times 0 = 0$
   c. $0 \times 1 = 0$
   d. $1 \times 1 = 1$ (p. 517)

16.5 Conversion from Decimal to Binary System

1. **Changing a number from decimal form to binary form:**
   a. Write the given number in decimal form.
   b. Divide it by 2.
   c. Write the quotient below the decimal and write the remainder at the right, even if it is 0.
   d. Divide the first quotient by 2 and write the remainder at the right; repeat the process until the final quotient is 0.
   e. Obtain the binary form of the decimal number by using the remainders from each step in order from bottom to top. (p. 518)

16.6 Conversion from Binary to Decimal System

1. **Changing a number from binary form to decimal form:** Find the place value of each digit of the given binary number and multiply it by its corresponding power of 2 or $\frac{1}{2}$. (p. 520)

16.7 Hexadecimal System

1. **Changing a number from hexadecimal form to decimal form:** Find the place value of each digit of the given binary number and multiply it by its corresponding power of 16. (p. 521)

2. **Changing a number from decimal form to hexadecimal form:**
   a. Write the given number in decimal form.
   b. Divide it by 16.
   c. Write the quotient below the decimal and write the remainder at the right, even if it is 0.
   d. Divide the first quotient by 16 and write the remainder at the right; repeat the process until the final quotient is 0.
   e. Obtain the hexadecimal form of the decimal number by using the remainders from each step in order from bottom to top. (p. 521)

16.8 Addition and Subtraction of Hexadecimal Numbers

1. **To add two hexadecimal numbers:** Using the hexadecimal table (Table 16.3), find the first digit in the first vertical column, find the second digit in the first horizontal row, and find their sum at the intersection. Continue this process until all pairs of digits have been added. (p. 522)
2. **To subtract two hexadecimal numbers:** This method uses the idea that subtraction can be written as an addition of the complement (negative) of a hexadecimal number. The following are *complementary pairs* of digits: 0 and F, 1 and E, 2 and D, 3 and C, 4 and B, 5 and A, 6 and 9, and 7 and 8. Change each digit in the number being subtracted to its complement and add the resulting hexadecimal numbers. *Note:* If the final result gives you more digits than you had when you started, add the extra digit 1 to the result to get the answer. (p. 524)

16.9 Binary to Hexadecimal Conversion

1. **Changing a number from binary form to hexadecimal form:** If the binary number is larger than any of the numbers in Table 16.4,
    a. insert enough zeros at the beginning of the binary number so that you have a total number of digits divisible by 4,
    b. mark the binary number in groups of 4, and
    c. use Table 16.4 to make the corresponding hexadecimal substitution for each group of 4 binary digits. (p. 525)
2. **Changing a number from hexadecimal form to binary form:** Replace each hexadecimal digit by its group of 4 binary equivalent from right to left. (p. 526)

REVIEW | CHAPTER 16

Change each binary number to decimal form:

1. 1101
2. 11001
3. 110100
4. 10110.11

Add the following binary numbers:

5. 11001
   1101
6. 1101001
   101101
7. 1001
   110
   101

Subtract the following binary numbers:

8. 10011
   1010
9. 11011
   1110

Multiply the following binary numbers:

10. 110
    11
11. 11011
    1001

Change each decimal number to binary form:

12. 36
13. 205
14. 1050

Change each hexadecimal number to decimal form:

15. E1
16. 2C
17. C1E

Change each decimal number to hexadecimal form:

18. 312
19. 52
20. 4624

Add the following hexadecimal numbers:

21. 4E + 35
22. 1A4 + EF
23. 6A12 + 7C2B

Change each binary number to hexadecimal form:

24. 10110
25. 100110
26. 101110011

Change each hexadecimal number to binary form:

27. 4C
28. 365
29. B2A
30. 4AA1

TEST | CHAPTER 16

Change each binary number to decimal form:

1. 10110
2. 10111.101

Add the following binary numbers:

3. 1110101
   11011
4. 11011
   10110
   1110

Subtract the following binary numbers:

5. 111011
   10010
6. 100010
   10111

7. Multiply the following binary numbers: 11011
   1101

Change each decimal number to binary form:

8. 407   **9.** 1142

Change each hexadecimal number to decimal form:

10. D3   **11.** 2F

Change each decimal number to hexadecimal form:

12. 628   **13.** 704

Add the following hexadecimal numbers:

14. 7D + 82   **15.** 36E + 15A   **16.** 23AC + 5F7

Change each binary number to hexadecimal form:

17. 101100   **18.** 1100011

Change each hexadecimal number to binary form:

19. 7F   **20.** 3BD

CUMULATIVE REVIEW | CHAPTERS 1–16

1. Round 2927.404 to **a.** the nearest hundredth and **b.** the nearest ten.

2. Perform the indicated operations and write the result in engineering notation rounded to three significant digits:
$$\frac{(612 \times 10^{-6})(15 \times 10^{-9})(2.7 \times 10^3)}{(82 \times 10^9)(8.16 \times 10^{-12})}$$

3. Convert 72°F to °C.

4. Use the rules for measurement to find the sum of the following set of measurements: 6128 km, 1520 km, 16.28 km, 225 km

5. Do as indicated and simplify:
$(2a^2 - 5a + 6) - (-2a^2 + 6a - 7) + (3a^2 - 7a - 2)$

6. Solve: $2(5y - 3) + 4(6y - 1) = 17(2y - 3) - 25y$

7. Solve (round the result to three significant digits):
$$\frac{x}{17} = \frac{14.6}{38.5}$$

8. Find the equation of a line having slope $-\frac{2}{3}$ and y intercept of 4.5.

9. The perimeter of a rectangle is 39.8 m. Its length is 10.1 m longer than its width. Find its length and width.

10. Factor completely: $30x^2 + 117x - 810$

11. Solve: $3x^2 - 5x = 2$

12. Solve: $7x^2 + 2x + 15 = 0$

13. Find the area of the trapezoid in Illustration 1.

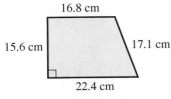

ILLUSTRATION 1

14. Find the volume of the sphere with radius 17.3 in.

15. Given $\cos B = 0.9128$, find angle B in degrees to nearest tenth of a degree.

16. Find side a in the right triangle in Illustration 2.

17. Find cos 256° rounded to four significant digits.

18. Find angle A in the triangle in Illustration 3.

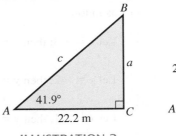 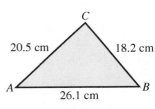

ILLUSTRATION 2   **ILLUSTRATION 3**

19. Find the mean of the following numbers: 20.2, 27.3, 35.1, 30.6, 29.6, 22.6.

20. Find the median of the data in Exercise 19.

21. Find the standard deviation of the data in Exercise 19.

22. Make a frequency distribution from the following data. The numbers of defective parts coming off an assembly line per eight-hour shift were as follows:

15, 12, 10, 9, 15, 22, 7, 23, 12, 8, 18, 22, 11, 30, 14, 18, 12, 20, 22, 35, 10, 8, 11, 19, 7, 23, 17, 15, 20, 16, 17, 18, 22, 15, 20, 13

23. Find the mean using grouped data from Exercise 22.

24. Find the median of the data in Exercise 22.

Do as indicated for the following binary numbers:

25. 1110101 + 10011   **26.** 110001 × 10111

27. Change the decimal number 612 to binary form.

28. Change the hexadecimal number E5 to decimal form.

29. Change the hexadecimal number 4AB to binary form.

30. Add the following hexadecimal numbers: 2B5 + 1A4D

APPENDIX A — Exponential Equations

Equations in the form $y = x^n$, where the exponent n is a constant and the base is a variable, are called *power equations*. For example, we graphed the simple power equation $y = x^2$ in Chapter 11.

Equations in the form $y = b^x$, where the exponent is a variable and the base b is a constant with $b > 0$ and $b \neq 1$, are called **exponential equations**. Two examples are $y = 2^x$ and $y = 3^{-x}$.

Example 1 Graph the exponential equation $y = 2^x$ by plotting points.

Set up a table, letting x equal some convenient values, and solve for each corresponding value of y as follows:

Let $x = -3$; then $y = 2^{-3} = \dfrac{1}{2^3} = \dfrac{1}{8}$.

Let $x = -2$; then $y = 2^{-2} = \dfrac{1}{2^2} = \dfrac{1}{4}$.

Let $x = -1$; then $y = 2^{-1} = \dfrac{1}{2^1} = \dfrac{1}{2}$.

Let $x = 0$; then $y = 2^0 = 1$.

Let $x = 1$; then $y = 2^1 = 2$.

Let $x = 2$; then $y = 2^2 = 4$.

Let $x = 3$; then $y = 2^3 = 8$.

x	-3	-2	-1	0	1	2	3
y	$\frac{1}{8}$	$\frac{1}{4}$	$\frac{1}{2}$	1	2	4	8

Then plot these points as in Figure A.1. It appears that the points follow the shape of a curve. To confirm, let's try some other points with x values that are not integers. To find the y values, you will need to find these powers with your calculator similar to the steps shown in Section 1.15. The first calculation is illustrated below.

Let $x = 0.5$; then $y = 2^{0.5} = 1.4$.

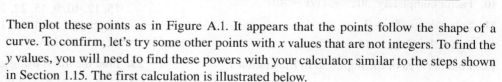

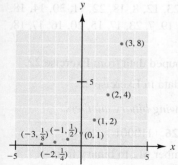

Figure A.1

APPENDIX A ♦ Exponential Equations 531

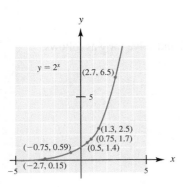

Figure A.2

Let $x = 0.75$; then $y = 2^{0.75} = 1.7$.
Let $x = 1.3$; then $y = 2^{1.3} = 2.5$.
Let $x = 2.7$; then $y = 2^{2.7} = 6.5$.
Let $x = -0.75$; then $y = 2^{-0.75} = 0.59$.
Let $x = -2.7$; then $y = 2^{-2.7} = 0.15$.

x	0.5	0.75	1.3	2.7	-0.75	-2.7
y	1.4	1.7	2.5	6.5	0.59	0.15

Next, plot these points as in Figure A.2. As you can see, these points lie along the same curve as outlined by using a few easy sample ordered pairs. ◆

In general, for $b > 1$, $y = b^x$ is an **increasing function**. That is, as x increases, y also increases.

Example 2 Graph the equation $y = \left(\dfrac{1}{3}\right)^x$.

The laws of exponents for powers of 10 in Section 2.5 also apply to powers of any base. Here, we can write $\dfrac{1}{3} = \dfrac{1}{3^1} = 3^{-1}$ and $\left(\dfrac{1}{3}\right)^x = (3^{-1})^x = 3^{-x}$. We can write this equation as $y = 3^{-x}$.

Set up a table, letting x equal some convenient values, and solve for y as follows:

Let $x = -3$; then $y = 3^{-(-3)} = 3^3 = 27$.
Let $x = -2$; then $y = 3^{-(-2)} = 3^2 = 9$.
Let $x = -1$; then $y = 3^{-(-1)} = 3^1 = 3$.
Let $x = 0$; then $y = 3^{-(0)} = 3^0 = 1$.
Let $x = 1$; then $y = 3^{-1} = \dfrac{1}{3}$.
Let $x = 2$; then $y = 3^{-2} = \dfrac{1}{3^2} = \dfrac{1}{9}$.
Let $x = 3$; then $y = 3^{-3} = \dfrac{1}{3^3} = \dfrac{1}{27}$.

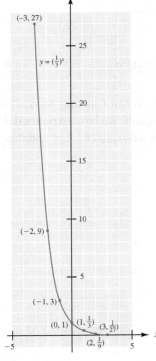

Figure A.3

x	-3	-2	-1	0	1	2	3
y	27	9	3	1	$\dfrac{1}{3}$	$\dfrac{1}{9}$	$\dfrac{1}{27}$

Then plot these points as in Figure A.3 and connect with a smooth curve. ◆

In general, for $0 < b < 1$, $y = b^x$ is a **decreasing function**. That is, as x increases, y decreases.

One special exponential equation describes a wide variety of natural growth and decay processes with its base e, which is approximately equal to 2.71828. Examples include population and bacteria growth, radioactive decay, electrical current and voltage analysis, and investment and savings growth. The fundamental equation is $y = e^x$.

APPENDIX A ♦ Exponential Equations

Example 3 Graph the equation $y = e^x$.

Again, set up a table, letting x equal some convenient values, and solve for y as follows:

Let $x = -3$; then $y = e^{-3} = 0.050$.

This first calculation with a calculator is illustrated below.

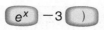

NOTE: The right parenthesis may or may not be needed for given calculators.

Let $x = -2$; then $y = e^{-2} = 0.14$.
Let $x = -1$; then $y = e^{-1} = 0.37$.
Let $x = 0$; then $y = e^0 = 1$.
Let $x = 1$; then $y = e^1 = 2.7$.
Let $x = 2$; then $y = e^2 = 7.4$.
Let $x = 3$; then $y = e^3 = 20.1$.

x	-3	-2	-1	0	1	2	3
y	0.050	0.14	0.37	1	2.7	7.4	20.1

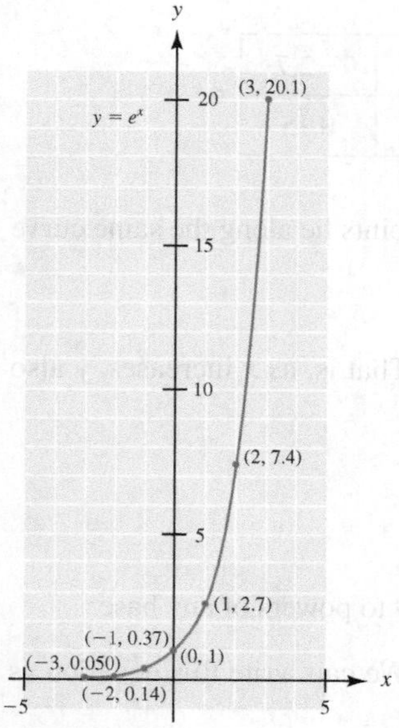

Then plot these points as in Figure A.4 and connect with a smooth curve. ♦

Figure A.4

There are many applications involving exponential equations. When a quantity *increases* so that the amount of increase is proportional to the amount present, we have **exponential growth**. One of the most commonly used exponential equations for exponential or continuous growth is

Exponential Growth

$$y = Ae^{rt}$$

where

y = the amount present at any given time t

A = the original amount

e = the natural number

r = the rate of growth and

t = the time the growth has occurred

This exponential growth equation can be used to find interest when compounded continuously, which is very common now. In Example 3 of Section 1.16, we found that $12,682.42 is owed at the end of 3 years if $10,000 is borrowed at 8% per year compounded quarterly and no payments are made on the loan.

Example 4

Find the amount of money owed at the end of 3 years if $10,000 is borrowed at 8% per year compounded continuously and no payments are made on the loan. Here,

$A = \$10{,}000,$

$r = 8\% = 0.08,$ and

$t = 3$ years.

$y = Ae^{rt},$

$y = (\$10{,}000)\, e^{(0.08)(3)}$

$y = \$12{,}712.49$

Using a scientific calculator, we have

10000 × e^x .08 × 3) =

```
12712.4915
```

NOTE: The e^x key usually provides the left parenthesis.

As you can see, interest accumulates more rapidly as the interest is compounded within shorter intervals of time.

When a quantity *decreases* so that the amount of decrease is proportional to the amount present, we have **exponential decay**. One of the most commonly used exponential equations for exponential or continuous decay is

Exponential Decay

$$y = Ae^{-rt}$$

where

$y = $ the amount present at any given time t

$A = $ the original amount

$e = $ the natural number

$r = $ the rate of decay and

$t = $ the time the decay has occurred

Example 5

The discharge current of a capacitor with an initial current of 3.60 A is given by the exponential equation $i = 3.6e^{-55t}$. Find the discharge current after 25 ms.

Here, $t = 25$ ms $= 0.025$ s. (Here, the time must be in seconds.) So, we have

$i = 3.60e^{-55t}$

$i = 3.60e^{(-55)(0.025)}$

$i = 0.91$ A

That is, after 25 ms, the discharge current has decreased to 0.91 A.

EXERCISES A

Graph each equation:

1. $y = 3^x$
2. $y = 4^x$
3. $y = 5^x$
4. $y = 10^x$
5. $y = \left(\dfrac{1}{2}\right)^x$
6. $y = \left(\dfrac{3}{4}\right)^x$
7. $y = \left(\dfrac{2}{3}\right)^x$
8. $y = \left(\dfrac{1}{4}\right)^x$
9. $y = 2^{-x}$
10. $y = 4^{-x}$
11. $y = 3^{2x}$
12. $y = 3^{-2x}$
13. $y = 2^{-4x}$
14. $y = 3^{5x}$
15. $y = 2^{2-x}$
16. $y = 3^{x+2}$
17. $y = e^{2x}$
18. $y = e^{-x}$
19. $y = e^{-3x}$
20. $y = e^{-2x}$

21. Mackenzie deposits $5000 in a savings account that pays $3\tfrac{3}{4}\%$ compounded continuously. Assuming she makes no more deposits, how much money will she have in her account after $2\tfrac{1}{2}$ years?

22. The amount of bacterial growth in a given culture is given by $N = N_0 e^{0.035t}$, where t is the time in hours, N_0 is the initial amount, and N is the amount after time t. Find how much bacteria we have after 4.00 h if we start with 7500 bacteria.

23. A city with a population of 65,000 has a growth rate of 3.5%. Find its expected population in 5 years.

24. The voltage in a circuit is increasing according to the equation $V = V_0 e^{0.025t}$. Find the voltage in 45.0 ms if the initial voltage is 275 mV. The result will be in mV.

25. Zachary retires with an annual pension of $75,000. With an inflation rate of 5%, what will be his purchasing power after 10 years? *Hint:* $r = -5\%$.

26. A given radioactive element decays according to the equation $y = y_0 e^{-0.0045t}$, where t is the time in seconds, y_0 is the initial amount, and y is the amount remaining after time t. How much of a 75.0-g radioactive element remains after 3.00 minutes?

27. A current flows in a given circuit according to the exponential equation $i = 0.025 e^{-0.175t}$, where t is in ms and i is in mA. Find the amount of current after 4.80 ms.

28. The voltage is discharged in an electric capacitor through a resistor according to the equation $E = E_0 e^{-t/(RC)}$, where E_0 is the original voltage in volts (V), t is the time t for E_0 to decrease to E, R is the resistance in ohms (Ω), and C is the capacitance in farads (F). Find the voltage after 0.500 s if the initial voltage is 25.0 mV, the resistance is 1.85 Ω, and the capacitance is 0.450 F.

Simple Inequalities

APPENDIX B

An **inequality** is a statement in one of the following forms:

1. $a < b$: One quantity is *less than* another; that is, a is *less than* b.
2. $a > b$: One quantity is *greater than* another; that is, a is *greater than* b.
3. $a \leq b$: One quantity is *less than or equal to* another; that is, a is *less than or equal to* b.
4. $a \geq b$: One quantity is *greater than or equal to* another; that is, a is *greater than or equal to* b.

The **sense** of an inequality refers to the direction (greater than or less than) in which the inequality sign points. The inequalities $a < b$ and $c < d$ and $m > n$ and $s > t$ have the *same sense*. The inequalities $a < b$ and $m > n$ have *opposite senses*.

We often find it more visual to draw a graph of an inequality on the number line.

Example 1   Draw a graph of the inequality $x > 3$ on the number line.

All real numbers greater than 3 are solutions of this inequality. Its graph is shown in Figure B.1.

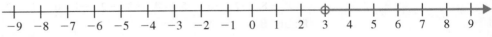

Figure B.1

NOTE: An open circle shows that the endpoint is *not* included.

Example 2   Draw a graph of the inequality $x \geq -4$ on the number line.

All real numbers greater than or equal to -4 are solutions of this inequality. Its graph is shown in Figure B.2.

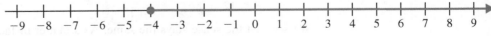

Figure B.2

NOTE: A closed circle shows that the endpoint *is* included.

536 APPENDIX B ♦ Simple Inequalities

Example 3 Draw a graph of the inequality $x < -1$ on the number line.

All real numbers less than -1 are solutions of this inequality. Its graph is shown in Figure B.3.

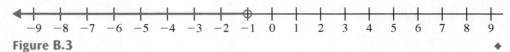

Figure B.3

Example 4 Draw a graph of the inequality $x \leq 6$ on the number line.

All real numbers less than or equal to 6 are solutions of this inequality. Its graph is shown in Figure B.4.

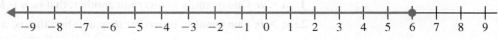

Figure B.4

Example 5 Write an inequality for the graph in Figure B.5.

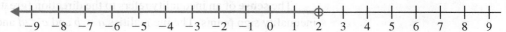

Figure B.5

The inequality is $x < 2$. Note the open circle does not include the endpoint.

Example 6 Write an inequality for the graph in Figure B.6.

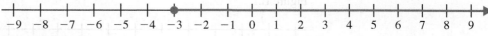

Figure B.6

The inequality is $x \geq -3$. Note the closed circle includes the endpoint.

Let's consider the inequality

$$3 < 7$$

What happens to the sense of this inequality if we add the same number, say 5, to both sides?

$$3 + 5 < 7 + 5$$
$$8 < 12$$

We see that the sense stays the same. As a matter of fact, **adding the same number to, or subtracting the same number from, both sides of an inequality does not change the sense.**

Let's consider the inequality

$$8 \geq 2$$

What happens to the sense of an inequality if we multiply both sides—say, 4—to both sides?

$$8(4) \geq 2(4)$$
$$32 \geq 8$$

We see that the sense stays the same.

But, what happens if we multiply both sides of that same inequality by a negative number, say −5?

$$8(-5) \geq 2(-5)$$
$$-40 \leq -10$$

We see that the sense of the inequality must be *reversed* for the statement to be true.

In general,

1. **Multiplying or dividing both sides of an inequality by a positive number does not change the sense.**
2. **Multiplying or dividing both sides of an inequality by a negative number does change the sense.**

We use the same principles for solving simple inequalities as we use for solving simple equations, except when we multiply or divide both sides of an inequality by a negative number, we must reverse its sense.

Example 7 Solve $x + 9 < 14$ and graph its solution on the number line.

$$x + 9 < 14$$
$$x + 9 - 9 < 14 - 9 \quad \text{Subtract 9 from both sides.}$$
$$x < 5$$

The graph of this solution is shown in Figure B.7.

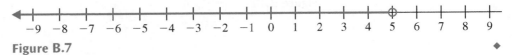

Figure B.7

Example 8 Solve $-6x \geq 30$ and graph its solution on the number line.

$$-6x \geq 30$$
$$\frac{-6x}{-6} \geq \frac{30}{-6} \quad \text{Divide both sides by −6.}$$
$$x \leq -5 \quad \text{Note the sense is reversed.}$$

The graph of this solution is shown in Figure B.8.

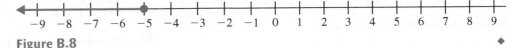

Figure B.8

Example 9 Solve $2x - 6 > 4$ and graph its solution on the number line.

$$2x - 6 > 4$$
$$2x - 6 + 6 > 4 + 6 \quad \text{Add 6 to both sides.}$$
$$2x > 10$$
$$x > 5 \quad \text{Divide both sides by 2.}$$

The graph of this solution is shown in Figure B.9.

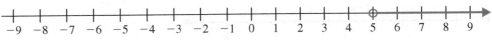

Figure B.9

APPENDIX B ♦ Simple Inequalities

Example 10 Solve $2(x + 4) \leq 6x - 12$ and graph its solution on the number line.

$$2(x + 4) \leq 6x - 12$$
$$2x + 8 \leq 6x - 12 \qquad \text{Remove parentheses.}$$
$$2x + 8 - 2x \leq 6x - 12 - 2x \qquad \text{Subtract } 2x \text{ from both sides.}$$
$$8 \leq 4x - 12$$
$$8 + 12 \leq 4x - 12 + 12 \qquad \text{Add 12 to both sides.}$$
$$20 \leq 4x$$
$$5 \leq x \qquad \text{Divide both sides by 4.}$$

Alternate solution:

$$2(x + 4) \leq 6x - 12$$
$$2x + 8 \leq 6x - 12 \qquad \text{Remove parentheses.}$$
$$2x + 8 - 6x \leq 6x - 12 - 6x \qquad \text{Subtract } 6x \text{ from both sides.}$$
$$-4x + 8 \leq -12$$
$$-4x + 8 - 8 \leq -12 - 8 \qquad \text{Subtract 8 from both sides.}$$
$$-4x \leq -20 \qquad \text{Divide both sides by } -4.$$
$$x \geq 5 \qquad \text{Note the sense is reversed.}$$

The graph of this solution is shown in Figure B.10.

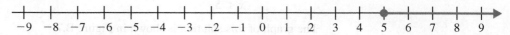

Figure B.10

EXERCISES B

Draw a graph for each inequality on the number line:

1. $x \leq 4$
2. $a \leq 1$
3. $x \geq 4$
4. $b \geq -5$
5. $m < 3$
6. $x < 7$
7. $a > -2$
8. $x > -6$
9. $x \geq 10$
10. $r \leq -12$
11. $x < -24$
12. $x < 36$

Write an inequality for each graph:

13.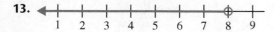

14.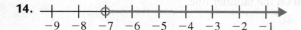

15.

16.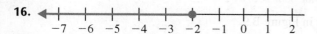

17.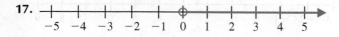

18.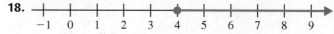

19.

20.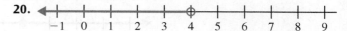

21.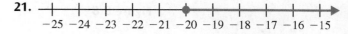

22.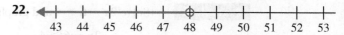

23.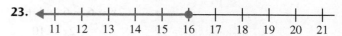

24.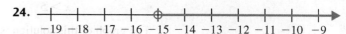

Solve each inequality and graph its solution on the number line:

25. $5x < 15$
26. $6x \geq 30$
27. $-4x \geq 20$
28. $-6x < -36$
29. $16 \leq x + 7$
30. $x + 13 < 1$
31. $x - 12 \geq 8$
32. $4 - x \geq -6$
33. $\dfrac{x}{2} > -4$
34. $-6 < \dfrac{x}{3}$
35. $-10 \leq \dfrac{x}{5}$
36. $\dfrac{x}{12} < 4$
37. $3x + 5 < 29$
38. $5x - 2 > 33$
39. $3 - 2x \geq -9$
40. $6 - 4x \leq -30$
41. $5x - 7 \leq 2x + 11$
42. $8x + 3 > 6x - 9$
43. $4(x - 3) \leq 6x - 14$
44. $-2(x - 4) > 4x - 10$
45. $3(2x + 4) < 2(x - 2)$
46. $-2(2 - x) \geq -(2x - 8)$
47. $\dfrac{9 + x}{4} > 10$
48. $\dfrac{3x + 2}{7} \geq x - 14$

APPENDIX C

Answers to Odd-Numbered Exercises and All Chapter Review and Cumulative Review Exercises

CHAPTER 1

Exercises 1.1 (pages 7–10)
1. 3255 **3.** 1454 **5.** 795,776 **7.** 5164 **9.** 26,008 **11.** 2820 **13.** 4195 Ω **15.** 224 **17.** 39 ft **19.** I: 1925 cm^3; O: 1425 cm^3; 500 cm^3 **21.** 27,216 **23.** 18,172,065 **25.** 35,360,000 **27.** 1809 **29.** 389 **31.** 844 r 40 **33.** 496 mi **35.** 325 cm^3 **37.** 13 km/L **39.** $86 **41.** 500 mi **43.** 4820 ft **45.** 9600 **47.** 67 in. from either corner; 48 in. above the floor **49.** 44 bu/acre **51. a.** 881 lb **b.** 15 lb **53.** 85 bu/acre **55.** $17,200 **57.** 36 **59.** 5 A **61.** 24 V **63.** 880 oz **65.** 4 **67.** 9 in. **69.** 19 in. **71.** 213,500 bd ft **73.** 320 **75.** 200 items **77. a.** $411 **b.** $137

Exercises 1.2 (page 13)
1. 2 **3.** 10 **5.** 18 **7.** 131 **9.** 73 **11.** 23 **13.** 16 **15.** 34 **17.** 102 **19.** 137 **21.** 230 **23.** 55 **25.** 0 **27.** 4 **29.** 0 **31.** 13 **33.** 19 **35.** 26 **37.** 21 **39.** 1 **41.** 102 **43.** 85

Exercises 1.3 (pages 15–17)
1. 96 yd^2 **3.** 307,500 ft^2 **5.** 13,943 in^2 **7.** 84 cm^2 **9.** 48 in^2 **11.** 52 in^2 **13.** 108 **15.** 32 gal **17. a.** $118,800 **b.** $257,840 **19.** 96 m^3 **21.** 720 cm^3 **23.** 3900 in^3 **25.** 600 cm^3 **27.** 11,520 in^3 **29.** 30 ft^3 **31.** 69,480 lb **33.** 14,880 lb **35.** $2750 **37. a.** 25 in. $\times$ 37 in., **b.** 925 in^2 **c.** 9 in^2 **39.** 128 ft^3 **41.** 48 ft^3

Exercises 1.4 (page 20)
1. 600 **3.** 11,250 **5.** 57,376 **7.** 38,400 **9.** 8 **11.** 5017 **13.** 40 in^2 **15.** 810 ft^2 **17.** 56 m^2 **19.** 648 ft^2 **21.** 32 ft^2 **23.** 1800 cm^2 **25.** 4500 cm^3 **27.** 172 **29.** 16

Exercises 1.5 (page 23)
1. a. yes **b.** no **3. a.** yes **b.** yes **5. a.** yes **b.** no **7.** prime **9.** not prime **11.** not prime **13.** not prime **15.** yes **17.** no **19.** no **21.** yes **23.** no **25.** yes **27.** yes **29.** no **31.** no **33.** yes **35.** no **37.** no **39.** yes **41.** yes **43.** yes **45.** $2 \cdot 2 \cdot 5$ **47.** $2 \cdot 3 \cdot 11$ **49.** $2 \cdot 2 \cdot 3 \cdot 3$ **51.** $3 \cdot 3 \cdot 3$ **53.** $3 \cdot 17$ **55.** $2 \cdot 3 \cdot 7$ **57.** $2 \cdot 2 \cdot 2 \cdot 3 \cdot 5$ **59.** $3 \cdot 3 \cdot 19$ **61.** $3 \cdot 5 \cdot 7$ **63.** $2 \cdot 2 \cdot 3 \cdot 3 \cdot 7$

Unit 1A Review (pages 23–24)
1. 241 **2.** 1795 **3.** 2,711,279 **4.** 620 **5.** 262 ft **6.** 254 bu **7.** 42 **8.** 7 **9.** 32 **10.** 499 in^2 **11.** 720 ft^3 **12.** 180 **13.** 10 **14.** 300 **15.** not prime **16.** prime **17.** yes **18.** no **19.** $2 \cdot 2 \cdot 2 \cdot 5$ **20.** $3 \cdot 3 \cdot 3 \cdot 5$

Exercises 1.6 (pages 28–29)
1. $\frac{3}{7}$ **3.** $\frac{6}{7}$ **5.** $\frac{3}{16}$ **7.** $\frac{1}{3}$ **9.** $\frac{4}{5}$ **11.** 1 **13.** 0 **15.** undefined **17.** $\frac{7}{8}$ **19.** $\frac{3}{4}$ **21.** $\frac{3}{4}$ **23.** $\frac{4}{5}$ **25.** $\frac{3}{10}$ **27.** $\frac{7}{8}$ **29.** $\frac{7}{9}$ **31.** $15\frac{3}{5}$ **33.** $9\frac{1}{3}$ **35.** $1\frac{1}{4}$ **37.** $9\frac{1}{2}$ **39.** $6\frac{1}{4}$ **41.** $\frac{23}{6}$ **43.** $\frac{17}{8}$ **45.** $\frac{23}{16}$ **47.** $\frac{55}{8}$ **49.** $\frac{53}{5}$ **51.** $4\frac{2}{3}$ pies

Exercises 1.7 (pages 37–41)
1. 16 **3.** 210 **5.** 48 **7.** $\frac{5}{6}$ **9.** $\frac{5}{32}$ **11.** $\frac{11}{28}$ **13.** $\frac{29}{64}$ **15.** $\frac{7}{20}$ **17.** $1\frac{3}{10}$ **19.** $\frac{37}{48}$ **21.** $\frac{13}{120}$ **23.** $\frac{67}{105}$ **25.** $\frac{1}{8}$ **27.** $\frac{1}{2}$ **29.** $\frac{4}{7}$ **31.** $\frac{1}{32}$ **33.** $7\frac{1}{4}$ **35.** $2\frac{5}{8}$ **37.** $4\frac{3}{4}$ **39.** $2\frac{5}{16}$ **41.** $13\frac{31}{45}$ **43.** $11\frac{5}{16}$ **45.** $9\frac{13}{24}$ **47.** $2237\frac{1}{4}$ ft **49. a.** $6\frac{1}{4}$ ft **b.** $1\frac{1}{2}$ ft

51. $35\frac{11}{20}$ gal **53.** $1\frac{1}{2}$ gal **55.** $\frac{5}{6}$ h **57.** $1\frac{31}{48}$ tons **59.** $8\frac{1}{4}$ in. **61. a.** $3\frac{5}{16}$ in. **b.** $29\frac{1}{16}$ in. **63. a.** $\frac{1}{4}$ in. **b.** $12\frac{5}{16}$ in.
65. $3\frac{1}{4}$ A **67.** $1\frac{43}{48}$ A **69.** $9\frac{5}{8}$ in. **71. a.** $10\frac{1}{2}$ in. **b.** $\frac{3}{4}$ in. **73. a.** $21\frac{13}{16}$ in. **b.** $\frac{7}{8}$ in. **75.** $14\frac{3}{8}$ in. **77.** $\frac{5}{64}$ in.
79. $1\frac{25}{32}$ in.; $1\frac{19}{32}$ in. **81.** $10\frac{13}{16}$ in.; $3\frac{3}{4}$ in. **83.** $52\frac{5}{32}$ in. **85. a.** $7\frac{3}{16}$ in. **b.** $1\frac{11}{16}$ in. **87.** $\frac{1}{2}$ acre **89.** $1\frac{1}{4}$ sticks **91.** $1\frac{1}{8}$ cups
93. $\frac{1}{8}$ bag

Exercises 1.8 (pages 45–48)

1. 12 **3.** 9 **5.** $\frac{35}{64}$ **7.** $\frac{2}{3}$ **9.** 10 **11.** $\frac{1}{8}$ **13.** $1\frac{1}{4}$ **15.** $\frac{2}{5}$ **17.** $\frac{18}{25}$ **19.** 18 **21.** 40 **23.** $1\frac{2}{33}$ **25.** $1\frac{43}{45}$ **27.** $2\frac{1}{4}$ **29.** $\frac{27}{32}$
31. $\frac{1}{126}$ **33.** $\frac{7}{12}$ **35.** $31\frac{1}{2}$ gal **37.** $45\frac{1}{2}$ in. **39.** $120\frac{3}{4}$ mi/h **41.** 33 ft **43.** 80 bd ft **45.** $1633\frac{1}{3}$ bd ft **47.** $3\frac{27}{32}$ in.
49. $1\frac{7}{8}$ in. **51.** $8\frac{5}{8}$ in. **53.** $8\frac{2}{3}$ min **55.** $2\frac{2}{3}$ ft^3 **57.** $1\frac{1}{4}$ h **59.** 2750 W **61.** 75 W **63.** $2\frac{2}{7}$ A **65. a.** 7 ft 3 in. **b.** 43 ft 9 in.
c. 21 ft $10\frac{1}{2}$ in. **67.** 40 lb **69.** 160 bu/acre **71.** $\frac{1}{2}$ oz **73.** $\frac{1}{2}$ **75.** $14\frac{1}{2}$ lb **77.** 24 **79.** $2\frac{1}{2}$ tsp **81. a.** $\frac{3}{4}$ in. **b.** 63 in^3
83. 4 Ω **85.** $3\frac{1}{5}$ Ω **87.** 75 white; 225 red **89.** 6 scoops **91.** 12 steaks **93.** $2\frac{5}{6}$ gal

Exercises 1.9 (pages 51–52)

1. 43 **3.** 83 **5.** 9 **7.** 9 **9.** 96 **11.** 6 **13.** 8 **15.** 5 **17.** $5\frac{1}{2}$ **19.** $3\frac{1}{2}$ **21.** $3\frac{1}{2}$ **23.** $30\frac{2}{3}$ **25.** 3520 **27.** $15\frac{5}{8}$
29. 6 ft 8 in. **31.** 153 in. **33.** 7 qt **35.** $\frac{66}{125}$ Ω **37.** 2520 in^2 **39. a.** 10,560 ft **b.** 3520 yd **41.** 48 oz **43.** 51 yd
45. $8\frac{1}{2}$ chains **47.** $410\frac{1}{10}$ gr **49.** 75 ft/min **51.** 72 mi/min **53.** $58\frac{2}{3}$ ft/s **55.** 120 ft/min **57.** 16 yd 1 ft 10 in. **59.** 9000 lb
61. 22 paces **63.** 28 qt **65.** $3\frac{3}{8}$ gal

Unit 1B Review (page 52)

1. $\frac{3}{5}$ **2.** $\frac{8}{9}$ **3.** $4\frac{1}{2}$ **4.** $\frac{17}{5}$ **5.** $1\frac{1}{2}$ **6.** $2\frac{23}{24}$ **7.** $\frac{4}{15}$ **8.** $\frac{6}{13}$ **9.** $3\frac{1}{4}$ **10.** 2 **11.** $\frac{7}{8}$ in. **12.** $17\frac{7}{8}$ in. **13.** $17\frac{5}{6}$ in. **14.** $16\frac{2}{3}$ in^2
15. 48 in. **16.** 8 yd **17.** 48 oz **18.** 5 gal **19.** 88 ft/s **20.** 5 ft 8 in.

Exercises 1.10 (pages 58–61)

1. four thousandths **3.** five ten-thousandths **5.** one and four hundred twenty-one hundred-thousandths **7.** six and ninety-two
thousandths **9.** 5.02; $5\frac{2}{100}$ or $5\frac{1}{50}$ **11.** 71.0021; $71\frac{21}{10,000}$ **13.** 43.0101; $43\frac{101}{10,000}$ **15.** 0.375 **17.** $0.7\overline{3}$ **19.** 0.34 **21.** $1.\overline{27}$
23. $18.\overline{285714}$ **25.** $34.\overline{2}$ **27.** $\frac{7}{10}$ **29.** $\frac{11}{100}$ **31.** $\frac{337}{400}$ **33.** $10\frac{19}{25}$ **35.** 150.888 **37.** 163.204 **39.** 86.6 **41.** 15.308
43. 8.68 **45.** 4.862 **47.** 10.0507 **49.** 1.45 ft × 4.2 ft **51.** 10.8 h **53.** 0.3125 in. **55.** $a = 4.56$ cm; $b = 4.87$ cm
57. 7.94 in. **59.** 4.8125 in. **61.** 2.605 A **63.** 1396.8 Ω **65.** 0.532 in. **67.** 0.22 in. **69.** 4.118 in. **71.** 3.8 billion
73. a. 1430.2 bbl **b.** 267.4 bbl **75.** 3.4 gal **77.** 6 lb

Exercises 1.11 (page 64)

1. a. 1700 **b.** 1650 **3. a.** 3100 **b.** 3130 **5. a.** 18,700 **b.** 18,680 **7. a.** 3.1 **b.** 3.142 **9. a.** 0.1 **b.** 0.057 **11. a.** 0.1
b. 0.070

	Hundred	Ten	Unit	Tenth	Hundredth	Thousandth
13.	600	640	636	636.2	636.18	636.183
15.	17,200	17,160	17,159	17,159.2	17,159.17	17,159.167
17.	1,543,700	1,543,680	1,543,679	—	—	—
19.	10,600	10,650	10,650	10,649.8	10,649.83	—
21.	600	650	650	649.9	649.90	649.900

23. 237,000 **25.** 0.0328 **27.** 72 **29.** 1,462,000 **31.** 0.0003376 **33.** 1.01

Exercises 1.12 (pages 68–71)
1. 0.555 **3.** 10.5126 **5.** 9,280,000 **7.** 30 **9.** 15 **11.** 248.23 **13.** 3676.47 **15.** 7.80 **17.** 6.59 **19.** $\frac{7}{12}$ **21.** $\frac{4}{5}$ **23.** 1.2 ft
25. 119 mi/h **27.** 27.7 mi/gal **29.** 9.682 in. **31. a.** 37.76 m **b.** 9.44 m **33.** 80 threads **35.** 2.95 in. **37.** 3000 sheets
39. $47,502 **41.** 51.20 in. **43.** 337.50 in^3 **45.** 0.5 L **47. a.** 0.186 in. **b.** 0.127 in. **49.** 240 gal **51.** 62.5¢ **53.** 94.2 Ω
55. 0.288 W **57.** 6.20 A **59.** 136.9 Ω **61.** 0.3 mg **63.** 5 **65.** 303.8 nautical mi **67.** 1290 lb **69.** 695,000 tons **71.** 42.7 ft^3
73. 3.75 bags

Exercises 1.13 (pages 75–76)
1. 0.27 **3.** 0.06 **5.** 1.56 **7.** 0.292 **9.** 0.087 **11.** 9.478 **13.** 0.0028 **15.** 0.00068 **17.** 0.0425 **19.** 0.00375 **21.** 54%
23. 8% **25.** 62% **27.** 217% **29.** 435% **31.** 18.5% **33.** 29.7% **35.** 519% **37.** 1.87% **39.** 0.29% **41.** 80% **43.** 12.5%
45. $16\frac{2}{3}$% **47.** $44\frac{4}{9}$% **49.** 60% **51.** 32.5% **53.** 43.75% **55.** 240% **57.** 175% **59.** $241\frac{2}{3}$% **61.** $\frac{3}{4}$ **63.** $\frac{4}{25}$ **65.** $\frac{3}{5}$
67. $\frac{93}{100}$ **69.** $2\frac{3}{4}$ **71.** $1\frac{1}{4}$ **73.** $\frac{43}{400}$ **75.** $\frac{107}{1000}$ **77.** $\frac{69}{400}$ **79.** $\frac{97}{600}$ **81.**

Fraction	Decimal	Percent
$\frac{3}{8}$	0.375	37.5%
$\frac{9}{20}$	0.45	45%
$\frac{9}{50}$	0.18	18%
$1\frac{2}{5}$	1.4	140%
$1\frac{2}{25}$	1.08	108%
$\frac{67}{400}$	0.1675	$16\frac{3}{4}$%

Exercises 1.14 (pages 80–83)
1. $P = 60$; $R = 25\%$; $B = 240$ **3.** $P = 108$; $R = 40\%$; $B = 270$ **5.** P = unknown; $R = 4\%$; $B = 28,000$ **7.** $P = 21$; $R = 60\%$; B = unknown **9.** $P = 2050$; $R = 6\%$; B = unknown **11.** $35,100 **13. a.** $137.96 **b.** $146.58 **15.** 50% **17.** $1600 **19.** 112.8
21. 8.96 V **23.** 74.6% **25.** 36.9% **27.** 6.7% **29.** 20% **31.** 766.7 ft **33.** 440 lb total; 352 lb active ingredients, 88 lb inert ingredients **35.** 850 gal; 35.7 gal butterfat **37.** 74% **39.** 7.5 mL **41.** 2.5% **43.** 53.3% **45.** 14.2% **47.** $45; $51 **49.** 1558 lb
51. 74.8% **53. a.** 35 deer/mi^2 **b.** 49 deer/mi^2 **55.** $16.65 **57.** $2005.97 **59.** $3523.90

Exercises 1.15 (page 86)
1. 225 **3.** 222 **5.** 0.00000661 **7.** 729 **9.** 562 **11.** 0.00483 **13.** 157 **15.** 2.96 **17.** 68.9 **19.** 42.4 **21.** 0.198

Exercises 1.16 (pages 91–92)
1. a. $300 **b.** $63.89 **3.** $9706.67 **5.** $23,152.64 **7.** $948.10 **9.** $78,276.71 **11. a.** $674.40; $24,278.40 **b.** $710.27; $25,569.72 Thus, choice **a** pays a total of $1291.32 less **13.** $59,931.28 **15.** $34,728.75 **17.** $34,854.67 **19.** $10,003.92
21. $1066.07 **23.** 56.4% **25.** 46.8% **27.** 36.9%

Unit 1C Review (pages 92–93)
1. 1.625 **2.** $\frac{9}{20}$ **3.** 10.129 **4.** 116.935 **5.** 5.854 **6.** 25.6 ft **7.** 160.2 ft **8. a.** 45.1 **b.** 45.06 **9. a.** 45.1 **b.** 45.06
10. 0.11515 **11.** 18.85 **12.** 6 cables; 2 in. left **13.** 0.25 **14.** 72.4% **15.** 69.3 **16.** 2000 **17.** 40% **18.** $17.49 **19.** 2110
20. 9.40

Chapter 1 Review (pages 96–98)
1. 8243 **2.** 55,197 **3.** 9,178,000 **4.** 226 r 240 **5.** 3 **6.** 43 **7.** 37 **8.** 11 **9.** 31.8 tsf **10.** 30 cm^3 **11.** 10 **12.** 3000
13. no **14.** $2 \cdot 3 \cdot 3 \cdot 3$ **15.** $2 \cdot 3 \cdot 5 \cdot 11$ **16.** $\frac{9}{14}$ **17.** $\frac{5}{6}$ **18.** $4\frac{1}{6}$ **19.** $6\frac{3}{5}$ **20.** $\frac{21}{8}$ **21.** $\frac{55}{16}$ **22.** 2 **23.** $1\frac{1}{2}$

24. $\frac{103}{180}$ **25.** $14\frac{29}{48}$ **26.** $1\frac{19}{24}$ **27.** $11\frac{3}{5}$ **28.** 5 **29.** $\frac{1}{4}$ **30.** 18 **31.** $\frac{1}{16}$ **32.** $\frac{3}{8}$ **33.** $13\frac{11}{25}$ **34.** $A: 3\frac{3}{8}$ in. $B: 5\frac{7}{16}$ in.
35. 105 **36.** 2016 **37.** 24 **38.** 63,360 yd **39.** 0.5625 **40.** $0.41\overline{6}$ **41.** $\frac{9}{20}$ **42.** $19\frac{5}{8}$ **43.** 168.278 **44.** 17.25 **45.** 68.665
46. 33.72 **47.** 3206.5 **48.** 1.9133 **49.** 3.18 **50.** 20.6 **51. a.** 200 **b.** 248.2 **c.** 250 **52. a.** 5.6 **b.** 5.65 **c.** 5.6491
53. 0.15 **54.** 0.0825 **55.** 6.5% **56.** 120% **57.** $1050

59. 38.1% **60.** 42.3% **61.** 48 tons **62.** 3 gal; $54.92
63. $\frac{5}{16}$ in. **64.** 30 in. × 41 in. **65.** 4020 **66.** 139
67. $15,615.04 **68.** $402.91

58.

Fraction	Decimal	Percent
$\frac{1}{4}$	0.25	25%
$\frac{3}{8}$	0.375	$37\frac{1}{2}$%
$\frac{5}{6}$	$0.83\frac{1}{3}$	$83\frac{1}{3}$%
$8\frac{3}{4}$	8.75	875%
$2\frac{2}{5}$	2.4	240%
$\frac{3}{2000}$	0.0015	0.15%

CHAPTER 2

Exercises 2.1 (page 104)
1. 3 **3.** 6 **5.** 4 **7.** 17 **9.** 15 **11.** 10 **13.** 7 **15.** −2 **17.** −12 **19.** 6 **21.** −9 **23.** 4 **25.** −3 **27.** −1 **29.** −4
31. −6 **33.** −2 **35.** 4 **37.** 9 **39.** 3 **41.** 4 **43.** 7 **45.** −7 **47.** 5 **49.** −20 **51.** −2 **53.** 0 **55.** −5 **57.** −19
59. −12 **61.** 7 **63.** 11 **65.** 4 **67.** 4 **69.** −14

Exercises 2.2 (pages 106–107)
1. −2 **3.** 11 **5.** 12 **7.** 6 **9.** 18 **11.** −6 **13.** −4 **15.** 0 **17.** 1 **19.** −10 **21.** −7 **23.** 15 **25.** −16 **27.** −10
29. −2 **31.** 14 **33.** 23 **35.** −15 **37.** 8 **39.** −4 **41.** 8 **43.** 2 **45.** −15 **47.** −3 **49.** −23 **51.** −4 **53.** −1 **55.** 2
57. −8 **59.** −10

Exercises 2.3 (page 109)
1. 24 **3.** −18 **5.** −35 **7.** 27 **9.** −72 **11.** 27 **13.** 0 **15.** 49 **17.** −300 **19.** −13 **21.** −6 **23.** 48 **25.** 21 **27.** −16
29. 54 **31.** −6 **33.** 24 **35.** 27 **37.** −63 **39.** −9 **41.** −6 **43.** 36 **45.** 30 **47.** −168 **49.** −162 **51.** 5 **53.** −9
55. 8 **57.** 2 **59.** 9 **61.** −4 **63.** 3 **65.** 17 **67.** 40 **69.** 15 **71.** 7 **73.** 4 **75.** −3 **77.** 10 **79.** 8

Exercises 2.4 (pages 113–114)
1. $-\frac{3}{16}$ **3.** $\frac{1}{16}$ **5.** $-12\frac{3}{20}$ **7.** $\frac{13}{18}$ **9.** $-\frac{1}{20}$ **11.** $1\frac{1}{16}$ **13.** $5\frac{3}{4}$ **15.** $-1\frac{3}{4}$ **17.** $-\frac{1}{63}$ **19.** 6 **21.** $-\frac{9}{10}$ **23.** $\frac{7}{24}$ **25.** $-\frac{9}{20}$
27. $\frac{3}{8}$ **29.** $1\frac{1}{4}$ **31.** $3\frac{1}{4}$ **33.** $\frac{1}{20}$ **35.** $-3\frac{1}{3}$ **37.** −48 **39.** −1 **41.** $-1\frac{1}{8}$ **43.** $\frac{1}{4}$ **45.** $-\frac{1}{4}$

Exercises 2.5 (page 117)
1. 10^{13} **3.** $\frac{1}{10^4}$ **5.** 10^3 **7.** 10^{12} **9.** $\frac{1}{10^{10}}$ **11.** 10^3 **13.** $\frac{1}{10^{12}}$ **15.** $\frac{1}{10^4}$ **17.** 10^2 **19.** $\frac{1}{10^6}$ **21.** 10^5 **23.** $\frac{1}{10^4}$
25. 10^{17} **27.** 10^6 **29.** 10^{14}

Exercises 2.6 (page 122)
1. 3.56×10^2 **3.** 6.348×10^2 **5.** 8.25×10^{-3} **7.** 7.4×10^0 **9.** 7.2×10^{-5} **11.** 7.1×10^5 **13.** 4.5×10^{-6}
15. 3.4×10^{-8} **17.** 6.4×10^5 **19.** 75,500 **21.** 5310 **23.** 0.078 **25.** 0.000555 **27.** 64 **29.** 960 **31.** 5.76 **33.** 0.0000064
35. 50,000,000,000 **37.** 0.00000062 **39.** 2,500,000,000,000 **41.** 0.000000000033 **43.** 0.0048 **45.** 0.00091 **47.** 0.00037

49. 0.0613 **51.** 0.0009 **53.** 1.0009 **55.** 0.00000000998 **57.** 0.000271 **59.** 2.4×10^{-15} **61.** 3×10^{24} **63.** 1×10^{-6}
65. 1.728×10^{18} **67.** 1.13×10^{-1} **69.** 1.11×10^{-25} **71.** 9×10^{-1} **73.** 6.67×10^{1} **75.** 7.46×10^{5} **77.** 1.17×10^{10}
79. 9.06×10^{-11} **81.** 2.66×10^{241}

Exercises 2.7 (page 125)
1. 28×10^{3} **3.** 3.45×10^{6} **5.** 220×10^{9} **7.** 6.6×10^{-3} **9.** 76.5×10^{-9} **11.** 975×10^{-3} **13.** 57,700 **15.** 4,940,000,000,000
17. 567,000,000 **19.** 0.000026 **21.** 0.000000005945 **23.** 0.00000000001064 **25.** 14.9×10^{18} **27.** 19.7×10^{-6}
29. 588×10^{12} **31.** 15.6×10^{-18} **33.** 339×10^{6} **35.** 123×10^{21} **37.** 8.97×10^{6} **39.** 1.31×10^{12}

Chapter 2 Review (pages 127–128)
1. 5 **2.** 16 **3.** 13 **4.** 3 **5.** -8 **6.** -3 **7.** -13 **8.** -3 **9.** -11 **10.** 19 **11.** -2 **12.** 0 **13.** -19 **14.** -24 **15.** 36
16. 72 **17.** -84 **18.** 6 **19.** -6 **20.** 5 **21.** $-\frac{1}{42}$ **22.** $\frac{1}{12}$ **23.** $-3\frac{1}{8}$ **24.** $-\frac{3}{2}$ or $-1\frac{1}{2}$ **25.** $\frac{1}{10^{5}}$ **26.** 10^{9} **27.** $\frac{1}{10^{12}}$ **28.** 1
29. 4.76×10^{5} **30.** 1.4×10^{-3} **31.** 0.0000535 **32.** 61,000,000 **33.** 0.00105 **34.** 0.06 **35.** 0.000075 **36.** 0.00183
37. 4.37×10^{-2} **38.** 2.8×10^{14} **39.** 2×10^{20} **40.** 2.025×10^{-15} **41.** 1.6×10^{37} **42.** 6.4×10^{7} **43.** 275×10^{3}
44. 32×10^{6} **45.** 450×10^{-6} **46.** 31,600,000 **47.** 0.746 **48.** 4.73×10^{6} **49.** 24.3×10^{-15} **50.** 46.1×10^{3}

Cumulative Review Chapters 1–2 (page 129)
1. 72 **2.** 51 cm² **3.** 3750 Ω **4.** No **5.** $2 \cdot 3 \cdot 3 \cdot 5 \cdot 7$ **6.** $3\frac{5}{9}$ **7.** 4480 ft² **8.** $\frac{13}{16}$ **9.** $1\frac{1}{16}$ **10.** $1\frac{7}{8}$ **11.** $\frac{1}{20}$ **12.** 83
13. a. 600 **b.** 615.3 **c.** 620 **d.** 615.288 **14.** 0.074 **15.** $3990 **16.** 662.5 **17.** 9.43% **18.** 6.25% **19.** 10 **20.** -432
21. $-2\frac{15}{16}$ **22.** $-\frac{25}{64}$ **23.** 3.1818×10^{5} **24.** 0.00213 **25.** $\frac{1}{10^{4}}$ **26.** 4.5×10^{3} **27.** 270×10^{-6} **28.** 0.000000281
29. 16,300,000 **30.** 7.02×10^{10} **31.** 4.75×10^{-3} **32.** 3.46×10^{-15} **33.** 2.07×10^{-3}

CHAPTER 3

Exercises 3.1 (page 133)
1. kilo **3.** centi **5.** milli **7.** mega **9.** h **11.** d **13.** c **15.** μ **17.** 65 mg **19.** 82 cm **21.** 36 μA **23.** 19 hL
25. 18 metres **27.** 36 kilograms **29.** 24 picoseconds **31.** 135 millilitres **33.** 45 milliamperes **35.** metre **37.** ampere
39. litre and cubic metre

Exercises 3.2 (pages 135–136)
1. 1 metre **3.** 1 kilometre **5.** 1 centimetre **7.** 1000 **9.** 0.01 **11.** 0.001 **13.** 0.001 **15.** 10 **17.** 100 **19.** cm
21. mm **23.** m **25.** km **27.** mm **29.** mm **31.** km **33.** m **35.** cm **37.** mm; mm **39.** *A*: 52 mm; 5.2 cm *B*: 11 mm; 1.1 cm *C*: 137 mm; 13.7 cm *D*: 95 mm; 9.5 cm *E*: 38 mm; 3.8 cm *F*: 113 mm; 11.3 cm **41.** 52 mm; 5.2 cm **43.** 79 mm; 7.9 cm
45. 65 mm; 6.5 cm **47.** 102 mm; 10.2 cm **49.** 0.675 km **51.** 1.54 m **53.** 0.65 m **55.** 730 cm **57.** 1.25 km **59.** 27.5 cm
61. 12.5 cm **63.** Answers vary.

Exercises 3.3 (pages 137–138)
1. 1 gram **3.** 1 kilogram **5.** 1 milligram **7.** 1000 **9.** 0.01 **11.** 1000 **13.** 1000 **15.** g **17.** kg **19.** g **21.** metric ton
23. mg **25.** g **27.** mg **29.** g **31.** kg **33.** metric ton **35.** 0.875 kg **37.** 85,000 mg **39.** 3600 g **41.** 0.27 g
43. 0.885 mg **45.** 0.375 mg **47.** 2500 kg **49.** 225 metric tons **51.** Answers vary.

Exercises 3.4 (page 141)
1. 1 litre **3.** 1 cubic centimetre **5.** 1 square kilometre **7.** 1000 **9.** 1,000,000 **11.** 100 **13.** 1000 **15.** L **17.** m² **19.** cm²
21. m³ **23.** mL **25.** ha **27.** m³ **29.** L **31.** m² **33.** ha **35.** cm² **37.** cm² **39.** 1.5 L **41.** 1,500,000 cm³ **43.** 85 mL
45. 0.085 km² **47.** 8.5 ha **49.** 500 g **51.** 0.675 ha

Exercises 3.5 (page 144)
1. 1 amp **3.** 1 second **5.** 1 megavolt **7.** 43 kW **9.** 17 ps **11.** 3.2 MW **13.** 450 Ω **15.** 1000 **17.** 0.000000001
19. 1,000,000 **21.** 0.000001 **23.** 350 mA **25.** 0.35 s **27.** 3 h 52 min 30 s **29.** 0.175 mF **31.** 1.5 MHz

Exercises 3.6 (page 146)
1. b **3.** c **5.** b **7.** c **9.** d **11.** 25 **13.** 617 **15.** 3.2 **17.** -26.7 **19.** -108.4

Exercises 3.7 (pages 149–150)

1. 3.63 **3.** 15.0 **5.** 366 **7.** 66.0 **9.** 81.3 **11.** 46.6 mi **13.** 10.8 mm **15.** 30.3 L **17. a.** 38 oz **b.** 1.08 kg **19. a.** 300 ft **b.** 91.4 m **21. a.** 12.7 cm **b.** 127 mm **23.** 2.51 m^2 **25.** 1260 ft^2 **27.** 116 cm^2 **29.** 1170 ft^2 **31.** 11.5 m^3 **33.** 279,000 mm^3 **35.** 2,380,000 cm^3 **37.** \$9/ft^2; \$810/frontage ft **39.** 34.4 acres **41.** 13.9 ha **43.** 0.619 acre **45.** 80 acres **47.** 9397 lb/acre; 168 bu/acre **49.** 0.606 acre **51.** 88 **53.** 363 lb/in^2 **55.** 29.1 m/s

Chapter 3 Review (page 151)

1. milli **2.** kilo **3.** M **4.** μ **5.** 42 mL **6.** 8.3 ns **7.** 18 kilometres **8.** 350 milliamperes **9.** 50 microseconds **10.** 1 L **11.** 1 MW **12.** 1 km^2 **13.** 1 m^3 **14.** 0.65 **15.** 0.75 **16.** 6100 **17.** 4.2×10^6 **18.** 1.8×10^7 **19.** 25,000 **20.** 25,000 **21.** 2.5 **22.** 6×10^5 **23.** 250 **24.** 22.2 **25.** -13 **26.** 0 **27.** 100 **28.** 81.7 **29.** 38.4 **30.** 142 **31.** 1770 **32.** 162 **33.** 176 **34.** 6.07 **35.** c **36.** a **37.** d **38.** d **39.** b **40.** b **41.** a **42.**

Prefix	Symbol	Power of 10	Sample unit	How many?	How many?
tera	T	10^{12}	m	10^{12} m = 1 Tm	1 m = 10^{-12} Tm
giga	G	10^9	W	10^9 W = 1 GW	1 W = 10^{-9} GW
mega	M	10^6	Hz	10^6 Hz = 1 MHz	1 Hz = 10^{-6} MHz
kilo	k	10^3	g	10^3 g = 1 kg	1 g = 10^{-3} kg
hecto	h	10^2	Ω	10^2 Ω = 1 hΩ	1 Ω = 10^{-2} hΩ
deka	da	10^1	L	10^1 L = 1 daL	1 L = 10^{-1} daL
deci	d	10^{-1}	g	10^{-1} g = 1 dg	1 g = 10^1 dg
centi	c	10^{-2}	m	10^{-2} m = 1 cm	1 m = 10^2 cm
milli	m	10^{-3}	A	10^{-3} A = 1 mA	1 A = 10^3 mA
micro	μ	10^{-6}	W	10^{-6} W = 1 μW	1 W = 10^6 μW
nano	n	10^{-9}	s	10^{-9} s = 1 ns	1 s = 10^9 ns
pico	p	10^{-12}	s	10^{-12} s = 1 ps	1 s = 10^{12} ps

CHAPTER 4

Exercises 4.1 (page 157)

1. 3 **3.** 4 **5.** 4 **7.** 4 **9.** 3 **11.** 4 **13.** 3 **15.** 3 **17.** 3 **19.** 4 **21.** 4 **23.** 5 **25.** 2 **27.** 6 **29.** 4 **31.** 6 **33.** 4 **35.** 2

Exercises 4.2 (page 160)

1. a. 0.01 A **b.** 0.005 A **3. a.** 0.01 cm **b.** 0.005 cm **5. a.** 1 km **b.** 0.5 km **7. a.** 0.01 mi **b.** 0.005 mi **9. a.** 0.001 A **b.** 0.0005 A **11. a.** 0.0001 W **b.** 0.00005 W **13. a.** 10 Ω **b.** 5 Ω **15. a.** 1000 L **b.** 500 L **17. a.** 0.1 cm **b.** 0.05 cm **19. a.** 10 V **b.** 5 V **21. a.** 0.001 m **b.** 0.0005 m **23. a.** $\frac{1}{3}$ yd **b.** $\frac{1}{6}$ yd **25. a.** $\frac{1}{32}$ in. **b.** $\frac{1}{64}$ in. **27. a.** $\frac{1}{16}$ mi **b.** $\frac{1}{32}$ mi **29. a.** $\frac{1}{9}$ in^2 **b.** $\frac{1}{18}$ in^2

Exercises 4.3A (pages 163–164)

1. 27.20 mm **3.** 63.55 mm **5.** 8.00 mm **7.** 115.90 mm **9.** 71.45 mm **11.** 10.25 mm **13.** 34.60 mm **15.** 68.45 mm **17.** 5.90 mm **19.** 43.55 mm **21.** 76.10 mm **23.** 12.30 mm

Exercises 4.3B (pages 166–167)

1. 1.362 in. **3.** 2.695 in. **5.** 0.234 in. **7.** 1.715 in. **9.** 2.997 in. **11.** 0.483 in. **13.** 1.071 in. **15.** 2.502 in. **17.** 0.316 in. **19.** 4.563 in. **21.** 2.813 in. **23.** 0.402 in.

Exercises 4.4A (pages 169–170)

1. 4.25 mm **3.** 3.90 mm **5.** 1.75 mm **7.** 7.77 mm **9.** 5.81 mm **11.** 10.28 mm **13.** 7.17 mm **15.** 8.75 mm **17.** 6.23 mm **19.** 5.42 mm

Exercises 4.4B (pages 171–172)

1. 0.238 in. **3.** 0.314 in. **5.** 0.147 in. **7.** 0.820 in. **9.** 0.502 in. **11.** 0.200 in. **13.** 0.321 in. **15.** 0.170 in. **17.** 0.658 in. **19.** 0.245 in.

Exercises 4.5 (pages 177–178)

1. a. 14.7 in. **b.** 0.017 in. **3. a.** 16.01 mm **b.** 0.737 mm **5. a.** 0.0350 A **b.** 0.00050 A **7. a.** All have the same accuracy. **b.** 0.391 cm **9. a.** 205,000 Ω **b.** 205,000 Ω and 45,000 Ω **11. a.** 0.04 in. **b.** 15.5 in. **13. a.** 0.48 cm **b.** 43.4 cm **15. a.** 0.00008 A **b.** 0.91 A **17. a.** 0.6 m **b.** All have the same precision. **19. a.** 500,000 Ω **b.** 500,000 Ω **21.** 18.1 m **23.** 94.8 cm **25.** 97,000 W **27.** 840,000 V **29.** 19 V **31.** 459 mm or 45.9 cm **33.** 126.4 cm **35.** 8600 mi **37.** 35 mm or 3.5 cm **39.** 65.4 g **41.** 0.330 in. **43.** 26.4 mm **45.** 12.7 ft **47.** 67 lb **49.** 1.3 gal **51.** 124 gal **53.** 10.66666 **55.** 9 lb

Exercises 4.6 (pages 180–181)

1. 4400 m² **3.** 1,230,000 cm² **5.** 901 m² **7.** 0.13 ΩA **9.** 7360 cm³ **11.** 4.7×10^9 m³ **13.** 35 A²Ω **15.** 2500 in² **17.** 40 m **19.** 340 V/A **21.** 2.1 km/s **23.** $3\overline{0}0$ V²/Ω **25.** 4.0 g/cm³ **27.** $9\overline{0}0$ ft³ **29.** 28 m **31.** 28.8 hp **33.** $1\overline{0}00$ m³ **35.** 270 ft³/min **37.** 4700 in³ **39.** 9.6 gal/h **41.** 1560 in² **43.** 19 ft³ **45.** 9 **47.** 581 ppm **49.** 2.3 mi³

Exercises 4.7 (pages 184–185)

1. 100 lb; 50 lb; 0.0357; 3.57% **3.** 1 rpm; 0.5 rpm; 0.000571; 0.06% **5.** 0.001 g; 0.0005 g; 0.00588; 0.59% **7.** 1 g; 0.5 g; 0.25; 25% **9.** 0.01 g; 0.005 g; 0.00225; 0.23% **11.** 0.01 kg; 0.005 kg; 0.005; 0.5% **13.** 0.001 A; 0.0005 A; 0.0122; 1.22% **15.** $\frac{1}{8}$ in.; $\frac{1}{16}$ in.; 0.00526; 0.53% **17.** 1 in.; 0.5 in.; 0.00329; 0.33% **19.** 13.5 cm **21.** 19.7 g **23.** Answer in text.

	Upper limit	Lower limit	Tolerance interval
25.	$6\frac{21}{32}$ in.	$6\frac{19}{32}$ in.	$\frac{1}{16}$ in.
27.	$3\frac{29}{64}$ in.	$3\frac{27}{64}$ in.	$\frac{1}{32}$ in.
29.	$3\frac{25}{128}$ in.	$3\frac{23}{128}$ in.	$\frac{1}{64}$ in.

	Upper limit	Lower limit	Tolerance interval
31.	1.24 cm	1.14 cm	0.10 cm
33.	0.0185 A	0.0175 A	0.0010 A
35.	26,000 V	22,000 V	4000 V
37.	10.36 km	10.26 km	0.10 km

39. $53,075 **41.** $1,567,020.60

Exercises 4.8 (pages 188–189)

1. 360 Ω; ±10% **3.** 830,000 Ω; ±20% **5.** 1,400,000 Ω, ±20% **7.** 70 Ω; ±5% **9.** 500,000 Ω; ±20% **11.** 10,000,000 Ω, ±20% **13.** yellow, gray, red **15.** violet, red, orange **17.** blue, green, yellow **19.** red, green, silver **21.** yellow, green, green **23.** violet, blue, gold **25. a.** 36 Ω **b.** 396 Ω **c.** 324 Ω **d.** 72 Ω **27. a.** 166,000 Ω **b.** 996,000 Ω **c.** 664,000 Ω **d.** 332,000 Ω **29. a.** 3.5 Ω **b.** 73.5 Ω **c.** 66.5 Ω **d.** 7 Ω

Exercises 4.9 (pages 192–194)

1. 6 V **3.** 6.4 V **5.** 0.4 V **7.** 1.4 V **9.** 40 V **11.** 230 V **13.** 7 Ω **15.** 12 Ω **17.** 35 Ω **19.** 85 Ω **21.** 300 Ω

Chapter 4 Review (pages 196–197)

1. 3 **2.** 2 **3.** 3 **4.** 3 **5.** 2 **6.** 3 **7.** 4 **8.** 5 **9. a.** 0.01 m **b.** 0.005 m **10. a.** 0.1 mi **b.** 0.05 mi **11. a.** 100 L **b.** 50 L **12. a.** 100 V **b.** 50 V **13. a.** 0.01 cm **b.** 0.005 cm **14. a.** 10,000 V **b.** 5000 V **15. a.** $\frac{1}{8}$ in. **b.** $\frac{1}{16}$ in. **16. a.** $\frac{1}{16}$ mi **b.** $\frac{1}{32}$ mi **17.** 42.35 mm **18.** 1.673 in. **19.** 11.84 mm **20.** 0.438 in. **21. a.** 36,500 V **b.** 9.6 V **22. a.** 0.0005 A **b.** 0.425 A **23.** 720,000 W **24.** $4\overline{0}0$ m **25.** 400,000 V **26.** 1900 cm³ **27.** 5.88 m² **28.** 1.4 N/m² **29.** 130 V²/Ω **30.** 0.0057; 0.57% **31.** 0.00032; 0.03% **32.** 2200 Ω; 1800 Ω **33.** 120,000 Ω; ±20% **34.** 0.85 Ω; ±5% **35.** 8.4 V **36.** 2.6 Ω

Cumulative Review Chapters 1–4 (page 198)

1. 27 **2. a.** 32,520 **b.** 32,518.61 **3.** 60 **4.** 102 **5.** $\frac{1}{10^5}$ **6.** 8.70×10^5 **7.** m **8.** 25 kg **9.** 250 microseconds **10.** 1 mega amp **11.** 120,000 m **12.** 2.5 m **13.** 0.05 kg **14.** 4.06 metric tons **15.** 186.8°F **16.** 10°C **17.** 1050 cm² **18.** 0.12 km **19.** 10,000 mL **20.** 2 **21.** 4 **22. a.** 0.01 mm **b.** 0.005 mm **23. a.** 0.1 lb **b.** 0.05 lb **24.** 77.75 mm

25. 3.061 in. **26.** 7.53 mm **27.** 0.537 in. **28.** 34,900 km **29.** 46.0 L **30.** 42,000 cm^2 **31.** 33.7 ft **32. a.** 0.001 cm **b.** 0.0005 cm **c.** 0.000234 **d.** 0.02% **33.** 165 V **34.** 110 Ω

CHAPTER 5

Exercises 5.1 (page 202)

1. 83 **3.** 8 **5.** 42 **7.** −6 **9.** −128 **11.** 1 **13.** −5 **15.** $\frac{3}{2}$ **17.** −24 **19.** $-9\frac{3}{5}$ **21.** −13 **23.** $-1\frac{3}{5}$ **25.** $\frac{1}{3}$ **27.** 331,776 **29.** −72 **31.** 50 **33.** 8 **35.** −78 **37.** −1 **39.** 25 **41.** 2

Exercises 5.2 (pages 205–206)

1. $a+b+c$ **3.** $a+b+c$ **5.** $a-b-c$ **7.** $x+y-z+3$ **9.** $x-y-z-3$ **11.** $2x+4+3y+4r$ **13.** $3x-5y+6z-2w+11$ **15.** $-5x-3y-6z+3w+3$ **17.** $2x+3y-z-w+3r-2s-10$ **19.** $-2x+3y-z-4w-4r+s$ **21.** $2b$ **23.** $3x^2+10x$ **25.** $3m$ **27.** $a+12b$ **29.** $6a^2-a+1$ **31.** $3x^2+3x$ **33.** $-1.8x$ **35.** $\frac{8}{9}x-\frac{1}{8}y$ **37.** $2x^2y-2xy+2y^2-3x^2$ **39.** $5x^3+3x^2y-5y^3+y$ **41.** 1 **43.** $3x+4$ **45.** $5-x$ **47.** $3y-7$ **49.** $4y+5$ **51.** 5 **53.** 6 **55.** 28 **57.** $\frac{5}{4}x-\frac{8}{3}$ **59.** $12x+36y$ **61.** $-36x^2+48y^2$ **63.** $2x+19$ **65.** $-8.5y-4$ **67.** −42 **69.** $4n+8$ **71.** $1.8x-7$ **73.** $-6n-2$ **75.** $-5x+6$ **77.** $-1.05x-8.4$

Exercises 5.3 (pages 208–209)

1. binomial **3.** binomial **5.** monomial **7.** trinomial **9.** binomial **11.** x^2-x+1; 2 **13.** $7x^2+4x-1$; 2 **15.** $5x^3-4x^2-2$; 3 **17.** $4y^3-6y^2-3y+7$; 3 **19.** $-7x^5-4x^4+x^3+2x^2+5x-3$; 5 **21.** $7a^2-10a+1$ **23.** $9x^2-5x$ **25.** $9a^3+4a^2+5a-5$ **27.** $4x+4$ **29.** $5x^2+18x-22$ **31.** $5x^3+13x^2-8x+7$ **33.** $8y^2-5y+6$ **35.** $4a^3-3a^2-5a$ **37.** $2x^2+6x+2$ **39.** $-3x^2+2x+2$ **41.** $a+3b$ **43.** $7a-4b-3x+4y$ **45.** x^2-3x+5 **47.** $7w^2-24w-6$ **49.** $x^2-4x-10$ **51.** $4x^2-4x+12$ **53.** $-6x^2+2x-10$ **55.** x^3+3x-3 **57.** $8x^5-18x^4+5x^2+1$

Exercises 5.4 (page 211)

1. $-15a$ **3.** $28a^3$ **5.** $54m^4$ **7.** $32a^8$ **9.** $-26p^2q$ **11.** $30n^3m$ **13.** $21a^4b$ **15.** $\frac{3}{8}x^3y^4$ **17.** $24a^3b^4c^3$ **19.** $\frac{3}{16}x^3y^5z^3$ **21.** $-371.64m^3n^2p^2$ **23.** $-255a^6b$ **25.** x^6 **27.** x^{24} **29.** $9x^8$ **31.** $-x^9$ **33.** x^{10} **35.** x^{30} **37.** $25x^6y^4$ **39.** $225m^4$ **41.** $15,625n^{12}$ **43.** $9x^{12}$ **45.** $8x^9y^{12}z^3$ **47.** $-32h^{15}k^{30}m^{10}$ **49.** −408 **51.** −144 **53.** 17,712 **55.** 16 **57.** −1728 **59.** 291,600 **61.** −1080 **63.** 324 **65.** −324 **67.** 16

Exercises 5.5 (page 213)

1. $4a+24$ **3.** $-18x^2-12y$ **5.** $4ax^2-6ay+a$ **7.** $3x^3-2x^2+5x$ **9.** $6a^3+12a^2-20a$ **11.** $-12x^3+21x^2+6x$ **13.** $-28x^3-12xy+8x^2y$ **15.** $3x^3y^2-3x^2y^3+12x^2y^2$ **17.** $-6x^3+36x^5-54x^7$ **19.** $5a^4b^2-5ab^5-5a^2b^3$ **21.** $\frac{28}{3}mn-8m^2$ **23.** $16y^2z^3-\frac{8}{35}yz^4$ **25.** $-5.2a^6-10a^3-4a$ **27.** $1334.4a^3+1668a^2$ **29.** $24x^4y-16x^3y^2+20x^2y^3$ **31.** $\frac{1}{2}a^3b^3-\frac{1}{3}a^2b^5+\frac{5}{9}ab^6$ **33.** $-19x^2+8x$ **35.** $3x^2y-2x^2y^2+2x^2y^3-7xy^3$ **37.** x^2+7x+6 **39.** $x^2+5x-14$ **41.** $x^2-13x+40$ **43.** $3a^2-17a+20$ **45.** $12a^2-10a-12$ **47.** $24a^2+84a+72$ **49.** $15x^2-4xy-4y^2$ **51.** $4x^2-12x+9$ **53.** $4c^2-25d^2$ **55.** $91m^2+32m-3$ **57.** $x^8-2x^5+x^2$ **59.** $10y^3-24y^2-32y+16$ **61.** $24x^2-78x-6y^2-39y$ **63.** $g^2-3g-h^2+9h-18$ **65.** $10x^7-3x^6-x^5+44x^4+x^3-x^2+16x-2$

Exercises 5.6 (page 215)

1. $3x^2$ **3.** $\frac{3x^8}{2}$ **5.** $\frac{6}{x^2}$ **7.** $\frac{2x}{3}$ **9.** x **11.** $\frac{5}{2}$ **13.** $\frac{5a^2}{b}$ **15.** $\frac{8}{mn}$ **17.** 0 **19.** $23p^2$ **21.** −2 **23.** $\frac{36}{r^2}$ **25.** $-\frac{23x^2}{7y^2}$ **27.** $\frac{8}{7a^3b^2}$ **29.** $\frac{4x^2z^4}{9y}$ **31.** $2x^2-4x+3$ **33.** x^2+x+1 **35.** $x-y-z$ **37.** $3a^4-2a^2-a$ **39.** $b^9-b^6-b^3$ **41.** $-x^3+x^2+x+4$ **43.** $12x+6x^2y-3$ **45.** $8x^3z-6x^2yz^2-4y^2$ **47.** $4y^3-3y-\frac{2}{y}$ **49.** $\frac{1}{2x^2}-3-2x^2$

Exercises 5.7 (pages 217–218)

1. $x + 2$ **3.** $3a + 3$ r 11 **5.** $4x - 3$ r -3 **7.** $y + 2$ r -3 **9.** $3b - 4$ **11.** $6x^2 + x - 1$ **13.** $4x^2 + 7x - 15$ r 5 **15.** $2x^2 - 2x - 12$ **17.** $2x^2 - 16x + 32$ **19.** $3x^2 + 10x + 20$ r 34 **21.** $2x^2 + 6x + 30$ r 170 **23.** $2x^3 + 4x + 6$ **25.** $4x^2 - 2x + 1$ r -2 **27.** $x^3 - 2x^2 + 4x - 8$ **29.** $3x^2 - 4x + 1$

Chapter 5 Review (page 219)

1. a **2.** 0 **3.** 1 **4.** -2 **5.** 50 **6.** 2 **7.** 1 **8.** 9 **9.** -30 **10.** $-\dfrac{9}{2}$ **11.** -12 **12.** 5 **13.** $6y - 5$ **14.** $6 - 8x$ **15.** $10x + 27$ **16.** $3x^3 + x^2y - 3y^3 - y$ **17.** binomial **18.** 4 **19.** $8a^2 + 5a + 2$ **20.** $9x^3 + 4x^2 + x + 2$ **21.** $10x^2 - 7x + 4$ **22.** $24x^5$ **23.** $-56x^5y^3$ **24.** $27x^6$ **25.** $15a^2 + 20ab$ **26.** $-32x^2 + 8x^3 - 12x^4$ **27.** $15x^2 - 11x - 12$ **28.** $6x^3 - 24x^2 + 26x - 4$ **29.** $\dfrac{7}{x}$ **30.** $5x^2$ **31.** $4a^2 - 3a + 1$ **32.** $3x - 4$ **33.** $3x^2 - 4x + 2$

CHAPTER 6

Exercises 6.1 (page 226)

1. 6 **3.** 17 **5.** $10\dfrac{1}{2}$ **7.** 19.5 **9.** 5.2 **11.** 301 **13.** 7 **15.** 20 **17.** $4\dfrac{1}{14}$ **19.** 0 **21.** 0 **23.** 16 **25.** 392 **27.** -4 **29.** -2 **31.** 2 **33.** -2 **35.** 32 **37.** -4 **39.** 18 **41.** $5\dfrac{2}{3}$ **43.** 42 **45.** $10\dfrac{1}{2}$ **47.** $-\dfrac{1}{3}$ **49.** -1 **51.** $\dfrac{77}{135}$ **53.** $1\dfrac{5}{7}$ **55.** $2\dfrac{12}{29}$ **57.** $-3\dfrac{1}{3}$ **59.** 25

Exercises 6.2 (page 228)

1. 8 **3.** 4 **5.** 4 **7.** 3 **9.** 6 **11.** 6 **13.** -5 **15.** 7 **17.** $2\dfrac{1}{5}$ **19.** -9 **21.** 3 **23.** 13 **25.** $\dfrac{3}{4}$ **27.** -1 **29.** 4

Exercises 6.3 (page 230)

1. 5 **3.** $\dfrac{2}{5}$ **5.** $-1\dfrac{1}{3}$ **7.** 6 **9.** -5 **11.** $7\dfrac{1}{5}$ **13.** -5 **15.** 0 **17.** $1\dfrac{1}{7}$ **19.** 9 **21.** 14 **23.** 45 **25.** 4 **27.** 22 **29.** -1 **31.** 10 **33.** 18 **35.** 2 **37.** -8 **39.** -2 **41.** -1 **43.** -7 **45.** $8\dfrac{1}{4}$ **47.** $-1\dfrac{1}{3}$ **49.** -3 **51.** 6 **53.** -2 **55.** 0 **57.** $\dfrac{19}{24}$ **59.** $6\dfrac{9}{44}$

Exercises 6.4 (page 234)

1. 8 **3.** 5 **5.** 3 **7.** 16 **9.** 1 **11.** 50 **13.** 6 **15.** 1 **17.** 24 **19.** 4 **21.** -56 **23.** 6 **25.** -3 **27.** $\dfrac{2}{3}$ **29.** $-\dfrac{1}{2}$ **31.** 1 **33.** $\dfrac{1}{5}$ **35.** 3 **37.** $\dfrac{2}{3}$ **39.** 5 **41.** $\dfrac{1}{7}$

Exercises 6.5 (pages 235–236)

1. $x - 20$ **3.** $\dfrac{x}{6}$ **5.** $x + 16$ **7.** $26 - x$ **9.** $2x$ **11.** $6x + 28 = 40$ **13.** $\dfrac{x}{6} = 5$ **15.** $5(x + 28) = 150$ **17.** $\dfrac{x}{6} - 7 = 2$ **19.** $30 - 2x = 4$ **21.** $(x - 7)(x + 5) = 13$ **23.** $6x - 17 = 7$ **25.** $4x - 17 = 63$

Exercises 6.6 (pages 239–240)

1. 10 in. **3.** 92 incandescent and 164 LED bulbs **5.** $825 to John; $1650 to Maria; $2475 to Betsy **7.** 10 cm × 20 cm **9.** 105 ft × 120 ft **11.** 42 ft; 42 ft; 38 ft **13.** $5\dfrac{1}{4}$ ft; $6\dfrac{3}{4}$ ft or 5 ft 3 in; 6 ft 9 in. **15.** 8 @ $6.50; 12 @ $9.50 **17.** $4500 @ 2.5%, $3000 @ 4% **19.** 20 L **21.** 320 mL of 30%; 480 mL of 80% **23.** 4 qt **25.** 558 ft³ **27.** $55,000 **29.** 30″ × 90″ **31.** 80% lean costs $2.40/lb; 90% lean costs $3.00/lb

Exercises 6.7 (page 243)

1. $\dfrac{E}{I}$ **3.** $\dfrac{F}{m}$ **5.** $\dfrac{C}{\pi}$ **7.** $\dfrac{V}{lh}$ **9.** $\dfrac{A}{2\pi r}$ **11.** $\dfrac{v^2}{2g}$ **13.** $\dfrac{Q}{I}$ **15.** vt **17.** $\dfrac{V}{I}$ **19.** $4\pi Er^2$ **21.** $\dfrac{1}{2\pi CX_C}$ **23.** $\dfrac{2A}{h}$ **25.** $\dfrac{QJ}{I^2 t}$ **27.** $\dfrac{5}{9}(F - 32)$ or $\dfrac{5F - 160}{9}$ **29.** $C_T - C_1 - C_3 - C_4$ **31.** $\dfrac{-By - C}{A}$ **33.** $\dfrac{Q_1 + PQ_1}{P}$ or $\dfrac{Q_1}{P} + Q_1$ **35.** $\dfrac{2A}{a + b}$ **37.** $\dfrac{l - a}{n - 1}$ **39.** $\dfrac{Ft}{V_2 - V_1}$ **41.** $\dfrac{Q}{w(T_1 - T_2)}$ **43.** $\dfrac{PV}{2\pi} - 3960$

Exercises 6.8 (pages 246–247)

1. a. $l = \dfrac{A}{w}$ **b.** 23.0 **3. a.** $h = \dfrac{3V}{\pi r^2}$ **b.** 20.0 **5. a.** $m = \dfrac{2E}{v^2}$ **b.** 2000 **7. a.** $t = \dfrac{v_f - v_i}{a}$ **b.** 9.90 **9. a.** $h = \dfrac{v_f^2 - v_i^2}{2g}$ **b.** 576
11. a. $r_1 = \dfrac{L - 2d}{\pi} - r_2$ **b.** 3.00 **13. a.** $r = \dfrac{Wv^2}{Fg}$ **b.** 1900 **15. a.** $b = \dfrac{2A}{h} - a$ **b.** 49.0 **17. a.** $d = \dfrac{2V}{lw} - D$ **b.** 3.00
19. a. $a = \dfrac{2S}{n} - l$ **b.** 16.6 **21.** 324 W **23.** 6.74 in. **25.** 6.72 ft **27.** 12.1 m **29.** 15.0 Ω **31.** 0.43 in. **33.** Speed would be 1.5 times the original speed. **35.** Momentum of truck is 3 times that of the automobile.

Exercises 6.9 (pages 249–250)

1. 4.80 Ω **3.** 18.0 Ω **5.** 40.0 Ω **7.** 2.50 cm **9.** 44.5 cm **11.** 9.00 Ω **13.** 318 Ω **15.** 3240 Ω **17.** 6.00 μF **19.** 1.91×10^{-6} F **21.** 1.74×10^{-8} F **23.** 219 Ω

Chapter 6 Review (page 251)

1. $1\frac{1}{2}$ **2.** −4 **3.** 57 **4.** 24 **5.** −7 **6.** 9 **7.** 8 **8.** 3 **9.** 6 **10.** $\dfrac{3}{5}$ **11.** 1 **12.** 6 **13.** 4 **14.** −2 **15.** 6 **16.** $\dfrac{1}{2}$ **17.** $2\dfrac{2}{3}$
18. $-7\dfrac{1}{2}$ **19.** 26 **20.** 2 **21.** 5 **22.** 6 in. × 18 in. **23.** 4.5 L of 100%; 7.5 L of 60% **24.** $g = \dfrac{F}{W}$ **25.** $A = \dfrac{W}{P}$
26. $t = 2L - 2A - 2B$ **27.** $m = \dfrac{2k}{v^2}$ **28.** $T_1 = \dfrac{P_1 T_2}{P_2}$ **29.** $v_0 = 2v - v_f$ **30.** 347 **31.** 19.5 **32.** 37 **33.** 30.0 Ω **34.** 70.6 μF

Cumulative Review Chapters 1–6 (page 252)

1. 2 · 2 · 2 · 3 · 29 **2.** 8.1% **3.** 0.0003015 **4.** 2.85×10^4 **5.** 50,000 **6.** 38.3° **7.** 43.4 **8. a.** 2 **b.** 1 **c.** 5
9. a. 55.60 mm **b.** 2.189 in. **10.** 0.428 in. **11.** 494,000 W **12.** $-8x + 8y$ **13.** $2y^3 + 4y^2 + 5y - 11$ **14.** $27y^9$
15. $-2x^3 + 6x^2 - 8x$ **16.** $12y^4 - 16y^3 + 3y^2 + 5y - 2$ **17.** $20x^2 - 7xy - 6y^2$ **18.** $\dfrac{43}{9xy^2}$ **19.** $-80x^6 y^8$
20. $x^2 - 3x + 4$ r -40 **21.** $6x^2 - 11xy + 15y^2$ **22.** $\dfrac{7}{2}$ **23.** 56 **24.** −6 **25.** $2\dfrac{2}{5}$ **26.** $\dfrac{11}{2}$ **27.** $a = 2C - b - c$
28. 11.1 m **29.** $7x = 250$ **30.** 5 ft × 10 ft

CHAPTER 7

Exercises 7.1 (pages 256–257)

1. $\dfrac{1}{5}$ **3.** $\dfrac{1}{3}$ **5.** $\dfrac{5}{3}$ **7.** $\dfrac{1}{5}$ **9.** $\dfrac{2}{1}$ **11.** $\dfrac{1}{32}$ **13.** $\dfrac{9}{14}$ **15.** $\dfrac{11}{16}$ **17.** $\dfrac{2}{1}$ **19.** $\dfrac{4}{1}$ **21.** $\dfrac{7}{5}$ **23.** $\dfrac{1}{4}$ **25.** 30 mi/gal **27.** 46 gal/h
29. 50 mi/h **31.** $\dfrac{3}{8}$ lb/gal **33.** $\dfrac{32}{5}$ or 6.4 to 1 **35.** 18 gal/min **37.** $\dfrac{1}{275}$ **39.** $\dfrac{12}{1}$ **41.** 36 lb/bu **43.** 25 gal/acre **45.** $\dfrac{1}{4}$
47. $1.69/ft **49.** $125/ft² **51.** $\dfrac{3}{4}$ **53.** $\dfrac{4}{25}$ **55.** 0.05 g/mL **57.** 38 drops/min **59.** 1 male cougar/150 mi² **61.** 50 drops/min
63. 45 drops/min **65.** $3\dfrac{1}{3}$ h **67.** $8\dfrac{1}{3}$ h **69.** $5.75/ft² **71.** $0.49/pencil **73.** $\dfrac{4}{3}$ **75.** $\dfrac{5}{1}$

Exercises 7.2 (pages 262–265)

1. a. 2, 3 **b.** 1, 6 **c.** 6 **d.** 6 **3. a.** 9, 28 **b.** 7, 36 **c.** 252 **d.** 252 **5. a.** 7, w **b.** x, z **c.** 7w **d.** xz **7.** yes; 30 = 30
9. no; 60 ≠ 90 **11.** yes; 12 = 12 **13.** 3 **15.** $5\dfrac{3}{5}$ **17.** 14 **19.** $3\dfrac{1}{3}$ **21.** 35 **23.** −7.5 **25.** 3.5 **27.** 0.5 **29.** 126
31. 20.6 **33.** $37\dfrac{1}{3}$ **35.** 38.2 **37.** 818 **39.** 9050 **41.** 21.3 **43.** 44 ft³ **45.** $228,000 **47.** $144 **49.** 10 gal **51.** 52.5 lb
53. 108 lb **55. a.** 12 bu/tree **b.** $4680 **57.** 1250 ft **59.** 595 turns **61.** 216 hp **63.** 26.4 gal **65.** 2 mL **67.** 6 mL
69. 5.9% **71. a.** 18% **b.** 117 lb **73.** 19 hL **75. a.** 13.3% **b.** 33.3% **c.** 53.3% **77.** 83 ft 4 in. **79.** 70 lb **81.** 35 turns
83. 4.5 mL **85.** 5 g **87.** 1.5 g **89.** 30 g **91.** 2.4 lb **93. a.** 5:2:1 **b.** 375 mL vegetable oil, 150 mL sherry vinegar, 75 mL salt **95.** $\dfrac{5}{8}$ cups margarine, $7\dfrac{1}{2}$ tbs sugar, $1\dfrac{1}{4}$ cups orange juice, 5 tsp lemon juice, $2\dfrac{1}{2}$ tbs grated orange rind, $1\dfrac{1}{4}$ tsp grated lemon rind, $1\dfrac{1}{4}$ cup toasted slivered almonds **97.** $1\dfrac{1}{8}$ cups butter, $1\dfrac{1}{8}$ cups all-purpose flour, $4\dfrac{1}{2}$ cups milk, $2\dfrac{1}{4}$ tsp salt, 36 oz sharp processed cheese, 18 egg yolks, 18 stiff beaten egg whites **99.** 12 sticks pie crust, 8 tbs soft butter, 4 eggs, 4 tsp salt, 4 tbs water, 2 tsp pepper

Exercises 7.3 (pages 269–270)

For all answers the allowable error is ±8 mi or ±8 km.

1. 48 mi **3.** 60 mi **5.** 86 mi **7.** 104 mi **9.** 96 mi **11.** 425 km **13.** 255 km **15.** 272 km **17.** 12.5 cm **19.** 100 cm²

21. 1 cm × 1 cm **23.** 6.5 cm **25.** 6.5 cm **27.** **29.**

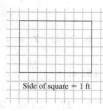

31. a. yes **b.** no **33.** 20 : 1 **35.** 9 : 1 **37.** 68 lb **39.** 238,500 lb **41.** 81 : 1 **43.** 25 : 1 **45.** 400 lb **47.** 400 lb **49.** 640 lb **51.** 225 mi **53.** 141

Exercises 7.4 (pages 272–274)

1. 36 rpm **3.** 80 rpm **5.** 22 cm **7.** 31 in. **9.** 520 rpm **11.** 50 in. **13.** 11 in. **15.** 10 cm; 17 cm **17.** 2080 rpm **19.** 11,800 rpm **21.** 160 teeth **23.** 1008 rpm **25.** 576 teeth **27.** 96 rpm **29.** 144 rpm **31.** 40 teeth **33.** 10 in. **35.** 120 lb **37.** 135 lb **39.** $\frac{1}{2}$ ton or 1000 lb **41.** 64 cm **43.** $25\frac{1}{3}$ in. **45.** 27 cm **47.** 3 A

Chapter 7 Review (page 275)

1. $\frac{1}{4}$ **2.** $\frac{3}{2}$ **3.** $\frac{2}{1}$ **4.** $\frac{11}{18}$ **5.** yes **6.** no **7.** 1 **8.** 30 **9.** 24 **10.** 40 **11.** 106 **12.** 41.3 **13.** 788 **14.** 529 **15.** $187.50 **16.** 1250 ft **17.** 216 h **18.** 300 lb **19.** Jones 58%; Hernandez 42% **20.** 8.8% **21. a.** direct **b.** inverse **22.** direct **23.** 362.5 mi **24.** 17.6 in. **25.** 562.5 rpm **26.** 75 rpm **27.** 50 lb **28.** 180 lb **29.** 2 A **30.** 108 h

CHAPTER 8

Exercises 8.1 (pages 282–283)

1. (3, 2) (8, −3)(−2, 7) **3.** (2, −1) (0, 5) (−2, 11) **5.** (0, −2) $\left(2, -\frac{1}{2}\right)$ (−4, −5) **7.** (5, 4) (0, 2) $\left(-3, \frac{4}{5}\right)$ **9.** (2, 4) (0, −5) (−4, −23) **11.** (2, 10) (0, 4) (−3, −5) **13.** (2, −3) (0, 7) (−4, 27) **15.** (3, 10) (0, 4) (−1, 2) **17.** (4, 14) (0, 4) (−2, −1) **19.** (1, 2) $\left(0, -\frac{5}{2}\right)$ (−3, −16) **21.** (2, 3) (0, 3) (−4, 3) **23.** (5, 4) (5, 0) (5, −2) **25.** $y = \frac{6 - 2x}{3}$ **27.** $y = \frac{7 - x}{2}$ **29.** $y = \frac{x - 6}{2}$ **31.** $y = \frac{2x - 9}{3}$ **33.** $y = \frac{2x + 6}{3}$ **35.** $y = \frac{-2x + 15}{3}$ **37.** $A(-2, 2)$ **39.** $C(5, -1)$ **41.** $E(-4, -5)$ **43.** $G(0, 5)$ **45.** $I(4, 2)$

47–65.

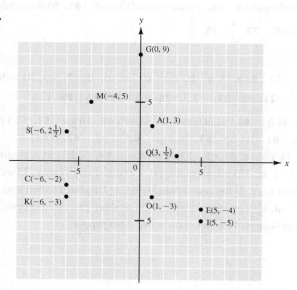

Exercises 8.2 (page 289)

1.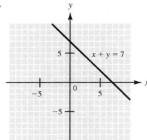
3.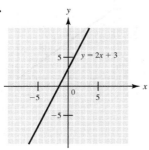
5.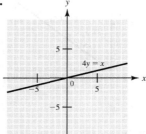
7.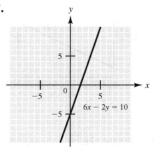
9.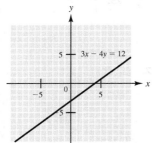
11.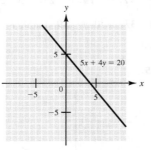
13.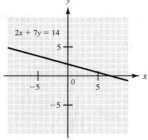
15.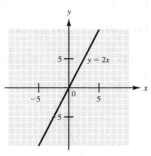
17.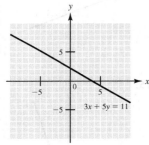
19.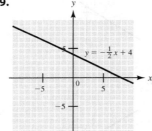
21.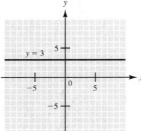
23.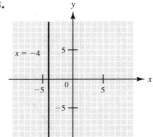
25.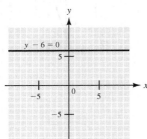
27.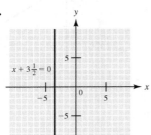
29.

	Independent	Dependent
31.	t	s
33.	V	R
35.	t	i
37.	t	v
39.	t	s

41. 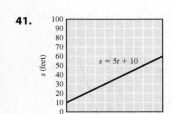 **43.** 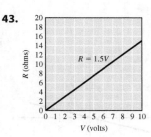 **45.**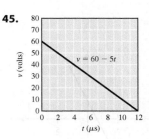

Exercises 8.3 (pages 294–295)

1. 1 **3.** 6 **5.** $-\dfrac{5}{4}$ **7.** 0 **9.** undefined **11.** $\dfrac{3}{5}$ **13.** -2 **15.** $\dfrac{3}{7}$ **17.** 0 **19.** 6 **21.** -5 **23.** $-\dfrac{3}{5}$ **25.** $\dfrac{1}{4}$ **27.** $\dfrac{5}{2}$
29. undefined **31.** parallel **33.** perpendicular **35.** perpendicular **37.** neither **39.** parallel

Exercises 8.4 (pages 299–300)

1. 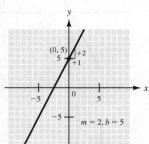 **3.** 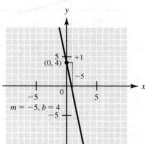 **5.** 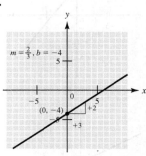 **7.**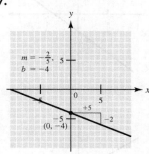

9. 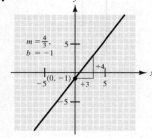 **11.** 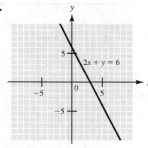 **13.** 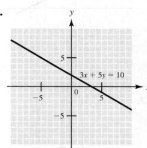 **15.**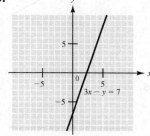

17. 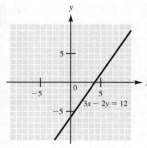 **19.**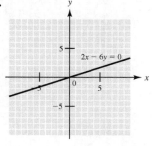

21. $y = 2x + 5$ **23.** $y = -5x + 4$ **25.** $y = \dfrac{2}{3}x - 4$ or $2x - 3y = 12$ **27.** $y = -\dfrac{6}{5}x + 3$ or $6x + 5y = 15$

29. $y = -\dfrac{3}{5}x$ or $3x + 5y = 0$

31. 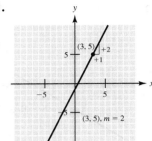 **33.** 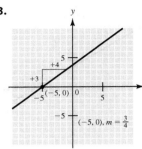 **35.** 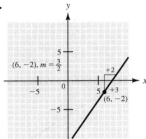 **37.**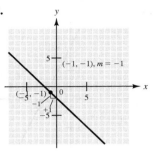

39. 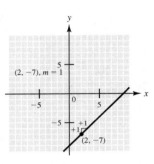 **41.** $2x - y = 1$ **43.** $3x - 4y = -15$ **45.** $3x - 2y = 22$ **47.** $10x + 3y = -33$ **49.** $x + y = 2$ **51.** $2x + 3y = 13$ **53.** $2x + y = 2$ **55.** $x - 3y = -3$ **57.** $x - 2y = -1$ **59.** $x + y = 6$

Chapter 8 Review (pages 301–302)

1. $\left(3, \frac{5}{2}\right)(0, 4)(-4, 6)$ **2.** $(3, -2)(0, -4)(-3, -6)$ **3.** $y = -6x + 15$ **4.** $y = \frac{3x + 10}{5}$ or $y = \frac{3}{5}x + 2$ **5.** $A(3, 5)$
6. $B(-2, -6)$ **7.** $C(2, -1)$ **8.** $D(4, 0)$ **9–12.** 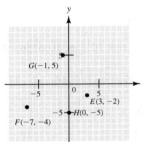 **13.**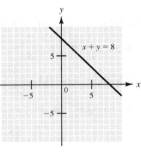

14. 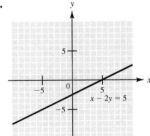 **15.** 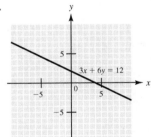 **16.** 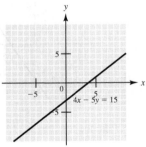 **17.**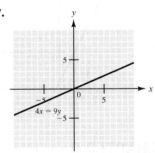

18. 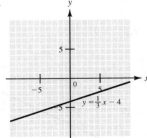 **19.** 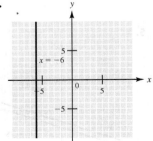 **20.**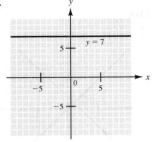

21. $\frac{9}{7}$ **22.** 1 **23.** 4 **24.** $-\frac{2}{5}$ **25.** $\frac{5}{9}$ **26.** perpendicular **27.** neither **28.** parallel **29.** perpendicular

554 APPENDIX C ♦ Selected Answers

30. 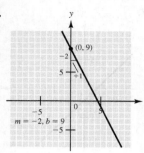 **31.** 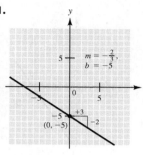 **32.** 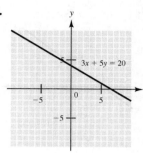 **33.**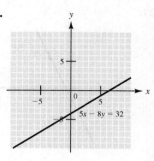

34. $y = -\dfrac{1}{2}x + 3$ or $x + 2y = 6$ **35.** $y = \dfrac{8}{3}x$ or $8x - 3y = 0$ **36.** $y = 0$

37. 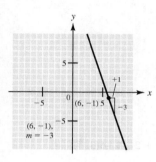 **38.**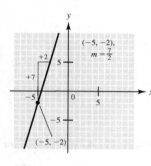

39. $x + y = 6$ **40.** $x + 4y = -20$
41. $x - y = 5$ **42.** $x - 2y = 12$

Cumulative Review Chapters 1–8   (page 303)

1. 5   **2.** 15   **3. a.** $4\dfrac{3}{16}$ in.   **b.** $2\dfrac{7}{8}$ in.   **4.** 98.4   **5.** 1.116×10^3 **6.** 0.061 m   **7.** 5   **8.** 7.82 mm   **9.** 350 m³   **10.** $7x - 11$
11. $-40x^4y^4$ **12.** $8x^2 - 6xy$ **13.** $-\dfrac{3}{5}$ **14.** $-\dfrac{52}{25}$ **15.** $V = \dfrac{3s - t}{2}$ **16.** $\dfrac{1}{13}$ **17.** $\dfrac{8}{27}$ **18.** 60   **19.** 41.0   **20.** 1450   **21.** $102.60
22. 170 mi   **23.** 10 teeth   **24.** (3, 2) (0, 4) (−3, 6)   **25.** $y = \dfrac{7 - 4x}{2}$

26. 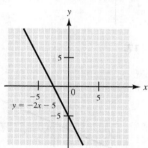 **27.**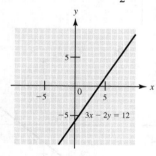

28. 1   **29.** $x - 2y = 10$ **30.** neither

CHAPTER 9

Exercises 9.1   (pages 309–310)

1. 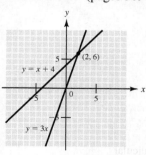 **3.** 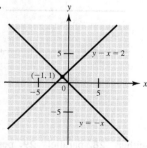 **5.** 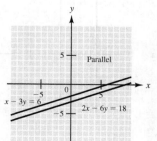 **7.**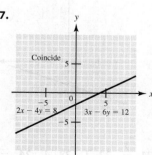

9.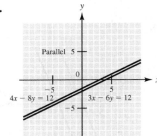
11.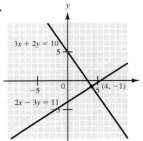
13.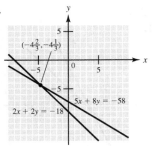
15.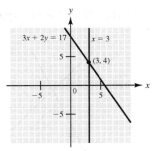

17.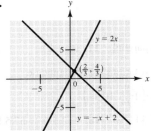
19.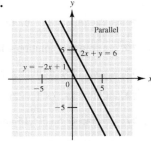
21.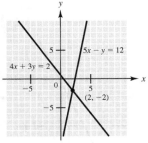
23.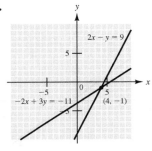

25.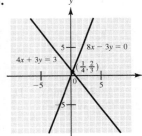
27.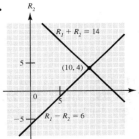
29. 4 ft³ of concrete; 16 ft³ of gravel
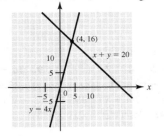

Exercises 9.2 (page 315)

1. (2, 1) **3.** (4, 2) **5.** (−2, 7) **7.** (3, 1) **9.** (−5, 5) **11.** (5, 1) **13.** (1, −2) **15.** (−1, 4) **17.** (7, −5) **19.** (4, −1)
21. (−2, 5) **23.** $\left(2, 1\frac{1}{4}\right)$ **25.** (−2, 4) **27.** coincide **29.** (2, −1) **31.** parallel **33.** $\left(\frac{1}{2}, 5\right)$ **35.** coincide **37.** parallel
39. (72, 30) **41.** $\left(\frac{1}{4}, \frac{2}{5}\right)$ **43.** (9, −2)

Exercises 9.3 (page 317)

1. $\left(\frac{12}{5}, \frac{36}{5}\right)$ or (2.4, 7.2) **3.** (10, 2) **5.** (6, 6) **7.** (4, 2) **9.** (−3, −1) **11.** (1, 4) **13.** (−2, 2) **15.** $\left(2\frac{1}{2}, -12\frac{1}{2}\right)$ **17.** (1, 3)
19. (9, −2) **21.** (3, 4) **23.** (6, 6) **25.** (2, 4)

Exercises 9.4 (pages 321–323)

1. 42 cm, 54 cm **3.** $2\frac{1}{2}$ h @ 180 gal/h; $3\frac{1}{2}$ h @ 250 gal/h **5.** 26 h @ $32; 22 h @ $41 **7.** 30 lb @ 5%; 70 lb @ 15%
9. 2700 bu corn; 450 bu beans **11.** 200 gal of 6%; 100 gal of 12% **13.** 5 @ 3 V; 4 @ 4.5 V **15.** 105 mL @ 8%; 35 mL @ 12%
17. 5 min @ 850 rpm; 9 min @ 1250 rpm **19.** 2 h @ setting 1; 3 h @ setting 2 **21.** 160 L of 3%; 40 L of 8% **23. a.** 6.8 A
b. 1.2 A **25.** 5 h @ 140 mL/h; 3 h @ 100 mL/h **27.** 31 of 2 mL; 11 of 5 mL **29.** 4 one-bedroom and 9 two-bedroom
31. 100 Ω, 450 Ω **33.** 30 cm, 90 cm **35.** 40 cm × 80 cm **37.** 150 mA **39.** 400 ft × 1000 ft **41. a.** 6 ft × 14 ft **b.** 8 ft × 12 ft
= 96 ft²; The second room is 12 ft² larger. **c.** 14% increase **43.** 30 ft × 75 ft; P: 4.5% increase; A: 24% increase
45. 76 in.; 68 in. **47.** 85 lb corn; 15 lb soybean meal **49.** $5000 @3.5%; $17,500 @8.5% **51.** 10 tables for 6; 11 tables for 8

Chapter 9 Review (page 324)

1.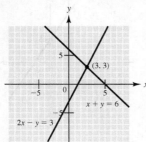
2.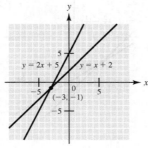
3.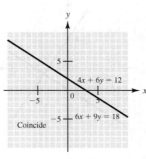
4.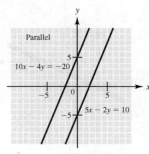
5.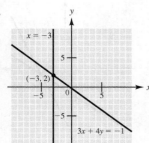
6.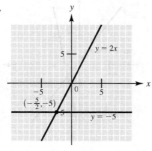

7. (3, 4) 8. (3, 1) 9. (4, 3)
10. no common solution—lines parallel 11. (6, −2)
12. (3, 1) 13. infinitely many solutions—lines coincide
14. (4, 8) 15. (−1, −2) 16. $\left(1\frac{9}{25}, 10\frac{11}{25}\right)$
17. 20 amp @ \$3.50; 15 amp @ \$11.50 18. 132.5 ft × 57.5 ft
19. 20 mH; 70 mH 20. 35 ft; 55 ft

CHAPTER 10

Exercises 10.1 (page 327)

1. $4(a + 1)$ 3. $b(x + y)$ 5. $5(3b − 4)$ 7. $x(x − 7)$ 9. $a(a − 4)$ 11. $4n(n − 2)$ 13. $5x(2x + 5)$ 15. $3r(r − 2)$
17. $4x^2(x^2 + 2x + 3)$ 19. $9a(a − x^2)$ 21. $10(x + y − z)$ 23. $3(y − 2)$ 25. $7xy(2 − xy)$ 27. $m(12x^2 − 7)$ 29. $12a(5x − 1)$
31. $13mn(4mn − 1)$ 33. $2(26m^2 − 7m + 1)$ 35. $18y^2(2 − y + 3y^2)$ 37. $3m(2m^3 − 4m + 1)$ 39. $-2x^2y^3(2 + 3y + 5y^2)$
41. $3abc(abc + 9a^2b^2c − 27)$ 43. $4xz^2(x^2z^2 − 2xy^2z + 3y)$

Exercises 10.2 (page 329)

1. $x^2 + 7x + 10$ 3. $6x^2 + 17x + 12$ 5. $x^2 − 11x + 30$ 7. $x^2 − 14x + 24$ 9. $2x^2 + 19x + 24$ 11. $x^2 + 4x − 12$
13. $x^2 − 19x + 90$ 15. $x^2 − 6x − 72$ 17. $8x^2 − 18x − 35$ 19. $8x^2 + 6x − 35$ 21. $14x^2 + 41x + 15$ 23. $3x^2 − 19x − 72$
25. $6x^2 + 47x + 35$ 27. $169x^2 − 104x + 16$ 29. $120x^2 + 54x − 21$ 31. $100x^2 − 100x + 21$ 33. $4x^2 − 16x + 15$
35. $4x^2 + 4x − 15$ 37. $6x^2 + 5x − 56$ 39. $6x^2 + 37x + 56$ 41. $16x^2 + 14x − 15$ 43. $2y^2 − 11y − 21$ 45. $6n^2 + 3ny − 30y^2$
47. $8x^2 + 26xy − 7y^2$ 49. $\frac{1}{8}x^2 − 5x + 48$

Exercises 10.3 (page 332)

1. $(x + 2)(x + 4)$ 3. $(y + 4)(y + 5)$ 5. $3(r + 5)^2$ 7. $(b + 5)(b + 6)$ 9. $(x + 8)(x + 9)$ 11. $5(a + 4)(a + 3)$
13. $(x − 4)(x − 3)$ 15. $2(a − 7)(a − 2)$ 17. $3(x − 7)(x − 3)$ 19. $(w − 6)(w − 7)$ 21. $(x − 9)(x − 10)$ 23. $(t − 10)(t − 2)$
25. $(x + 4)(x − 2)$ 27. $(y + 5)(y − 4)$ 29. $(a + 8)(a − 3)$ 31. $(c − 18)(c + 3)$ 33. $3(x − 4)(x + 3)$ 35. $(c + 6)(c − 3)$
37. $(y + 14)(y + 3)$ 39. $(r − 7)(r + 5)$ 41. $(m − 20)(m − 2)$ 43. $(x − 15)(x + 6)$ 45. $(a + 23)(a + 4)$ 47. $2(a − 11)(a + 5)$
49. $(a + 25)(a + 4)$ 51. $(y − 19)(y + 5)$ 53. $(y − 16)(y − 2)$ 55. $7(x + 2)(x − 1)$ 57. $6(x^2 + 2x − 1)$ 59. $(y − 7)(y − 5)$
61. $(a + 9)(a − 7)$ 63. $(x + 4)(x + 14)$ 65. $2(y − 15)(y − 3)$ 67. $3x(y − 3)^2$ 69. $(x + 15)^2$ 71. $(x − 9)(x − 17)$
73. $(x + 12)(x + 16)$ 75. $(x + 22)(x − 8)$ 77. $2b(a + 6)(a − 4)$ 79. $(y − 9)(y + 8)$

Exercises 10.4 (page 334)

1. $x^2 − 9$ 3. $a^2 − 25$ 5. $4b^2 − 121$ 7. 9991 or (10,000 − 9) 9. $9y^4 − 196$ 11. $r^2 − 24r + 144$ 13. $16y^2 − 25$
15. $x^2y^2 − 8xy + 16$ 17. $a^2b^2 + 2abd + d^2$ 19. $z^2 − 22z + 121$ 21. $s^2t^2 − 14st + 49$ 23. $x^2 − y^4$ 25. $x^2 + 10x + 25$
27. $x^2 − 49$ 29. $x^2 − 6x + 9$ 31. $a^2b^2 − 4$ 33. $x^4 − 4$ 35. $r^2 − 30r + 225$ 37. $y^6 − 10y^3 + 25$ 39. $100 − x^2$

Exercises 10.5 (page 336)

1. $(a + 4)^2$ 3. $(b + c)(b − c)$ 5. $(x − 2)^2$ 7. $(2 + x)(2 − x)$ 9. $(y + 6)(y − 6)$ 11. $5(a + 1)^2$ 13. $(1 + 9y)(1 − 9y)$
15. $(7 + a^2)(7 − a^2)$ 17. $(7x + 8y)(7x − 8y)$ 19. $(1 + xy)(1 − xy)$ 21. $(2x − 3)^2$ 23. $(R + r)(R − r)$ 25. $(7x + 5)(7x − 5)$

27. $(y - 5)^2$ **29.** $(b + 3)(b - 3)$ **31.** $(m + 11)^2$ **33.** $(2m + 3)(2m - 3)$ **35.** $4(x + 3)^2$ **37.** $3(3x + 1)(3x - 1)$
39. $a(m - 7)^2$

Exercises 10.6 (page 338)
1. $(5x + 2)(x - 6)$ **3.** $(2x - 3)(5x - 7)$ **5.** $(6x - 5)(2x - 3)$ **7.** $(2x + 9)(4x - 5)$ **9.** $(16x + 5)(x - 1)$ **11.** $4(3x + 2)(x - 2)$
13. $(5y + 3)(3y - 2)$ **15.** $(4m + 1)(2m - 3)$ **17.** $(7a + 1)(5a - 1)$ **19.** $(4y - 1)^2$ **21.** $(3x - 7)(x + 9)$ **23.** $(4b - 1)(3b + 2)$
25. $(5y + 2)(3y - 4)$ **27.** $(10 + 3c)(9 - c)$ **29.** $(3x - 5)(2x - 1)$ **31.** $(2y^2 - 5)(y^2 + 7)$ **33.** $(2b + 13)^2$ **35.** $(7x - 8)(2x - 5)$
37. $7x(2x + 5)^2$ **39.** $5a(2b + 7)(b - 5)$

Chapter 10 Review (pages 339–340)
1. $c^2 - d^2$ **2.** $x^2 - 36$ **3.** $y^2 + 3y - 28$ **4.** $4x^2 - 8x - 45$ **5.** $x^2 + 5x - 24$ **6.** $x^2 - 13x + 36$ **7.** $x^2 - 6x + 9$
8. $4x^2 - 24x + 36$ **9.** $1 - 10x^2 + 25x^4$ **10.** $6(a + 1)$ **11.** $5(x - 3)$ **12.** $x(y + 2z)$ **13.** $y^2(y + 18)(y - 1)$ **14.** $(y - 7)(y + 1)$
15. $(z + 9)^2$ **16.** $(x + 8)(x + 2)$ **17.** $4(a^2 + x^2)$ **18.** $(x - 9)(x - 8)$ **19.** $(x - 9)^2$ **20.** $(x + 15)(x + 4)$ **21.** $(y - 1)^2$
22. $(x - 7)(x + 4)$ **23.** $(x - 12)(x + 8)$ **24.** $(x + 11)(x - 10)$ **25.** $(x + 7)(x - 7)$ **26.** $(4y + 3x)(4y - 3x)$ **27.** $(x + 12)(x - 12)$
28. $(5x + 9y)(5x - 9y)$ **29.** $4(x - 13)(x + 7)$ **30.** $5(x + 12)(x - 13)$ **31.** $(2x + 7)(x + 2)$ **32.** $(4x - 1)(3x - 4)$
33. $(6x + 5)(5x - 3)$ **34.** $(12x - 1)(x + 12)$ **35.** $2(2x - 1)(x - 1)$ **36.** $(6x + 7y)(6x - 7y)$ **37.** $2(7x + 3)(2x + 5)$
38. $3(5x - 7)(2x + 1)$ **39.** $4x(x + 1)(x - 1)$ **40.** $25(y + 2)(y - 2)$

Cumulative Review Chapters 1–10 (pages 340–341)
1. 6 **2. a.** 746.8 **b.** 750 **3.** $-\dfrac{8}{3}$ **4. a.** 3.18×10^{-4} **b.** 318×10^{-6} **5.** 0.625 kg **6.** 64.6 ft² **7.** 70 V **8.** 95 cm³
9. $15x - 14$ **10.** $8a + c$ **11.** -60 **12.** 17 m; 12 m **13.** 19.1 **14.** 10 in. **15.** (3, 2), (0, 4), (−3, 6) **16.** $y = 3x - 5$
17. 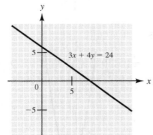 **18.** 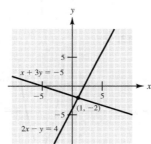 **19.** many solutions; lines coincide **20.** (2.3, 1.9)
21. (2, −4) **22.** no solution; lines parallel
23. $\left(\dfrac{1}{3}, -\dfrac{2}{5}\right)$ **24.** 6 days @ \$53.95; 10 days @ \$89.95
25. $6x^2 + x - 40$ **26.** $25x^2 - 70xy + 49y^2$
27. $15x^2 - 46x + 35$ **28.** $7x(x + 3)(x - 3)$
29. $4x^2(x + 3)$ **30.** $(2x + 1)(x - 4)$

CHAPTER 11

Exercises 11.1 (page 345)
1. −4, 3 **3.** −5, 4 **5.** 2, −1 **7.** 1, −1 **9.** 7, −7 **11.** −2, −3 **13.** 7, −3 **15.** 10, −4 **17.** 0, 9 **19.** 12, −9 **21.** −8, 2
23. $-\dfrac{5}{2}, -\dfrac{2}{5}$ **25.** $\dfrac{5}{2}, -\dfrac{5}{2}$ **27.** $\dfrac{4}{3}$ **29.** 0, −3 **31.** Base: 12 m; Height: 11 m **33.** 5 in. × 8 in.

Exercises 11.2 (pages 347–348)
1. $a = 1, b = -7, c = 4$ **3.** $a = 3, b = 4, c = 9$ **5.** $a = -3, b = 4, c = 7$ **7.** $a = 3, b = 0, c = -14$ **9.** −3, 2 **11.** −9, 1
13. $0, -\dfrac{2}{5}$ **15.** $\dfrac{5}{4}, -\dfrac{7}{12}$ **17.** 1.35, −1.85 **19.** $0, \dfrac{5}{3}$ **21.** $-1, \dfrac{3}{2}$ **23.** −1.38, −0.121 **25.** $-\dfrac{1}{4}, -1$ **27.** 2.38, −2.38 **29.** 15.5, −0.453
31. 4.15, −2.49 **33.** 0.206, −2.40

Exercises 11.3 (pages 350–352)
1. a. 4 s; 8 s **b.** 1.42 s; 10.6 s **c.** 16 s **3.** 3 ft × 7 ft **5.** 5 ft × 35 ft **7.** 20 m × 60 m **9.** 4.5 ft × 9 ft
11. a. 5 cm × 5 cm **b.** 7500 cm³ **13.** 2 ft **15. a.** 1400 ft² **b.** 2600 ft² **c.** $A = l(150 - l) = 150l - l^2$

d. Length (ft)	30	40	50	60	70	80	90	100	110	120	130	140
Area (ft²)	3600	4400	5000	5400	5600	5600	5400	5000	4400	3600	2600	1400

e. The maximum area given in the table is 5600 ft² when the length is 70 ft or 80 ft.

f.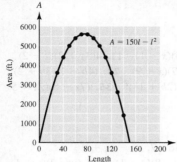

g. From the graph: When the length = the width = 75 ft, the maximum area is 5625 ft².

17. 65 ft × 141 ft

Exercises 11.4 (page 356)

1. 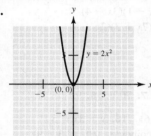 **3.** 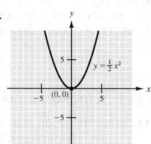 **5.** 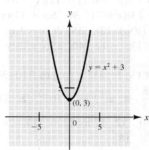 **7.**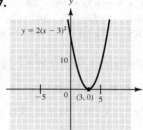

9. 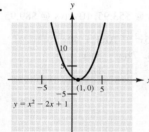 **11.** 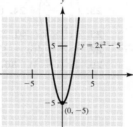 **13.** 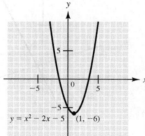 **15.**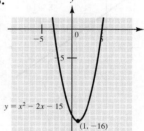

17. 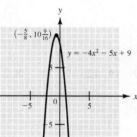 **19.**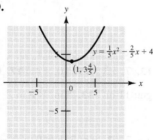

Exercises 11.5 (page 359)

1. $7j$ **3.** $3.74j$ **5.** $1.41j$ **7.** $7.48j$ **9.** $13j$ **11.** $5.20j$ **13.** $-j$ **15.** j **17.** $-j$ **19.** 1 **21.** -1 **23.** $-j$ **25.** two rational roots
27. two imaginary roots **29.** two irrational roots **31.** two imaginary roots **33.** two imaginary roots **35.** $3 + j; 3 - j$
37. $7 + 2j; 7 - 2j$ **39.** $-4 + 5j; -4 - 5j$ **41.** $-0.417 + 1.08j; -0.417 - 1.08j$ **43.** $1 + 1.15j; 1 - 1.15j$
45. $-0.8 + 0.4j; -0.8 - 0.4j$ **47.** $0.2, -3$ **49.** $-0.5 + 0.866j; -0.5 - 0.866j$

Chapter 11 Review (pages 360–361)

1. $a = 0$ or $b = 0$ **2.** $0, 2$ **3.** $2, -2$ **4.** $3, -2$ **5.** $0, \frac{6}{5}$ **6.** $7, -4$ **7.** $9, 5$ **8.** $6, -3$ **9.** $-\frac{8}{3}, -4$ **10.** $6, -\frac{2}{3}$
11. $0.653, -7.65$ **12.** $\frac{5}{2}, -3$ **13.** $4.45, -0.449$ **14.** $2.12, -0.786$ **15.** length 9 ft, width 4 ft **16. a.** $4 \,\mu s, 8 \,\mu s$
b. $6 \,\mu s$ **c.** $2.84 \,\mu s, 9.16 \,\mu s$

17. 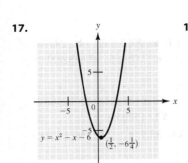 **18.**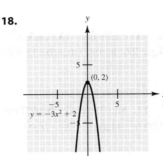

19. $6j$ **20.** $8.54j$ **21.** 1 **22.** $-j$
23. one rational root **24.** two imaginary roots
25. $2 + j, 2 - j$ **26.** $0.6 + 0.663j, 0.6 - 0.663j$
27. 2.5 ft $\times$ 8.5 ft **28.** 3 in.

CHAPTER 12

Exercises 12.1 (pages 367–369)
1. acute **3.** right **5.** acute **7.** obtuse **9.** right, perpendicular **11. a.** 1, 2; 2, 4; 3, 4; 1, 3; 5, 6; 6, 8; 7, 8; 5, 7
b. 1, 4; 2, 3; 5, 8; 6, 7 **13.** $\angle 2 = 57°; \angle 3 = 57°; \angle 4 = 123°$ **15.** 61° **17.** 52° **19. a.** Yes **b.** Yes **21.** 90° **23.** 15° **25.** 148°
27. 152° **29.** triangle **31.** hexagon **33.** quadrilateral **35.** heptagon

Exercises 12.2 (pages 371–373)
1. 60.0 cm; 225 cm² **3.** 32.0 m; 48 m² **5.** 53.5 m; $1\overline{0}0$ m² **7.** 89.4 in.; 411 in² **9.** 36.8 cm; 85 cm² **11.** 24.0 cm **13.** 13.4 ft
15. 352 in² **17. a.** 59.4 mi **b.** 197 mi² **19.** 16 in.; 48 in. **21.** 9 **23.** 36 ft **25.** 6 pieces **27.** 28 in² **29. a.** 208 in² **b.** $59\overline{0}$ in²
31. $40\overline{0},000$ acres **33.** 7 squares **35.** $927.35 **37.** 250 ft² **39.** 568.6 ft; $5401.70 **41.** 704 ft² **43.** $20\overline{0}$ ft² **45.** 105 ft

Exercises 12.3 (pages 377–381)
1. 10.0 m **3.** 25.0 m **5.** 17.0 m **7.** 13.5 cm **9.** 1460 km **11.** 350 ft **13.** 59.9 in. **15.** 33.4 in. **17.** 2.83 in. **19.** 16.6 in.
21. 116 V **23.** 21 A **25.** 127 Ω **27.** 2.6 Ω **29.** 158 m²; 60.8 m **31.** 15.6 cm²; 18.00 cm **33.** 395 m²; 108.7 m
35. 29.9 m²; 36.0 m **37.** 96.0 cm²; 48.0 cm **39.** 10.8 ft **41.** 74.3 mi **43.** 145 in² **45.** 81° **47.** 60° **49.** 5 bags
51. a. 9% **b.** 50.2 ft **53.** 71.1 acres

Exercises 12.4 (pages 383–385)
1. 5 **3.** Yes, all corresponding angles of $\triangle ABO$ and of $\triangle DCO$ are equal. **5.** $5\frac{1}{3}$ **7.** 120 ft **9. a.** 12.0 m **b.** 17.0 m **11.** 10.6 in.
13. $55\overline{0}$ ft **15.** 20.0 in. **17.** $6\overline{0}$ in. **19.** $A = 1.33$ ft; $B = 2.67$ ft; $C = 4.00$ ft; $D = 5.33$ ft; $E = 6.67$ ft; cross-piece is 10.0 ft
21. a. 3 in. = $\frac{3}{4}$ in.; 6 in. = $1\frac{1}{2}$ in.; 2 in. = $\frac{1}{2}$ in.; 1 in. = $\frac{1}{4}$ in.; 2.5 in. = $\frac{5}{8}$ in. **b.** 2.33 oz **23.** lengths: 36 in., 39 in., 53.1 in.; a 12-ft
board must be purchased

Exercises 12.5 (pages 388–392)
1. a. 31.4 in. **b.** 78.5 in² **3. a.** 28.9 mm **b.** 66.6 mm² **5. a.** 352 mi **b.** 9890 mi² **7.** 171° **9.** 43.4° **11.** 3.00 cm
13. 10.0 m **15.** 90° **17.** 12.6 ft² **19.** 3.98 in²; 7.07 in. **21.** 50.3 in. **23.** 2670 ft **25.** 13.4 in. **27.** 1.25 in. **29.** 1332 cm²
31. 24.42 in. **33. a.** 2.50 ft **b.** 3.54 ft **35.** 72° **37.** 45.2° **39.** 18,600 mi **41.** 171° **43.** 14° **45. a.** 120° **b.** 60°
c. 120° **47. a.** 60° **b.** 120° **c.** 60° **49. a.** 5.6 in² **b.** 4.13 in. **c.** 2.26 in. **d.** 8.4 oz **51.** 7050 gal **53.** Round tables require
2.5 ft² less space. **55. a.** 40 cups **b.** 20 sheets

Exercises 12.6 (page 396)
1. 180° **3.** 0.367 rad **5.** 60° **7.** $\frac{3\pi}{4}$ rad; 2.36 rad **9.** $\frac{4\pi}{3}$ rad; 4.19 rad **11.** $\frac{4\pi}{5}$ rad; 2.51 rad **13.** 31.4 cm **15.** 4.71 cm
17. 104 cm **19.** 2.80 rad **21.** 98.5° **23.** 8.83 cm **25.** $\frac{5}{\pi}$ rps or 1.59 rps **27.** 92.2 cm **29.** 25.8 cm **31. a.** 26.2 m **b.** 327 m²
c. 56 m² **33.** 14.1 in²

Exercises 12.7 (pages 400–402)
1. a. $44\overline{0}$ in² **b.** 622 in² **c.** 912 in³ **3.** 84 in²; 36 in³ **5.** $6\overline{0}$ ft³ **7. a.** 2490 m² **b.** 1230 m² **c.** 188 m³ **d.** $49\overline{0}0$ m²
9. 324 ft³ **11.** $3\overline{0}$ in. **13.** 12 ft 1 in. $\times$ 9 ft 11 in. **15.** $9\overline{0}$ in. $\times$ 113 in. **17.** 8 **19.** $36\overline{0}0$ ft³ **21. a.** 180 cups **b.** 80 cups
23. 2040 boxes

Exercises 12.8 (pages 405–407)

1. 13,600 mm^3 **3.** 8,280,000 L   **5.** 42.6 ft   **7.** 56.5 in^3 **9.** 1140 ft^3 **11.** 1.79 in.   **13.** 28.3 in^3 **15. a.** 2050 mm^2 **b.** 2470 mm^2 **17.** 101,000 cm^2 **19.** 18.0 in^3 **21.** 1.38 in^2 **23.** 6.1 gal   **25.** 25 kg   **27. a.** 68.0 in.   **b.** 13,800 lb   **c.** 2.08 gal   **d.** 88,400 lb   **29.** 15.5 ft   **31.** 10,100 bu   **33.** 318 cups   **35.** 114 cups

Exercises 12.9 (pages 411–413)

1. 147 in^3 **3.** 11$\overline{0}$0 m^3 **5.** 69,700 mm^3 **7.** 1440 mm^3 **9.** 312 ft^3 **11. a.** 1010 cm^3 **b.** 427 cm^2 **13.** 1520 bu   **15.** 225 in^2 **17.** 37,000 ft^3, 1$\overline{0}$0 truckloads   **19.** 15.0 m^3 **21.** 108,000 ft^3; 9420 ft^2 **23.** 28.0 in.   **25.** 33.9 ft^3 **27.** yes, 13.2%   **29.** 98.1 gal

Exercises 12.10 (pages 415–416)

1. a. 804 m^2 **b.** 2140 m^3 **3. a.** 4120 in^2 **b.** 24,800 in^3 **5.** 114,000 m^3 **7.** 1530 in^3 **9.** 7.56 in.   **11.** 933,000 gal   **13. a.** 0.375 or $\frac{3}{8}$ **b.** 0.25 or $\frac{1}{4}$ **c.** $\frac{3}{r}$ **15. a.** 2140 ft^2 **b.** 16,800 ft^3 **c.** 301,000 lb   **d.** 39 gal   **17.** 380 ft^2

Chapter 12 Review (pages 419–420)

1. acute   **2.** obtuse   **3.** $\angle 1 = 59°; \angle 2 = 121°; \angle 3 = 59°; \angle 4 = 59°$ **4.** adjacent or supplementary   **5.** 20°   **6. a.** quadrilateral   **b.** pentagon   **c.** hexagon   **d.** triangle   **e.** octagon   **7.** 36.00 cm; 60.0 cm^2 **8.** 360.4 m; 80$\overline{0}$0 m^2 **9.** 43.34 cm; 102 cm^2 **10.** 7.73 m   **11.** 1.44 cm   **12.** 159 m^2; 58.7 m   **13.** 796 m^2; 159.5 m   **14.** 4.73 cm^2; 10.93 cm   **15.** 41.9 m   **16.** 3.85 cm   **17.** 86°   **18.** 8.0 m   **19.** 423 cm^2; 72.9 cm   **20.** 12.1 cm   **21.** 216°   **22.** $\frac{2\pi}{15}$ rad or 0.419 rad   **23.** 10°   **24.** 42.2 cm   **25.** 1.55 rad   **26.** 291 cm   **27.** 1790 m^2; 2150 m^2 **28.** 5760 m^3 **29.** 1260 cm^3 **30.** 70.4 cm^3 **31. a.** 70.4 cm^2 **b.** 95.5 cm^2 **32.** 131 m^3 **33. a.** 57,300 m^3 **b.** 6380 m^2 **34. a.** 869 m^3 **b.** 44$\overline{0}$ m^2 **35.** 65,900 in^3; 6150 in^2 **36. a.** 13 gal; $520   **b.** 27$\overline{0}$0 lb

Cumulative Review Chapters 1–12 (pages 421–422)

1. -4 **2.** $13\frac{3}{4}$ in.   **3.** b   **4. a.** 100 L   **b.** 50 L   **5.** $2x^2 - 8x - 4$ **6.** $-8x^3 + 18x^2 + 23x - 15$ **7.** -7 **8.** 19.4   **9.** 34 lb/bu   **10.** 56 lb   **11.** $y = \frac{5x - 10}{8}$ **12.** $-\frac{7}{3}$ **13.** (4, 8)   **14.** 325 Ω, 975 Ω   **15.** $6x^2 + 11x - 35$ **16.** $16x^2 - 24x + 9$ **17.** $5x(x^2 - 3)$ **18.** $(x - 7)(x + 4)$ **19.** $\frac{7}{2}, -3$ **20.** $\frac{3}{2}, -1$ **21.** 0.4, -3 **22.** $-1.72, 4.22$ **23.** [graph of $y = 2x^2 - 3x - 2$ with vertex $V(\frac{3}{4}, -3\frac{1}{8})$]   **24.** $\angle 1 = \angle 2 = 82°; \angle 3 = \angle 5 = 98°$ **25.** 59°   **26.** 76.3 cm, 265 cm^2 **27.** 6.59 m, 1.86 m^2 **28.** 191 cm^2, 49.0 cm   **29.** 7.31 cm   **30.** 987 cm^3; 568 cm^2

CHAPTER 13

Exercises 13.1 (pages 427–428)

1. a **3.** c **5.** a **7.** B **9.** B **11.** 60.0 m   **13.** 55.2 mi   **15.** 17.7 cm   **17.** 265 ft   **19.** 35$\overline{0}$ m   **21.** 1530 km   **23.** 59,900 m   **25.** 0.5000   **27.** 0.5741   **29.** 0.5000   **31.** 0.7615   **33.** 2.174   **35.** 0.8686   **37.** 0.5246   **39.** 0.2536   **41.** 0.05328   **43.** 0.6104   **45.** 0.3929   **47.** 0.6800   **49.** 52.6°   **51.** 62.6°   **53.** 54.2°   **55.** 9.2°   **57.** 30.8°   **59.** 40.8°   **61.** 10.98°   **63.** 69.07°   **65.** 8.66°   **67.** 75.28°   **69.** 43.02°   **71.** 20.13°   **73.** sin A and cos B; cos A and sin B

Exercises 13.2 (page 430)

1. $A = 35.3°; B = 54.7°$ **3.** $A = 24.4°; B = 65.6°$ **5.** $A = 43.3°; B = 46.7°$ **7.** $A = 45.0°; B = 45.0°$ **9.** $A = 64.05°; B = 25.95°$ **11.** $A = 52.60°; B = 37.40°$ **13.** $A = 39.84°; B = 50.16°$ **15.** $A = 41.85°; B = 48.15°$ **17.** $A = 11.8°; B = 78.2°$ **19.** $A = 31.11°; B = 58.89°$ **21.** $A = 15.9°; B = 74.1°$ **23.** $A = 64.65°; B = 25.35°$

Exercises 13.3 (pages 431–432)

1. $b = 40.6$ m; $c = 54.7$ m   **3.** $b = 278$ km; $c = 365$ km   **5.** $a = 29.3$ cm; $b = 39.4$ cm   **7.** $a = 10.4$ cm; $c = 25.9$ cm   **9.** $a = 6690$ km; $c = 30,\overline{0}00$ km   **11.** $b = 54.92$ m; $c = 58.36$ m   **13.** $a = 161.7$ mi; $b = 197.9$ mi   **15.** $a = 4564$ m; $c = 13,170$ m   **17.** $b = 5950$ m; $c = 6270$ m   **19.** $a = 288.3$ km; $c = 889.6$ km   **21.** $a = 88.70$ m; $b = 163.0$ m   **23.** $a = 111$ cm; $b = 231$ cm

APPENDIX C ♦ Selected Answers 561

Exercises **13.4** (page 434)
1. $B = 39.4°$; $a = 37.9$ m; $b = 31.1$ m   **3.** $A = 48.8°$; $b = 234$ ft; $c = 355$ ft   **5.** $A = 22.6°$; $B = 67.4°$; $a = 30.0$ mi
7. $B = 37.9°$; $b = 56.1$ mm; $c = 91.2$ mm   **9.** $B = 21.2°$; $a = 36.7$ m; $b = 14.2$ m   **11.** $A = 26.05°$; $B = 63.95°$; $c = 27.33$ m
13. $B = 60.81°$; $a = 1451$ ft; $b = 2597$ ft   **15.** $A = 67.60°$; $B = 22.40°$; $c = 50.53$ m   **17.** $a = 319$ m; $b = 365$ m; $B = 48.9°$
19. $b = 263.8$ ft; $c = 1607$ ft; $A = 80.55°$ **21.** $a = 268.6$ m; $A = 44.90°$; $B = 45.10°$ **23.** $a = 14\overline{0}0$ ft; $c = 9470$ ft; $A = 8.5°$

Exercises **13.5** (pages 437–441)
1. 7.9 m   **3.** 5.2 m   **5.** 58.2°; 32.3 ft   **7.** 5°; 9.6%   **9.** 92.1 ft   **11.** 28.1°   **13.** 5.70 cm   **15.** 95.3 m   **17. a.** 430 Ω   **b.** 38°
c. 37°; $3\overline{0}0$ Ω   **19. a.** 53.7 V   **b.** 267 V   **c.** 236 V   **21.** 3.7529 in.   **23.** 9.5°   **25. a.** 3.46 in.   **b.** 2.04 in.   **27.** $A = 55.2°$;
$x = 22.8$ ft   **29.** $x = 4.50$ cm; $y = 3.90$ cm   **31.** $a = 12.4$ ft; $b = 30.6$ ft   **33. a.** 7.4 ft   **b.** 3.1 ft

Chapter **13** Review (pages 442–443)
1. 29.7 m   **2.** B **3.** Hypotenuse   **4.** 43.8 m   **5.** $\sin A$ **6.** Length of side adjacent to $\angle A$
7. $\dfrac{\text{length of side opposite } \angle B}{\text{length of side adjacent to } \angle B} = \dfrac{32.2 \text{ m}}{29.7 \text{ m}}$ **8.** 0.8070   **9.** 1.138   **10.** 0.4023   **11.** 45.5°   **12.** 10.4°   **13.** 65.8°   **14.** 39.2°
15. 50.8°   **16.** 7.17 m   **17.** 8.25 m   **18.** $b = 21.9$ m; $a = 18.6$ m; $A = 40.4°$ **19.** $b = 102$ m; $c = 119$ m; $B = 58.8°$
20. $A = 30.0°$; $B = 60.0°$; $b = 118$ mi   **21.** 4950 m   **22.** 5250 ft   **23.** 14.0°

CHAPTER 14

Exercises **14.1** (pages 450–451)
1. 0.6820   **3.** -0.4067 **5.** -0.4352 **7.** -0.9724 **9.** 0.7701   **11.** -0.4390
13. 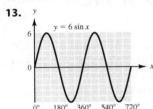   **15.** 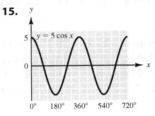   **17.** 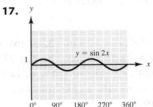   **19.**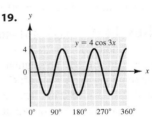

21. 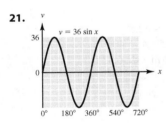   **23.**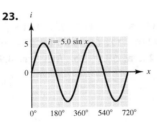

25. 25 V; -33 V   **27.** 3.5 A; -3.5 A   **29.** 2.0×10^{-4} s   **31.** 1.8 Hz   **33.** 8.8×10^9 Hz or 8.8 GHz   **35.** 2.25 m/s   **37.** 1.53 m

Exercises **14.2** (page 454)
1. 120°; 3   **3.** 60°; 8   **5.** 40°; 10   **7.** 720°; 6

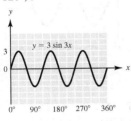   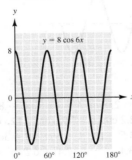   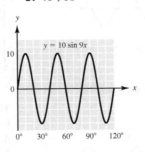   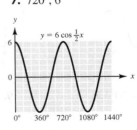

9. 540°; 3.5   **11.** 144°; 4   **13.** 360°; 1; 30° left   **15.** 360°; 2; 60° right

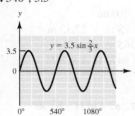 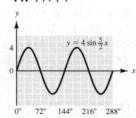 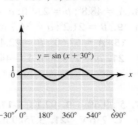 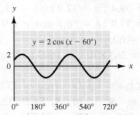

17. 120°; 4; 60° left   **19.** 90°; 10; 30° right   **21.** 720°; 5; 180° left   **23.** 1440°; 10; 720° left

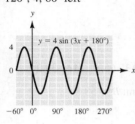 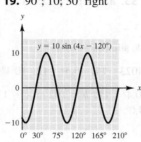 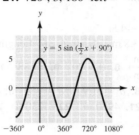 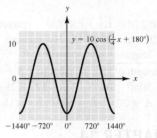

Exercises 14.3 (pages 456–457)

1. $c = 24.9$ m; $B = 41.5°$; $C = 70.5°$ **3.** $a = 231$ ft; $c = 179$ ft; $C = 42.9°$ **5.** $a = 14.4$ cm; $b = 32.9$ cm; $C = 130.3°$
7. $b = 32.5$ m; $c = 22.4$ m; $B = 82.9°$ **9.** $b = 79.3$ mi; $B = 145.0°$; $C = 14.9°$ **11.** $a = 6910$ m; $c = 5380$ m; $A = 74.9°$
13. $a = 26.4$ km; $A = 54.2°$; $C = 20.3°$ **15.** $c = 38\overline{0}$ ft; $B = 15.1°$; $C = 148.4°$ **17.** 63.5 m   **19.** 7.29 km   **21.** 214 ft

Exercises 14.4 (pages 461–462)

1. a. one solution   **b.** $B = 28.2°$; $C = 113.8°$; $c = 62.9$ m   **3. a.** two solutions   **b.** $B = 28.7°$; $C = 125.7°$; $c = 517$ m and $B = 151.3°$;
$C = 3.1°$; $c = 34.4$ m   **5.** no solution   **7. a.** two solutions   **b.** $A = 77.0°$; $B = 31.8°$; $b = 132$ cm and $A = 103.0°$; $B = 5.8°$; $b = 25.4$ cm
9. a. one solution   **b.** $A = 44.6°$; $C = 30.4°$; $c = 17.3$ mi   **11. a.** two solutions   **b.** $C = 34.3°$; $B = 114.2°$; $b = 656$ m and $C = 145.7°$; $B = 2.8°$; $b = 35.2$ m   **13.** no solution   **15. a.** two solutions   **b.** $C = 15.3°$; $A = 156.7°$; $a = 1280$ m and $C = 164.7°$; $A = 7.3°$; $a = 412$ m   **17.** 496 ft or 78.5 ft   **19.** 9.99 ft, 28.3°

Exercises 14.5 (pages 465–467)

1. $a = 21.3$ m; $C = 74.0°$; $B = 51.0°$ **3.** $c = 476$ ft; $B = 36.9°$; $A = 28.1°$ **5.** $A = 33.7°$; $B = 103.5°$; $C = 42.8°$
7. $A = 148.7°$; $C = 12.0°$; $b = 3070$ ft   **9.** $A = 30.7°$; $B = 107.3°$; $C = 42.0°$ **11.** 65.3 m   **13.** $C = 30.5°$; $A = 40.9°$
15. 61.0°; 30.0°; 89.0°   **17.** 132 mi   **19.** 147 in.   **21.** 43.6 ft; 19.7 ft   **23. a.** 59.5°   **b.** 70.5°   **c.** 4.92 m   **d.** 6.33 m
25. 114.3°   **27.** 36 yd   **29.** 31.5 ft   **31.** 134 m

Chapter 14 Review (pages 468–469)

1. −0.7536   **2.** −0.4970   **3.** 0.7361

4.   **5.**   **6.** 120°; 5   **7.** 90°; 3

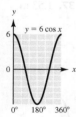 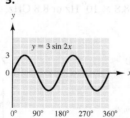 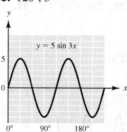 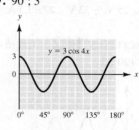

8. 360°; 4; 60° left   **9.** 180°; 6; 90° right

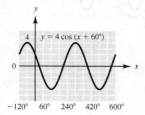 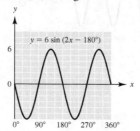

10. $c = 258$ m; $A = 42.5°$; $C = 84.8°$ **11.** $B = 43.2°$; $b = 62.2$ cm; $a = 79.6$ cm   **12.** $b = 129$ m; $C = 109.0°$; $A = 53.5°$ **13.** $A = 66.8°$; $B = 36.2°$; $C = 77.0°$ **14.** $c = 33.4$ m; $B = 33.1°$; $C = 117.4°$ and $B = 146.9°$; $C = 3.6°$; $c = 2.36$ m
15. $c = 3010$ m; $A = 20.5°$; $C = 141.0°$ and $c = 167$ m; $A = 159.5°$; $C = 2.0°$ **16.** $A = 27.3°$; $B = 59.7°$; $C = 93.0°$
17. no solution   **18. a.** 36.6°   **b.** 52.8 m   **19.** 9.40 in.
20. a. 23.5 m   **b.** 16.7 m   **c.** 15.1 m   **d.** 7.04 m

APPENDIX C ♦ Selected Answers

Cumulative Review Chapters 1–14 (pages 469–470)

1. 8 **2.** $-\dfrac{7}{30}$ **3.** 4.18×10^4 **4.** 198 **5.** 13 Ω **6. a.** 0.0001 mm **b.** 0.00005 mm **c.** 0.02778 **d.** 2.78% **7.** $6a^2 - 2a + 2$
8. 8 **9.** $\dfrac{26}{21}$ **10.** 102 m² **11.** $y = \dfrac{8 - 5x}{3}$ **12.** Graphs coincide; many solutions. **13.** $(x - 14)(x + 12)$ **14.** $3(x + 7)(x - 9)$
15. $5, -\dfrac{2}{3}$ **16.** 2.27; −1.77 **17.** 515 ft²; 126.3 ft **18.** 13.7 ft **19.** 6.38 m³ **20.** 2.379 **21.** 51.6° **22.** 55.8° **23.** 10.7 cm
24. 19.1 cm **25.** pitch $\dfrac{2}{3}$; 173 in. **26.** −0.9812 **27.** **28.** $P = 720°, A = 2$ **29.** 65.2° **30.** 25.8°

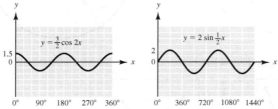

CHAPTER 15

Exercises 15.1 (pages 473–474)

1. $42,000 **3.** $47,000 **5.** $71,000 **7.** $82,000 **9.** $46,000 **11.** 1.8 million barrels **13.** United States **15.** 3.3 million barrels
17. 19 million barrels **19.** 2.4 million barrels
21. 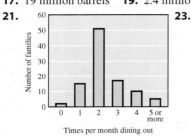 **23.**

25.
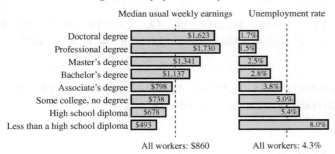

Exercises 15.2 (pages 476–477)

1. 94° **3.** 55° **5.** 270° **7.** 219° **9.** 15° **11.** 328°
13. United States Work-Related Deaths by Cause in 2013 **15.** Industrial Technology Credit Hour Requirements

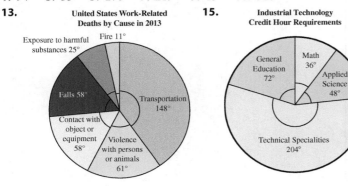

17.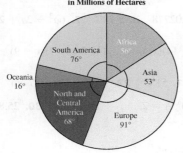

19.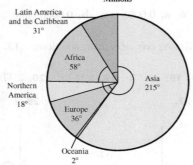

Exercises 15.3 (pages 478–479)

1. 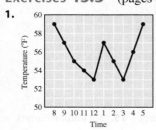 **3.**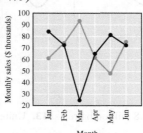

5. 29.52 in. Fri. near midnight **7.** 29.34 in. **9.** 98% **11.** 90° **13.** 95% **15. a.** 85 ppmv **b.** 1995

Exercises 15.4 (page 481)
1. 900 mW; 14 dB **3.** 85 mW **5.** 30 dB; 32 dB **7.** 16 dB **9.** 14 dB; 16 dB

Exercises 15.5 (pages 482–483)
1. 6911 **3.** 202$\overline{0}$ **5.** 1.096 **7.** 51 **9.** 4.58 **11.** 47.89 cm **13.** 0.2619 in. **15.** 30,292 mi **17.** 73° **19.** 161 kW **21.** 1059.8 million tons

Exercises 15.6 (page 485)
1. 6185.5 **3.** 2023 **5.** 1.102 **7.** 53 **9.** 4.17 **11.** 47.87 cm **13.** 0.2618 in. **15.** 28,985 mi **17.** 72.5° **19.** 169 kW **21.** 163 **23.** 107 **25.** 23 **27.** 2.85 mm **29.** 2 modes: 7291, 7285 **31.** 28 **33.** 6.2 in. **35.** 0.024 in. **37. a.** 9 **b.** 9 **c.** 9 **d.** 9

Exercises 15.7 (page 487)
1. 6061 **3.** 447 **5.** 0.208 **7.** 73 **9.** 8.54 **11.** 0.62 cm **13.** 0.0010 in. **15.** 19,393 mi **17.** 20°F **19.** 101 kW **21.** 1931 **23.** 15$\overline{0}$ **25.** 0.064 **27.** 25 **29.** 2.45 **31.** 0.26 cm **33.** 0.0003 in. **35.** 7221 mi **37.** 7°F **39.** 39 kW

Exercises 15.8 (pages 491–494)
1. 59 **3.** 7.7 months **5.** 36 h **7.** 42 lb **9.** 11 **11.** 4.2 days **13.** 99 strokes

15.

Interval	Midpoint	Frequency	Product
18.5–43.5	31	14	434
43.5–68.5	56	7	392
68.5–93.5	81	4	324
93.5–118.5	106	3	318
118.5–143.5	131	0	0
143.5–168.5	156	0	0
168.5–193.5	181	0	0
193.5–218.5	206	1	206
		29	1674

Mean = 58 players

APPENDIX C ♦ Selected Answers

Exercises 15.9 (page 496)
1. 9   **3.** 4 months   **5.** 5.2 h   **7.** 15 lb   **9.** 4   **11.** 3 days   **13.** 12 strokes   **15.** 38 players   **17. a.** 4 mm   **b.** 1.3 mm

Exercises 15.10 (pages 498–499)

1. a.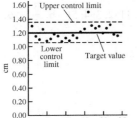

b. Out of control; in hour 16, the value was outside of control limits. And, there are nine values in a row on the same side of the target line.

3. a.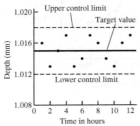

b. process in control

5. a.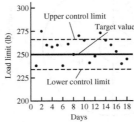

b. Out of control; days 2, 9, and 13 are outside the control limits.

Exercises 15.11 (pages 501–502)
1., **3.**, **5.**, **7.**, **9. a.** positive correlation   **11. a.**   **b.** no linear correlation

Exercises 15.12 (pages 504–505)
1. Approximately 95%   **3. a.** 63   **b.** 6.9   **c.** No   **5. a.** 3544   **b.** 1033   **c.** Yes   **7.** 68,000 blouses of sizes 10, 12, and 14
27,000 blouses of sizes 8 and 16
4700 blouses of sizes 4, 6, and 18
300 blouses of size 2

Exercises 15.13 (pages 506–507)
1. [A, 2, 3, 4, 5, 6, 7, 8, 9, 10, J, Q, K]   **3.** [J diamonds, Q diamonds, K diamonds, J hearts, Q hearts, K hearts]   **5.** [{red, red}, {red, white}, {red, white}]   **7.** $\frac{1}{13}$ **9.** $\frac{1}{6}$ **11.** 0   **13.** $\frac{3}{14}$ **15.** $\frac{1}{7}$ **17.** $\frac{13}{2500}$

Exercises 15.14 (pages 507–508)
1. $\frac{1}{16}$ **3.** $\frac{5}{32}$ **5.** $\frac{9}{25}$ **7.** $\frac{1}{12}$ **9.** $\frac{1}{28,561}$ **11.** $\frac{1}{216}$ **13.** $\frac{1}{49}$ **15.** $\frac{84}{6859}$

Chapter 15 Review (pages 509–510)

1. 126° **2.** 202° **3.** Students in fall 2016

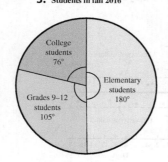

4.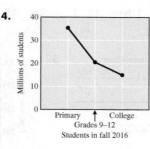

Students in fall 2016

5. 80°F **6.** 7.0036 mm **7.** 7.0036 mm **8.** 0.0001 mm **9. a.** 49.9 **b.** 18.3 **10.** 84.4 **11.** 83 **12.** 7.4 **13.** 19.2 **14.** 5.2 **15. a.** [1, 2, 3, 4, 5] **b.** $\frac{3}{5}$ **16.** $\frac{5}{324}$

CHAPTER 16

Exercises 16.1 (page 514)
1. 3 **3.** 6 **5.** 9 **7.** 51 **9.** 47 **11.** 78 **13.** 59 **15.** 156 **17.** 140 **19.** 63

Exercises 16.2 (page 515)
1. 1000 **3.** 1011 **5.** 1100 **7.** 10000 **9.** 100001 **11.** 100100 **13.** 1000111 **15.** 1000101 **17.** 1110101 **19.** 100100111 **21.** 100110 **23.** 1010001 **25.** 1000000 **27.** 1000011 **29.** 111111

Exercises 16.3 (page 517)
1. 100 **3.** 110 **5.** 10 **7.** 1 **9.** 1010 **11.** 11100 **13.** 10011 **15.** 10 **17.** 1001 **19.** 1000111 **21.** 10 **23.** −10 **25.** 10 **27.** −11 **29.** −11101 **31.** 10111010 **33.** −101110 **35.** 110000010

Exercises 16.4 (page 518)
1. 1111 **3.** 1100 **5.** 11001 **7.** 100100 **9.** 100001 **11.** 1001000 **13.** 1010100 **15.** 1001110011 **17.** 1111110 **19.** 100001001

Exercises 16.5 (page 519)
1. 1110 **3.** 111111 **5.** 1001000 **7.** 10100 **9.** 11000 **11.** 100000 **13.** 100101 **15.** 1100100 **17.** 1101111 **19.** 1110001

Exercises 16.6 (page 520)
1. 2.5 **3.** 2.25 **5.** 5.75 **7.** 4.125 **9.** 4.625 **11.** 12.75 **13.** 7.875 **15.** 26.5625 **17.** 4.8125 **19.** 3.8125

Exercises 16.7 (page 522)
1. 37 **3.** 293 **5.** 30 **7.** 197 **9.** 1031 **11.** 2594 **13.** 3010 **15.** 2859 **17.** 3294 **19.** 10,843 **21.** EB **23.** 3A **25.** D18 **27.** 15DA0 **29.** 8316

Exercises 16.8 (pages 524–525)
1. B **3.** 14 **5.** 17 **7.** A9 **9.** 13D **11.** 169 **13.** 77D **15.** 1247 **17.** 1678 **19.** CC3A **21.** CA51 **23.** 10011 **25.** B0EE **27.** E6CD **29.** 17AEC **31.** −C **33.** −3 **35.** −13 **37.** −26 **39.** −9 **41.** −91 **43.** −11F **45.** −367 **47.** −1EE **49.** −2B2 **51.** −A88 **53.** C15 **55.** −18AA **57.** 1B8F **59.** −342F

Exercises 16.9 (page 527)
1. 6 **3.** B **5.** 17 **7.** 24 **9.** 44 **11.** E6 **13.** 138 **15.** 73D **17.** CE7 **19.** CCC **21.** 6891 **23.** F1CA **25.** 110 **27.** 100100 **29.** 101010 **31.** 1001010001 **33.** 101000110010 **35.** 11111100100 **37.** 10011011101 **39.** 101011001101 **41.** 100101000111011 **43.** 1011110010101111

Chapter 16 Review (page 528)
1. 13 **2.** 25 **3.** 52 **4.** 22.75 **5.** 100110 **6.** 10010110 **7.** 10100 **8.** 1001 **9.** 1101 **10.** 10010 **11.** 11110011 **12.** 100100 **13.** 11001101 **14.** 10000011010 **15.** 225 **16.** 44 **17.** 3102 **18.** 138 **19.** 34 **20.** 1210 **21.** 83 **22.** 293 **23.** E63D **24.** 16 **25.** 26 **26.** 173 **27.** 1001100 **28.** 1101100101 **29.** 101100101010 **30.** 100101010100001

Cumulative Review Chapters 1–16 (page 529)

1. a. 2927.40 **b.** 2930 **2.** 37.0×10^{-9} **3.** 22.2°C **4.** 7890 km **5.** $7a^2 - 18a + 11$ **6.** $-\dfrac{41}{25}$ **7.** 6.45 **8.** $4x + 6y = 27$
9. 4.9 m $\times$ 15.0 m **10.** $3(5x - 18)(2x + 15)$ **11.** $-\dfrac{1}{3}, 2$ **12.** $-0.143 \pm 1.46j$ **13.** 306 cm^2 **14.** 21,700 in^3 **15.** 24.1°
16. 19.9 m **17.** -0.2419 **18.** 44.0° **19.** 27.6 **20.** 28.5 **21.** 5.5

22.

Interval	Midpoint	Frequency
5.5–10.5	8	7
10.5–15.5	13	11
15.5–20.5	18	10
20.5–25.5	23	6
25.5–30.5	28	1
30.5–35.5	33	1

23. 16 **24.** 15.5 **25.** 10001000 **26.** 10001100111
27. 1001100100 **28.** 229 **29.** 10010101011 **30.** 1D02

Appendix A (page 534)

1.   **3.** 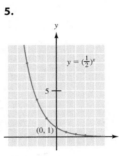 **5.** 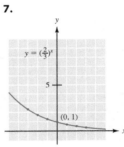 **7.** 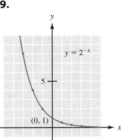 **9.**

11. 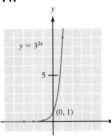 **13.** 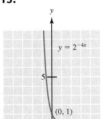 **15.** 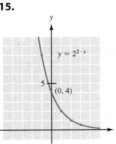 **17.** 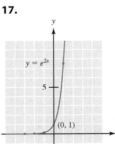 **19.**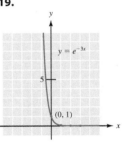

21. $5491.43 **23.** 77,400 **25.** $45,490 **27.** 0.0108 mA

Appendix B (pages 538–539)

1.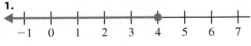

3.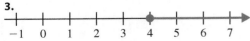

5.

7.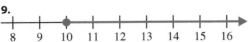

9.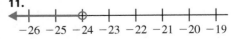

11.

13. $x < 8$ **15.** $b \geq -1$ **17.** $a > 0$ **19.** $x \leq 3$
21. $x \geq -20$ **23.** $a \leq 16$

25. $x < 3$

27. $x \leq -5$

29. $x \geq 9$

31. $x \geq 20$

33. $x > -8$

35. $x \geq -50$

37. $x < 8$

39. $x \leq 6$

41. $x \leq 6$

43. $x \geq 1$

45. $x < -4$

47. $x > 31$

INDEX

A

Absolute value, 101–102
Accuracy, of measurement, 155–156, 174–177
Acre, 148
Acute angle, 364, 454
Acute triangle, 374
Addition, 2, 200
    Associative Property for, 200
    binary numbers, 514–515
    Commutative Property for, 200
    decimal fractions, 56–58
    fractions, 29–32, 34–37, 43–45
    hexadecimal numbers, 522–523
    integers, 2
    measurements, 174–177
    method for solving pairs of linear equations, 310–315
    mixed numbers, 33
    polynomials, 207
    signed numbers, 101–103
Addition Property of Equality, 222
Additive Inverse, 200
Adjacent angles, 364
Algebra, 200–201
Algebraic expressions, 222
    evaluating, 201
    factoring
        finding binomial factors, 330–331
        finding monomial factors, 326–327
        trinomials, 332
    polynomials, 206–208, 211–213
    simplifying, 202–205
Algebraic symbols, translating words into, 235
Alternate angles, 365
Altitude, 369, 375
Amortization, 89
Ampere (A), 131, 142–143
Amplitude, 447
Angle, 363
    acute, 364, 454
    adjacent, 364
    alternate, 365
    complementary, 365
    corresponding, 365, 366
    exterior, 365, 366
    inscribed, 387
    interior, 365, 366
    measure of, 363–364, 392–396
    obtuse, 364, 454
    right, 364
    sides of, 363
    sum of angles of triangle, 377
    supplementary, 365
    using trigonometric ratios to find, 428–430
    vertex of, 363
    vertical, 365
Angle of depression, 434
Angle of elevation, 434
Annual depreciation, 6
Applications
    involving equations, 236–239
    involving pairs of linear equations, 317–320
    involving quadratic equations, 348–350
    percent, 86–91
    trigonometric ratios, 434–436
Approximate numbers, 154–155
Arcs, 387
Area, 13–14
    circle, 386
    metric system, 140–141
    parallelogram, 19, 370
    rectangle, 370
    rhombus, 370
    sector of circle, 394
    segment of circle, 395
    square, 370
    trapezoid, 19, 370
    triangle, 19, 376
Associative Property
    for Addition, 200
    for Multiplication, 200
Axis
    of cone, 408
    of cylinder, 402–403

B

Bar graph, 472–474
Base, 11
    cone, 408
    cylinder, 403
    percent, 76–77
    prism, 397
    pyramid, 407
    trapezoid, 19
    triangle, 19
Binary number system, 513–514
    addition, 514–515
    conversion from decimal to, 518–519
    conversion from hexadecimal to, 526
    conversion to decimal, 519–520
    conversion to hexadecimal, 525–526
    multiplication, 517
    subtraction, 516
Binomial factors, 330–331
Binomials, 206
    finding product of, mentally, 327–329
    square of, 333
Board foot, 43

C

Calculator
    adding and subtracting fractions using, 36–37
    adding and subtracting decimal fractions using, 58
    multiplying and dividing decimal fractions using, 67–68
    multiplying and dividing fractions using, 44–45
    reciprocal formulas using, 247–249
    scientific calculator, 6–7
    scientific calculator second function key, 28
Candela, 131
Cartesian coordinate system, 280–281
Cash discount, 90–91
Celsius scale, 144, 145
Center, of circle, 385
Centimetre (cm), 134
Chord, 387, 388
Circle, 385
    arcs, 387
    area, 386
    center, 385
    chord, 387, 388
    circumference, 385, 386
    diameter, 385
    intercepted arc, 387
    pi (π), 385
    radius, 385
    secant, 387
    sector, 394
    segment, 395
    tangent, 387, 388
Circle graph, 474–476
Circuit diagram, 3
Circular cylinder, 402–403
Circumference, circle, 385, 386
Color code of electrical resistors, 185–188
Common fraction, 24
    changing to decimal, 55–56
Commutative Property for Addition, 200
Commutative Property for Multiplication, 200
Complementary angles, 365
Complex numbers, 357
Compound interest, 87–88
Cone, 408
    axis of, 408
    frustum of, 410–411
    lateral surface area, 408
    right circular, 408
    slant height, 408
    total surface area, 408
    vertex of, 408
    volume, 408
Control charts, 496–497

570 INDEX

Conversion
  from binary to decimal, 519–520
  from binary to hexadecimal, 525–526
  from decimal to binary, 518–519
  from hexadecimal to binary, 526
  U.S./metric, 146–149
Conversion factors, 49
  metric system, 134–135
Coordinates, 281
Coordinate system, rectangular, 280–281
Corresponding angles, 365, 366
Cosine, 425
Cosine graph, 445–450
  period, 451
  phase shift, 452–453
Cube of a number, 84–85
Cube root, 85–86
Cubic centimetre, 138, 139
Cubic metre, 132, 138
Cubic units, 14
Current, 142–143
Curved line graph, 480
Cylinder, 403
  circular, 402–403
  lateral surface area, 404
  total surface area, 404
  volume, 403

D

Decimal fractions, 53
  addition, 56–58
  changing common fraction to, 55–56
  changing engineering notation to, 123–124
  changing percent to, 71–72
  changing scientific notation to, 119
  changing to percent, 72–73
  division, 65–68
  multiplication, 64–65
  place values for, 53–54
  repeating, 54–55
  subtraction, 56–58
  writing in engineering notation, 123
  writing in scientific notation, 118
Decimal number system, 513
  conversion from binary to, 519–520
  conversion from hexadecimal to, 521
  conversion to binary, 518–519
  conversion to hexadecimal, 521
Decreasing function, 531
Degrees, 392

Denominator, 24
  least common, 21, 30–31
Dependent variable, 288
Depreciation, 6
Diameter, circle, 385
Difference, 2
  of two squares, 335–336
Digital measuring instruments, precision of, 160
Digital micrometers, 173–174
Digital thermometers, 145
Digital vernier caliper, 166
Direct variation, 265–269
Discounts, 90–91
Discriminant, 357
Distributive Property, 200
Dividend, 5
Divisibility, 20–21
  by 2, 21–22, 23
  by 3, 22, 23
  by 5, 22, 23
  tests, 21–23
Division, 5–6, 200
  decimal fractions, 65–68
  engineering notation, 124
  fractions, 42
  measurements, 178–180
  by monomial, 213–214
  by polynomial, 215–217
  powers of 10, 114–115
  scientific notation, 120–121
  short, 21
  signed numbers, 108
  by zero, 26, 214
Division Property of Equality, 223
Divisor, 5
Drop factor, 255

E

Electrical resistors, color code of, 185–188
Electric circuit
  parallel, 4, 34–35
  series, 4
Electronics, units commonly used in, 143
Engineering notation, 123
  division, 124
  multiplication, 124
  powers of numbers in, 124
  square root of numbers in, 125
  writing decimal in, 123
  writing in decimal form, 123–124
English system of weights and measures, 13–14, 24, 48–51
  metric conversion, 146–149
Equations, 222
  applications involving, 236–239
  equivalent, 222

  exponential, 530–533
  of line, 295–296
    point-slope form, 298–299
    slope-intercept form, 296–298
  linear
    applications involving pairs of, 317–320
    graphing, 283–288
    solving by addition, 310–315
    solving by graphing, 305–309
    in two variables, 278–282
  power, 530
  quadratic
    applications involving, 348–350
    graphs of, 352–355
    solving by factoring, 343–345
    solving using quadratic formula, 345–347
  solving, 222–225
    containing fractions, 230–234
    containing parentheses, 228–230
    with variables in both members, 226–228
    translating words into algebraic symbols, 235
Equilateral triangle, 373–374
Equivalent equations, 222
Equivalent fractions, 24–25
  signed, 112–113
Evaluating an expression, 201
Even integers, 20
Exact number, 155
Exponential equations, 530–533
Exponential growth, 532
Exponents, 2, 11, 86, 200
  multiplying powers, 209–210
  raising power to a power, 210
  raising product to a power, 210
Exterior angles, 365, 366
Extremes, 258

F

Factoring
  algebraic expressions, 326–327, 330–331
  difference of two squares, 335–336
  general trinomials, 336–338
  perfect square trinomials, 334–335
  solving quadratic equations by, 343–345
  trinomials, 332
Factors, 5, 20
  binomial, 330–331
  monomial, 326–327
  prime, 20–21
  of special products, 334–336
Fahrenheit scale, 144, 145
Five, divisibility by, 22, 23

FOIL method, 328
Formulas, 18–19, 240–241
  amortization, 89
  area
    circle, 386
    parallelogram, 19, 370
    rectangle, 370
    rhombus, 370
    square of, 370
    trapezoid, 19, 370
    triangle, 19, 376
  base (percent), 77
  circumference of circle, 386
  compound interest, 88
  distance, 240
  exponential growth, 532
  geometry, 19
  lateral surface area
    cone, 408
    cylinder, 404
  part (percentage), 77
  perimeter
    parallelogram, 370
    rectangle, 370
    rhombus, 370
    square, 370
    trapezoid, 370
    triangle, 375
  power, 43
  pressure, 240
  Pythagorean theorem, 374–375, 424, 432–433
  quadratic, 345–347, 357
  rate (percent), 77
  reciprocal using calculator, 247–249
  simple interest, 87
  solving, 241–243
  subscripts in, 241
  substituting data in, 244–246
  surface area, sphere, 414
  velocity, 450
  volume
    cone, 408
    cylinder, 403
    prism, 399
    pyramid, 408
    sphere, 414
  work, 18
Fractions, 24. See also Decimal fractions
  addition, 29–32, 34–37, 43–45
  applications involving, 34–37, 43–44
  changing common fraction to decimal, 55–56
  changing improper fraction to mixed number, 26–28
  changing percent to, 74–75
  changing to percent, 73–74
  common, 24
  denominator, 24
  dividing, 42
  equal or equivalent, 24–25, 112–113

INDEX

finding least common
    denominator of, 30–31
improper, 26
multiplication, 41–42
numerator, 24
proper, 26
reducing to lowest terms, 25–26
signed, 110–113
simplifying, 25–26
solving equations with,
    230–234
subtraction, 32–33, 34–37,
    43–45
Frequency, 449
Frequency distribution, 488, 490
Frustum
    of cone, 410–411
    of pyramid, 409–410
Fulcrum, 272

G

Gear system relationship, 271
Geometry
    angles, 363–366
    circles, 385–388
    cones, 408–409, 410–411
    cylinders, 402–405
    plane, 363
    polygons, 366–367
    prisms, 397–400
    pyramids, 407–408, 409–410
    quadrilaterals, 369–371
    radian measure, 392–396
    similar polygons, 381–382
    solid, 397
    spheres, 414
    triangles, 373–377
Gram (g), 136
Graphing
    inequalities, 535–538
    linear equations, 283–288
    quadratic equations, 352–355
    sine and cosine, 445–450
    solving pairs of linear
        equations by, 305–309
Graphs
    bar, 472–474
    circle, 474–476
    curved line, 480
    histogram, 499
    line, 477
    run chart, 500
    scattergram, 500–501
    semilogarithmic, 480
Greatest common factor (GCF),
    of polynomial, 326–327
Greatest possible error, 158–160
Grouped data, 488–491
    standard deviation for,
        494–496
Grouping symbols, 11

H

Hectare, 140, 141, 148
Heptagon, 367
Hexadecimal number system,
    520–522
    addition, 522–523
    for color codes, 526
    conversion from binary to,
        525–526
    conversion from decimal to, 521
    conversion to binary, 526
    conversion to decimal, 521
    subtraction, 523–524
Hexagon, 367
Histogram, 499
Horizontal line, 287
    slope of, 291
Hydraulic press, 267–268
Hypotenuse, 374, 424

I

Imaginary numbers, 356–359
Improper fraction, 26
Increasing function, 531
Independent events, 507
Independent variable, 288
Individual reading chart, 496
Inequality, 535
    graphing, 535–538
    sense of, 535
Inscribed angle, 387
Integers
    even, 20
    negative, 101, 102
    odd, 20
    positive, 2, 20, 101, 102
Intercepted arc, 387
Interest, 86
    amortization formula, 89
    cash discount, 90–91
    compound, 87–88
    simple, 86–87
Interior angles, 365, 366
International System of Units
    (SI), 131
Intersect, 364
Interval, 488
Inverse variation, 271–272
Irrational numbers, 101, 102
Isosceles triangle, 373, 374

J

Joule, 132

K

Kelvin (K), 131, 144
Kilogram (kg), 131, 137
Kilometre (km), 134
Kitchen ratio, 261–262

L

Lateral faces, 397
Lateral surface area
    cone, 408
    cylinder, 404
    prism, 397
Law of cosines, 462–465
Law of sines, 454–456, 457–461
Least common denominator
    (LCD), 21, 30–31
Length, 133–134
Lever arms, 272
Lever principle relationship, 272
Like terms, 203–204
Linear equations
    graphing, 283–288
    pairs of
        applications involving,
            317–320
        solving by addition, 310–315
        solving by graphing, 305–309
        solving by substitution,
            316–317
    in two variables, 278–282
Line graph, 477
Line of symmetry, 354
Lines
    equation of, 295–296
        point-slope form, 298–299
        slope-intercept form, 296–298
    graphing, 283–288
    horizontal, 287, 291
    intersecting, 364
    parallel, 293, 364, 366
    perpendicular, 293, 364
    secant, 387
    slope, 289–292, 295
    tangent, 387, 388
    transversal, 365, 366
    vertical, 288, 292
    x intercept of, 285–287
    y intercept of, 285–287, 295
Litre (L), 132, 138, 139

M

Mass, 136–137
Mathematical principles, 200
Mathematics at Work, 1, 100,
    130, 153, 199, 221, 253,
    304, 325, 342, 362, 423,
    444, 471, 512
Mean control chart, 496–497
Mean measurement, 481–482
Means, 258
Measurement, 49, 153, 154
    accuracy, 155–156, 174–177
    addition and subtraction,
        174–177
    approximate numbers vs.
        exact numbers, 154–155
    area, 140–141
    greatest possible error, 158–160
    length, 133–134
    mass, 136–137
    mean, 481–482
    median, 483–484
    metric system, 131–133
    micrometer caliper, 167–174
    multiplication and division,
        178–180
    nonuniform scales, 191–192
    percent of error, 182
    precision, 157–158, 160,
        174–177
    random errors, 184
    relative error, 182
    significant digits, 155–156
    single vs. multiple, 183–184
    systematic errors, 184
    time, 142
    tolerance, 183
    uniform scales, 189–191
    U.S. system of weights and
        measures, 48–51
    vernier caliper, 161–167
    volume, 138–140
    weight, 137
Mechanical advantage (MA),
    267–268
Median chart, 496
Median measurement, 483–484
Metre (m), 131, 133
Metre per second, 132
Metric micrometer, 168
Metric ruler, 134
Metric system, 13–14, 131
    area, 140–141
    basic units, 131
    conversion factors, 134–135
    current, 142–143
    length, 133–134
    mass, 136–137
    prefixes, 132
    temperature, 144–146
    time, 142
    units, 132
    U.S./English conversion,
        146–149
    volume, 138–140
    weight, 137
Metric ton, 137
Micrometer caliper, 167–174
Micrometer with vernier scale, 172
Millimetre (mm), 134
Mixed numbers, 26
    addition, 33
    changing improper fraction to,
        26–28
    changing percent containing,
        to fraction, 75
    subtraction, 33–34
Mode, 484

Mole, 131
Monomial factors, 326–327
Monomials, 206
    division by, 213–214
    multiplication, 209–211
Mortgage, 89
Multimeter, 192
Multiplication, 5, 200
    Associative Property for, 200
    binary numbers, 517
    binomials, 327–329
    Commutative Property for, 200
    decimal fractions, 64–65
    engineering notation, 124
    fractions, 41–42
    measurements, 178–180
    monomials, 209–211
    polynomials, 211–213
    powers, 209–210
    powers of 10, 114
    scientific notation, 120
    signed numbers, 107–108
Multiplication Property of Equality, 223
Multiplicative Inverse, 200

N

Negative integers, 101, 102
Negative power of 10, 116
Newton (N), 132, 137
Nonagon, 367
Nonuniform scales, 191–192
Normal distribution, 502–504
nth percentile, 484
Number line, 101
Number plane, 281
    plotting points in, 281–282
Numbers
    absolute value of, 101–102
    binary system of, 513–514
    complex, 357
    cube of, 84–85
    decimal system of, 513
    hexadecimal, 520–522
    imaginary, 356–359
    irrational, 101, 102
    prime, 20
    rational, 101, 102
    real, 101, 102, 357
    reciprocal of, 247–249
    rounding, 61–63
    signed, 101–103, 105–106, 107–108
    square of, 83–84
    square root of, 84
    whole, 2
Numerator, 24
Numerical coefficient, 203

O

Oblique triangle, 454–456
    solving, 462–465
Obtuse angle, 364, 454
Obtuse triangle, 374
Octagon, 367
Odd integers, 20
Ohm's law, 6, 44, 241
Ordered pairs, 278, 281
Order of operations, 11–13, 200
Origin, 280, 281
Out of phase, 452

P

Parabola
    graphing, 352–354
    line of symmetry, 354
    vertex of, 354–355
Parallel circuit, 4, 34–35
Parallel lines, 293, 364, 366
Parallelogram, 369
    area of, 19, 370
    perimeter of, 370
Parentheses, 11
    nested, 12
    removing, 202–203
    solving equations with, 228–230
Part (percentage), 76–80
Payment amount, 87–88, 89–90
Pentagon, 367
Pentagonal prism, 397
Percent, 71
    applications involving, 86–91
    base, 76–77
    changing decimal to, 72–73
    changing fraction to, 73–74
    changing to decimal, 71–72
    changing to fraction, 74–75
    finding the part, 76–80
    rate, 76–77
Percentiles, 484
Percent of error, 182
Perfect squares, 84
Perfect square trinomials, 333, 334–335
Perimeter, 5, 35–36
    parallelogram, 370
    rectangle, 370
    rhombus, 370
    square, 370
    trapezoid, 370
    triangle, 375
Period, of sine or cosine graph, 448, 451
Perpendicular lines, 293, 364
Phase shift, 452–453
pi (π), 385
Place value
    for decimals, 53–54
    rounding numbers to particular, 62

Plane geometry, 363
Plotting points, 281–282
Point-slope form, 298–299
Polygons, 366–367, 397. *See also specific types*
    regular, 367
    similar, 381–382
Polynomials, 206. *See also Binomials; Monomials; Trinomials*
    addition and subtraction, 206–208
    division by, 215–217
    greatest common factor of, 326–327
    multiplication, 211–213
Positive integers, 2, 20, 101, 102
Power, 11, 43, 86
Power equations, 530
Powers
    multiplication, 209–210
    raising power to a power, 210
    raising product to a power, 210
Powers of 10, 2
    division, 114–115
    multiplication, 114
    negative, 116
    raising to a power, 115–116
    zero, 116
Precision, 157–158, 160, 174–177
Prime factorization, 20–23
Prime factors, 20–21
Prime number, 20
Prism, 397
    base, 397
    lateral faces, 397
    lateral surface area, 397
    pentagonal, 397
    right, 397
    total surface area, 397
    triangular, 397
    volume, 399
Probability, 505–506
Problem solving, 244
Product, 5
    to a power, 210
    of sum and difference of two terms, 333
    of two binomials, finding mentally, 327–329
Proper fraction, 26
Proportion, 257–261
Protractor, 363
Pulley system relationship, 271
Pyramid, 407
    frustum of, 409–410
    square, 407
    triangular, 407
    volume, 408
Pythagorean theorem, 374–375, 424, 432–433

Q

Quadrants, 281
Quadratic equations, 343
    applications involving, 348–350
    graphs of, 352–355
    solving
        by factoring, 343–345
        using quadratic formula, 345–347
Quadratic formula, 345–347, 357
Quadrilateral, 367, 369–371
Quotient, 5, 6

R

Radian (rad), 132, 363, 392–396
Radius, circle, 385
Random errors, 184
Range, 485
Rate, 255
    percent, 76–77
Ratio, 254–256, 425
    direct variation, 265–269
    kitchen, 261–262
    trigonometric, 424–427
Rational numbers, 101, 102
Real number line, 101
Real numbers, 101, 102, 357
Reciprocal of number, 247–249
Rectangle, 369
    area of, 370
    perimeter of, 370
Rectangular coordinate system, 280–281
Regular polygon, 367
Relative error, 182
Remainder, 6
Repeating decimals, 54–55
Resistor, value of, 186
Rhombus, 369, 370
    area of, 370
    perimeter of, 370
Right angle, 364
Right circular cone, 408
Right circular cylinder, 403
Right prism, 397
Right triangle, 374, 424
    solving, 432–433
    using trigonometric ratios to find angles, 428–430
    using trigonometric ratios to find sides, 430–431
Roots
    cube, 85–86
    equations, 222
    square, 84, 125
Rounding numbers, 61–63
Run chart, 500

INDEX

S

Sample space, 505
Scale drawing, 265–267
Scalene triangle, 373, 374
Scales
    nonuniform, 191–192
    uniform, 189–191
Scattergram, 500–501
Scientific notation, 118
    comparing numbers using, 119–120
    division, 120–121
    multiplication, 120
    powers of numbers in, 121–122
    writing decimal in, 118
    writing in decimal form, 119
Secant, 387
Second (s), 131, 142
Sector, 394
Segment, of circle, 395
Semilogarithmic graph, 480
Sense, of inequality, 535
Series circuit, 4
Short division, 21
SI (International System of Units), 131
Signed numbers, 102
    adding and subtracting combinations of, 105–106
    addition, 101–103
    division, 108
    fractions, 110–113
    multiplication, 107–108
    subtraction, 105–106
Significant digits, 62–63, 155–156
Similar polygons, 381–382
Simple interest, 86–87
Simplifying algebraic expression, 202–205
Sine, 425
Sine graph, 445–450
    period, 451
    phase shift, 452–453
Slant height, 408
Slope-intercept form, 296–298
Slope of line, 289–292, 295
Solid geometry, 397
Solution, 222
Solving equations, 222–225
    containing fractions, 230–234
    containing parentheses, 228–230
    pairs of linear
        by addition, 310–315
        applications for, 317–320
        by graphing, 305–309
        by substitution, 316–317
    with variables in both members, 226–228
Solving oblique triangles, 462–465
Solving right triangles, 432–433

Special products, 332–333
    finding factors of, 334–336
Sphere, 414
    surface area, 414
    volume, 414
Square, 369, 370
    area of, 370
    perimeter of, 370
Square centimetre, 140
Square metre, 132, 140
Square millimetre, 140
Square of a number, 83–84
Square root, 84, 125
Squares
    of binomials, 333
    factoring difference of two, 335–336
    perfect, 84
Square units, 14
Standard deviation, 485–487, 494–496
Standard unit, 49
Statistical data, 499–501
Statistical process control, 496–498
Statistics, 472
    bar graph, 472–474
    circle graph, 474–476
    grouped data, 488–491
    independent events, 507
    line graph, 477
    mean, 481–482
    median, 483–484
    mode, 484
    normal distribution, 502–504
    percentiles, 484
    probability, 505–506
    range, 485
    standard deviation, 485–487
        for grouped data, 494–496
Straight-line depreciation, 6
Subscripts, in formulas, 241
Substitution, method for solving pairs of linear equations, 316–317
Subtraction, 2–3, 200
    binary numbers, 516
    decimal fractions, 56–58
    fractions, 32–33, 34–37, 43–45
    hexadecimal numbers, 523–524
    measurements, 174–177
    mixed numbers, 33–34
    polynomials, 208
    signed numbers, 105–106
Subtraction Property of Equality, 223
Sum, 2
Supplementary angles, 365
Surface area
    lateral, 397, 404, 408
    sphere, 414
    total, 397, 404, 408
Systematic errors, 184

T

Tachometer, 154–155
Tangent, 387, 388
Temperature, 144–146
Ten, powers of, 2
Terms, 203
    like terms, 203–204
    unlike terms, 203
Three, divisibility by, 22, 23
Tolerance, 183
Tolerance interval, 183
Total surface area
    cone, 408
    cylinder, 404
    prism, 397
Transversal, 365, 366
Trapezoid, 369, 370
    area of, 19, 370
    perimeter of, 370
Triangle, 367, 373
    acute, 374
    altitude, 375
    area of, 19, 376
    equilateral, 373–374
    isosceles, 373, 374
    oblique, 454–456, 462–465
    obtuse, 374
    perimeter, 375
    Pythagorean theorem, 374–375
    right, 374, 424
    scalene, 373, 374
    sum of angles of, 377
Triangular prism, 397
Triangular pyramid, 407
Trigonometric ratios, 424–427
    applications involving, 434–436
    using to find angles, 428–430
    using to find sides, 430–431
Trigonometry
    law of cosines, 462–465
    law of sines, 454–456, 457–461
    sine and cosine graphs, 445–450
    solving right triangles, 432–433
    trigonometric ratios, 424–427
        using to find angles, 428–430
        using to find sides, 430–431
Trinomials, 206
    factoring, 332
    factoring general, 336–338
    perfect square, 333, 334–335
Two, divisibility by, 21–22, 23

U

Uniform scales, 189–191
Unlike terms, 203
U.S. micrometer, 170

U.S. system of weights and measures, 13–14, 24, 48–51
    metric conversion, 146–149

V

Variables, 200, 222
    dependent, 288
    independent, 288
    linear equations in two, 278–282
Variation
    direct, 265–269
    inverse, 271–272
Velocity, 450
Vernier caliper, 161–167
Vertex
    of angle, 363
    of cone, 408
    of parabola, 354–355
    of pyramid, 407
Vertical angles, 365
Vertical line, 288
    slope of, 292
Volt-ohm meter (VOM), 189–190, 192
Volume, 14–15, 138–140
    cone, 408
    cylinder, 403
    prism, 399
    pyramid, 408
    sphere, 414

W

Watt, 132
Wavelength, 449
Wave velocity, 450
Weight, 137
Whole numbers
    basic operations with, 2–10
    defined, 2
Work, 18

X

x axis, 280
x intercept, 285–287

Y

y axis, 280
y intercept, 285–287, 295

Z

Zero, division by, 26, 214
Zero power of 10, 116

Applications Symbols Used in This Text

- Agriculture and Horticulture
- Allied Health
- Auto/Diesel Service
- Aviation
- Business and Personal Finance
- CAD/Drafting
- Culinary Arts
- Electronics
- HVAC
- Industrial and Construction Trades
- Manufacturing
- Natural Resources
- Welding

U.S. Weights and Measures

Length

Standard unit: inch (in. or ″)

12 inches = 1 foot (ft or ′)
3 feet = 1 yard (yd)
5½ yards or 16½ feet = 1 rod (rd)
5280 feet = 1 mile (mi)

Weight

Standard unit: pound (lb)

16 ounces (oz) = 1 pound
2000 pounds = 1 ton

Volume

Liquid

3 teaspoons (tsp) = 1 tablespoon (tbs)
16 tablespoons = 1 cup
2 cups = 1 pint (pt)
16 fluid ounces (fl oz) = 1 pint (pt)
2 pints = 1 quart (qt)
4 quarts = 1 gallon (gal)

Dry

2 pints (pt) = 1 quart (qt)
8 quarts = 1 peck (pk)
4 pecks = 1 bushel (bu)

Metric System Prefixes

Multiple or Submultiple* Decimal Form	Power of 10	Prefix	Prefix Symbol	Pronunciation	Meaning
1,000,000,000,000	10^{12}	tera	T	těr′ă	one trillion times
1,000,000,000	10^{9}	giga	G	jĭg′ă	one billion times
1,000,000	10^{6}	mega	M	měg′ă	one million times
1,000	10^{3}	kilo**	k	kĭl′ō or kēl′ō	one thousand times
100	10^{2}	hecto	h	hěk′tō	one hundred times
10	10^{1}	deka	da	děk′ă	ten times
0.1	10^{-1}	deci	d	děs′ĭ	one tenth of
0.01	10^{-2}	centi**	c	sěnt′ĭ	one hundredth of
0.001	10^{-3}	milli**	m	mĭl′ĭ	one thousandth of
0.000001	10^{-6}	micro	μ	mī′krō	one millionth of
0.000000001	10^{-9}	nano	n	năn′ō	one billionth of
0.000000000001	10^{-12}	pico	p	pē′kō	one trillionth of

*Factor by which the unit is multiplied.
**Most commonly used prefixes.

As an example, the prefixes are used below with the metric standard unit of length, metre (m).

1 *tera*metre (Tm) = 1,000,000,000,000 m
1 *giga*metre (Gm) = 1,000,000,000 m
1 *mega*metre (Mm) = 1,000,000 m
1 *kilo*metre (km) = 1,000 m
1 *hecto*metre (hm) = 100 m
1 *deka*metre (dam) = 10 m
1 *deci*metre (dm) = 0.1 m
1 *centi*metre (cm) = 0.01 m
1 *milli*metre (mm) = 0.001 m
1 *micro*metre (μm) = 0.000001 m
1 *nano*metre (nm) = 0.000000001 m
1 *pico*metre (pm) = 0.000000000001 m

1 m = 0.000000000001 Tm
1 m = 0.000000001 Gm
1 m = 0.000001 Mm
1 m = 0.001 km
1 m = 0.01 hm
1 m = 0.1 dam
1 m = 10 dm
1 m = 100 cm
1 m = 1,000 mm
1 m = 1,000,000 μm
1 m = 1,000,000,000 nm
1 m = 1,000,000,000,000 pm

Metric and U.S. Conversion

Length

U.S.	Metric
1 inch (in.)	= 2.54 cm
1 foot (ft)	= 30.5 cm
1 yard (yd)	= 91.4 cm
1 mile (mi)	= 1610 m
1 mi	= 1.61 km
0.0394 in.	= 1 mm
0.394 in.	= 1 cm
39.4 in.	= 1 m
3.28 ft	= 1 m
1.09 yd	= 1 m
0.621 mi	= 1 km

Weight

U.S.	Metric
1 ounce (oz)	= 28.3 g
1 pound (lb)	= 454 g
1 lb	= 0.454 kg
0.0353 oz	= 1 g
0.00220 lb	= 1 g
2.20 lb	= 1 kg

Capacity

U.S.	Metric
1 gallon (gal)	= 3.79 L
1 quart (qt)	= 0.946 L
0.264 gal	= 1 L
1.06 qt	= 1 L
1 metric ton	= 1000 kg

Area

U.S.
- 1 ft² = 144 in²
- 1 yd² = 9 ft²
- 1 rd² = 30.25 yd²
- 1 acre = 160 rd²
- = 4840 yd²
- = 43,560 ft²
- 1 mi² = 640 acres
- = 1 section

Metric
- 1 m² = 10,000 cm² or 10⁴ cm²
- = 1,000,000 mm² or 10⁶ mm²
- 1 cm² = 100 mm²
- = 0.0001 m²
- 1 km² = 1,000,000 m²
- 1 hectare (ha) = 10,000 m²
- = 1 hm²

U.S. — Metric
- 1 in² = 6.45 cm²
- = 645 mm²
- 1 ft² = 929 cm²
- = 0.0929 m²
- 1 yd² = 8361 cm²
- = 0.8361 m²
- 1 rd² = 25.3 m²
- 1 acre = 4047 m²
- = 0.004047 km²
- = 0.4047 ha
- 1 mi² = 2.59 km²

Metric — U.S.
- 1 m² = 10.76 ft²
- = 1550 in²
- = 0.0395 rd²
- = 1.196 yd²
- 1 cm² = 0.155 in²
- 1 km² = 247 acres
- = 1.08 × 10⁷ ft²
- = 0.386 mi²
- 1 ha = 2.47 acres

Volume

U.S.
- 1 ft³ = 1728 in³
- 1 yd³ = 27 ft³

Metric
- 1 m³ = 10⁶ cm³
- 1 cm³ = 10⁻⁶ m³
- = 10³ mm³
- 1 cm³ = 1 mL

U.S. — Metric
- 1 in³ = 16.39 cm³
- 1 ft³ = 28,317 cm³
- = 0.028317 m³
- 1 yd³ = 0.7646 m³

Metric — U.S.
- 1 cm³ = 0.06102 in³
- 1 m³ = 35.3 ft³
- = 1.31 yd³

Temperature

$$C = \frac{5}{9}(F - 32°) \qquad F = \frac{9}{5}C + 32°$$

Formulas from Geometry

	Perimeter	Area		Volume	Lateral Surface Area
Rectangle	$P = 2(b + h)$	$A = bh$	Prism	$V = Bh$	
Square	$P = 4b$	$A = b^2$	Cylinder	$V = \pi r^2 h$	$A = 2\pi r h$
Parallelogram	$P = 2(a + b)$	$A = bh$	Pyramid	$V = \frac{1}{3}Bh$	
Rhombus	$P = 4b$	$A = bh$	Cone	$V = \frac{1}{3}\pi r^2 h$	$A = \pi r s$
Trapezoid	$P = a + b + c + d$	$A = \frac{(a+b)h}{2}$	Sphere	$V = \frac{4}{3}\pi r^3$	$A = 4\pi r^2$
Triangle	$P = a + b + c$	$A = \frac{1}{2}bh$			
Circle	$C = 2\pi r$ or $C = \pi d$	$A = \pi r^2$			